GROUNDWATER RESOURCE EVALUATION

McGraw-Hill Series in Water Resources and Environmental Engineering

VEN T. CHOW, ROLF ELIASSEN, and RAY K. LINSLEY *Consulting Editors*

Walton: GROUNDWATER RESOURCE EVALUATION

GROUNDWATER RESOURCE EVALUATION

WILLIAM C. WALTON
Professor of Geology and Geophysics
Director, Water Resources Research Center
University of Minnesota

McGraw-Hill Book Company
NEW YORK ST. LOUIS SAN FRANCISCO
DÜSSELDORF LONDON MEXICO PANAMA SYDNEY TORONTO

To ELLEN, CYNTHIA, and LOUISE

GROUNDWATER RESOURCE EVALUATION

Library of Congress Catalog Card Number 71-78959

ISBN 07-068051-5

4567890 MAMM 7987654

This book was set in Times Roman and printed on permanent paper and bound by The Maple Press Company. The designer was Edward Zytko; the drawings were done by John Cordes, J. & R. Technical Services, Inc. The editors were Roland S. Woolson and Sally Anderson. William P. Weiss supervised the production.

PREFACE

During the last decade it has been more fully realized that refined quantitative answers are needed concerning available groundwater resources. Utilization of aquifers continues to accelerate to meet the needs for irrigation, industrial, urban, and suburban water supplies. As groundwater development intensifies, well owners become more interested in the response of aquifers to heavy pumping, whereas initially they were concerned largely with the delineation and exploration of aquifers. Competition for available resources has brought about an awareness that one of the principal problems confronting groundwater hydrologists is resource management. Before groundwater resources can be managed, they must be quantitatively appraised. In ever-increasing numbers, engineers, geologists, and others are being called upon to estimate how much groundwater is available for development and what will be the consequences of development. Groundwater users are continually asking for suggestions as to how available resources can be properly managed.

Proper planning for the development and management of groundwater resources requires the testing of all possible schemes and the appraising of the relative merits of various alternatives. Groundwater resource development and management personnel are concerned with the sustained yields of wells and aquifers, the interference between wells and well fields, the interrelation between surface water and groundwater, and the quality of groundwater. Questions pertaining to the use of groundwater resources require that pumping be related to water-level changes with reference to time and space. The hydrogeologic properties and dimensions of aquifers and existing aquitards and the boundaries of aquifers are of utmost importance in relating cause and effect. Cause cannot be related to effect until hydrogeologic maps are available. These maps must encompass all nonhomogeneous and irregular hydrogeologic conditions. Electric analog and digital computer techniques that allow the rapid study of cause-and-effect relationships involving complex aquifer conditions have recently been developed.

Along with the "information explosion" in the scientific world during recent years, there has been rapid and enormous progress in the science and technology concerned with groundwater. This progress has stimulated a great deal of interest, and correspondingly a large amount of research, among a widely diversified group of scientists. The ever-increasing scientific research on groundwater has resulted in a large body of knowledge which, for the most part, can only be found scattered through various scientific journals and technical reports, but in general is not easily accessible to many inquirers who wish to view all aspects of groundwater resource evaluation. It is highly desirable that the voluminous material in the groundwater resource evaluation field be assembled and briefed in order that engineers, geologists, and others actively engaged in quantitative studies can have available a ready reference.

My experience in teaching groundwater courses to engineers and geologists during the past several years has indicated the urgent need for a text which presents both the basic principles of groundwater resource evaluation and the practical application of techniques to well and aquifer problems. Derivation of equations as well as numerous illustrative case histories of analyses based on actual field data are presented in this book. Numerous practical problems which may be solved with available equations and methods of analysis are given. A comprehensive bibliography is presented containing references to the literature germane to groundwater resource evaluation that may be used to expand the reader's understanding of subject matter. The equations and analytical methods available to groundwater hydrologists are almost unlimited in number, and the discussion of all of them would necessitate several volumes and unwarranted duplications. A selection has therefore been made to include equations and analytical methods most frequently applied to actual groundwater resource evaluation field problems. The subject matter includes equations and methods used to quantitatively appraise the hydrogeologic parameters affecting the water-yielding capacity of wells and aquifers, and equations and methods used to quantitatively appraise the response of wells and aquifers to heavy pumping.

This book, in addition to serving as a text for groundwater courses, should aid in the systematic appraisal of groundwater resource problems by practicing engineers and geologists. Graduate students should be stimulated by the broad perspective of groundwater resource evaluation.

WILLIAM C. WALTON

CONTENTS

CHAPTER 1

INTRODUCTION

Groundwater is one of the earth's most widely distributed and most important resources. Groundwater exists wherever water penetrates beneath the surface, the rocks beneath the surface are permeable enough to transmit this water, and the rate of infiltration is sufficient that the rocks are saturated to an appreciable thickness. These conditions are met and groundwater exists at least intermittently in a very large part of the United States. Groundwater becomes a usable resource when the rocks in the zone of saturation are permeable enough to yield useful supplies of water to wells, springs, or streams; when the zone of saturation is perennial, or at least persists long enough each season to allow practical exploitation; and when the mineral substances dissolved by the water as it percolates through the soil and rocks do not reach such concentrations as to make the water unfit for the desired use. At least the small quantities of potable water needed for domestic supply are available very widely. Only where impermeable or saline rocks or permanently frozen ground extend to great depths or where the climate is exceedingly dry, or on an island too small or too permeable to retain useful supplies of water, is potable groundwater entirely or practically absent; even in large regions having these general characteristics, groundwater may be available locally.

While the pattern of surface-water flow is generally known and the potentials of surface-water reservoirs on principal rivers have been extensively appraised, there has been only limited study of the nation's groundwater resources. Groundwater studies to date have been predominantly in problem areas and have been largely concerned with falling groundwater levels, failure of wells, or impairment of groundwater quality. Recent studies have added to the knowledge of the geographical distribution of the nation's groundwater resources and have provided the base for rough estimates of the magnitude of undeveloped potentials. More detailed studies in limited areas have clarified a few of the complex features of groundwater reservoirs. The determination and appraisal of groundwater resources is slow and time-consuming work. Although the general geologic and groundwater conditions throughout the United States are known, the detailed information necessary for meaningful quantitative estimates of groundwater resources is known only for a small part of the country.

Groundwater's importance to the nation is not fully appreciated. Groundwater accounts for only a fifth to a sixth of the total amount of water withdrawn from all sources. But, its place in the nation's economy is not be be measured solely by the ratio of groundwater withdrawn to surface water withdrawn. For example, groundwater is more widely and easily available than surface water, and most groundwater is unpolluted and relatively safe from pollution. About one-third of the conterminous United States is underlain by groundwater reservoirs generally capable of yielding at least 50 gallons per minute of water per well. There are large areas where hundreds or even thousands of gallons per minute can be obtained from wells. Groundwater generally averages out to be a little harder and more mineralized than surface water in the same locality, but its quality is more uniform during the year. It is rarely if ever necessary to consider removal of sediment from groundwater. Where groundwater is polluted, treatment to make it safe is simple and relatively inexpensive. Groundwater is the practical source of emergency supplies in the event of a national nuclear disaster. The temperature of groundwater, like its chemical quality, is relatively uniform throughout the year. This makes it preferable for cooling purposes in the summer, when surface water is warmer. Groundwater enabled the settlement and continued occupation of our agricultural areas and furnished public and industrial water for most of the towns and cities that grew up around the country. Groundwater is still an underdeveloped resource in large areas, is contributing substantially to our industrial growth, and is the most practical source for supplemental irrigation.

Groundwater reservoirs accept water; they filter it to remove sediment and disease-causing organisms; they store it in quantities vastly exceeding those which are or conceivably could be stored in all natural and artificial surface-water bodies put together; they even out its temperature and chemical quality; they transport it from areas of replenishment to areas of need; and they slow down its natural discharge to the surface so that it makes up the bulk of the dry-weather flow of streams. To an increasing extent they will be used to store surplus surface water through artificial recharge. This practice will have a benefit in reducing floodflows and flood damage, but its greatest value will be in reducing the waste of good-quality water which otherwise serves no useful purpose.

Water and the materials over and through which groundwater flows constitute a vast, complex dynamic system in which any change in the operation of one part is reflected in more or less substantial changes in other parts. In analyzing groundwater systems, consideration must be given to the effect groundwater withdrawals will have on such things as the water levels in nearby wells or in a wetland, the flow of contaminants or salt water toward wells, and the reduction of flow in a nearby stream. The pumping may serve a desirable purpose, but it may be desirable also to maintain the flow of the stream to dilute wastes, to preserve the wetlands as a habitat for wildlife, or to protect the well itself or other wells from salt-water encroachment. The need is to understand cause-and-effect relationships well enough to be able to describe a series of possible alternative development and managment choices and the consequences of each. For full development of groundwater reservoirs, groundwater's place in the hydrologic cycle and in hydrologic systems must be evaluated and the groundwater resource must be described quantitatively rather than just in qualitative terms. Adequate data and imagination, uncircumscribed by disciplinary boundaries, can lead to vast improvement in the efficiency of groundwater use. Conjunctive use and management of surface water and groundwater multiplies the number of alternatives available for water management because it multiplies the times, places, and means for water development and the number of circumstances under which development can be accomplished.

1.1 Hydrologic Cycle

The chain of events describing the history of water is called the *hydrologic cycle*. The cycle involves the total earth system: atmosphere, hydrosphere, and lithosphere. The atmosphere is the gaseous envelope above the hydrosphere, the hydrosphere consists of the bodies of water that cover the surface of the earth, and the lithosphere is the solid rock environment below the hydrosphere. The cycle of activities of water extend through these three parts of the earth system from an average depth of at least a half mile in the lithosphere to a height of about 10 miles in the atmosphere.

In the hydrologic cycle, water evaporates from the oceans and the land and becomes a part of the atmosphere. The evaporated moisture is lifted and carried in the atmosphere until it precipitates to the earth, either on land or in the oceans. The precipitated water may be intercepted or transpired by plants, may run over the ground surface and into streams to oceans, or may infiltrate into the ground. Much of the intercepted and transpired water and some of the surface runoff return to the air through evaporation and transpiration. The infiltrated water may percolate downward to be temporarily stored as groundwater which later flows out of rocks as springs, or seeps into streams as runoff to oceans, or evaporates into the atmosphere to complete the cycle. Thus, the hydrologic cycle undergoes various complicated processes of evaporation, precipitation, interception, transpiration, infiltration, percolation, storage, and runoff. Figure 1.1 illustrates the hydrologic cycle.

The hydrologic cycle reaches into the atmosphere and traverses the domains of hydrometeorology, meteorology, and climatology. In the hydrosphere, it crosses, or embodies, the domains of potamology (surface streams), limnology (lakes), cryology (snow and ice), glaciology, and oceanology. In the lithosphere, the hydrologic cycle relates to agronomy, hydrogeology (emphasizing geologic aspects), geohydrology (emphasizing hydrologic aspects), and geomorphology. As water affects plants as well as animals, the hydrologic cycle extends itself into plant ecology, silviculure, biohydrology (emphasizing hydrologic aspects), and hydrobiology (emphasizing biologic aspects). The cycle has its important influence in agriculture, forestry, geography, watershed management, political science (water law and policy), economics (hydroeconomics), and sociology; and it has practical applications in structural design, water supply, waste-water disposal and treatment, irrigation, drainage,

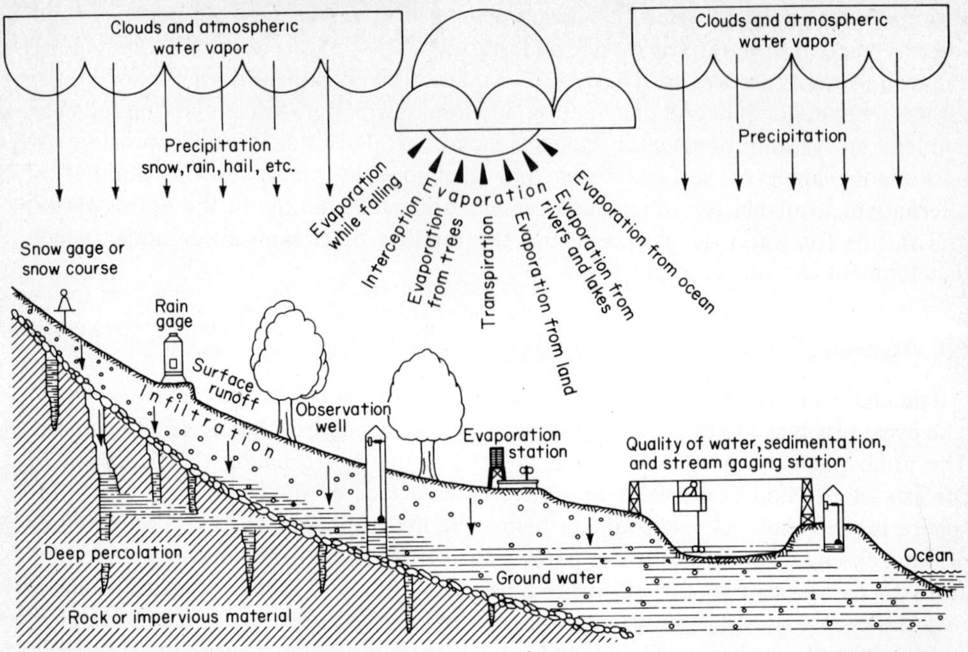

Fig. 1.1 Diagrammatic representation of the hydrologic cycle. (*Adapted from ASCE Hydrology Handbook, 1949.*)

hydropower, flood control, navigation, erosion and sediment control, salinity control, pollution abatement, recreational use of water, fish and wildlife reservation, insect control, and coastal works.

The amounts of evaporation, precipitation, runoff, and other hydrologic cycle quantities are not evenly distributed on the earth, either geographically or time wise. In the conterminous United States, annual rainfall averages 1,430 cubic miles; evaporation is about 1,000 cubic miles; discharge of water to the oceans from surface streams is about 390 cubic miles; and about 40 cubic miles of water is discharged directly into the oceans from groundwater reservoirs (Nace, 1965). Several times 40 cubic miles of water passes through inland groundwater reservoirs and reaches stream channels to be discharged to the oceans via surface streams.

That part of the hydrologic cycle that occurs beneath the land surface, and that is therefore invisible, is surrounded with misunderstanding and misconceptions. Groundwater is regarded by a great many people as being no less mysterious than it appeared to our forebears of the Middle Ages, who often explained it by means of supernatural phenomena. Many still feel that the occurrence of groundwater, its replenishment, movement, and discharge, is so strange as to defy scientific explanation; they feel that groundwater can be understood and located only by persons who are supposedly possessed of occult powers (water dowsers) or by the use of instruments whose operations are as mysterious as their operators. The puzzling and apparently mysterious complexity of the occurrence and movement of groundwater is largely the result of the complex geologic history of our earth. It results from the fact that our continent, which now appears so stable and permanent, has been subjected during eons past to uplift, then erosion, sedimentation and folding and faulting, deep-seated intrusions of molten rocks and surficial outpourings of lava, invasions of the land areas by the sea, and burial under huge masses of glacial ice advancing from the mountains or the frigid north.

The geologic history through which any particular area has passed very largely influences the hydrologic cycle in that area. If the history has been such as to form layers of sand, the soil probably will readily absorb rainfall, and there will be a minimum of runoff over the land surface. The streams will reach flood stages only during very heavy storms and they will tend to have a uniform, steady flow even during dry-weather periods. Drilled or dug wells probably will produce large amounts of water. On the other hand, if the history has been such that subsurface materials are dominantly clay or silt, the soil will absorb water with difficulty. Storms of relatively small magnitude will cause substantial overland runoff. The streams will flood frequently and will tend to dry up during even moderate droughts. Wells will produce relatively little water, and in some places none at all. Inasmuch as the geologic history is infinitely varied in different parts of the country, the number of variations to the simple examples cited is likewise infinite.

1.2 History of Groundwater Resource Evaluation

Many centuries were required to establish the hydrologic cycle concept. The solution of the problem of the origin of groundwater contributed to the development and slow acceptance of the concept (see Parizek, 1963). According to Keilhack (1912), the origin of groundwater was mentioned in Book 21 of Homer's *Iliad*, written about 1000 B.C., as follows: "with Zeus neither does the mighty river Achelous fight nor the mighty strength of the deep-flowing Oceanus, from which flows all rivers and every sea and all springs and deep wells." Herodotus (484–425 B.C.) reasoned that the sun "attracts the water. After attracting it [the sun] again repels it into the upper regions, where the winds lay hold of it, scatter it, and reduce it to a vapor." Before the sun again returns to its summer position in the sky "the other rivers run big, from the quantity of rainwater which they bring down from countries where so much moisture falls that all the land is cut into gullies; but in summer, when the showers fail, and the sun attracts their water, they become low" (Herodotus, *Persian Wars*).

Anaxagoras (500–428 B.C.) recognized the importance of evaporation and rainfall as a source of water for rivers, but thought in terms of great reservoirs probably of the Tartarus vast-cavern type. Anaxagoras also recognized the importance of rainfall, recharge, and underground storage even though he incorrectly understood the nature of this storage.

Plato wrote in *Critias*, around 350 B.C., about the rainfall-runoff segment of the cycle, the infiltration of surface water and underground storage segments, and made correct statements concerning the origins of springs and streams. Aristotle (384–322 B.C.) recognized the cyclic path of water between land and air, evaporation and condensation and their importance in rainfall, and that some rain is discharged by streams, while some percolates into the earth and reappears in springs (*Meteorologica*). The importance of evaporation, rainfall, and infiltration to the origin of groundwater was clearly and correctly stated by Vitruvius, who lived at the time of Christ (Vitruvius, *Architecture*).

Leonardo da Vinci recognized artesian underground systems. Bernard Palissy (1509–1589) correctly explained the relationship between wells and rivers, the origin of springs and rivers, the hydrologic cycle, and changes in water levels in wells (Palissy, *Discours Admirables*). Pierre Perrault (1608–1680), a French scientist, demonstrated that rainfall is more than adequate to account for the discharge of rivers and springs. His work was extended some years later by another French scientist, Mariotte (1620–1684), who correlated seepage into a cellar with rainfall and showed that the flow of springs varied according to rainfall and that infiltration of rainfall supplied springs and wells with their water. In 1693 Edmund Halley (1656–1742), a British astronomer, demonstrated that evaporation was ample to supply the quantity of water returned to the sea by rivers flowing into it. Bernardino Ramazzini gave an accurate account of the origin of artesian pressure in 1691. Vallesiere (1715) recognized the importance of an impermeable stratum to serve as a confining layer to an artesian system. La Metherie in 1791 explained that a part of the water from rain and snow flows off directly, a second part moistens the soil and evaporates or feeds the plants, and a third part penetrates to subsurface reservoirs at greater depths, from which it gradually issues as springs.

The series of developments between 1856 and 1955 which helped to establish the principles of groundwater resource evaluation as a quantitative science were described by Ferris and Sayre (1955). Darcy (1856) did experimental work on the flow of water in sands and derived a formula known as Darcy's law, which expresses the relationship between velocity of percolation, permeability of water-yielding materials, and the hydraulic gradient. Darcy's law serves as the basis for numerous quantitative methods in groundwater resource evaluation. Dupuit (1863) was the first scientist to develop a steady-state formula for the flow of water into a well. A. Thiem (1870) developed equations for flow toward wells and galleries. Forchheimer (1886) introduced the concepts of conformal mapping and the construction of flow nets, the method of images, and the theory of functions of the complex variable. He was the first to solve the problem of groundwater flow in a semi-infinite water-yielding formation bounded by a perennial stream and the problem of a well discharging from a water-yielding formation that is supplied by uniform recharge. Badon Ghyben (1889) established the laws of equilibrium between sea water and fresh groundwater. Slichter (1898) showed the relationship between grain size of rocks and permeability, and the motion of groundwater. Chamberlin (1885) defined the requisite geologic controls necessary to the occurrence of artesian wells. Detailed information on the relation of head to groundwater movement was presented by King (1892).

Turneaure and Russell (1901) published a textbook on public water supplies, which included their counterpart of Thiem's and Slitcher's equations for radial flow toward a discharging well. G. Thiem (1906) developed a field method for determining permeability of water-yielding formations and the rate of flow of water, using a pumping well and the resultant drawdown in observation wells. Horton (1906) extended the steady-state formulas of Slichter to provide a method for determining the proper depth and spacing of tile drains. Meinzer (1923) evaluated early studies on the framework of principles and methodology for investigations of the occurrence and distribution of groundwater and provided the first manual for groundwater hydrologists.

The water-logging of lands was investigated by Gardner et al. (1928), Israelsen and McLaughlin (1935), and Kirkham (1945), who analyzed the hydraulics of flow toward drains overlying artesian water-yielding formations. The work of Hopf and Trefftz (1921), Pavlovsky (1922), Kozeny (1933), Lindquist (1933), Hamel (1934), Muskat (1935), Verdernikov (1934), Meleschenko (1936), and Vreedenburgh and Stevens (1936) greatly expanded the development of solutions for two- and three-dimensional flow toward canals and drains.

De Glee (1930) provided a solution to the problem of a steady flow toward a well in a leaky artesian water-yielding formation replenished by an overlying water-yielding formation. Lewis (1932) provided the first practical solution for steady radial flow to a well in a water-yielding formation receiving steady, uniform recharge. The distribution of potential about a partially penetrating well was analyzed by Forchheimer (1930), Kozeny (1932), Muskat (1932), and Hantush (1964). Methods for determining the entrance losses in a pumped well were described by Kano (1939), Jacob (1947), Rorabaugh (1953), and Engelund (1953). Methods for estimating groundwater recharge or discharge from changes in groundwater storage were devised by Smith (1924), White (1932), Meinzer (1928), and Williams and Lohman (1949). One of the most important milestones in the development of groundwater resource evaluation was Theis's (1935) introduction of an equation for the non-steady-state flow to a well.

Sand-tank models were applied to radial-flow problems by Wyckoff et al. (1932), Babbitt and Caldwell (1948), Hansen (1952), Childs et al. (1953), Kozeny (1932), Muskat (1935), Vaidhianathan et al. (1934), Baumann (1951), and Day and Luthin (1954). Spaced-plate or viscous-flow models were used by Dietz (1944), Moore (1949), and Zanger (1953). Electric analog models were developed for the solution of complex flow patterns by Muskat (1935), Babbitt and Caldwell (1948), Wyckoff et al. (1933), and Zanger (1953).

Streamflow records were analyzed to determine recharge to the groundwater reservoir and discharge of groundwater to streams by Mendenhall (1905), Piper (1939), Houk (1921), and Meinzer (1928). Muskat (1937) published a comprehensive text on fluid flow through porous media. The use of the method of images and Green's functions were developed and solutions were presented for semi-infinite, quarter-infinite, and infinite-strip water-yielding formations. The hydraulics of groundwater flow toward wells in reentrants, buried channels, horst and graben structures, and other geologic situations was clearly developed.

Theis (1937) formulated the first transient-state flow of groundwater to a discharging ditch or drain. Kirkham (1945), Childs (1945), and Engelund (1951) provided thorough analyses of the flow of groundwater and soil water toward drains.

Hubbert (1940) derived from the Navier-Stokes equations Darcy's law and introduced the concept of force potential in his mathematical derivation. Hubbert (1940) pointed out the physical principles of the fresh-water–salt-water relationship for the hydrodynamic condition. Jacob (1940) clarified quantitatively the storage coefficient as it applies to the elastic artesian water-yielding formation. Several useful corollary equations were developed from the Theis formula by Theis (1935) and Jacob (1950). Guyton (1941) demonstrated that drawdowns computed with the Theis formula checked closely with actual field observations. Mathematical analyses of groundwater flow on a basin-wide scale were made by Hubbert (1940), Jacob (1945), Steggewentz (1933), Werner and Noren (1951), and Ferris (1951). Subsidence of land due to groundwater withdrawals was investigated by Tolman and Poland (1940). Methods for treating river-water infiltration to streamside wells were presented by Theis (1941), Rorabaugh (1951), Hantush (1964), Kazmann (1948), and Walton (1963). The application of Theis's formula and the method of images for locating water-yielding formation boundaries was described by Ferris (1948). Methods for analyzing data on water-level interference resulting from operation of several wells were developed by Wenzel and Greenlee (1943) and Cooper and Jacob (1946). More comprehensive analyses of the leaky artesian problem were added by Jacob (1946) and Hantush (1964). Jacob (1946) formulated transient-state solutions for a well in a leaky artesian situation and for a well of constant drawdown in an elastic artesian situation.

Principles of cation exchange were described by Piper et al. (1953) and Love (1944). Base exchange in groundwaters was studied by Renick (1925) and Poland et al. (1948). Quantitative methods for the study of the influx of mineralized waters in areas of intensive pumping were developed by Hill (1942), Langelier and Ludwig (1942), and Piper (1944).

Detailed discussions of the use of numerical methods for solving problems involving flow in aquifers were given by Yang (1949), Boulton (1951), Kashef et al. (1952), Luthin and Scott (1952), Shaw and Southwell (1941), and Stallman (1955).

The quantitative relationship between water quality and electrical resistivity was developed and used by Jones and Buford (1951). Hansen (1952) and Zanger (1953) used mechanical (elastic) models to solve groundwater problems. Stuart et al. (1954) analyzed groundwater problems associated with iron-ore mining. Hubbert (1953) presented new concepts relating to the entrapment of petroleum and groundwater conditions. Ahrens (1957) presented well-design criteria. Hem (1959) summarized methods for interpretation of chemical characteristics of natural water. Henry (1959) and Bear and Dagen (1963) derived equations concerning salt-water intrusion into fresh-water aquifers. Kohout (1960) presented data on the cyclic flow of salt water in coastal water-yielding formations. Hydrodynamic dispersion and mixing of water in aquifiers was analyzed by Bear (1961).

Koenig (1960) analyzed economic aspects of water-well stimulation. Rasmussen and Andreasen (1959) and Schicht and Walton (1961) prepared hydrologic budgets for basins. Back (1961) presented techniques for mapping hydrochemical facies. Skibitzke (1961) applied electric analog techniques to solving complex groundwater system problems. Walton and Prickett (1963) extended the application of analog techniques in solving complex groundwater system problems. Walton and Neill (1961) analyzed groundwater problems with a digital computer. Poland (1961) described the coefficient of storage in a region of major subsidence caused by compaction of water-yielding formations. Lusczynski (1961) derived equations for the head and flow of groundwater of variable density. Patten and Bennett (1963) summarized applications of electrical and radioactive well logging to groundwater hydrology. Sniegocki (1963) summarized problems in artificial recharge through wells. Toth (1963) provided a theoretical analysis of groundwater flow in small drainage basins. Walton and Neill (1963) analyzed specific-capacity data to determine the hydraulic properties of water-yielding formations. Boulton (1963) devised a method for analyzing pumping tests under water-table conditions which was extended by Prickett (1965). Cooper and Rorabaugh (1963) analyzed groundwater movements and bank storage due to flood stages in surface streams. Hantush (1964) provided solutions for many well-hydraulics problems, including those associated with leaky artesian conditions, partially penetrating wells, induced infiltration of streamflow, and horizontal collectors. Walton et al. (1967) analyzed with electric analog computers recharge from induced streambed infiltration under varying groundwater-level and stream-stage conditions.

Forced by the necessity of solving pressing groundwater supply problems, most recent advances in groundwater resources evaluation have dealt with water-well hydraulics and well-field designs. The most striking trend in 1968 was the renewed interest in Darcy's law and all aspects of regional groundwater flow, as opposed to the previous preoccupation with hydraulics of water wells and well fields. In 1968, groundwater hydrologists were concerned with whether Darcy's concept of groundwater motion can be extended and how much further it can be applied. This trend manifests itself in two ways: on the one side are studies pertaining to the explicit validity of Darcy's law, in which special attention is given to the exact meaning of the permeability constant, and the meaning of hydraulic head and of hydraulic gradient. On the other side are attempts to apply Darcy's law in a much wider sense, whereby it is assumed that the law is valid for all natural conditions of groundwater flow. In this respect particular attention is being paid to the laws that govern the regional distribution of hydraulic head. Finally, the question of whether there are other laws that govern groundwater motion is being investigated.

During the 1960's, there has been a tremendous increase in knowledge with regard to hydrodynamic flow systems. Groundwater–surface-water relationships received more attention as hydrologists increasingly adopted the "systems analysis" point of view in attempting to visualize the effects throughout whole hydrologic systems of various water development and management projects. Some fundamental work was accomplished on induced infiltration from streams to wells, on filling and emptying of "bank storage" as a result of natural fluctuations in stream storage, on analysis of water-level fluctuations in wells resulting from tidal and other changes in stream storage in terms of water-yielding formation characteristics, and on interchange of fresh and saline water between coastal water-yielding formations and the ocean.

Geophysical methods are being used increasingly in groundwater studies. Prospecting by the electrical-resistivity and seismic methods has been supplemented on a small scale by magnetic and gravity surveys. "Borehole geophysics," however, was the principal tool in 1968. The chief advances have been in neutron and gamma-ray logging, in the hydraulics of multiple-aquifer wells, and in textural and geochemical interpretations from resistivity and spontaneous-potential data. Borehole geophysics is expected to be a principal tool in getting more reliable information at practical cost from wells and test holes in the construction of analog models of groundwater systems.

There has been an increasing emphasis on regional or basin analysis of groundwater systems. Electrical analog models employing the analogies between resistors and permeability and between capacitors and storage coefficient have been developed to the stage where they can be used routinely on areal studies as simulation models of groundwater systems. New techniques for automated data processing using the digital computer now offer the opportunity for organized storage and retrieval of the vast quantities of groundwater data that are presently available, and, in addition, they provide means for computing and compiling the data into printout tables, graphs, and maps. The use of simulation models, either analog or digital, for analysis of groundwater systems has required field studies attempting to quantify geology by trying to relate various geologic factors to such characteristics as the magnitude and distribution of permeability.

The subjects of effects of urbanization, salt-water encroachment, brackish groundwater, corrosion of well casings, use of specific-capacity data for estimating transmissibility, subsidence, and artificial recharge were receiving considerable attention in 1968.

1.3 Needed Research

Much valuable groundwater research has been accomplished and is now underway. However, intensified interest and demand for knowledge of groundwater resources calls for continuing additional effort. Several professional society committees have studied and reported on groundwater research needs. Some of the subjects mentioned in committee reports are listed in the hope that this will stimulate thought and additional research in groundwater resource evaluation.

1. The jurisdiction, powers, and type of organization best s[illegible] groundwater activities of an area for optimum eco[illegible]d s[illegible] all who are affected, and the proper interrela[illegible] existing agencies.
2. The alterations in the nature of recharge water, p[illegible]ty, necessary to maximize rates of recharge.
3. The physics, chemistry, and biology of the effluent from oxidation ponds and septic tanks during its movement to the water table.
4. The nature of groundwater law which will be continuously responsive to the public interest in a dynamic society.
5. Better methods for: (*a*) determining or estimating hydrogeologic properties of the aquifers and aquitards from geologic, geophysical, and hydraulic evidence; (*b*) determining rate and direction of groundwater movement; (*c*) determining storage capacity, including improvement of direct field methods for measuring moisture content, porosity, and other properties; (*d*) relating laboratory results to field data and applying them to field conditions, and (*e*) operating and maintaining groundwater recharge facilities to promote and prolong very high rates of infiltration (reduce or prevent clogging due to sedimentation and chemical and biological activity).
6. The economic effects of alternate depletion rates and pumping patterns in a groundwater mining situation.
7. The economic effects of alternate forms for organizing groundwater systems for providing quantity and quality management.
8. The economic effects of using alternative criteria for operating integrated surface-water–groundwater systems in terms of quantity and quality of water service.
9. Approaches to make groundwater recharge facilities or areas multipurpose, usable, for example, as recreational areas and wildlife habitats and for aesthetic improvement of the environment.
10. The process of interflow (or subsurface flow distinguished from surface runoff and groundwater or base flow) with special attention to the role of soil structure, physiography, and geology.
11. The factors governing the amount of water absorbed and transmitted by substrates with high specific surface (e.g., clays) under various conditions. The effects of solute quality, quantity, and distribution, of thermal and electric gradients, and of other factors on the movement of groundwater.
12. The changes occurring in the properties of porous media as water moves through media. Chemical changes such as solution or deposition and physical changes including compaction and resulting subsidence.
13. Fluctuations of streams and lakes which have a pronounced influence on nearby groundwater bodies and vice versa (time, space, and volume relations).
14. Man's use of land, such as urbanization, drainage, irrigation, cultivation, deforestation and affordation, and grazing and highway construction, and its

influence on water yields, water quality, stream regimen, groundwater recharge, and erosion, flooding, and attendant sedimentation.

15. The role of ephemeral frost, permafrost, and the active layer on surface runoff, the occurrence and movement of groundwater, and the quality of water.

16. Physical properties of porous media that influence the occurrence, storage, and movement of groundwater; including small-scale (including microscopic) factors such as grain size and shape, intermediate factors such as facies change, folds, and faults, and large-scale factors such as geologic terrains, major physiographic and climatic provinces.

17. Principles of multiphase flow and diffusion, including contiguous salt water and fresh water under natural conditions and as modified by man.

18. Transient hydraulics of partial or entire aquifer system by flow nets, analogs, mathematical, and other models.

19. Electrochemical, mathematical, and analytical systems and reactions of water and earth materials that cause changes from place to place and from time to time in the quality of the water.

There is an appalling lack of understanding of all facets of groundwater resource development and management on the part of many, both laymen and professional, who are in policy-making positions. The full value of research efforts will be obtained only if the results and their significance are fully communicated to all concerned. This is particularly true as regards the policy-making levels of government, the Congress, state legislatures, and political subdivisions.

1.4 Groundwater Investigations

Many organizations make information on groundwater available to the public through groundwater investigations. The public consists of individuals who use water as well as those who are concerned with exploration, development of supplies, distribution, and chemical quality. By information is meant: (1) raw data, such as observations on groundwater levels; (2) interpretations, such as on quantity, distribution, sources of replenishment, and discharge in areas of concern; and (3) knowledge of fundamental principles and relationships. It is necessary to keep track of groundwater withdrawals, to inventory the storage, and, in places of need, to know the physical facts concerning the volume of groundwater in storage, its occurrence, replenishment, relation to adjoining surface water, and its state of contamination, if any. A critical examination of all information should be made before it is used in analyses (see Pettyjohn, 1967).

On a nationwide basis, the Water Resources Division, Geological Survey of the U.S. Department of the Interior, is the most active organization in the field of groundwater investigations. Federal organizations concerned with groundwater investigations are listed in Fig. 1.2. Other organizations that are concerned with groundwater investigations are: state and local agencies, private organizations, and educational institutions.

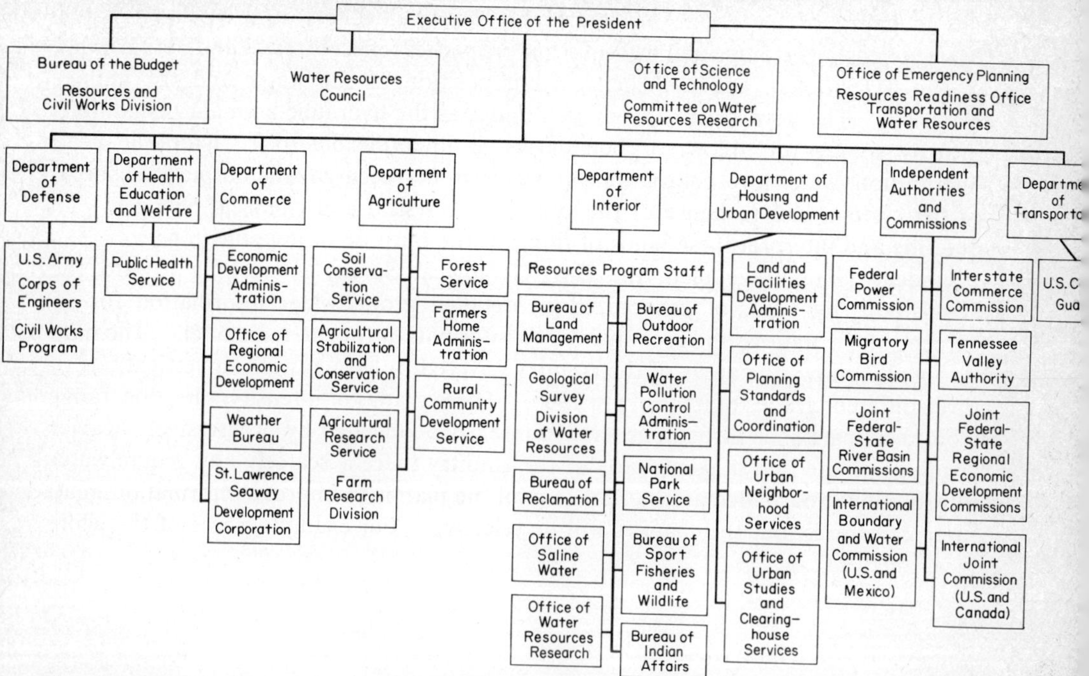

Fig. 1.2 Federal organizations concerned with groundwater investigations in 1967. (*Drawn by W. C. Walton.*)

In 1964, the Federal Bureau of the Budget issued Circular A-67. This circular prescribes guidelines for coordinating federal activities in acquiring data on streams, lakes, reservoirs, estuaries, and groundwater. It provides also for the planning of a national network to assure that the needed data are collected in a timely and efficient manner. Responsibility for implementing the circular is vested in the Department of the Interior with the Geological Survey as the lead agency (see Upson, 1967). In ensuing years, the increased need for and interest in water data indicated by the circular will result in continued strengthening of national data programs, including that of the Geological Survey for acquiring groundwater data. In general, a sizable and fairly sharp increase in data collection is expected. This will consist of more than simply maintaining records of groundwater level fluctuations and well yields. Temperature and dissolved constituents, including radioactive substances and pesticides, will be of increasing importance. Inasmuch as raw data by themselves are of very limited value, by far the greater amount of data collection will be in conjunction with areal investigations.

By investigations is meant the collection, synthesis, and interpretation of data so that for various areas one can define the environment in which groundwater occurs and the flow system of the groundwater body or bodies. The environment includes the geometry of the aquifer system (area, depth, thickness, and extent of water-bearing beds), its storage and transmission properties (effective yield and permeability of the formations), and the chemical composition of the water and the aquifer materials. The flow system covers such things as the hydraulic boundary conditions (nearby sources of inflow, or boundaries of impermeable rock), hydraulic head distribution (water-level contour maps), water input (recharge) and output (discharge) of the system, and response of the system to physical and chemical changes. To describe and interpret these kinds of things is the purpose of investigations.

Under a long-range plan, the U.S. Geological Survey is working toward the acquisition, by 1973, of generalized or detailed groundwater information for 75 percent of the nation, and reconnaissance information for the remainder. The map in Fig. 1.3 shows the areas of the country covered by groundwater investigations of three different grades of intensity in 1965. These grades of intensity, or scope, range from reconnaissance through general to intensive, indicated by progressively heavier shading. Note that in some parts of the country there is scarcely any groundwater information (nonshaded areas). In general the plan is to improve information in all three categories, but the changes will be selective. For example, some of the white

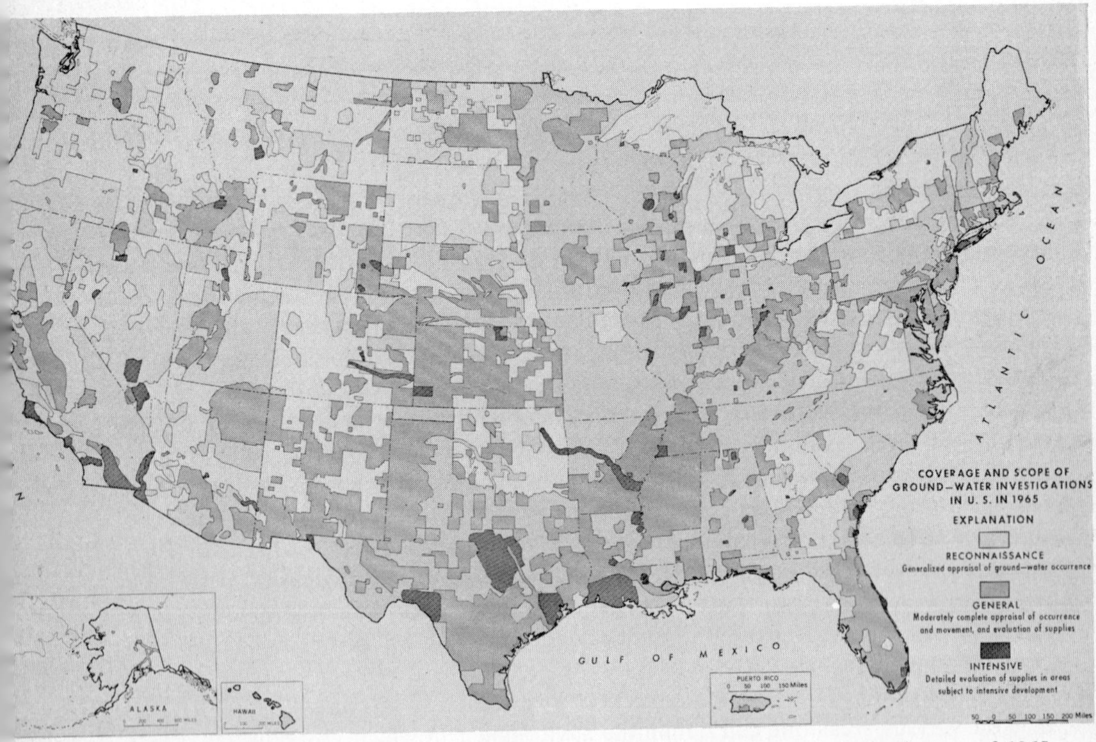

Fig. 1.3 Map of the United States showing coverage by groundwater investigations as of 1965 (*From Upson, 1967.*)

areas will be changed to light shading through reconnaissance work. Some of the lightly shaded areas will be changed to moderately shaded, though some will remain moderately shaded for some time, at least until population growth or other development makes more information necessary. Probably the largest change, measured in man-years of effort, will be in shifting selected areas from light shading to moderate shading. At the same time, there will be an increase in the number of intensive investigations (heaviest shading) for special problem areas. Although these may not require so much manpower, they may take the most money because of the cost of special testing.

1.5 Educational Facilities

The great and continued expansion in the utilization of groundwater has resulted in an ever-increasing need for professional people to enter the field of groundwater resource evaluation. Prominent scientists and educators have indicated that the small number of graduates each year is inadequate to meet the current needs, much less the future demands, for groundwater geologists and engineers.

Statistical data on the educational facilities in groundwater geology and hydrology in the United States and Canada were presented by Walton (1964). Courses in groundwater are offered in both engineering and geology departments and the instructional emphasis differs from school to school and from department to department. In 1964, of the 67 departments offering or planning to offer course work in groundwater geology or hydrology, 46 geology and 15 engineering departments indicated that they offered formal course work in this area, and 4 geology, 1 soil and water science, and 1 earth science departments planned to offer this course work in the near future. In most departments that do not offer formal course work in groundwater geology or hydrology, these subjects are included as a significant part of the course work in allied subjects.

Among the institutions offering curricula with broad coverage in groundwater geology and hydrology in 1964 were: Agricultural and Mechanical College of Texas, University of Arizona, University of California, Colorado State University, Columbia University, University of Idaho, Indiana University, University of Illinois, University of Nevada, New Mexico Institute of Mining and Technology, Pennsylvania State University, Purdue University, University of Saskatchewan, University of Southern California, Stanford University, University of Utah, and Virginia Polytechnic Institute. Courses offered at most of these institutions, selected to show the range in curricula approaches, are listed in Table 1.1. The program at the University of Arizona was described in detail by Harshbarger and Ferris (1963).

Graduate degrees (M.S. or Ph.D.) with a major in groundwater geology or hydrology were granted at 5 engineering and 16 geology departments in 1964. A full list of the institutions that provided formal course work in groundwater geology or hydrology in 1964 is given in Table 1.2.

TABLE 1.1 COURSE WORK IN GROUNDWATER GEOLOGY AND HYDROLOGY AT SELECTED INSTITUTIONS IN UNITED STATES IN 1964

Educational institution, department, and course title	*Credit hours or units*	*Instructional level*
Agricultural and Mechanical College of Texas		
Department of Geology and Geophysics		
Geology of groundwater	3	Graduate
Advanced groundwater geology	3	Graduate
University of Arizona		
Department of Geology and Hydrology		
Geology of groundwater	3	Undergraduate, graduate
Quantitative determination of aquifer performance	3	Graduate
Appraisal and development of groundwater supplies	3	Graduate
Land mass hydrology	3	Graduate
Analog model analysis of hydrologic systems	3	Graduate
University of California		
Department of Civil Engineering		
Hydrology	2	Undergraduate
Groundwater hydrology	3	Graduate
Department of Irrigation		
Water supply	3	Undergraduate
Advanced groundwater problems	2	Graduate
Flow in porous media	2	Graduate
Physics of soil water	2	Graduate
Colorado State University		
Departments of Civil Engineering and Geology		
Groundwater development	3	Graduate
Groundwater hydraulics	3	Graduate
Fluid mechanics of porous media	3	Graduate
Hydrogeology	2	Undergraduate, graduate
Groundwater geology	4	Undergraduate, graduate
University of Idaho		
Departments of Geology and Geography		
Fundamentals of groundwater	2	Undergraduate
Groundwater systems	3	Graduate
University of Illinois		
Departments of Geology and Civil Engineering		
Groundwater geology	1	Graduate
Groundwater geology seminar	1	Graduate
Groundwater	1	Graduate

TABLE 1.1 COURSE WORK IN GROUNDWATER GEOLOGY AND HYDROLOGY AT SELECTED INSTITUTIONS IN UNITED STATES IN 1964 (*Continued*)

Educational institution, department, and course title	*Credit hours or units*	*Instructional level*
Indiana University		
Department of Geology		
Principles of hydrology	3	Undergraduate, graduate
Groundwater geology	3	Undergraduate, graduate
Seminar in hydrogeology	1–3	Graduate
University of Nevada		
Department of Geology		
Groundwater	3	Undergraduate, graduate
Hydrogeology 1	3	Graduate
Hydrogeology 2	3	Graduate
New Mexico Institute of Mining and Technology		
Section of Groundwater Hydrology, Graduate School		
Theory of groundwater motion	3	Graduate
Applied groundwater hydrology	3	Graduate
Prospecting for groundwater	3	Graduate
Groundwater laboratory	2	Graduate
Principles of hydrochemistry	2	Graduate
Pennsylvania State University		
Departments of Geology and Geophysics		
Introduction of hydrogeology	3	Undergraduate, graduate
Groundwater hydrology	3	Graduate
Advanced seminar in hydrogeology	3	Graduate
Flow of fluids through porous media	3	Graduate
Purdue University		
School of Civil Engineering		
Groundwater and seepage	3	Graduate
Transient flow of groundwater	3	Graduate
University of Saskatchewan		
Department of Geological Sciences		
Sedimentary aquifers		Undergraduate, graduate
Groundwater geology		Undergraduate, graduate
Groundwater hydrology		Graduate
University of Southern California		
Departments of Civil Engineering and Geology		
Groundwater	3	Graduate
Groundwater hydrology	3	Graduate

TABLE 1.1 COURSE WORK IN GROUNDWATER GEOLOGY AND HYDROLOGY AT SELECTED INSTITUTIONS IN UNITED STATES IN 1964 (*Continued*)

Educational institution, department, and course title	*Credit hours or units*	*Instructional level*
Stanford University		
Departments of Geology and Civil Engineering		
Seminar	2	Graduate
Hydrogeology	5	Undergraduate, graduate
Development of groundwater resources	3	Graduate
Physics of underground fluids	5	Undergraduate, graduate
Flow in permeable media	2	Graduate
University of Utah		
Department of Geology		
Groundwater geology	4	Undergraduate
Advanced groundwater geology	3	Graduate
Groundwater hydrology	3	Graduate
Advanced groundwater hydrology	3	Graduate
Virginia Polytechnic Institute		
Civil Engineering Department		
Groundwater and seepage	3	Graduate
Groundwater hydraulics	3	Graduate

Institutions offer a variety of courses in groundwater study. The titles of the courses most commonly offered are: groundwater geology, geology of groundwater, hydrogeology, groundwater hydrology, groundwater flow and seepage, geology of water resources, and flow in porous media. The titles of courses in which groundwater geology or hydrology is most likely to be included as a significant part are: engineering geology, economic geology, engineering hydrology, water supply and treatment, physical geology, advanced hydrology, irrigation principles and practice, public water supply, and water resources and mineral resources.

Case histories and areal studies are used as teaching aids in many of the institutions offering courses in groundwater. In most instances, the instructor uses case histories and areal studies based on both published material and personal field experience. The textbooks most widely used in formal courses in groundwater geology and hydrology are listed below:

Tolman, C. F., 1937. "Ground Water," McGraw-Hill Book Company, New York.
Meinzer, O. E., 1942. "Hydrology," McGraw-Hill Book Company, New York.
Jacob, C. E., 1947. Flow of Groundwater, in "Engineering Hydraulics," chap. 5, John Wiley & Sons, Inc., New York.
Thomas, H. E., 1951. "The Conservation of Ground Water," McGraw-Hill Book Company, New York.
Butler, S. S., 1957. "Engineering Hydrology," Prentice-Hall, Inc., Englewood Cliffs, N.J.

TABLE 1.2 COLLEGES AND UNIVERSITIES OFFERING FORMAL COURSES IN GROUNDWATER GEOLOGY OR HYDROLOGY IN UNITED STATES IN 1964

Geology departments

University of Alabama	Montana State College
University of Arizona (M)[a]	Montana State University
Arizona State University	University of Nevada (M)
Boston College	University of North Dakota
Boston University	Ohio State University
Brown University	Ohio University
University of Colorado	University of Oklahoma
Colorado State University (M)	Pennsylvania State University (M)
Columbia University (M)	Rensselaer Polytechnic Institute
University of Florida	San Jose State College
Florida State University	University of Saskatchewan
University of Hawaii	South Dakota School of Mines and Technology (M)
University of Idaho	University of Southern California (M)
University of Illinois (M)	Stanford University (M)
Indiana University (M)	University of Texas (M)
University of Iowa	Agricultural and Mechanical College of Texas
University of Kansas (M)	University of Utah (M)
Kansas State University (M)	University of Virginia
Kent State University	Washington State University
Lamar State College of Technology	Western Reserve University
University of Michigan	University of Wisconsin (M)
Michigan State University	University of Wyoming
University of Missouri	
Missouri School of Mines (M)	

Engineering departments

University of California, Berkeley (M)	Northwestern University
University of California, Davis	Princeton University (M)
California Institute of Technology (M)	Purdue University (M)
Colorado State University	Rensselaer Polytechnic Institute
University of Illinois	Stanford University
University of Nebraska	University of Southern California
New Mexico Institute of Mining and Technology (M)	Virginia Polytechnic Institute
North Carolina State College of Agriculture and Engineering	

[a] (M) indicates those institutions offering a major or thesis in groundwater geology or hydrology.

Todd, D. K., 1959. "Ground Water Hydrology," John Wiley & Sons, Inc., New York.
Harr, M. E., 1962. "Ground Water and Seepage," McGraw-Hill Book Company, New York.
Hantush, M. S., 1964. Hydraulics of Wells, in "Advances in Hydroscience," Academic Press Inc., New York.
De Wiest, R. J. M., 1965. "Geohydrology," John Wiley & Sons, Inc., New York.
Davis, S. N., and R. J. M. De Wiest, 1966. "Hydrogeology," John Wiley & Sons, Inc., New York.

These are supplemented by many texts, technical papers, and governmental publications that include: "Wisler & Brater's Hydrology"; "Applied Hydrology," by Linsley et al.; journals of the American Geophysical Union (AGU), American Society of Civil Engineers (ASCE), American Water Works Association (AWWA), and National Water Well Association (NWWA); U.S. Geological Survey bulletins, circulars, and Water Supply Papers; and state publications.

A total of 78 instructors reported that in 1964 they either taught formal course work in groundwater or included groundwater geology or hydrology as a significant part of course work in allied subjects. Of a total of 51 instructors in geology departments, 50 taught a formal course; of a total of 27 instructors in engineering departments, 17 taught a formal course. The professional training in groundwater of 57 percent of the instructors includes only field experience, whereas 26 percent of instructors lack both formal education and field experience. About 4 percent of the instructors reported academic instruction in groundwater as their only professional training and 13 percent reported both field experience and academic instruction as their professional training. Experience of instructors was gained through employment by governmental organizations, educational institutions, and private firms. The U.S. Geological Survey employed about 17 percent of the instructors; the percentages of instructors reporting employment with other governmental agencies, private consulting firms, and institutions were 22, 27, and 34, respectively. Papers in groundwater geology and hydrology have been published by about 50 percent of the instructors.

Many institutions providing course work in groundwater offer students part-time employment in groundwater geology or hydrology. The organizations and firms providing employment are: U.S. Geological Survey, other governmental agencies, and private consulting firms. Student employment is mostly by governmental agencies.

The growing involvement of many skills in the business of developing and managing groundwater resources creates some special requirements for students. The student must, first of all, be well grounded in the basis of groundwater geology and hydrology. To this should be added some knowledge of the related disciplines which must be brought to bear on complicated groundwater resource problems. Top positions go to those who not only possess depth in their chosen functional field but who also are able and willing to look at and appreciate the overall picture. Because he will be working with others in many disciplines, the successful career man is one who can communicate, both orally and in writing. This is often found to be a serious deficiency in the ranks of those who enter the field of groundwater resources. He must be able to develop a concise, complete report and recommendation in summarizing a study or operation. He must be able to tell representatives of other agencies what his organization is doing and why. At the same time, he is required to bring back reliable information regarding the activities of others. He must be able to appreciate an opposing viewpoint and to adjust or compromise when it is clearly desirable. Finally, there is an urgent need for people with initiative, imagination, and dedication. It is not too difficult to employ individuals who can extrapolate from the past. And their skills are needed. However, equally important are those who can create, improvise, and think in bold new terms.

1.6 Professional Employment

Professional people who concern themselves with groundwater resource evaluation problems represent a variety of skills. All types of engineers are active in the field—agricultural, chemical, civil, electronic and electrical, hydraulic, mechanical, mining, sanitary, etc. Those qualified in the field organize and make quantitative surveys on water-yielding formations. They appraise the dependable groundwater supplies of the nation.

Basic studies by physicists, chemists, and even biochemists are needed particularly in relation to the problems of pollution and how and why pollutants move as they do. Theoretical physicists serve as consultants to research personnel in the increasingly important field known as radiohydrology. The need to express groundwater processes in nature using mathematical models provides opportunities for mathematicians and statisticians to work effectively on problems, often using computers.

In the earth sciences field, one finds skills related to geology, geomorphology geophysics, geochemistry, glaciology, soil science, meteorology, and oceanography. Geologists study the earth's crust and map the groundwater resources. Long-range water plans require accurate knowledge of the geology of areas and its relation to water supply. Geomorphologists, geophysicists, and geochemists also are necessary to study erosion and sedimentation, to explore new groundwater supplies, and to examine the relations of groundwater chemistry to the geology and other characteristics of groundwater basins.

Soil scientists are concerned with interactions of water and soil in the hydrologic cycle. This involves infiltration, seepage, and evaporation from the soil. Meteorologists chart the rainfall and consider the possibilities of increasing groundwater supply by weather modification. Oceanographers can solve such problems as salt-water intrusion in fresh-water supplies and hydraulic relations between sea water and groundwater by means of underwater mapping, photography, sediment sampling, and sonar measurements.

There is increasing concern with land-water-plant relationships. Since the hydrologic cycle includes the use of water by living vegetation, the services of agronomists, biologists, botanists, limnologists, foresters, and ecologists are required. A sizable portion of the water which falls on the earth as precipitation returns to the atmosphere by transpiration from growing plants. This process must be better understood. Studies are made of the relation between the distribution of plants and elements of the environment, such as soil moisture and texture, low flow of streams, and floods. There are investigations to be conducted on the effects of the biota upon the quality of groundwater, past trends in climate, floodplain deposition and erosion, and pollen deposits.

Of recent origin is the realization that the skills of the social and behavioral sciences are necessary to determine how and why individual citizens, and public and private institutions, perceive and react to groundwater problems. The ways in which people and institutions regard groundwater resources—in the values assigned to groundwater, the uses of groundwater, their recognition of and attitudes toward solutions offered—will be the final factors determining the manner in which groundwater resources are conserved and developed. Plans for river-basin development, or for the best use of groundwater in industry, require informed appraisals and projections of population, income, and industrial and urban development.

Understanding of the economic implications of various alternatives of groundwater development is far from adequate. More effective administrative and political institutions for groundwater planning, development, and management are required. Economists, political scientists, sociologists, geographers, ecologists, anthropologists, and historians all have a part to play. They are involved in the growing public interest in recreation, quality of environment, and other social objectives which are not easy to evaluate.

Groundwater law and regulation is an interesting area of specialization for those in the legal profession. Many schools of law have elective courses in natural resources, thus recognizing that the lawyer is essential in any formulation of imaginative legal solutions to groundwater problems. Our present groundwater laws are often inadequate, confusing, or difficult to apply. Because groundwater crosses state lines, there is often conflict between states, and between state and federal agencies, in regard to control over groundwater resources. Groundwater law must respond to scientific and social changes. Often legislation is responsive only to a particular problem with the eventual result of a hodgepodge of statutes pertaining to groundwater resources. There is great need for updating the legislation on the books, and many states have made progress in this direction.

The field of groundwater resource evaluation requires more trained administrators and executives. For the most part, these positions are open only to those employees who have been a part of the operation for some time. However, the fact that there are "openings at the top" at all levels should be recognized and appreciated. The special skill required is an ability to formulate plans and to supervise operations while taking into account all of the physical, political, and socioeconomic constraints and limitations found in the management of groundwater resources. Qualities of leadership, in addition to technical competency, are a prerequisite.

The opportunities for careers in groundwater resource evaluation are almost without limit. There is a shortage of qualified personnel. Employment opportunities include work in data collection and analysis, groundwater resource planning, drought and flood analysis, groundwater quality investigations, groundwater studies, and research and teaching, to name a few. The opportunities extend from government employment in local, state, or federal agencies to employment in industries, as research, groundwater supply, or pollution specialists. Private consulting firms in such fields as engineering, groundwater geology, irrigation development, and systems analysis offer tremendous challenges. For the people with interest in pursuing their university training beyond the bachelor's degree, teaching and research offer a rewarding career. Those who desire to work abroad will find that there are many openings with federal agencies, private firms, or international agencies. Foreign-aid programs employ numerous groundwater resource specialists. Some of the primary needs of underdeveloped countries lie in the area of groundwater supply and irrigation.

1.7 Technical Societies and Periodicals

Numerous scientific, engineering, and geological societies are concerned with groundwater in the United States. The most important ones are: Geological Society of America, American Geophysical Union, American Society of Civil Engineers, American Society of Agricultural Engineers, American Water Works Association, National Water Well Association, and American Water Resources Association.

Groundwater papers and articles are currently published in the United States mostly in the following periodicals: water resources research of the American Geophysical Union; transactions, proceedings, and journals of divisions of the American Society of Civil Engineers and the American Society of Agricultural Engineers; journals of the American Water Works Association; and a journal on groundwater published by the National Water Well Association.

References

Ahrens, T. P., 1957. Well Design Criteria, *Natl. Water Well Assoc. J.*

Aristotle, "Meteorologica," translated by E. W. Webster, in W. D. Ross (ed.), "The Oxford Translation of the Complete Works of Aristotle," 1923–1955, vol. 3, Clarendon Press, Oxford, 1923–1955.

Babbitt, H. E., and D. H. Caldwell, 1948. "The Free Surface around, and Interference between, Gravity Wells," *Univ. Ill. Bull.*, vol. 45, no. 30.

Back, W., 1961. "Techniques for Mapping Hydrochemical Facies," U.S. Geol. Surv. Profess. Paper 424-D.

Baumann, Paul, 1951. Ground-water Movement Controlled through Spreading, *Proc. Am. Soc. Civil Engrs.*, vol. 77, separate no. 86.

Bear, J., 1961. Some Experiments in Dispersion, *Geophys. Res.*, vol. 66.

——— and G. Dagan, 1963. Some Exact Solutions of Interface Problems by Means of the Hodograph Method, *J. Geophys. Res.*, vol. 69

Boulton, N. S., 1951. The Flow Pattern near a Gravity Well in a Uniform Water-bearing Medium, *J. Inst. Civil Engrs.*, vol. 10.

———, 1963. Analysis of Data from Nonequilibrium Pumping Tests Allowing for Delayed Yield from Storage, *Proc. Inst. Civil Engrs.*, vol. 26, no. 6693.

Chamberlin, T. C., 1885. "The Requisite and Qualifying Conditions of Artesian Wells," U.S. Geol. Surv. 5th Ann. Rept.

Childs, E. C., 1945. The Water Table, Equipotentials, and Streamlines in Drained Land, *Soil Sci.*, vol. 59.

———, A. H. Cole, and D. H. Edwards, 1953. The Measurement of the Hydraulic Permeability of Saturated Soil in Situ, pt. 2, *Proc. Roy. Soc.* (*London*), ser. A, vol. 216, no. 1124.

Cooper, H. H., Jr., and C. E. Jacob, 1946. A Generalized Graphical Method for Evaluating Formation Constants and Summarizing Well-field History, *Trans. Am. Geophys. Union*, vol. 27.

——— and M. I. Rorabaugh, 1963. "Ground-water Movements and Bank Storage due to Flood Stages in Surface Streams," U.S. Geol. Surv. Water Supply Paper 1536-J.

Darcy, H., 1956. "Les fontaines publiques de la ville de Dijon," Paris.

Day, P. R., and J. N. Luthin, 1954. Sand-model Experiments on the Distribution of Water-pressure under an Unlined Canal, *Soil Sci. Soc. Am. Proc.*, vol. 18, no. 2.

de Glee, G. J., 1930. "On Groundwater Currents through Draining by Means of Wells," thesis, Technische Hogeschool te Delft, Delft.

Dietz, D. N., 1944. Ervaringen met modelonderzoek in de hydrologie, *Water*, no. 3.

Dupuit, J., 1863. "Etudes theoriques et practiques sur le mouvement des eaux," 2d ed., Paris.

Engelund, F., 1951. "Mathematical Discussion of Drainage Problems," Danish Acad. Tech. Sci. Trans. Bull. 3.

———, 1953. "On the Laminar and Turbulent Flows of Groundwater through Homogeneous Sand," Danish Acad. Tech. Sci. Trans. Bull. 4.

Ferris, J. G., 1948. "Groundwater Hydraulics As a Geophysical Aid," Mich. Dept. Conserv., Geol. Surv. Div. Tech. Rept. 1.

———, 1951. Cyclic Fluctuations of Water Level As a Basis for Determining Aquifer Transmissibility, Union Geodesique et Geophysique Internationale, Association Internationale d'Hydrologie Scientifique Assemblee Generale de Bruxelles, vol. 2.

——— and A. N. Sayre, 1955. The Quantitative Approach to Groundwater Investigations, *Economic Geology*, 50th anniversary volume.

Forchheimer, P., 1886. Uber die Ergiebigkeit von Brunnenanlagen und Sickerschlitzen, *Z. Architekt. Ing.-Verein*, Hannover, vol. 32, no. 7.

———, 1930. "Hydraulik," B. G. Teubner Verlagsgesellschaft, Berlin.

Gardner, W., O. W. Israelsen, and W. W. McLaughlin, 1928. The drainage of Land Overlying Artesian Basins, *Soil Sci.*, vol. 26.

Ghyben, W. (Badon), 1889. Nota in verband met de voorgenomen putboring nabij Amsterdam, *K. Inst. Ing. Tydschr.*, the Hague.

Guyton, W. F., 1941. Applications of Coefficients of Transmissibility and Storage to Regional Problems in the Houston District, Texas, *Trans. Am. Geophys. Union*, 21st Ann. Meeting, pt. 3.

Hamel, G., 1934. Uber Grundwasserstromung, *Z. Angew. Math. Mech.*, vol. 14.

Hansen, V. E., 1952. Complicated Well Problems Solved by the Membrane Analogy, *Trans. Am. Geophys. Union*, vol. 33, no. 6.

Hantush, M. S., 1964. Hydraulics of Wells, in "Advances in Hydroscience," Academic Press Inc., New York.

Harshbarger, J. W., and J. G. Ferris, 1963. Interdisciplinary Training Program in Scientific Hydrology, *Ground Water*, vol. 1, no. 2.

Hem, J. D., 1959. "Study and Interpretation of the Chemical Characteristics of Natural Water," U.S. Geol. Surv. Water Supply Paper 1473.

Henry, H. R., 1959. Salt Intrusion into Fresh-water Aquifers, *J. Geophys. Res.*, vol. 64, no. 11.

Herodotus, "Persian Wars," translated by G. Rawlinson, in F. R. B. Godolphin (ed.), "The Greek Historians, Complete and Unabridged Historical Works of Herodotus," vol. 1, Random House, Inc., New York, 1942.

Hill, R. A., 1942. Salts in Irrigation Water, *Trans. Am. Soc. Civil Engrs.*, vol. 107.

Hopf, L., and E. Trefftz, 1921. Grundwasserstromung in einem abfallenden Gelande mit Abfanggraben, *Z. Angew. Math. Mech.*, vol. 1.

Horton, R. E., 1906. Surface Drainage of Land by Tile, *Mich. Engr.*

Houk, I. E., 1921. "Rainfall and Runoff in the Miami Valley," state of Ohio, Miami Conserv. Dist. Tech. Repts., pt. VIII.

Hubbert, M. K., 1940. The Theory of Groundwater Motion, *J. Geol.*, vol. 48, no. 8.

———, 1953. Entrapment of Petroleum under Hydrodynamic Conditions, *Am. Assoc. Petrol. Geol. Bull.*, vol. 37, no. 8.

Israelsen, O. W., and W. W. McLaughlin, 1935. "Drainage of Land Overlying an Artesian Groundwater Reservoir," Utah Agr. Exp. Sta. Bull. 259.

Jacob, C. E., 1940. On the Flow of Water in an Elastic Artesian Aquifer, *Trans. Am. Geophys. Union*, 21st Ann. Meeting.

———, 1945. "Correlation of Groundwater Levels and Precipitation on Long Island, N.Y.," New York Dept. Conserv., Water Power and Control Comm. Bull. GW-14.

———, 1946. Radial Flow in a Leaky Artesian Aquifer, *Trans. Am. Geophys. Union*, vol. 27.

———, 1947. Drawdown Test to Determine Effective Radius of Artesian Well, *Trans. Am. Soc. Civil Engrs.*, vol. 112.

———, 1950. Flow of Groundwater, in "Engineering Hydraulics," chap. 5, John Wiley & Sons, Inc., New York.

Jones, P. H., and T. B. Buford, 1951. Electric Logging Applied to Groundwater Exploration, *Geophysics*, vol. 6, no. 1.

Kano, T., 1939. Frictional Loss of Head in the Wall of a Well, *Japan. J. Astron. Geophys.*, vol. 17, no. 1.

Kashef, A. I., Y. S. Toulouklan, and R. E. Fadum, 1952. "Numerical Solutions of Steady-state and Transient Flow Problems," Purdue Univ. Exp. Sta. Bull., vol. 36, Res. Ser. 117.

Kazmann, R. G., 1948. River Infiltration As a Source of Groundwater Supply, *Trans. Am. Soc. Civil Engrs.*, vol. 113.

Keilhack, K., 1912. "Groundwasser and Quellen Kinde, " 1st ed., Berlin.

King, F. H., 1892. "Observations and Experiments on the Fluctuations in Level and the Rate of Movement of Groundwater on the Wisconsin Agricultural Experiment Station Farm and at Whitewater, Wis.," U.S. Weather Bur. Bull. 5.

Kirkham, D., 1945. Artificial Drainage of Land, Streamline Experiments, the Artesian Basin—III, *Trans. Am. Geophys. Union*, vol. 26, no. 3.

Koenig, L., 1960. Economic Aspects of Water Well Stimulation, *J. Am. Water Works Assoc.*, vol. 52, no. 5.

Kohout, F. A., 1960. Cyclic Flow of Salt Water in the Biscayne Aquifer of Southeastern Florida, *J. Geophys. Res.*, vol. 65, no. 7.

Kozeny, J., 1932. Hydrologische Grundlagen des Draversuchs, *Commun. Trans. Intern. Soc. Soil Sci.*, 6th, Groningen.

———, 1933. Theorie und Berechnung der Brunnen, *Wasserkraft Wasserwirtschaft*, vol. 28.

Langelier, W. F., and H. F. Ludwig, 1942. Graphical Methods for Indicating the Mineral Character of Natural Waters, *J. Am. Water Works Assoc.*, vol. 34.

Lewis, M. R., 1932. Flow of Groundwater As Applied to Drainage Wells, *Trans. Am. Soc. Civil Engrs.*, vol. 96.

Lindquist, E., 1933. "On the Flow of Water through Porous Soil," Repts. to the First Cong. of Large Dams, Stockholm.

Love, S. K., 1944. Cation Exchange in Groundwater Contaminated with Seawater near Miami, Fla., *Trans. Am. Geophys. Union*, 25th Ann. Meeting, pt. 6.

Lusczynski, N. J., 1961. Head and Flow of Ground Water of Variable Density, *J. Geophys. Res.*, vol. 66, no. 12.

Luthin, J. N., and V. H. Scott, 1952. Numerical Analysis of Flow through Aquifers towards Wells, *Agr. Eng.*, vol. 33, no. 5.

Meinzer, O. E., 1923. "The Occurrence of Groundwater in the United States," U.S. Geol. Surv. Water Supply Paper 489.

——— and N. D. Stearns, 1928. "A Study of Groundwater in the Pomperaug Basin, Connecticut, with Special Reference to Intake and Discharge," U.S. Geol. Surv. Water Supply Paper 597-B.

Meleschenko, N. T., 1936. Analysis of Groundwater Movement under Structures Equipped with Drainage Openings, *Izv. Nauchn. Issled. Inst. Gidrothn.*, vol. 19.

Mendenhall, W. C., 1905. "The Hydrology of San Bernardino Valley, California," U.S. Geol. Surv. Water Supply Paper 142.

Moore, A. D., 1949. Fields from Fluid Flow Mappers, *J. Appl. Phys.*, vol. 20.

Muskat, M., 1932. Potential Distributions in Large Cylindrical Disks with Partially Penetrating Electrodes, *Physics*, vol. 2.

———, 1935. Seepage of Water through Dams with Vertical faces, *Physics*, vol. 6.

———, 1937. "The Flow of Homogeneous Fluids through Porous Media," McGraw-Hill Book Company, New York.

Nace, R. L., 1965. Global Thirst and the International Hydrogical Decade, *J. Am. Water Works Assoc.*, vol. 57, no. 7.

Palissy, B., "Discours Admirables," translated by A. La Rocque, in "The Admirable Discourses of Bernard Palissy," The University of Illinois Press, Urbana, Ill.

Parizek, R. R., 1963. The Hydrologic Cycle Concept, *Mineral Ind.*, Penn. State Univ., vol. 32, no. 7.

Patten, E. P., Jr., and G. D. Bennett, 1963. "Application of Electrical and Radioactive Well Logging to Groundwater Hydrology, U.S. Geol. Surv. Water Supply Paper 1544-D.

Pavlovsky, N. N., 1922. "Teoriia dvizheniia gruntovykh vod pod gidrotekhnicheskini seeruzeniaimi," Leningrad.

Pettyjohn, W. A., 1967. Evaluation of Basic Data and a Variety of Techniques Needed in Hydrologic Systems Analysis, in "System Approach to Water Quality in the Great Lakes," *Proc. Ann. Symp. Water Resources Res. Ohio State Univ. Water Resource Center*, 3d.

Piper, A. M., 1944. A Graphic Procedure in the Geochemical Interpretation of Water-analyses, *Trans. Am. Geophys. Union*, 25th Ann. Meeting, pt. 6, discussion by R. A. Hill, W. F. Langelier, and A. M. Piper.

——— et al., 1939. "Geology and Groundwater Hydrology of the Mokelumne Area, California," U.S. Geol. Surv. Water Supply Paper 780.

——— et al., 1953. "Native and Contaminated Ground Waters in the Long Beach–Santa Ana Area, California," U.S. Geol. Surv. Water Supply Paper 1136.

Plato, "Critias" and "Timaeus," translated by R. G. Bury, in "Plato's Works in the Loeb Classical Library," vol. 7, Harvard University Press, Cambridge, Mass., 1929.

Poland, J. F., 1961. "The Coefficient of Storage in a Region of Major Subsidence Caused by Compaction of an Aquifer System," U.S. Geol. Surv. Profess. Paper 424-B.

———, A. A. Garrett, and A. Sinnotte, 1948. "Geology, Hydrology, and Chemical Character of the Ground Waters in the Torrance–Santa Monica Area, Los Angeles County, California," U.S. Geol. Surv. mimeographed rept.

Prickett, T. A., 1965. Type-curve Solution to Aquifer Tests under Water-table Conditions, *Ground Water*, vol. 3, no. 3.

Rasmussen, W. C., and G. E. Andreasen, 1959. "Hydrologic Budget of the Beaverdam Creek Basin, Maryland," U.S. Geol. Surv. Water Supply Paper 1472.

Renick, B. C., 1925. "Base Exchange in Groundwater by Silicates As Illustrated in Montana," U.S. Geol. Surv. Water Supply Paper 520-D.

Rorabaugh, M. I., 1951. Stream-bed Percolation in Development of Water Supplies, Union Geodesique et Geophysique Internationale, Association Internationale d'Hydrologie Scientifique, Assemblee Generale de Bruxelles, vol. II.

———, 1953. Graphical and Theoretical Analysis of Step Drawdown Test of Artesian Well, *Proc. Am. Soc. Civil Engrs.*, vol. 79.

Schicht, R. J., and W. C. Walton, 1961. "Hydrologic Budgets for Three Small Watersheds in Illinois," Illinois State Water Surv., Rept. of Invest. 40.

Shaw, F. S., and R. V. Southwell, 1941. Relaxation Methods Applied to Engineering Problems, pt. VII, Problems Relating to the Percolation of Fluids through Porous Materials, *Proc. Roy. Soc. (London)*, ser. A, vol. 178.

Skibitzke, H. E., 1961. "Electronic Computers As an Aid to the Analysis of Hydrologic Problems," Intern. Assoc. of Sci. Hydrology Publ. 52.

Slichter, C. S., 1898. "Theoretical Investigations of the Motion of Groundwaters," U.S. Geol. Surv. 19th Ann. Rept., pt. 2.

Smith, G. E. P., 1924. The Effect of Transpiration of Trees on the Groundwater Supply (abstract), *J. Wash. Acad. Sci.*, vol. 14.

Sniegocki, R. T., 1963. "Problems in Artificial Recharge through Wells in the Grand Prairie Region Arkansas," U.S. Geol. Surv. Water Supply Paper 1615-E.

Stallman, R. W., 1955. Numerical Analysis of Regional Water Levels to Define Aquifer Hydrology (abstract), *Trans. Am. Geophys. Union*, vol. 36, no. 3.

Steggewentz, J. H., 1933. "De invloed van de getijbeweging van zeeen en getijrivieren op de stijghoogte van grondwater," thesis, Technische Hogeschool te Delft, Delft.

Stuart, W. T., E. A. Brown, and E. C. Rhodehamel, 1954. "Ground Water Investigations of the Marquette Iron-mining District," Mich. Dept. Conserv., Geol. Surv. Div. Tech. Rept. 3.

Theis, C. V., 1935. The Relation between the Lowering of the Piezometric Surface and the Rate and Duration of Discharge of a Well Using Ground-water Storage, *Trans. Am. Geophys. Union*, 16th Ann. Meeting, pt. 2.

———, 1937. Amount of Ground-water Recharge in the Southern High Plains, *Trans. Am. Geophys. Union*, 18th Ann. Meeting, pt. 2.

———, 1941. The Effect of a Well on the Flow of a Nearby Stream, *Trans. Am. Geophys. Union*, 22nd Ann. Meeting, pt. 3.

Thiem, A., 1870. Die Ergiebigkeit artesischer Bohrlocher, Schachtbrunnen, und Filtergallerien, *J. Gasbeleuchtung Wasserversorgung* (*Munich*), vol. 14.

Thiem, G., 1906. "Hydrologische Methoden," J. M. Gephardt, Leipzig.

Tolman, C. F., and J. F. Poland, 1940. Ground-water, Salt-water Infiltration and Ground-surface Recession in Santa Clara Valley, Santa Clara County, Calif., *Trans. Am. Geophys. Union*, 21st Ann. Meeting, pt. 1.

Toth, J. A., 1963. Theoretical Analysis of Groundwater Flow in Small Drainage Basins, *J. Geophys. Res.*, vol. 68.

Turneaure, F. E., and H. L. Russell, 1901. "Public Water Supplies," 1st ed., John Wiley & Sons, Inc., New York.

Upson, J. E., 1967. Plans of the U.S. Geological Survey, Water Resources Division for Research Investigations, and Data Collection in Groundwater, *Ground Water*, vol. 5, no. 2.

Vaidhianathan, V. I., H. R. Luthra, and N. K. Bose, 1934. A Hydrodynamical Investigation of the Subsoil Flow from Canal Beds by means of Models, *Proc. Indian Acad. Sci.*, sec. A, vol. 1.

Verdernikov, V. V., 1934. Versickerung aus Kanalen, *Wasserkraft Wasserwirtsch.*, vols. 11–13.

Vreedenburgh, C. G. J., and O. M. Stevens, 1936. Electric Investigation of Underground Water Flow Nets, *Proc. Intern. Conf. Soil Mech. Found. Eng.*, vol. 1.

Walton, W. C., 1963. Estimating the Infiltration Rate of a Streambed by Aquifer-test Analysis, Intern. Assoc. Sci. Hydrology, Gen. Assembly Berkeley.

———, 1964. Education Facilities in Groundwater Geology and Hydrology in the United States and Canada, 1963, *Ground Water*, vol. 2, no. 3.

——— and J. C. Neill, 1961. "Analyzing Groundwater Problems with Mathematical Models and a Digital Computer," Intern. Assoc. of Sci. Hydrology Publ. 52.

——— and ———, 1963. Statistical Analysis of Specific-capacity Data for a Dolomite Aquifer, *J. Geophys. Res.*, vol. 68.

——— and T. A. Prickett, 1963. Hydrogeologic Electric Analog Computers, *Hydrology Div., Am. Soc. Civil Engrs.*, vol. 89, no. HY6.

———, D. L. Hills, and G. M. Grundeen, 1967. "Recharge from Induced Streambed Infiltration under Varying Groundwater-level and Stream-stage Conditions," Minn. Water Resources Res. Center Bull. 6.

Wenzel, L. K., 1936. "The Thiem Method for Determining Permeability of Water-bearing Materials and Its Application to the Determination of Specified Yield, Results of Investigations in the Plattee River Valley, Nebr.," U.S. Geol. Surv. Water Supply Paper 679-A.

——— and A. L. Greenlee, 1943. A Method for Determining Transmissibility and Storage Coefficients by Tests of Multiple Well-systems, *Trans. Am. Geophys. Union*, 24th Ann. Meeting, pt. 2.

Werner, P. W., and D. Noren, 1951. Progressive Waves in Non-artesian Aquifers, *Trans. Am. Geophys. Union*, vol. 32, no. 2.

White, W. N., 1932. "A Method of Estimating Groundwater Supplies Based on Discharge by Plants and Evaporation from Soil, Results of Investigations in Escalante Valley, Utah," U.S. Geol. Surv. Water Supply Paper 659-A.

Williams, C. C., and S. Lohman, 1949. "Geology and Ground-water Resources of a Part of Southern-central Kansas, with Special Reference to the Wichita Municipal Water Supply," Kansas Geol. Surv. Bull. 79.

Wyckoff, R. D., H. G., Botset, and M. Muskat, 1932. Flow of Liquids through Porous Media under the Action of Gravity, *Physics*, vol. 3.

———, ———, and ———, 1933. The Mechanics of Porous Flow Applied to Water Flooding Problems, *Trans. Soc. Petrol. Engrs. AIME*, vol. 103.

Yang, St. T., 1949. "Seepage toward a Well Analyzed by the Relaxation Method," doctoral thesis, Harvard Univ.

Zanger, C. N., 1953. "Theory and Problems of Water Percolation," U.S. Bur. Recl. Eng. Mon. 8.

CHAPTER 2

GROUNDWATER MOVEMENT, STORAGE, EXPLORATION, AND DATA

Most rocks contain numerous open spaces, called *interstices*, in which water may be stored and through which water can move. Water that exists in the interstices of rocks is called *subsurface water* (Meinzer, 1959); that part of subsurface water in interstices completely saturated with water is called *groundwater*. Subsurface water in interstices above the zone of saturation in the zone of aeration where interstices are only partially saturated with water is called *vadose water*. The zone of aeration is subdivided into the soil water zone, the intermediate zone, and the capillary zone. The *soil water zone* consists of soil and other materials near the surface which discharge water into the atmosphere by evapotranspiration. The *capillary zone* extends immediately above the zone of saturation to the limit of capillary rise of water. The *intermediate zone* lies between the soil water and capillary zones (*capillary fringe*). This book is primarily concerned with groundwater.

An *aquifer* is a saturated bed, formation, or group of formations which yields water in sufficient quantity to be of consequence as a source of supply. An *aquitard* is a saturated bed, formation, or group of formations which yields inappreciable quantities of water to drains, wells, springs, and seeps compared to an aquifer but through which appreciable leakage of water is possible. An *aquiclude* is a saturated bed, formation, or group of formations which yields inappreciable quantities of water to drains, wells, springs, and seeps and through which there is inappreciable movement of water. A formation may be classified as an aquifer in one area but only as an aquitard or aquiclude in a different area depending upon the availability of groundwater. An aquifer serves as a transmission conduit and storage reservoir. It transports water from recharge areas to surface bodies of water, wetlands, springs, areas of evapotranspiration and wells and other water-collecting devices. As a storage reservoir, an aquifer provides reserve water for use during periods when withdrawals exceed recharge. The quantities of water available in storage in some productive aquifers are so great that in places large withdrawals over a period of years fail to produce marked evidence of depletion.

An *unconfined aquifer* is one in which groundwater possesses a free surface open to the atmosphere. The upper surface of the zone of saturation is called the *water table*. Changes in the stage of the water table correspond to changes in the thickness of the zone of saturation; when the water table declines, gravity drainage of interstices occurs. In most places there is only one water table, but in some localities because of the presence of aquitards or aquicludes there may be perched aquifers with additional water tables.

An *artesian aquifer* is one in which groundwater is confined under pressure by overlying and underlying aquitards or aquicludes and water levels in wells rise above the top of the aquifer. Artesian aquifers are classified as leaky or nonleaky depending upon whether groundwater in the aquifer is confined by aquitards or aquicludes. Water levels in wells tapping artesian aquifers sometimes rise above the surface to cause wells to flow. Rises and falls of the water levels in wells tapping artesian aquifers correspond to changes in the pressure of the water. The imaginary surface to which water rises in wells tapping artesian aquifers is called the *piezometric surface*. When water is released from storage by the compaction of the aquifer and its associated beds and by expansion of the water itself, while the interstices remain saturated, the piezometric surface is lowered. An artesian aquifer becomes an unconfined aquifer when the piezometric surface declines below the top of the aquifer. The piezometric surface may rise to stages above or below water tables depending upon whether the vertical movement of water is from or into the aquifer. Examples of unconfined and artesian aquifers are shown in Fig. 2.1.

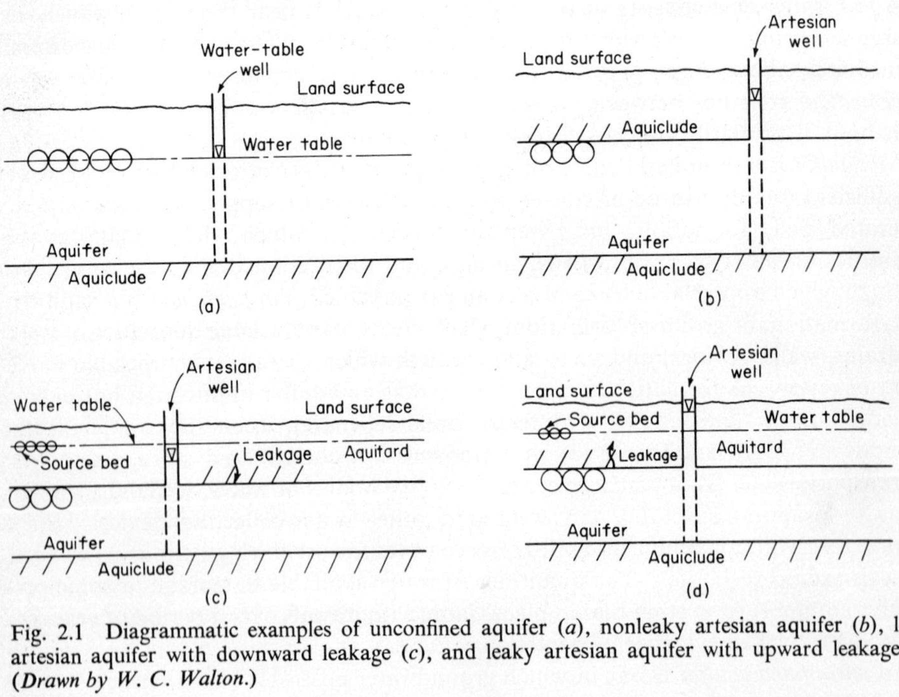

Fig. 2.1 Diagrammatic examples of unconfined aquifer (*a*), nonleaky artesian aquifer (*b*), leaky artesian aquifer with downward leakage (*c*), and leaky artesian aquifer with upward leakage (*d*). (*Drawn by W. C. Walton*.)

2.1 Interstices

Differences in the number, size, shape, interconnection, and arrangement of the interstices of aquifers, aquitards, and aquicludes result from the great diversity of geologic processes by which rocks were produced and later modified. Most interstices are small and interconnected, some are cavernous in size, and others are small and largely isolated so that there is little opportunity for movement of water from one interstice to another. The nature of interstices is determined by the geologic framework of rocks, and an orderly description of the geology and geologic history of an area is essential for an understanding of the movement and storage of groundwater.

The interstices of rocks may be divided into two groups, original and secondary (Meinzer, 1959). *Original interstices* came into existence when the rocks were formed and can be subdivided into those of sedimentary origin and those of igneous origin. *Secondary interstices* are the result of processes by which rocks were modified after coming into existence and largely comprise joints and other fracture and solution openings. Original interstices consist of the spaces between adjacent fragments of sedimentary rock and small cavities or inclusions, within crystals and small intercrystal spaces developed in igneous rocks during their congealing. Most consolidated rocks are broken by joints cutting the rocks in various directions and extending to varying distances and depths. These secondary interstices, produced chiefly by shrinkage, pressure, and deformation of rocks, commonly vary in number and size. Joints often intersect one another, frequently have no regularity of spacing, and tend to become tighter and spaced farther apart with depth. Secondary interstices are also produced by the chemical decomposition of rocks and the solution and subsequent removal of the soluble products, or by the solution and removal of soluble rocks. The removal of the calcareous cement from the original interstices of a sandstone or the removal of soluble material such as limestone results in abundant secondary openings in many areas. In some consolidated sedimentary rocks, the original interstitial spaces are likely to decrease with depth; most deep wells have encountered few interstices below a depth of a mile. In many igneous rocks, most interstices are found within a few hundred feet of the surface. Several types of interstices are shown in Fig. 2.2.

The shape of interstices and the degree of their interconnection are extremely complex. In order to lay the foundation for a mental picture of interstitial space, the interstitial fabric of the simple geometry of aggregates of well-rounded fragments of rock of the same diameter will be described. No rocks consist exclusively of spheres of one size, but some well-sorted sand deposits probably develop interstitial patterns similar to those developed by packing of perfect spheres. The theoretical treatment of such aggregates is worthwhile if the profound modifications of interstices caused by irregularity in shape and degree of assortment of grains of rocks are kept in mind.

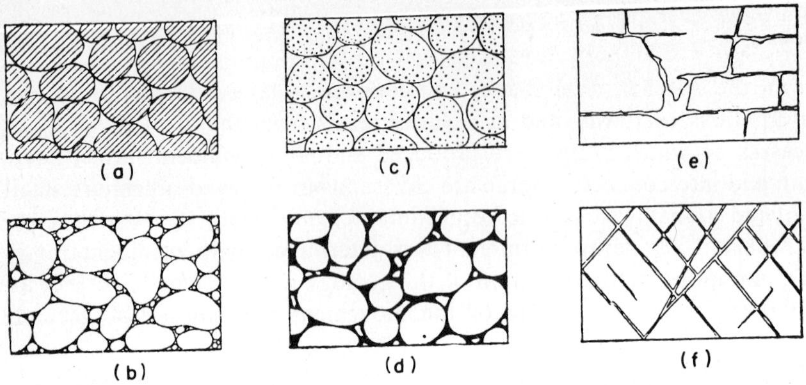

Fig. 2.2 Several types of interstices and the relation of rock texture to porosity. (*a*) Well-sorted sedimentary deposit having high porosity; (*b*) poorly sorted sedimentary deposit having low porosity; (*c*) well-sorted sedimentary deposits consisting of fragments of rock that are themselves porous, so that the deposit has a very high porosity; (*d*) well-sorted sedimentary deposit whose porosity has been diminished by the deposition of mineral matter in interstices; (*e*) rock rendered porous by solution; and (*f*) rock rendered porous by fracturing. (*From Meinzer, 1959.*)

It has been shown by Graton and Fraser (1935) that the least compact arrangement of spheres is that of a cubical array of spheres and the most compact arrangement is that of a rhombohedral array. The porosity of the cubical array is 47.6 percent and the porosity of the rhombohedral array is 26.0 percent. Figure 2.3 shows the packing of spheres in the two arrays and the resulting interstitial volumes.

The most stable array of spheres is rhombohedral. It might be expected that under moderate agitation spheres would assume the most stable array and hence the assemblage of least porosity. However, under natural conditions the required perfection in packing is confined to a limited number of grains, and beyond this point random packing results. In natural assemblages agitated to induce compact packing, groups of spheres packed in orderly arrays are separated by boundaries in which random packing occurs and where the porosity is even higher than that for cubic packing. These zones are maintained because of the bridging of groups of particles under pressures less than the crushing strength of the particles.

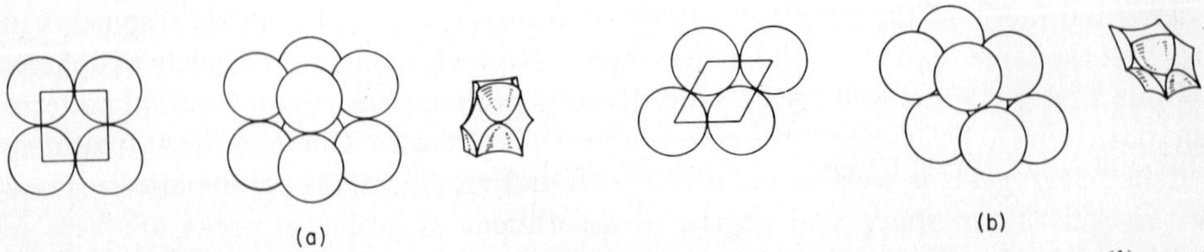

Fig. 2.3 Packing of spheres and interstitial volumes. (*a*) Least compact arrangement; (*b*) most compact arrangement. (*Drawn by W. C. Walton.*)

2.2 Porosity

The *porosity* of a rock is a measure of the interstitial space of the rock and is expressed quantitatively as the percentage of the total volume of rock occupied by interstices (Meinzer, 1959). Figure 2.2 shows several types of interstices and their relation to rock texture. Representative porosity ranges for selected rocks are given in Table 2.1. In general, a porosity greater than 20 percent is considered large, a porosity between 5 and 20 percent is considered medium, and a porosity less than 5 percent is considered small. The highest porosity known is 80 to 95 percent, which has been reported for freshly deposited alluvium of the Mississippi delta.

TABLE 2.1 REPRESENTATIVE POROSITY RANGES FOR SELECTED ROCKS

Rocks	*Porosity*, %
Clay	45–55
Sand	35–40
Gravel	30–40
Sand and gravel	20–35
Sandstone	10–20
Shale	1–10
Limestone	1–10

The porosity of a sedimentary deposit depends chiefly on the shape and arrangement of its constituent particles, the degree of assortment of its particles, the cementation and compacting to which it has been subjected since its deposition, the removal of mineral matter through solution by percolating waters, and the fracturing of the rock, resulting in joints and other interstices. In well-sorted and well-rounded deposits, the size of grains has no influence on porosity; thus, a deposit of boulders may have the same porosity as a deposit of clay. The porosity of many deposits is increased by the irregular angular shapes of its constituent grains. Porosity decreases with increases in the variety of size of grains; small grains fill interstices between large grains. Meinzer (1959) outlined several methods for determining porosity.

2.3 Specific Yield

Not all the water in the interstices of a saturated rock can be withdrawn through wells, drains, springs, or seeps. A part of the water is retained in interstices largely by the forces of molecular attraction, adhesion, and cohesion. The attraction between the walls of interstices and the adjacent molecules of water is called *adhesion;* the attraction between adjacent molecules of water is called *cohesion*. The walls of interstices retain a film of water by adhesion and some water is retained in detached

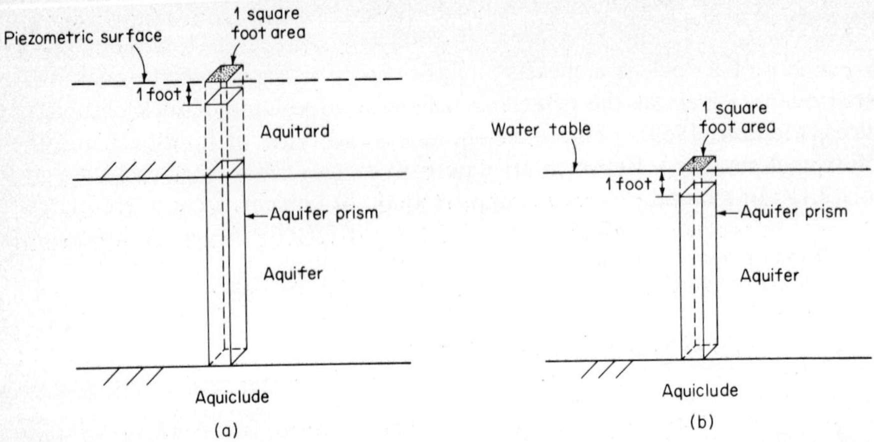

Fig. 2.4 Diagrammatic representation of coefficient of storage. (*a*) Artesian condition; (*b*) water-table condition. (*After Ferris et al., 1962.*)

interstices. The amount of water retained varies directly as the aggregate surface of the interstices and indirectly as the size of the interstices; thus retention is greatest in rocks having small interstices. The amount of water retained in interstices also depends on the time of drainage; on the temperature and mineral composition of groundwater which affect its surface tension, viscosity, and specific gravity; and on various physical relations of the rock.

The *specific yield* of a rock is a measure of the water-yielding capacity of the rock and is expressed quantitatively as the percentage of the total volume of rock occupied by the ultimate volume of water released from or added to storage in a water-table aquifer per unit (horizontal) area of aquifer and per unit decline or rise of the water table (see Fig. 2.4*b*). The *specific retention* of a rock is a measure of the water-retaining capacity of the rock and is expressed quantitatively as the percentage of the total volume of rock occupied by groundwater that will be retained in interstices against the force of gravity. Representative specific-yield ranges for selected rocks

TABLE 2.2 REPRESENTATIVE SPECIFIC YIELD RANGES FOR SELECTED ROCKS

Rocks	*Specific yield,* %
Clay	1–10
Sand	10–30
Gravel	15–30
Sand and gravel	15–25
Sandstone	5–15
Shale	0.5–5
Limestone	0.5–5

are given in Table 2.2. The porosity of a saturated rock is equal to the sum of specific yield and the specific retention. Meinzer (1959) outlined several methods for determining specific yield.

For most rocks, gravity drainage of interstices is not instantaneous and the water-yielding capacity increases at a diminishing rate as the time of drainage increases, gradually approaching the specific yield. *Gravity yield* has been defined as the percentage of the total volume of rock occupied by groundwater that will drain under the action of gravity during a given period of drainage (see Rasmussen and Andreasen, 1959).

2.4 Coefficients of Storage, Permeability, and Transmissibility

The *coefficient of storage* S of an aquifer has been defined as the volume of water the aquifer releases from or takes into storage per unit surface area of the aquifer per unit decline or rise of head. In Fig. 2.4, the volume of water released from storage in the aquifer prism divided by the product of the prism's cross-sectional area and the change in head results in a dimensionless number which is the coefficient of storage. Under water-table conditions, the coefficient of storage is equal to the specific yield, provided gravity drainage is complete. The coefficient of storage of water-table aquifers ranges from about 0.02 to 0.30. Although rigid limits cannot be established, the storage coefficients of artesian aquifers may range from about 0.00001 to 0.001.

Permeability is a measure of the ease of movement of groundwater through aquifers and aquitards. The *field coefficient of permeability* of an aquifer, P, has been defined as the rate of flow of water, in gallons per day, through a cross-sectional area of 1 square foot of the aquifer (see opening A, Fig. 2.5) under a hydraulic gradient of 1 foot per foot at the prevailing temperature of the water. A related term, the *coefficient of transmissibility* T, indicates the capacity of an aquifer to transmit water through its entire thickness and is equal to the coefficient of permeability multiplied by the saturated thickness of the aquifer, m, in feet. The coefficient of transmissibility is defined as the rate of flow of water, in gallons per day (gpd), through a vertical strip of the aquifer (see opening B, Fig. 2.5) 1 foot wide and extending the full saturated thickness under a hydraulic gradient of 1 foot per foot at the prevailing temperature of the water. The rate of vertical leakage of groundwater through an aquitard is dependent upon the permeability of the aquitard. The *aquitard field coefficient of permeability* P' is defined as the rate of vertical flow of water, in gpd, through a horizontal cross-sectional area of 1 square foot of the aquitard (see opening C, Fig. 2.5) under a vertical hydraulic gradient of 1 foot per foot at the prevailing temperature of the water. Some authors use the terms *hydraulic conductivity* ($K = k\gamma/\mu$), *permeability* (k), *transmissivity* ($T = KD$), and *storativity* (ζ).

Laboratory coefficients of permeability P_L and P'_L may be computed by adjusting the field coefficients to equivalent values for the standard temperature of 60°F. The

In basins bordered by mountain ranges occur tremendous volumes of unconsolidated deposits including sand and gravel. The sand and gravel derived by erosion of mountains underlying intermontane valleys yield more than half of all the water pumped from wells in the conterminous United States. There is some recharge directly from precipitation, but generally most recharge is by seepage from streams into the alluvial fans lying between the base of mountains and the basin floors. Subsurface flow through the walls of the mountains into intermontane valley deposits is also appreciable. The storage of these aquifers is very large in comparison to annual recharge in the arid regions.

About 30 percent of the land surface of the earth has been covered by glacial ice during the past million years. At present, 10 percent of the land surface is covered. Glacial till covers several million square miles of the earth's surface, yet relatively few wells draw their supplies directly from the till. Of the wells drawing water from till, most of them probably obtain their water from joints or small sand lenses within the till. A number of wells in mountain valleys probably draw water from coarse-grained till.

Eolian deposits (formed by the wind) are far less abundant than either stream or glacial deposits. The actual volume of eolian material encountered in subsurface exploration is hard to evaluate because distinctive textural and structural features are seldom recognized in drilling cuttings. Almost certainly many silts and sands assumed to be alluvial or lacustrine are actually eolian. Eolian deposits can be divided into two types, loess and dune sand. Loess is not commonly an aquifer because of its low permeability and because where its permeability is the highest it is usually in high topographic positions where subsurface drainage is efficient. If loess overlies impermeable soils or consolidated rocks, stock or farm wells can be developed that will give good service except during long periods of dry weather. Aquifers of dune sand are not widely utilized because wells that prevent entrance of the loess sand are difficult to construct by standard practices. In some places active dune areas are favorable for water development because of high recharge rates, good water quality, and moderately high permeabilities.

Limestones and dolomites, constituting between 5 and 10 percent of all sedimentary rocks in the conterminous United States, vary widely in permeability. The most important aquifers are those in which secondary solution openings occur. In regions where limestone forms the land surface and extends to considerable depth, groundwater dissolves the rock and waters are hard because of dissolved mineral matter. The ultimate development of limestone terraces is a karst region with such features as sinkholes, lost rivers, large springs, and subsurface drainage. Solution openings are irregularly distributed both horizontally and vertically. Some wells may penetrate deep into limestone without yielding appreciable quantities of water; nearby wells may have large yields.

Gypsum is sufficiently soluble to develop a high permeability. Water which is hard and high in sulfate is pumped from gypsum in places.

The porosity of sandstone and conglomerate is generally much less than that of their equivalents, sand and gravel, because of the presence of cementing materials between grains. The interstices of quartzites generally are almost completely closed by cementation, and the rock yields water from joints or other secondary openings. Many sandstones are only partly cemented and yield most water from the original interstices between grains and from joints. Sandstone wells seldom have yields as high as the yields of wells in unconsolidated sand and gravel.

Shale, silt, clay, glacial till, weathered residual rocks, and poorly sorted alluvium often have high porosities but retain much water by molecular attraction and have a low permeability. They furnish small supplies to domestic wells at many places. Siliceous shale, some claystones, and most argillites will develop closely spaced joints if the rocks are near the surface. Also, if these rocks are involved in faulting, fractures that stay open at considerable depths may develop. The joints and fractures may yield small amounts of water to wells. Most commonly, however, the fine-grained rocks will be barriers to the movement of groundwater.

The volcanic rock basalt has about the same range of permeability as limestone. Permeable zones include flow breccias, porous zones between successive lava beds, lava tubes, and cracks and joints. Joints caused by cooling, lava tubes, vesicles that intersect, tree molds, fractures caused by buckling of partly congealed lava, and voids left between successive flows are some of the features that give recent andesite and basalt its high permeability. Sediments interbedded with the lava will greatly increase the average porosity of large volumes of rock that are predominantly volcanic. Both the permeability and porosity of volcanic rock tend to decrease slowly with geologic time. Pyroclastic rocks associated with lava flows are generally porous but not very permeable. Exceptions are blockly, coarse material near volcanic vents and tuffs which have been reworked by water. Pyroclastic materials generally have medium permeabilities.

Generally crystalline and metamorphic rocks, commonly occurring in mountainous areas, are poor aquifers, especially in areas where these rocks are buried beneath other rocks. Where crystalline and metamorphic rocks are the uppermost bedrock, they yield small but reliable amounts of water to wells in the upper decayed portions of these rocks extending to depths up to 100 feet. Some water occurs at depths up to 300 feet in joints to shear planes. Most wells in crystalline and metamorphic rocks yield supplies sufficient only for domestic use. The average permeability of metamorphic and plutonic igneous rocks decreases rapidly with depth.

A study of the surface and subsurface distribution of rocks and of their character, thickness, and depth below land surface is prerequisite to an understanding of the occurrence and movement of groundwater at any locality. The earth's crust consists of layers of rocks laid down one upon another and underlain or intersected in places by massive or foliated rocks. Most sedimentary rocks and some igneous and metamorphic rocks are stratified to some degree, whereas most igneous rocks form massive bodies that were intruded into or extruded through the stratified rocks. Rock formations, distinct units of the earth's crust consisting of rocks of one or more kinds, may range from a few feet to thousands of feet in thickness and occur at the surface or may be buried beneath other rocks. Formations may extend over thousands of square miles or may be limited in areal extent to less than a square mile. There may be important differences in the same formation at different horizons and in different localities.

Water wells penetrate formations deposited in a region during the geologic ages of the past; generally the youngest formation is encountered first and then successively older formations are passed through. At places, because of the occurrence of intrusive rocks, folds, faults, and thrusts, older formations may rest upon younger ones. In most places, the lowest known rocks are crystalline igneous or metamorphic rocks (basal complex) which have an eroded upper surface and are overlain by younger formations. There is great variety in the succession of conformable beds overlying the basal complex; commonly a coarse-grained formation rests unconformably on older rocks. Generally, beds of a sedimentary series become increasingly fine-grained and calcareous from the bottom upward. The thickness of most sedimentary formations consisting of layers of rocks is very small in comparison with the areal extent of the formations. The formations above or below an aquifer may be distinctive and persistent.

Formations generally are stratified; individual beds may differ in thickness, composition, and compactness. Aquifers may be interbedded with aquitards or aquicludes. Due to differences in the conditions under which deposition occurred, a stratified formation generally changes gradually in thickness and character of rock from place to place. As a rule, formations are found to become thinner and finer grained at increasing distances from the source of the sediments. These lateral gradations are of great significance with respect to the occurrence and movement of groundwater.

There is often a rapid lateral gradation in unconsolidated deposits; lenses, layers, and stringers of sand and gravel occur irregularly and give way abruptly to clays. Radical changes in thickness and character are to be expected from place to place. Extreme local variations in glacial deposits and river alluvial deposits are in striking contrast to the relatively uniform conditions of some sedimentary bedrock aquifers. Glacial till has a very chaotic structure and consists of clayey materials with highly irregular interbedded lenses or layers of sand and gravel. Outwash deposits of sand and gravel made by streams that flowed from the melting glacial ice often persist over fairly large areas and sometimes change gradually in thickness and character from place to place. Alluvium tends to be irregular in structure but often has fewer local irregularities than glacial till. Lake deposits are much better stratified than either glacial drift or alluvium. The composition of glacial deposits and alluvium and lake deposits depends largely upon the nature of the rocks which were eroded to supply the material out of which they were formed.

Rock formations are rarely horizontal; dips may be due to deposition in a sloping position or to deformation after deposition. Alluvium generally dips downstream, lake deposits dip away from shore, and lava beds dip away from vents. Because of deformation, formations may have slight inclination, or they may turn into a vertical position, or overturn. Knowledge of the dip of the formation underlying a region is necessary for the determination of the distribution of the aquifers, aquitards, and aquicludes and for the forecast of the depths to aquifers. The determination of the distribution and depth of aquifers would be relatively simple if formations everywhere had the same angle and direction of dip. However, many formations have been warped to form folds including anticlines, synclines, simple sills or monoclines, or irregular flexures. Anticlines may form ridges or structural domes; synclines may be troughs or structural basins. Formations dip in all directions away from a common center in a structural dome and from all directions toward a common center in a structural basin.

At places, rocks are separated from one another by unconformities representing intervals of time during which existing rocks eroded and perhaps were tilted, warped, or broken and practically no sediments were deposited. An unconformity shows the approximate topography of a region just before the deposition of overlying formations. The distribution of an aquifer or aquitard in formations either overlying or underlying an unconformity may be controlled largely by the intervening unconformity.

Fractures result chiefly from compression during earth movements and from rock shrinkage due to drying of sediments or cooling of igneous rocks. A joint is a natural rock fracture; if blocks of rock on opposite sides of a fracture are dislocated with reference to each other, the fracture is called a *fault*. Faults are of two kinds, normal and thrust. They differ greatly in their lateral extent, in depth, and in amount of displacement. Large faults may affect the distribution and position of aquifers and act as subsurface dams or as conduits through which deep-seated waters may escape. At places there is a fault zone called *fault breccia* containing many small parallel faults or masses of broken rock. The raised side of a fault may produce an escarpment with radical differences in the altitude and topography of the surface on opposite sides of the fault. Deposition of coarse materials may result on the downthrown side from rapid erosion of the exposed rocks on the raised side. So much erosion may occur with the lapse of geologic ages that the escarpment is obliterated. A fault may displace alternating permeable and impermeable beds so that the impermeable beds face permeable beds. There may be clayey gouge along the fault planes. The opposite sides of many faults are not everywhere pressed together; few faults are single, clear-cut breaks.

2.6 Characteristics of Aquifers in the Conterminous United States

Meinzer (1959) divided the conterminous United States into 21 groundwater provinces; later Thomas (1952) consolidated and rearranged Meinzer's 21 provinces into 10 groundwater regions. Thomas's regions are used herewith; his descriptions of the characteristics and potentialities of aquifers in the regions together with those of McGuinness (1963) are briefly summarized in the following sections. The regions (see Fig. 2.6) are: (1) Western Mountain Ranges, (2) Arid Basins, (3) Columbia Lava

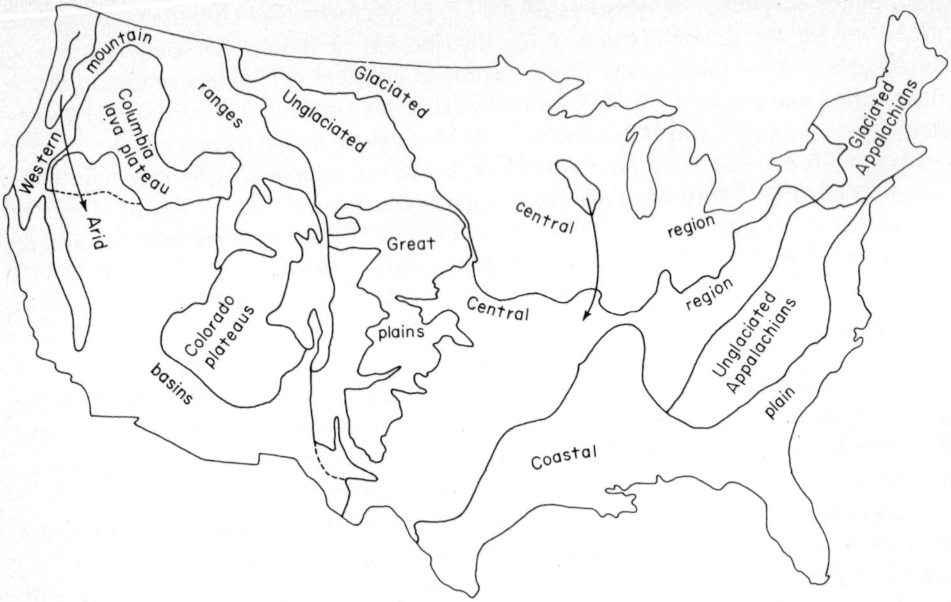

Fig. 2.6 Groundwater regions of the conterminous United States. (*From Thomas, 1952.*)

Plateau, (4) Colorado Plateau, (5) Great Plains, (6) Unglaciated Central, (7) Glaciated Central, (8) Unglaciated Appalachian, (9) Glaciated Appalachian, and (10) Atlantic and Gulf Coastal Plain.

The Western Mountain Ranges groundwater region includes the Sierra Nevada of California, the Cascade and Coast Ranges farther north, and the Rocky Mountains from Montana to New Mexico as shown in Fig. 2.7. Most of the rocks in the region are igneous rocks, metamorphic rocks, and well-indurated sedimentary rocks. In the extensive areas underlain by these rocks, wells yield only small supplies from joints and other fractures and from fault zones and other openings in the otherwise well-consolidated rocks. Small quantities of groundwater are obtained from weathered material and loose mantle overlying the consolidated rocks. Limestones and sandstones are important aquifers in some parts of the region, such as in the Uinta Mountains of northeastern Utah and in the Bighorn Basin, Wyoming. The groundwater reservoirs having the best potentialities for development in the region are those in the loose gravel and sand of the intermontane valleys.

The Arid Basins groundwater region includes the heavily populated part of California, the Great Basin in California, Nevada, Oregon, Idaho, Utah, and Arizona, the drainage basins of the Gila River and Rio Grande in Arizona, New Mexico, and west Texas, and the lowland extending southward from Puget Sound through the Willamette Valley in Oregon as shown in Fig. 2.8. The Arid Basins flank the mountain ranges of the West and consist of vast depressed areas dotted with isolated mountain ranges.

The basins, sunken blocks of the earth's crust that intervene between uplifted mountain blocks, are filled with rock particles eroded from mountains. The aquifers of the Arid Basins are almost exclusively the unconsolidated sediment valley fill underlying the valleys. In most valleys, the valley fill has been accumulating for a long period of geologic history and now forms a flat or undulating surface. The valley fill is known to be hundreds of feet thick in places; thicknesses of several thousand feet are possible. The great bulk of the valley fill is saturated with water. Most of the valley fill is silt and clay; stringers and sheets of sand and gravel are abundant enough to be excellent aquifers. Large-capacity wells generally obtain water from aquifers of sand and gravel which are common near the mouths of the

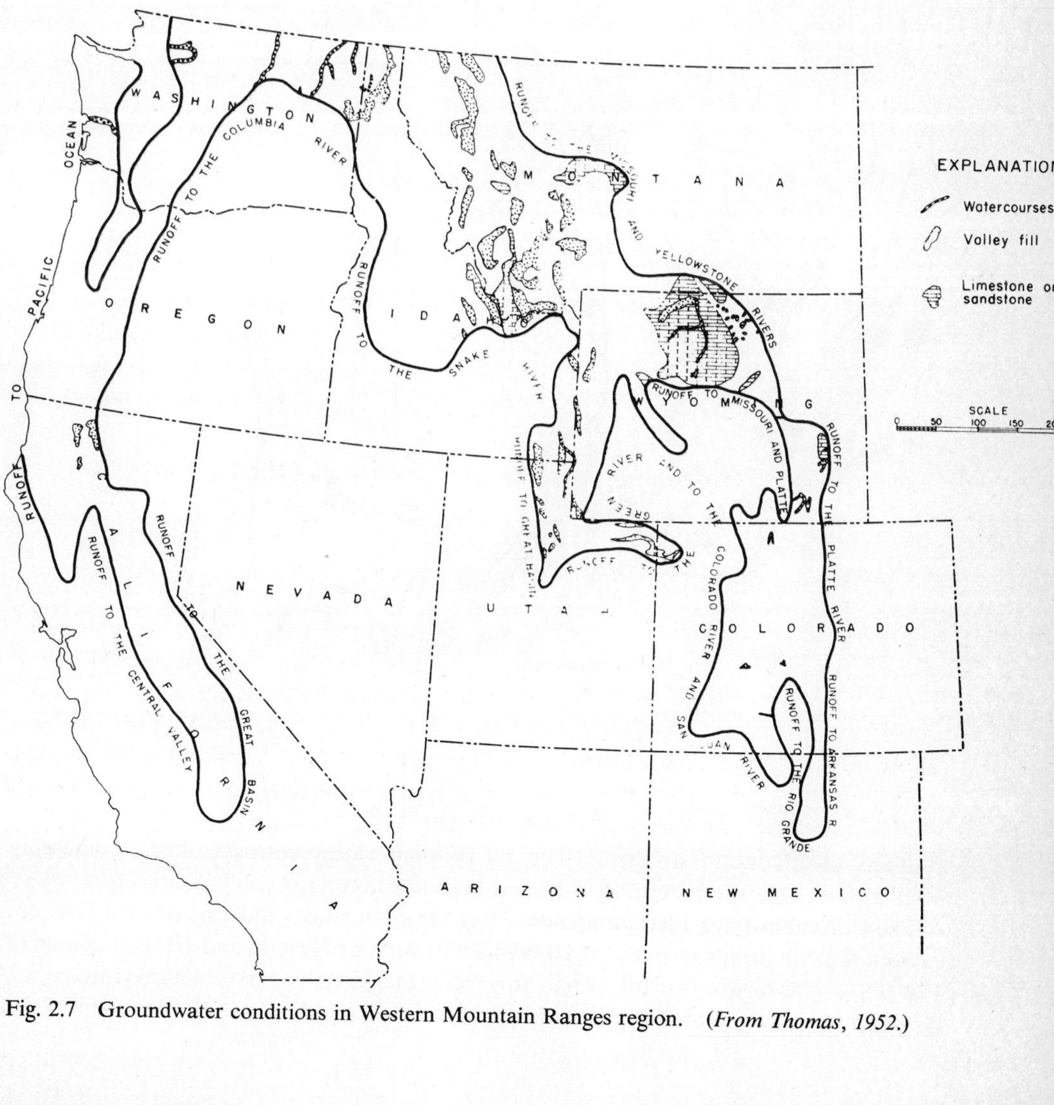

Fig. 2.7 Groundwater conditions in Western Mountain Ranges region. (*From Thomas, 1952.*)

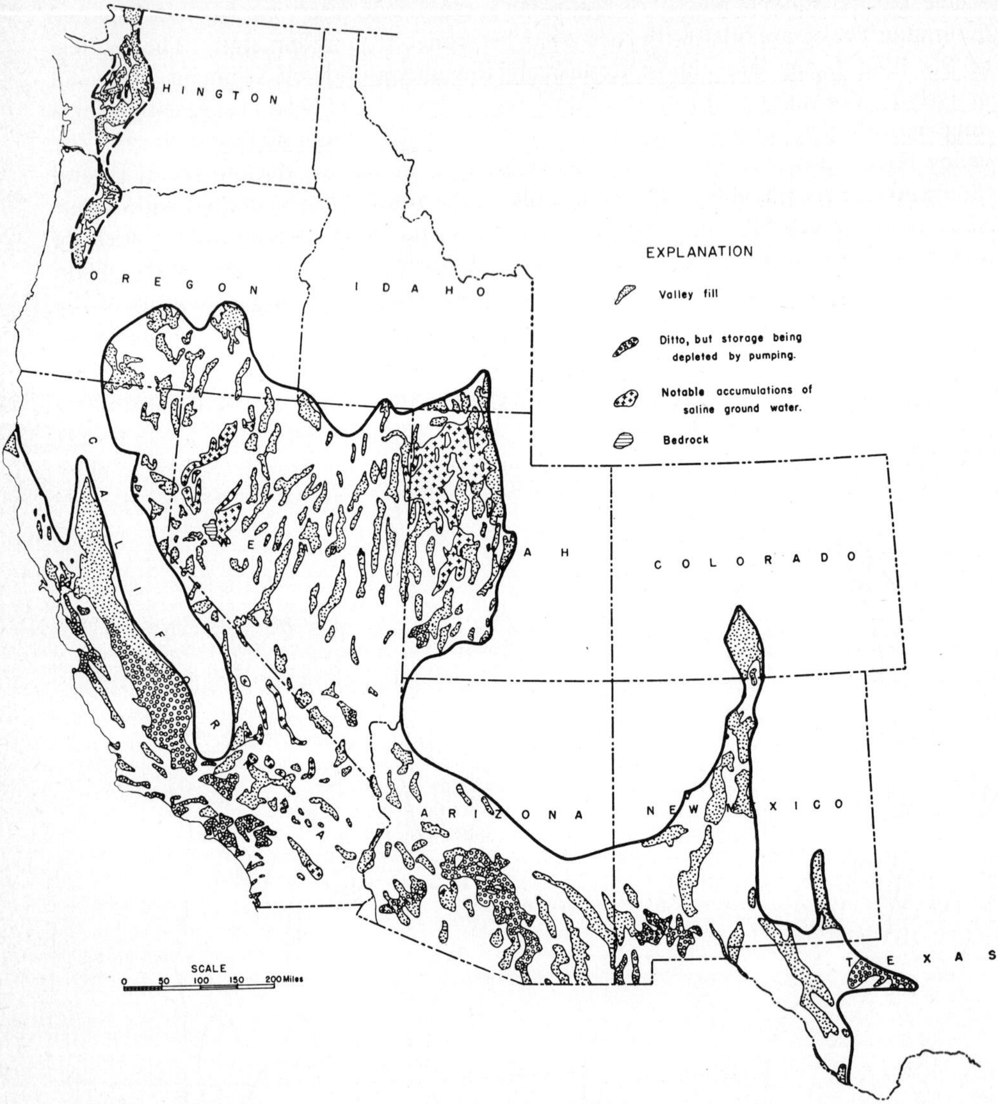

Fig. 2.8 Groundwater conditions in the Arid Basins region. (*From Thomas, 1952.*)

canyons of mountain streams. The fill of each valley consists of the coalescing alluvial fans deposited from all the surrounding canyons.

The Columbia Lava Plateau groundwater region includes most of eastern Oregon, eastern Washington, southern Idaho, and small parts of Nevada and Utah as shown in Fig. 2.9. The region is a large area formed principally by extrusive volcanic rocks, largely lava flows, which are interbedded with or overlain by alluvium and lake sediments. The volcanic materials range from less than a hundred feet thick near the borders of the region to hundreds or even thousands of feet in thickness in many places. These great thicknesses are produced by many individual lava flows commonly 20 to 500 feet thick. The lava rocks include many permeable zones formed by shrinkage cracks and, especially, by the covering over of the rough surfaces of lava flows by the chilled bottoms of the next flows, which did not fill in the irregularities. completely. Gas bubbles, lava tubes, or small caverns add to the pore space. Permeable sand and gravel beds also occur in the volcanic sequence. The massive lava flows and intrusive rocks such as dikes, sills, or plugs are generally impermeable. Clay beds interbedded with the basalts at places are also impermeable.

Groundwater is also found within the region in unconsolidated deposits interbedded with lava flows or lying in valleys cut into these rocks. The most productive unconsolidated deposits are glacial outwash sand and gravel along present streams or in abandoned melt-water channels.

The Colorado Plateau groundwater region lies west of the Rocky Mountains and includes parts of the states of Wyoming, Colorado, Utah, New Mexico, and Arizona as shown in Fig. 2.10. Most of the rocks of the region are sedimentary strata, chiefly interbedded sandstone and shale. These sediments in general are nearly horizontal, but in places are tilted or folded and in others are broken by faults into large blocks forming plateaus that stand at different levels. Most of the aquifers are beds of sandstone, although limestone and alluvium yield water in a few places. The average productivity of the rocks is very low; the most important aquifer is the

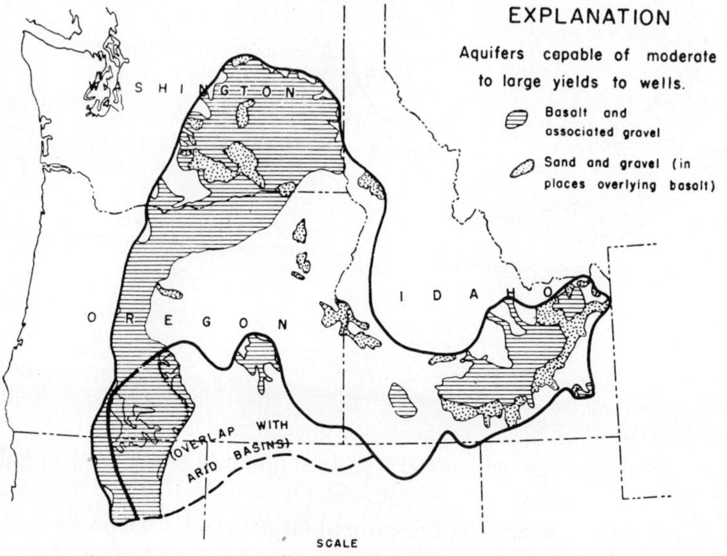

Fig. 2.9 Groundwater conditions in Columbia Lava Plateau region. (*From Thomas, 1952.*)

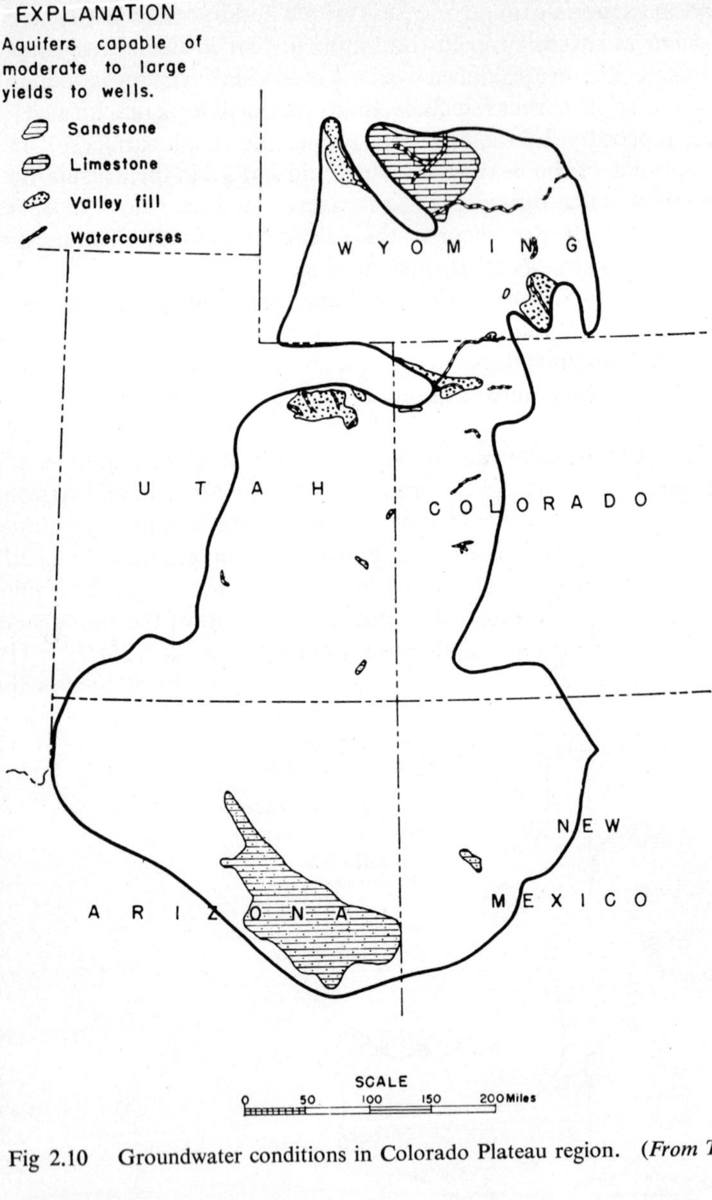

Fig 2.10 Groundwater conditions in Colorado Plateau region. (*From Thomas, 1952.*)

Coconino Sandstone of Permian age, which yields several hundred gallons per minute to some wells.

Alluvial deposit aquifers are scattered and few are of large extent. Deposits in the Uinta Basin at the foot of the Uinta Range in northeastern Utah and at the foot of the Wind River and Medicine Bow Mountains in Wyoming are the principal ones.

The Great Plains groundwater region, extensive semiarid to subhumid plains and plateaus east of the Rockies, extends from South Dakota to western Texas and eastern New Mexico as shown in Fig. 2.11. The groundwater region is a very sizable remnant of a gigantic alluvial apron formed in late Tertiary time and which formerly extended uninterrupted frcm the foot of the Rockies and the mountains to the south in New Mexico to an unknown terminus probably some hundreds of miles east of its present edge. The alluvium was deposited on the eroded, eastward sloping surface of a thick body of stratified sandstone, shale, and limestone of Paleozoic, Mesozoic, and Tertiary age.

Unconsolidated sediments containing beds of sand and gravel underlie the region,

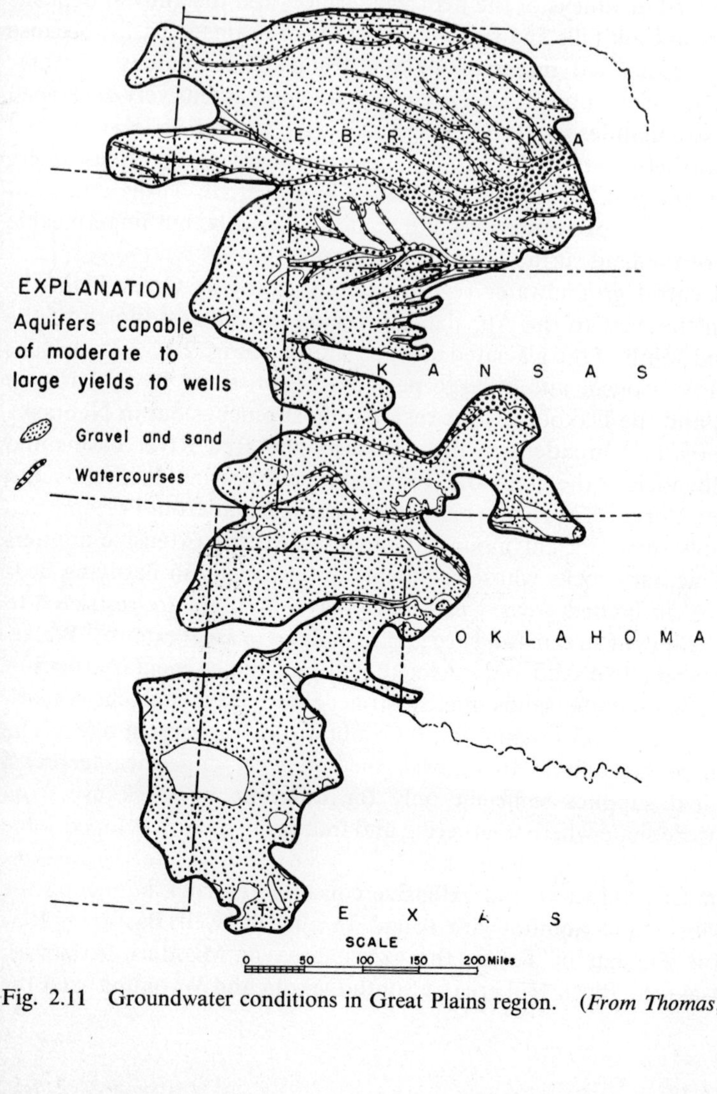

Fig. 2.11 Groundwater conditions in Great Plains region. (*From Thomas, 1952.*)

essentially the area of the High Plains, to considerable depth in many places. Groundwater is also found in consolidated sedimentary rocks, in stream channels or terrace deposits, and in glacial drift. Generally the deposits of the Great Plains region have a higher proportion of sand, silt, and fine gravel, and a better degree of sorting, than the valley fill of the western basins.

Aquifers in most places are capable of yielding moderate to large quantities of water to wells and occupy more extensive areas than do those in any of the groundwater regions farther west. The aquifers are mostly stream deposits of gravel, sand, silt, and clay in various degrees of sorting, and they have a wide range in permeability. These aquifers contain abundant irregularly distributed lenses and stringers of sand and gravel. The thickness of the alluvial deposits is quite variable, as much as 700 feet of alluvium was deposited in valleys of the bedrock surface, and the alluvial deposits are thin or absent over ancient hills and ridges or in areas of drainage basins because of erosion. The unconsolidated materials hold a tremendous amount of water generally under water-table conditions; they are either the most intensively developed or have the greatest potentialities for development in Texas, Oklahoma, Kansas, and Nebraska. In the northern part of the region there are many sandy areas where aquifers are readily recharged. South of the Canadian River, in Texas and New Mexico, there are small areas of sand dunes and some sandy soils, but impermeable material under most of the land surface impedes recharge.

The Unglaciated Central groundwater region extends from the Great Plains or Rocky Mountains on the west to the Appalachian region on the east, north of the Gulf Coastal Plain and south of the glaciated area, as shown in Fig. 2.12. It includes the Driftless area of Wisconsin and a large part of the Missouri River basin in Montana, Wyoming, and the Dakotas. The region almost pinches out in Nebraska and in southern Illinois and broadens in the Arkansas and Red River basins and farther east to include much of the Ohio and Tennessee River basins; it extends into Pennsylvania and New York. The region has rocks ranging in age from Precambrian to Recent and includes some ancient mountain ranges. All the extensive aquifers are consolidated sedimentary rocks whose sediments are generally in flat-lying beds but are steeply inclined in limited areas. Sand and gravel aquifers are restricted to the valleys or former valleys of streams and are greatly limited in areal extent. Watercourses are the most productive sand and gravel aquifers. The sedimentary rocks of the region include: conglomerate, sandstone, siltstone, shale, clay, coal, chert, marl, limestone, and dolomite gypsum ranging from Cambrian to Tertiary in age. The rocks vary greatly in permeability. In general, the relatively impermeable rocks, clays and shales, yield supplies sufficient only for domestic or stock use from joints and zones of brittle shale where weathering and fracturing have developed some interstices.

Limestone is the most productive and extensive consolidated rock aquifer in the region. Important limestone aquifers are found in the Roswell Basin in New Mexico; the Edwards Plateau in Texas; the Ozark area in Missouri, Arkansas, Oklahoma, and Kansas; the Black Hill area in South Dakota and Wyoming; and the

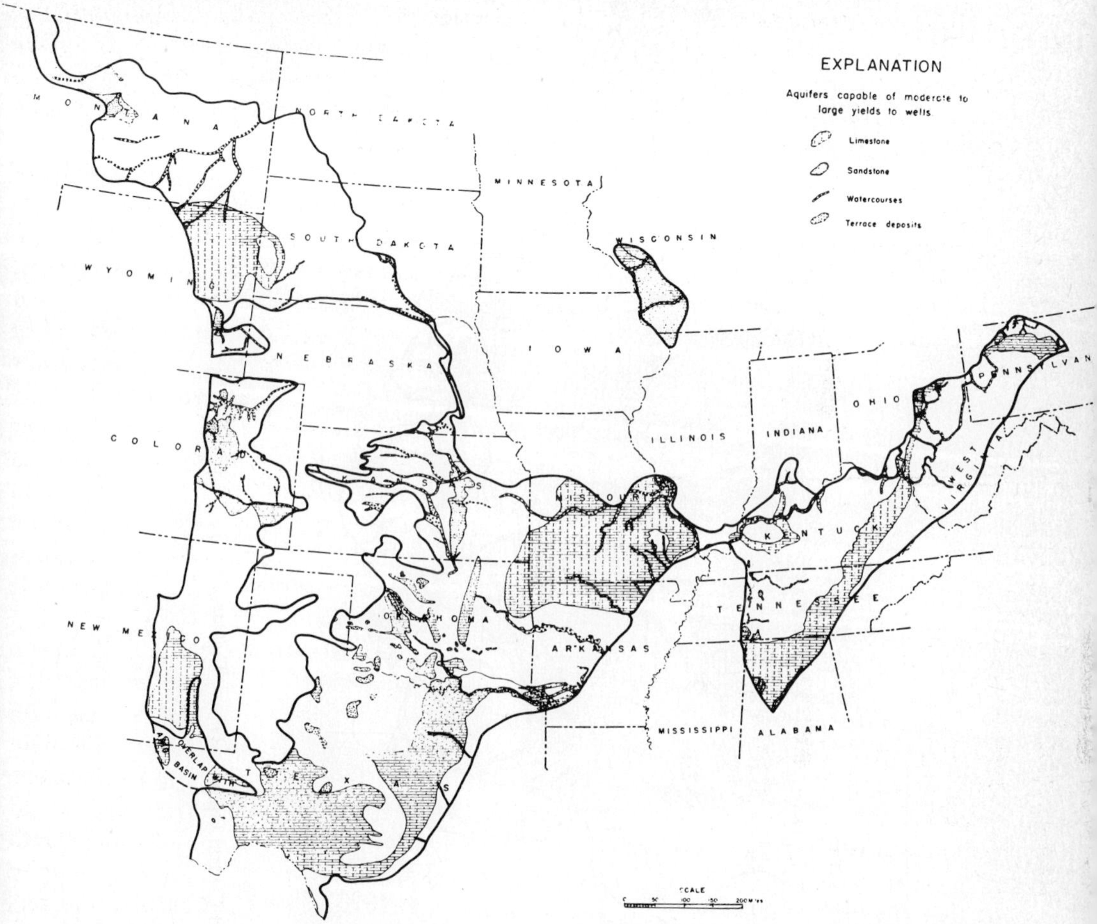

Fig. 2.12 Groundwater conditions in Unglaciated Central region. (*From Thomas, 1952.*)

Tennessee Valley in Tennessee and Alabama. Sandstones are also important consolidated rock aquifers; the Dakota sandstone is a well-known deep artesian aquifer of the Dakotas and several other states. Other sandstone aquifers occur in parts of Texas, Oklahoma, Kansas, Colorado, Wisconsin, Pennsylvania, and Ohio. Many of the consolidated rock aquifers in the region contain water that is too highly mineralized for use. Most sedimentary rocks are of marine origin and in places they contain fresh water only within a few hundred feet of the surface.

Sand and gravel of stream valleys are relatively small in volume but may be highly productive especially with recharge by induced infiltration of surface water in streams. Most alluvium yields sufficient amounts of water for domestic or stock use; aquifers capable of yielding large quantities of water constitute only a minor proportion of

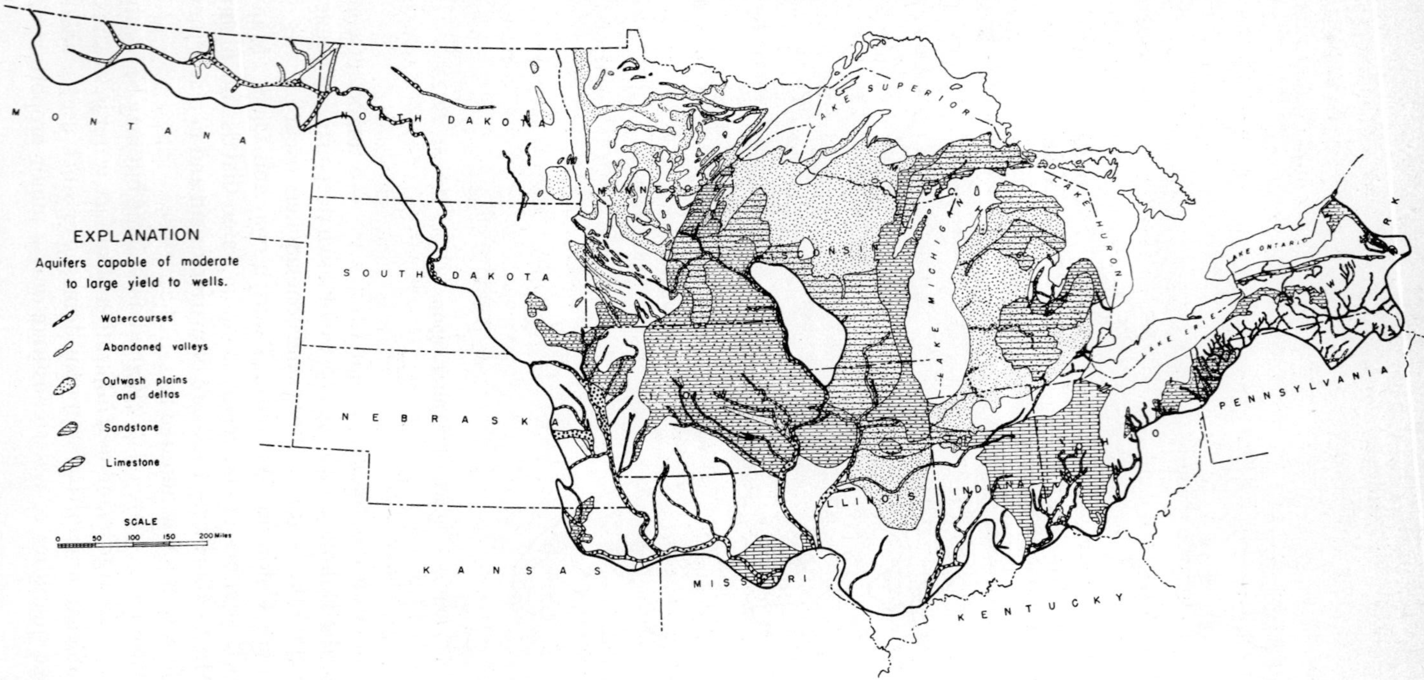

Fig. 2.13 Groundwater conditions in Glaciated Central region. (*From Thomas, 1952.*)

the total alluvium in most streams valleys. Sand and gravel have accumulated in considerable quantities near the north boundary of the unglaciated region as glacial outwash. The outwash is also found in many valleys beyond the limit of glaciation. Some sand and gravel beds deposited by ancient streams form terraces above present stream channels. Terrace deposits constitute important aquifers in Oklahoma and Texas.

The Glaciated Central groundwater region includes the glaciated part of the central lowland extending from the Canadian border southward to approximately the present channel of the Missouri River from Great Falls to St. Louis and that of the Ohio River from Pittsburgh to its mouth as shown in Fig. 2.13. The preglacial surface was formed by consolidated sedimentary rocks including sandstones and limestones. Several of the rocks beneath the glacial drift form important aquifers. The Dakota Sandstone extends into Minnesota, Iowa, Montana, and Nebraska. Sandstone aquifers extend also through large parts of Iowa, eastern Minnesota, Wisconsin, northern Illinois, and Michigan. These aquifers are deeply buried in places, and water is confined in them under artesian pressure. Limestones are important sources of water in some areas. The limestone contains permeable zones in the coral reefs or along joints and fractures. In the vicinity of Lake Superior, sedimentary rocks are absent and crystalline or metamorphic rocks crop out or are buried beneath glacial drift. The yields of these rocks are generally sufficient only for domestic and stock use.

The glacial drift is very thin in some places and hundreds of feet thick in others; it is thickest in the southern peninsula of Michigan and the adjacent northern three-fifths of Indiana, and nearly as thick in western Ohio, northeastern Illinois, north-central Iowa, eastern and northern Wisconsin, and a substantial part of Minnesota. The bulk of the glacial drift is till, deposited directly by the melting ice, and entirely unassorted. Where the till is composed chiefly of clay-size particles, it is relatively impermeable; in places the till is composed chiefly of sandy material and is fairly permeable. The sand and gravel deposits of streams that issued from the ice occur in abundant irregular lenses and stringers intermingled with the till, in deltas formed in glacial lakes, in outwash aprons, and in valley trains. The sand and gravel intermingled with the till yield moderate to small water supplies. The sediments deposited in glacial lakes are predominately of fine texture; however, sandy deposits are found near shorelines. The outwash aprons are generally well-sorted sand and gravel deposits having large areal extents and large yields.

The valley-train deposits of sand and gravel, occupying watercourses of present streams of valleys no longer occupied by streams, are the aquifers that produced the greatest quantities of water. The valley trains extend for many miles from the ice front and many go beyond the limits of the glaciated region. Many of the cities of the glaciated region and adjacent parts of the unglaciated region are along watercourses where sand and gravel beds yield large quantities of water to wells. Sand and gravel deposits in valleys that were occupied by rivers draining the ice sheet but have since been abandoned by their streams yield moderate to large supplies of water.

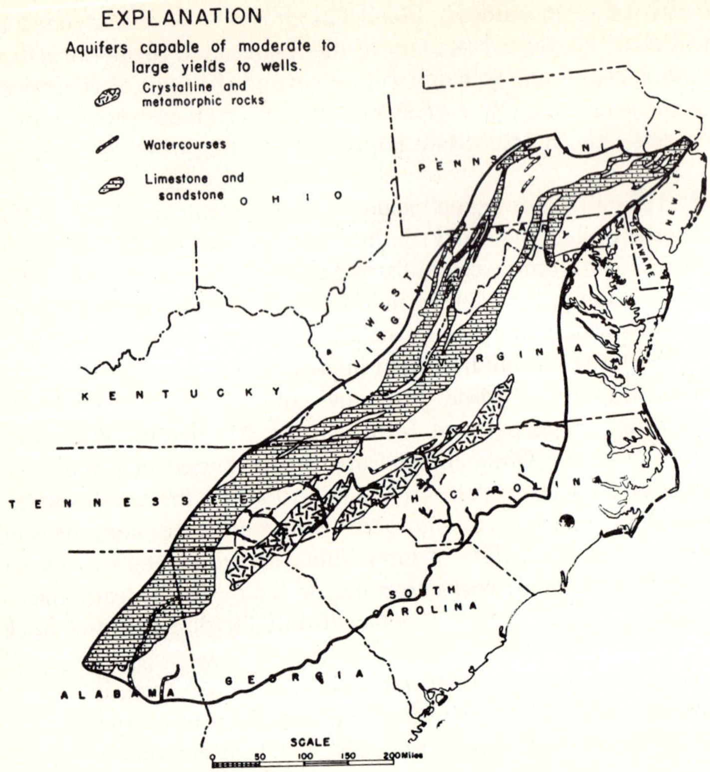

Fig. 2.14 Groundwater conditions in Unglaciated Appalachian region. (*From Thomas, 1952.*)

In substantial areas where the glacial drift is not productive, the bedrock is capable of furnishing moderate quantities of water to wells.

The Unglaciated Appalachian groundwater region covers the unglaciated parts of the Appalachian Valley and Piedmont provinces and includes parts of Alabama, Georgia, Tennessee, South and North Carolina, Virginia, Maryland, Delaware, Pennsylvania, and New Jersey as shown in Fig. 2.14.

The dominant bedrocks are crystalline or metamorphic or well-consolidated sedimentary rocks. These rocks, including chiefly granite, gneiss, and schist, are the principal aquifers of the region; water occurs in joints and fractures that occur in the fresh rock to considerable depths or in the unconsolidated weathered surficial materials. Some limestones contain large and extensive solution cavities and channels. These rocks furnish water for practically all the domestic supplies and many of the small industrial and public supplies. Topography and rock type are important guides to the productivities of wells. Granitic rocks commonly have parting planes usually parallel to the rock surface on hills and in valleys. Wells in sheeted granitic terranes usually have higher yields on the lowlands than on uplands. Interstices are most abundant near the surface and little water is obtained at depths greater than a few hundred feet. Gneiss produces some water particularly in North and South Carolina; schist is generally the least permeable of the crystalline rocks. Metamorphic and sedimentary rocks yield water chiefly from fractures; sandstones are important aquifers in New Jersey and eastern Pennsylvania. Limestones and dolomites yield moderate to large quantities of water in many parts of the region. For large quantities of water the best prospects are the sand and gravel deposits in the alluvium of streams. Many watercourses contain sand and gravel aquifers.

The Glaciated Appalachian groundwater region includes the glaciated part of the Appalachian Mountains and practically all of New England, plus the eastern part of New York and the northern parts of Pennsylvania and New Jersey as shown in Fig. 2.15. Except for small scattered areas in the Connecticut Valley in Connecticut and Massachusetts, in northern New Jersey and adjacent New York, along the Hudson section of the Valley and Ridge province, and at the south edge of the Adirondacks, underlain by limestone or sandstone of Paleozoic age, bedrock aquifers are not highly productive. Most of New England is underlain by metamorphosed sediments which generally yield small quantities of water chiefly from joints and fractures. Ancient crystalline rocks crop out from beneath the flanking sedimentary strata in the middle of the mountains.

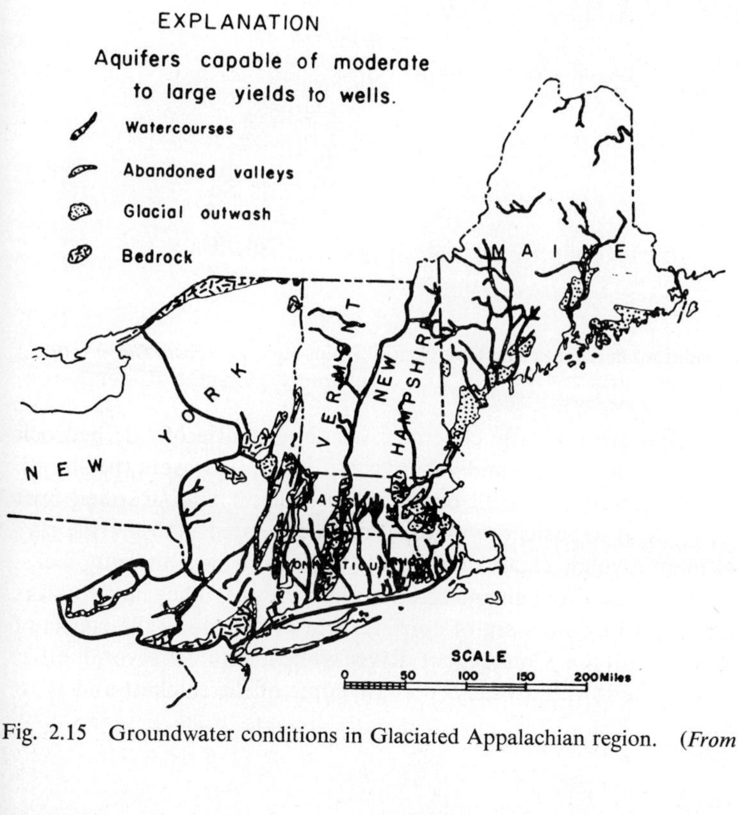

Fig. 2.15 Groundwater conditions in Glaciated Appalachian region. (*From Thomas, 1952.*)

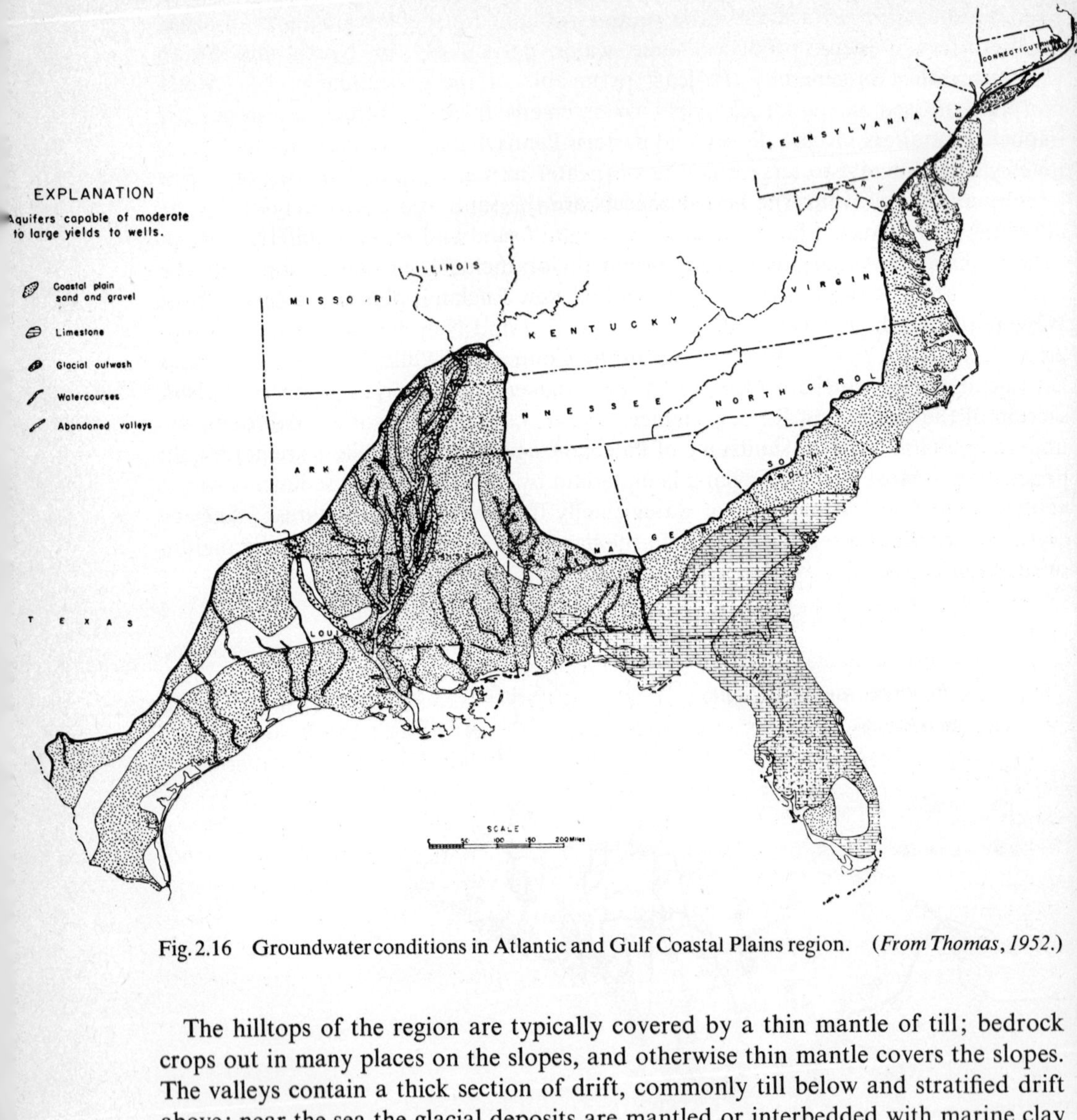

Fig. 2.16 Groundwater conditions in Atlantic and Gulf Coastal Plains region. (*From Thomas, 1952.*)

The hilltops of the region are typically covered by a thin mantle of till; bedrock crops out in many places on the slopes, and otherwise thin mantle covers the slopes. The valleys contain a thick section of drift, commonly till below and stratified drift above; near the sea the glacial deposits are mantled or interbedded with marine clay and silt. The glacial mantle yields small supplies of water to drilled and dug wells, and sand and gravel aquifers are often encountered in the valleys. The chief aquifers are the glacial outwash deposits which are of considerable extent along the coasts of Maine and Massachusetts, in the Connecticut River valley, and in several other places in New England. The lowland valleys contain some of the thickest and most permeable sand and gravel deposits and therefore the most productive aquifers. Abandoned or buried valleys containing extensive sand and gravel deposits are numerous in the region.

The Atlantic and Gulf Coastal Plain groundwater region extends from Long Island and the sandy islands and capes of Massachusetts to Texas, and includes the Mississippi Embayment as far north as Cairo, Illinois, as shown in Fig. 2.16. Most of the region is underlain by a thick succession of loose or semiconsolidated sediments that dip gently seaward. Sand and gravel aquifers interbedded with less permeable clayey deposits occur in the sediments. The most heavily developed sand and gravel aquifers are in eastern Arkansas, southwestern Louisiana, and southeastern Texas. The Houston-Galveston and Baltimore, Maryland, areas are other areas of intensive pumping. Some areas have had difficulties with salt-water encroachment. Limestones in Florida and southeast Georgia are very permeable aquifers. Glacial outwash deposits are the most important aquifers of Long Island. The sediments of the watercourses are prevailingly of fine texture, and permeable sand and gravel aquifers are rare.

The region is a huge seaward thickening wedge of generally unconsolidated but locally consolidated sedimentary rocks lying on Precambrian, Paleozoic, and Mesozoic bedrocks. The sediments are mainly clay, silt, sand, gravel, marl, and limestone in alternating and intertonguing strata. The thicknesses of the deposits vary from as little as a few hundred feet in places in the northeasternmost part of the region to tens of thousands of feet in the general vicinity of the mouth of the Mississippi River. The region, expecially the southern part of it, contains some of the most extensive and productive aquifers in the conterminous United States. Only a broad strip running northeastward through Texas into Louisiana is underlain by relatively impermeable rocks of Tertiary age; a smaller crescent-shaped strip in Mississippi and Alabama and small scattered areas in Texas, Louisiana, and Florida are underlain by aquifers in which all the groundwater is saline. There is a very wide range in the number and depth of permeable sand and gravel beds in the sediments. There is also a wide variation in the extent to which different beds have been flushed of salt water. Some of the beds are thick and extend over hundreds or thousands of square miles; others are thin and small in areal extent and intertongue complexly with other beds of different kinds.

GROUNDWATER EXPLORATION

The geologist utilizes petrography, stratigraphy, structural geology, and geomorphology in the search for groundwater. Logs of wells and excavations, either artificial or natural, give geologic sections of the earth's crust. Successive deposits are examined as they exist in place in natural outcrops. Surface and subsurface geophysical techniques are useful in the interpretation of geologic structure and stratigraphy.

By well construction, mining, tunneling, and other excavations, man has added hundreds of thousands of places where he can collect data concerning aquifers beneath the land surface. Although these potential observation points are numerous, they represent in the aggregate a negligible proportion of the total volume of rock materials through which groundwater moves, and in large areas concepts of the occurrence and movement of groundwater are based on inferences. Nevertheless, the collected data form the basis for a good working knowledge of the physical principles governing the movement and storage of groundwater.

The geologist maps the surface extent of the various lithologic units and their water-yielding properties. This is accomplished with the aid of aerial photographs, regional geologic maps, and investigations in the field. Modern mapping techniques are described by Compton (1962). Hydrogeologic maps do not necessarily correspond to conventional geologic maps. In preparing hydrogeologic maps, emphasis is placed on the description of aquifers and aquitards.

The position and thickness and continuity of aquifers, aquitards, and aquicludes are determined with stratigraphic tools. Structural geology is used in conjunction with stratigraphy to locate rocks which have been displaced by earth movements. Geomorphological studies of land forms can yield important lithologic, stratigraphic, and structural information. Rock types with distinctive water-yielding properties are identified through petrographic studies.

The correlation of formations depends on either tracing beds from place to place or identfying the same beds in different localities. Identification can often be made by studying characteristic physical properties. Determination of the relative age of formations in different localities is sometimes made with radioactive dating methods. Correlation on the basis of characteristic physical properties is often impossible or at least uncertain because of the lateral changes in formations and the similarity of different beds. Fossils often can be used to correlate widely separated sections. In outcrop areas, boundaries of formations can be mapped from field observations; where formations are concealed by other rocks this is impossible and the boundary must be inferred from data on the dip of the formation and the altitude of the land surface. Dips of rocks are often determined by correlating the formations in outcrops and well sections. Unconformities are characterized by great irregularities; it is often possible to construct a rough approximation of a contour map of a buried unconformity from observation of natural and artificial exposures and from well logs.

The method most commonly used to correlate aquifers, aquitards, and aquicludes is based on a study of well records and examination of samples of drilling cuttings. Driller's logs furnish valuable information on the thickness and location of formations, diameters of holes and casings, fluctuations of water level, and depth, discharge, and head of water-yielding beds. Samples of the drilling cuttings of wells taken at intervals not exceeding 10 feet and every change in formation afford a lithologic record and section inferior in value only to an exposure of the edges of the formations in an outcrop. Such reliable data are available for many wells. To be of optimium value, samples should be placed, unwashed, in proper containers in the field and plainly and accurately labeled in ink with the location and owner of the well, the date, and the depth from which each was taken. Drilling cuttings studied petrographically aid in identifying, from well to well, the formations from which they come. A systematic outline of procedures for examination of well samples was given by Meinzer (1959).

Subsurface conditions may differ considerably from conditions indicated by surficial geologic maps, particularly in glaciated areas. Therefore, it is advantageous to use a series of lithofacies maps as a sequence of overlays plotted on transparent paper (Pettyjohn and Randich, 1966). Such maps permit direct visual determination of the lithology and areal extent of aquifers and indicate areas of possible stream losses or gains. They indicate areas and boundaries of confining layers and, in some places, they may be used to delineate areas of water-table and artesian conditions. They are especially useful in determining buried channels in glaciated areas. The porosity of rock units making up an aquifer may be plotted on lithofacies maps, and the product of the volume of each lithologic type and the porosity provides an estimate of the total water in storage. In addition, if the water-yielding properties are known, the quantity of water moving through an aquifer and the amount available to wells may be determined. Lithofacies maps also aid in the explanation of complex changes in water quality.

Lithofacies maps are constructed as a sequence of overlay maps. Each map represents a unit thickness above or below a datum chosen to suit conditions. The unit thickness for each map is determined by the overall thickness and complexity of the aquifer. Lithofacies maps of clastic sediments may provide a useful tool in groundwater studies.

Geologic and hydrologic exploration should proceed together for the best results. The hydrologist maps water-level conditions in various lithologic units. Water levels are measured in wells and other excavations. The location and magnitude of springs and discharge areas give a good indication of general groundwater conditions. Aerial photography and direct aerial observation can be used under favorable conditions as a reconnaissance method of mapping areas of groundwater discharge. In exploration programs, the quantities of groundwater discharged by springs, evapotranspiration, streams, and water-collecting systems are measured and studied. Water samples are collected from wells and natural discharge areas and are analyzed to provide useful data on the occurrence and movement of groundwater.

2.7 Water-level and Pumpage Data

Water-level data provide records of short-term changes and long-term trends of fluctuations of storage within groundwater reservoirs. Recognized uses of data on groundwater levels include the following: (1) to identify areas of detrimentally low or high water levels; (2) to facilitate prediction of the groundwater supply outlook for the future by showing the time-rate of change in groundwater storage; (3) to provide information for evaluating the water-yielding properties of groundwater reservoirs; (4) to appraise the relationship between water-level fluctuations and pumping, precipitation, and other factors; (5) to indicate the status of groundwater in transit; (6) to aid in estimating the base flow of streams; (7) to furnish information for use in research.

Water levels in wells are almost constantly fluctuating and decline or rise a fraction of an inch or many feet within a relatively short time. Water levels in wells in artesian aquifers under natural conditions generally fluctuate to a much greater extent than water levels in water-table aquifers and are sensitive to such factors as earthquakes and earth tides and changes in atmospheric pressure, surface-water stage, and surface loading. The water levels in artesian wells also are influenced by recharge from precipitation in the water-table portions of aquifers, although the effects of recharge are sometimes not noticeable immediately, and by withdrawals from wells and springs. Water levels in water-table aquifers are affected by direct recharge from precipitation, evapotranspiration, withdrawals from wells, discharge to streams, and sometimes changes in atmospheric pressure.

Fluctuations in water levels indicate both changes in the actual quantity of water stored in aquifers and movement of groundwater. A continual decline in water levels results when discharge exceeds recharge; water levels usually rise when recharge is greater than discharge. The amount of water taken from or added to storage per unit change in water levels under water-table conditions is generally many times larger than under artesian conditions (that is, $S \ll S_y$).

During the growing season, the water table fluctuates in response to evapotranspiration losses. Groundwater storage is reduced as the water requirements of plants are met and evaporation from the soil occurs. The drawdown of water levels is greatest during sunlight period when evapotranspiration losses are high; the maximum rate of drawdown occurs during the midday hours. Acceleration of the drawdown of water levels generally begins about 8 A.M. each day; the rate of drawdown begins to decrease about 6 P.M. The decreased rate of drawdown during the hours of darkness appears as recovery, but water levels do not rise above the stages at daybreak. The magnitude of fluctuations of the water table in response to evapotranspiration losses is relatively small and depends in part upon weather conditions. Water-level fluctuations caused by evapotranspiration are negligible after killing frosts and begin with the appearance of foliage.

The water levels in artesian wells fluctuate in response to changes in atmospheric pressure; water levels in some water-table wells fluctuate, to a very small degree, because of atmospheric-pressure changes. Water levels in wells recede when the atmospheric pressure increases and rise when the atmospheric pressure decreases. Diurnal and semidiurnal atmospheric-pressure changes, culminating in two maximums and two minimums during the course of 24 hours, are recorded at places by barometers. Diurnal pressure changes produce maximums during the coldest hours and minimums during the warmest hours. Barometric records show semidiurnal pressure changes culminating in maximums at about 10 A.M. and 10 P.M. and minimums at about 4 A.M. and 4 P.M., depending in part upon the season, elevation, and weather conditions. These diurnal and semidiurnal atmospheric-pressure changes persist through all seasons and are produced by daily temperature changes.

The application of an additional load on an artesian aquifer in the form of a passing railroad train compresses the aquifer and increases the hydrostatic pressure. A rapid rise in water level results, and a maximum water level is recorded at about the time the locomotive is opposite the well. The water level then slowly recovers to its original position, usually within a few minutes. Water-level fluctuations due to passing railroad trains usually appear as straight lines on normal recording gage charts because of the compressed time scale.

Water levels in wells near lakes and streams fluctuate in response to changes in surface-water stages. Under water-table conditions the fluctuations are largely due to actual movement of water into and out of the aquifer; the response of artesian wells is due to the changing load on the aquifer. Wells which respond to changes in surface-water stage because of the loading effect should also respond to changes in atmospheric pressure. In general, fluctuations from the effect of surface-water–stage changes decrease with distance from the surface-water body.

Earthquakes affect water levels in some artesian wells (see Vorhis, 1955). These fluctuations probably result from compression and expansion of elastic artesian aquifers with the passage of earthquake waves. The speed of these waves is approximately 125 miles per minute (mi/min) (Parker and Stringfield, 1950); thus, water-level fluctuations may occur within an hour after the most distant earthquakes.

Water levels in some wells also fluctuate in response to earth tides in a regular semidiurnal cycle. Fluctuations caused by earth tides are generally very small (Robinson, 1939). Minor water-level changes are sometimes caused by wind blowing over the tops of wells. Air pressure within the well is suddenly lowered with the passing of a gust of wind, causing the water level to rise. After the gust passes, the atmospheric pressure rises, and the water level returns to its initial position.

Drawdown and recovery occur each time pumps are operated. Weekday-to-weekend variations in nearby pumping are reflected in water-level trends in many wells within pumping centers.

Groundwater levels are influenced by seasonal cycles in such factors as recharge from precipitation, evapotranspiration, and discharge from wells and show a seasonal pattern of fluctuations. These seasonal variations in water levels are superimposed on secular fluctuations extending over periods of several years or more. Water levels respond to alternating series of wet and dry years in which recharge from precipitation is above or below the mean. In overdeveloped areas, groundwater levels may continue in a downward trend for many years as discharge exceeds recharge.

The water table in many places under natural conditions recedes in late spring, summer, and early fall when discharge by evapotranspiration and by groundwater runoff to streams is greater than recharge from precipitation. Water levels begin to recover in wells late in the fall when evapotranspiration losses are small and conditions are favorable for the infiltration of rainfall, first to replenish depleted soil moisture and later to percolate to the water table. The rise of water levels is especially pronounced in the wet spring months when the groundwater reservoir receives most of its annual recharge. The high and low points of the annual cycle of water levels occur at different times of the year, depending in large part upon the seasonal and areal distribution and intensity of rainfall. Groundwater recharge is greatest in spring months of heavy rainfall and during wet years, and is least in summer and fall months and during dry years. Hydrographs for humid areas show that during most years downward trends of water levels can persist almost without interruption for periods exceeding four months, usually from June through October. Minimum and maximum water levels are recorded at various times of the year depending upon antecedent water-level conditions in addition to the precipitation conditions during the year of record. Extended dry periods have pronounced adverse effects on water levels; groundwater storage is considerably reduced during these periods.

Discharge of groundwater to streams is greatest during periods when the water table is high and is least during periods when the water table is low. A large portion of streamflow is groundwater runoff; therefore, streamflow is considerably influenced by groundwater levels.

Water levels in areas remote from pumping centers furnish a good index of the long-term trends of the water table. The stage of the water table correlates with the precipitation; in general, water levels rise during periods of high precipitation and decline during periods of drought. Water levels are at low stages during dry periods; however, as precipitation increases, water levels return to high stages. No long-term trends of general rise or decline of the water table are discernible according to information on long-term water-level measurements made in various parts of the conterminous United States for periods ranging from about 20 to 50 years (Fishel, 1956).

In heavily pumped areas, changes in water levels caused by pumping are superimposed on seasonal and secular fluctuations due to natural phenomena. Water levels are influenced by seasonal fluctuations in pumping rate. During the late spring and summer, increased pumpage of groundwater for cooling and irrigation results in a local or regional decline in water levels. During the fall, winter, and early spring months, decreases in air temperature result in a reduction in pumpage and a recovery or a decrease in the time-rate of decline of water levels.

In heavily pumped areas a downward trend of water levels may continue for many years because of continual increases in pumpage or withdrawals in excess of recharge, or both. In some instances, large developments of groundwater have caused pronounced and serious declines of water levels. There are, however, many areas of groundwater development where serious water-level declines have not occurred.

Water levels in wells are measured by chalked tape, electric probes, air lines, and the reflection of sound (see Davis and DeWiest, 1966). Various types of electrically actuated pressure transducers are available and can measure changes in water level of less than 0.001 foot. Mechanically actuated drum recorders attached to floats in wells are one of the oldest and most reliable types of water-level measuring devices. With proper gear ratios, water-level fluctuations of less than 0.001 foot can be recorded.

Data on groundwater pumpage provide records of short-term changes and long-term trends of fluctuations in withdrawals from aquifers. Pumpage data are useful in the following ways: (1) to identify areas of detrimentally high groundwater withdrawals; (2) to facilitate prediction of future groundwater supplies by showing the time-rate of change in groundwater pumpage; and (3) to appraise the relation between pumpage and water-level declines.

The reliability of pumpage data varies greatly. Municipal pumpage is usually metered in cities, but many smaller villages and subdivisions operate without meters. A few of the larger institutions meter their pumpage. Only a small part of industrial and irrigation pumpage is metered. Pumpage of unmetered public supplies is usually estimated from the population and number of services, number and types of industries served, number of hours pumps are in operation, and the capacity of the pumps. Estimates of pumpage of industrial supplies are often made from the use classification of the plant, number of hours per day and days per week the industry operates, number of hours the pumps run, and the capacities of the pumps. Pumpage for domestic supplies may be estimated on the basis of detailed use surveys of a few selected typical areas. For all these reasons, it is difficult to ascertain exact pumpage figures.

Pumpage-use data are generally classified according to four main categories. These are: (1) public, including (*a*) municipal and (*b*) institutional; (2) industrial; (3) rural nonirrigation; (4) irrigation, including (*a*) form and (*b*) golf courses and cemeteries. Most water-supply systems furnish water for several types of use. For example, a public supply commonly includes water used for drinking and other domestic purposes, manufacturing processes, and lawn sprinkling. The water supplies for industries, as well as those for golf courses and cemeteries, may also be used for drinking. In all cases, the total pumpage may be known approximately, but the final use of the water cannot always be determined.

2.8 Surface Geophysical Methods

Earth-resistivity and seismic and gravity prospecting often can be used to locate aquifers and production wells (see Dudley et al., 1964; Foster and Buhle, 1951; McDonald and Wantland, 1961; McGinnis and Kempton, 1961). Geophysical methods based on measurements at the earth's surface do not yield as complete information about the subsurface as does test drilling, and they will never even under favorable conditions replace test drilling. However, they can in many cases reduce substantially the amount of test drilling required and aid in a more intelligent selection of test-hole sites. The most common use of earth-resistivity methods is in determining the thickness of shallow sand and gravel deposits. Seismic prospecting provides fair estimates of layer depth, and gravity prospecting may be used successfully in detecting broad, deeply buried valleys.

The success of surface geophysical methods depends on measurements of variations of density, electrical conductivity, electrical potential, elasticity, and other physical properties of the earth. If the contrasts in properties are slight and/or the spatial distribution of geologic units is too complex, measurements cannot be interpreted geologically.

Earth-resistivity prospecting involves the passage of an electric current into the ground through a pair of current electrodes and the measurement of the resulting voltage drop at the earth's surface between a pair of potential electrodes. A quantity called the *apparent resistivity* may be derived from the ratio of the measured voltage to the impressed current. The apparent resistivity will vary as the positioning and spacing of the four electrodes change, and variations in resistivity depend on the true resistivities and distribution of the subsurface materials. The ability of the resistivity methods to detect sand and gravel deposits depends upon the thickness of the deposit relative to its depth of burial and on the resistivity contrast with the overlying material. The wrong combination of these factors can make a sand and gravel deposit undetectable.

Various electrode configurations are employed in earth-resistivity exploration. The most commonly used is one in which the electrodes are evenly spaced along a line, the outer pair being the current electrodes and the inner the potential electrodes. The distance between two adjacent electrodes is roughly equivalent to the depth of penetration. Data are obtained either by keeping the central point of the electrode array fixed and taking current and potential readings for a series of electrode spacings, or only one or two spacings may be used at any one location and a larger area covered. The former method, called *depth-profiling*, is most suited for quantitative treatment of the data while the latter method, called the *lateral-profiling* method, may be used to delineate the areas in which anomalies occur. Depth-profiling data may be interpreted qualitatively or quantitatively; in either case, it is usual to plot the apparent resistivity against increasing values of the electrode spacing. A qualitative interpretation may be accomplished by studying the depth-profiles.

Various quantitative methods for interpretation of depth-profiling data exist. Great difficulty is encountered in analyzing data quantitatively largely because of the differences between actual and the idealized geology assumed in theory. The apparent-resistivity curve that would be obtained over an earth consisting of horizontal, homogeneous, and isotropic layers for any combination of layer thicknesses and resistivities can be derived theoretically. However, the unique identification of an observed resistivity curve with one specific combination of layer thickness and resistivities is not as easily solved except in the two-layer case. A number of empirical methods exist which base their depth determinations on the electrode spacings at which points of inflection, or maxima and minima, are observed in the resistivity curve. In the curve-fitting method, the field curve obtained using the electrode configuration mentioned earlier is compared to the members of a set of curves computed for a number of discrete resistivity and layer-thickness ratios, and for the two-, three-, or four-layer cases. Over 2,000 of these curves are available for comparison but they are insufficient to match all the possible combinations. This necessitates interpolation between the curves with respect to both resistivity and layer thickness and causes great difficulty in solving problems. Another method involves matching the initial part of the field curve to a simple layer situation and then gradually adding deeper layers and adjusting the properties of the upper layers so that the field data are matched over a greater and greater range of electrode spacings.

The earth-resistivity method is useful for the detection and delineation of near-surface aquifers and for outlining the courses of buried valleys. The method has less success in the detection of deeper aquifers. The standard quantitative interpretative methods employed in earth resistivity fail or are subject to misinterpretation if such factors as inhomogeneities, dipping formations, or anisotropy assume importance.

If a shock is generated at or near the earth's surface by exploding a small dynamite charge or by the impact of a heavy instrument, the resulting shock wave will spread out through the earth in all directions from the source. The nature of subsurface materials influences the paths taken by and the speed of travel of the seismic waves. Thus, seismic prospecting often can be used to locate aquifers.

The normal field technique used in seismic exploration is reverse shooting. Detectors or geophones are equally spaced along a linear spread and explosive charges are fired successively at the two ends of the spread. A geophone is a very sensitive electromagnetic device capable of picking up and converting into an electric signal slight movements of the ground beneath it. A seismograph records for each geophone the elapsed time between the firing of one of the charges and the arrival of the resulting shock wave at the geophone. A seismogram is thus obtained for each of the two shots fired. The graphs of arrival times plotted against distances from the shot points determine the values of the seismic wave velocities in formations. The true velocities, apparent dips, and depths to bedrock at any point along the length of the spread may be calculated. Spread length is determined by the desired depth of penetration and by velocity contrasts.

Most seismic equipment is multichannel equipment and records the ground movements at six, twelve, or more geophone locations simultaneously. Inexpensive single-channel seismic equipment which may be very useful in near-surface geophysical prospecting has become available. The multichannel and some single-channel instruments provide continuous records of all seismic energy arriving at each geophone. In all common arrangements with multichannel equipment, the geophones are spaced at equal intervals along a straight line, the intervals being determined by the type of prospecting. In seismic reflection, the spacings between geophones are small compared to the depths of investigation. In seismic refraction, geophone spacings are more comparable to the depths of investigations; the spread length is always greater than the maximum depth of penetration. Seismic-refraction prospecting is most suited to shallow exploration, because the length of the spread will not then become prohibitively large. The reflection technique, on the other hand, is not suited to very shallow exploration because the reflected waves from shallow horizons reach the surface so soon after the onset of the direct and refracted waves that their effects are confused with and overshadowed by the still vigorous ground vibrations excited by these earlier arrivals.

Interpretation of seismic-refraction results is based on the time-distance graph, a plot of the observed arrival time for the first arrival at each geophone against the distance of the geophone from the point of origin of the seismic energy. The combination of low-velocity contrast and deep bedrock leads to difficulty in seismic-refraction prospecting. The principal limitation for conventional seismic-refraction equipment is any combination of low-velocity contrast and deep bedrock that makes the methods so unwieldy because of the special spread arrangements required that the expense and time involved in a single seismic depth determination are prohibitive.

Small compact seismic timers can be used for shallow refraction surveys. If penetration of 50 feet or less is desired, a sledgehammer can be used to create the needed vibrations. This eliminates the use of high explosives.

Gravity is an attractive force that acts between any two masses. It is directly proportional to the product of the masses and inversely proportional to the square of their separation. Because of the gravitational force any body at the earth's surface is attracted by the mass of the earth as a whole. Local variations in the force of attraction are observed, however, and these are a consequence of the variations in density of the materials within the earth's crust, the materials closest to the body being attracted having the greatest effect.

The major part of the earth's gravitational field is a result of the attraction of the earth's mantle and core, and the variations caused by crustal variations in density are very rarely much greater than 1/10,000 of the total field. It is these small variations that are of interest in gravity prospecting, so that instruments used for this purpose must be extremely sensitive. This requirement becomes even more stringent in groundwater exploration because the anomalies sought are some two orders of magnitude smaller.

The gravity meter, or gravimeter, is the instrument most commonly used for geophysical exploration. Field operation consists essentially of the "weighting" of a known object with great precision at a number of selected stations. The difference between the "weights" at two different gravity stations is a measure of the gravity difference between the two stations. The elevation of each gravity station is an essential part of the data because of the corrections that have to be applied to the observations. Periodic observations are required at one or more base stations as a check on the instrumental drift.

The reduction of gravity measurements consists of the application of a series of corrections designed to reduce the observed value at each station to an equivalent value that would have been obtained if the station had been at sea level and at the equator, on a nonrotating earth and surrounded by a completely featureless terrain.

Interpretations of the corrected observations may vary from a simple search for anomalous areas to attempts to explain the sizes and trends of the observed anomalies, making reasonable assumptions concerning the subsurface geology.

Shallow narrow buried valleys are probably detectable by this method. For this reason, gravity exploration may be useful in areas where unconsolidated deposits are thinner but where there is a lack of contrast between layer velocities. Gravity prospecting may prove successful in the detection of relatively broad and deeply buried valleys. Local changes in the anomaly associated with a buried valley may be due to the presence of permeable deposits and permit the direct detection of sand and gravel.

2.9 Subsurface Geophysical Methods

Subsurface geophysical methods of investigation that yield data interpretable in terms of aquifer depth, thickness, continuity, areal extent, structure, porosity, permeability, and degree of saturation and on the chemical quality and temperature of the contained water are now in general use (see Patten and Bennett, 1963). Spontaneous potential logs, resistivity logs, radioactive logs, acoustic logs, temperature logs, caliper logs, and fluid-conductivity and fluid-velocity logs are most useful in groundwater exploration. Many subsurface geophysical methods are restricted to use in uncased wells, although surveys such as gamma radioactivity and caliper can be used with casing in the hole. Electrical logs are most useful for correlating subsurface geologic formations and for determining the occurrence of fresh groundwater. Information concerning subsurface geophysical methods given by Patten and Bennett (1963) is briefly summarized in the following sections.

In cable-tool holes the examination of drilling cuttings permits the obtaining of logs which are reasonably accurate. This is not so when rotary drilling is used, as the samples taken contain drilling cuttings from several feet of hole and also debris resulting from the erosion of the exposed formation by the mud stream. Electric logging, because of its accuracy, reliability, and simplicity, is generally used to obtain information from rotary-drilled wells.

Electric logs can only be recorded in open holes. Conventional electric logs consist of a potential curve and one or more resistivity curves recorded side by side as a graph. The potential curve is the continuous recording in terms of depth of the natural electric potential in millivolts. It is sometimes called the SP curve (spontaneous potential). The potential in clay (or shale) is generally used as a reference, for convenience. Potential readings in aquifers vary in direction and amplitude according to the respective salinities of the drilling mud and of the formation water.

A resistivity curve is obtained by lowering one or several electrodes and recording, also in terms of depth, appropriate electrical measurements. As the logging tool travels in the hole, the recorded variations of formation resistivity constitute a resistivity curve. When only one electrode is used, the graph obtained is called a *single-electrode resistivity curve*. The graph obtained when the downhole tool carries several electrodes is called a *normal* or *lateral* resistivity curve depending upon the electrode arrangement.

Basically, an electric logger consists of a downhole tool, sometimes called a logging electrode or *sonde*, connected to a reel-mounted logging cable, a depth-measuring device, a control panel, and an automatic recorder. One end of the logging cable is connected to the surface control panel and the recorder. The measuring device is mechanically connected to the chart drive mechanism so that the chart moves in synchronism with the sonde. Two recording pens moving back and forth, according to the amplitude of the signals (potential and resistivity) received from the sonde, draw the corresponding curves. "Two-curve" loggers are the simplest and least expensive. They permit recording a potential and a single-electrode resistivity curve and are satisfactory for usual water-well drilling applications. "Multiple-electrode" loggers permit recording a potential and one or more resistivity curves. The latter loggers are somewhat more complicated and expensive but they permit obtaining quantitative information regarding the fluid content and other characteristics of the formation penetrated by the drill.

The spontaneous-potential logging device consists of a recording galvanometer connected between a ground terminal at the surface and a moving electrode on an insulated conductor in the borehole. The galvanometer measures the potential difference between the earth at the surface and the point in the mud- or water-filled borehole occupied by the moving electrode, and it records this potential difference as a function of depth. The result is a log in which the deflections indicate potential differences between points along the axis of the well. The qualitative recognition of zones containing highly mineralized water is an important aspect of interpretation of spontaneous potential. In water-filled boreholes, recognition of the highly mineralized zones is possible as long as the borehole contains water that is chemically different from that in the zone in question. The spontaneous-potential log is a valuable qualitative method in groundwater studies. It has great value as a means of lithologic correlation, and it can sometimes aid in the identification of internal flow or in the location of zones of highly mineralized water. The assumptions involved in quantitative interpretation of spontaneous potential, however, are seldom satisfied in groundwater work. The use of the method as a quantitative tool is, therefore, likely to produce erroneous results.

When resistivity logging is done by the single-point resistance method, the measurement made indicates the resistance of a segment of borehole and earth extending to a radius of approximately 10 times the radius of the electrode. Much of the zone of investigation of the point-resistance device lies within the open borehole. The zone of investigation of the short normal includes a greater proportion of the surrounding formation and that of the long normal includes an even greater proportion.

The resistivity of a rock is generally inversely proportional to its porosity and to the ionic concentration of its interstitial fluid. Empirical relations between rock resistivity, interstitial fluid, resistivity, texture, and porosity are used to determine the *formation-resistivity factor* of a rock. The formation-resistivity factor is defined as the constant ratio of true rock resistivity to the resistivity of the water saturating the rock. In problems in groundwater the saturating fluid is normally fresh or slightly salty water, and large amounts of clay minerals in an aquifer may cause the formation-resistivity factor to vary. The presence of clay minerals can lead to many mistaken interpretations of electric logs, both lithologic and hydrologic.

The electrical properties of a rock can be expected to change wherever lithology changes significantly. As the logging assembly is lowered through a thick, homogeneous, and isotropic formation the apparent resistivity recorded by each device reflects the resistivity of the formation and the relative effect of the borehole within the zone of investigation of the device. The large zone of investigation of the long normal serves to minimize the borehole effect so that the response of this device is generally close to the true rock resistivity in a thick homogeneous formation. At the same time, however, this large zone of investigation makes it impossible for this device to indicate lithologic detail in zones of thin, alternating beds. The single-point device, on the other hand, because of its relatively small zone of investigation, defines thin beds in detail and identifies geologic contacts with greater accuracy, but it generally gives a rather poor approximation of the true rock resistivity.

During the logging of water wells, the borehole is usually filled with drilling mud. If the drilling mud has invaded the formation to a certain radius and if contamination of the drilling mud by ionic diffusion is negligible, the borehole will be surrounded by a region of low resistivity. A low short-normal resistivity is not a reliable indication of invasion of the formation, and differences between the normal resistivity curves cannot always be used for the identification of permeable zones on the assumption that only permeable zones will be invaded.

If the bed is not invaded by drilling mud, the difference between the apparent resistivities shown by the long- and short-normal curves is very small and is due only to the greater effect of the borehole in the zone of investigation of the short normal. Because the resistivity contrast of the drilling mud and formation water is very small and differences in the apparent resistivities are greatly affected by other factors, a quantitative form of interpretation to calculate the true formation water resistivity is subject to a high degree of uncertainty.

These situations illustrate the inability of resistivity logging to supply complete information, even qualitative information, regarding groundwater conditions. Thus, before the interpretation of resistivity data is attempted, all available evidence from other sources, including other types of well logs, should be studied.

The familiar features of secondary porosity in crystalline and dense limestone rocks, fractures, joints, shattered fault zones, and solution openings are the water-yielding zones of the rocks and are characterized by a relatively low electrical resistivity. Features of secondary porosity can be indicated by resistivity logging in dense limestone and crystalline rocks; in a sandstone-shale sequence, however, resistivity logging is of little help in identifying zones of this type. The effect of clay minerals in decreasing the resistivity of shale generally overshadows any resistivity variation due to changes in porosity. Thus, the difference in resistivity between a fracture zone in a shale interval and the unfractured shale itself is usually too small to be indicated by resistivity logging.

Differences in the intensity of natural radioactivity emitted by rocks of the earth's crust make possible the use of radiation measurements for well logging. The radioactivity of rocks other than those containing radioactive ores is extremely small, and highly sensitive instruments are required for its measurement. Inasmuch as gamma rays have high penetrating power and can easily pass through steel casings, they are useful in well logging to indicate the radioactivity of rocks. The basic instrument used in radioactive logging consists of a detecting unit, which may be one of the following: (1) ionization chamber, (2) Geiger-Müller counter, (3) scintillation counter, or (4) neutron detector. Because neutrons are largely affected only by material containing hydrogen, for example water, a measure of the dissipation of neutron energy in the material adjacent to a borehole constitutes the basis of a useful well logging method.

Radiation logs can be divided into two classes: (1) radioactivity logs (generally gamma ray), and (2) neutron logs. The most common type of rock that can be readily identified on radioactivity logs is shale, which has a medium radioactivity intensity, and clayey or shaly sand or sandstone, which have intermediate intensities. Limestone and dolomite beds have intensities in the same range with sandstone, and a knowledge of the local lithology is necessary for their differentiation. Difficulties in the interpretation of rock type from radioactivity data are indicated by the fact that the ranges in radioactivity of several types of rock overlap. However, there is usually sufficient difference to give the logs great utility after the general correlation of lithology with radioactivity has been established for a given locality.

Radioactivity logs are useful for identification of beds, for accurate determination of tops and bottoms of beds, and for structural and stratigraphic studies involving logs of several boreholes in a locality. Although they give no direct indication of the water content or salinity of water in an aquifer, they have been used with supplementary data to locate fresh-water sands penetrated by unsuccessful or abandoned oil-test wells. The principal advantage of radioactivity logging is that it provides the only means by which information on the lithology of beds penetrated by cased holes may be obtained at any time. The principal deficiency of the radioactivity log is that it does not provide information on the porosity of the rocks. To obtain porosity data, the neutron logging methods were devised.

The two neutron logging methods in common use differ in that one measures the intensity of the gamma radiation of neutron captures, whereas the other measures the number of slow neutrons that reach a detecting element in unit time. Both phenomena are indirect quantitative functions of the same condition, namely, the fluid content (porosity) of the rock forming the borehole wall. In reality, it is the abundance of hydrogen atoms that is measured.

The neutron-neutron logging method, like the neutron–gamma-ray method, utilizes a radium-beryllium source below a counter which is sensitive only to slow neutrons. The principal advantage of the neutron-neutron logging method is that it is not affected by the natural or artificially induced gamma rays. Because of that feature, the neutron-neutron log indicates the fluid content of the rock material more accurately than the neutron–gamma-ray log. The combined radioactivity and neutron log provides interpretive power comparable to that of the electric log. Interpretive procedures for utilizing gamma-ray logs have, for the most part, been restricted to qualitative identification of lithology and formational contacts.

The acoustic log measures the speed of sound in the rocks near a borehole. Inasmuch as the travel time of the sound is strongly dependent on the porosity, the acoustic log is widely used to estimate porosity. The present techniques have been applied most widely to moderately porous, well-compacted sandstones, limestones, and dolomites.

As a result of low thermal conductivity there is a rather steep depth-temperature gradient in the zone of rock penetrated by exploratory boreholes. The thermal gradient with depth in any locality is mainly a function of the thermal conductivity of the rocks that underlie it. The reference temperature at shallow depth (50 to 100 feet) is approximately the average annual air temperature. The temperature change with depth below the land surface and the zone of seasonal fluctuation is generally expressed in terms of the reciprocal gradient, that is, depth per degree of temperature change. Reciprocal gradients in the crust within 20,000 feet of the surface generally range from 50 to 100 ft/°F. Since the thermal gradient is in part a function of the textural properties of rocks, temperature logs may provide information on the lithologic units penetrated by a well.

A temperature log is a record of the apparent temperature of the rocks penetrated by a borehole. Temperature logging is readily accomplished by adapting conventional electric logging circuits to measure change of resistance of a temperature-sensitive metallic conductor lowered into the hole on the logging cable. One of the first devices of this type employed a platinum wire about 20 inches long recessed in a groove in the rubber wall of the electrode plummet or sonde. Some temperature logging units utilize a glass-insulated semiconductor (thermistor) which has a high negative-temperature coefficient of resistance.

A differential temperature surveying method involving measurement of very small temperature differences between two temperature-sensitive elements spaced 6 feet apart in the borehole has recently been devised. This is called the *delta log*, and its principal advantage is that it responds passively to normal thermal gradients and actively to abnormal gradients. Logging rates of 50 to 100 feet per minute (fpm) yield excellent detail, the record being obtained on the downward traverse. The sensitivity of the equipment is very high, differentials of 0.001°F being measurable. For accurate determination of the relative thermal conductivity of the beds penetrated, the log must be made several days after drilling has stopped, when thermal equilibrium between formations and borehole fluid has been reached.

In many groundwater investigations, temperature measurements are obtained only for water being discharged from flowing or pumped wells. Measurements of water temperature in wells have been made with thermometers inserted in bottles and lowered down the wells. The most precise way to measure the temperature of water in place in wells is with an electrical thermometer. Such thermometers, accurate to 0.1°F, are available from several manufacturers.

The definition of zones having a constant thermal gradient is a function of thickness of the bed and of the contract in their thermal conductivity. An alternating sequence of thin beds of sandstone and shale, or of any other rock types, gives no appreciable inflection to the depth-temperature curve on the conventional temperature log, and no characteristic straight-line segments. The log of a borehole penetrating dense igneous or metamorphic rock beneath a sedimentary rock mantle generally shows a marked inflection of the curve at the contact.

Temperature logs of wells penetrating several aquifers at different depths can be used to identify the aquifer or aquifers tapped by the wells. Measurements of the temperature of water from a flowing or pumped well can be used to estimate the depth of the aquifer tapped, if the depth-temperature gradient locally has been determined for equilibrium conditions. If a well taps two aquifers, the temperature of the water pumped from the well may be used to estimate the relative rate of withdrawal from each aquifer.

The rocks that form the walls of a borehole are worn away as the hole is deepened, partly by abrasion from drilling tools and partly by erosion and hydration effects of the drilling mud. The extent of diameter variation with depth depends upon: (1) the action of the drilling tools; (2) the relative hardness, competence, and fluid content and permeability of the rock formations penetrated; and (3) the degree to which the rocks dissolve in or become hydrated by the drilling mud.

Rock formations that are dense, hard, and competent and insoluble in the mud fluid drill slowly, and the diameter of the borehole closely approximates bit size. Deposits of thick-bedded limestone or dolomite, quartzite, schist, hard shale or slate, and other massive formations only slightly soluble in water form smooth borehole walls when drilled with water-base fluid. Anhydrite or gypsum, rock salt, and deposits that contain thin beds of these and other water-soluble minerals cave when drilled with water-base fluid but stand up well if oil-base fluid is used. Shales vary widely in their tendency to cave, and not all the factors involved are well understood.

Borehole enlargement by drilling-fluid erosion, or by swelling and collapse, is common in sandy or silty clay and shale. Soft fissile or massive shale and compacted clay sometimes tend to become hydrated when drilled with water-base mud fluid. The rock swells, becomes incompetent, and sloughs into the hole, exposing fresh rock for further hydration. Most permeable granular rock formations drill rapidly, and, where they are penetrated at depths greater than several hundred feet, the diameter of the borehole is seldom appreciably greater than bit size. At shallower depths, beds of sand sometimes cave while beds of clay or shale do not. The hole size through beds of sand or sandstone at depth generally is approximately equal to bit size as a consequence of mud invasion and the deposition of mud residue (filter cake) on the wall of the hole. Beds of sandstone, sand, and gravel often become evident by constriction of the borehole.

The variation of borehole diameter with depth is measured mechanically and recorded electrically by making a traverse of the hole with a caliper tool. The caliper tool is a chrome-plated device, 3 inches in diameter and 6 feet long, with four arms hinged to it. A shielded electric cable is used to lower the tool to the well bottom, where the arms are released and press against the wall of the borehole. As the caliper is slowly withdrawn from the well, the spring-loaded arms follow every change in the diameter of the hole. These changes are transmitted over the suspending cable and recorded at the surface on photographic film. Hole diameters are automatically and continuously plotted in this manner.

Different types of rock may produce similar patterns of diameter change when drilled under analogous conditions. However, there generally is sufficient variation among rock types so that, with a general knowledge of the sequence to be expected, certain identifications are possible. Cavities indicated on borehole-diameter logs of wells drilled in consolidated rock may be caused by local zones of fractures or solution openings. Because groundwater in consolidated rock commonly occurs in such zones, the cavity may indicate a water-yielding zone.

Flow-meter studies in uncased wells tapping consolidated rock are likely to give more accurate information if measured velocities are adjusted for the hole diameter. Quantitative interpretation of electric logs of thin beds requires hole-diameter data for computation of true resistivity, and the interpretation of temperature logs of boreholes in which thermal equilibrium has not been reached is facilitated by the borehole-diameter log.

The fluid-conductivity log of a water-filled borehole often can give a general indication of water quality and can sometimes provide information on the hydraulics of a well. It is useful in determining the depth of salt-water leaks in cased artesian wells and the depth and relative artesian head of salt-water aquifers penetrated by uncased wells. The conductivity of the water in the well at a given depth depends upon the concentrations, mobilities, and valences of the ions in the water at that depth. It is, therefore, controlled partly by the processes of ionic diffusion and dispersion and ion transport by moving fluids.

Fluid-conductivity logging requires the use of a pair of closely spaced electrodes which are lowered into the well on two-conductor insulated cables connected to a battery or generator and an ammeter or ohmmeter at the land surface. The insulated cable is lowered into the hole over a calibrated sheave, so that the depth of the electrodes is known at all times. Values of fluid conductivity or conductance (or their reciprocals, resistivity or resistance) at selected depths are recorded and plotted on coordinate paper with depth as the ordinate.

The interpretation of fluid-conductivity data for wells penetrating a single aquifer is controlled by the following principles: If the log shows no gradient, the conductivity of the water in the aquifer is probably the same from top to bottom and is equal to the conductivity shown on the log; if the log displays a gradient, the conductivity of the water in the aquifer differs from top to bottom although perhaps not in exactly the same way as does the water in the borehole. Experience in a particular area or with a particular aquifer will often enable the log interpreter to estimate water quality from the conductivity log and to decide whether a laboratory analysis is necessary.

The flow-velocity log serves to identify and evaluate: (1) the aquifers tapped by cased wells having multiple screens, (2) leaks in cased wells, and (3) permeable zones penetrated by uncased wells. It is useful also in the interpretation of the spontaneous-potential curve of the electric log of a cable-tool well.

A flow-velocity meter is a record of the magnitude of the velocity and direction of movement of water in a well at all depths. The log may be made while the well is discharging water at the land surface, while it is receiving water from the land surface, or while there is no flow at the land surface. Flow-velocity logging requires the use of a velocity-sensitive instrument, a system of cables and conductors for lowering the device into a well, a depth-measuring device by which the position of the flow meter can be determined at all times, a cable reel and well-head equipment for lowering or raising the instrument, and a recording device at the land surface. The flow meter in wide use in water-well studies is the so-called Au Current meter. A recently developed meter for measuring fluid velocities in water wells is based upon the relation of fluid velocity to the dissipation of heat from an electrically heated element. It is much smaller than the Au Current meter and often can be lowered in wells without removing pumps and pump columns. The resistance of the element varies with temperature. The direction of flow of water in the well is easily determined by slowly raising and lowering the meter through a 20- to 30-foot section of hole and recording velocity measurements during both traverses. If the velocity measured is greater during the downward traverse than during the upward traverse, the direction of flow is upward, and vice versa.

Analysis of flow-velocity logs is facilitated by replotting the logs in terms of the reciprocal velocity gradient (depth interval per unit of velocity). Changes in flow rate in wells due to leakage through casing, or gains from or losses to aquifers, cause marked anomalies on the reciprocal velocity gradient log.

For quantitative analysis of hydraulic conditions in two or more aquifers tapped by a well it is necessary to obtain flow-velocity logs for the well at two or more different flowing or pumping rates. A flow-velocity log made while the well is pumped at a moderate rate or if the artesian head is sufficient and is allowed to flow will identify the zones contributing to the discharge and provide information on their thickness and relative yield at that discharge rate.

A closed-circuit television system for use in water wells has been developed that can be operated in boreholes 8 inches in diameter and larger. The equipment is well adapted for locating and examining fractures and solution openings, and for identifying changes in lithology in uncased holes drilled in consolidated rock. It is useful also for examining the position and condition of screens, casings, and obstructions in cased holes. The equipment was designed to be as simple as possible and to operate at a maximum hydrostatic pressure of 1,000 feet of water.

The system uses a high-resolution Kin Tel camera and monitor modified for use in boreholes. The monitor and camera provide a full 500-line resolution, which is about twice that of home television receivers. A 14-inch aluminized picture tube with a Polaroid filter mounted in front provides a bright clear picture even in high ambient light. The most difficult problem of design has been adequate illumination in the viewed area when the camera is below water level in a well. The most satisfactory lights used are 3-inch quartz iodide lamps rated at 600 watts each, wired in parallel. In larger holes, three lamps directed ahead at a 45° angle are placed 8 to 10 inches to the side of the camera.

Packers have been used for many years in testing and completing oil wells, but they have not been used extensively in testing water wells. Recent developments in design now make the use of inflatable straddle packers feasible for testing sandstone and siltstone aquifers. The packer elements are inflated against the wall of the test hole by water under a pressure of about 850 to 1,250 pounds per square inch (psi). Valve ports above, below, or between the packer elements are opened or closed by raising or lowering the suspension pipe. This permits water-level measurements, sampling, or pumping of the isolated zone between the two packer elements or of the composite zone above the top element or below the bottom element.

2.10 Case History of Groundwater Exploration

During 1964, the Saskatchewan Research Council conducted an extensive test-hole drilling program aimed at delineating the sand and gravel aquifers in the buried Missouri River and Yellowstone River valleys west, northwest, and north of Estevan in southeastern Saskatchewan, Canada (see Walton, 1966). The results of the exploratory investigations were later used to determine the potential yield of the aquifers with selected schemes of development.

The test-hole drilling program was designed taking into consideration information in two reports (Meneley et al., 1957; Christiansen and Parizek, 1961) and the results of studies on water-level conditions in the Estevan area (Walton, 1964). According to Meneley et al. (1957), the valley of the preglacial Missouri River can be traced from North Dakota into southeastern Saskatchewan as shown in Fig. 2.17. The preglacial Missouri River was jointed by the preglacial Yellowstone River at a point about 15 miles north of the international border. Downstream from the junction, the preglacial Missouri River flowed north of a highland in the vicinity of Estevan before entering Manitoba, where it joined the Red River to drain into the Hudson Bay.

The most pronounced topographic features of the southeastern corner of Saskatchewan are the Missouri Coteau in the southwestern part and Moose Mountain in the northeastern part. Between the Missouri Coteau, which rises to an altitude of 2,400 feet, and Moose Mountain, which crests at an altitude of 2,700 feet, lies the Weyburn Plain, at a general altitude of 1,950 feet. A large sag, nearly 10 miles in width across the Missouri Coteau, in the northeastern corner of Montana marks the

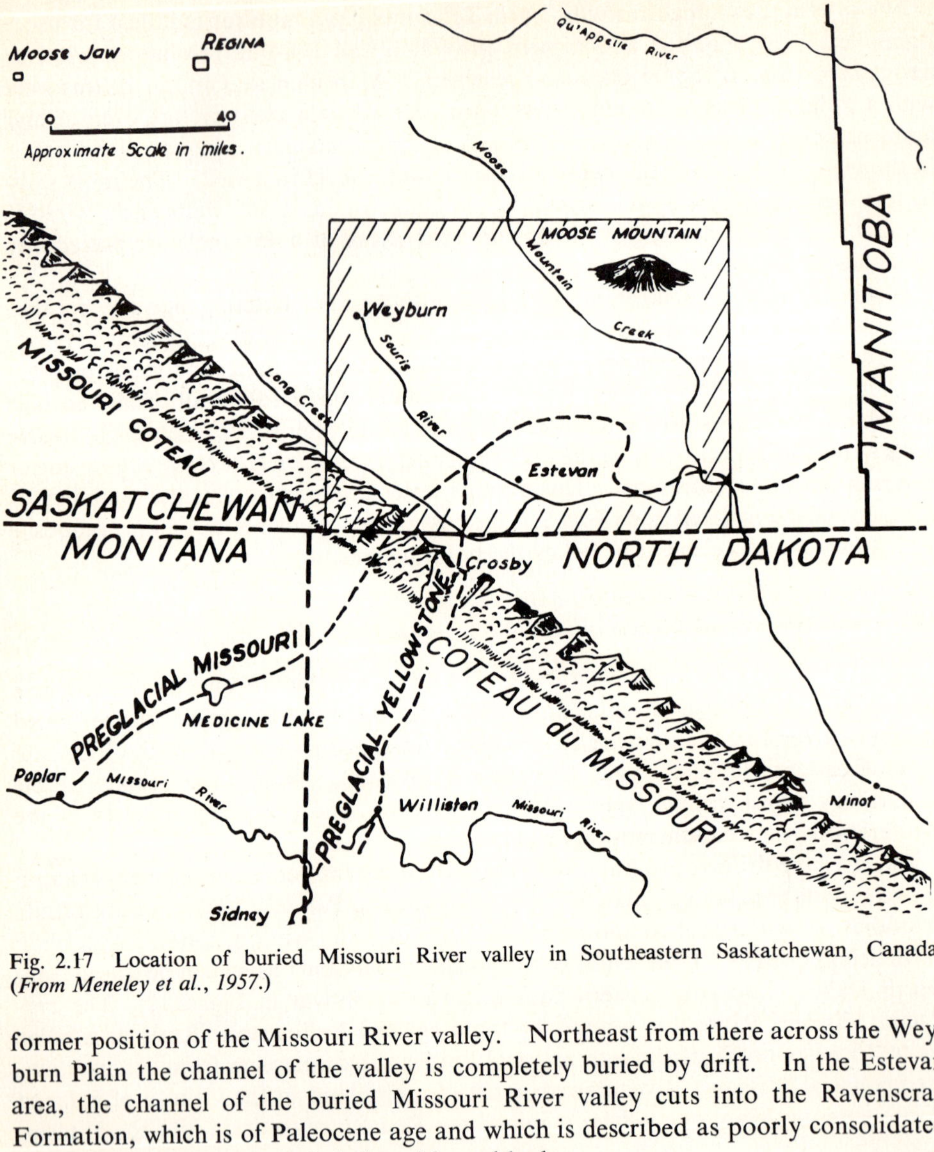

Fig. 2.17 Location of buried Missouri River valley in Southeastern Saskatchewan, Canada. (*From Meneley et al., 1957.*)

former position of the Missouri River valley. Northeast from there across the Weyburn Plain the channel of the valley is completely buried by drift. In the Estevan area, the channel of the buried Missouri River valley cuts into the Ravenscrag Formation, which is of Paleocene age and which is described as poorly consolidated interbedded sandstones and shales with coal beds.

A preglacial bedrock topography map prepared by Meneley et al. (1957) from subsurface information obtained from oil wells, shallow test wells, coal-mining operations, and water wells indicates that the buried Missouri River valley is about 12 miles wide and the average gradient of its floor is about 2.1 feet per mile. The glacial drift, which almost completely covers the bedrock surface, is more than 400 feet thick in places. The drift includes some alluvium, coluvium, etc.; it may also include gravels and other materials deposited by preglacial streams. Outside the preglacial valleys, the drift consists predominantly of till with some stratified sediments.

Based on the findings of Meneley et al. (1957), a reconnaissance investigation of the sand and gravel deposits in the buried Missouri and Yellowstone River valley was conducted in 1959 and 1960 by the Saskatchewan Research Council (see Christiansen and Parizek, 1961). Numerous resistivity studies were conducted to locate more precisely the permeable drift deposits within the buried valleys. Based on these studies, three test holes were drilled north and west of Estevan. The results of this reconnaissance test-hole drilling program suggested that the deepest parts of the channels of the preglacial Missouri and Yellowstone Rivers were at lower altitudes than anticipated by Meneley et al. A permeable sand and aquifer in the lower part of the drift was encountered in the test holes. Geological and geophysical evidence suggested the possibility that the aquifer had large areal extent and that large quantities of groundwater could be developed from the aquifer. Christiansen and Parizek (1961) recommended further exploratory test-hole drilling to determine more precisely the areal extent and character of the permeable sands and gravels in the deeper parts of the buried valley. Available information suggested that the most permeable sand and gravel deposits might occur in a narrow bedrock channel at lower altitudes than penetrated by existing test holes.

Before designing an extensive test-hole drilling program, the Saskatchewan Research Council decided to study water-level conditions in the Estevan area, reasoning that water-level data might shed additional light on the location of permeable sand and gravel deposits. The Geological Survey of the Federal Department of Mines, Canada, made an extensive study of groundwater conditions in the Estevan area in 1935, when groundwater development was inappreciable. Water levels in several hundred shallow and deep wells in glacial drift and bedrock were measured.

Previous studies suggested the possibility that permeable sand and gravel deposits of large areal extent might occur in two zones, between the depths of 150 and 250 feet and 250 and 450 feet. Water-level data for wells terminating in these two zones were studied for clues concerning the areal extent and continuity of sand and gravel deposits. It was reasoned that water-level contours might warp around positions of the buried-valley system where permeable aquifers occur.

Figure 2.18 is a water-level-contour map, prepared from data for wells terminating in bedrock or glacial drift at depths between 150 and 250 feet. Valleys in the piezometric surface correspond with the Souris River and its tributaries. Contours are also warped to some extent in the vicinity of the buried-valley system especially north and west of Estevan. The general pattern of flow of groundwater at depths between 150 and 250 feet in 1935 was slow movement in all directions toward the Souris River and its tributaries and toward the buried valley. The fact that piezometric contours are deflected only slightly in the vicinity of the buried valley suggested that the sand and gravel deposits between the depths of 150 and 250 feet were of limited areal extent and/or permeability and thickness. It was decided that test-hole drilling would have

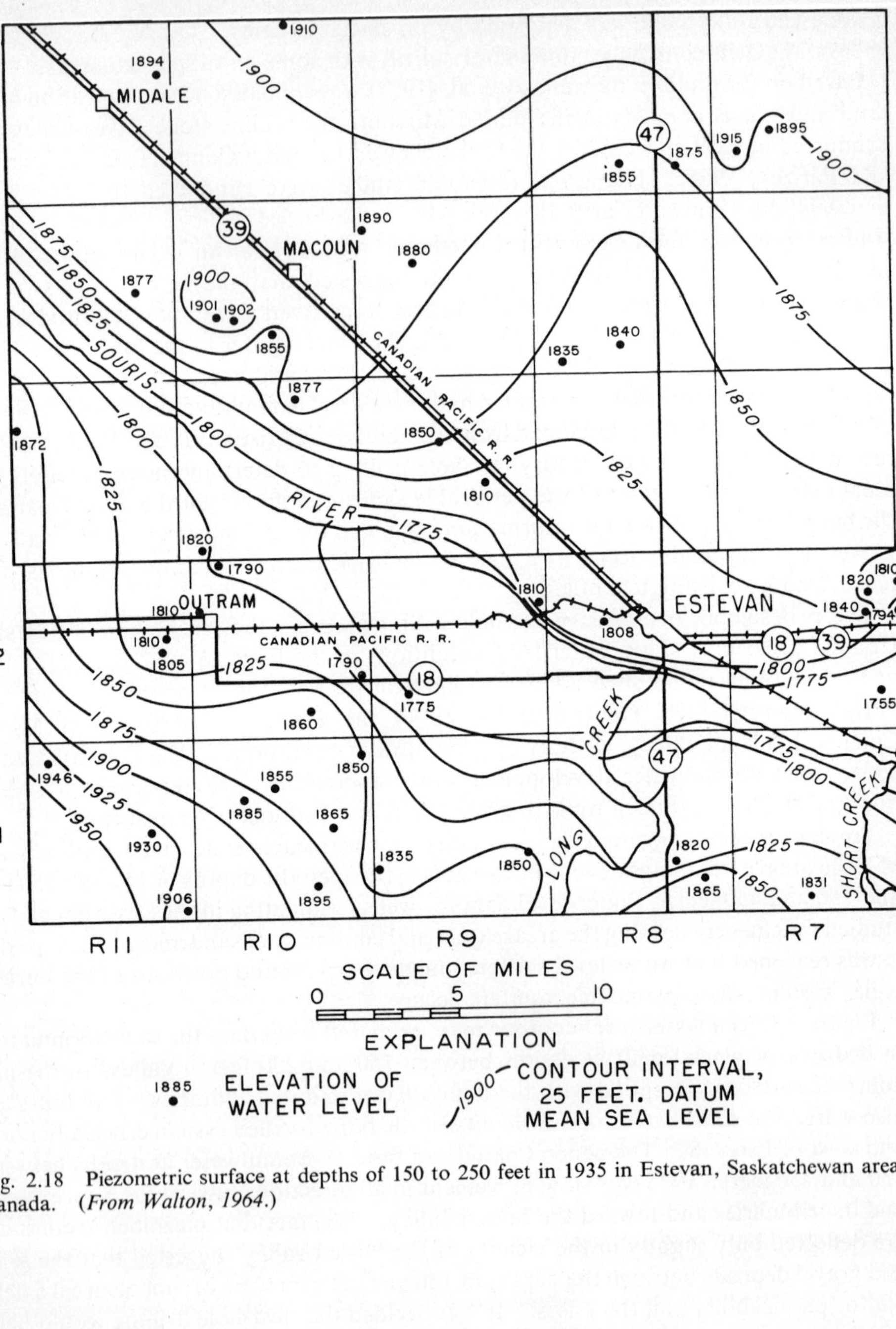

Fig. 2.18 Piezometric surface at depths of 150 to 250 feet in 1935 in Estevan, Saskatchewan area, Canada. (*From Walton, 1964.*)

to be extended to depths greater than 250 feet in search of an aquifer capable of yielding large quantities of water to wells.

Figure 2.19 is a water-level-contour map, prepared from data for wells terminating in bedrock or glacial drift at depths between 250 and 450 feet. Valleys in the piezometric surface correspond to the Souris River and its tributaries. Contours are also warped to a great extent around the buried-valley system especially north, northwest, and west of Estevan. The large and deep valley in the piezometric surface over the buried valley suggests that permeable sand and gravel deposits of large areal extent and thickness exist at depths exceeding 250 feet. The piezometric-surface map further indicates that the sand and gravel deposits pinch out in both the Missouri and Yellowstone River buried valleys south of Outram.

Based on these water-level studies it was decided that: (1) the extensive test-hole drilling program should concentrate on locating sand and gravel deposits at depths between 250 and 500 feet, (2) test holes should be drilled north and northwest of the Estevan along three lines at right angles to the center lines of the buried-bedrock valley and the valley in the piezometric surface, and (3) test holes should be drilled west and southeast of Outram along two lines at right angles to the center lines of the buried-bedrock valleys to test the validity of the conclusion that the sand and gravel deposits pinch out in the Outram area. It was estimated that at least 24 test holes might have to be drilled, assuming that the deepest channel in the bedrock containing the most permeable deposits was narrow and had a highly irregular course.

The test-hole drilling program started in March, 1964, and was completed the last of July, 1964. Test holes 1 through 29 shown in Fig. 2.20 were drilled using the hydraulic-rotary method. Drilling cutting samples were collected at 10-foot intervals and at each significant change in formation character. These samples were examined visually to determine the thickness, lithology, and stratigraphy of the formations encountered. Information on the color and oxidation of samples, grain-size characteristics, mode of sediment deposition, predominating rock types, and calcareous versus noncalcareous reaction of samples was noted. These features were correlated with driller's logs and also with electric logs obtained from a Widco electric logger. A professional geologist supervised the test-hole drilling program and had direct control over the distribution of drilling, the drilling methods employed, and the methods and procedures of sampling. Test hole 30 was drilled by the hydraulic-rotary method to a depth of 355 feet; a cable tooling was used to drill the lower 103 feet of stratified sediments including sand and gravel deposits. Bailer samples of drilling cuttings were taken at frequent intervals between the depths of 355 and 458 feet. Mechanical analyses of the bailer samples were made. Generalized logs for the test holes are shown in Figs. 2.21, 2.22, and 2.23.

Approximately 300 feet of predominantly glacial till overlies a distinctly stratified zone averaging 190 feet thick that in turn rests directly on bedrock. The most permeable sand and gravel deposits occur in the deepest parts of the buried-valley channel as a belt averaging 3,600 feet wide and 75 feet thick. These deposits pinch out west and southeast of Outram as suggested by water-level data and extend in

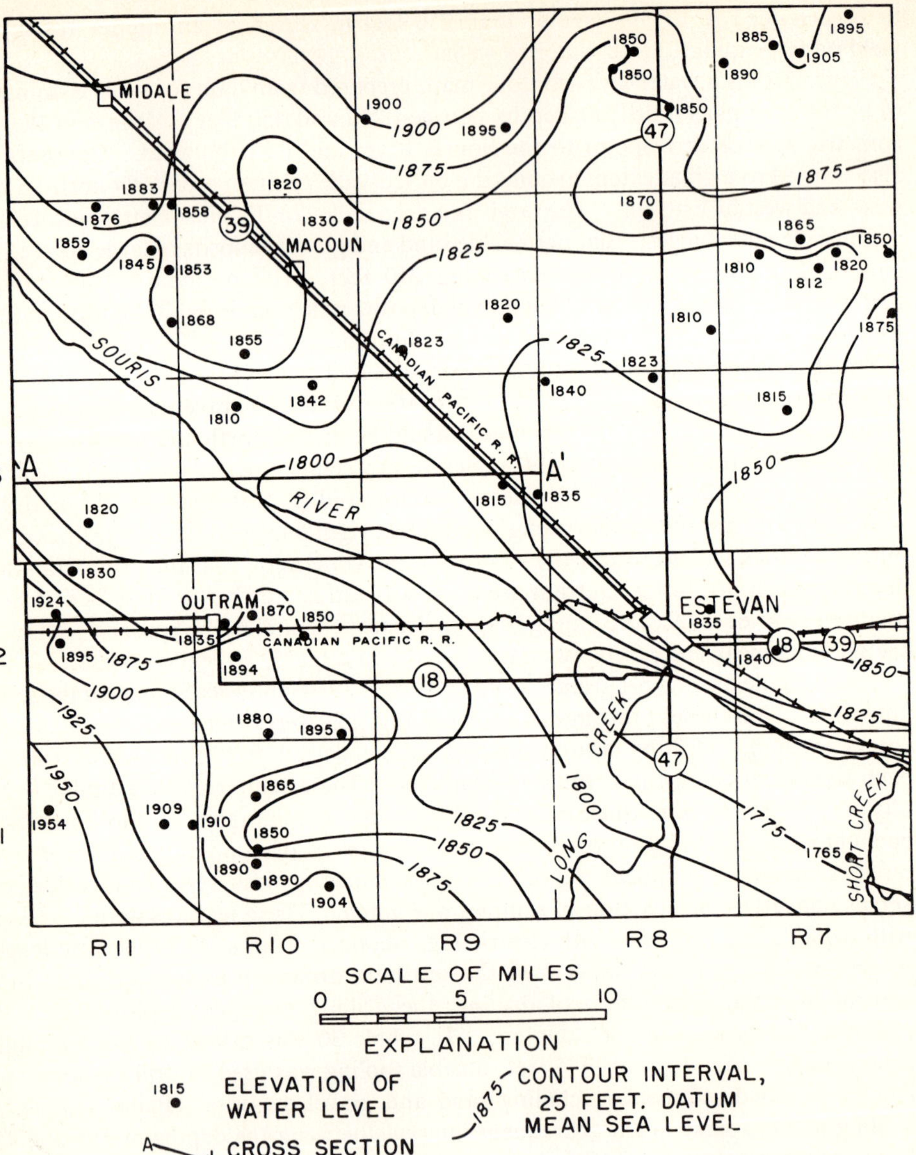

Fig. 2.19 Piezometric surface at depths of 250 to 450 feet in 1935 in Estevan, Saskatchewan, area, Canada. (*From Walton, 1964.*)

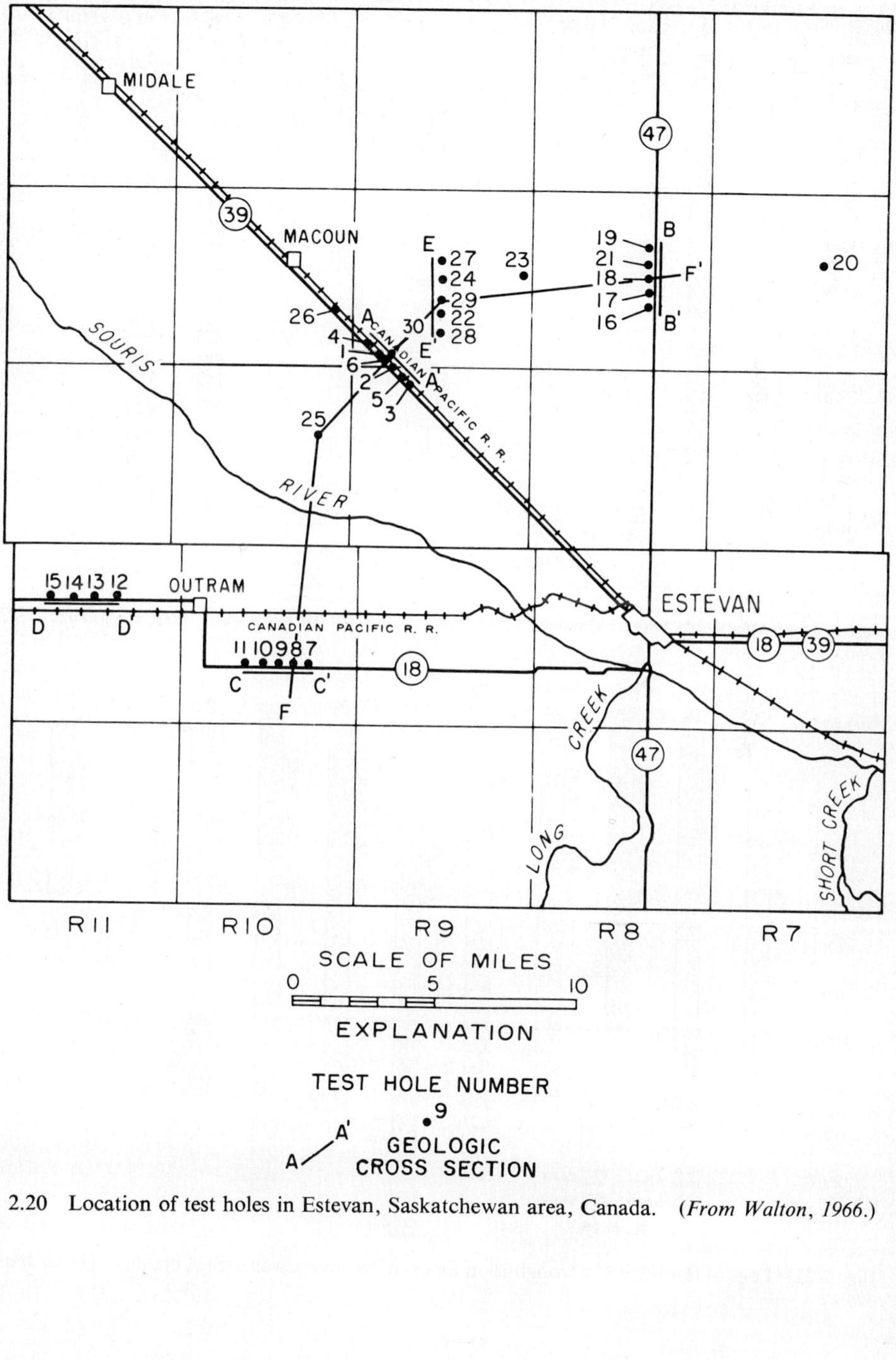

Fig. 2.20 Location of test holes in Estevan, Saskatchewan area, Canada. (*From Walton, 1966.*)

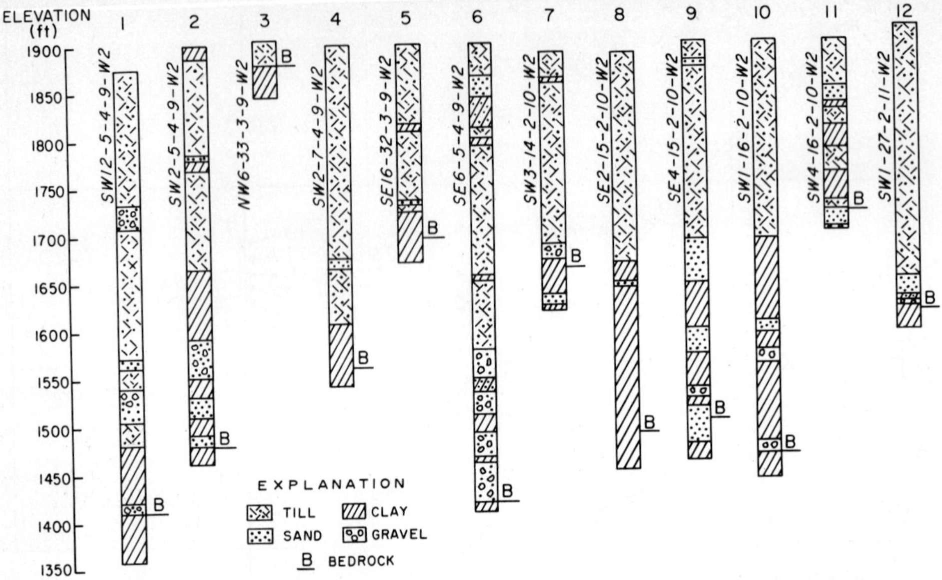

Fig. 2.21 Logs of test holes 1 through 12 in Estevan, Saskatchewan area, Canada. (*From Walton, 1966.*)

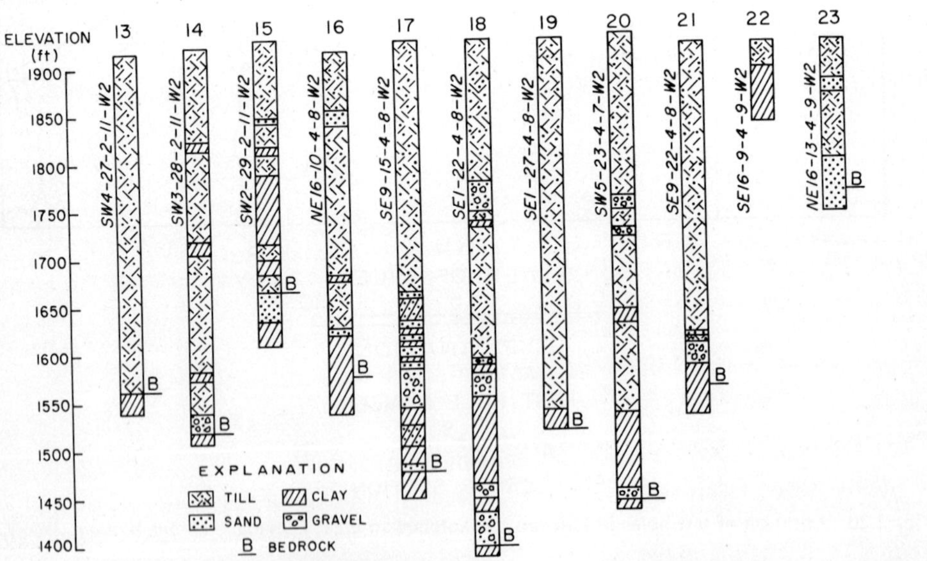

Fig. 2.22 Logs of test holes 13 through 23 in Estevan, Saskatchewan area, Canada. (*From Walton, 1966.*)

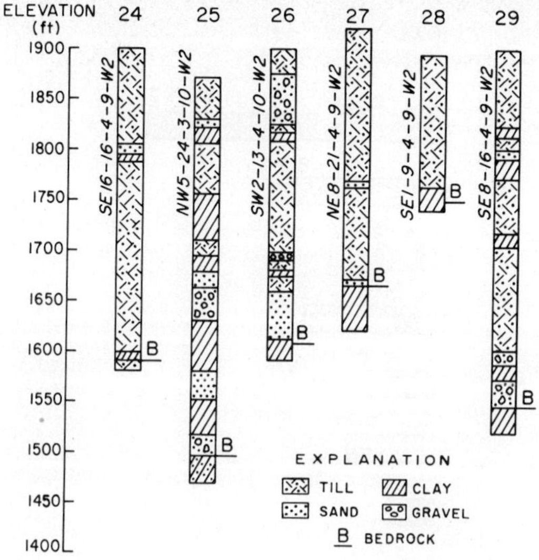

Fig. 2.23 Logs of test holes 24 through 29 in Estevan, Saskatchewan area, Canada. (*From Walton, 1966.*)

a northeasterly direction from Outram for at least 35 miles. The sand and gravel deposits are overlain by clayey materials averaging 25 feet thick and a fine-grained sand and silt zone averaging 8,000 feet wide and 90 feet thick.

Gravels in the deepest part of the buried-valley channel are clean, coarse, exceptionally well rounded, and highly permeable. Quartz, chert, quartzite, brown and maroon quartzitic sandstones, argillite, ultrabasic pebbles, and petrified wood are common lithologic constituents of this zone. The stratified sand silts and clays above this zone and below the upper till zone were probably laid down either as fine-grained lacustrine deposits in a preglacial lake or as preglacial quiet-water floodplain silts and clays. The predominantly till zone, which extends from the ground surface down to a depth of about 300 feet, contains randomly distributed inclusions of stratified drift.

The results of the test-hole drilling program, which cost about $31,000 (in Canadian dollars), indicate that large quantities of groundwater may be developed from the sand and gravel deposits occurring between the depths of 400 and 500 feet in the deepest parts of the buried Missouri River valley channel.

PRESENTATION OF HYDROGEOLOGIC DATA

Hydrogeologic maps record the status of groundwater resources at some point in time by defining the container (the geology of the area), its contents (the water), and its rate of filling and emptying (recharge and discharge). A principal clue to the groundwater conditions of a locality is its geologic section. The character, thickness, and succession of the underlying formations give important data as to existing aquifers, aquitards, and aquicludes and the depths at which they can be tapped.

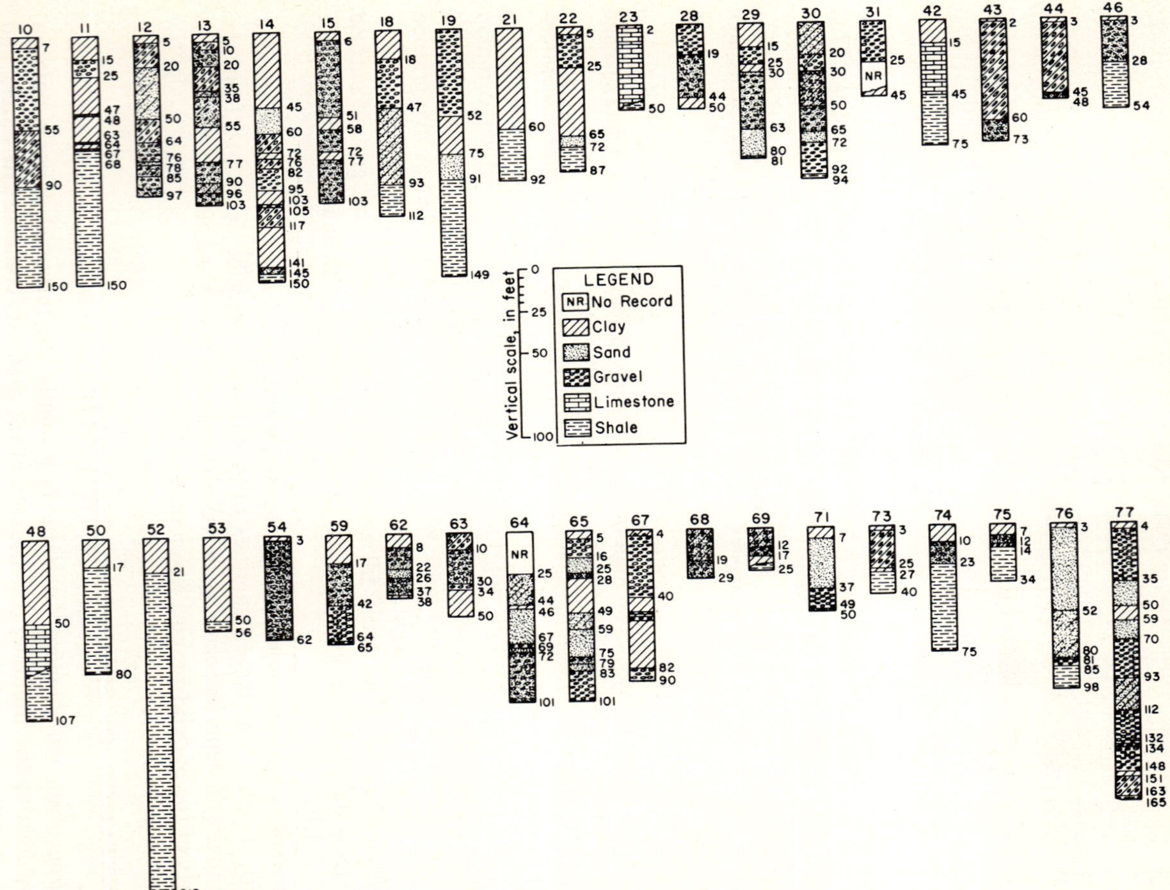
LEGEND
NR No Record
Clay
Sand
Gravel
Limestone
Shale
Vertical scale, in feet
0
25
50
100

(a)

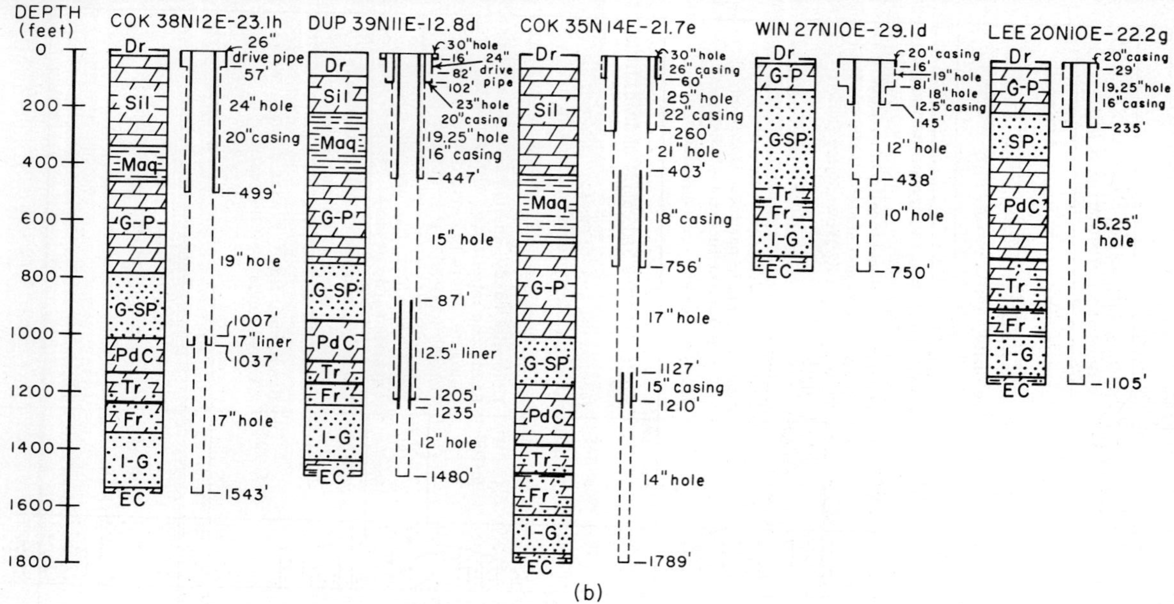

Fig. 2.24 Logs of selected wells in Fairborn area, Ohio (*a*) and in northeastern Illinois (*b*). (*From Walton and Scudder, 1960; Walton and Csallany, 1962.*)

SYSTEM	SERIES	GROUP OR FORMATION	GEOHYDROLOGIC UNITS		LOG	THICKNESS (FT)	DESCRIPTION
QUATERNARY	PLEISTOCENE		Glacial drift aquifers			0-200±	Unconsolidated glacial deposits-pebbly clay (till), silt, sand and gravel Alluvial silts and sands along streams
DEVONIAN						Fissure Fillings	Shale, sandy, brown to black
SILURIAN	NIAGARAN	Racine Waukesha Joliet	Niagaran aquifer	Silurian dol. aquifer		0-170	Dolomite, very pure to highly argillaceous, silty, cherty; reefs in upper part Dolomite, shaly, and shale, dolomitic; maroon, green, pink
SILURIAN	ALEXANDRIAN	Kankakee Edgewood	Alexandrian aquifer	Silurian dol. aquifer		0-90	Dolomite, glauc.; thin grn. shale partings Dolomite, argillaceous, silty and/or sandy, cherty
ORDOVICIAN	CINCINNATIAN	Neda	Confining beds of the Maquoketa Formation			0-20	Shale, red; oolites
ORDOVICIAN	CINCINNATIAN	Maquoketa	Confining beds of the Maquoketa Formation			85-230	Shale, silty, dolomitic, greenish gray, weak (Upper unit) Dolomite and limestone, white, light gray, interbedded shale (Middle unit) Shale, dolomitic, brown, gray (Lower unit)
ORDOVICIAN	MOHAWKIAN	Galena Decorah Platteville	Galena-Platteville	Cambrian-Ordovician aquifer		300-350	Dolomite, and/or limestone, cherty Dolomite, shale partings, speckled Dolomite and/or limestone, cherty, sandy at base
ORDOVICIAN	MOHAWKIAN	Glenwood	Glenwood-St. Peter	Cambrian-Ordovician aquifer		200-375	Sandstone, fine and coarse grained; little dolomite; shale at top Sandstone, fine to medium grained; locally cherty red shale at base
ORDOVICIAN	CHAZYAN	St. Peter	Glenwood-St. Peter	Cambrian-Ordovician aquifer			
ORDOVICIAN	PRAIRIE DU CHIEN	Shakopee New Richmond Oneota	Prairie du Chien	Cambrian-Ordovician aquifer		0-200	Dolomite, sandy, cherty (oolitic); sandstone Sandstone interbedded with dolomite Dolomite, white to pink, coarse grained cherty (oolitic), sandy at base
CAMBRIAN	CROIXAN	Trempealeau	Trempealeau	Cambrian-Ordovician aquifer		80-190	Dolomite, white, fine grained; geodic quartz; sandy at base
CAMBRIAN	CROIXAN	Franconia	Franconia	Cambrian-Ordovician aquifer		70-100	Dolomite, sandstone and shale, glauconitic, green to red, micaceous
CAMBRIAN	CROIXAN	Ironton Galesville	Ironton-Galesville	Cambrian-Ordovician aquifer		175-200	Sandstone, fine to coarse grained, well sorted; upper part dolomitic
CAMBRIAN	CROIXAN	Eau Claire	Confining beds of the Eau Claire Formation (upper and middle beds)			300-400	Shale and siltstone, dolomitic, glauconitic; sandstone, dolomitic, glauconitic
CAMBRIAN	CROIXAN	Mt. Simon	Eau Claire (lower beds) and Mt. Simon Formations	Mt. Simon aquifer		2,000±	Sandstone, coarse grained, white, red in lower half; lenses of shale and siltstone, red, micaceous
Precambrian							

Fig. 2.25 Stratigraphic section, hydrogeologic units, water-yielding properties of the rocks, and

DRILLING AND CASING CONDITIONS	WATER YIELDING PROPERTIES	CHEMICAL QUALITY OF WATER	WATER TEMPERATURE
Boulders, heaving sand locally; sand and gravel wells usually require screens and development; casing required in wells into bedrock	Sand and gravel, permeable Some wells yield more than 1000 gpm; specific capacities range from 1.0 to 40.7 gpm/ft, av. 13.8 gpm/ft	Hardness from 387 to 596 ppm, av. 485	46° min. 52° av. 54° max.
Upper part usually weathered and broken; extent of crevicing varies widely Chert layers slow drilling rate	Niagaran aquifer more productive than Alexandrian aquifer Basal units of Niagaran aquifer locally may retard recharge Specific capacities from 0.5 to 530 gpm/ft Coefficient of transmissibility averages 114,000 gpm/ft	Variable hardness < 200 to >1000 ppm Iron > 0.5 ppm in 60% of analyses	49° min. 52° av. 59° max.
Shale requires casing Dolomite units creviced			
Top of Galena usually selected for hole reduction and seating of casing	Development and yields of crevices are small	Regionally hardness < 100 ppm H_2S often present High alkalinity > 350 ppm	54° to 55°
Lower cherty shales cave and usually are cased Friable sand may slough	Small to moderate quantities of water Coefficient of transmissibility probably about 15% of that of Cambrian-Ordovician aquifer	Water similar in quality or slightly harder than that in Ironton-Galesville Sandstone	53° to 56°
Crevices encountered locally in the dolomite, especially in Trempealeau Casing not required	Crevices in dolomite and sandstone generally yield small to moderate quantities of water; Trempealeau locally well creviced and partly responsible for exceptionally high yields of several deep wells Coefficient of transmissibility probably averages about 30% of that of total Cambrian-Ordovician aquifer		
Amount of cementation variable Lower parts more friable Sometimes sloughs	Most productive unit of Cambrian-Ordovician aquifer Coefficient of transmissibility probably averages about 50% of that of total Cambrian-Ordovician aquifer	Iron usually < 0.4 ppm Hardness av. 200 ppm	55° to 58°
Casing usually not necessary; locally weak shales may require casing	Shales, generally not water-yielding; act as a confining bed between Ironton-Galesville and Mt. Simon		
Casing not required	Moderate amounts of water; permeability between that of Glenwood-St. Peter and Ironton-Galesville	Hardness from 247 to 544 ppm, av. 352 Chlorides increase at rate of 400 ppm each additional 25' depth below elevation -1275'	66° at elev. -1300', increasing 1° with each additional 100' depth
crystalline rocks			

character of groundwater in DuPage County, northern Illinois. (*From Zeizel et al., 1962.*)

Maps have their beginning in a base map. The utility of any hydrogeologic map is enhanced by a precise base map at a suitable scale. The base map is selected for readability, accuracy, density of data, degree of quantitative measurement desired, ease of compilation, and economy of preparation.

Water levels measured in wells are presented by means of water-level-contour maps, water-level-change maps, depth-to-water maps, water-level profiles, and hydrographs. Water-level-contour maps are classified as water-table maps and piezometric-surface maps. The general direction of groundwater flow is shown on water-level-contour maps; water-level-change maps indicate the change of water levels in wells during a given span of time. Water-level profiles are useful in depicting differences in water levels at various depths and show the hydraulic gradients associated with the vertical movement of groundwater. Long- and short-term water-level hydrographs give some indication of the rates of discharge and recharge to aquifers and changes in storage of water within aquifers and aquitards. Natural and artificial areas of recharge and discharge, springs, wells, well fields, and pumping centers are commonly recorded on hydrogeologic maps.

Well log records (Fig. 2.24.) are often included in groundwater reports. The general sequence, lithologic character, water-yielding properties, structure, and distribution of rocks may be illustrated as shown in Fig. 2.25.

A detailed report on the groundwater resources of the Chicago region in northeastern Illinois (Suter et al., 1959) contains several kinds of hydrogeologic maps. Maps include the thickness of the unconsolidated deposits (Fig. 2.26); bedrock topography (Fig. 2.27); areal geology of the bedrock surface and major structures (Fig. 2.28); and elevation of the top and thickness of a major aquifer, the Ironton-Galesville Sandstone (Fig. 2.29). One method of presenting information on the alluvial and glacial deposits is illustrated in Fig. 2.30.

The saturated thickness of an aquifer may be expressed as shown in Fig. 2.31. Examples of water-table and piezometric-surface maps are given in Figs. 2.32 and 2.33. In areas where both water-table and artesian conditions exist, it is often desirable to prepare a map similar to Fig. 2.34. Selected water-level-change maps are shown in Fig. 2.35. Geologic sections and water-level profiles are commonly combined as shown in Figs. 2.36 and 2.37.

The three-dimensional fence diagram and isometric block diagram in Figs. 2.38 and 2.39 are easily read and represent a high degree of integration and interpretation of data on the geology and hydrology of areas.

Hydrographs of a well in Illinois in Fig. 2.40*a* illustrate the daily cycle of water-level fluctuations caused by evapotranspiration. The hydrograph of another well in Illinois in Fig. 2.40*b* illustrates atmospheric fluctuations in wells. Water-level fluctuations on an expanded time scale due to an earthquake are shown in Fig. 2.41*a*. Railroad water-level fluctuations in a well in Illinois are illustrated in Fig. 2.41*b*. Illinois River stage changes and corresponding changes in water levels in a well 50 feet from the river's edge are shown in Fig. 2.42*a*. The effects of passing barges are also illustrated. Figure 2.42*b* shows water-level fluctuations in a well in the Chicago, Illinois, area near a well field where pumping is intermittent.

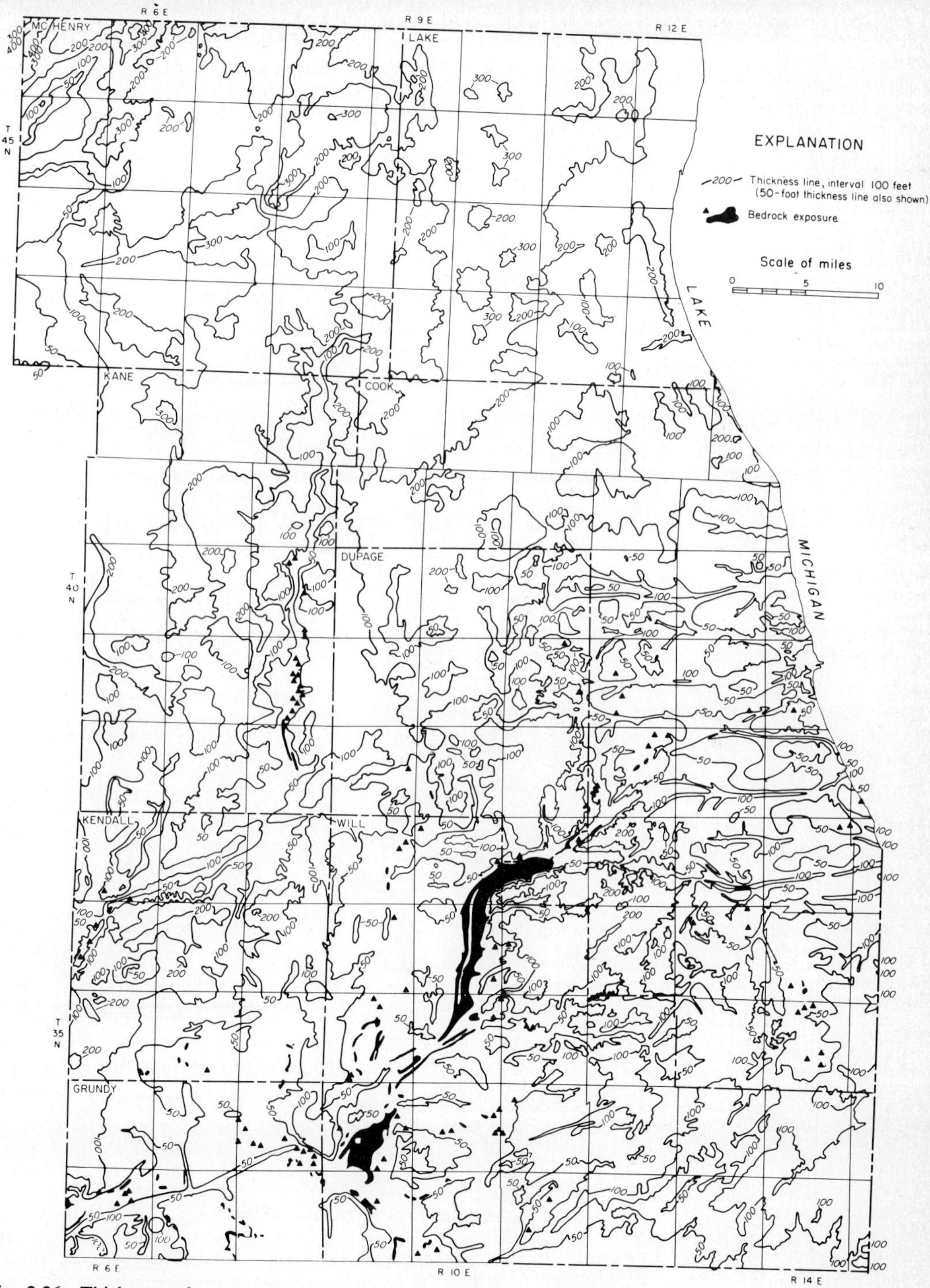

Fig. 2.26 Thickness of unconsolidated deposits overlying bedrock in Chicago region, Illinois. (*From Suter et al., 1959.*)

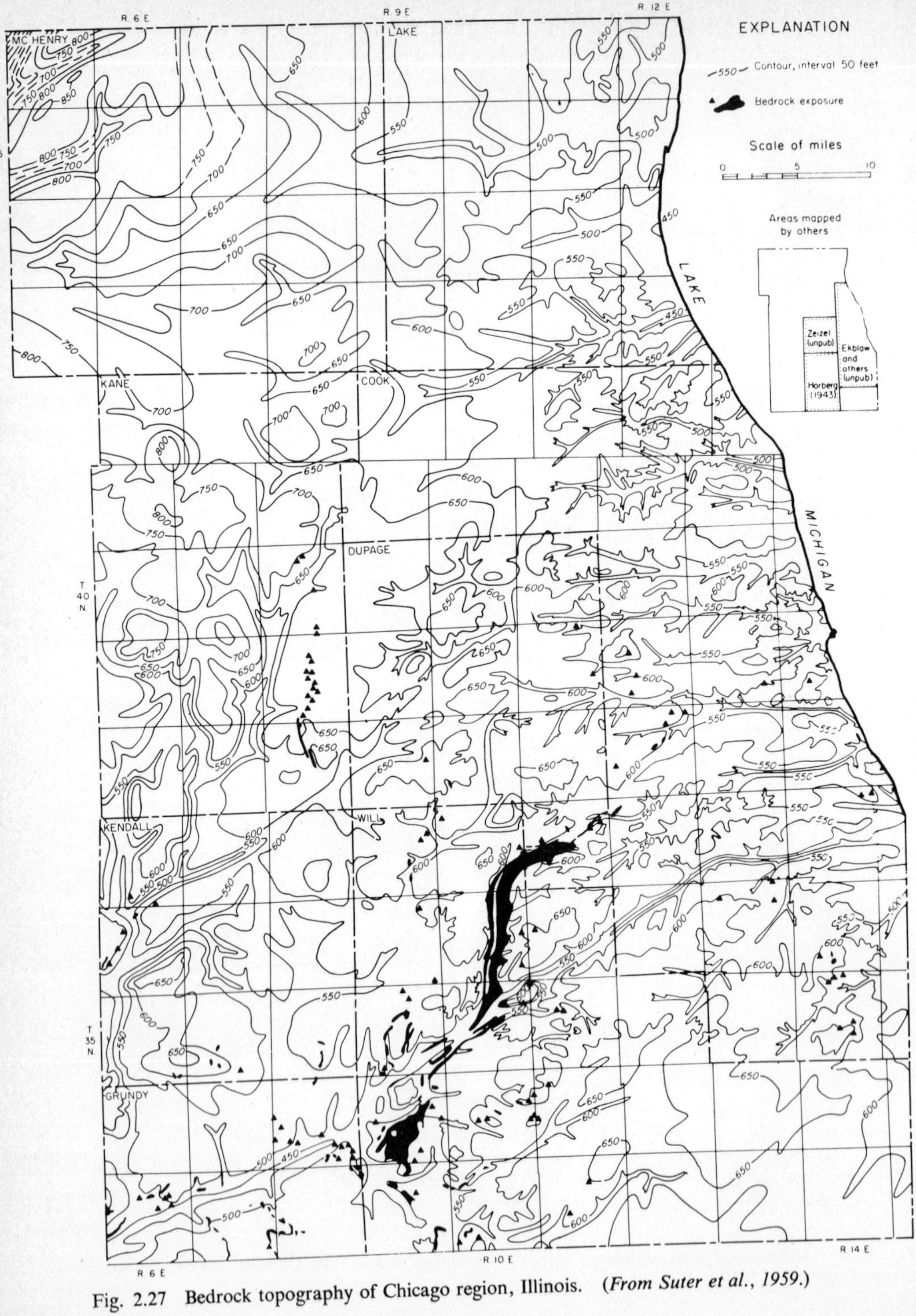

Fig. 2.27 Bedrock topography of Chicago region, Illinois. (*From Suter et al., 1959.*)

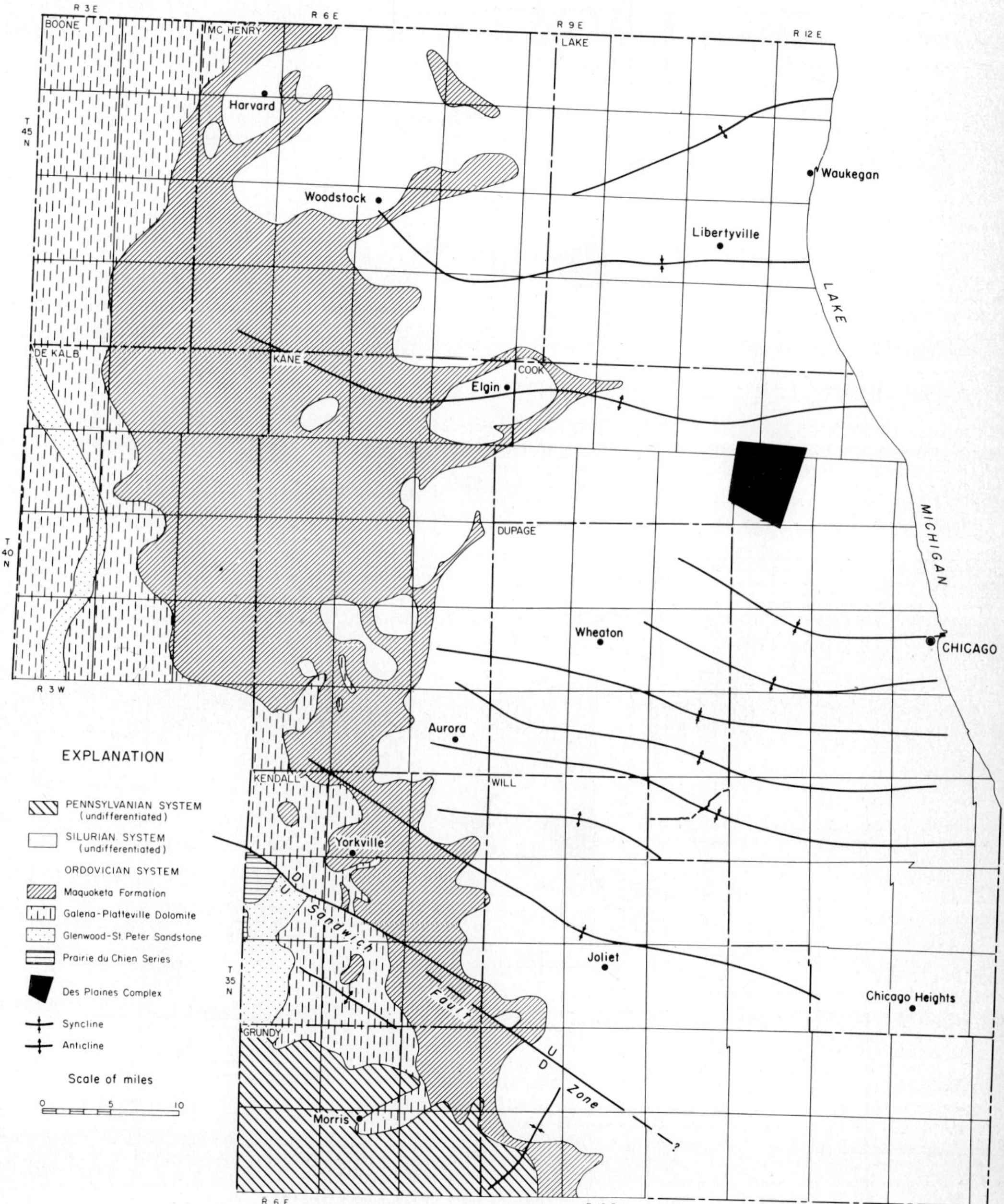

Fig. 2.28 Areal geology of bedrock surface and major structures in Chicago region, Illinois. (*From Suter et al., 1959.*)

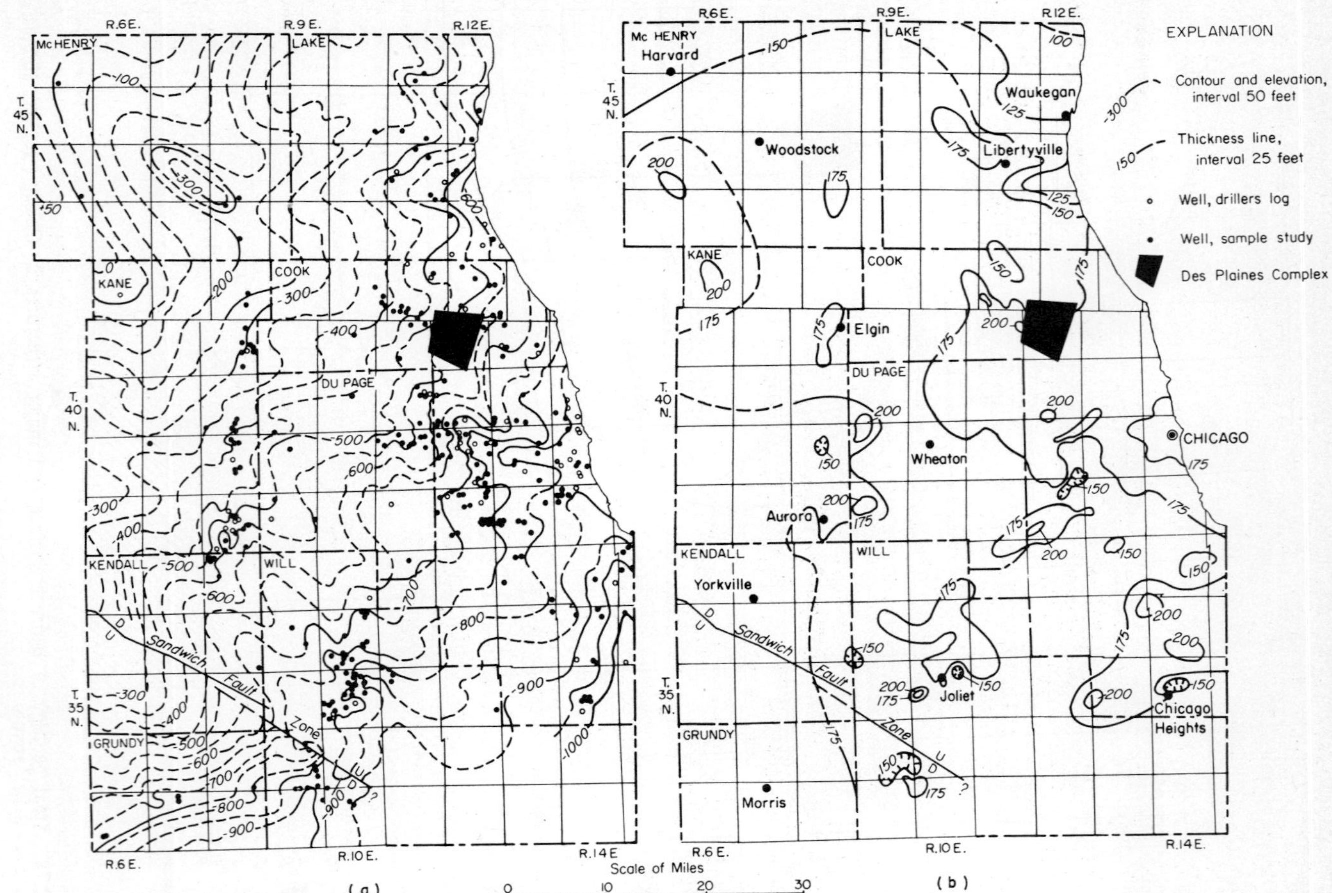

Fig. 2.29 Elevation of top (*a*) and thickness (*b*) of Ironton-Galesville Sandstone in Chicago region, Illinois. (*From Suter et al., 1959.*)

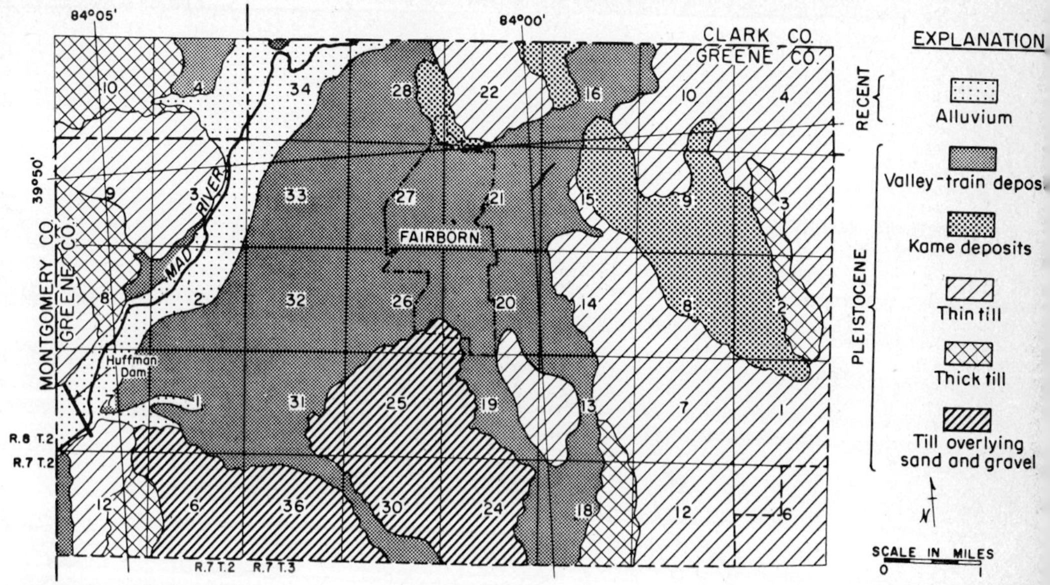

Fig. 2.30 Map of alluvial and glacial deposits of Fairborn area, Ohio. (*From Walton and Scudder, 1960.*)

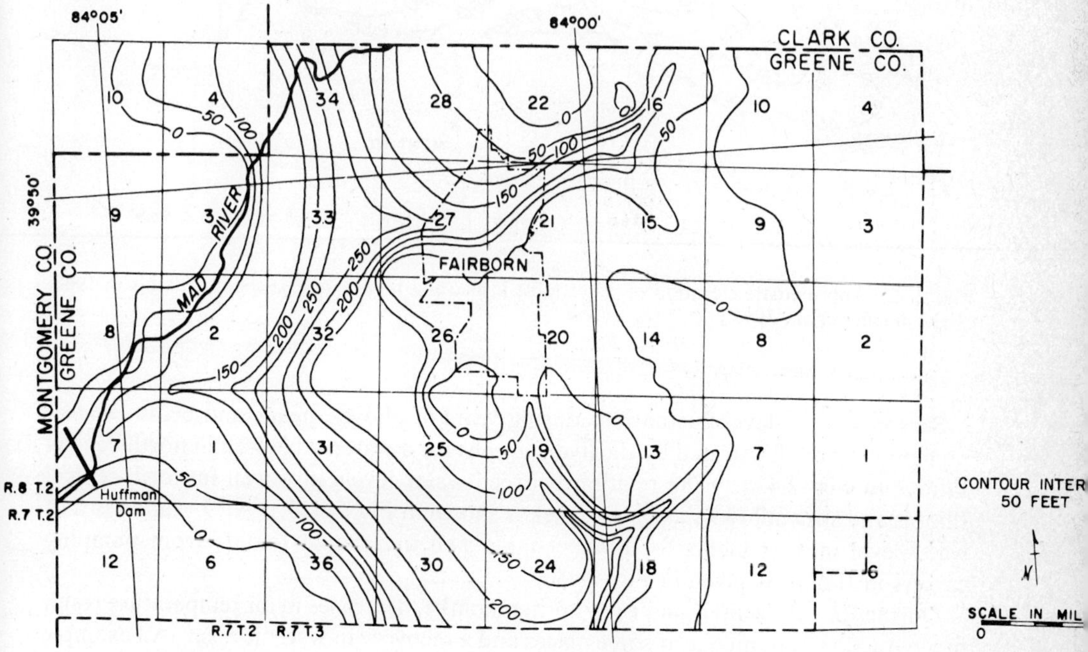

Fig. 2.31 Map showing saturated thickness of valley-train deposits in Fairborn area, Ohio. (*From Walton and Scudder, 1960.*)

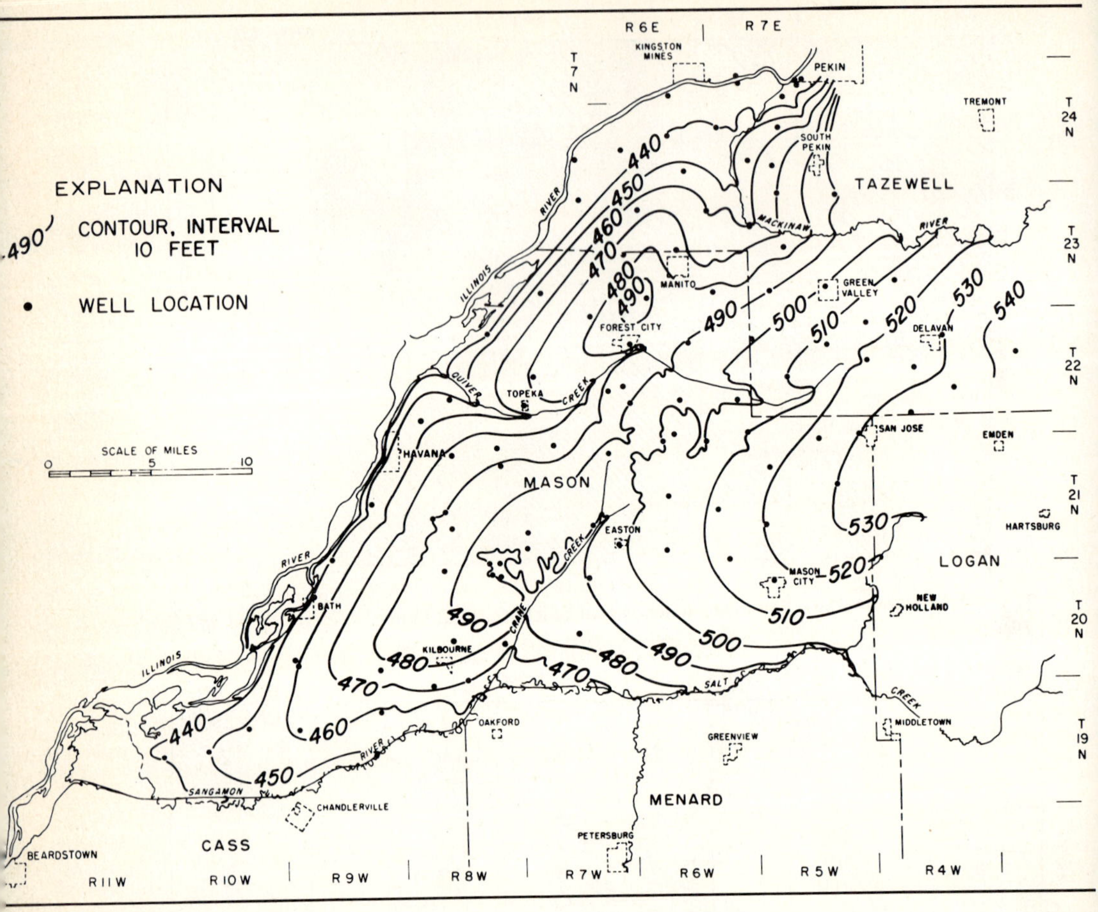

Fig. 2.32 Approximate elevation of water table in Havana region in west-central Illinois in 1960. (*From Walker et al., 1965.*)

Seasonal water-level fluctuations during years of above-, near-, and below-normal precipitation are illustrated by the hydrographs for a water-table well in north-central Illinois in Fig. 2.43*a*. The relation between water levels in a well in north-central Illinois and streamflow in a nearby creek is shown in Fig. 2.43*b*. A 19-year record of water-level measurements for a water-table well in an area remote from pumping centers in Illinois is given in Fig. 2.44*a*.

During the fall, winter, and early spring months, decreases in air temperature result in a reduction in pumpage at some places and a recovery in water levels. An example of such fluctuations is shown in the hydrograph for a well in a heavily pumped area in Chicago in Fig. 2.44*b*. In heavily pumped areas, a downward trend of water levels

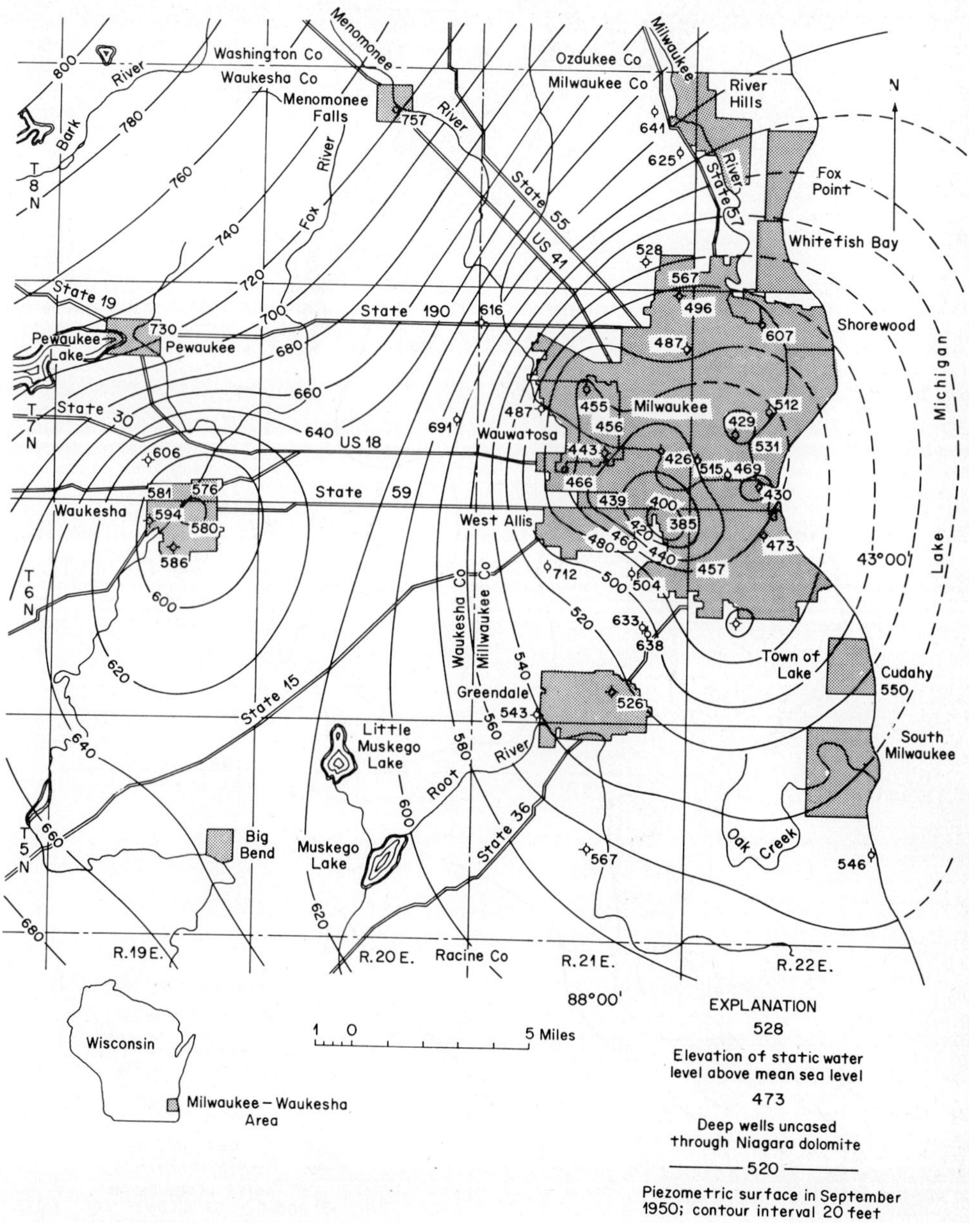

Fig. 2.33 Piezometric surface of sandstone aquifer in Milwaukee-Waukesha area, Wisconsin, September, 1950. (*From Foley et al., 1953.*)

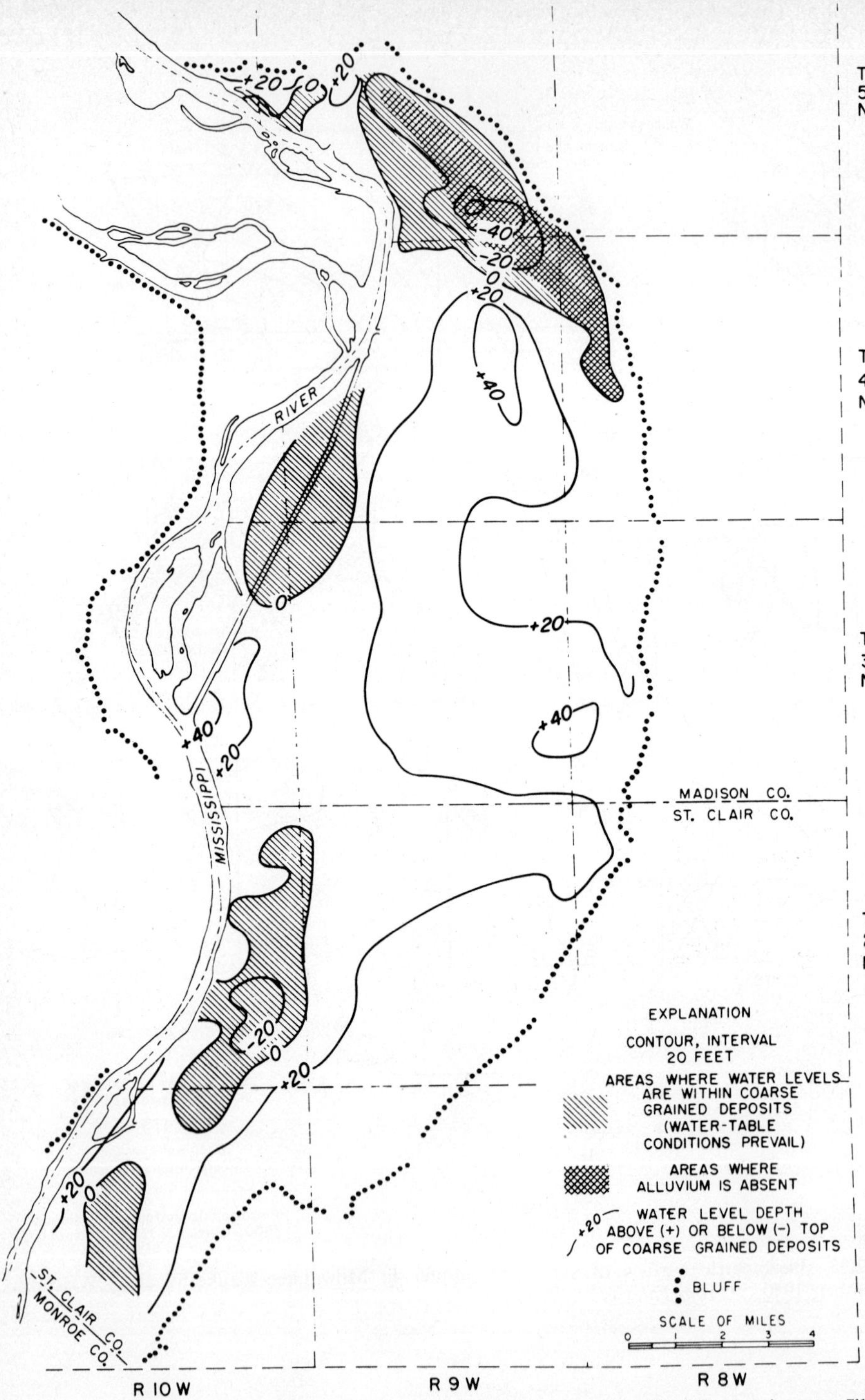

Fig. 2.34 Location of areas where water-table conditions prevail in East St. Louis area, Illinois. (*From Schicht 1965.*)

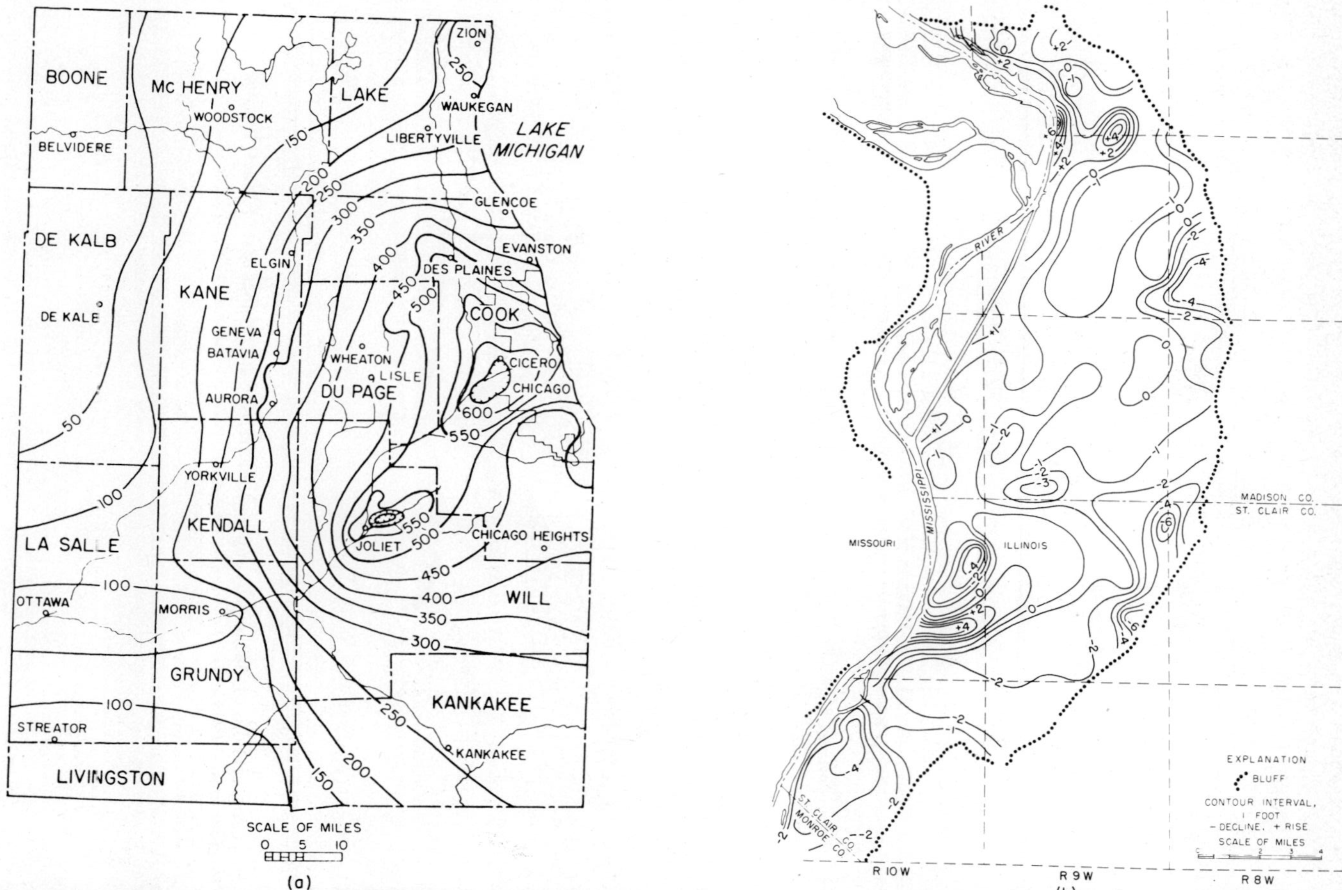

Fig. 2.35 Decline of artesian pressure in Cambrian-Ordovician Aquifer in northeastern Illinois, 1864–1958 (*a*), and estimated change in water levels in East St. Louis area, Illinois, June to November, 1961 (*b*). (*From Suter et al., 1959; Schicht, 1965.*)

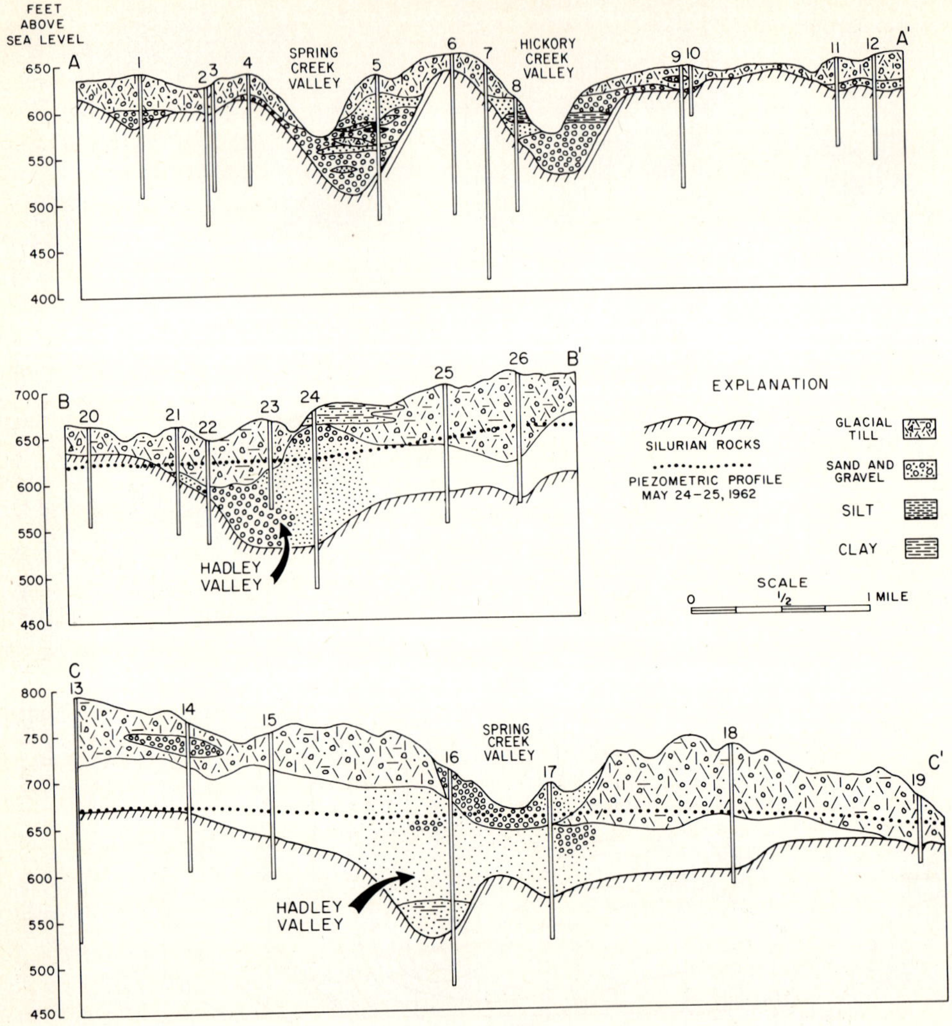

Fig. 2.36 Cross sections of glacial drift and piezometric-surface profiles east of Joliet, Illinois. (*From Prickett et al., 1964.*)

may continue for many years because of continual increases in pumpage or withdrawals in excess of recharge or both. Such is the case as illustrated by the hydrographs of deep sandstone wells in the Chicago, Illinois, area shown in Fig. 2.44.

The distribution of groundwater pumpage is commonly illustrated as shown in Fig. 2.45. The time-rate of growth in groundwater pumpage may be presented as shown in Fig. 2.46*a*. Useful methods for showing production-well locations and information on individual wells are illustrated in Fig. 2.46*b* and *c*. Data on the yields of wells can be conveniently presented as shown in Fig. 2.47. Areas influenced by withdrawals of groundwater may be shown as illustrated in Fig. 2.48*a*. Information concerning recharge rates is commonly expressed as shown in Fig. 2.48*b*.

The international legend for hydrogeological maps, given in Fig. 2.49, is the result of work and meetings by study groups and general meetings of the International Association of Scientific Hydrology (IASH) and the International Association of Hydrogeologists (IAH). These two associations, the first forming part of the International Union of Geodesy and Geophysics, and the second, part of the International Geological Congress, International Union of Geological Sciences, have for some years been studying the setting up of this legend. The legend should facilitate the drawing up, in a relatively standardized form, of specialized maps having topography and geology as a basis (specialized topographical and geological maps) upon which groundwater and surface-water resources could be represented with all their characteristics.

Problems

2.1 Using as many of the references mentioned in McGuinness (1963) as may be available to the reader, assemble data on porosity; specific yield; and coefficients of storage, permeability, and transmissibility for the major aquifers of the 10 groundwater provinces within the conterminous United States. Prepare tables showing the range of the values for the water-yielding properties of various types of aquifers (sand and gravel, sandstone, limestone, etc.). Prepare a report summarizing the results of the above-mentioned studies.

2.2 Using as source material as many of the references mentioned in this book as may be available to the reader, compile a list of case histories of the use of surface and subsurface geophysical methods in groundwater exploration. Evaluate the effectiveness of these methods in groundwater studies.

2.3 Using as source material as many of the references mentioned in this book as may be available to the reader, prepare a short report summarizing the various techniques used in geologic and hydrologic methods for groundwater exploration.

2.4 Using information available in the files of state and federal water resources agencies, prepare a set of hydrogeologic maps for an area within your state. Also prepare a hydrogeologic profile showing groundwater conditions of a major aquifer within your state.

2.5 Obtain well log data from a water-well contractor and prepare hydrogeologic maps and profiles for a selected area within your state.

2.6 Using information available in the files of state and federal water resources agencies, prepare hydrographs for selected wells and a report describing the significance of water-level fluctuations in a selected area within your state.

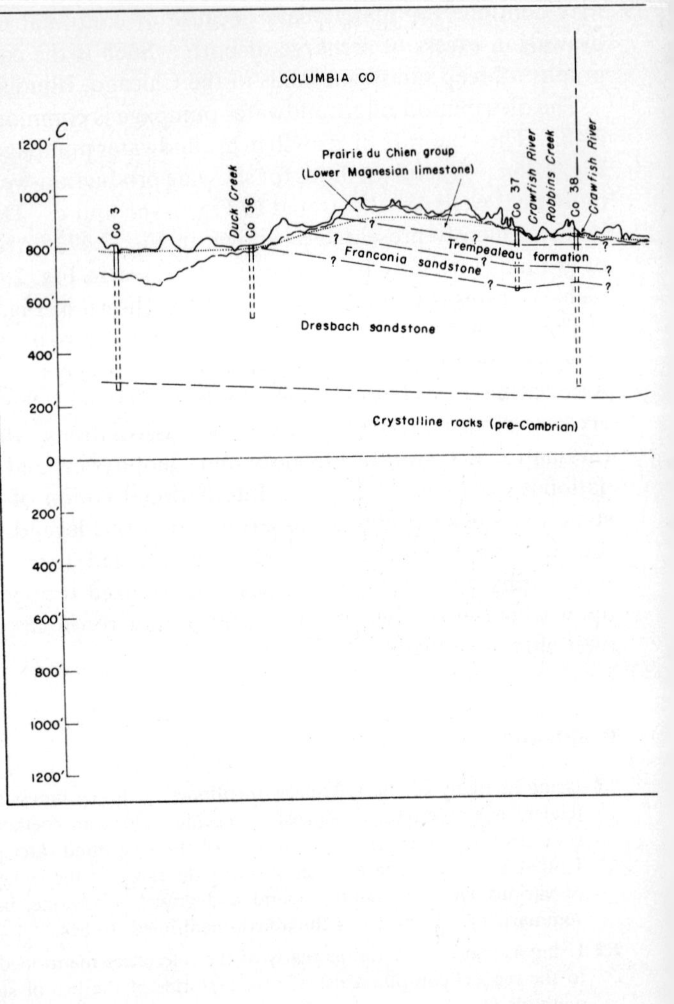

Fig. 2.37 Geologic cross section from Portage to Milwaukee, Wisconsin, and profiles of piezometric surface of Niagara Dolomite and sandstone aquifer. (*From Foley et al., 1953.*)

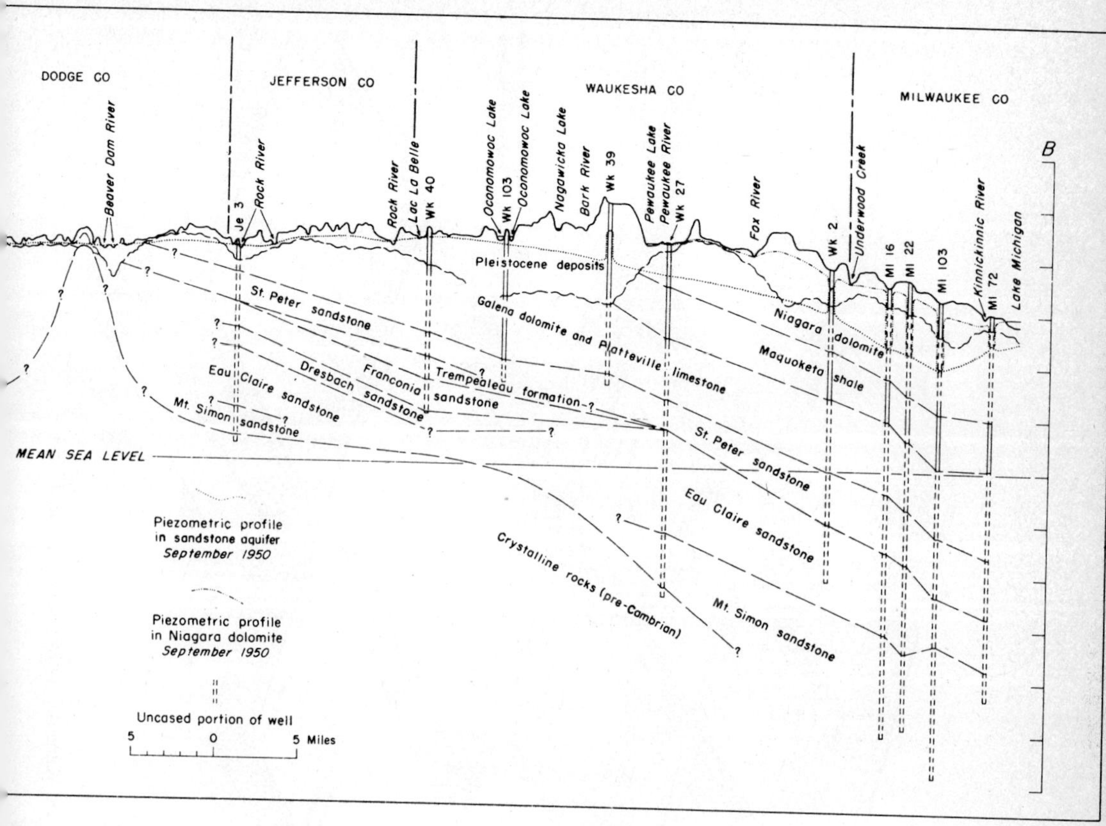
DODGE CO
JEFFERSON CO
WAUKESHA CO
MILWAUKEE CO
B
Beaver Dam River
Rock River
Je 3
Rock River
Lac La Belle
Wk 40
Oconomowoc Lake
Wk 103
Oconomowoc Lake
Nagawicka Lake
Bark River
Wk 39
Pewaukee Lake
Pewaukee River
Wk 27
Fox River
Wk 2
Underwood Creek
Ml 16
Ml 22
Ml 103
Kinnickinnic River
Ml 72
Lake Michigan
Pleistocene deposits
St. Peter sandstone
Galena dolomite and Platteville limestone
Niagara dolomite
Maquoketa shale
Franconia sandstone
Trempealeau formation
Dresbach sandstone
Eau Claire sandstone
Mt. Simon sandstone
MEAN SEA LEVEL
St. Peter sandstone
Eau Claire sandstone
Crystalline rocks (pre-Cambrian)
Mt. Simon sandstone
Piezometric profile in sandstone aquifer September 1950
Piezometric profile in Niagara dolomite September 1950
Uncased portion of well
5 0 5 Miles

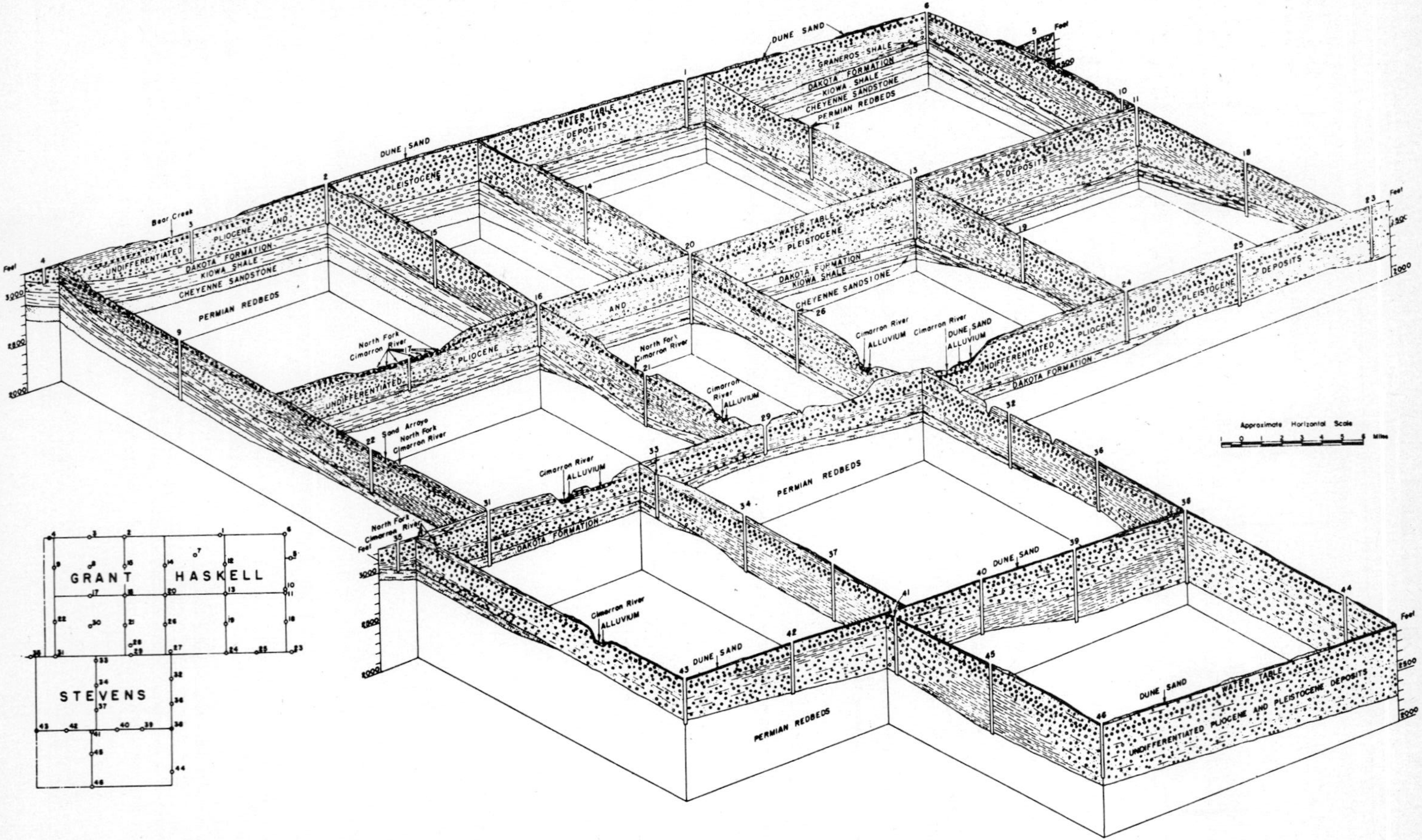

Fig. 2.38 Fence diagram for Grant, Haskell, and Stevens Counties, Kansas. (*After McLaughlin, 1946.*)

Fig. 2.39 Isometric block diagram showing water-bearing formations of Defiance Uplift, Chuska Mountains and vicinity, Navajo County, Arizona–Utah–New Mexico. (*After Harshbarger et al., 1953.*)

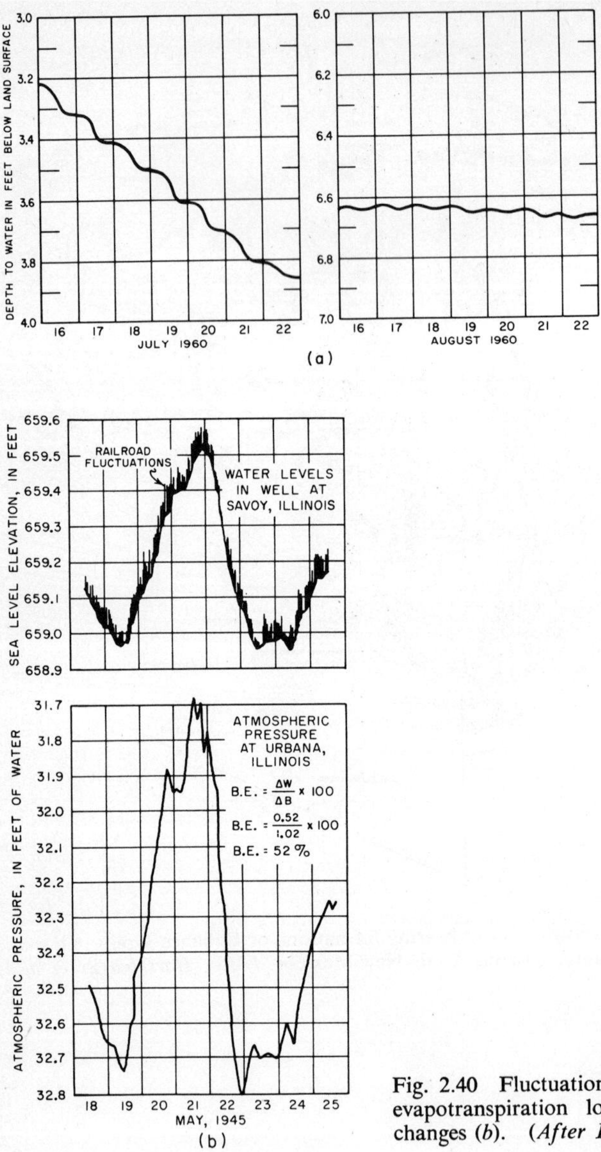

Fig. 2.40 Fluctuations of water levels in wells caused by evapotranspiration losses (*a*) and atmospheric-pressure changes (*b*). (*After Roberts and Romine, 1947.*)

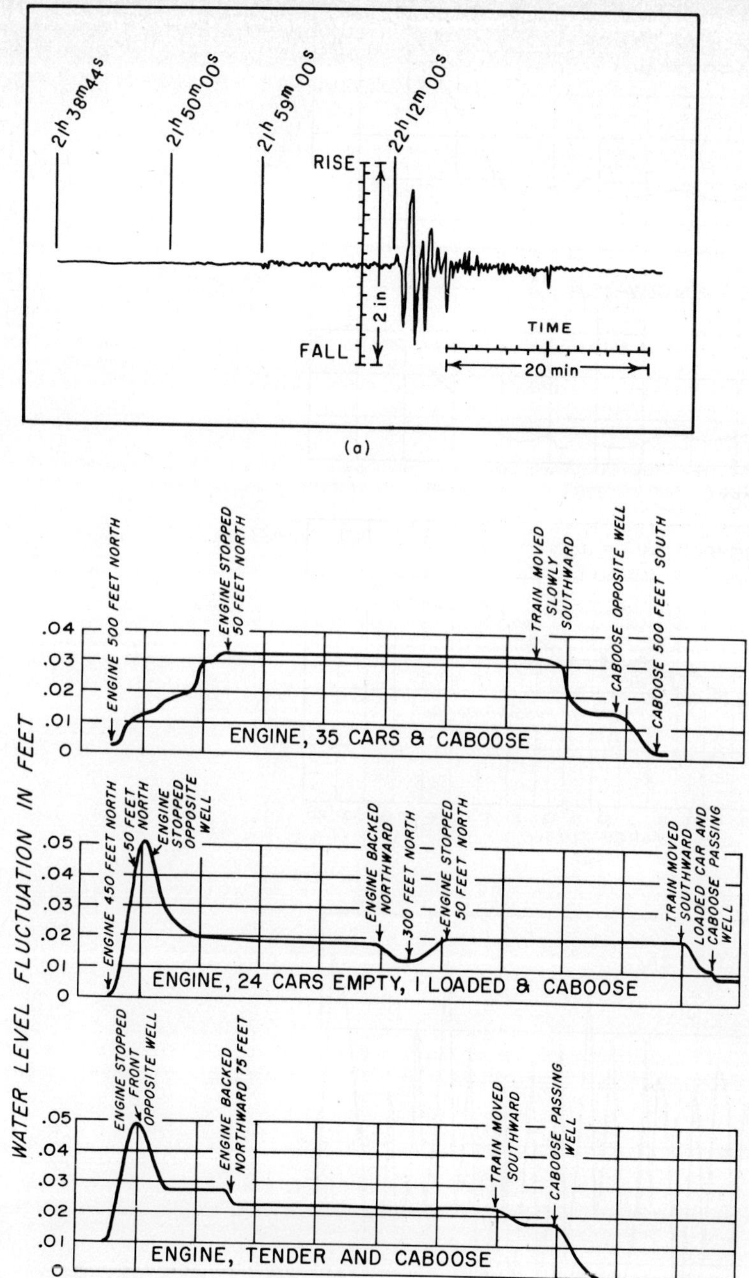

Fig. 2.41 Fluctuations of water levels in wells caused by an earthquake (*a*) and a passing railroad train (*b*). (*From Russell, 1963.*)

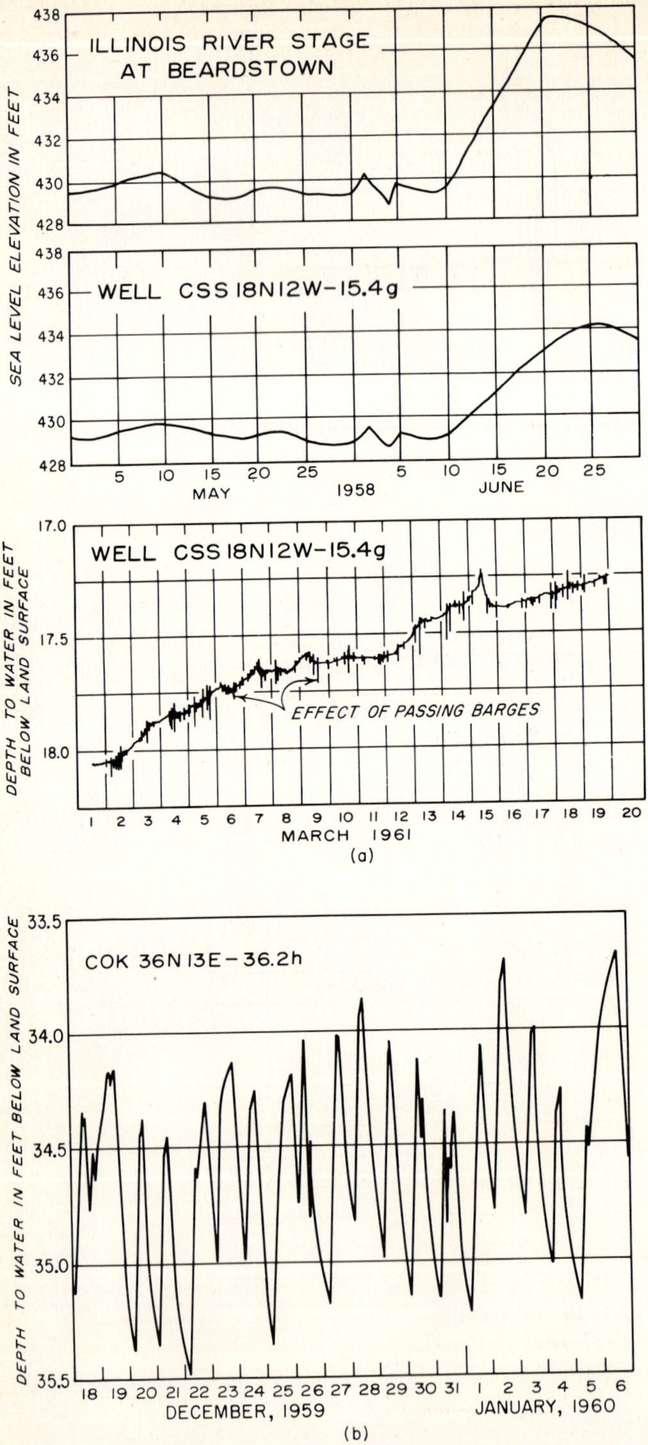

Fig. 2.42 Fluctuations of water levels caused by changes in river stages (*a*) and intermittent pumpage (*b*). (*From Russell, 1963.*)

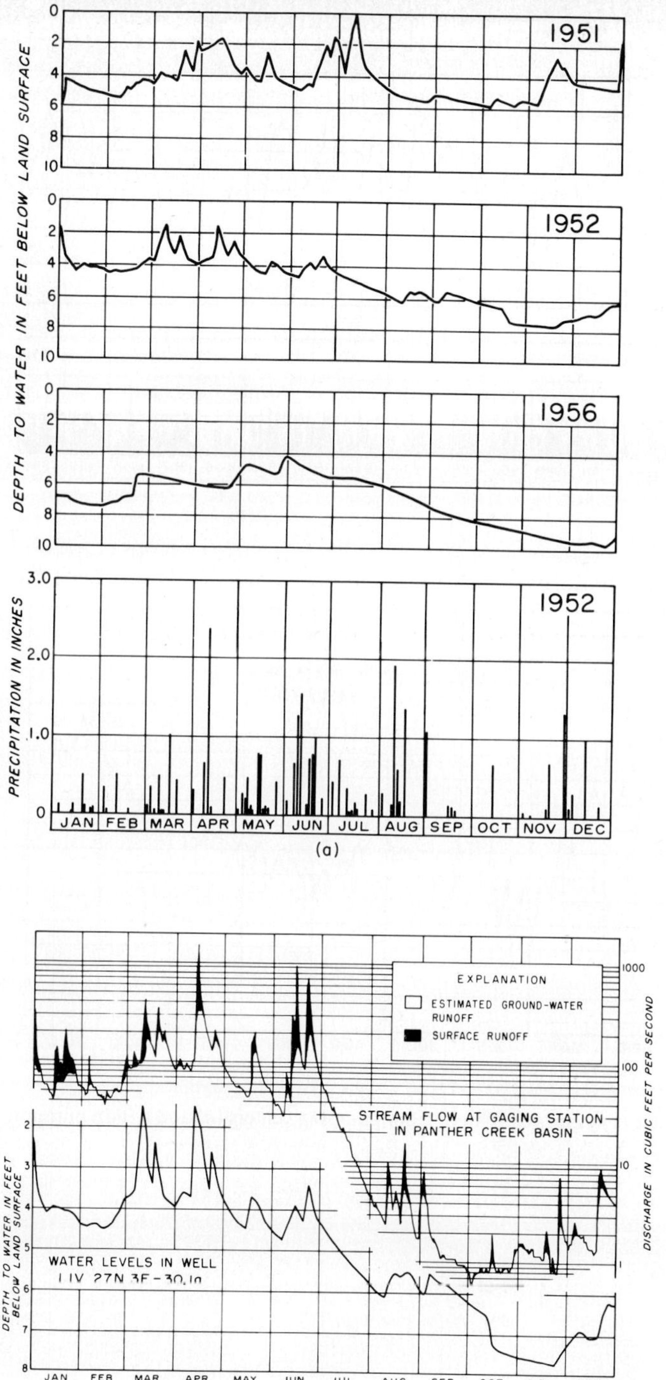

Fig. 2.43 Fluctuations in water levels caused by recharge from precipitation (*a*) and the relation between groundwater levels and streamflow (*b*). (*After Schicht and Walton, 1961.*)

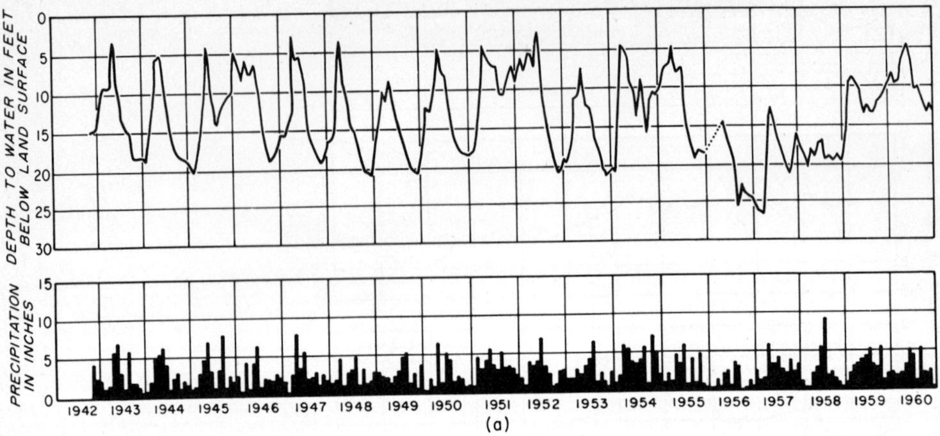

(a)

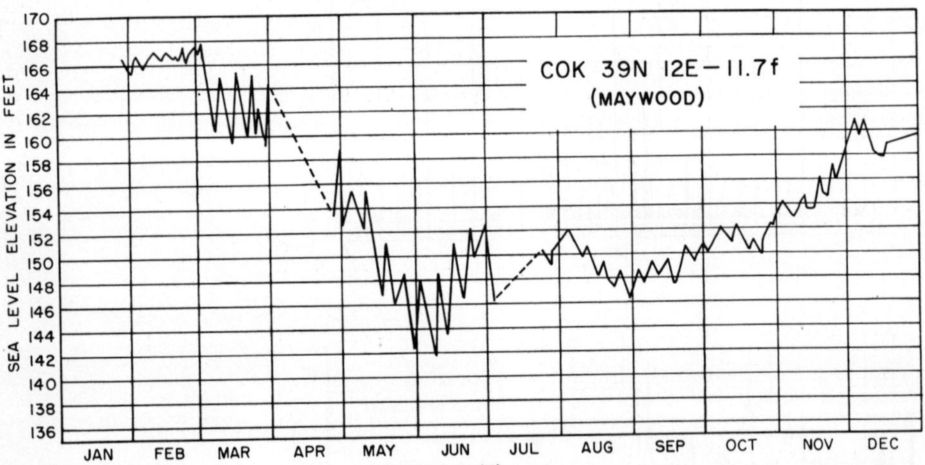

(b)

Fig. 2.44 Fluctuations in water levels in well remote from pumping centers (*a*) and within pumping

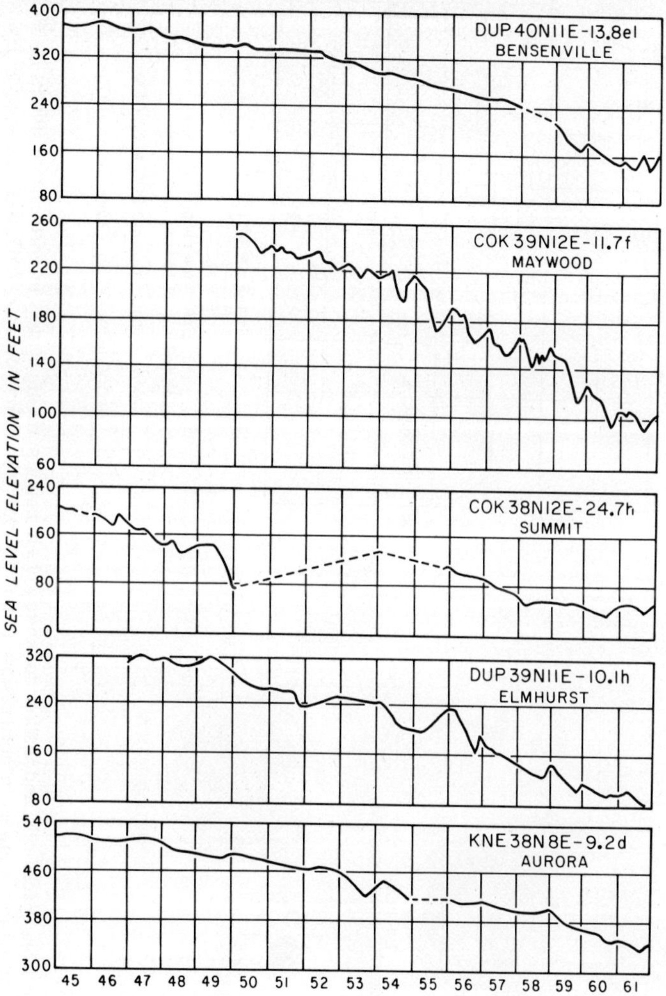

(c)

centers (*b* and *c*). (*From Russell, 1963.*)

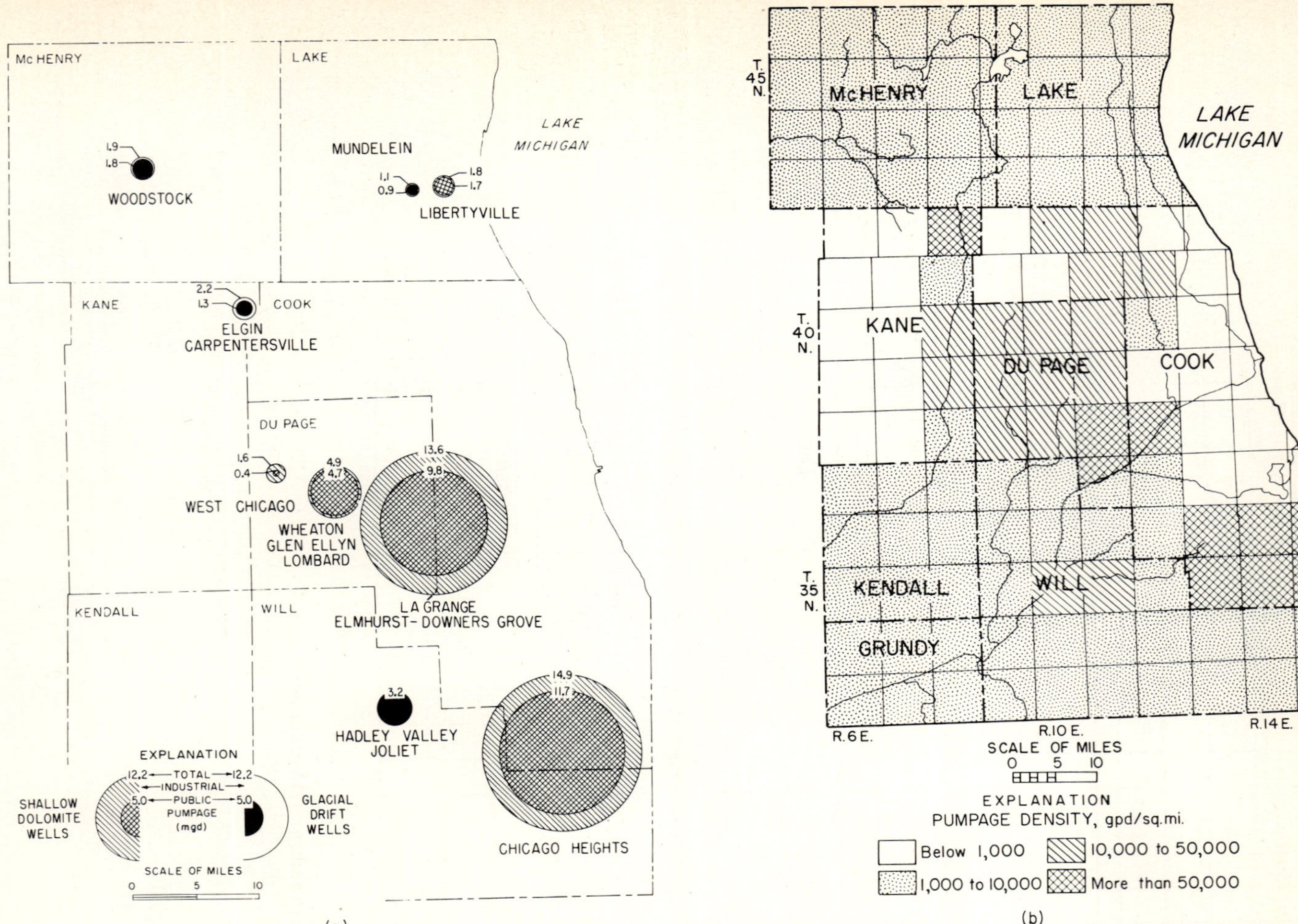

Fig. 2.45 Distributions of pumpage from glacial drift and shallow dolomite wells in major pumping centers in northeastern Illinois in 1962 (*a*) and pumpage per square mile from shallow aquifers in northeastern Illinois in 1959 (*b*). (*From Sasman, 1965; Suter et al., 1959.*)

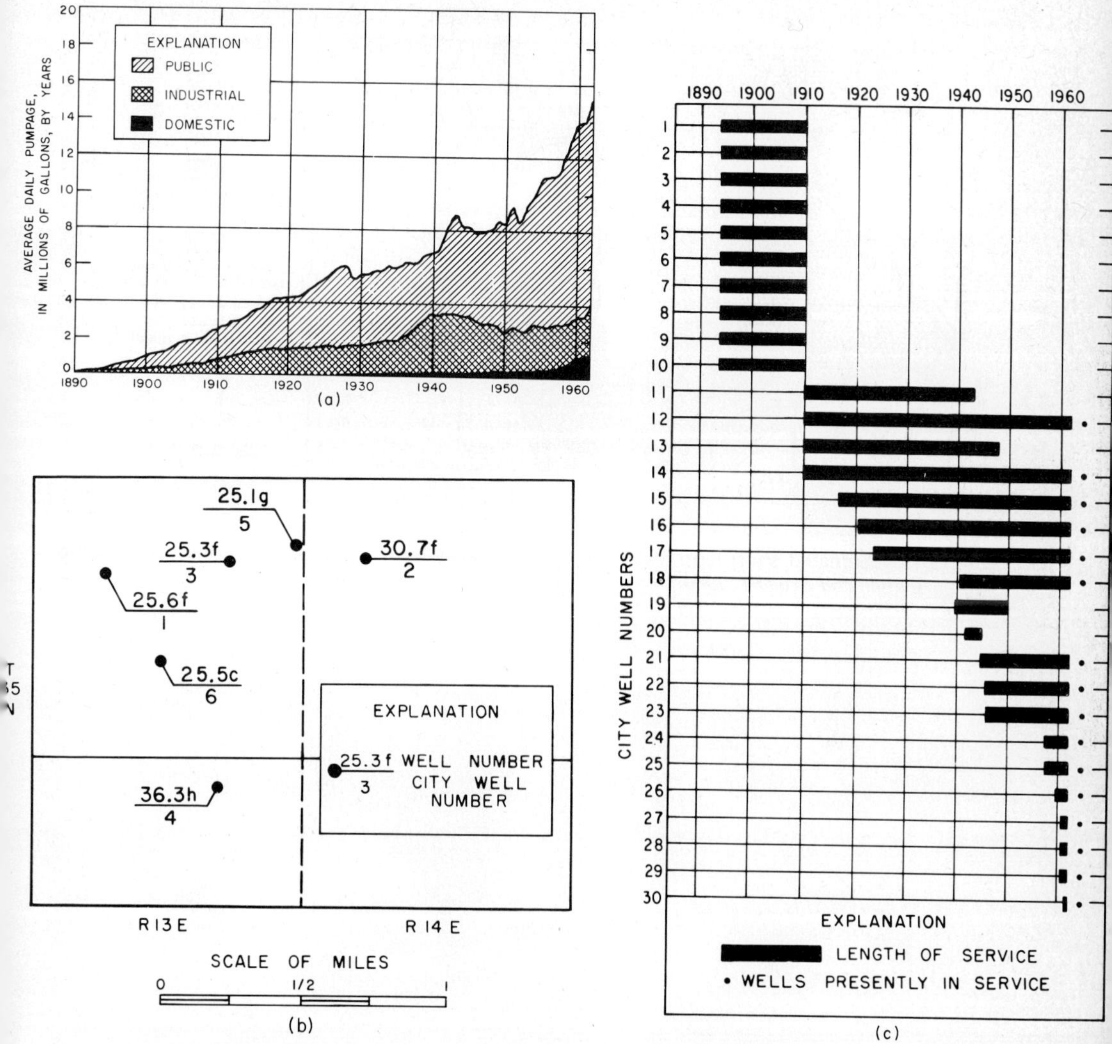

Fig. 2.46 Estimated groundwater pumpage from Silurian dolomite aquifer in Chicago Heights area, Illinois, 1890–1962 (*a*), location of municipal wells at Park Forest, Illinois (*b*), and length of service of municipal wells at Chicago Heights, Illinois (*c*). (*From Prickett et al., 1964.*)

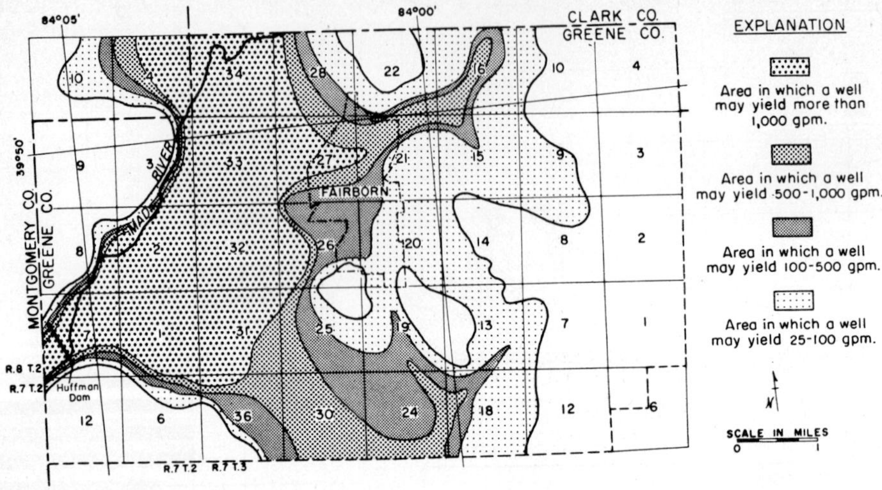

Fig. 2.47 Estimated short-term yields of wells in valley-train deposits of Fairborn area, Ohio. (*From Walton and Scudder, 1960.*)

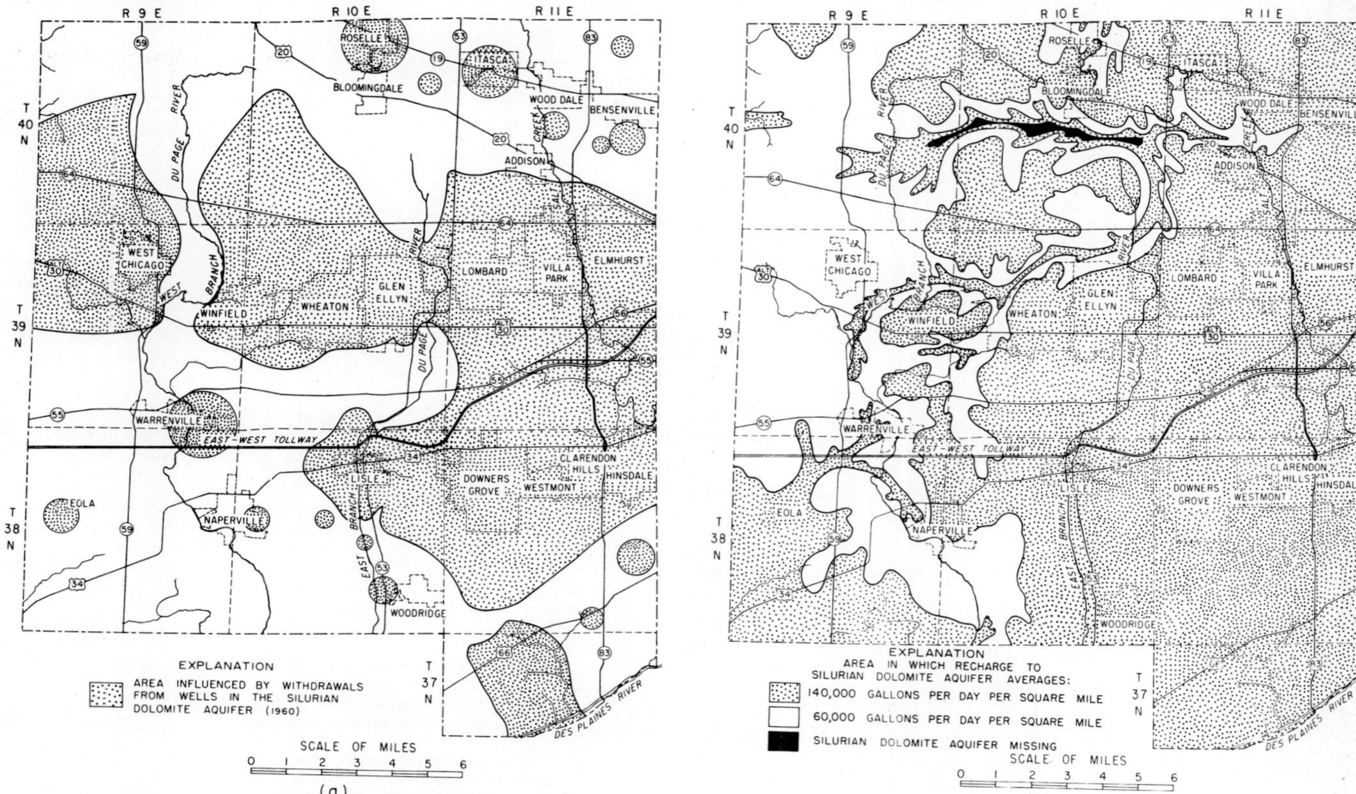

Fig. 2.48 Areas influenced by withdrawals from wells in Silurian dolomite aquifer in DuPage County, Illinois (*a*), and estimated recharge rates for Silurian dolomite aquifer in DuPage County, Illinois (*b*). (*From Zeizel et al., 1962.*)

Subject	Recommended Symbol	
A. TOPOGRAPHY		
Topography	Symbols conform as far as possible with international usage (grey)	
B. GEOLOGY		
1. Geological formation	Only if age is essential to hydrogeological understanding, colours should be used and conform as far as possible with international geological usage.	
2. Stratigraphy	Letters, symbols and patterns should conform with international usage	(BLACK)
3. Height or depth of formation (top or base) relative to the national reference level	Contour line, broken where uncertain 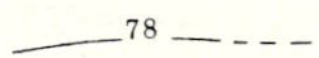	(BLACK)
4. Contact between permeable and impermeable or semi-permeable formations	Line of contact 	(BLACK)
5. Strike and dip		(BLACK)
6. Axis of anticline, with direction of axial plunge	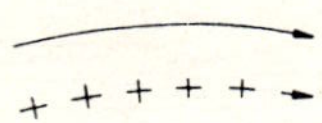 (or)	(BLACK)
7. Axis of syncline, with direction of axial plunge	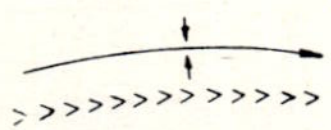	(BLACK)
8. Flexure, with direction of downthrow side	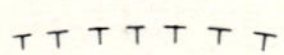	(BLACK)
9. Flexure, not affecting covering layers	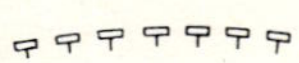	(BLACK)
10. Fault, with direction of downthrow side		(BLACK)
11. Fault, not affecting covering layers		(BLACK)
12. Overthrust fault	(Teeth on upper plate)	(BLACK)

Subject	Recommended Symbol	
11. Chemical properties of the formation	The lithological symbols may also indicate chemical properties	
D. HYDROGRAPHY		
All natural waters in blue		
1. Perennial stream with direction of flow		(BLUE)
2. Perennial stream, highly polluted	See F. 7	
3. Perennial stream, with high chloride content	See F. 8	
4. Seasonal stream, with driection of flow	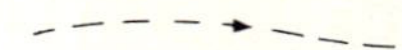	(BLUE)
5. Intermittent stream, with direction of flow	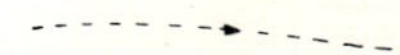	(BLUE)
6. Disappearance point of stream		(BLUE)
7. Gauging station, with yearly average flow and area of catchment	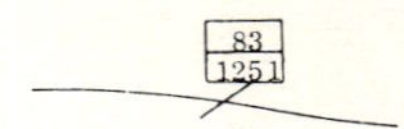	(BLUE)
8. Marsh, seasonal marsh		(BLUE)
9. Flood-stage area, area inundated during floods	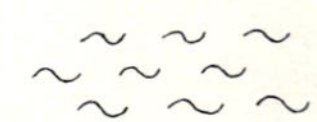	(BLUE)
10. Surface water divide		(BLUE)
11. Spring	 The inside of the symbol should be reserved for hydrochemical data (in colours according to F.3 and F.5), the outside for hydrodynamical data. The example given shows one of ths possi-	(BLUE OR DARK BLUE)

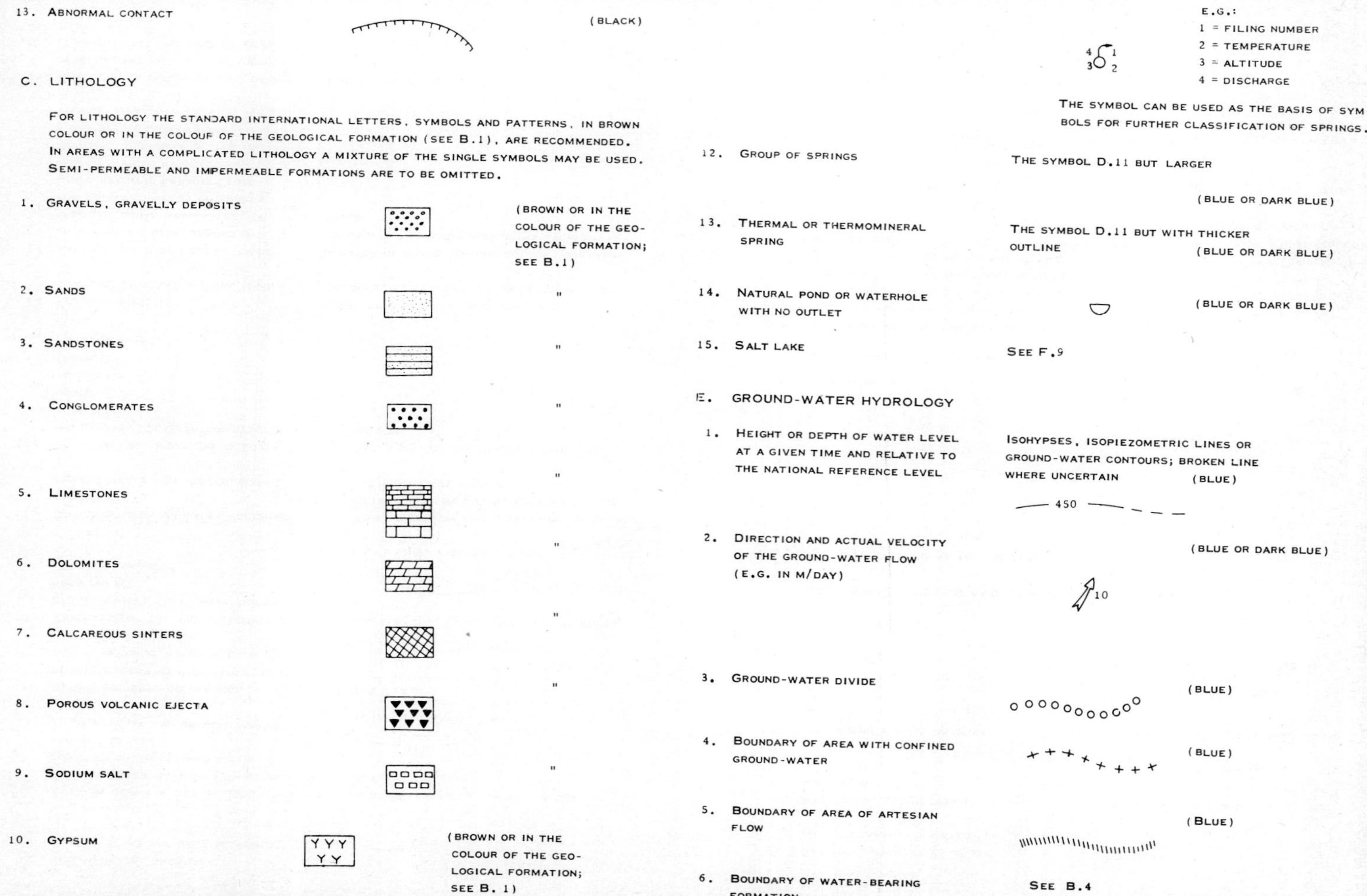

Fig. 2.49 International legend for hydrological maps. (*From La Moreaux, 1966.*)

Subject	Recommended Symbol
7. Ground-water barrier (Large-scale and special maps)	 (Blue)
8. Average depth of top of saturated part of water-bearing formation, confined or unconfined below ground surface (Large-scale and special maps)	Contour lines e.g. in the colour of the formation 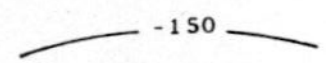
9. Height or depth of top and/or base of water-bearing formation relative to the national reference level	See B.3
10. Thickness of the water-saturated bed at a given time, with thickness (in m) (Special maps)	Isopachytes, lines of equal thickness  (Blue line, figures in red)
11. Different ground-water horizons (aquifers) (Large-scale and special maps)	To be shown by cross sections or planimetrically (by colour left to the discretion of the author)
12. Infiltration conditions of covering layers. Qualitative description, e.g. -good -moderate -poor (Special maps)	Patterns, at the discretion of the author
13. Transmissibility (Large-scale and special maps)	Lines of equal transmissibility or colours, at the discretion of the author
14. Average yield of wells. Order of magnitude represented by areas of equal well yield, or for selected wells of approximate specific capacity (discharge divided by drawdown or by total discharge of the wells for a specific drawdown). (Large-scale and special maps)	A range of shades of one colour, greater intensity of colour indicating greater yield
15. Exploitable yield per unit of the development area of the aquifer. (Large-scale and special maps)	A range of shades of blue
16. Annual rainfall (depth in mm)	Isohyets (Blue)

Subject	Recommended Symbol
Not determined	⊖
6. Chemical properties of the water-bearing formation	See C.11
7. Highly polluted stream (organic pollution)	(Blue line with grey shading on each side) 
8. Stream with high chloride content	(Blue line with violet shading on each side) 
9. Salt lake	(Blue line with violet shading along margin of lake) 

G. Boreholes, Wells and Other Works

All artificial works are indicated in red.

Subject	Recommended Symbol	
1. Borehole	•	(Red)
2. Dug well *	○	(Red)
3. Dug well, dry	●	(Red)
4. Drilled well *		(Red)

* The inside of the symbol should be reserved for hydrochemical data (in colours according to F.3 and F.5). The outside for hydrodynamical data. The example given shows one of the possibilities.

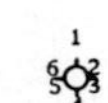

E.g.:
1 = Number
2 = Static level
3 = Depth
4 = Temperature
5 = Drawdown
6 = Yield

Subject	Recommended Symbol	
5. Drilled well, dry		(Red)
6. Artesian well, flowing *		(Red)
7. Artesian well, non-flowing *		(Red)
8. Recharge well		(Red)

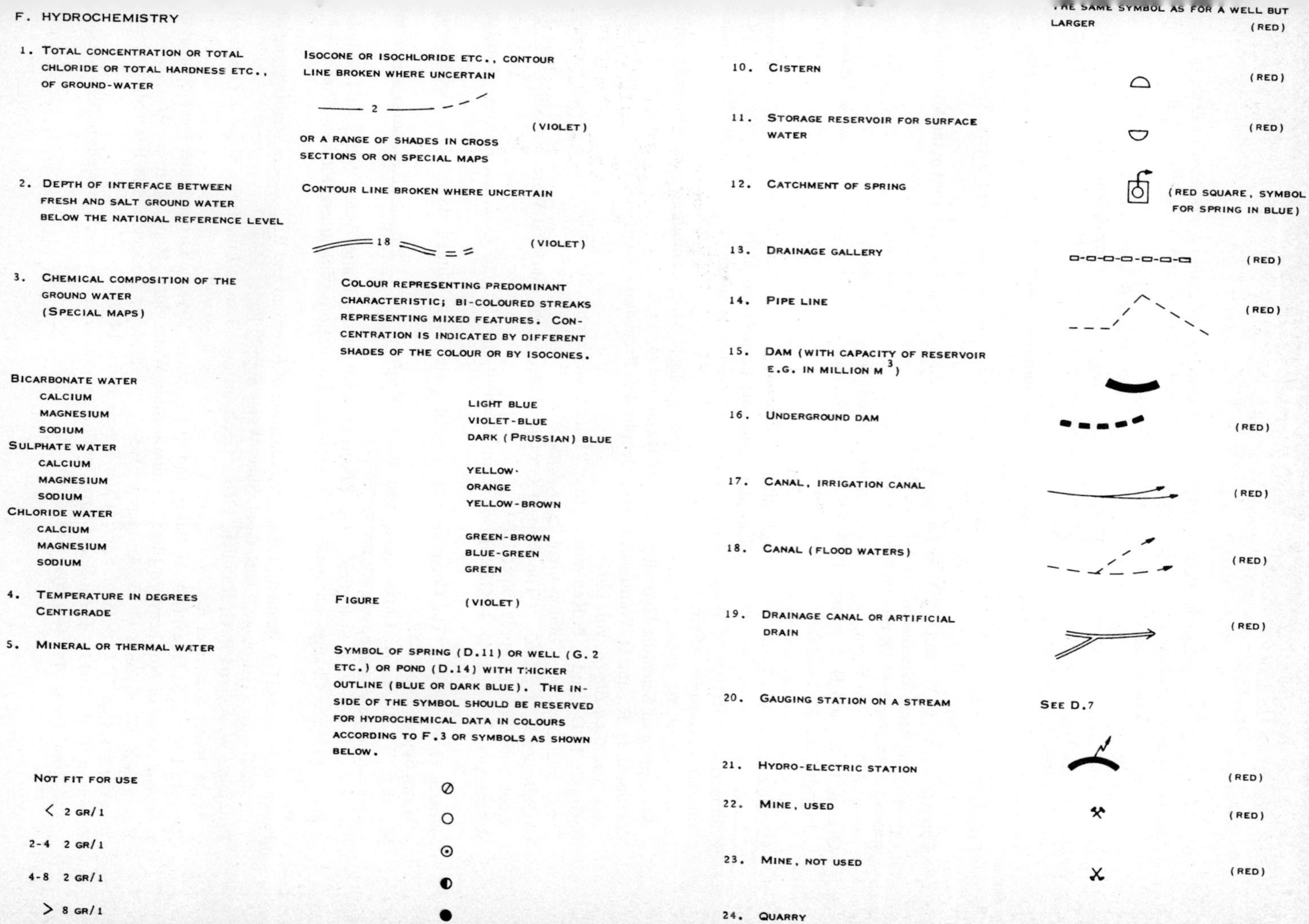

Fig. 2.49 (*Continued*) International legend for hydrological maps. (*From La Moreaux, 1966.*)

References

Christiansen, E. A., and R. R. Parizek, 1961. "A Summary of Studies Completed to Date of the Ground-water Geology and Hydrology of the Buried Missouri and Yellowstone Valleys near Estevan," Saskatchewan Res. Council, Geol. Div. Circ. 1.

Compton, R. R., 1962. "Manual of Field Geology," John Wiley & Sons, Inc., New York.

Davis, S. N., and R. J. M. DeWiest, 1966. "Hydrogeology," John Wiley & Sons, Inc., New York.

Dudley, W. W., Jr., et al., 1964. "Geophysical Studies in Nevada Relating to Hydrogeology," Desert Res. Inst., Univ. Nevada, Tech. Rept. 2.

Ferris, J. G., D. B. Knowles, R. H. Brown, and R. W. Stallman, 1962. "Theory of Aquifer Tests," U.S. Geol. Surv. Water Supply Paper 1536-E.

Fishel, V. C., 1956. Long-term Trends of Ground-water Levels in the United States, *Trans. Am. Geophys. Union.*, vol. 37.

Foley, F. C., W. C. Walton, and W. J. Drescher, 1953. "Ground-water Conditions in the Milwaukee-Waukesha Area, Wisconsin," U.S. Geol. Surv. Water Supply Paper 1229.

Foster, J. W., and M. B. Buhle, 1951. An Integrated Geophysical Investigation of Aquifers in Glacial Drift near Champaign-Urbana, Illinois, *Econ. Geol.*, vol. 46.

Graton, L. C., and H. J. Fraser, 1935. Systematic Packing of Spheres with Particular Relation to Porosity and Permeability, and Experimental Study of the Porosity and Permeability of Clastic Sediments, *J. Geol.*, vol. 43.

Harshbarger, J. W., C. A. Repenning, and J. T. Callahan, 1953. The Navajo Country, Arizona–Utah–New Mexico, in "The Physical and Economic Foundation of Natural Resouces," pt. IV, Subsurface Facilities of Water Management and Patterns of Supply-type Area Studies, House Interior and Insular Affairs Comm. U.S. Congress.

La Moreaux, P. E., 1966. Activities of the Hydrogeologic Map Committee of the International Association of Scientific Hydrology, *Ground Water*, vol. 4, no. 4.

McDonald, H. R., and D. Wantland, 1961. Geophysical Procedures in Ground Water Study, *Trans. Am. Soc. Civil Engrs.*, vol. 126.

McGinnis, L. D., and J. P. Kempton, 1961. "Integrated Seismic, Resistivity and Geologic Studies of Glacial Deposits," Illinois Geol. Surv. Circ. 323.

McGuinness, C. L., 1963. "The Role of Ground Water in the National Water Situation," U.S. Geol. Surv. Water Supply Paper 1800.

McLaughlin, T. G., 1946. "Geology of Grant, Haskell, and Stevens Counties, Kansas," Kansas Geol. Surv. Bull. 61.

Meinzer, O. E., 1959. "The Occurrence of Ground Water in the United States," U.S. Geol. Surv. Water Supply Paper 489.

Meneley, W. A., E. A. Christiansen, and W. O. Kupsch, 1957. Preglacial Missouri River in Saskatchewan, *J. Geol.*, vol. 65, no. 4.

Parker, G. G., and V. T. Stringfield, 1950. Effects of Earthquakes, Trains, Tides, Winds, and Atmospheric Pressure Changes on Water in the Geologic Formations of Southern Florida, *Econ. Geol.*, vol. 45.

Patten, E. P., and G. D. Bennett., 1963. "Application of Electrical and Radioactive Well Logging to Ground-water Hydrology," U.S. Geol. Surv. Water Supply Paper 1544-D.

Pettyjohn, W. A., and P. G. Randich, 1966. Geohydrologic Use of Lithofacies Maps in Glaciated Areas, *Water Resources Res.*, vol. 2, no. 4.

Prickett, T. A., L. R. Hoover, W. H. Baker, and R. T. Sasman, 1964. "Groundwater Development in Several Areas of Northeastern Illinois," Illinois State Water Surv., Rept. of Invest. 47.

Rasmussen, W. C., and G. E. Andreasen, 1959. "Hydrologic Budget of the Beaverdam Creek Basin, Maryland," U.S. Geol. Surv. Water Supply Paper 1472.

Robinson, T. W., 1939. Earth-tides Shown by Fluctuations of Water-levels in Wells in New Mexico and Iowa, *Trans. Am. Geophys. Union.*, vol. 20.

Russell, R. R., 1963. "Ground-water Levels in Illinois through 1961," Illinois State Water Surv. Rept. of Invest. 45.

Sasman, R. T., 1965. "Ground-water Pumpage in Northeastern Illinois through 1962," Illinois State Surv., Rept. of Invest. 50.

Schicht, R. J., 1965. "Ground-water Development in East St. Louis Area, Illinois," Illinois State Water Surv., Rept. of Invest. 51.

——— and W. C. Walton, 1961. "Hydrologic Budgets for Three Small Watersheds in Illinois," Illinois State Water Surv., Rept. of Invest. 40.

Suter, Max, R. E. Bergstrom, H. F. Smith, G. H. Emrich, W. C. Walton, and T. E. Larson, 1959. "Preliminary Report on Ground-water Resources of the Chicago Region, Illinois," Illinois State Water Surv. and Geol. Surv. Cooperative Ground-water Rept. 2.

Thomas, H. E., 1952. Ground-water Regions of the United States—Their Storage Facilities, in "The Physical and Economic Foundation of Natural Resources," vol. 3, House Interior and Insular Affairs Comm., U.S. Congress.

Todd, D. K., 1959. "Ground Water Hydrology," John Wiley & Sons, Inc., New York.

Vorhis, R. C., 1955. Interpretation of Hydrologic Data Resulting from Earthquakes, *Geol. Rundschau*, vol. 43.

Walker, W. H., R. E. Bergstrom, and W. C. Walton, 1965. "Preliminary Report on the Ground-water Resources of the Havana Region in West-central Illinois," Illinois State Water Surv. and Geol. Surv. Cooperative Ground-water Rept. 3.

Walton, W. C., 1964. "Potential Yield of a Sand and Gravel Aquifer in a Buried Valley near Estevan, Saskatchewan—Water-level Conditions," typewritten report prepared for Saskatchewan Research Council.

———, 1966. "Potential Yield of a Sand and Gravel Aquifer in a Buried Valley near Estevan, Saskatchewan—Results of Test Hole Drilling Program," typewritten report prepared for Saskatchewan Research Council.

——— and S. Csallany, 1962. "Yields of Deeps Sandstone Wells in Northern Illinois," Illinois State Water Surv., Rept. of Invest. 43.

——— and G. D. Scudder, 1960. "Ground-water Resources of the Valley-train Deposits in the Fairborn Area, Ohio Div. of Water Tech. Rept. 3.

Zeizel, A. J., W. C. Walton, R. T. Sasman, and T. A. Prickett, 1962. "Ground-water resources of DuPage County, Illinois," Illinois State Water Surv. and Geol. Surv. Cooperative Ground-water Rept. 2.

CHAPTER 3

BASIC PRINCIPLES AND FUNDAMENTAL EQUATIONS

Groundwater movement and storage are governed by established hydraulic principles. Hagen (1839) and Poiseville (1846) showed that the velocity of flow in capillary tubes is proportional to the hydraulic gradient. Darcy (1856) confirmed the applicability of this principle to the laminar flow of water in uniform sands. He observed that the velocity of flow through porous media is proportional to the first power of the hydraulic gradient. This statement is known universally as *Darcy's law* and serves as the foundation of quantitative groundwater resource evaluation studies. General flow equations of groundwater are derived from Darcy's law and the equation of continuity.

Darcy's law states that the flow rate through porous media is proportional to the head loss and inversely proportional to the length of flow path. In a useful form, this may be expressed as

$$Q_m = \frac{P(h_1 - h_2)A_d}{L_s} \tag{3.1}$$

in which Q_m is the flow rate, P is the coefficient of permeability, $h_1 - h_2$ is the head loss over a distance L_s in the direction of flow, and A_d is the cross-sectional area of the porous medium across which the flow occurs.

3.1 Differential Form and Range of Validity of Darcy's Law

On a microscopic scale the velocity of movement of a particle of water through the interstices of an aquifer is invariably random. The velocity changes continuously as water moves through interconnected interstices whose space dimensions vary. The head and its gradient are unevenly distributed along the direction of flow. The *head* h is the sum of elevation of a point above an arbitrarily chosen datum of elevation and the height of water in a noncapillaric pipe open at the considered point. Thus, the head of flow $h(x,y,z,t)$ at the point (x,y,z) is defined as (Hantush, 1964)

$$h(x,y,z,t) = \frac{p}{\gamma} + z + f \tag{3.2}$$

where p is the hydrostatic pressure at the point; γ is the unit weight of water; x, y, z are rectangular coordinates; f is the elevation of the xy plane above an arbitrarily chosen datum of elevation; and t is the time of observation.

On a macroscopic scale, the head and its gradient may be assumed uniform and equal to the average of the heads and gradients at many points over small volumes of the aquifer containing a great number of interstices. Consequently, the flow per unit area may also be assumed uniform and equal to the average of the flows through many small elemental cross-sectional areas cutting a great number of interstices. In this way the flow system is replaced by a mathematical continuum in which the flow can be described by Darcy's law in terms of uniform average heads.

The *specific discharge* Q_m/A_d, from Eq. (3.1) (volume of water passing through a unit cross section of the aquifer per unit time), is a hypothetical macroscopic velocity and should not be confused with the actual or field velocity (volume of water passing through a unit area of the interstitial space, in which flow of water occurs, per unit time).

In laminar flow through aquifers the inertial forces may be considered negligible in comparison with viscous forces. The only macroscopic force exerted on the water by the aquifer skeleton is the viscous resistance to flow. The viscous force must be in equilibrium with the weight of the water and the net force due to water pressure in the direction of flow.

Consider the elemental volume $\Delta V = \Delta A\ \Delta d$ of the aquifer oriented in the direction of flow and inclined to the horizontal plane. ΔA and Δd are, respectively, the elemental flow cross-sectional area and the length of the elemental volume in the direction of flow. The pressure acting on one end of the elemental volume is p and the pressure on the other end is $p + (\partial p/\partial d)\,\Delta d$. The net force f_p acting on any plane within the elemental volume in the direction of flow is equal to $-\theta\,\Delta A(\partial p/\partial d)\,\Delta d$, in which θ is the porosity of the elemental volume of the aquifer (Hantush, 1964).

The weight of water in the elemental volume of aquifer is equal to $\gamma\theta\,\Delta A\,\Delta d$. The component of force f_g due to the weight of the water parallel to the direction of flow is equal to $-\gamma\theta\,\Delta A\,\Delta d(\partial z/\partial d)$ in which $\partial z/\partial d$ is the sine of the angle which Δd makes with its projection on the horizontal plane. The viscous force f_u resisting flow and acting parallel to the direction of flow is equal to $-\mu a_s\,\Delta A\,\Delta d\,v_s$ (that is, linearly proportional to v_s), in which μ is the dynamic viscosity of water, a_s is a factor that depends on the geometry of the interstices in the direction of flow, and v_s is the bulk velocity in the direction of flow. It is assumed that the total surface area of the grains within the elemental volume is proportional to $\Delta A\,\Delta d$ (Hantush, 1964).

As stated earlier, the forces f_p, f_g, and f_u must be in equilibrium. Thus,

$$\theta\,\Delta A\,\frac{\partial p}{\partial d}\,\Delta d - \gamma\theta\,\Delta A\,\Delta d\,\frac{\partial z}{\partial d} - \mu a_s\,\Delta A\,\Delta d\,v_s = 0$$

or

$$-\theta\,\Delta A\,\Delta d\left(\frac{\partial p}{\partial d} + \gamma\,\frac{\partial z}{\partial d} + \frac{\mu a_s}{\theta}\,v_s\right) = 0$$

and $$-\left(\frac{\partial p}{\partial d} + \gamma \frac{\partial z}{\partial d} + \frac{\mu a_s}{\theta} v_s\right) = 0$$

In terms of the bulk velocity v_s, the above equation becomes (Hantush, 1964)

$$v_s = -\frac{\gamma P_s}{\mu} \frac{\partial(p/\gamma + z)}{\partial d} \tag{3.3}$$

where P_s, a constant characteristic of the aquifer, is substituted for θ/a_s. P_s has the dimension of length square, and is usually known as the *intrinsic permeability* of the aquifer. From Eq. (3.2), $\partial h/\partial d = \partial(p/\gamma + z)/\partial d$ since f is a constant. Substitution of $\partial h/\partial d$ for $\partial(p/\gamma + z)/\partial d$ in Eq. (3.3) results in the following equation

$$v_s = -P\frac{\partial h}{\partial d} \tag{3.4}$$

where $P = (\gamma/\mu)P_s$ and is known as the *coefficient of permeability.*

Equation (3.4) is the differential form of Darcy's law, which states that the velocity of flow through porous media is proportional to the first power of the hydraulic gradient. The negative sign signifies that the fluid moves in the direction of decreasing head. The components of velocity in the three rectangular-coordinate directions may be written as (Hantush, 1964)

$$v_x = -P\frac{\partial h}{\partial x} \qquad v_y = -P\frac{\partial h}{\partial y} \qquad v_z = -P\frac{\partial h}{\partial z} \tag{3.5}$$

Since $P = (\gamma/\mu)P_s$, and $P_s = \gamma/\mu(\theta/a_s)$, P depends upon the properties of the water flowing through the aquifer, the porosity of the aquifer, the range and distribution of grain sizes, the shape of the grains and their orientation and arrangement, and the mean grain diameter. The ratio of the specific weight to the viscosity varies from temperature to temperature. However, the temperature in many problems of groundwater flow varies inappreciably, and P may be regarded as constant in the equation of flow.

Darcy's law represents the statistical macroscopic equivalent of the Navier-Stokes equations for the viscous flow of groundwater. Viscous effects are accounted for completely by Darcy's law.

For Darcy's law to be applicable, the flow must be viscous and irrotational. The flow through individual interstices is rotational; however, on a macroscopic scale when a volume of aquifer containing many interstices is considered, the rotations within individual interstices balance out and the resultant flow is irrotational.

In deriving Darcy's law, the flow was assumed laminar. By analogy to flow through pipes, the *Reynolds number* is used as an index to determine the tendency of flow to be laminar. The Reynolds number, when adapted to flow through porous media, is given by (Hantush, 1964)

$$R = \rho V \frac{d_m}{\mu} \tag{3.6}$$

where ρ is the density of the fluid, V is the bulk velocity, d_m is the mean grain diameter, and μ is the dynamic viscosity of the fluid. The mean grain diameter was suggested by Hazen (1893) as the diameter of sand grain such that 10 percent of the aquifer material (by weight) is of smaller grains and 90 percent is of larger grains.

Experiments on natural and artificial sands of approximately uniform spherical grains show that departure from laminar flow starts at values of R between 1 and 10 (Todd, 1959), depending upon the range of grain size and shape. There is no known lower limit of applicability of Darcy's law (Meinzer and Fishel, 1934). However, the law of flow in material of extremely fine grains, such as colloidal clays or sands that are not completely saturated with water, may be somewhat different from Darcy's law. From both theory and experiment, almost all natural groundwater flow has Reynolds numbers less than unity, and Darcy's law is applicable. In some rock aquifers, such as dolomite, containing cracks and crevices, the flow may deviate from Darcy's law. Moreover, flows in the immediate vicinity of a heavily pumped well with steep hydraulic gradients may violate Darcy's law. Ordinarily it may be assumed that the flow is laminar, and the analyses that follow will be based on Darcy's law with modifications to account for turbulence in the immediate vicinity of production wells.

3.2 Volume Elasticity of Aquifers

Meinzer (Meinzer and Hard, 1925), while studying the relation between declines in artesian head and withdrawals of water from the Dakota Sandstone in North Dakota, came to the conclusion that withdrawals could not be accounted for by the compressibility of the water alone, but might be accounted for by the compressibility of the aquifer. Thus, the compressibility and elasticity of aquifers have long been recognized. The volume elasticity of aquifers is relatively important in the storage of groundwater only under artesian conditions; under water-table conditions, the compressibility of aquifers is relatively unimportant with respect to the specific yield associated with the gravity drainage of interstices.

Consider the forces acting on the aquifer-aquitard or aquifer-aquiclude interface (see Ferris et al., 1962). The forces due to the weight of the formations overlying the aquifer and other loads such as atmospheric pressure are borne by the water contained in the interstices of the aquifer and by the structural skeleton of the aquifer. If the aquifer is compressible and elastic and the total load on the interface remains constant, a reduction of water pressure due to pumping from a well results in an increase in the load borne by the skeleton of the aquifer. The aquifer is compacted, reducing the interstitial space. At the same time, the water expands to the extent permitted by its elasticity. In response to the slight expansion of the water itself and a small decrease in porosity due to the slight compaction of the skeleton of the aquifer, a certain amount of water will be released from storage within the aquifer. Conversely, in response to cessation of pumping, the water pressure gradually increases, resulting in compression of the water and expansion of the skeleton of the aquifer. This releases interstitial space for storing water moving into the part of the aquifer affected by the previous compression. If the aquifer is perfectly elastic and water levels in recharge and discharge areas do not change, the original artesian head will ultimately be restored.

Consider the elemental volume $\Delta V = \Delta A \, \Delta z$ in a compressible aquifer. Assuming that water pressure acts throughout the elemental volume, the compressive stress of the skeleton of the aquifer acts over the horizontal area ΔA, the lateral deformations of the elemental volume are negligible, and the atmospheric pressure is constant, the following equation may be written provided that the arching action of the aquitard overlying the aquifer is neglected (Hantush, 1964)

$$w = p + k = \text{const} \tag{3.7}$$

where w is the weight per unit area of the atmosphere and of the material overlying the elemental volume of the aquifer, p is the pressure of the water in the elemental volume, and k is the compressive stress of the skeleton of the elemental volume. By differentiation

$$dk = -dp \tag{3.8}$$

Thus, an increase in water pressure must be accompanied by a corresponding decrease in the compressive stress. The density of the water ρ increases as p increases in accordance with

$$d\rho = \beta\rho \, dp \tag{3.9}$$

in which β is the reciprocal of the bulk modulus of elasticity of the water and ρ is the density of the water. According to Young's modulus of elasticity, when ΔV is compressed, Δz varies in accordance with (Jacob, 1950)

$$d(\Delta z) = -\alpha \, \Delta z \, dk \tag{3.10}$$

or from Eq. (3.8)

$$d(\Delta z) = \alpha \, \Delta z \, dp \tag{3.11}$$

in which α is the reciprocal of the modulus of elasticity of the skeleton of the aquifer.

The volume of solid aquifer skeleton may be assumed constant and is given by $\Delta V_s = (1 - \theta) \, \Delta A \, \Delta z$. By differentiation

$$\frac{d(\Delta V_s)}{\Delta A} = (1 - \theta) \, d(\Delta z) - \Delta z \, d\theta = 0$$

or from Eq. (3.11)

$$d\theta = \alpha(1 - \theta) \, dp \tag{3.12}$$

The mass of water in the elemental volume of the aquifer is $\Delta M = \rho\theta \, \Delta A \, \Delta z$. If ΔA remains constant, ΔM will vary with the density of the water, the porosity of the aquifer, and the vertical dimension of the elemental volume. By differentiation, the change of the mass of water is given by (Hantush, 1964)

$$\frac{d(\Delta M)}{\Delta V} = \theta\, d\rho + \rho\, d\theta + \frac{\rho\theta\, d(\Delta z)}{\Delta z} \tag{3.13}$$

Substituting from Eqs. (3.8), (3.11), and (3.12),

$$\frac{d(\Delta H)}{\Delta V} = \theta\beta\rho\, dp + \rho\alpha(1-\theta)\, dp + \rho\theta\alpha\, dp \tag{3.14}$$

or

$$\frac{d(\Delta M)}{\rho\,\Delta V} = (\theta\beta + \alpha)\, dp \tag{3.15}$$

From Eq. (3.2) it can be seen that the pressure p in an elemental volume, at constant elevation within the aquifer, changes directly with the head h, that is, $dp = \gamma\, dh$, and

$$\frac{d(\Delta M)}{\rho\,\Delta V} = \gamma\theta\beta\left(1 + \frac{\alpha}{\theta\beta}\right) dh = S_s\, dh \tag{3.16}$$

The coefficient S_s of dimension L^{-1} has been termed the *specific storage* of the aquifer (Hantush, 1964) and is defined as the volume of water which a unit volume of the aquifer releases from storage because of expansion of water and compression of the aquifer under a unit decline in the average head within the unit volume of the aquifer. In artesian aquifers the coefficient of storage S based on Eq. (3.16) is given by (Jacob, 1950)

$$S = \gamma\theta\beta m\left(1 + \frac{\alpha}{\theta\beta}\right) \tag{3.17}$$

where m is the saturated thickness of the aquifer. In water-table aquifers the coefficient of storage is given by (see Hantush, 1964)

$$S = S_y + \gamma\theta\beta m\left(1 + \frac{\alpha}{\theta\beta}\right) \tag{3.18}$$

where S_y is the specific yield of the aquifer and m is the saturated thickness of the aquifer. Usually $\gamma\theta\beta m(1 + \alpha/\theta\beta)$ is negligible in comparison with S_y, and S under water-table conditions for all practical purposes is the specific yield. The fractions of storage attributable to expansibility of the water and to the compressibility of the skeleton of the aquifer are, respectively, $\theta\gamma m\beta$ and $\gamma m\alpha$.

Equation (3.17) was derived by assuming a hypothetical isotropic elastic artesian aquifer bounded by impermeable planes. In the case where the aquifer has included clay beds and the release of stored water from the clay is virtually instantaneous, the coefficient may be given as (see Jacob, 1940)

$$S = \gamma\theta m\left(\frac{1}{E_w} + \frac{1}{\theta E_s} + \frac{c_b}{E_c}\right) \tag{3.19}$$

where E_w is the bulk modulus of elasticity of water, E_s is the modulus of compression of the skeleton of the aquifer, E_c is the modulus of compression of the clay, and c_b is a dimensionless quantity that depends largely on the thickness, configuration, and distribution of the clay beds.

Water levels in artesian aquifers respond to changes in atmospheric pressure. Water-level fluctuations in a well due to atmospheric-pressure changes are an index of the elasticity of the aquifer. Assuming that changes in atmospheric-pressure loading on an aquitard overlying the aquifer are transmitted through it undiminished in magnitude, the following equation may be written to express the relation between forces acting on the aquifer-aquitard interface (Hantush, 1964)

$$p + k = p_a + \text{const} \tag{3.20}$$

in which p_a is the atmospheric pressure. By differentiation

$$dp + dk = dp_a \tag{3.21}$$

or

$$dp - dp_a = -dk \tag{3.22}$$

The weight of the column of water in the well above the top of the artesian aquifer plus the weight of the atmosphere is balanced by the pressure of the water in the aquifer. Therefore, the change of water pressure is given by

$$dp = dp_a + \gamma\, dh \tag{3.23}$$

in which h is the elevation of the water surface in the well. From Eqs. (3.23) and (3.22) the ratio of the change in water level to the change in atmospheric-pressure head producing it is (Hantush, 1964)

$$\frac{dh}{dp_a/\gamma} = \frac{-dk}{dp + dk} = \frac{-1}{1 + dp/dk} \tag{3.24}$$

The volume of water ΔV_w in an elemental aquifer volume ΔV is equal to $\theta\, \Delta V$. Assuming the grains of the skeleton of the aquifer to be incompressible, as the aquifer is compressed, the change in volume $d(\Delta V)$ is equal to the change in water volume $d(\Delta V_w)$ and

$$\frac{d(\Delta V_w)}{\Delta V_w} = \frac{d(\Delta V)}{\theta\, \Delta V} \tag{3.25}$$

From the definition of the bulk modulus of elasticity, the left-hand member of Eq. (3.25) equals $-\beta\, dp$ and the right-hand member equals $-\alpha\, dk/\theta$. The negative sign indicates a decrease in volume accompanying an increase in stress. Thus,

$$\frac{dp}{dk} = \frac{\alpha}{\theta\beta} \tag{3.26}$$

Substituting Eq. (3.26) into Eq. (3.24) gives (Jacob, 1950)

$$\frac{dh}{dp_a/\gamma} = \frac{-1}{1 + \alpha/\theta\beta} \tag{3.27}$$

in which

$$\frac{dh}{dp_a/\gamma}$$

is the barometric efficiency (BE) of the aquifer.

From Eq. (3.17)

$$S = \gamma\theta\beta m \frac{1}{\text{BE}} \tag{3.28}$$

Water levels in artesian aquifers respond to tidal fluctuations. Water-level rises in artesian wells near the ocean, lakes, or streams due to stage rises reflect increased pressure head in the aquifer caused by the additional loading. A well in an artesian aquifer that responds to tidal fluctuations should also respond to atmospheric-pressure changes because it is the same mechanism in the aquifer that produces both types of response. The relation between forces acting on the aquifer-aquitard interface in the case of tidal fluctuations can be expressed as (Jacob, 1950)

$$p + k = \gamma H + \text{const} \tag{3.29}$$

where H is the stage of the tide (corrected for density if necessary). By differentiation

$$dp + dk = \gamma\, dH \tag{3.30}$$

The change of the pressure of the water in the aquifer is given by

$$dp = \gamma\, dh \tag{3.31}$$

From Eqs. (3.31) and (3.30), the ratio of a change of water level to a change in the stage of the tide is (Jacob, 1950)

$$\frac{dh}{dH} = \frac{dp}{\gamma\, dH} = \frac{dp/dk}{1 + dp/dk} \tag{3.32}$$

since

$$\frac{dp}{dk} = \frac{\alpha}{\theta\beta} \qquad \frac{dh}{dH} = \frac{\alpha/\theta\beta}{1 + \alpha/\theta\beta} = \text{TE} \tag{3.33}$$

where dh/dH is the tidal efficiency (TE) of the aquifer. The sum of the barometric and tidal efficiencies equals unity, that is,

$$\text{BE} + \text{TE} = 1 \tag{3.34}$$

3.3 Differential Equations Governing Groundwater Flow

Consider the elemental volume ΔV of an aquifer situated in the field of flow. From the law of conservation of matter, the net inward flux through ΔV must equal the rate at which matter is accumulating within ΔV (see DeWiest, 1965). The mass flows per unit time through the three pairs of faces of the elemental volume $\Delta V = \Delta x\, \Delta y\, \Delta z$ are approximately

$$-\left(\frac{\partial(\rho v_s)}{\partial x}\right) \Delta x\, \Delta y\, \Delta z \qquad -\left(\frac{\partial(\rho v_y)}{\partial y}\right) \Delta y\, \Delta x\, \Delta z \qquad \text{and} \qquad -\left(\frac{\partial(\rho v_z)}{\partial z}\right) \Delta z\, \Delta x\, \Delta y$$

Summing (Jacob, 1950)

$$\text{Net inward flux} = -\left(\frac{\partial(\rho v_x)}{\partial x} + \frac{\partial(\rho v_y)}{\partial y} + \frac{\partial(\rho v_z)}{\partial z}\right) \Delta x\, \Delta y\, \Delta z \tag{3.35}$$

The rate at which matter is accumulating within ΔV is $\partial(\Delta M)/\partial t$. The mass ΔM of water within ΔV is $\rho\theta\, \Delta x\, \Delta y\, \Delta z$. Assuming that the skeleton of the aquifer and the water contained therein are compressible, changes in the lateral dimensions of the elemental volume are negligible and, therefore, the density of the water, the porosity of the aquifer, and the vertical dimensions of the elemental volume are variable, and the rate of change of the mass of water is, from Eq. (3.15),

$$\frac{\partial(\Delta M)}{\partial t} = \left[\theta\rho\left(\beta + \frac{\alpha}{\theta}\right)\frac{\partial p}{\partial t}\right] \Delta x\, \Delta y\, \Delta z \tag{3.36}$$

From the law of conservation of matter, net inward flux equals $\partial(\Delta M)/\partial t$ or

$$-\left(\frac{\partial(\rho v_x)}{\partial x} + \frac{\partial(\rho v_y)}{\partial y} + \frac{\partial(\rho v_z)}{\partial z}\right) \Delta x\, \Delta y\, \Delta z = \theta\rho\left(\beta + \frac{\alpha}{\theta}\right)\frac{\partial p}{\partial t} \Delta x\, \Delta y\, \Delta z \tag{3.37}$$

Dividing through by $\Delta x\, \Delta y\, \Delta z$

$$-\frac{\partial(\rho v_x)}{\partial x} - \frac{\partial(\rho v_y)}{\partial y} - \frac{\partial(\rho v_z)}{\partial z} = \theta\rho\left(\beta + \frac{\alpha}{\theta}\right)\frac{\partial p}{\partial t} \tag{3.38}$$

or
$$-\rho\left(\frac{\partial v_x}{\partial x} + \frac{\partial v_y}{\partial y} + \frac{\partial v_z}{\partial z}\right) - \left(v_x \frac{\partial p}{\partial x} + v_y \frac{\partial p}{\partial y} + v_z \frac{\partial p}{\partial z}\right) = \theta\rho\beta\left(1 + \frac{\alpha}{\theta\beta}\right)\frac{\partial p}{\partial t} \tag{3.39}$$

From Eqs. (3.9) and (3.2)

$$\frac{\partial p}{\partial x} \approx \rho\beta \frac{\partial p}{\partial x} = \rho^2\beta\gamma \frac{\partial h}{\partial x}$$

$$\frac{\partial p}{\partial y} \approx \rho\beta \frac{\partial p}{\partial y} = \rho^2\beta\gamma \frac{\partial h}{\partial y}$$

$$\frac{\partial p}{\partial z} \approx \rho\beta \frac{\partial p}{\partial z} = \rho^2\beta\gamma \left(\frac{\partial h}{\partial z} - 1\right)$$

and
$$\frac{\partial p}{\partial t} = \gamma \frac{\partial h}{\partial t}$$

Substituting in Eq. (3.39) the above results and expressing the velocity components in terms of hydraulic gradients as given by Eq. (3.5)

$$P\left(\frac{\partial^2 h}{\partial x^2} + \frac{\partial^2 h}{\partial y^2} + \frac{\partial^2 h}{\partial z^2}\right) + P\beta\gamma\left[\left(\frac{\partial h}{\partial x}\right)^2 + \left(\frac{\partial h}{\partial y}\right)^2 + \left(\frac{\partial h}{\partial z}\right)^2 - \frac{\partial h}{\partial z}\right] = \theta\beta\gamma\left(1 + \frac{\alpha}{\theta\beta}\right)\frac{\partial h}{\partial t} \tag{3.40}$$

The second term on the left-hand side of Eq. (3.40) is usually small in comparison to the first term, especially when $\partial h/\partial z \ll 1$. Neglecting this term, Eq. (3.40) becomes

$$P\left(\frac{\partial^2 h}{\partial x^2} + \frac{\partial^2 h}{\partial y^2} + \frac{\partial^2 h}{\partial z^2}\right) = \theta\beta\gamma\left(1 + \frac{\alpha}{\theta\beta}\right)\frac{\partial h}{\partial t} \tag{3.41}$$

or, in terms of the specific storage S_s [Eq. (3.16)],

$$\frac{\partial^2 h}{\partial x^2} + \frac{\partial^2 h}{\partial y^2} + \frac{\partial^2 h}{\partial z^2} = \frac{S_s}{P}\frac{\partial h}{\partial t} \tag{3.42}$$

which is the approximate partial differential equation governing the unsteady-state flow of groundwater in a homogeneous and isotropic aquifer. Equation (3.42) is of the same form as the fundamental equation of applied physics known as the *diffusion equation*. Equations of the same form as Eq. (3.42) appear in the theories of unsteady-state flow of heat and electricity from which solutions to many groundwater flow problems may be obtained by analogy.

Under *steady-state* flow conditions, the velocity and therefore the pressure distribution does not change with time, and Eq. (3.42) reduces to

$$\frac{\partial^2 h}{\partial x^2} + \frac{\partial^2 h}{\partial y^2} + \frac{\partial^2 h}{\partial z^2} = 0 \tag{3.43}$$

Equation (3.43), which is of the same form as the fundamental equation of applied physics known as *Laplace's equation*, is the approximate differential equation governing the steady-state flow of groundwater (or of an incompressible fluid) in a homogeneous and isotropic aquifer.

For the special case of horizontal (two-dimensional) flow in aquifers of essentially uniform thickness m, Eq. (3.42) becomes (Jacob, 1950)

$$\frac{\partial^2 h}{\partial x^2} + \frac{\partial^2 h}{\partial y^2} + \frac{\partial^2 h}{\partial z^2} = \frac{S}{T}\frac{\partial h}{\partial t} \tag{3.44}$$

in which the relations $S = S_s m$ and $T = Pm$ have been used.

The derivation of Eq. (3.42) presented here is essentially the original version given by Jacob (1950) and widely used (see Hantush, 1964, and DeWiest, 1965) until recently (personal communication, Papadopulos, Harza Engineering Co., Chicago).

DeWiest (1966) questioned Jacob's (1950) derivation of Eq. (3.42), indicating that "in one side of the equation the net inward mass flux was calculated for a volume element without deformation while in the other side of the equation, to compute the rate of change of the mass inside the volume element, the element itself was deformed." He derived the equation of groundwater flow (DeWiest, 1966) by using a control volume whose boundaries are fixed in space.

Although he obtained the same approximate equation as Jacob (1950), in the process he redefined the specific storage S_s (DeWiest, 1966; Davis and DeWiest, 1966) as

$$S_s = \rho g[(1 - \eta)\alpha + \theta\beta]$$

"whereby the relation between specific storage and storage coefficient is no longer the thickness of the aquifer but the result of the integration of the equation of flow over the thickness of the aquifer."

Cooper (1966), however, noted that in his derivation DeWiest (1966) "did not . . . distinguish between the rate of flow relative to the moving grains of the medium and the rate of flow across the fixed boundaries of his control volume while the material is deforming." He proceeded to rederive the equation of flow (Cooper, 1966), in terms of both fixed and deforming coordinates, showing that Jacob's (1950) expressions for the specific storage [Eqs. (3.16) and (3.42)] are correct if one considers the vertical coordinate z to be a deforming coordinate.

The fact that z should be a deforming coordinate has little effect in most practical applications of Eq. (3.42), as Cooper (1966) concludes: "In field and laboratory investigations the piezometers or other instruments for measuring heads do not ordinarily remain fixed in space as the material deforms; more often than not they move with the grains of the medium. Observed gradients are therefore more likely to be in terms of deforming coordinates than fixed coordinates. The difference between the two is negligible in most aquifer materials and can be appreciable only in clays and similar materials that compact substantially when the pressure of the fluid in them is reduced."

APPLICATION OF DIFFERENTIAL EQUATIONS TO WELL HYDRAULICS

Having been introduced, the basic principles and equations governing groundwater movement and storage can now be applied to problems of flow to and from production wells. To obtain analytical solutions for particular aquifer and well conditions, assumptions must be made regarding permeability, release of water from storage, type of flow, and boundary conditions. Idealization of the aquifer and of boundary conditions is necessary; aquifers are assumed to be homogeneous, isotropic, and infinite in areal extent; if boundaries are present, they are often represented as straight-line demarcations. Although such solutions only approximate field conditions, when applied with discretion they provide practical answers to well-hydraulics problems. Analytical solutions for several aquifer and well conditions are given in succeeding sections.

3.4 Unsteady-state Radial Flow in Isotropic Nonleaky Artesian Aquifier with Fully Penetrating Wells and Constant-discharge Conditions

Figure 3.1 shows wells fully penetrating an artesian aquifer overlain and underlain by aquicludes. The aquifer is homogeneous, isotropic, infinite in areal extent, and is of the same thickness throughout. The wells completely penetrate the aquifer and flow is radial throughout. The production well is pumped at a constant rate Q.

When a production well in such an aquifer is pumped, water is continuously withdrawn from storage within the aquifer as the cone of depression progresses radially outward from the well. Because of the absence of a source of recharge in the form of vertical leakage or a recharge boundary, there can be no stabilization of water levels and the head in the aquifer will continue to decline provided the aquifer is effectively infinite in areal extent. However, the rate of decline of head continuously decreases as the cone of depression spreads. Water is released from storage by the compaction of the aquifer and its associated beds and by the expansion of the water itself. Assuming that water is released instantaneously with a decline in head, the differential equation governing the unsteady-state radial flow in the nonleaky artesian aquifer from Eq. (3.44) in polar-coordinate notation is (Jacob, 1950)

$$\frac{\partial^2 h}{\partial r^2} + \frac{1}{r}\frac{\partial h}{\partial r} = \frac{S}{T}\frac{\partial h}{\partial t} \tag{3.45}$$

where r is the distance from the production well to the point of observation.

To obtain a simple mathematical solution, the production well is replaced in the analysis by a mathematical sink of constant strength and is assumed to have an

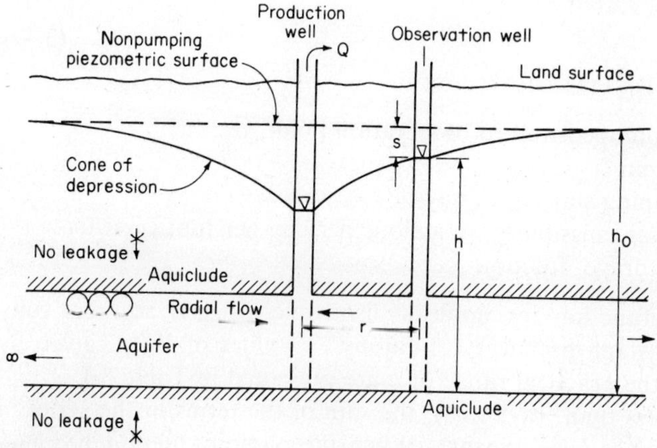

Fig. 3.1 Nonleaky artesian aquifer with fully penetrating wells. (*Drawn by W. C. Walton.*)

infinitesimal diameter. For the boundary conditions $h \to h_0$ as $r \to \infty$ for $t \geq 0$

$$\lim_{r \to 0} \left(r \frac{\partial h}{\partial r} \right) = \frac{Q}{2\pi T}$$

and the initial conditions $h(r,0) = h_0$ for $t \leq 0$, the solution to the problem given by Eq. (3.45) is as follows (see Jacob, 1950):

$$h_0 - h = \frac{Q}{4\pi T} \int_{r^2 S/4Tt}^{\infty} \frac{e^{-u}}{u} \, du \tag{3.46}$$

Theis (1935) first applied this equation to well hydraulics and this equation is known as the *nonequilibrium equation*. The integral is a function of the lower limit u and is tabulated as the exponential integral under the symbol $-E_i(-u)$. Expanding the exponential integral in a convergent series, the drawdown, $s = h_0 - h$, in an observation well at a distance r and at a time t for a constant discharge becomes

$$s = \frac{Q}{4\pi T} \left(-0.5772 - \ln u + u - \frac{u^2}{2.2!} + \frac{u^3}{3.3!} - \frac{u^4}{4.4!} + \cdots \right) \tag{3.47}$$

where $u = r^2 S/4Tt$. The drawdown at the production-well face is found from Eq. (3.47) for r equal to the effective radius of the well.

With the exponential integral expressed symbolically as $W(u)$, that is,

$$W(u) = \int_u^{\infty} \frac{e^{-u}}{u} \, du$$

Eq. (3.46) may be rewritten in the gallon-day-foot system of units in abbreviated form as

$$s = \frac{114.6Q}{T} W(u) \tag{3.48}$$

where $u = \dfrac{1.87 r^2 S}{Tt}$ (3.49)

s = drawdown, in feet
r = distance from pumped well to observation point, in feet
Q = discharge, in gpm
t = time after pumping started, in days
T = coefficient of transmissibility, in gallons per day per foot (gpd/ft)
S = coefficient of storage, fraction

$W(u)$ is read as the "well function for nonleaky isotropic artesian aquifers fully penetrated by wells and constant-discharge conditions." Values of $W(u)$, given by Wenzel (1942), in terms of the practical range of u are presented in Table 3.1.

When u becomes small (less than, say, 0.01), the sum of the terms in the series in Eq. (3.47) beyond $\ln u$ becomes insignificant. When the pumping period becomes large or when the distance r is small, values of u will be small. When $u \leq 0.01$, then

TABLE 3.1 VALUES OF $W(u)$ OR $W(u_{xy})$. (*After Wenzel, 1942.*)

N \ u or u_{xy}	$N\times10^{-15}$	$N\times10^{-14}$	$N\times10^{-13}$	$N\times10^{-12}$	$N\times10^{-11}$	$N\times10^{-10}$	$N\times10^{-9}$	$N\times10^{-8}$	$N\times10^{-7}$	$N\times10^{-6}$	$N\times10^{-5}$	$N\times10^{-4}$	$N\times10^{-3}$	$N\times10^{-2}$	$N\times10^{-1}$	N
1.0	33.9616	31.6590	29.3564	27.0538	24.7512	22.4486	20.1460	17.8435	15.5409	13.2383	10.9357	8.6332	6.3315	4.0379	1.8229	0.2194
1.5	33.5561	31.2535	28.9509	26.6483	24.3458	22.0432	19.7406	17.4380	15.1354	12.8328	10.5303	8.2278	5.9266	3.6374	1.4645	0.1000
2.0	33.2684	30.9658	28.6632	26.3607	24.0581	21.7555	19.4529	17.1503	14.8477	12.5451	10.2426	7.9402	5.6394	3.3547	1.2227	0.04890
2.5	33.0453	30.7427	28.4401	26.1375	23.8349	21.5323	19.2298	16.9272	14.6246	12.3220	10.0194	7.7172	5.4167	3.1365	1.0443	0.02491
3.0	32.8629	30.5604	28.2578	25.9552	23.6526	21.3500	19.0474	16.7449	14.4423	12.1397	9.8371	7.5348	5.2349	2.9591	0.9057	0.01305
3.5	32.7088	30.4062	28.1036	25.8010	23.4985	21.1959	18.8933	16.5907	14.2881	11.9855	9.6830	7.3807	5.0813	2.8099	0.7942	0.006970
4.0	32.5753	30.2727	27.9701	25.6675	23.3649	21.0623	18.7598	16.4572	14.1546	11.8520	9.5495	7.2472	4.9482	2.6813	0.7024	0.003779
4.5	32.4575	30.1549	27.8523	25.5497	23.2471	20.9446	18.6420	16.3394	14.0368	11.7342	9.4317	7.1295	4.8310	2.5684	0.6253	0.002073
5.0	32.3521	30.0495	27.7470	25.4444	23.1418	20.8392	18.5366	16.2340	13.9314	11.6280	9.3263	7.0242	4.7261	2.4679	0.5598	0.001148
5.5	32.2568	29.9542	27.6516	25.3491	23.0465	20.7439	18.4413	16.1387	13.8361	11.5330	9.2310	6.9289	4.6313	2.3775	0.5034	0.0006409
6.0	32.1698	29.8672	27.5646	25.2620	22.9595	20.6569	18.3543	16.0517	13.7491	11.4465	9.1440	6.8420	4.5448	2.2953	0.4544	0.0003601
6.5	32.0898	29.7872	27.4846	25.1820	22.8794	20.5768	18.2742	15.9717	13.6691	11.3665	9.0640	6.7620	4.4652	2.2201	0.4115	0.0002034
7.0	32.0156	29.7131	27.4105	25.1079	22.8053	20.5027	18.2001	15.8976	13.5950	11.2924	8.9899	6.6879	4.3916	2.1508	0.3738	0.0001155
7.5	31.9467	29.6441	27.3415	25.0389	22.7363	20.4337	18.1311	15.8280	13.5260	11.2234	8.9209	6.6190	4.3231	2.0867	0.3403	0.0000658
8.0	31.8821	29.5795	27.2769	24.9744	22.6718	20.3692	18.0666	15.7640	13.4614	11.1589	8.8563	6.5545	4.2591	2.0269	0.3106	0.0000376
8.5	31.8215	29.5189	27.2163	24.9137	22.6112	20.3086	18.0060	15.7034	13.4008	11.0982	8.7957	6.4939	4.1990	1.9711	0.2840	0.0000216
9.0	31.7643	29.4618	27.1592	24.8566	22.5540	20.2514	17.9488	15.6462	13.3437	11.0411	8.7386	6.4368	4.1423	1.9187	0.2602	0.0000124
9.5	31.7103	29.4077	27.1051	24.8025	22.4999	20.1973	17.8948	15.5922	13.2896	10.9870	8.6845	6.3828	4.0887	1.8695	0.2387	0.0000071

Cooper and Jacob, 1946)

$$\int_u^\infty \frac{e^{-u}}{u}\,du = (-0.5772 - \ln u)$$

and

$$s = \frac{Q}{4\pi T}(-0.5772 - \ln u) \tag{3.50}$$

For a particular time, Eq. (3.50) is the equation of a straight-line plot of s versus $\log r$. Drawdowns s_1 and s_2 on the straight line at distances r_1 and r_2, respectively, are from Eq. (3.50).

$$s_1 = \frac{Q}{4\pi T}(-0.5772 - \ln u_1) = \frac{Q}{4\pi T}\left[\ln\left(\frac{1}{u_1}\right) - 0.5772\right] \tag{3.51}$$

$$s_2 = \frac{Q}{4\pi T}(-0.5772 - \ln u_2) = \frac{Q}{4\pi T}\left[\ln\left(\frac{1}{u_2}\right) - 0.5772\right] \tag{3.52}$$

Thus,

$$s_1 - s_2 = \frac{Q}{4\pi T}\ln\frac{4Ttr_2^2S}{4Ttr_1^2S} \tag{3.53}$$

Converting to logarithms to the base 10, Eq. (3.53) may be written in the gallon-day-foot system of units as

$$s_1 - s_2 = \frac{527.7Q}{T}\log\frac{r_2}{r_1} \tag{3.54}$$

For a log cycle, $\log (r_2/r_1) = 1$ and Eq. (3.54) becomes

$$T = \frac{527.7Q}{\Delta s} \tag{3.55}$$

where T = coefficient of transmissibility, in gpd/ft
Q = discharge, in gpm
Δs = drawdown difference per log cycle, in feet

For a particular distance from the pumped well, Eq. (3.50) is the equation of a straight-line plot of s versus $\log t$. Drawdowns s_1 and s_2 on the straight line at times t_1 and t_2, respectively, are given in Eqs. (3.51) and (3.52). The difference between s_2 and s_1 is

$$s_2 - s_1 = \frac{Q}{4\pi T}\left[\ln\frac{u_2}{u_1}\right] = \frac{Q}{4\pi T}\left[\ln\frac{t_2}{t_1}\right] \tag{3.56}$$

Converting to logarithms to the base 10, Eq. (3.56) may be written in the gallon-day-foot system of units as

$$s_2 - s_1 = \frac{264Q}{T}\log\frac{t_2}{t_1} \tag{3.57}$$

For a log cycle, $\log (t_2/t_1) = 1$ and Eq. (3.57) becomes

$$T = \frac{264Q}{\Delta s} \tag{3.58}$$

where T = coefficient of transmissibility, in gpd/ft
Q = discharge, in gpm
Δs = drawdown difference per log cycle, in feet

The straight-line plots of distance-drawdown and time-drawdown may be extrapolated to their intersections with the zero-drawdown axis. At the zero-drawdown intercepts, $s = 0$ and from Eq. (3.50)

$$0.5772 = \ln \frac{1}{u} = \ln \frac{4Tt_0}{r^2 S} \tag{3.59}$$

and

$$0.5772 = \ln \frac{4Tt}{r_0^2 S} \tag{3.60}$$

Converting to logarithms to the base 10, Eqs. (3.59) and (3.60) may be rewritten in the gallon-day-foot system of units as

$$S = \frac{0.3Tt_0}{r^2} \tag{3.61}$$

$$S = \frac{0.3Tt}{r_0^2} \tag{3.62}$$

where T = coefficient of transmissibility, in gpd/ft
S = coefficient of storage, fraction
t_0 = intersection of time-drawdown semilog straight line with zero-drawdown axis, in days
t = time after pumping started, in days
r_0 = intersection of distance-drawdown semilog straight line with zero-drawdown axis, in feet
r = distance from pumped well to observation point, in feet

3.5 Unsteady-state Flow in Anisotropic Nonleaky Artesian Aquifer with Fully Penetrating Wells and Constant-discharge Conditions

Figure 3.1 shows wells penetrating an artesian aquifer overlain and underlain by aquicludes. Suppose that the aquifer is anisotropic (the permeability varies in two directions), homogeneous, infinite in areal extent, and of the same thickness throughout. The wells completely penetrate the aquifer and the production well is pumped at a constant rate Q.

Assuming that water is released instantaneously with a decline in head, the distribution of drawdown around the production well is described by the following boundary-value problem (Papadopulos, 1965)

$$T_{xx}\frac{\partial^2 s}{\partial x^2} + 2T_{xy}\frac{\partial^2 s}{\partial x\,\partial y} + T_{yy}\frac{\partial^2 s}{\partial y^2} + QD_\delta(x)D_\delta(y) = S\frac{\partial s}{\partial t} \tag{3.63}$$

$$s(x,y,0) = 0 \qquad s(\pm\infty,y,t) = 0 \qquad s(x,\pm\infty,t) = 0$$

where T_{xx}, T_{yy}, and T_{xy} are the components of the transmissibility tensor, D_δ is the Dirac delta function, and x and y are the coordinates of an arbitrary set of orthogonal axes with the origin at the production well.

In anisotropic aquifers, the velocity vector $\mathbf{V} = -P$ grad h (according to Darcy's law) and the hydraulic gradient vector are generally not parallel (Papadopulos, 1965) and P is a symmetric tensor of the second rank (Ferrandon, 1948; Scheidegger, 1954; Liakopoulos, 1962) called the *permeability tensor*, which transforms the components of the hydraulic gradient into those of velocity. The velocity and hydraulic gradient have the same direction only along one of three orthogonal axes called the *principal axes* of the permeability tensor. The anisotropy of an aquifer can be defined by the orientation of the principal axes and the magnitudes of the components of permeability along them.

For the case where the permeability varies in only two directions, the *transmissibility tensor* (product of the two-dimensional permeability tensor and the thickness of the aquifer) is used. In matrix notation the two-dimensional symmetric transmissibility tensor T can be written as (Papadopulos, 1965)

$$T = \begin{vmatrix} T_{xx} & T_{xy} \\ T_{xy} & T_{yy} \end{vmatrix}$$

where x and y are an arbitrary set of orthogonal axes. For the principal axes X and Y the above equation reduces to

$$T = \begin{vmatrix} T_{XX} & 0 \\ 0 & T_{YY} \end{vmatrix}$$

where T_{XX} and T_{YY} are the maximum and minimum transmissibilities, respectively.

The theory of integral transforms was used by Papadopulos (1965) to solve Eq. (3.63). The solution to Eq. (3.63) is

$$s = \frac{Q}{4\pi\sqrt{T_{xx}T_{yy} - T_{xy}^2}}\int_{u_{xy}}^{\infty}\frac{e^{-u_{xy}}}{u_{xy}}\,du \tag{3.64}$$

where

$$u_{xy} = \frac{S}{4t}\,\frac{T_{xx}y^2 + T_{yy}x^2 - 2T_{xy}xy}{T_{xx}T_{yy} - T_{xy}^2} \tag{3.65}$$

If the coordinate axes x and y coincide with the principal axes X and Y of the transmissibility tensor, Eq. (3.64) reduces to (Papadopulos, 1965)

$$s = \frac{Q}{4\pi\sqrt{T_{XX}T_{YY}}}\int_{u_{XY}}^{\infty}\frac{e^{-u_{XY}}}{u_{XY}}\,du \tag{3.66}$$

where T_{XX} and T_{YY} are the principal transmissibilities and

$$u_{XY} = \frac{S}{4t} \frac{T_{XX}Y^2 + T_{YY}X^2}{T_{XX}T_{YY}} \tag{3.67}$$

Expanding the exponential integral in Eq. (3.64) in a convergent series, the drawdown becomes

$$s = \frac{Q}{4\pi\sqrt{T_{xx}T_{yy} - T_{xy}^2}}\left(-0.5772 - \ln u_{xy} + u_{xy} - \frac{u_{xy}^2}{2.2!} + \frac{u_{xy}^3}{3.3!} - \frac{u_{xy}^4}{4.4!} + \cdots\right)$$

With the exponential integral expressed symbolically as $W(u_{xy})$, that is,

$$W(u_{xy}) = \int_{u_{xy}}^{\infty} \frac{e^{-u_{xy}}}{u_{xy}}\, du$$

Eq. (3.64) may be rewritten in the gallon-day-foot system of units in abbreviated form as

$$s = \frac{114.6QW(u_{xy})}{\sqrt{T_{xx}T_{yy} - T_{xy}^2}} \tag{3.68}$$

where

$$u_{xy} = \frac{1.87S}{t} \frac{T_{xx}Y^2 + T_{yy}X^2 - 2T_{xy}XY}{T_{xx}T_{yy} - T_{xy}^2} \tag{3.69}$$

s = drawdown, in feet
x and y = coordinates of observation point, in feet
Q = discharge, in gpm
t = time after pumping started, in days
S = coefficient of storage, fraction
T_{xy}, T_{xx}, and T_{yy} = components of the second-rank symmetric tensor of transmissibility, in gpd/ft

$W(u_{xy})$ is read as the "well function for nonleaky anisotropic artesian aquifers fully penetrated by wells and constant-discharge conditions." Values of $W(u_{xy})$ in terms of the practical range of u_{xy} are presented in Table 3.1.

Lines of equal drawdown around a production well in an anisotropic aquifer have the form of concentric ellipses with transverse axes along the maximum transmissibility axis and conjugate axes along the minimum transmissibility axis.

3.6 Unsteady-state Radial Flow in Isotropic Nonleaky Artesian Aquifer with Fully Penetrating Wells and Constant-drawdown Conditions

Suppose the well in Fig. 3.1 is a flowing well. Initially the well is capped and the drawdown is zero throughout the aquifer. The cap is removed, the water level in the well drops instantaneously to a lower stage, and a constant drawdown is maintained thereafter, resulting in a variable discharge governed by Eq. (3.45).

TABLE 3.2 VALUES OF $W(\lambda)$. (*From Jacob and Lohman, 1952.*)

N \ λ	$N\times 10^{-4}$	$N\times 10^{-3}$	$N\times 10^{-2}$	$N\times 10^{-1}$	$N\times 1$	$N\times 10$	$N\times 10^{2}$	$N\times 10^{3}$	$N\times 10^{4}$	$N\times 10^{5}$	$N\times 10^{6}$	$N\times 10^{7}$	$N\times 10^{8}$	$N\times 10^{9}$	$N\times 10^{10}$	$N\times 10^{11}$
1	56.9	18.34	6.13	2.249	0.985	0.534	0.346	0.251	0.1964	0.1608	0.1360	0.1177	0.1037	0.0927	0.0838	0.0764
2	40.4	13.11	4.47	1.716	0.803	0.461	0.311	0.232	0.1841	0.1524	0.1299	0.1131	0.1002	0.0899	0.0814	0.0744
3	33.1	10.79	3.74	1.477	0.719	0.427	0.294	0.222	0.1777	0.1479	0.1266	0.1106	0.0982	0.0883	0.0801	0.0733
4	28.7	9.41	3.30	1.333	0.667	0.405	0.283	0.215	0.1733	0.1449	0.1244	0.1089	0.0968	0.0872	0.0792	0.0726
5	25.7	8.47	3.00	1.234	0.630	0.389	0.274	0.210	0.1701	0.1426	0.1227	0.1076	0.0958	0.0864	0.0785	0.0720
6	23.5	7.77	2.78	1.160	0.602	0.377	0.268	0.206	0.1675	0.1408	0.1213	0.1066	0.0950	0.0857	0.0779	0.0716
7	21.8	7.23	2.60	1.103	0.580	0.367	0.263	0.203	0.1654	0.1393	0.1202	0.1057	0.0943	0.0851	0.0774	0.0712
8	20.4	6.79	2.46	1.057	0.562	0.359	0.258	0.200	0.1636	0.1380	0.1192	0.1049	0.0937	0.0846	0.0770	0.0709
9	19.3	6.43	2.35	1.018	0.547	0.352	0.254	0.198	0.1621	0.1369	0.1184	0.1043	0.0932	0.0842	0.0767	0.0706
10	18.3	6.13	2.25	0.985	0.534	0.346	0.251	0.196	0.1608	0.1360	0.1177	0.1037	0.0927	0.0838	0.0764	0.0704

The well is replaced by a mathematical sink of variable strength. With the boundary condition $h \to h_0$ as $r \to \infty$ for $t > 0$ and $h = h_0 - s_w$ for $r = r_w$ and for $t > 0$, where s_w is the drawdown in the flowing well and r_w is the effective radius of the flowing well, and the initial condition $h = h_0$ for $t \leq 0$ and for all values of r_w, the solution to the problem given by Eq. (3.45) is (Jacob and Lohman, 1952)

$$Q = 2\pi T s_w \frac{4\lambda}{\pi} \int_0^\infty x e^{-\lambda z} \left[\frac{\pi}{2} + \tan^{-1} \frac{Y_0(x)}{J_0(x)}\right] dx \tag{3.70}$$

where $\lambda = Tt/Sr_w^2$, $J_0(x)$ is the zero-order Bessel function of the first kind, and $Y_0(x)$ is the zero-order Bessel function of the second kind.

With the integral expressed symbolically as $W(\lambda)$, that is,

$$W(\lambda) = \int_0^\infty x e^{-\lambda x^2} \frac{\pi}{2} + \tan^{-1} \frac{Y_0(x)}{J_0(x)}\, dx$$

Eq. (3.70) may be rewritten in the gallon-day-foot system of units in abbreviated form as

$$Q = \frac{T s_w}{229} W(\lambda) \tag{3.71}$$

where $$\lambda = \frac{1.34 \times 10^{-1} Tt}{r_w^2 S} \tag{3.72}$$

Q = discharge of flowing well, in gpm
s_w = drawdown, in feet
r_w = effective radius of well, in feet
t = time after discharge started, in days
T = coefficient of transmissibility, in gpd/ft
S = coefficient of storage, fraction

$W(\lambda)$ is read as the "well function for nonleaky isotropic artesian aquifers fully penetrated by wells and constant-drawdown conditions." Values of $W(\lambda)$ in terms of the practical range λ, given by Jacob and Lohman (1952), are presented in Table 3.2.

3.7 Unsteady-state Flow in Nonleaky Artesian Aquifer with Partially Penetrating Wells

Figure 3.2 shows wells partially penetrating an artesian aquifer overlain and underlain by aquicludes. The wells are screened between the depth m_d and the base of the aquifer. The aquifer is homogeneous, isotropic, infinite in areal extent, and is of the same thickness throughout. The production and observation wells are screened in the same zone. Assuming that $r < 1.5m$ and that water is released instantaneously with a decline in head, the differential equation governing the unsteady-state flow to the

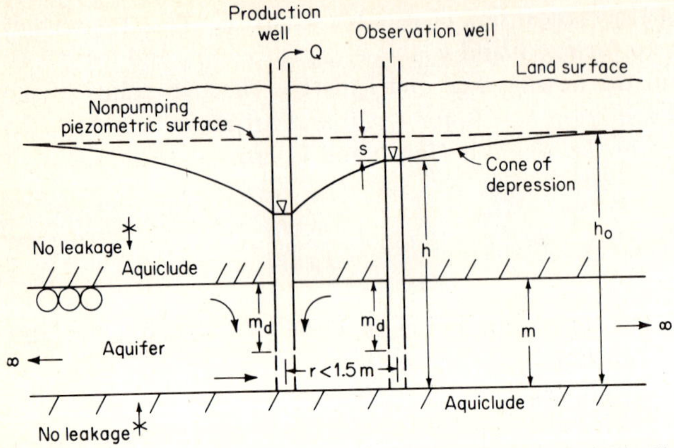

Fig. 3.2 Nonleaky artesian aquifer with partially penetrating wells. (*Drawn by W. C. Walton.*)

partially penetrating well in the nonleaky artesian aquifer in polar-coordinate notation is (Hantush, 1964)

$$\frac{\partial^2 s}{\partial r^2} + \frac{1}{r}\frac{\partial s}{\partial r} + \frac{\partial^2 s}{\partial z^2} = \frac{S}{T}\frac{\partial s}{\partial t} \tag{3.73}$$

With the boundary conditions

$$s(r,z,0) = s(\infty,x,t) = 0$$

$$\frac{\partial s(r,0,t)}{\partial z} = \frac{\partial s(r,m,t)}{\partial z} = 0$$

$$\lim_{r\to 0}\left[(m - m_d)r\frac{\partial s}{\partial r}\right] = 0 \qquad 0 < z < d$$

$$= -\frac{Q}{2\pi P} \qquad m_d < z < m$$

The solution to the problem given by Eq. (3.73) is (Adrian Visocky, Illinois State Water Survey, personal communication)

$$s = \frac{Q}{4\pi T}\left[\int_u^\infty \frac{e^{-u}}{u}\,du + \frac{2m^2}{\pi^2(m - m_d)^2}\sum_{n=1}^{\infty}\frac{1}{n^2}\sin\left(\frac{n\pi m_d}{m}\right) \times \int_u^\infty \frac{1}{y}\exp\left(-y - \frac{\left(\frac{n\pi r}{m}\right)^2}{4y}\right)dy\right] \tag{3.74}$$

where $u = r^2S/4Tt$. Evaluating the integrals, the drawdown at a distance r and a time t for a constant discharge becomes (see Hantush, 1964)

$$s = \frac{Q}{4\pi T}\Bigg\{\left(-0.5772 - \ln u + u - \frac{u^2}{2.2!} + \frac{u^3}{3.3!} - \frac{u^4}{4.4!}\cdots\right)$$

$$+ \frac{2m^2}{\pi^2(m-m_d)^2}\sum_{n=1}^{\infty}\frac{1}{n^2}\left[\sin\left(\frac{n\pi m}{m}\right) - \sin\left(\frac{n\pi m_d}{m}\right)\right]^2$$

$$\times\left[2K_0\left(\frac{n\pi r}{m}\right) - I_0\left(\frac{n\pi r}{m}\right)\right]\left[-E_i\left(-\frac{(n\pi r/m)^2}{4u}\right)\right] + \left[\exp\left(-\frac{(n\pi r/m)^2}{4u}\right)\right]$$

$$\times\Bigg\{0.5772 + \ln u + [-E_i(-u)] - u + u\left[\frac{I_0(n\pi r/m) - 1}{(n\pi r/m)^2}\right]$$

$$- u^2\sum_{n=1}^{\infty}\sum_{m=1}^{\infty}\frac{(-1)^{n+m}(n-m+1)!}{(n+2)!^2}\left(\frac{n\pi r}{2m}\right)^{2m}u^{n-m}\Bigg\}\Bigg\} \qquad (3.75)$$

where $K_0(x)$ is the modified Bessel function of the second kind and zero order and $I_0(x)$ is the modified Bessel function of the first kind and zero order.

With

$$\int_u^\infty \frac{e^{-u}}{u}\,du + \frac{2m^2}{\pi^2(m-m_d)^2}\sum_{n=1}^{\infty}\frac{1}{n^2}\sin^2\left(\frac{n\pi m_d}{m}\right)\int_u^\infty \frac{1}{y}\exp\left(-y - \frac{(n\pi r/m)^2}{4y}\right)dy$$

expressed symbolically as $W(u, r/m, \gamma)$, Eq. (3.74) may be rewritten in the gallon-day-foot system of units in abbreviated form as

$$s = \frac{114.6Q}{T}W\left(u, \frac{r}{m}, \gamma\right) \qquad (3.76)$$

where $u = \dfrac{1.87r^2S}{Tt}$ (3.77)

s = drawdown, in feet
r = distance from pumped well to observation point, in feet
Q = discharge, in gpm
t = time after pumping started, in days
T = coefficient of transmissibility, in gpd/ft
S = coefficient of storage, fraction
m = thickness of aquifer, in feet
m_d = distance from top of aquifer to top of screen, in feet
$\gamma = \dfrac{m - m_d}{m}$

$W(u, r/m, \gamma)$ is read as the "well function for nonleaky isotropic artesian aquifers partially penetrated by wells." Values of $W(u, r/m, \gamma)$, computed by Adrian Visocky, Illinois State Water Survey (personal communication), in terms of the practical

TABLE 3.3 VALUES OF $W(u, r/m, \gamma)$.

$\gamma = 0.75$

u	$r/m = 0.1$	0.01	0.001
10^{-6}	13.8767	15.2580	16.7637
10^{-5}	11.5741	12.9554	14.2530
10^{-4}	9.2716	10.6478	11.3995
10^{-3}	6.9699	8.1392	8.3991
10^{-2}	4.6712	5.2967	5.3635
10^{-1}	2.2597	2.4103	2.4193
1	0.2823	0.2898	0.2898
2	0.0634	0.0643	0.0645
3	0.0167	0.0169	0.0169

$\gamma = 0.50$

u	$r/m = 0.5$	0.2	0.1	0.03	0.01	0.001
10^{-6}	13.5665	14.4689	15.4989	17.6358	19.7506	24.2954
10^{-5}	11.2639	12.1663	13.1963	15.3332	17.4498	21.1506
10^{-4}	8.9614	9.8638	10.8938	13.0307	15.1224	17.0340
10^{-3}	6.6597	7.5621	8.5921	10.6994	11.9812	12.5845
10^{-2}	4.3661	5.2685	6.2757	7.4555	7.8851	8.0462
10^{-1}	2.1511	2.8822	3.2620	3.5305	3.6050	3.6304
1	0.3384	0.3986	0.4185	0.4319	0.4349	0.4353
2	0.0808	0.0910	0.0942	0.0964	0.0966	0.0968
3	0.0223	0.0247	0.0252	0.0254	0.0254	0.0255

$\gamma = 0.25$

u	$r/m = 1.00$	0.75	0.20	0.10	0.03	0.01	0.001
10^{-6}	13.3385	13.9367	16.2123	18.9845	25.1707	31.4176	44.9718
10^{-5}	11.0359	11.6341	13.9097	16.6837	22.8681	29.1150	40.7960
10^{-4}	8.6334	9.3316	11.6072	14.3794	30.5656	26.7666	33.5338
10^{-3}	6.4317	7.0299	0.3055	12.0777	18.2045	22.6026	24.9428
10^{-2}	4.1381	4.7363	7.0119	9.7382	13.8971	15.3684	15.9702
10^{-1}	1.9231	2.5213	4.4451	5.7545	6.8298	7.1101	7.1913
1	0.2981	0.4959	0.7160	0.7856	0.8493	0.8549	0.8531
2	0.0806	0.1271	0.1675	0.1794	0.1900	0.1875	0.1893
3	0.0245	0.0366	0.0454	0.0472	0.0501	0.0481	0.0481

range of u and r/m and values of γ equal to 25, 50, and 75 percent are presented in Table 3.3.

For observation wells at distances greater than about 1.5m the drawdown is described by Eq. (3.48).

The drawdown in Eq. (3.74) is the sum of the drawdown described by Eq. (3.46) for fully penetrating conditions and the drawdown s_p due to the effects of partial penetration. The drawdown s_p due to the effects of partial penetration is given by the following equation (Adrian Visocky, Illinois State Water Survey, personal communication)

$$s_p = \frac{Q}{4\pi T}\left[\frac{2m^2}{\pi(m - m_d)^2}\sum_{n=1}^{\infty}\frac{1}{n^2}\sin^2\left(\frac{n\pi m_d}{m}\right)\int_u^{\infty}\frac{1}{y}\exp\left(-y - \frac{(n\pi r^2/m)^2}{4y}\right)dy\right] \tag{3.78}$$

where $u = r^2S/4Tt$. With

$$\frac{2m^2}{\pi(m - m_d)^2}\sum_{n=1}^{\infty}\frac{1}{n^2}\sin^2\frac{n\pi m_d}{m}\int_u^{\infty}\frac{1}{y}\exp\left(-y - \frac{(n\pi r^2)/m)^2}{4y}\right)dy$$

expressed symbolically as $W_p(u, r/m, \gamma)$, Eq. (3.78) may be written in the gallon-day-foot system of units in abbreviated form as

$$s_p = \frac{114.6Q}{\cdot T} W_p\left(u, \frac{r}{m}, \gamma\right) \tag{3.79}$$

Values of $W_p(u, r/m, \gamma)$, computed by Adrian Visocky, Illinois State Water Survey, personal communication, for the practical range of u and r/m and for values of γ equal to 25, 50, and 75 percent are presented in Table 3.4.

If the aquifer is anisotropic, that is, $P_h \neq P_v$ where P_h is the horizontal permeability of the aquifer and P_v is the vertical permeability of the aquifer, the differential equation governing the unsteady-state flow to the partially penetrating well in a nonleaky artesian aquifer in polar-coordinate notation is (Nathan Columbus, Illinois State Water Survey, personal communication)

$$\frac{P_h}{P_v}\left(\frac{\partial^2 s}{\partial r^2} + \frac{1}{r}\frac{\partial s}{\partial r}\right) + \frac{\partial^2 s}{\partial z^2} = \frac{S}{T}\frac{\partial s}{\partial t} \tag{3.80}$$

where $T = P_v m$. The solution to the problem given by Eq. (3.80) is (Nathan Columbus, Illinois State Water Survey, personal communication)

$$s = \frac{Q}{4\pi T}\left[\int_u^{\infty}\frac{e^{-u}}{u}du + \frac{2m^2}{\pi^2(m - m_d)^2}\sum_{n=1}^{\infty}\frac{1}{n^2}\sin^2\left(\frac{n\pi m_d}{m}\right)\right.$$
$$\left.\times\int_u^{\infty}\frac{1}{y}\exp\left(\frac{-y - (P_v/P_h)(n\pi r/m)^2}{4y}\right)dy\right] \tag{3.81}$$

TABLE 3.4 VALUES OF $W_p(u, r/m, \gamma)$.

$\gamma = 0.25$

u	$r/m = 0.01$	0.03	0.1	0.2	0.5	1.0
$<10^{-5}$	18.18	11.93	5.75	2.97	0.70	0.10
10^{-5}	18.18	11.93	5.75	2.97	0.70	0.10
10^{-4}	18.13	11.93	5.75	2.97	0.70	0.10
10^{-3}	16.27	11.87	5.75	2.97	0.70	0.10
10^{-2}	11.33	9.86	5.70	2.97	0.70	0.10
10^{-1}	5.29	5.01	3.93	2.62	0.70	0.10
1	0.63	0.63	0.57	0.50	0.28	0.08
2	0.14	0.14	0.13	0.12	0.08	0.03

$\gamma = 0.50$

u	$r/m = 0.01$	0.03	0.1	0.2	0.5	1.0
$<10^{-5}$	6.51	4.40	2.26	1.23	0.33	0.05
10^{-5}	6.51	4.40	2.26	1.23	0.33	0.05
10^{-4}	6.49	4.40	2.26	1.23	0.33	0.05
10^{-3}	5.65	4.37	2.26	1.23	0.33	0.05
10^{-2}	3.85	3.42	2.24	1.23	0.33	0.05
10^{-1}	1.78	1.71	1.44	1.06	0.33	0.05
1	0.22	0.21	0.20	0.18	0.12	0.01

$\gamma = 0.75$

u	$r/m = 0.01$	0.1	0.2	0.5
$<10^{-5}$	2.02	0.64	0.33	0.08
10^{-5}	2.02	0.64	0.33	0.08
10^{-4}	2.01	0.64	0.33	0.08
10^{-3}	1.81	0.64	0.33	0.08
10^{-2}	1.26	0.64	0.33	0.08
10^{-1}	0.59	0.44	0.30	0.08
1	0.07	0.06	0.05	0.03

where $u = r^2S/4Tt$. Evaluating the integrals, the drawdown at a distance r and a time t for a constant discharge becomes (see Hantush, 1964)

$$s = \frac{Q}{4\pi T}\Bigg\{\left(-0.5772 - \ln u + u - \frac{u^2}{2.2!} + \frac{u^3}{3.3!} - \frac{u^4}{4.4!}\cdots\right)$$

$$+ \frac{2m^2}{\pi^2(m-m_d)^2}\sum_{n=1}^{\infty}\frac{1}{n^2}\left[\sin\left(\frac{n\pi m}{m}\right) - \sin\left(\frac{n\pi m_d}{m}\right)\right]^2\Bigg\{2K_0\left(\sqrt{\frac{P_v}{P_h}}\right)\left(\frac{n\pi r}{m}\right)$$

$$- I_0\left(\sqrt{\frac{P_v}{P_h}}\right)\left(\frac{n\pi r}{m}\right)\left[-E_i\left(-\frac{(P_v/P_h)(n\pi r/m)^2}{4u}\right)\right] + \left[\exp\left(-\frac{(P_v/P_h)(n\pi r/m)^2}{4u}\right)\right]$$

$$\times\Bigg\{0.5772 + \ln u + [-E_i(-u)] - u + \frac{u[I_0(\sqrt{P_v/P_h})(n\pi r/m) - 1]}{(P_v/P_h)(n\pi r/2m)^2}$$

$$- u^2\sum_{n=1}^{\infty}\sum_{m=1}^{\infty}\frac{(-1)^{n+m}(n-m+1)!}{(n+2)!^2}\left(\frac{P_v}{P_h}\frac{n\pi r}{2m}\right)^{2m}u^{n-m}\Bigg\}\Bigg\} \qquad (3.82)$$

with

$$\int_u^{\infty}\frac{e^{-u}}{u}\,du + \frac{2m^2}{\pi^2(m-m_d)^2}\sum_{n=1}^{\infty}\frac{1}{n^2}\sin^2\left(\frac{n\pi m_d}{m}\right)\int_u^{\infty}\frac{1}{y}\exp\left[\frac{-y-(P_v/P_h)(n\pi r/m)^2}{4y}\right]dy$$

Expressed symbolically as $W(u, \sqrt{P_v/P_h}\cdot r/m, \gamma)$, Eq. (3.81) may be rewritten in the gallon-day-foot system of units in abbreviated form as

$$s = \frac{114.6Q}{T}\,W\left(u, \sqrt{P_v/P_h}\cdot\frac{r}{m}, \gamma\right) \qquad (3.83)$$

$W(u, \sqrt{P_v/P_h}\cdot r/m, \gamma)$ is read as the "well function for nonleaky anisotropic artesian aquifers partially penetrated by wells."

3.8 Unsteady- and Steady-state Radial Flow in Isotropic Leaky Artesian Aquifer with Fully Penetrating Wells without Water Released from Storage in Aquitard and Constant-discharge Conditions

Figure 3.3 shows wells penetrating an artesian aquifer overlain by an aquitard and underlain by an aquiclude. Overlying the aquitard are deposits (source bed) in which there is a water table. The aquifer is homogeneous, isotropic, infinite in areal extent, and is of the same thickness throughout. The wells completely penetrate the aquifer, and flow in the aquifer is radial throughout. Flow in the aquifer is augmented by vertical leakage through the aquitard. The flow lines are assumed to be refracted a full right angle as they cross the aquitard-aquifer interface. The aquitard is assumed to be more or less incompressible so that water released from storage therein is negligible. The water table is not influenced appreciably by pumping.

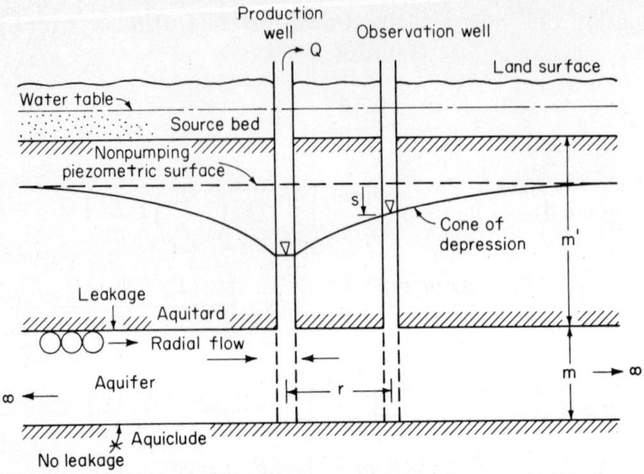

Fig. 3.3 Leaky artesian aquifer with fully penetrating wells. (*Drawn by W. C. Walton.*)

The discharge of water from a well in such an aquifer is supplied from storage within the aquifer, as well as from leakage through the aquitard. Because of the presence of recharge in the form of leakage, water levels will stabilize when the entire discharge of the well is derived from leakage. The rate of vertical leakage is proportionate to the difference in head between the water table and the piezometric surface of the aquifer. If P' is the coefficient of permeability of the aquitard, m' is the thickness of the aquitard, and $B^2 = T/(P'/m')$

$$\frac{\partial^2 s}{\partial r^2} + \frac{1}{r}\frac{\partial s}{\partial r} - \frac{s}{B^2} = \frac{S}{T}\frac{\partial s}{\partial t} \tag{3.84}$$

Equation (3.84) is the differential equation in polar-coordinate notation governing the unsteady-state radial flow in a leaky artesian aquifer with fully penetrating wells without water released from storage in the aquitard (Jacob, 1946).

To facilitate the solution, the production well is replaced by a mathematical sink of constant strength. With the boundary conditions $s \to 0$ as $r \to \infty$ for $t \geq 0$

$$\lim_{r\to 0} r\frac{\partial s}{\partial r} = \frac{-Q}{2\pi T}$$

and the initial condition $s = 0$ for $t < 0$, and for all values of r, the solution to the problem given by Eq. (3.84) is (Hantush and Jacob, 1955)

$$s = \frac{Q}{4\pi T}\int_u^\infty \frac{1}{y}\exp\left(-y - \frac{r^2}{4B^2 y}\right) dy \tag{3.85}$$

Evaluating the integral, the drawdown at a distance r and a time t for a constant discharge becomes

$$s = \frac{Q}{4\pi T}2K_0\left(\frac{r}{B}\right) - I_0\left(\frac{r}{B}\right)\left[-E_i\left(-\frac{r^2}{4B^2u}\right)\right] + \exp\left(-\frac{r^2}{4B^2u}\right)$$
$$\times\left\{0.5772 + \ln u + [-E_i(-u)] - u + \frac{u[I_0(r/B) - 1]}{r/4B^2}\right.$$
$$\left. - u^2\sum_{n=1}^{\infty}\sum_{m=1}^{\infty}\frac{(-1)^{n+m}(n-m+1)!}{(n+2)!^2}\left(\frac{r^2}{4B^2}\right)^m u^{n-m}\right\}$$

where
$$u = \frac{r^2S}{4Tt} \qquad B = \sqrt{\frac{T}{P'/m'}}$$

With the integral in Eq. (3.85) expressed symbolically as $W(u, r/B)$, that is,

$$\int_u^{\infty}\frac{1}{u}\exp\left(-y - \frac{r^2}{4B^2y}\right)dy = W\left(u, \frac{r}{B}\right)$$

Eq. (3.85) may be rewritten in the gallon-day-foot system of units in abbreviated form as

$$s = \frac{114.6Q}{T}W\left(u, \frac{r}{B}\right) \tag{3.86}$$

where $u = \dfrac{1.87r^2S}{Tt}$ (3.87)

$$\frac{r}{B} = \frac{r}{\sqrt{T/(P'/m')}} \tag{3.88}$$

s = drawdown, in feet
r = distance from pumped well to observation point, in feet
Q = discharge, in gpm
t = time after pumping started, in days
T = coefficient of transmissibility, in gpd/ft
S = coefficient of storage, fraction
P' = coefficient of permeability of aquitard, in gpd/sq ft
m' = saturated thickness of aquitard, in feet

$W(u, r/B)$ is read as the "well function for leaky artesian aquifers with fully penetrating wells without water released from storage in aquitard and constant-discharge conditions." Values of $W(u, r/B)$ in terms of the practical range of u and r/B given by Hantush (1956) are presented in Table 3.5.

When discharge is balanced by leakage and water levels stabilize at permanent stages, a steady state exists. The differential equation in polar-coordinate notation governing the steady-state radial flow in the leaky artesian aquifers is (Jacob, 1946)

$$\frac{\partial^2 s}{\partial r^2} + \frac{1}{r}\frac{\partial s}{\partial r} - \frac{s}{B^2} = 0 \tag{3.89}$$

TABLE 3.5 VALUES OF $W(u, r/B)$ OR $W(u'', r/B)$. (*After Hantush, 1956.*)

u or u'' \ r/B	0.01	0.015	0.03	0.05	0.075	0.10	0.15	0.2	0.3	0.4	0.5	0.6	0.7	0.8	0.9	1.0	1.5	2.0	2.5
0.000001																			
0.000005	9.4413																		
0.00001	9.4176	8.6313																	
0.00005	8.8827	8.4533	7.2450																
0.0001	8.3983	8.1414	7.2122	6.2282	5.4228														
0.0005	6.9750	6.9152	6.6219	6.0821	5.4062	4.8530													
0.001	6.3069	6.2765	6.1202	5.7965	5.3078	4.8292	4.0595	3.5054											
0.005	4.7212	4.7152	4.6829	4.6084	4.4713	4.2960	3.8821	3.4567	2.7428	2.2290									
0.01	4.0356	4.0326	4.0167	3.9795	3.9091	3.8150	3.5725	3.2875	2.7104	2.2253	1.8486	1.5550	1.3210	1.1307					
0.05	2.4675	2.4670	2.4642	2.4576	2.4448	2.4271	2.3776	2.3110	1.9283	1.7075	1.4927	1.2955	1.2955	1.1210	0.9700	0 8409			
0.1	1.8227	1.8225	1.8213	1.8184	1.8128	1.8050	1.7829	1.7527	1.6704	1.5644	1.4422	1.3115	1.1791	1.0505	0.9297	0.8190	0.4271	0.2278	
0.5	0.5598	0.5597	0.5596	0.5594	0.5588	0.5581	0.5561	0.5532	0.5453	0.5344	0.5206	0.5044	0.4860	0.4658	0.4440	0.4210	0.3007	0.1944	0.1174
1.0	0.2194	0.2194	0.2193	0.2193	0.2191	0.2190	0.2186	0.2179	0.2161	0.2135	0.2103	0.2065	0.2020	0.1970	0.1914	0.1855	0.1509	0.1139	0.0803
5.0	0.0011	0.0011	0.0011	0.0011	0.0011	0.0011	0.0011	0.0011	0.0011	0.0011	0.0011	0.0011	0.0011	0.0011	0.0011	0.0011	0.0010	0.0010	0.0009

The production well is replaced by a mathematical sink of constant strength. With boundary conditions $s \to 0$ as $r \to \infty$ for $t \geq 0$

$$\lim_{r \to 0} r \frac{\partial s}{dr} = \frac{-Q}{2\pi T}$$

and the initial conditions $s = 0$ for $t < 0$, and for all values of r, the solution to the problem given by Eq. (3.89) is (Steggewentz and Van Nes, 1939)

$$s = \frac{Q}{2\pi T} K_0 \frac{r}{B} \tag{3.90}$$

Equation (3.90) may be written in the gallon-day-foot system of units in abbreviated form as

$$s = \frac{229Q}{T} K_0 \frac{r}{B} \tag{3.91}$$

where $\dfrac{r}{B} = \dfrac{r}{\sqrt{T/(P'/m')}}$

s = drawdown, in feet
r = distance from pumped well to observation point, in feet
Q = discharge, in gpm
T = coefficient of transmissibility, in gpd/ft
P' = coefficient of permeability of aquitard, in gpd/sq ft
m' = thickness of aquitard, in feet

Values of $K_0(r/B)$ in terms of the practical range of r/B are presented in Table 3.6. The coefficient of storage cannot be computed under steady-state conditions because, under such conditions of flow, the entire yield of the well is derived from only leakage sources.

TABLE 3.6 VALUES OF $K_0(r/B)$. (*After Hantush, 1956.*)

N	$r/B = N \times 10^{-3}$	$N \times 10^{-2}$	$N \times 10^{-1}$	N
1.0	7.0237	4.7212	2.4271	0.4210
1.5	6.6182	4.3159	2.0300	0.2138
2.0	6.3305	4.0285	1.7527	0.1139
2.5	6.1074	3.8056	1.5415	0.0623
3.0	5.9251	3.6235	1.3725	0.0347
3.5	5.7709	3.4697	1.2327	0.0196
4.0	5.6374	3.3365	1.1145	0.0112
4.5	5.5196	3.2192	1.0129	0.0064
5.0	5.4143	3.1142	0.9244	0.0037
5.5	5.3190	3.0195	0.8466	
6.0	5.2320	2.9329	0.7775	0.0012
6.5	5.1520	2.8534	0.7159	
7.0	5.0779	2.7798	0.6605	0.0004
7.5	5.0089	2.7114	0.6106	
8.0	4.9443	2.6475	0.5653	
8.5	4.8837	2.5875	0.5242	
9.0	4.8266	2.5310	0.4867	
9.5	4.7725	2.4776	0.4524	

3.9 Unsteady-state Radial Flow in Isotropic Leaky Artesian Aquifer with Fully Penetrating Wells without Water Released from Storage in Aquitard and Constant-drawdown Conditions

Suppose the well in Fig. 3.3 has a constant drawdown and is discharging at a variable rate by natural flow. The variation of discharge of the well can be determined from Eq. (3.84). Initially the well is capped and the drawdown is zero throughout the aquifer. The cap is removed and the water level in the well drops instantaneously to a lower stage, and a constant drawdown is maintained thereafter.

The well is replaced by a mathematical sink of variable strength. With the boundary condition $s \rightarrow 0$ as $r \rightarrow \infty$ for $t \geq 0$ and $s = s_w$ for $r = r_w$ for $t \geq 0$, where s_w is the drawdown in the well and t_w is the effective radius of the well, and the initial condition $s = 0$ for $t \leq 0$, and for all values of r_w, the solution to the problem given by Eq. (3.84) is (Hantush, 1959)

$$Q = 2\pi T s_w \left\{ \frac{r_w}{B} \frac{K_1(r_w/B)}{K_0(r_w/B)} + \frac{4}{\pi^2} \exp\left[-\lambda\left(\frac{r_w}{B}\right)^2\right] \times \int_u^\infty \frac{u \exp(-\lambda u^2)}{J_0^2(u) + Y_0^2(u)} \frac{du}{u^2 + (r_w/B)^2} \right\} \tag{3.92}$$

where $K_1(X) =$ first-order modified Bessel function of the second kind, $J_0(X) =$ zero-order Bessel function of the first kind, $Y_0(X) =$ zero-order Bessel function of the second kind, and $u = r_w\sqrt{1/\exp(i\pi)B^2}$. With

$$\frac{r_w}{B} \frac{K_1(r_w/B)}{K_0(r_w/B)} + \frac{4}{\pi^2} \exp\left[-\lambda\left(\frac{r_w}{B}\right)^2\right] \int_0^\infty \frac{u \exp(-\lambda u^2)}{J_0^2(u) + Y_0^2} \frac{du}{u^2 + (r_w/B)^2}$$

expressed symbolically as $W(\lambda, r/B)$, Eq. (3.92) may be rewritten in the gallon-day-foot system of units in abbreviated form as

$$Q = \frac{T s_w}{229} W\left(\lambda, \frac{r}{B}\right) \tag{3.93}$$

where
$$\lambda = \frac{1.34 \times 10^{-1} T t}{r_w^2 S} \tag{3.94}$$

$$\frac{r_w}{B} = \frac{r_w}{\sqrt{T/(P'/m')}} \tag{3.95}$$

$Q =$ discharge, in gpm
$s_w =$ drawdown, in feet
$r_w =$ effective radius of well, in feet
$t =$ time after discharge started, in days
$T =$ coefficient of transmissibility, in gpd/ft
$S =$ coefficient of storage, fraction
$P' =$ coefficient of permeability of aquitard, in gpd/sq ft
$m' =$ thickness of aquitard, in feet

$W(\lambda, r/B)$ is read as the "well function for leaky artesian aquifers with fully penetrating wells without water released from storage in aquitard and constant-drawdown conditions." Values of $W(\lambda, r/B)$ in terms of the practical range of λ and r/B given by Hantush (1959) are presented in Table 3.7.

3.10 Unsteady-state Radial Flow in Isotropic Leaky Artesian Aquifer with Fully Penetrating Wells with Water Released from Storage in Aquitard

Figure 3.3 shows a well penetrating an artesian aquifer overlain by an aquitard and underlain by an aquiclude. Overlying the aquitard are permeable deposits (source bed) in which there is a water table. The aquifer is homogeneous, isotropic, infinite in areal extent, and is of the same thickness throughout. The well completely penetrates the aquifer, and flow in the aquifer is radial throughout. Flow in the aquifer is augmented by vertical leakage through the aquitard. The flow lines are assumed to be refracted a full right angle as they cross the aquitard-aquifer interface. Suppose that water released from storage within the aquitard is appreciable and that the water table is not influenced appreciably by pumping.

The discharge of water from a well in such an aquifer is supplied from storage within the aquifer and aquitard, as well as from leakage through the aquitard. Because of the presence of recharge in the form of leakage, water levels will stabilize when the entire discharge of the well is derived from leakage within the aquitard.

Assuming that water is released from storage instantaneously with a decline in head, the differential equation governing the unsteady-state flow in the leaky artesian aquifer in polar-coordinate notation is (Hantush, 1964)

$$\frac{\partial^2 s}{\partial r^2} + \frac{1}{r}\frac{\partial s}{\partial r} + \frac{P'}{T}\frac{\partial s'}{\partial z} = \frac{S}{T}\frac{\partial s}{\partial t} \tag{3.96}$$

With the boundary conditions

$$s(r,0) = 0$$

$$s(\infty,t) = 0$$

$$\lim_{r \to 0} r\frac{\partial s}{\partial r}(r,t) = \frac{-Q}{2\pi T}$$

$$s_c(r,z,0) = 0$$

$$s_c(r,m,t) = s(r,t)$$

$$s_c(r, m + m', t) = 0$$

where s_c is the drawdown in the aquitard.

TABLE 3.7 VALUES OF $W(\lambda, r_w/B)$. (*From Hantush, 1959.*)

λ \ r_w/B	0	1×10^{-5}	2×10^{-5}	4×10^{-5}	6×10^{-5}	8×10^{-5}	10^{-4}	2×10^{-4}	4×10^{-4}	6×10^{-4}	8×10^{-4}	10^{-3}	2×10^{-3}	4×10^{-3}	6×10^{-3}	6×10^{-3}	10^{-2}
1×10^{2}	0.346																0.346
5×10^{2}	0.274												0.274	0.274	0.275	0.275	0.276
1×10^{3}	0.251												0.251	0.252	0.252	0.254	0.255
5×10^{3}	0.210												0.210	0.212	0.215	0.218	0.222
1×10^{4}	0.196											0.196	0.197	0.200	0.204	0.209	0.216
5×10^{4}	0.170											0.170	0.173	0.181	0.192		
1×10^{5}	0.161								0.161	0.162	0.162	0.162	0.167	0.178			
5×10^{5}	0.143							0.143	0.143	0.144	0.145	0.148	0.161				
1×10^{6}	0.136						0.136	0.137	0.138	0.139	0.141	0.144	0.159				
5×10^{6}	0.123						0.123	0.124	0.128	0.133							
1×10^{7}	0.118						0.118	0.120	0.127								
5×10^{7}	0.108																
1×10^{8}	0.104		0.104	0.104	0.105	0.106	0.108										
5×10^{8}	0.0958	0.0958	0.0966	0.0989													
1×10^{9}	0.0927	0.0930	0.0943	0.0980													
5×10^{9}	0.0864	0.0880	0.0916														
1×10^{10}	0.0838	0.0867	0.0914														
5×10^{10}	0.0785																
10×10^{10}	0.0764	0.0860	0.0914	0.0976	0.102	0.105	0.107	0.116	0.126	0.133	0.138	0.142	0.158	0.177	0.191	0.202	0.212

Two asymptotic solutions to the problem given by (3.96) are possible. For the condition $t \geq m'S'/74.8P'$ the solution of Eq. (3.96) is (Hantush, 1964)

$$s = \frac{Q}{4\pi T} \int_u^\infty \frac{e^{-y}}{y} \operatorname{erfc}\left[\frac{\psi\sqrt{u}}{\sqrt{y(y-u)}}\right] dy \tag{3.97}$$

where $u = \dfrac{r^2 S}{Tt}$

$$\psi = \frac{r}{4}\sqrt{\frac{S'P'}{TSm'}}$$

S' = coefficient of storage of aquitard

erfc (x) is the complementary error function defined by

$$\operatorname{erf}(x) = 1 - \operatorname{erfc}(x) = \frac{2}{\sqrt{\pi}} \int_0^x \exp(-y^2)\, dy$$

With the integral expressed symbolically as $W(u,\psi)$, that is,

$$W(u,\psi) = \int_u^\infty \frac{e^{-y}}{y} \operatorname{erfc} \frac{\psi\sqrt{u}}{\sqrt{y(y-u)}}\, dy$$

Eq. (3.97) may be rewritten in the gallon-day-foot system of units in abbreviated form as

$$s = \frac{114.6Q}{T} W(u,\psi) \tag{3.98}$$

where $$u = \frac{1.87r^2 S}{Tt} \tag{3.99}$$

$$\psi = \left(\frac{r}{4}\right)\sqrt{\frac{S'P'}{TSm'}} \tag{3.100}$$

s = drawdown, in feet
r = distance from pumped well to observation point, in feet
Q = discharge, in gpm
t = time after pumping started, in days
T = coefficient of transmissibility of aquifer, in gpd/ft
S = coefficient of storage of aquifer, fraction
P' = coefficient of permeability of aquitard, in gpd/sq ft
m' = thickness of aquitard, in feet
S' = coefficient of storage of aquitard, fraction

$W(u,\psi)$ is read as the "well function for leaky artesian aquifers with fully penetrating wells with water released from storage in aquitard when $t \leq m'S'/74.8P'$." Values of $W(u,\psi)$, given by Hantush (1964) in terms of the practical range of u and ψ, are presented in Table 3.8.

TABLE 3.8 VALUES OF $W(u,\psi)$. *(From Hantush, 1964.)*[a]

ψ / u	(−3)			(−2)			(−1)		
	1	2	5	1	2	5	1	2	5
1(−6)	11.9842	11.4237	10.5908	9.9259	9.2469	8.3395	7.6497	6.9590	6.0463
5(−6)	10.8958	10.4566	9.7174	9.0866	8.4251	7.5284	6.8427	6.1548	5.2459
1(−5)	10.3739	9.9987	9.3203	8.7142	8.0657	7.1771	6.4944	5.8085	4.9024
5(−5)	9.0422	8.8128	8.3171	7.8031	7.2072	6.3523	5.6821	5.0045	4.1090
1(−4)	8.4258	8.2487	7.8386	7.3803	6.8208	5.9906	5.3297	4.6581	3.7700
5(−4)	6.9273	6.8375	6.6024	6.2934	5.8561	5.1223	4.4996	3.8527	2.9933
1(−3)	6.2624	6.1969	6.0193	5.7727	5.4001	4.7290	4.1337	3.5045	2.6650
5(−3)	4.6951	4.6649	4.5786	4.4474	4.2231	3.7415	3.2483	2.6891	1.9250
1(−2)	4.0163	3.9950	3.9334	3.8374	3.6669	3.2752	2.8443	2.3325	1.6193
5(−2)	2.4590	2.4502	2.4243	2.3826	2.3040	2.1007	1.8401	1.4872	0.9540
1(−1)	1.8172	1.8116	1.7949	1.7677	1.7157	1.5768	1.3893	1.1207	0.6947
5(−1)	0.5584	0.5570	0.5530	0.5463	0.5333	0.4969	0.4436	0.3591	0.2083
1(0)	0.2189	0.2184	0.2169	0.2144	0.2097	0.1961	0.1758	0.1427	812(−4)
5(0)	115(−5)	114(−5)	114(−5)	112(−5)	110(−5)	104(−5)	934(−6)	763(−6)	423(−6)
10(0)	415(−8)	414(−8)	411(−8)	407(−8)	399(−8)	375(−8)	339(−8)	277(−8)	153(−8)

[a] The numbers in parenthesis are powers of 10 by which the other numbers are multiplied, e.g.,

Equation (3.98) may completely describe the drawdown distribution when the aquitard is thick and relatively impermeable. For the condition $t \geq 0.27m'S'/P'$, the solution to the problem given by Eq. (3.96) is

$$s = \frac{Q}{4\pi T}\int_{u''}^{\infty} \frac{e^{-y-r^2/4B^2y}}{y}\,dy \tag{3.101}$$

With the integral expressed symbolically as $W(u'', r/B)$, that is,

$$W\left(u'', \frac{r}{B}\right) = \int_{u''}^{\infty} \frac{e^{-y-r^2/4B^2y}}{y}\,dy$$

Eq. (3.101) may be rewritten in the gallon-day-foot system of units in abbreviated form as

$$s = \frac{114.6Q}{T}\,W\left(u'', \frac{r}{B}\right) \tag{3.102}$$

where $$u'' = \frac{1.87r^2S}{Tt}\,\frac{1+S'}{3S} \tag{3.103}$$

$$\frac{r}{B} = \frac{r}{\sqrt{T/(P'/m')}} \tag{3.104}$$

$W(u'', r/B)$ is read as the "well function for leaky artesian aquifers with fully penetrating wells with water released from storage in aquitard when $t \geq 0.27m'S'/P'$." Values of $W(u'', r/B)$, given by Hantush (1964) in terms of the practical range of u'' and r/B, are presented in Table 3.5.

ψ	(0)			(1)			(2)		
u	1	2	5	1	2	5	1	2	5
1(−6)	5.3575	4.6721	3.7756	3.1110	2.4671	1.6710	1.1361	0.6879	0.2698
5(−6)	4.5617	3.8836	3.0055	2.3661	1.7633	1.0574	0.6256	0.3091	787(−4)
1(−5)	4.2212	3.5481	2.6822	2.0590	1.4816	0.8285	0.4519	0.1978	388(−4)
5(−5)	3.4394	2.7848	1.9622	1.3943	0.8994	0.4024	0.1685	494(−4)	405(−5)
1(−4)	3.1082	2.4658	1.6704	1.1359	0.6878	0.2698	963(−4)	222(−4)	107(−5)
5(−4)	2.3601	1.7604	1.0564	0.6252	0.3089	787(−4)	166(−4)	169(−5)	129(−7)
1(−3)	2.0506	1.4776	0.8271	0.4513	0.1976	388(−4)	590(−5)	361(−6)	
5(−3)	1.3767	0.8915	0.4001	0.1677	493(−4)	403(−5)	205(−6)	228(−8)	
1(−2)	1.1122	0.6775	0.2670	955(−4)	221(−4)	106(−5)	274(−7)		
5(−2)	0.5812	0.2923	755(−4)	160(−4)	164(−5)	126(−7)			
1(−1)	0.3970	0.1789	359(−4)	552(−5)	340(−6)				
5(−1)	0.1006	325(−4)	288(−5)	151(−6)	171(−8)				
1(0)	365(−4)	993(−5)	547(−6)	151(−7)					
5(0)	167(−6)	309(−7)							
10(0)									

488(−4) = 0.0488.

Equation (3.102) may completely describe the drawdown distribution when the aquitard is thin and relatively permeable:

$$t \to \infty \qquad u'' \to 0$$

and

$$s = \frac{229Q}{T} K_0 \frac{r}{B} \tag{3.105}$$

Thus, when discharge is balanced by leakage and water levels stabilize at permanent stages, a steady state exists and Eq. (3.105) describes the drawdown distribution.

3.11 Unsteady-state Flow in Isotropic Water-table Aquifer with Fully Penetrating Wells and Constant-discharge Conditions

Figure 3.4 shows wells fully penetrating a water-table aquifer underlain by an aquiclude. The aquifer is homogeneous, isotropic, infinite in areal extent, and is of the same thickness throughout. The wells completely penetrate the aquifer and water is unconfined. Drawdown is assumed to be very small in comparison to the original saturated thickness of the aquifer.

When a well in such an aquifer is pumped, water is continuously withdrawn from storage within the aquifer as the cone of depression progresses radially outward from the well. Because of the absence of a source of recharge, there can be no steady-state flow, and the head in the aquifer will continue to decline as long as the aquifer is effectively infinite. However, the rate of decline of head continuously decreases as the cone of depression spreads. Water is released from storage by the gravity drainage of the interstices in the portion of the aquifer drained by the pumping and

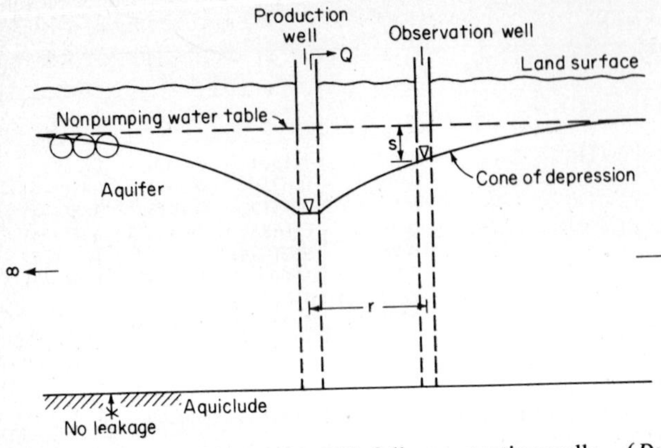

Fig. 3.4 Water-table aquifer with fully penetrating wells. (*Drawn by W. C. Walton.*)

by the compaction of the aquifer and the expansion of the water. The gravity drainage of water through stratified sediments is not immediate; the coefficient of storage appears to vary and to increase at a diminishing rate with the time of pumping.

Unconfined stratified sediments often react to pumping for a short time after discharge starts as would an artesian aquifer. Gravity drainage of water through stratified sediments is not instantaneous but water is released instantaneously from storage by the compaction of the aquifer and by the expansion of the water itself. After a short initial period, the cone of depression slows in its rate of expansion as water released from storage by gravity drainage of sediments reaches the cone. Finally, the rate of expansion of the cone of depression increases and the cone continues to expand as gravity drainage keeps pace with declining water levels (see Walton, 1962).

If the delayed gravity drainage were absent, the differential equation in polar-coordinate notation governing the unsteady-state flow in the water-table aquifer would be (Boulton, 1954)

$$T\left(\frac{\partial^2 s}{\partial r^2} + \frac{1}{r}\frac{\partial s}{\partial r}\right) = S\frac{\partial s}{\partial t} \tag{3.106}$$

The amount of water derived from storage, due to an increment of drawdown Δs between times τ and $\tau + \Delta\tau$ after pumping started, consists of two components (Boulton, 1963): (1) a volume of water instantaneously released from storage per unit horizontal area, $\Delta s S$ (S is the coefficient of storage as defined for artesian conditions), and (2) a delayed yield from storage, per unit horizontal area, at any time t ($t > \tau$) from the start of pumping, $\Delta s\ D_i S_y e^{-D_i(t-\tau)}$ (D_i is an empirical constant). The total volume of delayed gravity drainage per unit area per unit drawdown is

$$D_i S_y \int_\tau^\infty e^{-D_i(t-\tau)}\, dt = S_y \tag{3.107}$$

The effective coefficient of storage under water-table conditions is $S + S_y = NS$ and the time-rate of delayed gravity drainage per unit area at time t from Eq. (3.107) is

$$D_i S_y \int_0^t \frac{\partial s}{\partial \tau} e^{-D_i(t-\tau)} \, d\tau \tag{3.108}$$

Then, the differential equation governing the unsteady-state flow to the fully penetrating well in the water-table aquifer in polar-coordinate notation from Eqs. (3.106) to (3.108) is (Boulton, 1963)

$$T\left(\frac{\partial^2 s}{\partial r^2} + \frac{1}{r}\frac{\partial s}{\partial r}\right) = S\frac{\partial s}{\partial t} + D_i S_y \int_0^t \frac{\partial s}{\partial t} e^{-D_i(t-\tau)} \, d\tau \qquad (t = \tau) \tag{3.109}$$

With the boundary conditions $s \to 0$ as $r \to \infty$ for $t \geq 0$ and the initial condition $s = 0$ for $t < 0$, and for all values of r, the solution to the problem given by Eq. (3.109) is (Boulton, 1963)

$$s = \frac{Q}{4\pi T}\int_0^\infty \frac{2}{x}\left\{1 - e^{-u_1}\left[\cosh u_2 + \frac{D_i N(1 - x^2)t}{2u_2} \sinh u_2\right]\right\} J_0\left(\frac{r}{\nu D_t} x\right) dx \tag{3.110}$$

where $$u_1 = \frac{D_i t N(1 - x^2)}{2}$$

$$u_2 = \frac{D_i t \sqrt{N^2(1 + x^2)^2 - 4Nx^2}}{2}$$

$$\nu = \sqrt{\frac{N - 1}{N}} = \sqrt{\frac{S_y}{S + S_y}}$$

$$D_t = \sqrt{\frac{T}{D_i S_y}}$$

To cover the useful field of finite N values would require very extensive tabulation. For the case when N is large and tends to infinity, Eq. (3.110) can be reduced to (Boulton, 1963)

$$s = \frac{Q}{4\pi T}\int_0^\infty 2J_0\left(\frac{r}{D_t} x\right)\left[1 - \left(\frac{1}{x^2 + 1}\right) \exp\left(-\frac{D_i t x^2}{x^2 + 1}\right) - F\right]\frac{dx}{x} \tag{3.111}$$

where $$F = \frac{x^2}{x^2 + 1} \exp\left[-D_i N t(x^2 + 1)\right]$$

The function F vanishes when $t > 0$, but is finite as t approaches zero and Nt approaches a finite value. Thus, for small values of t, Eq. (3.111) reduces to (Boulton, 1963)

$$s = \frac{Q}{4\pi T}\int_0^\infty 2J_0\left(\frac{r}{D_t} x\right)\left(\frac{x^2}{x^2 + 1}\right)\{1 - \exp\left[-D_i N t(x^2 + 1)\right]\}\frac{dx}{x} \tag{3.112}$$

TABLE 3.9 VALUES OF $W(u_{ay}, r/D_t)$. (*From Boulton, 1963.*)

$1/u_a = N_n \times 10^n$

$r/D_t = 0.01$			$r/D_t = 0.1$			$r/D_t = 0.2$			$r/D_t = 0.316$			$r/D_t = 0.4$			$r/D_t = 0.6$		
N	n	$W(u_a, r/D_t)$	N	n	$W(u_a, r/D_t)$	N	n	$W(u_a, r/D_t)$	N	n	$W(u_a, r/D_t)$	N	n	$W(u_a, r/D_t)$	N	n	$W(u_a, r/D_t)$
1	1	1.82	1	1	1.80	5	0	1.19	1	0	0.216	1	0	0.213	1	0	0.206
1	2	4.04	5	1	3.24	1	1	1.75	2	0	0.544	2	0	0.534	2	0	0.504
1	3	6.31	1	2	3.81	5	1	2.95	5	0	1.153	5	0	1.114	5	0	0.996
5	3	7.82	2	2	4.30	1	2	3.29	1	1	1.655	1	1	1.564	1	1	1.311
1	4	8.40	5	2	4.71	5	2	3.50	5	1	2.504	5	1	2.181	2	1	1.493
1	5	9.42	1	1	4.83	1	3	3.51	1	2	2.623	1	2	2.225	5	1	1.553
1	6	9.44	1	4	4.85				1	3	2.648	1	3	2.229	1	2	1.555

$r/D_t = 0.8$			$r/D_t = 1.0$			$r/D_t = 1.5$			$r/D_t = 2.0$			$r/D_t = 2.5$			$r/D_t = 3.0$		
N	n	$W(u_a, r/D_t)$	N	n	$W(u_a, r/D_t)$	N	n	$W(u_a, r/D_t)$	N	n	$W(u_a, r/D_t)$	N	n	$W(u_a, r/D_t)$	N	n	$W(u_a, r/D_t)$
5	−1	0.046	5	−1	0.0444	5	−1	0.0394	3.33	−1	0.0100	5	−1	0.0271	5	−1	0.0210
1	0	0.197	1	0	0.1855	1	0	0.1509	5	−1	0.0335	1	0	0.0803	1	0	0.0534
2	0	0.466	2	0	0.421	1.25	0	0.199	1	0	0.114	1.25	0	0.0961	1.25	0	0.0607
5	0	0.857	5	0	0.715	2	0	0.301	1.25	0	0.144	2	0	0.1174	2	0	0.0681
1	1	1.050	1	1	0.819	5	0	0.413	2	0	0.194	5	0	0.1247	5	0	0.0695
2	1	1.121	2	1	0.841	1	1	0.427	5	0	0.227	1	1	0.1247	1	1	0.0695
5	1	1.131	5	1	0.842	2	1	0.428	1	1	0.228						

$1/u_y = N_n \times 10^n$

$r/D_t = 0.01$			$r/D_t = 0.1$			$r/D_t = 0.2$			$r/D_t = 0.316$			$r/Dt = 0.4$			$r/D_t = 0.6$		
N	n	$W(u_y, r/D_t)$	N	n	$W(u_y, r/D_t)$	N	n	$W(u_y, r/D_t)$	N	n	$W(u_y, r/D_t)$	N	n	$W(u_y, r/D_t)$	N	n	$W(u_y, r/D_t)$
4	2	9.45	4	0	4.86	4	−1	3.51	4	−1	2.66	1	−1	2.23	4.44	−1	1.586
4	3	9.54	4	1	4.95	4	0	3.54	4	0	2.74	1	0	2.26	2.22	0	1.707
4	4	10.23	4	2	5.64	2	1	3.69	4	1	3.38	5	0	2.40	4.44	0	1.844
4	5	12.31	4	3	7.72	4	1	3.85	4	2	5.42	1	1	2.55	1.67	1	2.448
4	6	14.61	4	4	10.01	1.5	2	4.55	4	3	7.72	3.75	1	3.20	4.44	1	3.255
						4	2	5.42				1	2	4.05			

$r/D_t = 0.8$			$r/D_t = 1.0$			$r/D_t = 1.5$			$r/D_t = 2.0$			$r/D_t = 2.5$			$r/D_t = 3.0$		
N	n	$W(u_y, r/D_t)$	N	n	$W(u_y, r/D_t)$	N	n	$W(u_y, r/D_t)$	N	n	$W(u_y, r/D_t)$	N	n	$W(u_y, r/D_t)$	N	n	$W(u_y, r/D_t)$
2.5	−2	1.133	4	−2	0.844	7.11	−2	0.444	4	−2	0.239	2.56	−2	0.1321	1.78	−2	0.0743
2.5	−1	1.158	4	−1	0.901	3.55	−1	0.509	2	−1	0.283	1.28	−1	0.1617	8.89	−2	0.0939
1.25	0	1.264	4	0	1.356	7.11	−1	0.587	4	−1	0.337	2.56	−1	0.1988	1.78	−1	0.1189
2.5	0	1.387	4	1	3.140	2.67	0	0.963	1.5	0	0.614	9.6	−1	0.3990	6.67	−1	0.2618
9.37	0	1.938				7.11	0	1.569	4	0	1.111	2.56	0	0.7977	1.78	0	0.5771
2.5	1	2.704															

Equation (3.112) is identical (except for notation) to Eq. (3.85) for the leaky artesian aquifer without water released from storage in the aquitard.

With the integral in Eq. (3.111) expressed symbolically as $W(u_{ay}, r/D_t)$, that is,

$$W\left(u_{ay}, \frac{r}{D_t}\right) = \int_0^{\infty} 2J_0\left(\frac{r}{D_t}x\right)\left[1 - \frac{1}{x^2+1}\exp\left(-\frac{D_i t x^2}{x^2+1}\right) - F\right]\frac{dx}{x}$$

Eq. (3.111) may be rewritten in the gallon-day-foot system of units in abbreviated form as (Prickett, 1965)

$$s = \frac{114.6Q}{T} W\left(u_{ay}, \frac{r}{D_t}\right) \tag{3.113}$$

where $u_a = \dfrac{1.87r^2S}{Tt}$ (applicable for small values of t) (3.114)

$$u_y = \frac{1.87r^2S_y}{Tt} \quad \text{(applicable for large values of } t\text{)} \tag{3.115}$$

$$\frac{r}{D_t} = \frac{2.73r}{\sqrt{T/D_iS_y}} \tag{3.116}$$

$$D_i = \frac{(r/D_t)^2\, 1/u_y}{4t} \tag{3.117}$$

$$u_y = \frac{u_a}{N-1} \tag{3.118}$$

$$N = \frac{S+S_y}{S} \tag{3.119}$$

s = drawdown, in feet
r = distance from pumped well to observation point, in feet
Q = discharge, in gpm
t = time after pumping started, in days
T = coefficient of transmissibility, in gpd/ft
S = coefficient of storage under artesian conditions, fraction
S_y = coefficient of storage associated with gravity drainage of interstices, fraction (commonly referred to as the *specific yield*)
D_i = reciprocal of delay index, in days^{-1}

$W(u_{ay}, r/D_t)$ is read as the "well function for water-table aquifers with fully penetrating wells." Values of $W(u_{ay}, r/D_t)$, given by Boulton (1963), in terms of the practical range of u_a, u_y, and r/D_i, are presented in Table 3.9.

APPLICATION OF IMAGE-WELL THEORY TO HYDROGEOLOGIC BOUNDARIES

The development of the equations discussed in the preceding sections was predicted in part on the assumption of infinite areal extent of the aquifer. The influence of hydrogeologic boundaries on groundwater movement and storage may be determined by means of the image-well theory described by Ferris (1959). The *image-well theory* may be stated as follows: the effect of a barrier boundary on the drawdown in a well, as a result of pumping from another well, is the same as though the aquifer were infinite and a like discharging well were located across the real boundary on a perpendicular thereto and at the same distance from the boundary as the real pumping well. For a recharge boundary the principle is the same except that the image well is assumed to be discharging the aquifer instead of pumping from it.

Thus, the effects of hydrogeologic boundaries on the drawdown in a well can be simulated by use of hypothetical image wells. Hydrogeologic boundaries are replaced for analytical purposes by imaginary wells which produce the same disturbing effects as the boundaries. Boundary problems are thereby simplified to consideration of an infinite aquifer in which real and image wells operate simultaneously.

3.12 Barrier Boundary

Consider an aquifer bounded on one side by an impervious formation (barrier boundary). The impervious formation cannot contribute water to the pumped well; water cannot flow across a line that defines the effective limit of the aquifer. The problem is to create a hypothetical infinite hydraulic system that will satisfy the boundary conditions dictated by the finite aquifer system.

Consider the cone of depression that would exist if the barrier boundary were not present, as shown in Fig. 3.5*a*. If a boundary is placed across the cone of depression, as shown in Fig. 3.5*b*, the hydraulic gradient cannot remain as it was because it would cause flow across the boundary. An imaginary discharging well placed across the boundary perpendicular to and equidistant from the boundary would produce a hydraulic gradient from the boundary toward the image well equal to the hydraulic gradient from the boundary toward the pumped well. A groundwater divide would exist at the boundary as shown in Fig. 3.5*c*, and this would be true everywhere along the boundary. The condition of no flow across the boundary line has been fulfilled. Therefore, the imaginary hydraulic system of a well and its image counterpart in an infinite aquifer satisfies the boundary conditions dictated by the field geology. The resultant real cone of depression is the summation of the components of both the real- and image-well depression cones as shown in Fig. 3.5*d*. The constant profile of the cone of depression is flatter on the side of the real well toward the boundary and steeper on the opposite side away from the boundary than it would be if no boundary were present. A generalized-plan view of the flow net in the vicinity of a discharging well near a barrier boundary is shown in Fig. 3.6.

Under barrier boundary conditions, water levels in wells will decline at an initial rate under the influence of the production well only. When the cone of depression of the image well appreciably affects the real well, the time-rate of drawdown will change. It will be increased in this instance because the total rate of withdrawal

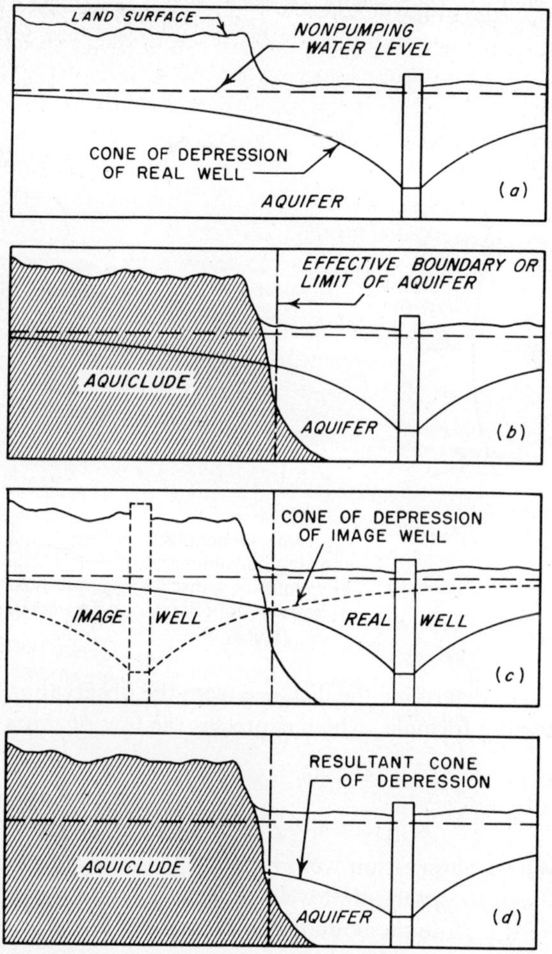

Fig. 3.5 Diagrammatic representation of image-well theory as applied to a barrier boundary. (*From Walton, 1955.*)

from the aquifer is now equal to that of the production well plus that of the discharging image well. Thus, the drawdown curve of the real well is deflected downward.

For a given aquifer the times of occurrence of equal drawdown vary directly as the squares of the distances from an observation well to production wells of equal discharge. This principle is analogous to the *law of times* defined by Ingersoll et al. (1948). The law of times is

$$\frac{t_1}{r_1^2} = \frac{t_2}{r_2^2} = \cdots = \frac{t_n}{r_n^2} \tag{3.120}$$

It follows that, if the time intercept of a given drawdown in an observation well caused by pumping a well at a given distance is known, and if the time intercept of an equal amount of divergence of the time-drawdown curve caused by the effect of the

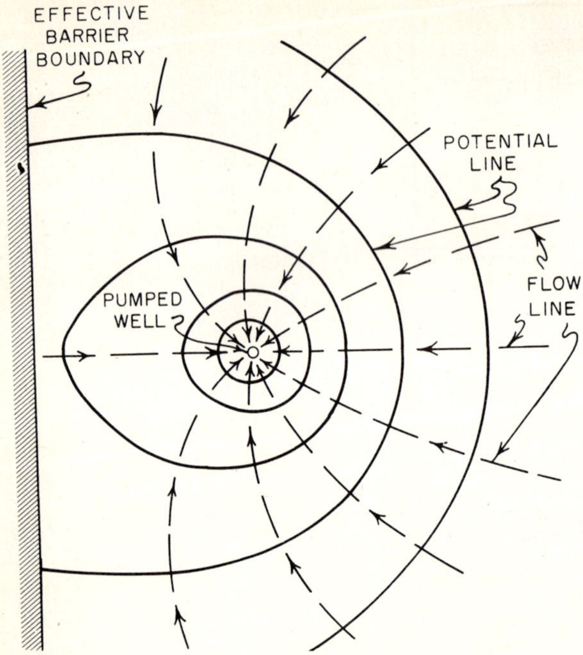

Fig. 3.6 Generalized flow net showing flow lines and potential lines in vicinity of a discharging well near a barrier boundary. (*After Ferris et al., 1962.*)

image well is also known, it is possible to determine the distance from the observation well to the image well using the following formula, which expresses the law of times

$$r_i = r\sqrt{\frac{t_i}{t_p}} \tag{3.121}$$

where r_i = distance from image well to observation well, in feet

r = distance from pumped well to observation well, in feet

t_p = time after pumping started, before the boundary becomes effective, for a particular drawdown to be observed, in days

t_i = time after pumping started, after the boundary becomes effective, when the divergence of the time-drawdown curve caused by the influence of the image well is equal to the particular value of drawdown at t_p, in days

3.13 Recharge Boundary

Recharge boundaries can also be analyzed by methods similar to those pertaining to a barrier boundary problem. Consider an aquifer bounded on one side by a recharge boundary as shown in Fig. 3.7. The cone of depression cannot spread beyond the stream. The condition is established that there shall be no drawdown along an effective line of recharge somewhere offshore. The imaginary hydraulic system of a production well and its image counterpart in an infinite aquifer shown in

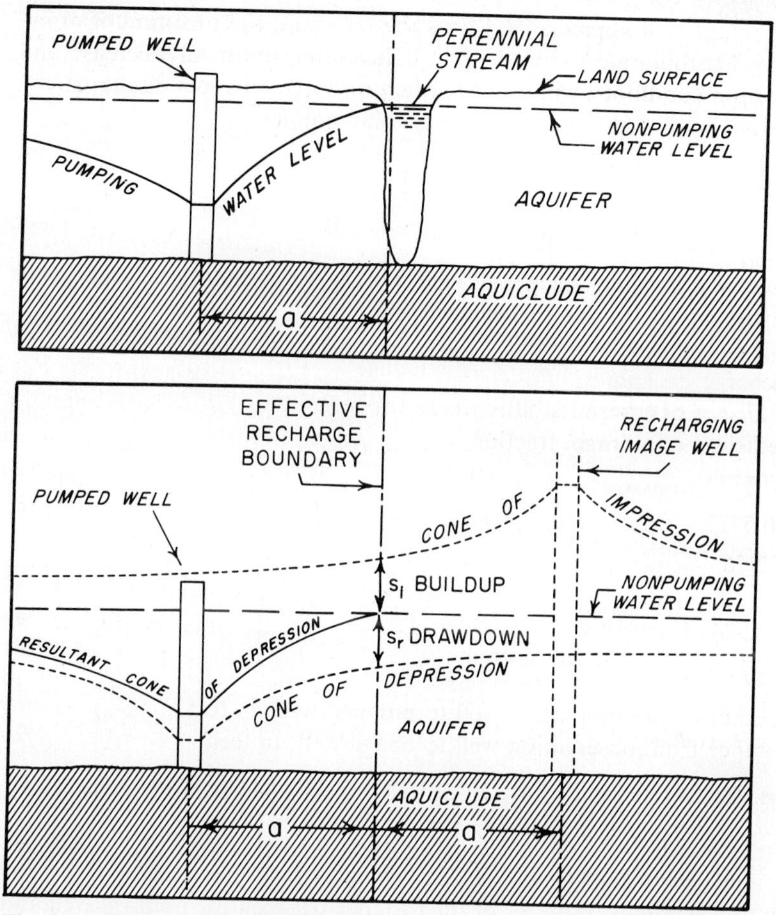

Fig. 3.7 Diagrammatic representation of image-well theory as applied to a recharge boundary. (*After Ferris et al., 1962.*)

Fig. 3.7 satisfy the foregoing boundary condition. An imaginary recharge well has been placed directly opposite and at the same distance from the stream as the real well. The recharge image well operates simultaneously and at the same rate as the real well. The resultant real cone of depression is the arithmetic summation of the components of the real-well cone of depression and the image-well cone of impression as shown in Fig. 3.7. The resultant profile of the real cone of depression is steeper on the river side of the real well and flatter on the land side of the real well than it would be if no boundary were present. A generalized-plan view of the flow net in the vicinity of a discharging well near a recharge boundary is shown in Fig. 3.8.

Under recharge boundary conditions, water levels in wells will draw down at an initial rate under the influence of the production well only. When the cone of impression of the image well appreciably affects the real well, the time-rate of drawdown will change. The time-rate of drawdown will thereafter continually decrease, and eventually equilibrium conditions will prevail, when recharge balances discharge.

Under equilibrium conditions in a nonleaky artesian aquifer

$$s_r = s - s_i \tag{3.122}$$

or

$$s_r = \frac{114.6Q}{T}[W(u) - W(u_i)] \tag{3.123}$$

where s_r = drawdown in observation well near a recharge boundary, in feet
s = drawdown due to pumped well, in feet
s_i = buildup due to image well, in feet
Q = discharge, in gpm
T = coefficient of transmissibility, in gpd/ft
S = coefficient of storage, fraction

$$W(u_i) = -0.5772 - \ln u_i \tag{3.124}$$

$$W(u) = -0.5772 - \ln u \tag{3.125}$$

$$u = \frac{1.87r^2S}{Tt} \tag{3.126}$$

$$u_i = \frac{1.87r_i^2S}{Tt} \tag{3.127}$$

r = distance from observation well to pumped well, in feet
r_i = distance from observation well to image well, in feet

Substituting Eqs. (3.124) to (3.127) into Eq. (3.123) results in the expression

$$s_r = \frac{528Q}{T}\log\frac{r_i}{r} \tag{3.128}$$

This equation was expressed in terms of the distance between the pumped well and the line of recharge by Rorabaugh (1956) as

$$s_r = \frac{528Q\log(\sqrt{4a^2 + r^2 - 4ar\cos B_r}/r)}{T} \tag{3.129}$$

where a = distance from pumped well to recharge boundary, in feet
B_r = angle between a line connecting the pumped and image wells and a line connecting the pumped and observation wells

For the particular case where the observation well is on a line parallel to the recharge boundary, the following equation (Rorabaugh, 1956) applies

$$s_r = \frac{528Q\log(\sqrt{4a^2 + r^2}/r)}{T} \tag{3.130}$$

For the particular case where the observation well is on a line perpendicular to the recharge boundary and on the river side, the following equation (Rorabaugh, 1956)

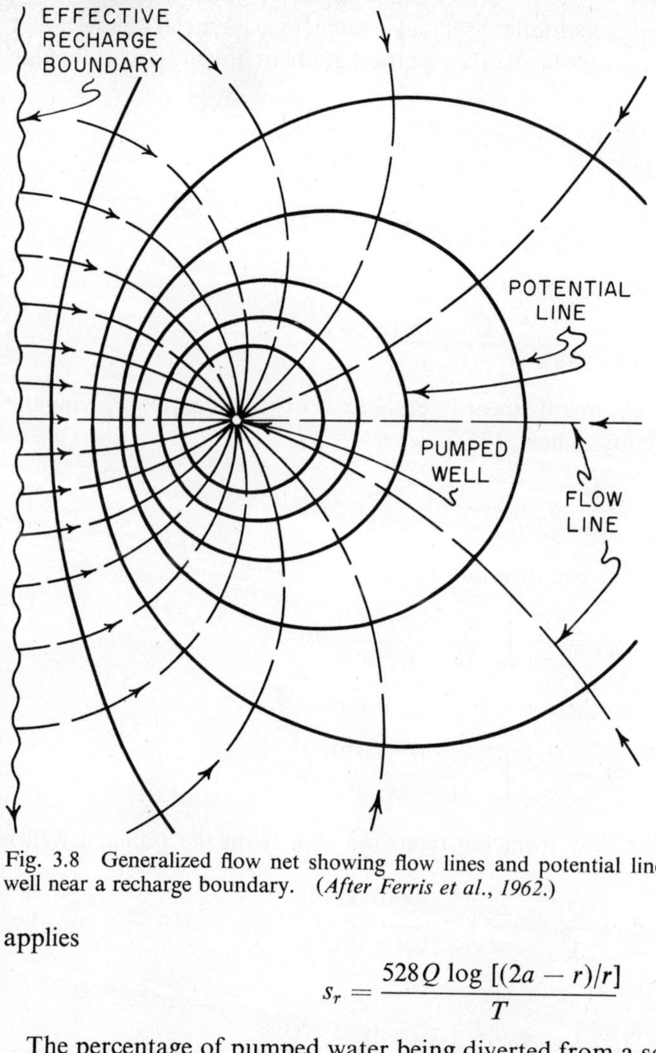

Fig. 3.8 Generalized flow net showing flow lines and potential lines in vicinity of a discharging well near a recharge boundary. (*After Ferris et al., 1962.*)

applies

$$s_r = \frac{528Q \log [(2a - r)/r]}{T} \tag{3.131}$$

The percentage of pumped water being diverted from a source of recharge depends upon the hydraulic properties of the aquifer, the distance from the pumped well to the recharge boundary, and the pumping period.

At a distance x along the line of recharge, measured from the perpendicular joining the pumped and image wells, the hydraulic gradient caused by the real well is, from Eq. (3.46) (Theis, 1941),

$$\frac{ds}{dr} = \frac{Q}{2\pi Tr} e^{-r_r^2 S/4Tt}$$

$$\frac{ds}{dr} = \left(\frac{Q}{2\pi T}\sqrt{a^2 + x^2}\right) e^{-(a^2+x^2)S/4Tt}$$

The image well will produce a hydraulic gradient similar in amount at the point x. By vectorial addition of the two gradients, the resultant gradient directly perpendicular from the river is

$$\frac{ds}{dy} = \frac{2Q}{2\pi T}\sqrt{a^2 + x^2}\,\frac{a}{\sqrt{a^2 + x^2}}\,e^{-(a^2+x^2)S/4Tt}$$

$$\frac{ds}{dy} = \frac{Qa}{\pi T\,(a^2 + x^2)}\,e^{-(a^2+x^2)S/4Tt}$$

Let $F_f = a^2S/4Tt$

then
$$\frac{ds}{dy} = \frac{Q}{\pi Ta\,[1 + (x^2/a^2)]}\,e^{-F_f[1+(x^2/a^2)]}$$

The flow from the river, Q_t, in any linear interval dx is obtained by multiplying the gradient by the transmissibility (Theis, 1941)

$$\frac{dQ_r}{dx} = T\frac{ds}{dy} = \frac{Q}{\pi a[1 + (x^2/a^2)]}\,e^{-F_f[1+(x^2/a^2)]}$$

the total flow from the river in one direction is

$$Q_t = \frac{Q}{\pi a}\int_0^\infty \frac{e^{-F_f[1+(x^2/a^2)]}}{[1 + (x^2/a^2)]}\,dx$$

and the grand-total flow from the river, for both directions, is

$$Q_t = \frac{2Q}{\pi a}\int_0^\infty \frac{e^{-F_f[1+(x^2/a^2)]}}{[1 + (x^2/a^2)]}\,dx$$

The ratio P_r between the flow from the river and that from the pumped well is (Theis, 1941)

$$P_r = \frac{Q_t}{Q} = \frac{2}{\pi a}\int_0^\infty \frac{e^{-F_f[1+(x^2/a^2)]}}{[1 + (x^2/a^2)]}\,dx$$

Letting $z = x/a$

$$P_r = \frac{2}{\pi}\int_0^\infty \frac{e^{-F_f(1+z^2)}}{1 + z^2}\,dz$$

and (de Han, 1939)

$$P_r = 1 - \left(\frac{2e^{-F_f/2}}{\sqrt{\pi}}\right)\sqrt{\sum_1^\infty \frac{F_f n}{n!}\sum_1^n \frac{(-1)^{m-1}}{2m - 1}} = 1 - e^{-F_f/2}$$

$$(1.273239F_f + 0.424413F_f^2 + 0.183912F_f^3 + 0.038399F_f^4 + 0.008859F_f^5 + 0.001316F_f^6 + 0.000207F_f^7 + 0.000024F_f^8 + \cdots)^{\frac{1}{2}}$$

This series converges rapidly for fractional values of F_f. For larger values of F_f, the integral can be more easily evaluated graphically or by Simpson's rule in the following form: Let $z = \tan u$, then $dz = \sec^2 u\, du$, and $1 + z^2 = 1 + \tan^2 u = \sec^2 u$. Changing the limits of integration (Theis, 1941)

$$P_r = \frac{2}{\pi} \int_0^{\pi/2} e^{-F_f \sec^2 u}\, du \tag{3.132}$$

where $u = \tan^{-1} \dfrac{r_r}{a}$

$$F_f = \frac{1.87a^2S}{Tt} \tag{3.133}$$

P_r = percentage of pumped water being diverted from a source of recharge
a = distance from pumped well to recharge boundary, in feet
S = coefficient of storage, fraction
T = coefficient of transmissibility, in gpd/ft
t = time after pumping started, in days
r_r = distance along recharge boundary measured from the perpendicular joining the real and image well, in feet

Figure 3.9 gives values of P_r for various values of F_f and shows therefore the percentage of the pumped water diverted from a source of recharge.

Considerable time elapses before a cone of depression stabilizes, water is no longer taken from storage within the aquifer, and a new state of approximate equilibrium is established in an aquifer with a recharge boundary. The time required to reach approximate equilibrium may be computed with the following equation (see Foley et al., 1953).

$$t_e = \frac{a^2S}{112T\epsilon_e \log (2a/r)^2} \tag{3.134}$$

where t_e = time required to reach approximate equilibrium, in years
a = distance from pumped well to recharge boundary, in feet
r = distance from pumped well to observation point, in feet
S = coefficient of storage, fraction
T = coefficient of transmissibility, in gpd/ft
ϵ_e = deviation from absolute equilibrium (generally arbitrarily assumed to be 0.05)

The drawdowns caused by a well discharging, under equilibrium conditions, from an aquifer having a recharge boundary of finite length can be determined from a formula developed by Theis (1963). Assume that the recharge boundary of length $2c_r$ lies along the x axis of a cartesian coordinate system and is bisected by the y axis, i.e., the boundary extends from $-c_r$ to $+c_r$ on the x axis. The pumped wel lis located at the point x_0, y_0, and the observation well is located at the point (x,y). With c_r as the basic unit of reference, the coordinates of these points are $(x_0/c_r, y_0/c_r)$ and $(x/c_r, y/c_r)$, respectively. The cartesian system of coordinates may be transformed

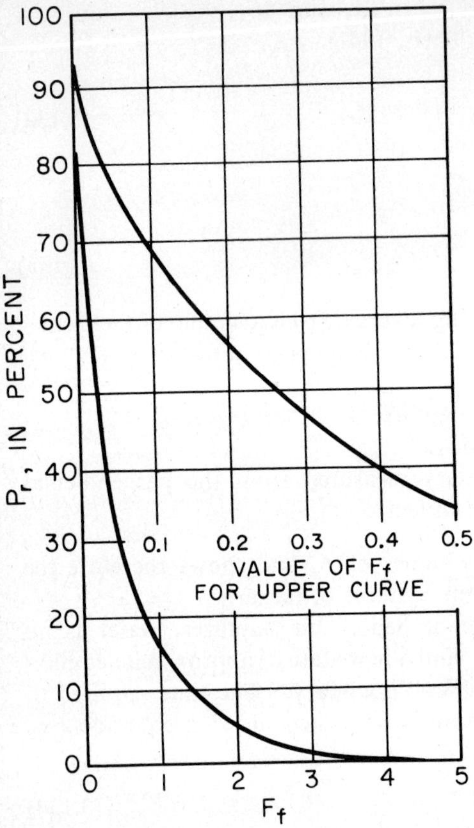

Fig. 3.9 Graph for determination of percentage of pumped water being diverted from a source of recharge. (*From Theis, 1941.*)

by the equations (Theis, 1963)

$$\frac{x_0}{c_r} = \cosh \epsilon_0 \cos \eta_0$$

$$\frac{x}{c_r} = \cosh \epsilon_r \cos \eta_r$$

$$\frac{y_0}{c_r} = \sinh \epsilon_0 \sin \eta_0$$

$$\frac{y}{c_r} = \sinh \epsilon_r \sin \eta_r$$

The resulting system of orthogonal coordinates represents the flow lines and the equipotential lines from the line source. Figure 3.10 gives values of ϵ_0 and η_0 corresponding to values of x_0/c_r and y_0/c_r and values of ϵ_r and η_r corresponding to values of x/c_r and y/c_r; ϵ_0, ϵ_r, η_0, and η_r are in radian measure. Tables of cosh ϵ_0

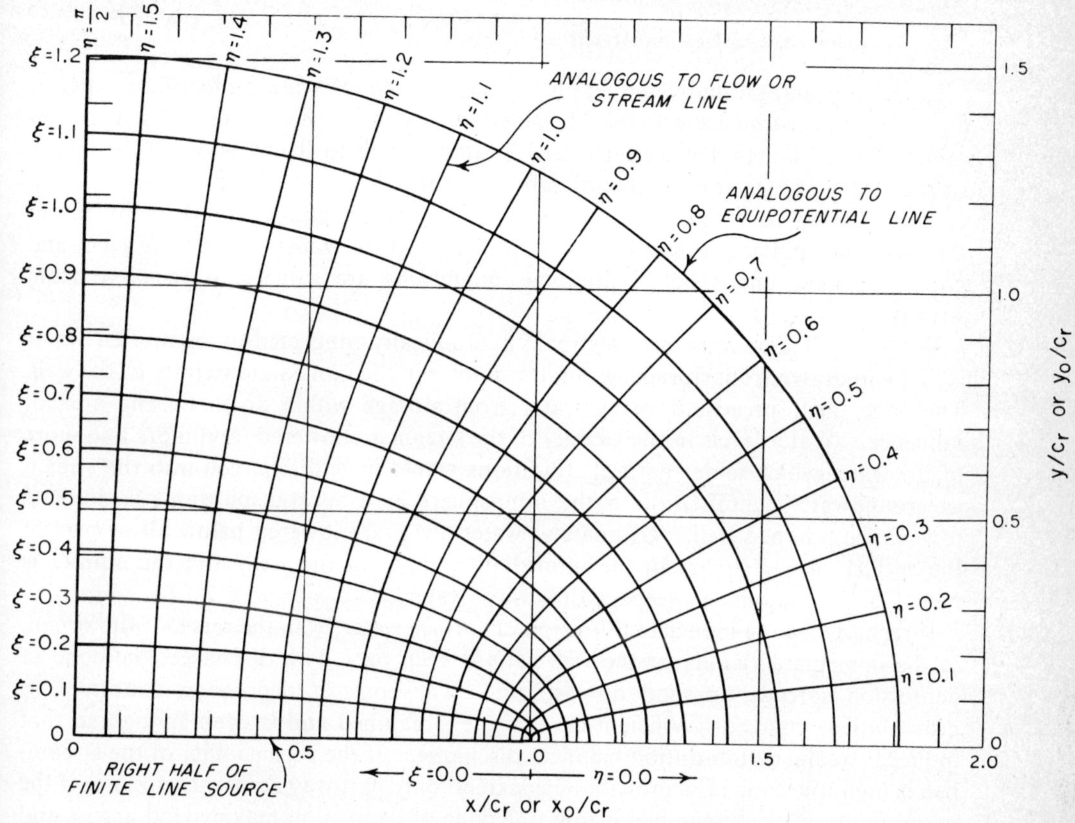

Fig. 3.10 Diagram for effecting transformation from rectangular coordinates to a form of hyperbolic coordinates. (*From Theis, 1963.*)

or ϵ_r, cos η_0 or η_r, and sin η_0 or η_r can be found in a handbook of mathematical tables.

The equation for the required drawdown is (Theis, 1963)

$$s_f = \frac{264Q}{T} \log \left[\frac{\cosh(\epsilon_r + \epsilon_0) - \cos(\eta_r - \eta_0)}{\cosh(\epsilon_r - \epsilon_0) - \cos(\eta_r - \eta_0)} \frac{\cosh(\epsilon_r + \epsilon_0) - \cos(\eta_r + \eta_0)}{\cosh(\epsilon_r - \epsilon_0) - \cos(\eta_r + \eta_0)} \right] \tag{3.135}$$

where s_f = drawdown in observation well near a finite recharge boundary, in feet
Q = discharge, in gpm
T = coefficient of transmissibility, in gpd/ft

In Eq. (3.135) the hyperbolic cosine is always positive [cosh x = cosh $(-x)$] and the natural cosine is positive for any angle in the first and fourth quadrants and negative for any angle in the second and third quadrants

$$\cos x = \cos(-x) \qquad \text{and} \qquad \cos(\pi - x) = -\cos x$$

3.14 Induced Streambed Infiltration

In recent years, groundwater supplies using induced infiltration of surface water as a source of recharge have been extensively developed especially in areas in north-central United States that were covered by or adjacent to the Pleistocene ice sheets. In the glaciated areas or closely adjacent to them, valleys cut into the bedrock often contain limited to extensive sand and gravel outwash aquifers. At many places the outwash pattern coincides with present stream valleys, and the sand and gravel aquifers under heavy pumping conditions are subject to recharge from streams.

When a well is pumped near a stream hydraulically connected to an aquifer, water is first withdrawn from storage within the aquifer in the immediate vicinity of the well. The cone then spreads, drawing water from storage within an increasing area of influence. Water levels in the vicinity of the stream are lowered, and more and more of the water which under natural conditions would have discharged into the stream as groundwater runoff, or into the atmosphere as evapotranspiration, is diverted toward the pumped well. Ultimately, water levels are lowered below all or part of the surface of the stream in the immediate vicinity of the well, and the aquifer is recharged by the influent seepage of surface water.

If recharge by the induced infiltration of surface water from the reach of the stream in the immediate vicinity of the well cannot keep pace with discharge, the cone of depression spreads up- and downstream. The cone of depression continues to grow until it intercepts sufficient area of the streambed and is deep enough so that induced streambed infiltration balances discharge. If the permeability of the streambed is high, the cone of depression may extend only part way across the stream; if the permeability of the streambed is low, the cone of depression may expand across and beyond the stream.

Based on Darcy's law, the infiltration rate of a streambed is a direct function of permeability and head. Recharge from streamflow is directly proportional to the permeability of the streambed and to the difference between water levels in the aquifer immediately below the streambed and the surface of the stream, until the water table has declined below the streambed.

Recharge by the influent seepage of surface water takes place over an area of the streambed. However, to make flow problems associated with induced streambed infiltration amenable to mathematical treatment, the area is replaced by a line source. The assumption is made that water levels in the aquifer will behave the same way whether recharge occurs over an area or along a line.

The stabilized cone of depression of a well near a recharge boundary can be described and the area of influence of pumping can be delineated with Eq. (3.129), provided the distance a and the coefficient of transmissibility are known. These basic factors can often be determined from aquifer test data.

The coefficient of transmissibility can often be evaluated from distance-drawdown data for observation wells on a line parallel to the recharge boundary. Provided that the observation wells are not too close to the recharge boundary and not too far from the pumped well, the hydraulic gradient of the profile of the cone of depression through the wells will not be distorted to any great degree. The wells are approximately equidistant from the image well associated with the recharge boundary, and the effects of induced streambed infiltration on drawdowns are approximately equal at all wells. Thus, the distance-drawdown data for wells on a line parallel to the stream closely describe the hydraulic gradient toward the pumped well that would exist if the aquifer were infinite in areal extent. A plot of drawdowns in the observation wells versus the logarithm of the distances from the pumped well will yield a straight-line graph. The slope of the straight line is substituted into Eq. (3.55) to compute the coefficient of transmissibility.

Although the hydraulic gradient is not distorted, the total values of the drawdowns in the observation wells are much less than they would be under infinite aquifer conditions because of the effects of recharge, and the coefficient of storage of the aquifer cannot be determined from the distance-drawdown graph.

If T is known, the distance a can be computed on the basis of the total drawdown in each observation well on a line parallel to the stream with Eq. (3.130). The ratio of the drawdowns in any two observation wells on a line parallel to the stream (see Rorabaugh, 1956) may also be used to determine a.

Equation (3.130) assumes that the cone of depression has stabilized, water is no longer taken from storage within the aquifer, and equilibrium conditions prevail. The pumping period required to reach approximate equilibrium conditions can be computed by using Eq. (3.134).

The effects of slow gravity drainage of interstices under water-table conditions and leakage through a confining bed under leaky artesian conditions closely resemble the effects of recharge by induced streambed infiltration. The mere fact that the cone of depression stabilizes is not positive proof that induced streambed infiltration is occurring. The decrease in the time-rate of drawdown can be attributed either to the effects of slow gravity drainage, the effects of leakage through a confining bed, or the effects of induced streambed infiltration if the effects of partial penetration are excluded. Rorabaugh (1956) gave methods for proving whether or not water levels stabilize because of the effects of slow gravity drainage.

Equations (3.123) to (3.127) and the computed values of T and a can be used to determine the coefficient of storage. Several assumed values of the coefficient of storage are substituted into Eqs. (3.126) and (3.127) for computation of u and u_i. Corresponding values of $W(u)$ and $W(u_i)$ are obtained from Table 3.1 and are substituted into Eq. (3.123) to determine drawdowns. The computed drawdowns are then compared with actual drawdowns in observation wells, and that coefficient of storage which was used to compute drawdowns equal to actual drawdowns is assigned to the aquifer.

If the hydraulic properties of the aquifer and the distance a are known, the percentage of pumped water being diverted from a stream can be computed with Eq. (3.132). The amount of recharge by induced infiltration is then given by the following equation:

$$Q_r = \frac{QP_r}{100} \tag{3.136}$$

where Q_r = amount of induced infiltration, in gpm
Q = discharge of pumped well, in gpm
P_r = percentage of pumped water being diverted from the source of recharge

The area of the streambed through which influent seepage of surface water is occurring can be determined with Eq. (3.123). The pumped well, the image well associated with the recharge boundary, and the streambed are drawn to scale on a map. Several points within the streambed that are equidistant up- and downstream from the pumped well are selected for computations of drawdown. The distances between the selected points, the pumped well, and the image well are scaled from the map. Values of drawdown at the points are then computed with Eqs. (3.123) to (3.127). The reach of the streambed, L_r, within the area of influence of pumping is ascertained by noting the location of the points up- and downstream where drawdown is negligible (say, $\leq$0.01 ft). The *area of induced infiltration* A_r is then the product of L_r and the average distance between the shoreline and the recharge boundary or the average width of the streambed, depending upon the position of the recharge boundary.

When Q_r and A_r are known, the *infiltration rate of the streambed* can be computed with the following equation (Walton, 1963)

$$I_a = \frac{6.3 \times 10^7 Q_r}{A_r} \tag{3.137}$$

where I_a = average infiltration rate of streambed, in gallons per day per acre (gpd/acre)
Q_r = amount of induced infiltration, in gpm
A_r = streambed area of infiltration, in square feet

The average *head loss* h_r due to the vertical percolation of water through the streambed can be determined from data for observation wells installed within the streambed area of infiltration at depths just below the streambed. In many cases the installation of observation wells in the stream channel is impractical, and h_r must be estimated with Eq. (3.123) and the hydrogeologic properties and position of the recharge boundary determined by aquifer test analysis. The drawdowns at many points within the streambed area of infiltration are computed with Eq. (3.123). The average of the computed drawdowns is the approximate average h_r within the streambed area of infiltration. The average infiltration rate of the streambed per unit area per foot of head loss can be estimated by use of the following equation

$$I_h = \frac{I_a}{h_r} \tag{3.138}$$

where I_h = average infiltration rate of streambed, in gallons per day per acre of streambed per foot of head loss (gpd/acre/ft)
I_a = average infiltration rate of streambed, in gpd/acre
h_r = average head loss within streambed area of infiltration, in feet

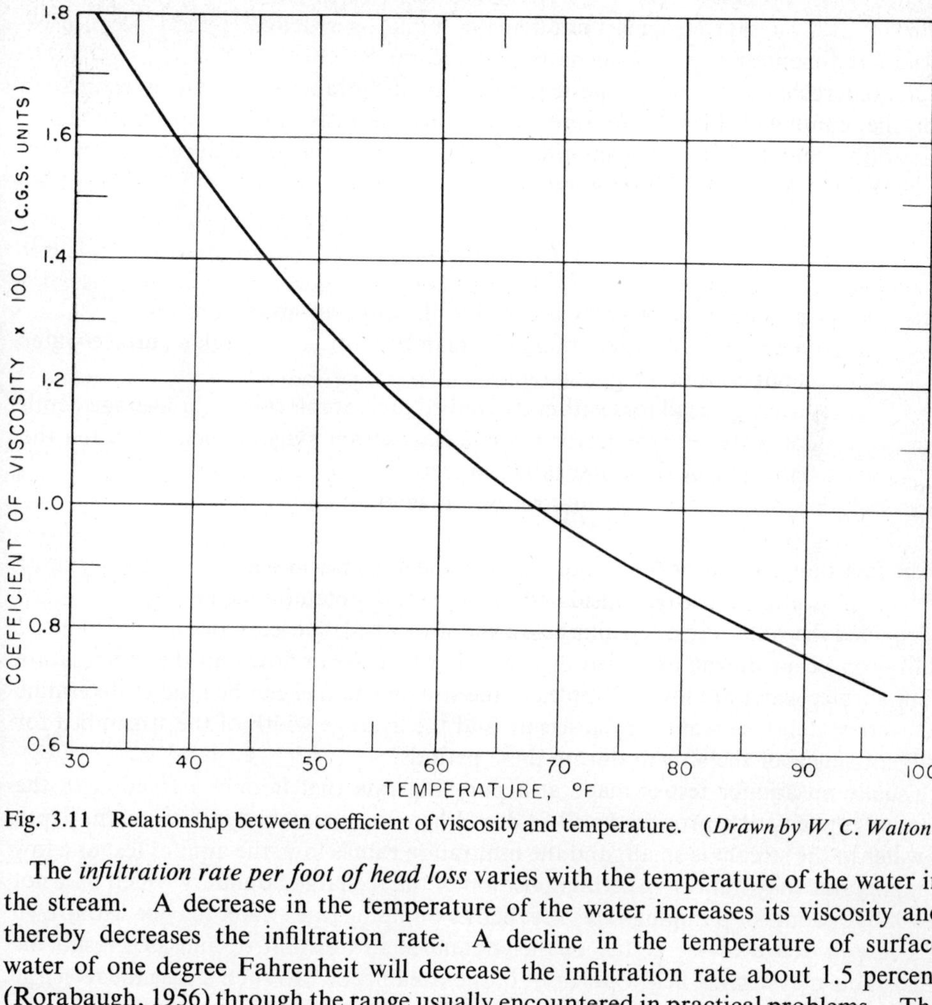

Fig. 3.11 Relationship between coefficient of viscosity and temperature. (*Drawn by W. C. Walton.*)

The *infiltration rate per foot of head loss* varies with the temperature of the water in the stream. A decrease in the temperature of the water increases its viscosity and thereby decreases the infiltration rate. A decline in the temperature of surface water of one degree Fahrenheit will decrease the infiltration rate about 1.5 percent (Rorabaugh, 1956) through the range usually encountered in practical problems. The infiltration rate for any particular surface-water temperature can be computed with Fig. 3.11 and the following equation

$$I_t = I_h \frac{\mu_a}{\mu_t} \tag{3.139}$$

where I_t = average infiltration rate of streambed for a particular surface-water temperature, in gpd/acre/ft

I_h = average infiltration rate of streambed determined from aquifer test results, in gpd/acre/ft

μ_a = coefficient of dynamic viscosity at temperature of surface water during aquifer test, in centimeter-gram-second (cgs) units

μ_t = coefficient of dynamic viscosity at a particular temperature of surface water, in cgs units

The infiltration rate also may not remain stable over a long period because of alternate sedimentation and scouring by the stream.

Potential recharge by the induced infiltration of surface water can be estimated from the computed infiltration rate, streamflow records, and temperature data. Infiltration cannot exceed streamflow. Potential recharge by induced infiltration can be determined by substituting data in the following equation

$$R_I = I_t h_r A_r \tag{3.140}$$

where R_I = potential recharge by induced infiltration, in gpd
I_t = average infiltration rate of streambed for a particular surface-water temperature, in gpd/acre/ft
h_r = average head loss within streambed area of infiltration or average depth of water in stream for a particular stream stage, depending upon the position of the water table, in feet
A_r = streambed area of infiltration, in acres

The fact that the average depth of water in the stream will decrease as the result of induced infiltration must be considered in estimating potential recharge.

Provided the water table remains below the streambed, the least amounts of induced infiltration occur during extended dry periods when streamflow and the temperature of the surface water are low. Profiles of the stream channel can be used to determine the average depth of water in the stream and the average width of the streambed for different stages of the stream during these periods.

Usually an aquifer test is made at a pumping rate that is only a fraction of the planned capacity of a proposed well field. Where the streambed is narrow, the depth of water in the stream is small, and the infiltration rate is low, the aquifer test at a low pumping rate may indicate a certain position of the recharge boundary which may not be valid for a higher pumping rate. At higher pumping rates water may be withdrawn at a rate in excess of the ability of the streambed to transmit it, and as a result the water table may decline below portions of the streambed. In such a case the recharge boundary moves away from the pumped well as maximum infiltration occurs in the reach of the stream in the immediate vicinity of the pumped well, and the cone of depression spreads further up- and downstream.

The position of the recharge boundary at higher pumping rates can be determined by the process of trial and error. Several positions of the recharge boundary are assumed, and drawdowns beneath the streambed and streambed areas of infiltration are computed with Eq. (3.123). Values of R_I are then computed with Eq. (3.140), keeping in mind that h_r is either the average head loss within the streambed area of infiltration or the average depth of water in the streambed area of infiltration, depending upon the stage of the water table. The position of the recharge boundary which results in R_I balancing the pumping rate is judged to be the correct position for the higher pumping rate.

3.15 Multiple-boundary Systems

Aquifers are often delimited by two or more boundaries. Two converging boundaries delimit a *wedge-shaped aquifer*; two parallel boundaries form an *infinite-strip aquifer*; two parallel boundaries intersected at right angles by a third boundary form a *semi-infinite–strip aquifer*; and four boundaries intersecting at right angles form a *rectangular aquifer*. The image-well theory may be applied to such cases by taking into consideration successive reflections on the boundaries.

A number of image wells are associated with a pair of converging boundaries. A primary image well placed across each boundary will balance the effect of the pumped well at each boundary. However, each primary image well produces an unbalanced effect at the opposite boundary. Secondary image wells must be added at appropriate positions until the effects of the real and image wells are balanced at both boundaries.

Although image-well systems can be divised regardless of the wedge angle involved, simple solutions of closed image systems are preferred. The actual aquifer wedge angle W_e is approximated as equal to one of certain aliquot parts of 360°. These particular values of W_e were specified by Ferris et al. (1962) as follows: If the aquifer wedge boundaries are of like character, W_e must be an aliquot part of 180°. If the boundaries are not of like character, W_e must be an aliquot part of 90°. Under the foregoing conditions and from the solution of the wedge problem as given by Carslaw and Jaeger (1959), it is evident that the exact number of image wells N_i depends upon the angle W_e included between the boundaries and is given by the expression

$$N_i = \frac{360°}{W_e} - 1 \tag{3.141}$$

The locus of image-well locations is a circle whose center is at the apex of the wedge and whose radius is equal to the distance from the pumped well to the wedge apex.

If the aquifer wedge boundaries are not of like character, the character of each image well, recharge or discharge, can be ascertained by balancing the image-well system, considering each boundary separately (see Walton, 1962). A primary image well placed across a barrier boundary is discharging in character, and a primary image well placed across a recharge boundary is recharging in character. A secondary image well placed across a barrier boundary has the same character as its parent image well, and a secondary image well placed across a recharge boundary has the opposite character of its parent image well. Image-well systems for several wedge-shaped aquifers are shown in Fig. 3.12.

If the arrangement of two boundaries is such that they are parallel to each other, analysis by the image-well theory requires the use of an image-well system extending to infinity (Ferris et al., 1962). Each successively added secondary image well produces a residual effect at the opposite boundary. However, in practice it is only necessary to add pairs of image wells until the next pair has negligible influence on the sum of all image-well effects out to the point. Image-well systems for several parallel-boundary situations are shown in Figs. 3.13 and 3.14.

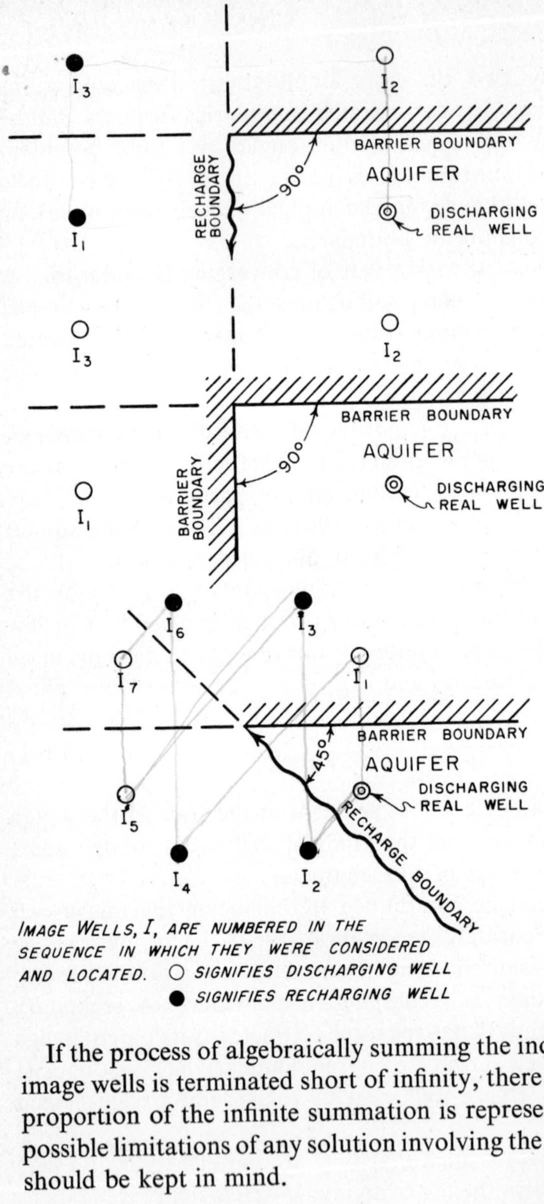

Fig. 3.12 Plans of image-well systems for several wedge-shaped aquifers. (*After Ferris et al., 1962.*)

If the process of algebraically summing the individual effects of an infinite system of image wells is terminated short of infinity, there is no simple way of determining what proportion of the infinite summation is represented by the partial summation. The possible limitations of any solution involving the use of an infinite system of image wells should be kept in mind.

APPLICATION OF DIFFERENTIAL EQUATIONS TO FLOW FROM AND TO STREAMS AND DITCHES

A number of investigators have presented equations for groundwater flow from and to streams and ditches. Equations have been applied to the determination of the

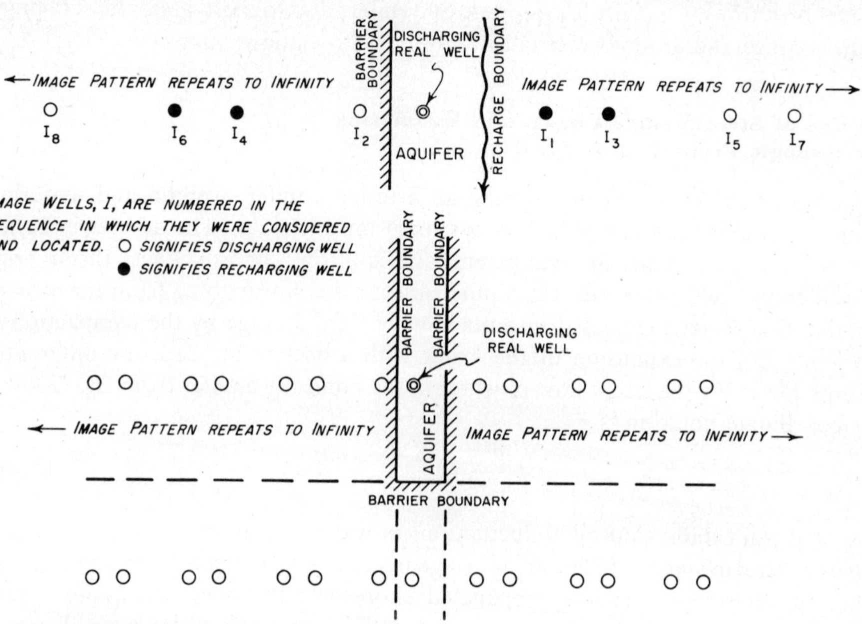

Fig. 3.13 Plans of image-well systems for selected parallel-boundary situations. (*After Ferris et al., 1962.*)

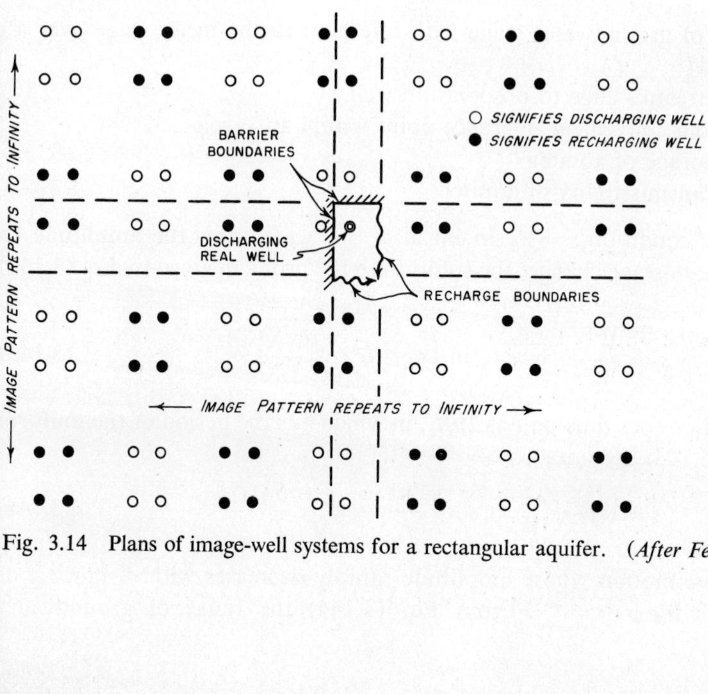

Fig. 3.14 Plans of image-well systems for a rectangular aquifer. (*After Ferris et al., 1962.*)

coefficients of transmissibility and storage of aquifers and to the evaluation of changes in bank storage and groundwater contribution to streamflow.

3.16 Use of Stream-stage Changes in Estimating Hydrogeologic Properties of Aquifers

Figure 3.15 shows a well penetrating an artesian aquifer overlain and underlain by aquicludes and fully penetrated by a stream on one side. The aquifer is homogeneous, isotropic, infinite in areal extent, and is of the same thickness throughout. The well completely penetrates the aquifer and flow is linear to or from the stream. Assuming that water is released instantaneously from storage by the compaction of the aquifer and the expansion of the water with a decline in head, the differential equation governing the linear flow of water in the nonleaky aquifer from Eq. (3.44) in polar-coordinate notation is

$$\frac{\partial^2 s}{\partial x^2} = \frac{S}{T}\frac{\partial s}{\partial t} \tag{3.142}$$

The well will exhibit sinusoidal fluctuations of water level in response to correlative changes of stream stage. If the stream stage fluctuates as a simple harmonic motion, a train of sinusoidal waves is propagated shoreward through the aquifer. With increasing distance from the stream, the amplitude of the transmitted wave decreases and the time lag of a given maximum or minimum increases (Ferris, 1953). In terminology adapted to the sinusoidal stream-stage changes and associated water-level fluctuations,

s = net rise or fall of groundwater stage with reference to the mean stage over an observed period
x = distance from stream's edge to observation well
t = time elapsed from convenient reference point within any cycle
S = coefficient of storage of aquifer
T = coefficient of transmissibility of aquifer

For the boundary condition $s = s_0 \sin \omega_s t$ at $x = 0$ where s_0 is the amplitude or half range of the stream-stage change, the solution to the problem given by Eq. (3.142) is (Ferris, 1951)

$$s = s_0 e^{-x\sqrt{\omega_s S/2T}} \sin\left(\omega_s t - x\sqrt{\frac{\omega_s S}{2T}}\right) \tag{3.143}$$

Expressing ω_s in radians per time unit as $2\pi/t_s$, in which t_s is the period of the uniform stage change, Eq. (3.143) becomes

$$s = s_0 e^{-x\sqrt{\pi S/t_s T}} \sin\left(\frac{2\pi t}{t_s} - x\sqrt{\frac{\pi S}{t_s T}}\right) \tag{3.144}$$

which defines a wave motion whose amplitude rapidly decreases with distance x as given by the factor $s_0 e^{-x\sqrt{\pi S/t_s T}}$. From Eq. (3.144) the range of groundwater

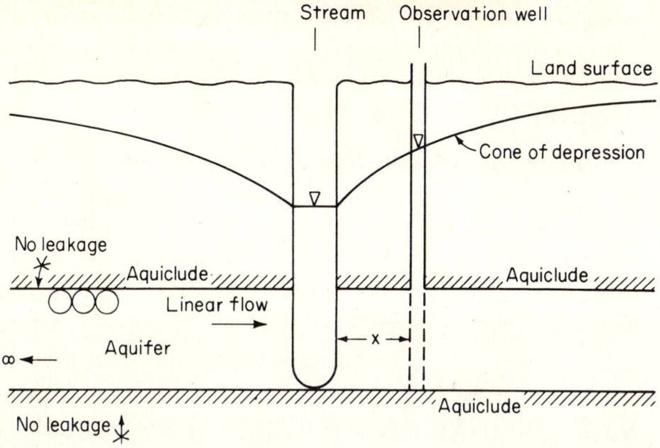

Fig. 3.15 Nonleaky artesian aquifer with fully penetrating stream. (*Drawn by W. C. Walton.*)

fluctuation at an observation well a distance x from the stream's edge is (Ferris, 1951)

$$s_s = 2s_0 e^{-x\sqrt{\pi S/t_s T}} \tag{3.145}$$

The lag in time of occurrence of a given maximum or minimum groundwater stage, t_m, is

$$t_m = x\sqrt{\frac{t_s S}{4\pi T}} \tag{3.146}$$

The apparent velocity of transmission of the wave through the aquifer, V_{ap}, is

$$V_{\text{ap}} = \frac{x}{t_m} = \sqrt{\frac{4\pi T}{t_s S}} \tag{3.147}$$

and the wavelength W_L is

$$W_L = V_{\text{ap}} t_s = \sqrt{\frac{4\pi t_s T}{S}} \tag{3.148}$$

The quantity of flow into or from the aquifer per half cycle, $Q_{\frac{1}{2}c}$, from Darcy's law is

$$Q_{\frac{1}{2}c} = TIL \tag{3.149}$$

where L is the length of the cross section of the aquifer through which flow occurs and I is the hydraulic gradient of the piezometric surface, ds/dx. I as given by Ferris (1951) is

$$I = s_0 e^{-x\sqrt{\omega_s S/2T}} \left(-\frac{\omega_s S}{2T}\right)\left[\sin\left(\omega_s t - x\sqrt{\frac{\omega_s S}{2T}}\right) + \cos\left(\omega_s t - x\sqrt{\frac{\omega_s S}{2T}}\right)\right] \tag{3.150}$$

The gradient is not in phase with s. From Eq. (3.150), at $x = 0$ the gradient is zero at $t = -\pi/4\omega_s = -(t_s/8)$, reaches a minimum at $t = \pi/4\omega_s = t_s/8$, and returns to 0 at $t = 3\pi/4\omega_s = 3t_s/8$.

The quantity of flow per unit length of aquifer is (Ferris, 1951)

$$\frac{Q_L}{L} = -T \int_{-t_s/8}^{3t_s/8} \left(\frac{\partial s}{\partial x}\right)_{x=0} dt = s_0 \sqrt{\frac{2t_s ST}{\pi}} \tag{3.151}$$

Equation (3.145) may be rewritten in the gallon-day-foot system of units as

$$\frac{s_s}{2s_0} = e^{-4.8x\sqrt{S/Tt_s}} \tag{3.152}$$

and

$$\log \frac{s_s}{2s_0} = -2.1x\sqrt{\frac{S}{t_s T}}$$

thus

$$2.1\sqrt{\frac{S}{t_s T}} = \frac{-\log (s_s/2s_0)}{x} \tag{3.153}$$

Equation (3.153) expressed in terms of T becomes

$$T = \frac{4.4[-\log (s_s/2s_0)]^2 S}{t_s} \tag{3.154}$$

where t_s is the period of the stream-stage change, in days; T is the coefficient of transmissibility, in gpd/ft; S is the coefficient of storage, fraction; and the logarithmic quantity $s_s/2s_0$ is in effect the ratio of the range of groundwater stage to the range of stream stage.

As indicated by Eq. (3.154), it is necessary to know S in order to evaluate T. $s_s/2s_0$ approaches unity as x approaches zero. Thus, the negative x value associated with the $s_s/2s_0$ ratio of unity represents the effective distance offshore to the sub-outcrop.

Equation (3.146) may be rewritten in the gallon-day-foot system in terms of T as

$$T = \frac{0.60x^2 S t_s}{t_m{}^2} \tag{3.155}$$

in which t_m is the time lag in occurrence of a given maximum or minimum groundwater stage following the occurrence of a similar stream-stage change, in days; x is the distance from edge of stream to observation point, in feet.

At the suboutcrop, where $x = 0$, the time lag is zero. The negative value of x at $t_m = 0$ is the effective distance offshore to the suboutcrop.

In 1938 Theis (in Wenzel and Sand, 1942) developed a formula for determining the hydrogeologic properties of a nonleaky artesian aquifer with data on the decline in artesian head at any distance from a drain discharging at a uniform rate. The formula can be written nondimensionally as (Ferris et al., 1962)

$$s = \frac{Q_b x}{2T}\left(\frac{e^{-u^2}}{u\sqrt{\pi}} - 1 + \frac{2}{\sqrt{\pi}} \int_0^{x/2\sqrt{Tt/S}} e^{-u^2}\, du\right) \tag{3.156}$$

where

$$u = x\sqrt{\frac{S}{4Tt}}$$

With the quantity in brackets expressed symbolically as $W(u_q)$, that is,

$$W(u_q) = \frac{e^{-u^2}}{u\sqrt{\pi}} - 1 + \frac{2}{\sqrt{\pi}} \int_0^{x/2\sqrt{Tt/S}} e^{-u^2}\, du$$

Eq. (3.156) may be written in the gallon-day-foot system of units in abbreviated form as

$$s = \frac{720 Q_b x W(u_q)}{T} \tag{3.157}$$

where $$u_q^2 = \frac{1.87 x^2 S}{Tt} \tag{3.158}$$

s = drawdown at any point in the vicinity of the drain discharging at a constant rate, in feet
Q_b = constant discharge of the drain, in gpm per lineal foot of drain
x = distance from the drain to the observation point, in feet
t = time after the drain started discharging, in days
T = coefficient of transmissibility, in gpd/ft
S = coefficient of storage, fraction

and $W(u_q)$ is read as the "well function for nonleaky artesian aquifers with a constant-discharging drain." Values of $W(u_q)$, given by Ferris et al. (1962) in terms of the practical range of u_q, are presented in Table 3.10.

Ferris et al. (1962) gave a formula for determining the hydrogeologic properties of a nonleaky artesian aquifer with data on the decline in artesian head at any distance from a drain whose stage has been suddenly changed and whose discharge is variable. The formula can be written nondimensionally as

$$s = s_d\left(1 - \frac{2}{\sqrt{\pi}} \int_0^{x/2\sqrt{Tt/S}} e^{-u^2}\, du\right) \tag{3.159}$$

where $$u = x\sqrt{\frac{S}{4Tt}} \tag{3.160}$$

With the quantity in brackets expressed symbolically as $W(u_h)$, that is,

$$W(u_h) = 1 - \frac{2}{\sqrt{\pi}} \int_0^{x/2\sqrt{Tt/S}} e^{-u^2}\, du$$

Eq. (3.159) may be written in the gallon-day-foot system of units in abbreviated form as

$$s = s_d W(u_h) \tag{3.161}$$

where $$u_h^2 = \frac{1.87 x^2 S}{Tt} \tag{3.162}$$

s = drawdown at any point in the vicinity of the drain, in feet
s_d = abrupt change in drain stage, in feet
x = distance from the drain to the observation point, in feet
t = time after the abrupt drain-stage change, in days
T = coefficient of transmissibility, in gpd/ft
S = coefficient of storage, fraction

TABLE 3.10 VALUES OF $W(u_q)$. *(From Ferris et al., 1962.)*

u	u^2	$W(u_q)$
0.0500	0.0025	10.32
0.0600	0.0036	8.468
0.0700	0.0049	7.109
0.0800	0.0064	6.130
0.0900	0.0081	5.331
0.1000	0.010	4.714
0.1140	0.013	4.008
0.1265	0.016	3.532
0.1414	0.020	3.079
0.1581	0.025	2.657
0.1732	0.030	2.354
0.1871	0.035	2.109
0.2000	0.040	1.943
0.2236	0.050	1.658
0.2449	0.060	1.441
0.2646	0.070	1.282
0.3000	0.090	1.049
0.3317	0.110	0.8810
0.3605	0.130	0.7598
0.4000	0.160	0.6284
0.4359	0.190	0.5324
0.4796	0.230	0.4384
0.5291	0.280	0.3517
0.5745	0.330	0.2895
0.6164	0.380	0.2434
0.6633	0.440	0.2008
0.7071	0.500	0.1837
0.7616	0.580	0.1345
0.8124	0.660	0.1094
0.8718	0.760	0.0864
0.9486	0.900	0.0623
1.0000	1.000	0.0507

and $W(u_h)$ is read as the "well function for nonleaky artesian aquifers with a drain having an abrupt change in stage." Values of $W(u_h)$, given by Ferris et al. (1962), in terms of the practical range of u_h, are presented in Table 3.11.

The relation expressing the discharge from the aquifer, per unit length of stream channel, Q_b, resulting from the abrupt change in stream stage taking into account water contribution from both sides of the drain may be expressed nondimensionally as (Ferris et al., 1962)

$$Q_b = \frac{2s_d}{\sqrt{\pi t}} \sqrt{ST} \tag{3.163}$$

TABLE 3.11 VALUES OF $W(u_h)$. *(From Ferris et al., 1962.)*

u	u^2	$W(u_h)$
0.03162	0.0010	0.9643
0.04000	0.0016	0.9549
0.05000	0.0025	0.9436
0.06325	0.0040	0.9287
0.07746	0.0060	0.9128
0.08944	0.0030	0.8994
0.1000	0.010	0.8875
0.1265	0.016	0.8580
0.1581	0.025	0.8231
0.2000	0.040	0.7730
0.2449	0.060	0.7291
0.2828	0.080	0.6892
0.3162	0.10	0.6548
0.4000	0.16	0.5716
0.5000	0.25	0.4795
0.6325	0.40	0.3711
0.7746	0.60	0.2733
0.8944	0.80	0.2059
1.000	1.00	0.1573
1.140	1.30	0.1069
1.265	1.60	0.0736
1.378	1.90	0.0513
1.483	2.20	0.0359
1.581	2.50	0.0254
1.643	2.70	0.0202
1.732	3.00	0.0143
1.789	3.20	0.0114

or in the gallon-day-foot systems of units as

$$Q_b = (2.15 \times 10^{-3}) s_d \sqrt{\frac{ST}{t}} \tag{3.164}$$

where Q_b is the discharge of the drain, in gpm per foot of drain length.

Equations (3.161) and (3.164) afford a means for evaluating the hydrogeologic properties of aquifers, inasmuch as the ratio S/T is determined from Eq. (3.161) and the product ST is obtained from Eq. (3.164).

3.17 Estimating Groundwater Contribution to Streamflow and Bank Storage due to Flood Stages

Figure 3.16 shows a water-table aquifer bounded on two sides by fully penetrating parallel streams of infinite length. The aquifer is homogeneous and isotropic and is recharged at a constant and uniform rate W_R per unit area. Flow in the aquifer is

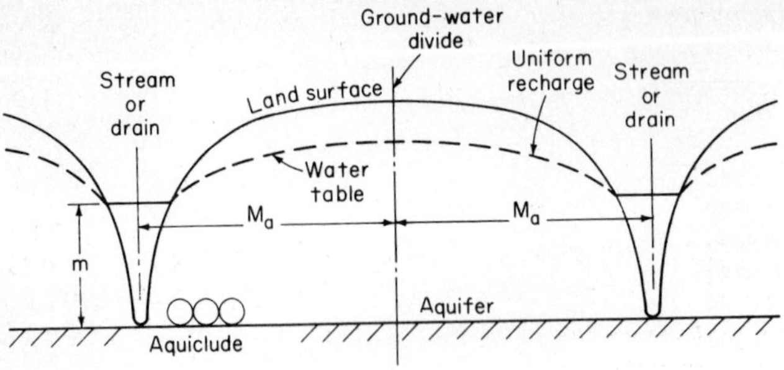

Fig. 3.16 Flow to parallel streams with uniform recharge. (*Drawn by W. C. Walton.*)

one-dimensional and a groundwater divide is created at a distance M_a midway between the streams. It is assumed that the saturated thickness of the aquifer below the streams is great compared to the maximum height of the water table above stream level, so that the saturated thickness of the aquifer may be regarded as constant. Let h_0 be the elevation of the water table with respect to the mean stream level. At any point x the rate of lateral flow per unit normal thickness is (Jacob, 1943)

$$q = -T\frac{dh_0}{dx} = W_R(x - M_a) \tag{3.165}$$

Integrating Eq. (3.165) after multiplying through with dx

$$Th_0 = W_R\left(M_a \frac{x - x^2}{2}\right) + C \tag{3.166}$$

To satisfy the condition $h_0 = 0$ at $x = 0$, the constant integration C must be zero. The equation of the steady-state profile is (Jacob, 1943)

$$h_0 = \frac{W_R}{T} M_a \frac{x - x^2}{2} = \frac{M_a{}^2 W}{2T}\left(\frac{2x}{M_a} - \frac{x^2}{M_a{}^2}\right) \tag{3.167}$$

This is the particular solution of the fundamental differential equation

$$\frac{\partial^2 h_0}{\partial x^2} = -\frac{W_R}{T} \tag{3.168}$$

obtained by differentiating Eq. (3.165), with boundary conditions $h_0 = 0$ at $x = 0$ or at $x = 2M_a$ and $dh_0/dx = 0$ at $x = M_a$.

Equation (3.167) may be written in the gallon-day-foot system of units as (see Ferris et al., 1962)

$$W_R = \frac{T}{1.71 \times 10^{-3}\,[(M_a x/h_0) - (x^2/2h_0)]} \tag{3.169}$$

where W_R = constant rate of recharge to water table, in inches per year (in./yr)
T = coefficient of transmissibility, in gpd/ft
M_a = distance from stream to groundwater divide, in feet
x = distance from stream to observation point, in feet
h_0 = elevation of water table, at observation point, with respect to mean stream level, in feet

Recharge must equal the discharge to the stream, provided changes in storage are insignificant. Q_b, the groundwater contribution to streamflow in gpm per foot of stream channel, can be expressed as (Ferris et al., 1962)

$$Q_b = \frac{W_R M_a}{4.22 \times 10^5} \tag{3.170}$$

Changes in groundwater levels, groundwater contribution to streamflow, and bank storage are caused by flood waves in surface streams. The shapes of many flood-wave–stage hydrographs may be approximated by one of the family of asymmetric curves defined by (Cooper and Rorabaugh, 1963)

$$W_\psi(t) = \begin{cases} N_c h_a e^{-\delta t(1 - \cos wt)} & \text{when} \quad 0 \le t \le \tau_w \\ 0 \quad \text{when} \quad t \ge \tau_w \end{cases} \tag{3.171}$$

where h_a is the maximum rise in stage, t is the time since the beginning of the flood wave, τ_w is the duration of the wave, $\omega = 2\pi/\tau_w$, $\delta = \omega \cot (\omega t_c/2)$ is a constant that determines the degree of asymmetry, t_c is the time of the flood crest, and

$$N_c = \frac{1}{e^{-\delta t_c}(1 - \cos \omega t_c)} = \frac{\delta^2 + \omega^2}{2\omega^2} e^{2\delta/\omega} \arctan \frac{\omega}{\delta} \tag{3.172}$$

is a constant that serves to make the curves of the family peak at the same height h_a. Curves corresponding to $\delta = w$ and $\delta = 0$ are shown in Fig. 3.17. For $\delta = 0$ the curve is sinusoidal.

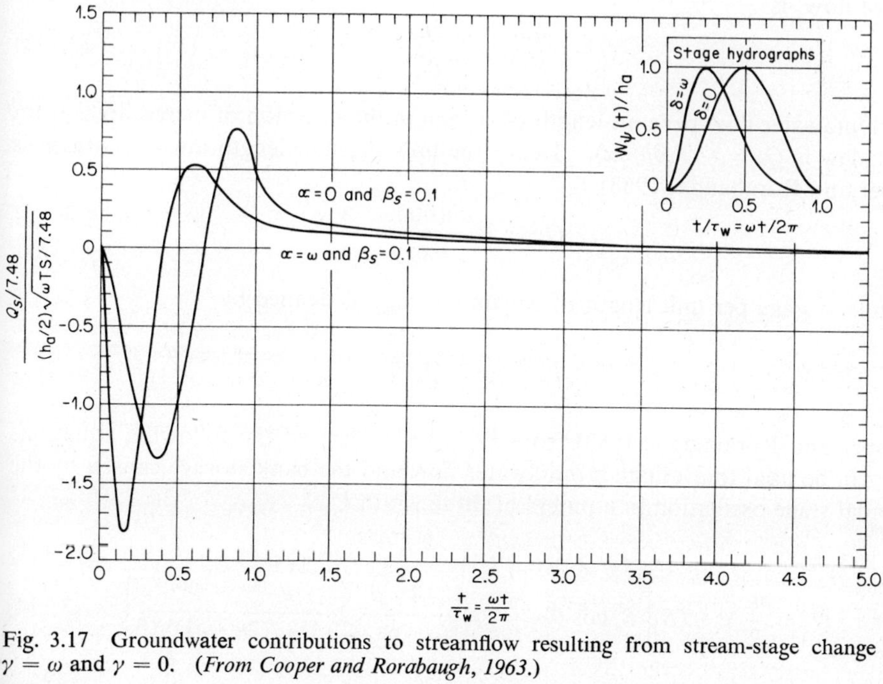

Fig. 3.17 Groundwater contributions to streamflow resulting from stream-stage change when $\gamma = \omega$ and $\gamma = 0$. (*From Cooper and Rorabaugh, 1963.*)

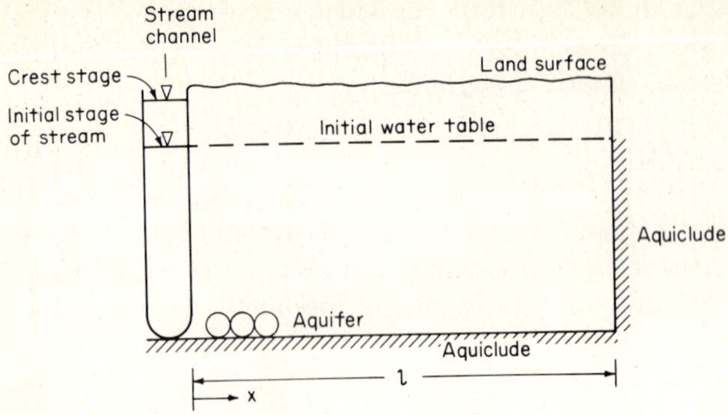

Fig. 3.18 Water-table aquifer bounded by a stream and a barrier boundary. (*Drawn by W. C. Walton.*)

Consider an aquifer of finite width l bounded on one side by a fully penetrating infinite stream with vertical banks and on the other side by an infinite vertical barrier boundary as shown in Fig. 3.18. It is assumed that the slope and fluctuations of the water table are small in comparison to the thickness of the saturated zone, the stream does not overflow its banks, the initial water table is horizontal, and groundwater level changes are due solely to changes in the stage of the stream.

The differential equation governing the linear flow of water in the aquifer, provided the variations in the height of the water table are small in relation to the average depth of flow, is

$$\frac{\partial^2 h}{\partial x^2} = \frac{S}{T}\frac{\partial h}{\partial t} \tag{3.173}$$

The groundwater flow per unit length of stream in the direction of increasing x from Darcy's law is $Q_s = -T(\partial h/\partial x)$. Hence the flow per unit length into the stream is (Cooper and Rorabaugh, 1963)

$$Q_s = T\frac{\partial h(0,t)}{\partial x}$$

The bank storage per unit length of stream at time t is defined by

$$V_s = -\int_0^t Q_s \, dt$$

Cooper and Rorabaugh (1963) from Eq. (3.173) derived the following equations which can be used to evaluate groundwater flow and the bank storage caused by the sinusoidal stage oscillation in aquifers of finite width l:

When $t \leq \tau_w$

$$Q_s = \frac{h_a}{2}\sqrt{\omega TS}\left[B_s \cos(\omega t + \varphi) - \frac{4\sqrt{\beta_s}}{\pi}\sum_{n=1}^{\infty}\frac{e^{-(2n-1)^2\beta_s\omega t}}{1 + (2n-1)^4\beta_s^2}\right] \tag{3.174}$$

and

$$V_s = \frac{h_a}{2}\sqrt{\frac{TS}{\omega}}\left[\frac{\pi}{2\sqrt{\beta_s}} - B_s \sin(\omega t + \varphi) - \frac{4\sqrt{\beta_s}}{\pi}\sum_{n=1}^{\infty}\frac{e^{-(2n-1)^2\beta_s\omega t}}{(2n-1)^2\beta_s + (2n-1)^6\beta_s^3}\right] \tag{3.175}$$

When $t \geq \tau_w$

$$Q_s = \frac{h_a}{2}\sqrt{\omega TS}\,\frac{4\sqrt{\beta_s}}{\pi}\sum_{n=1}^{\infty}\frac{(e^{-(2n-1)^2 2\pi\beta_s} - 1)e^{-(2n-1)^2\beta_s\omega t}}{1 + (2n-1)^4\beta_s^2} \tag{3.176}$$

and

$$V_s = \frac{h_a}{2}\sqrt{\frac{TS}{\omega}}\,\frac{4\sqrt{\beta_s}}{\pi}\sum_{n=1}^{\infty}\frac{(e^{-(2n-1)^2 2\pi\beta_s} - 1)e^{-(2n-1)^2\beta_s\omega t}}{(2n-1)^2\beta_s + (2n-1)^6\beta_s^3} \tag{3.177}$$

where T = coefficient of transmissibility
S = coefficient of storage

$$B_s = \left[\frac{\cosh(\pi/\sqrt{2\beta_s}) - \cos(\pi/\sqrt{2\beta_s})}{\cosh(\pi/\sqrt{2\beta_s}) + \cos(\pi/\sqrt{2\beta_s})}\right]^{\frac{1}{2}}$$

$$\varphi = \arctan\left[\frac{\sinh(\pi/\sqrt{2\beta_s}) + \sin(\pi/\sqrt{2\beta_s})}{\sinh(\pi/\sqrt{2\beta_s}) - \sin(\pi/\sqrt{2\beta_s})}\right]$$

$$\beta_s = \frac{\pi\tau_w T}{8l^2 S}$$

Q_s = groundwater flow per width length into stream
V_s = bank storage per unit length of stream

In the case where the aquifer is semi-infinite, i.e., there is no barrier boundary, and $\delta = 0$, Cooper and Rorabaugh (1963) from Eq. (3.173) derived the following equations which can be used to evaluate groundwater flow and the bank storage caused by the sinusoidal stage oscillation:

When $t \leq \tau_w$

$$Q_s = \frac{h_a\sqrt{\omega TS}}{2}\sqrt{\frac{2}{\pi}}\left\{\frac{1}{\sqrt{2\omega t}}\left[-\frac{(2\omega t)^2}{3} + \frac{(2\omega t)^4}{3\cdot 5\cdot 7} - \frac{(2\omega t)^6}{3\cdot 5\cdot 7\cdot 9\cdot 11} + \cdots\right]\right\} \tag{3.178}$$

and

$$V_s = \frac{h_a}{2}\sqrt{\frac{TS}{\omega}}\sqrt{\frac{2}{\pi}}\sum_{n=1}^{\infty}\frac{(-1)^{n-1}(2\omega t)^{(4n+1)/2}}{\prod_{m=1}^{2n}(2m+1)} \tag{3.179}$$

When $t \geq \tau_w$

$$Q_s = \frac{h_a\sqrt{\omega TS}}{2}\sqrt{\frac{2}{\pi}}\left\{\frac{1}{\sqrt{2\omega t}}\left[-\frac{(2\omega t)^2}{3} + \frac{(2\omega t)^4}{3\cdot 5\cdot 7} - \frac{(2\omega t)^6}{3\cdot 5\cdot 7\cdot 9\cdot 11} + \cdots\right]\right.$$
$$\left. - \frac{1}{\sqrt{2\omega t - 4\pi}}\left[-\frac{(2\omega t - 4\pi)^2}{3} + \frac{(2\omega t - 4\pi)^4}{3\cdot 5\cdot 7} - \frac{(2\omega t - 4\pi)^6}{3\cdot 5\cdot 7\cdot 9\cdot 11} + \cdots\right]\right\} \tag{3.180}$$

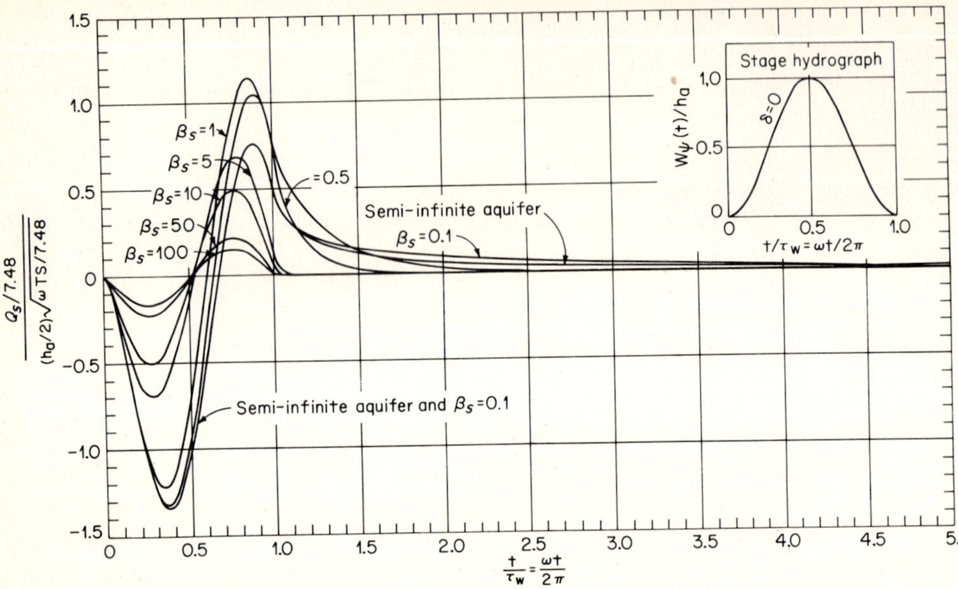

Fig. 3.19 Groundwater contribution to stream flow resulting from stream-stage change $\gamma = 0$. (*From Cooper and Rorabaugh, 1963.*)

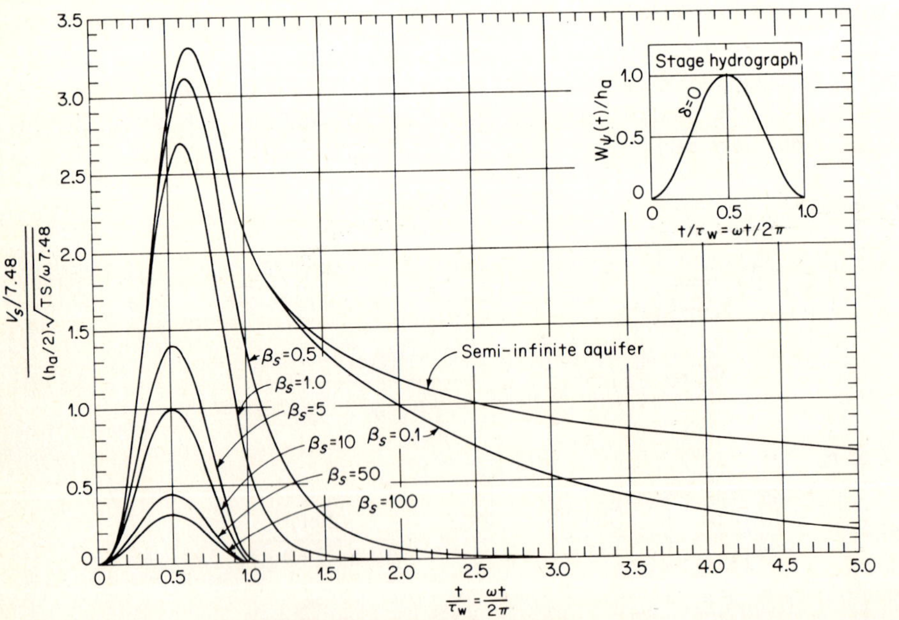

Fig. 3.20 Bank storage resulting from stream-stage change when $\gamma = 0$. (*From Cooper and Rorabaugh, 1963.*)

and
$$V_s = \frac{h_a}{2}\sqrt{\frac{TS}{\omega}}\sqrt{\frac{2}{\pi}}\sum_{n=1}^{\infty}\frac{(-1)^{n-1}[(2\omega t)^{(4n+1)/2} - (2\omega t - 4\pi)^{(4n+1)/2}]}{\prod_{m=1}^{2n}(2m+1)} \tag{3.181}$$

Curves (Cooper and Rorabaugh, 1963) computed from Eqs. (3.173) to (3.177) representing, respectively, the groundwater flow and the bank storage caused by the sinusoidal stage oscillation described by Eq. (3.171) when $\delta = 0$ are given in Figs. 3.19 and 3.20. The curves for $\beta_s > 0$, corresponding to aquifers of finite widths, were computed from Eqs. (3.174) and (3.175) for $t \leq \tau_w$ and from Eqs. (3.176) and (3.177) for $t \geq \tau_w$. Those for the semi-infinite aquifer ($\beta_s = 0$) were computed from Eqs. (3.178) and (3.179) for $t \leq \tau_w$ and from Eqs. (3.180) and (3.181) for $t \geq \tau_w$. From these curves the groundwater flow and bank storage can be computed for any combination of values of the parameters h_a, T, S, w, and l that corresponds to one of the seven values of β_s for which curves are given, including the limiting case of the semi-infinite aquifer for which $\beta_s = 0$. To illustrate the effect of asymmetry in the stage oscillation, curves representing the groundwater flow and bank storage for $\delta = 0$, $\beta_s = 0.1$ are compared with those for $\delta = 0$, $\beta_s = 0.1$ in Figs. 3.18 and 3.21 (Cooper and Rorabaugh, 1963). The asymmetry causes the rate of flow out of the stream to be larger during more rapid rise in stage, but, because it also shortens the duration of this flow, it lessens the maximum bank storage.

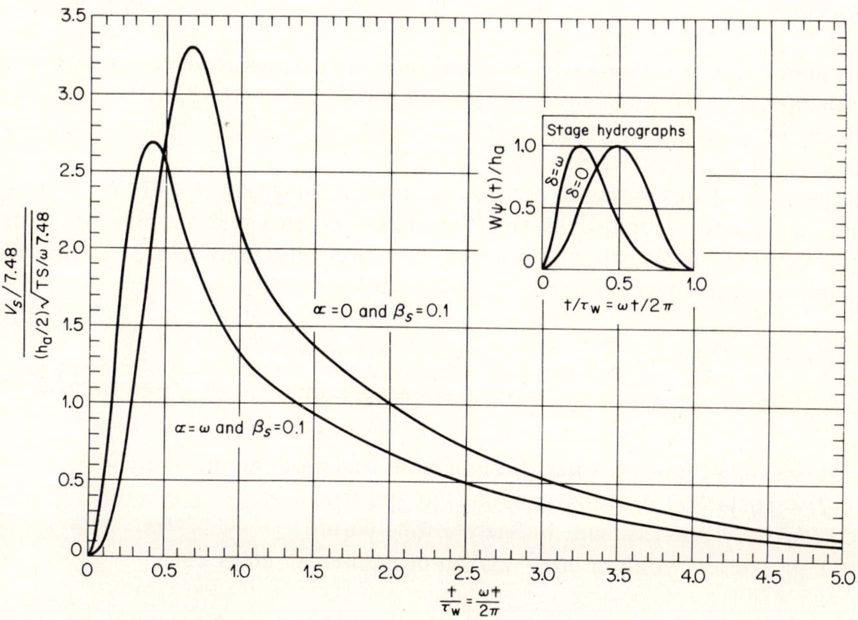

Fig. 3.21 Bank storage resulting from stream-stage change when $\gamma = 0$ and $\gamma = \omega$. *(From Cooper and Rorabaugh, 1963.)*

In Figs. 3.18 through 3.21,

Q_s = groundwater flow per unit length into stream at time t, in gpd/ft
h_a = rise in stream stage at time t, in feet
$\omega = \dfrac{2\pi}{\tau_w}$, in days,
τ_w = duration of flood wave, in days
T = coefficient of transmissibility, in gpd/ft
S = coefficient of storage, fraction
$\beta_s = \dfrac{\pi\tau_w T}{59.8l^2S} = \dfrac{\alpha}{\omega}$
l = width of aquifer, in feet
t = time after beginning of flood wave, in days
V_s = bank storage per unit length of stream, in gallons per foot (gal/ft), at time t
δ = a constant that imparts asymmetry $= \omega \cot \dfrac{\omega t_c}{2}$ where t_c is the time of the flood crest, in days

3.18 Flow Nets

Groundwater movement and storage may be quantitatively appraised by graphical analysis of flow nets. A *flow net* is composed of two families of lines or curves; one family of curves represents *flow lines*, paths followed by particles of water as they move through an aquifer in the direction of decreasing head. Intersecting the flow lines at right angles is a family of curves termed *equipotential lines*, which represent piezometric-surface or water-table contours.

If the quantity of groundwater percolating through a given cross section (*flow channel*) of an aquifer delimited by two flow lines and two piezometric-surface or water-table contours is known, the coefficient of transmissibility can be estimated from the following modified form of the Darcy equation

$$T = \frac{Q_N}{IL} \tag{3.182}$$

where Q_N = rate of flow of water through cross section of aquifer, in gpd
T = coefficient of transmissibility, in gpd/ft
I = hydraulic gradient, in feet per mile (ft/mi)
L = average width of cross section of aquifer, in miles

L is obtained from the piezometric-surface or water-table map with a map measurer. The hydraulic gradient can be calculated by using the following formula (see Foley et al., 1953)

$$I = \frac{c_i}{W_a} \tag{3.183}$$

where I = hydraulic gradient, in ft/mi

c_i = contour interval of piezometric-surface or water-table map, in feet

$W_a = \dfrac{A'}{L}$, where A' is the area between two limiting flow lines and piezometric-surface or water-table contours, in square miles; and L is the average length of the piezometric-surface or water-table contours between the two limiting flow lines, in miles

The coefficient of storage can be estimated by studying the difference in quantities of water flowing across successive water-table or piezometric-surface contours between two limiting flow lines provided there is no leakage from above or below into the aquifer with the following equation

$$S = \frac{Q_2 - Q_1}{\Delta h_t A_L (2.1 \times 10^8)} \tag{3.184}$$

where $Q_2 - Q_1$ = difference in quantity of water crossing successive contour lines between limiting flow lines, in gpd

Δh_t = average rate of water-level decline or rise in area between limiting flow lines and successive contours, in feet per day (fpd)

A_L = area between limited flow lines and successive contours, in square miles

S = coefficient of storage, fraction

If the coefficient of vertical permeability of an aquitard, the thickness of the aquitard, the area of the aquitard through which leakage occurs, and the difference between the head in the aquifer and in the source bed above the aquitard are known, the quantity of leakage through the aquitard can be computed from the following modified form of the Darcy equation

$$Q_c = \frac{P' \, \Delta h A_c}{m'} \tag{3.185}$$

where P' = coefficient of vertical permeability of aquitard, in gpd/sq ft

Q_c = leakage through aquitard, in gpd

m' = thickness of aquitard through which leakage occurs, in feet

A_c = area of aquitard through which leakage occurs, in square feet

Δh = difference between the head in the aquifer and in the source bed above the aquitard, in feet

If the quantity of leakage through an aquitard into an aquifer, the thickness of the aquitard, the area of the aquitard through which leakage occurs, and the difference between the head in the aquifer and that in the source bed above the aquitard are known, the coefficient of vertical permeability can be computed from the following modified form of the Darcy equation

$$P' = \frac{Q_c m'}{\Delta h A_c} \tag{3.186}$$

The velocity of the flow of groundwater from the equation of continuity is

$$V_g = \frac{Q_N}{7.48 S_y A} \tag{3.187}$$

where V_g = velocity of the flow of groundwater, in fpd
Q_N = rate of flow of water through cross section of aquifer, in gpd
A = cross-sectional area of the aquifer, in square feet
S_y = specific yield of aquifer, fraction

The specific yield converts the total cross-sectional area of the aquifer to the effective area of the pore openings through which flow actually occurs.

Q_N can be calculated from Darcy's equation as the product of the coefficient of permeability, hydraulic gradient, and cross-sectional area of aquifer, or

$$Q_N = PIA \tag{3.188}$$

Substitution of Eq. (3.187) in Eq. (3.188) results in the expression (see Butler, 1957)

$$V_g = \frac{PI}{7.48 S_y} \tag{3.189}$$

where P = coefficient of permeability, in gpd/sq ft

Recharge by the vertical leakage of water through an aquitard can be estimated by studying the difference in the quantity of water crossing successive water-table or piezometric-surface contours between two limiting flow lines with the following equation

$$Q_c = \frac{(Q_2 - Q_1) \pm \Delta h_t S A_L (2.1 \times 10^8)}{A_L} \tag{3.190}$$

where Q_c = rate of vertical leakage, in gpd/sq mi
$Q_2 - Q_1$ = difference in quantity of water crossing successive contours, in gpd
Δh_t = average rate of water-level decline or rise in area between limiting flow lines and successive contours, in fpd
A_L = area between limiting flow lines and successive contours, in square miles
S = coefficient of storage, fraction

The + sign in Eq. (3.190) is used when there is a water-level rise and the − sign is used when there is a water-level decline.

Ferris et al. (1962) present a graphical solution of flow nets to analyze problems of groundwater flow. A flow net is constructed using only a few flow lines and equipotential lines such that the total quantity of flow lines and the total drop in head across the system are evenly divided between adjacent pairs of equipotential lines.

Since the direction of water movement is everywhere synonymous with paths that are normal to the equipotential lines, the system of flow lines must be drawn orthogonal to the system of equipotential lines.

The flow net is a pattern of rectangles in which the ratio of the mean dimensions of each rectangle is constant. In flow fields involving curved paths of flow, the elemental geometric forms in the net are curvilinear and are not true rectangles; however, the corners of each rectangle are right angles and the mean distances between the two pairs of opposite sides are equal.

The discharge through any path of the flow net may be obtained from Eq. (3.188). Considering a unit width or thickness of the aquifer, measured normal to the direction of flow, Eq. (3.188) may be rewritten as (Ferris et al., 1962)

$$\Delta q = PIb_s \tag{3.191}$$

where Δq is the flow occurring between a pair of adjacent flow lines, and b_s is the spacing of flow lines. If L_s represents the spacing and Δh the drop in head between the equipotential lines, Eq. (3.191) becomes

$$\Delta q = P\,\Delta h\,\frac{b}{L_s} \tag{3.192}$$

If the flow net was constructed so that the sides of each rectangle are equal, the ratio b/L_d is equal to unity and the same incremental flow Δq occurs between each pair of adjacent flow lines. If there are n_f flow channels, the total flow q through a unit thickness of the aquifer is (Ferris et al., 1962)

$$q = n_f\,\Delta q \tag{3.193}$$

If there are n_d potential drops, the total drop in head, h_T, is given by (Ferris et al., 1962)

$$h = n_d\,\Delta h \tag{3.194}$$

Substitution of Eqs. (3.193) and (3.194) in Eq. (3.191) results in the equation

$$q = \frac{n_f}{n_d}\,Ph \tag{3.195}$$

The total flow through the full thickness of the aquifer becomes (Ferris et al., 1962)

$$Q_N = \frac{n_f}{n_d}\,Ph_Tm \tag{3.196}$$

where Q_N = flow through full thickness of aquifer, in gpd
n_f = number of flow channels
n_d = number of potential drops
P = coefficient of permeability, in gpd/sq ft
m = saturated thickness of aquifer, in feet
h_T = total potential drop, in feet

The preceding discussion of flow nets concerns steady-state two-dimensional flow fields in homogeneous and isotropic aquifers.

3.19 Numerical Analysis of Water Levels

The equations presented in the preceding sections of this chapter for describing groundwater movement and storage involve application of the rules of calculus for integrating an infinite set of infinitesimal changes in water level between two or more points in an aquifer. In numerical analysis, solutions to differential equations are found by substituting finite entities for the differential forms that appear in fundamental differential equations (Stallman, 1956). Approximations are introduced in that an area is considered to be composed of small but finite parts $(\Delta x)(\Delta y)$, instead of an infinite set of infinitesimal areas $(dx)(dy)$. A sensibly small and finite set of component parts are added instead of the infinite set of infinitesimal areas postulated by the orthodox calculus.

The differential equation describing two-dimensional flow in a homogeneous and isotropic aquifer infinite in areal extent subject to recharge at a steady rate of accretion W_R is (Stallman, 1956)

$$\frac{\partial^2 h}{\partial x^2} + \frac{\partial^2 h}{\partial y^2} = \frac{S}{T}\frac{\partial h}{\partial t} - \frac{W_R}{T} \tag{3.197}$$

The equation states that the net change in the rate at which water moves past the point x, y equals the rate of recharge at that point.

The differential lengths dx and dy are expanded to an equivalent finite length a, and dt is considered equivalent to Δt. Thus, a point is visualized as a small area of length and width such that it is insignificant in size compared with the size of the aquifer. The rectangular grid in Fig. 3.22 is superposed over the aquifer. The grid lines are equally spaced in both directions. The first two differentials in Eq. (3.197)

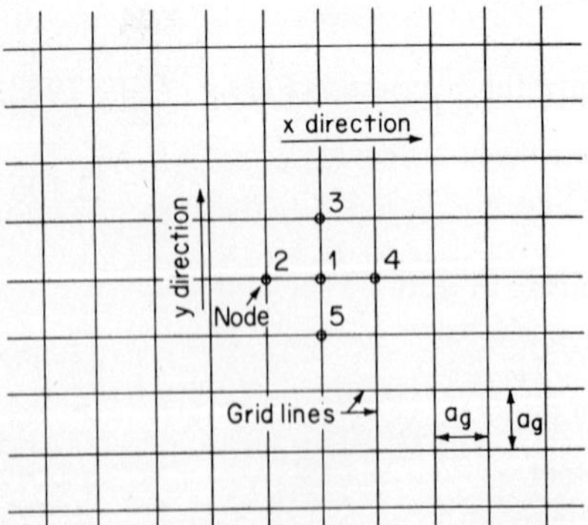

Fig. 3.22 Finite grid nomenclature used in numerical analysis. (*Drawn by W. C. Walton.*)

may be expressed in terms of the head values at the nodes as (Southwell, 1946)

$$\frac{\partial^2 h}{\partial x^2} \approx \frac{h_2 + h_4 - 2h_1}{a_g^{\,2}} \tag{3.198}$$

$$\frac{\partial^2 h}{\partial y^2} \approx \frac{h_3 + h_5 - 2h_1}{a_g^{\,2}} \tag{3.199}$$

$\partial h/\partial t$ may be considered equivalent to $\Delta h_{1t}/\Delta t$ (Stallman, 1956) where Δh_{1t} is the change in head which occurs at node through the time interval Δt. The finite-difference form of Eq. (3.197) is then

$$h_2 + h_3 + h_4 + h_5 - 4h_1 = \left(\frac{S}{T}\frac{\Delta h_{1t}}{\Delta t} - \frac{W_R}{T}\right) a_g^{\,2} \tag{3.200}$$

Equation (3.200) may be used to compute the ratios S/T and W_R/T or the head distribution in time and space, in an aquifer of given size and shape.

The head distribution can be found either by relaxation or iteration methods of numerical analysis. The aim of the computing methods is to find the particular head values at all nodal points such that the finite-difference equation is satisfied at all nodes simultaneously.

The differential equation describing two-dimensional flow in a nonhomogeneous aquifer infinite in areal extent subject to recharge at a steady rate of accretion W_R is (Stallman, 1956)

$$T\left(\frac{\partial^2 h}{\partial x^2} + \frac{\partial^2 h}{\partial y^2}\right) + \frac{\partial T}{\partial x}\frac{\partial h}{\partial x} + \frac{\partial T}{\partial y}\frac{\partial h}{\partial y} = S\left(\frac{\partial h}{\partial t}\right) - W_R \tag{3.201}$$

It is assumed that all parameters vary in space and that W_R varies both in space and in time.

Regional trends of variations of T in space can be estimated by assuming that T varies as a semiarbitrary function of X and Y which can be expressed as (Stallman, 1956)

$$\frac{T_{(x,y)}}{T_{(0,0)}} = 1 + f(X, Y) \tag{3.202}$$

Specified unknowns in Eq. (3.202) are solved using a group of finite-difference equations constructed from water-level data and Eq. (3.200). Storage and accretion factors must be taken into account. If these factors must be evaluated by the configuration of the piezometric surface, two methods are available for eliminating them from the computations of transmissibility variations. Under certain conditions

$$S\frac{\partial h}{\partial t} - W_R = 0 \tag{3.203}$$

For any case (Stallman, 1956)

$$S\frac{\partial h}{\partial t} - W_R = f(h, T, X, Y) \tag{3.204}$$

Use of Eqs. (3.201) to (3.203) for analyzing the water-level data permits the evaluation of the manner of variation of T in space. Use of Eqs. (3.201), (3.202), and (3.204) permits individual evaluation of all hydrologic parameters as a function of space, if the values of $(\partial h/\partial t)_{x,y}$ are known.

Data for well arrays spanning such hydraulic controls as recharging or discharging wells or surface-water bodies hydraulically connected with the aquifer cannot be used in numerical analysis.

3.20 Salt-water Encroachment

Under natural conditions, fresh groundwater is discharged to the sea or ocean from artesian and water-table coastal aquifers. Heavy pumping in coastal areas reduces or reverses the natural hydraulic gradient sloping toward the ocean. At places sea water enters and penetrates for considerable distances inland in aquifers. This phenomenon is called *salt-water encroachment*. Production wells may become contaminated with salt water, resulting in abandonment of well fields.

Assuming fresh and salt water to be immiscible fluids of different densities, a sharp interface is formed with a shape governed by a hydrodynamic balance wherever the fluids are in contact. A coastal cross section of a water-table aquifer is shown in Fig. 3.23. According to Ghyben (1888–1889) and Herzberg (1901), the cross sections of coastal line, interface, ocean level, and water table meet at a point. In Fig. 3.23, the total hydrostatic pressure at A is (Todd, 1959)

$$p_A = \rho_s g h_s \tag{3.205}$$

where ρ_s is the salt-water density, g is the acceleration of gravity, and h_s is the difference between the elevation of the surface of the ocean and the elevation of the interface. At B (Todd, 1959)

$$p_B = \rho_f g h_f + \rho_f g h_s \tag{3.206}$$

where ρ_f is the fresh-water density and h_f is the difference between the elevation of the surface of the ocean and the elevation of the water table directly above B.

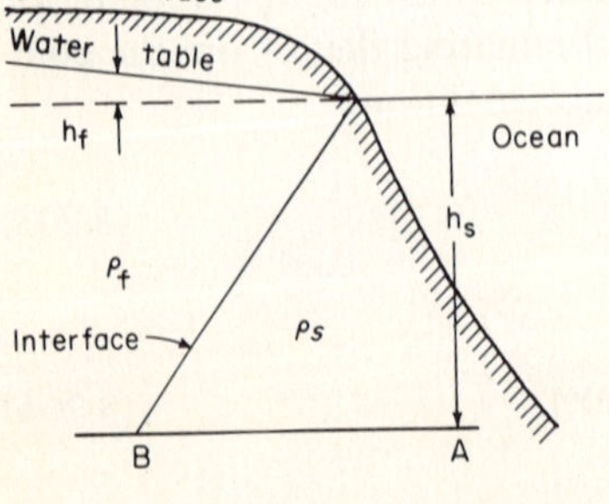

Fig. 3.23 Fresh-water and salt-water distribution in a water-table coastal aquifer according to the Ghyben-Herzberg relation. (*Drawn by W. C. Walton.*)

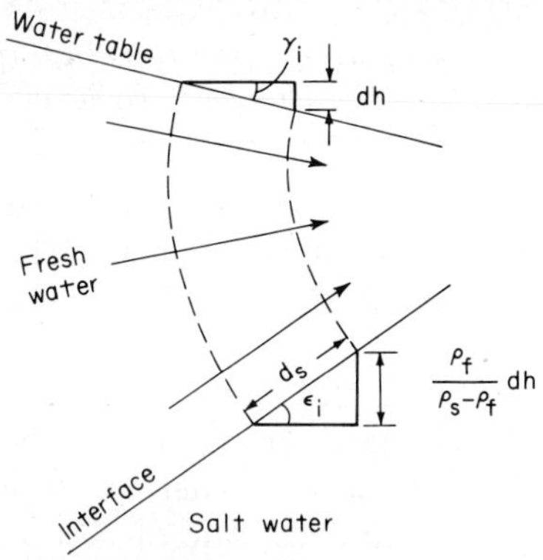

Fig. 3.24 Relation between slopes of the water-table and the fresh-water–salt-water interface. (*Drawn by W. C. Walton.*)

Equating Eqs. (3.205) and (3.206) results in the Ghyben-Herzberg relation

$$h_s = \frac{\rho_f}{\rho_s - \rho_f} h_f \tag{3.207}$$

Taking $\rho_s = 1.025$ g/cm³ and $\rho_f = 1.000$ g/cm³

$$h_s = 40h_f \tag{3.208}$$

In the Ghyben-Herzberg relation, the interface slopes downward from the coast. From Darcy's law

$$\sin \gamma_i = \frac{dh}{ds} = \frac{V_g}{P} \tag{3.209}$$

where γ_i is the slope of the water table, V_g is the velocity of the flow of groundwater, and P is the permeability. The slope of the interface, ϵ_i, as shown in Fig. 3.24, is given by (Todd, 1959)

$$\sin \epsilon_i = \frac{\rho_f}{\rho_s - \rho_f} \frac{V_g}{P} \tag{3.210}$$

Equations for the interface under several sets of boundary conditions were derived by Henry (1959).

The Ghyben-Herzberg relation describes a salt-water wedge (see Fig. 3.25). From Darcy's law, the seaward fresh-water flow for a confined aquifer is given by

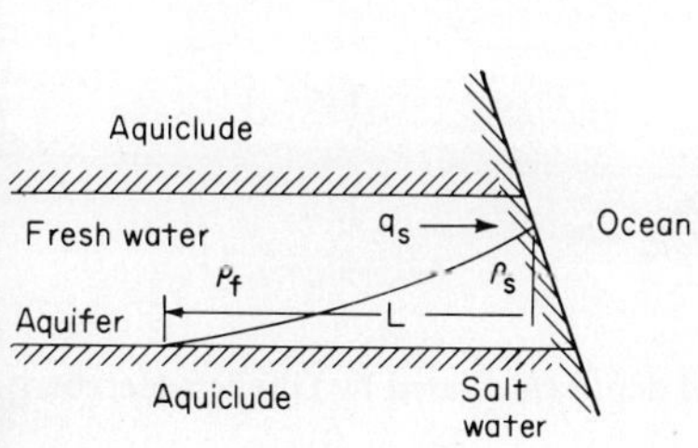

Fig. 3.25 Salt-water wedge in a nonleaky artesian aquifer. (*Drawn by W. C. Walton.*)

(Todd, 1959)

$$q_s = \frac{1}{2}\frac{\rho_s - \rho_f}{\rho_f}\frac{Pm^2}{L} \tag{3.211}$$

where q_s is the seaward fresh-water flow per foot of ocean front, m is the saturated thickness of the aquifer, and L is the width of the cross section of the aquifer. The length of the intruded wedge is inversely proportional to the fresh-water flow. The interface represents a flow line.

The limitations of the Ghyben-Herzberg relation were pointed out by DeWiest (1965). If static conditions existed, the water table would have zero slope and the interface would become horizontal with fresh water overlying salt water by mere density difference. In addition, the fresh water is in a continuous state of motion and there is a dynamic rather than a hydrostatic equilibrium of the fresh-water–salt-water interface (see Hubbert, 1940, and Wentworth, 1942). Hubbert (1940) pointed out the discrepancy between the actual depth of salt water and the depth based on the Ghyben-Herzberg equation.

Referring to Fig. 3.26, the slope of the interface is given by (DeWiest, 1965)

$$\sin \epsilon_i = \frac{\partial z}{\partial s} = -\left[\frac{1}{P_f}\left(\frac{\rho_f}{\rho_f - \rho_s}\right)V_{f,s} - \frac{1}{P_s}\left(\frac{\rho_s}{\rho_f - \rho_s}\right)V_{s,s}\right] \tag{3.212}$$

where the subscripts f and s indicate fresh water and salt water respectively and $V_{f,s}$ and $V_{s,s}$ are the components of the specific discharge respectively of the fresh water and the salt water along the trace of the interface.

If the salt water were stagnant and the fresh water were flowing over it, $V_{s,s} = 0$ and $V_{f,s} \neq 0$. In this case, because $\rho_f < \rho_s$, $\sin \epsilon_i > 0$ and ϵ_i will increase with $V_{f,s}$.

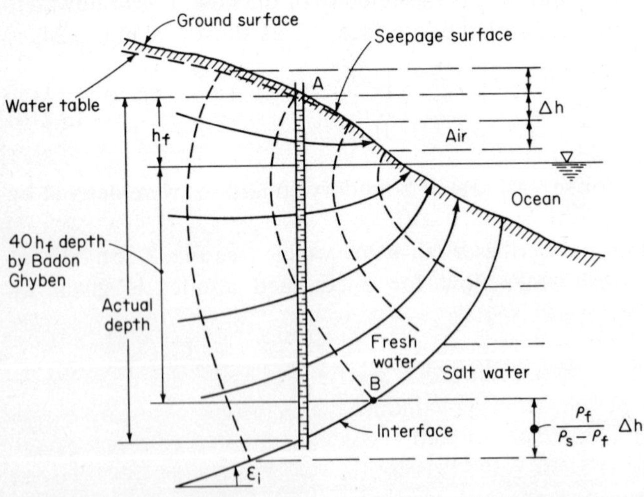

Fig. 3.26 Discrepancy between actual depth to salt water and depth calculated by Ghyben-Herzberg relation. (*After Hubbert, 1940.*)

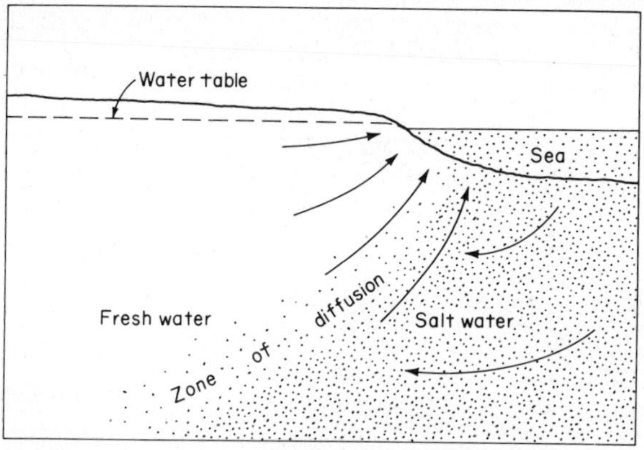

Fig. 3.27 Circulation of salt water from sea to zone of diffusion and back to sea. (*From Cooper, 1959.*)

Equation (3.207) is valid between the points of intersection of any equipotential line of the flowing fresh water as indicated by points A and B in Fig. 3.26. Because the equipotential line is curved, the vertical distance between the water table and the interface is greater than that given by Eq. (3.207).

The width of the gap through which fresh water escapes to the ocean is given approximately by (DeWiest, 1965)

$$X_w = -\frac{Q_N}{2P(\gamma_s - \gamma_f)/\gamma_f} \tag{3.213}$$

where X_W is the width of gap through which fresh water escapes, Q_N is the flow of fresh water, P is permeability, γ_s is the unit weight of salt water, and γ_f is the unit weight of fresh water.

Cooper (1959) advanced the hypothesis that under dynamic conditions salt water is not static but flows perpetually in a cycle from the floor of the sea into a zone of diffusion (a zone of substantial thickness in which there is a gradation of salt content from that of fresh water to that of salt water) and back to the sea; this flow tends to lessen the extent to which the salt water occupies the aquifer. There is a continuous discharge of salty water from the zone of diffusion into the sea as shown in Fig. 3.27. Figure 3.28 shows the zone of diffusion in the Cutler area, near Miami, Florida. Salts are transported from the floor of the sea through the aquifer into the zone of diffusion and hence back to the sea. Salts are transported largely by a hydraulic flow of the salt water with a consequent loss of head in the salt-water environment. Sea water and fresh water become intimately mixed in the zone of diffusion; diluted sea water, being less dense than fresh water, rises along a seaward path. Salts that are introduced into the fresh-water environment are carried back to the sea by the flow of the fresh-water system. Circulation is induced by the transfer of salts out of the salt-water environment. The zone of diffusion is caused by the reciprocative motion of the salt-water front that occurs as the result of ocean tides and of the rise and fall of the water table due to variations in recharge and other forces, including pumping (Cooper, 1959). Both convection and molecular diffusion are important

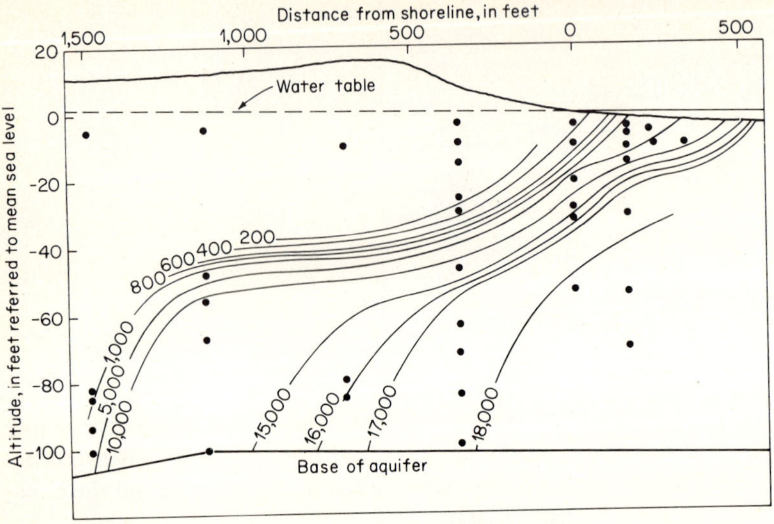

Fig. 3.28 Section through Cutler area, near Miami, Florida, showing zone of diffusion, 1958. Lines represent chloride content in parts per million. (*From Cooper, 1959.*)

parts of the dispersion process (the process by which two miscible liquids interfuse about their boundary when the boundary is caused to move by hydraulic flow), convection in producing large transfers and molecular diffusion in completing the blending.

Assuming that the coefficient of dispersion in the direction of flow (longitudinal dispersion) is proportional to the first power of the mean interstitial velocity (Cooper, 1959)

$$D_D = \frac{4M_d A_m}{t_t} \tag{3.214}$$

where D_D is the coefficient of dispersion, M_d is a constant whose magnitude depends on the properties of the aquifer, A_m is the amplitude, and t_t is the period of the displacement of the water in the aquifer caused by ocean tides.

The amplitude of the tide-produced displacement of water is related approximately to the amplitude of the tide and the distance from the shoreline (Cooper, 1959). The tide-produced change in the artesian head with reference to its mean in a semi-infinite artesian aquifer is (Ferris, 1951)

$$h = h_a \exp\left(-x\sqrt{\frac{\pi S}{t_t T}}\right) \sin\left(\frac{2\pi t}{t_t} - x\sqrt{\frac{\pi S}{t_t T}}\right) \tag{3.215}$$

where h_a is the amplitude of the tide, x is the distance from the shoreline, t is the time referred to the beginning of a tidal cycle, t_t is the period of the tide cycle, S is the coefficient of storage, and T is the coefficient of transmissibility.

The gradient producing the displacement will be (Cooper, 1959)

$$\frac{\partial h}{\partial x} = -h_a \sqrt{\frac{\pi S}{t_t T}} \exp\left(-x\sqrt{\frac{\pi S}{t_t T}}\right)\left[\sin\left(\frac{2\pi t}{t_t} - x\sqrt{\frac{\pi S}{t_t T}}\right) + \cos\left(\frac{2\pi t}{t_t} - x\sqrt{\frac{\pi S}{t_t T}}\right)\right] \tag{3.216}$$

The displacement is zero at a time t when $\partial h/\partial x$ is a maximum, or when the quantity in parentheses in Eq. (3.216) is $\pi/4 + n\pi$. It is extreme at a time t_z when $\partial h/\partial x = 0$, or when the quantity in parentheses in Eq. (3.216) is $3\pi/4 + n\pi$. Thus, with the substitutions (Cooper, 1959)

$$u = \frac{2\pi t}{t_t} - x\sqrt{\frac{\pi S}{t_t T}}$$

$$du = \frac{2\pi\, dt}{t_t}$$

the amplitude of the displacement is (Cooper, 1959)

$$A_m = \frac{P}{\theta}\int_{t_1}^{t_2} \frac{\partial h}{\partial x}\, dt$$

$$A_m = \frac{Ph_a}{\theta}\sqrt{\frac{t_t S}{4\pi T}} \exp\left(-x\sqrt{\frac{\pi S}{t_t T}}\right)\int_{\pi/4}^{3\pi/4} (\sin u + \cos u)\, du$$

$$A_m = \frac{Ph_a}{\theta}\sqrt{\frac{t_t S}{2\pi T}} \exp\left(-x\sqrt{\frac{\pi S}{t_t T}}\right) \tag{3.217}$$

where P is the permeability and θ is the effective porosity of the aquifer.

With Eqs. (3.214) and (3.217) the coefficients of tide-produced dispersion at given distances from the shoreline in a typical coastal aquifer can be estimated. Equation (3.217) was derived for artesian aquifers but probably will give a fair approximation of the amplitude of the displacement in a water-table aquifer if an appropriate value of its coefficient of storage is used. The coefficient of dispersion in an aquifer of non-uniform permeability may be considerably greater than that for a homogeneous aquifer (Cooper, 1959).

Groundwater of uniform density moves in the direction of decreasing fluid potential. Where the density varies from point to point, as along the coast, measurements of head do not indicate the direction of movement directly (Kohout, 1960). For wells that contain salty water, the equivalent head of fresh water must be computed. Conversion of environmental salt-water head to fresh-water head in a given well is accomplished by application of the hydrostatic-pressure equation (Kohout, 1960)

$$p = \rho g l_e \tag{3.218}$$

where p is the pressure at casing terminus, ρ is the density of water in the casing, g is the acceleration due to gravity, and l_c is the measured length of water column above the casing terminus. Equating the right term of the above equation for fresh-water and salt-water columns (Kohout, 1960)

$$\rho_f g l_f = \rho_s g l_s$$

$$l_f = \frac{\rho_s l_s}{\rho_f} \tag{3.219}$$

where the subscripts f and s refer to fresh water and salt water, respectively. The density of fresh water may be assumed to be 1.000.

Lusczynski (1961) introduced the concepts of point-water head, fresh-water head, and environmental-water head to compute the velocity distribution in the zone of dispersion and to determine the front of the dispersed water in the fresh water.

Problems

3.1 The barometric efficiency of a well in an artesian aquifer infinite in areal extent is 60 percent. If the thickness of the aquifer is 150 feet and the porosity of the aquifer is 0.31, estimate the coefficient of storage of the aquifer.
Answer: 1.1×10^{-4}.

3.2 The tidal efficiency of a well in an artesian aquifer infinite in areal extent overlain by an extensive body of tidewater is 42 percent. If the thickness of the aquifer is 200 feet and the porosity of the aquifer is 0.35, estimate α, the reciprocal of the modulus of elasticity of the aquifer skeleton.
Answer: 8.4×10^{-7} in.^{2}/lb.

3.3 The coefficient of storage of an artesian aquifer infinite in areal extent is 2.8×10^{-4}. If the thickness of the aquifer is 200 feet and the porosity of the aquifer is 0.35, estimate the fraction of the coefficient of storage attributable to expansibility of water and that attributable to compressibility of the aquifer skeleton.
Answer: 36 percent expansibility; 64 percent compressibility.

3.4 The coefficients of transmissibility and storage of a nonleaky artesian aquifer infinite in areal extent are 1.15×10^4 gpd/ft and 1.6×10^{-4}, respectively. A fully penetrating production well has been discharging water at a constant rate of 60 gpm for a period of 30 minutes. Compute the drawdowns at distances of 50, 175, and 500 feet from the production well and estimate the virtual radius of the cone of depression.
Answer: 3.1 ft; 1.6 ft; 0.5 ft; about 2,000 ft.

3.5 The coefficients of transmissibility and storage of a nonleaky artesian aquifer infinite in areal extent are 1.7×10^4 gpd/ft and 4×10^{-4}, respectively. A fully penetrating production well is discharging water at a constant rate of 300 gpm. Compute the drawdowns at distances of 10, 100, and 1,000 feet from the production well for pumping periods of 60 minutes and 1 day.
Answer: 60 min—17.4 ft, 8.1 ft, and 0.4 ft; 1 day—23.8 ft, 14.4 ft, and 5.2 ft.

3.6 The coefficients of transmissibility and storage of a nonleaky artesian aquifer infinite in areal extent are 1×10^5 gpd/ft and 3×10^{-4}, respectively. A fully penetrating production well has been discharging water at a constant rate of 1,000 gpm for a period of 1 year. Compute the drawdowns at a distance of 10,000 feet from the production well for pumping periods of 10 days, 50 days, and 1 year and estimate the virtual radius of the cone of depression for pumping period of 1 year.
Answer: 2.7 ft; 4.4 ft; 6.8 ft; about 630,000 ft.

3.7 The coefficients of transmissibility and storage of a nonleaky artesian aquifer infinite in areal extent are 110 gpd/ft and 8×10^{-3}, respectively. A 6-inch-diameter fully penetrating flowing production well was capped for several hours before testing. The water level in the well stabilized at a stage of 100 feet above the top of the casing. The well was uncapped and allowed to flow for a period of 1 day. Compute the discharge at the end of the test.
Answer: 8.6 gpm.

3.8 The coefficients of transmissibility and storage of a nonleaky artesian aquifer infinite in areal extent and 50 feet thick are 5×10^{4} gpd/ft and 4×10^{-4}, respectively. A production well penetrating 50 percent of the aquifer has been discharging water at a constant rate of 300 gpm for a period of 12 hours. Compute the drawdown in an observation well 25 feet from the production well.
Answer: 7.4 ft.

3.9 Using data in Problem 3.8, compute the drawdown due to the effects of partial penetration.
Answer: 0.2 ft.

3.10 The coefficients of transmissibility and storage of a leaky artesian aquifer infinite in areal extent and 50 feet thick are 5.98×10^{4} gpd/ft and 3.4×10^{-4}, respectively. The coefficient of vertical permeability of an overlying aquitard 20 feet thick is 0.10 gpd/sq ft. A fully penetrating production well has been discharging at a constant rate of 1,043 gpm for a period of 1,000 minutes. Compute the drawdowns at distances of 100, 500, and 2,000 feet from the production well and estimate the coefficient of permeability of the aquifer.
Answer: 14.4 ft; 8.1 ft; 3.0 ft; 1,200 gpd/sq ft.

3.11 The coefficients of transmissibility and storage of a leaky artesian aquifer infinite in areal extent and 67 feet thick are 1.9×10^{5} gpd/ft and 1.2×10^{-3}, respectively. The coefficient of vertical permeability of an overlying aquitard 30 feet thick is 1.17 gpd/sq ft. A fully penetrating production well has been discharging at a constant rate of 1,130 gpm. Compute the drawdowns at a distance of 1,770 feet from the production well for pumping periods of 40, 100, and 1,000 minutes. Estimate the leakage coefficient of the aquitard and the coefficient of permeability of the aquifer.
Answer: 0.1 ft, 0.3 ft, and 0.7 ft; 3.9×10^{-2} gpd/cu ft and 2.8×10^{3} gpd/sq ft.

3.12 The coefficients of transmissibility and permeability of a water-table aquifer infinite in areal extent and 274 feet thick are 2.8×10^{5} gpd/ft and 1.02×10^{3} gpd/sq ft, respectively. $S = 7.3 \times 10^{-3}$, $S_y = 1.44 \times 10^{-1}$, $N = 20.8$, and the delay index = 82.6 minutes. A fully penetrating production well has been discharging water at a constant rate of 1,000 gpm. Compute the drawdowns at a distance of 50 feet from the production well for pumping periods of 10, 100, and 1,000 minutes and estimate the time after pumping started at which the effects of delayed gravity drainage are negligible.
Answer: 0.9 ft; 1.3 ft; 2.2 ft; 265 min.

3.13 The coefficients of transmissibility and storage of a water-table aquifer infinite in areal extent and 42 feet thick are 3.24×10^{5} and 3.6×10^{-2}, respectively. A fully penetrating production well has been discharging at a constant rate of 1,100 gpm for a period of 1,147 minutes. Compute the drawdowns at distances of 22 and 1,125 feet from the production well.
Answer: 3.3 ft; 0.3 ft.

3.14 The coefficients of transmissibility and storage of a nonleaky artesian aquifer are 3.85×10^{5} gpd/ft and 3.4×10^{-4}, respectively. The aquifer is bounded on one side by a barrier boundary. A fully penetrating production well has been discharging at a constant rate of 1,485 gpm. The drawdown in an observation well 300 feet from the pumped well due to the effects of the barrier boundary for a pumping period of 100 minutes is 0.61 foot. Compute the distance from the observation well to the discharging image well associated with the barrier boundary.
Answer: 2,630 ft.

3.15 A production well has been discharging water at a constant rate for 10 years. The well is in a nonleaky artesian aquifer bounded on one side by a source of recharge. The coefficients of transmissibility and storage of the aquifer are 1.7×10^4 gpd/ft and 4.0×10^{-4}, respectively. The distance from the production well to the recharge boundary is 25 miles. Compute the percentage of the discharge being diverted from the source of recharge.
Answer: 56 percent.

3.16 A production well has been discharging water at a constant rate from a water-table aquifer bounded on one side by a source of recharge. The coefficients of transmissibility and storage of the aquifer are 1.0×10^5 gpd/ft and 0.2, respectively. The radius of the well is 1 foot. The recharge boundary is 1 mile from the well. Compute the time that must elapse before water levels in the well stabilize.
Answer: 1.3 yr.

3.17 An aquifer having a coefficient of transmissibility of 2×10^4 gpd/ft is bounded on one side by a recharge boundary of finite length of 8 units. A production well has been discharging water at a constant rate of 1,000 gpm. The coordinates of the production well and an observation well are $X_0 = -3$, $Y_0 = -1$, and $X = -3.25$, $Y = -0.5$, respectively. Compute the drawdown in the observation well under equilibrium conditions.
Answer: 13 ft.

3.18 An aquifer is delimited by two converging barrier boundaries. The aquifer wedge angle is 20°. Compute the number of image wells associated with the multiple-boundary system.
Answer: 17.

3.19 The coefficients of transmissibility and storage of a water-table aquifer hydraulically connected to a stream are 2.6×10^5 gpd/ft and 0.20, respectively. The ratios of recorded ranges of groundwater stages in observation wells located 42 and 252 feet from the stream as compared to the stage ranges of the stream are 0.72 and 0.30 foot, respectively. The period of the stream fluctuation is 1 day. Compute the river efficiency of a well 106 feet from the stream and the effective distance offshore to the suboutcrop.
Answer: 54 percent; 40 ft.

3.20 The coefficients of transmissibility and storage of a nonleaky artesian aquifer fully penetrated by a drain are 5×10^4 gpd/ft and 4×10^{-4}, respectively. The drain has been discharging at a constant rate of 10 gpm per lineal foot of drain. Compute the drawdown in an observation well 500 feet from the drain at a time of 1 day after the drain started to discharge.
Answer: 1.2 ft.

3.21 A water-table aquifer is bounded on one side by a barrier boundary and on the other side by a fully penetrating stream. The width of the aquifer is 1 mile and the coefficients of transmissibility and storage of the aquifer are 1.0×10^5 gpd/ft and 0.20, respectively. During a period of 2 days the stream stage rises 5 feet and returns to its original stage. $\delta = 0$. Assuming that the initial water table was horizontal, compute the groundwater contribution to streamflow 2 days after the beginning of the flood wave.
Answer: 825 gpd/ft.

3.22 With the aquifer-stream situation described in Problem 3.21, compute the bank storage resulting from the stage change.
Answer: 1110 gal/ft.

3.23 The piezometric-surface map of a nonleaky artesian aquifer with an average coefficient of transmissibility of 5×10^4 gpd/ft in the vicinity of pumping center A is shown in the following illustration. Compute the quantity of water moving through the aquifer into pumping center A.
Answer: 7.5 million gallons per day (mgd).
If the total withdrawal from pumping center B is 8.5 mgd, compute the average coefficient of transmissibility of the aquifer in the vicinity of pumping center B.
Answer: 3.5×10^4 gpd/ft.

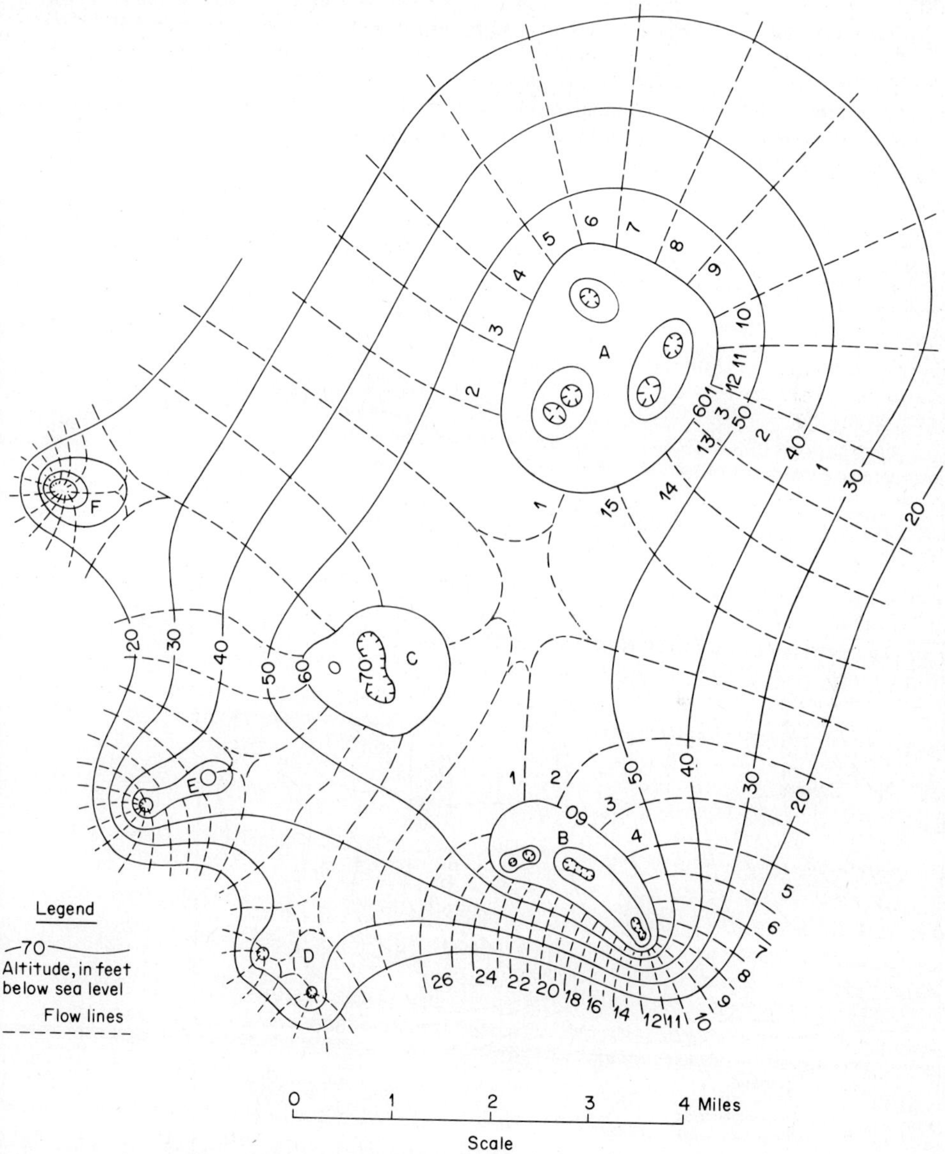
A
B
C
D
E
F
Legend
70
Altitude, in feet below sea level
Flow lines
0 1 2 3 4 Miles
Scale

Fig. P. 3.23

3.24 The coefficients of transmissibility of flow cross sections 1 and 2, 3 and 4 in the following illustration are 2.2×10^4 and 1.98×10^4 gpd/ft, respectively. Compute the quantities of water flowing through the flow cross sections.
Answer: 0.5 mgd; 5.3 mgd; 5.8 mgd; 6.5 mgd.
If the total withdrawals from pumping center A are 11.2 mgd, compute the average coefficient of transmissibility of the aquifer in the vicinity of pumping center A.
Answer: 16,600 gpd/ft.

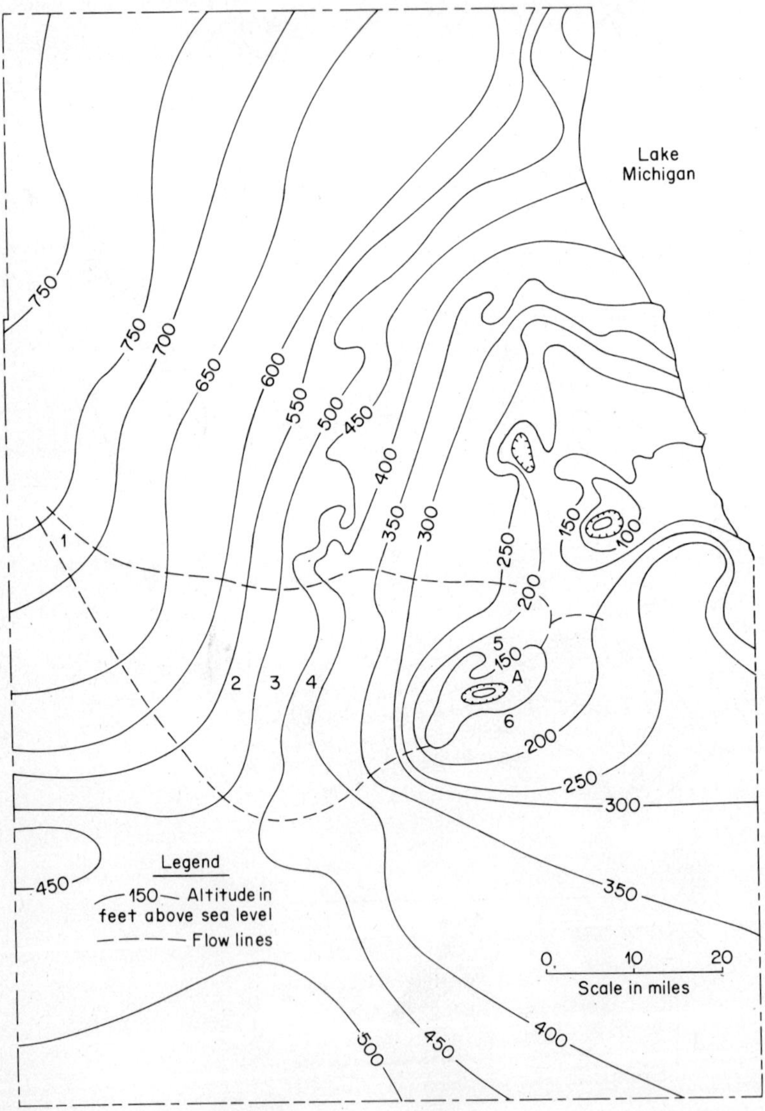

Fig. P. 3.24

References

Boulton, N. S., 1954. "Unsteady Radial Flow to a Pumped Well Allowing for Delayed Yield from Storage," Intern. Assoc. Sci. Hydrology, Gen. Assembly Rome, vol. 2, Publ. 37.
———, 1963. Analysis of Data from Nonequilibrium Pumping Tests Allowing for Delayed Yield from Storage, *Proc. Inst. Civil Engrs.* (London), vol. 26, no. 6693.
———, 1964. Discussions on the Analysis of Data from Nonequilibrium Pumping Tests Allowing for Delayed Yield from Storage, *Proc. Inst. Civil. Engrs.* (London), vol. 28.
Butler, S. S., 1957. "Engineering Hydrology," Prentice-Hall, Inc., Englewood Cliffs, N.J.
Carslaw, H. S., and J. C., Jaeger, 1959. "Conduction of Heat in Solids," Oxford University Press, New York.
Cooper, H. H., Jr., 1959. A Hypothesis Concerning the Dynamic Balance of Fresh Water and Salt Water in a Coastal Aquifer, *J. Geophys. Res.*, vol. 64, no. 4.
———, 1966. The Equation of Groundwater Flow in Fixed and Deforming Coordinates, *J. Geophys. Res.*, vol. 71, no. 20.
——— and M. I. Rorabaugh, 1963. "Groundwater Movements and Bank Storage due to Flood Stages in Surface Streams," U.S. Geol. Surv. Water Supply Paper 1536-J.
Darcy, Henri, 1856. "Les fontaines publiques de la Ville de Dijon," Dalmont, Paris.
Davis, S. N., and R. J. M. DeWiest, 1966. "Hydrogeology," John Wiley & Sons, Inc., New York.
de Han, Bierens, 1939. "Nouvelles Tables d'Intégrales de finies," table 91, no. 13, G. E. Stechert & Company, New York.
DeWiest, R. J. M., 1965. "Geohydrology," John Wiley & Sons, Inc., New York.
———, 1966. On the Storage Coefficient and the Equations of Groundwater Flow, *Geophys. Res.* vol. 71, no. 4.
Ferris, J. G., 1951. "Cyclic Fluctuations of Water Levels As a Basis for Determining Aquifer Transmissibility," Intern. Assoc. Sci. Hydrology, Gen. Assembly Brussels, Publ. 33, vol. 2.
———, 1959. Ground Water, in C. O. Wisler and E. F. Brater (eds.), "Hydrology," chap. 7, John Wiley & Sons, Inc., New York.
———, D. B. Knowles, R. H. Brown, and R. W. Stallman, 1962. "Theory of Aquifer Tests," U.S. Geol. Surv. Water Supply Paper 1536-E.
Foley, F. C., W. C. Walton, and W. J. Drescher, 1953. "Ground-water Conditions in the Milwaukee-Waukesha area, Wisconsin," U.S. Geol. Surv. Water Supply Paper 1229.
Ghyben, W. (Badon), 1888–1889. Nota in verband met de voorgenomen putboring nabij Amsterdam, *Tijdschr. Koninkl. Inst. Ing.*, The Hague.
Glover, R. E., 1959. The pattern of Fresh-water Flow in a Coastal Aquifer, *J. Geophys. Res.*, vol. 64, no. 4.
Hagen, G., 1839. Bervegung des Wassers in engen cylindrischen Rohren, *Pogg. Ann.*, vol. 47.
Hantush, M. S., 1956. Analysis of Data from Pumping Tests in Leaky Aquifers, *Trans. Am. Geophys. Union*, vol. 37, no. 6.
———, 1959. Non-steady Flow to Flowing Wells in Leaky Aquifers, *J. Geophys. Res.*, vol. 64, no. 8.
———, 1964. Hydraulics of Wells, in "Advances in Hydroscience," Academic Press, Inc., New York.
——— and C. E. Jacob, 1955. Non-steady Radial Flow in an Infinite Leaky Aquifer, *Trans. Am. Geophys. Union*, vol. 36, no. 1.
Hazen, A., 1893. "Some Physical Properties of Sands and Gravels with Special Reference to Their Use in Filtration," 24th Ann. Rept., Mass. State Bd. Health, Boston.
Henry, H. R., 1959. Salt Intrusion into Fresh-water Aquifers, *J. Geophys. Res.*, vol. 64, no. 11.
Herzberg, B., 1901. Die Wasserversorgung einiger Nordseebader, *J. Gasbeleuchtung Wasserversorgung*, vol. 44.
Hubbert, M. K., 1940. The Theory of Ground-water Motion, *J. Geol.*, vol. 48.
———, 1956. Darcy's Law and the Field Equations of the Flow of Underground Fluids, *Trans. AIME*, vol. 207.

Ingersoll, L. R., O. J. Zobel, and A. C. Ingersoll, 1948. "Heat Conduction with Engineering and Geological Applications," McGraw-Hill Book Company, New York.

Jacob, C. E., 1940. On the Flow of Water in an Elastic Artesian Aquifer, *Trans. Am. Geophys. Union*, pt. 2.

———, 1943. Correlation of Ground-water Levels and Precipitation on Long Island, New York, *Trans. Am. Geophys. Union*, pt. 1.

———, 1946. Radial Flow in a Leaky Artesian Aquifer, *Trans. Am. Geophys. Union*, vol. 27, no. 2.

———, 1950. Flow of Ground Water, in "Engineering Hydraulics," chap. 5, John Wiley & Sons, Inc., New York.

——— and S. W. Lohman, 1952. Nonsteady Flow to a Well of a Constant Drawdown in an Extensive Aquifer, *Trans. Am. Geophys. Union*, vol. 33, no. 4.

Kohout, F. A., 1960. Cyclic Flow of Salt Water in the Biscayne Aquifer of Southeastern Florida, *J. Geophys. Res.*, vol. 65, no. 7.

Lusczynski, N. J., 1961. Head and Flow of Ground Water of Variable Density, *J. Geophys. Res.*, vol. 66, no. 12.

Meinzer, O. E., and V. C. Fishel, 1934. Tests of Permeability with Low Hydraulic Gradients, *Trans. Am. Geophys. Union.*, vol. 15.

——— and H. A. Hard, 1925. "The Artesian Water Supply of the Dakota Sandstone in North Dakota, with Special Reference to the Edgeley Quadrangle," U.S. Geol. Surv. Water Supply Paper 520.

Papadopulos, I. S., 1965. Nonsteady Flow to a Well in an Infinite Anisotropic Aquifer, Symp. Intern. Assoc. Sci. Hydrology, Dubrovinik.

Poiseville, J. M., 1846. Experimental Investigations on the Flow of Liquids in Tubes of Very Small Diameters, *Proc. Roy. Acad. Sci. Inst. (France), Math. Phys. Sci. Mem.*

Prickett, T. A., 1965. Type-curve Solution to Aquifer Tests under Water-table Conditions, *Ground Water*, vol. 3, no. 3.

Rorabaugh, M. I., 1956. "Ground-water Resources of the Northeastern Part of the Louisville Area, Kentucky," U.S. Geol. Surv. Water Supply Paper 1360-B.

Stallman, R. W., 1956. Numerical Analysis of Regional Water Levels to Define Aquifer Hydrology, *Trans. Am. Geophys. Union*, vol. 37.

Steggewentz, J. H., and J. L. Van Nes, 1939. Calculating the Yield of a Well, Taking into Account Replenishment of the Ground Water above, *Water Water Eng.*, vol. 41.

Theis, C. V., 1935. The Relation between the Lowering of Piezometric Surface and the Rate and Duration of Discharge of a Well Using Ground-water Storage, *Trans. Am. Geophys. Union*, 16th Ann. Meeting, pt. 2.

———, 1941. The Effect of a Well on the Flow of a Nearby Stream, *Trans. Am. Geophys. Union*, pt. 3.

———, 1963. Drawdowns Caused by a Well Discharging under Equilibrium Conditions from an Aquifer Bounded on a Finite Straight-line Source, in "Shortcuts and Special Problems in Aquifer Tests," U.S. Geol. Surv. Water Supply Paper 1545-C.

Todd, D. K., 1959. "Ground-water hydrology," John Wiley & Sons, Inc., New York.

Walton, W. C., 1955. Ground-water Hydraulics As an Aid to Geologic Interpretation, *Ohio J. Sci.*, vol. 55, no. 1.

———, 1962. "Selected Analytical Methods for Well and Aquifer Evaluation," Illinois State Water Surv. Bull. 49.

———, 1963. "Estimating the Infiltration Rate of a Streambed by Aquifer-test Analysis," Intern. Assoc. Sci. Hydrology, Gen. Assembly Berkeley.

Wentworth, C. K., 1942. Storage Consequences of the Ghyben-Herzberg Theory, *Trans. Am. Geophys. Union*, vol. 23.

Wenzel, L. K., 1942. "Methods of Determining Permeability of Water-bearing Materials, with Special Reference to Discharging-well Methods," U.S. Geol. Surv. Water Supply Paper 887.

——— and H. H. Sand, 1942. "Water Supply of the Dakota Sandstone in the Ellentown-Jamestown Area, North Dakota," U.S. Geol. Surv. Water Supply Paper 889-A.

CHAPTER 4

AQUIFER TEST AND FLOW-NET ANALYSIS

The hydrogeologic properties of aquifers (T and S) and aquitards (P' and S') may be determined by means of *aquifer tests*, wherein the effect of pumping a well (e.g. at a known constant rate) is measured in the pumped well and in observation wells penetrating the aquifer. Graphs of drawdown versus time after pumping started, and/or of drawdown versus distance from the pumped well, are used to solve equations presented in Chap. 3 which express the relation between the hydrogeologic properties of an aquifer and its aquitard, if present, and the lowering of water levels in the vicinity of a pumped well. Aquifer tests generally are at least 8 hours in duration and may extend for periods of one or more weeks. Piezometric-surface and water-table maps also may be used to evaluate the hydrogeologic properties of aquifers and aquitards by means of flow-net analysis.

4.1 Drawdown

To determine *drawdown*, the water-level trend before pumping started is extrapolated through the pumping period, and differences between extrapolated stages of the water level that would have been observed if the well had not been pumped and water levels measured during the pumping period are computed. Drawdown is not determined from the water level that was measured just prior to the start of the aquifer test. Ideally, water levels should be measured for a period prior to pumping equal to the pumping period.

Before water-level data are used to determine hydrogeologic properties, they must be adjusted for any pumping-rate changes in nearby production wells. Drawdown data must also be adjusted for atmospheric-pressure changes before they are used to determine hydrogeologic properties. Changes in the time-rate of drawdown due to atmospheric-pressure fluctuations may be mistaken for evidence of hydrogeologic boundary conditions. A time interval during which water levels are not affected by pumping changes is selected. Barometer readings, converted to feet of water (inches of mercury $\times$ 1.13 = feet of water), are inverted and plotted on plain coordinate paper, together with water-level data for the pumped and observation wells. Prominent barometric fluctuations are used to compare changes in water level caused by changes in atmospheric pressure. The amount of rise in water level as a result of a decrease in atmospheric pressure and the amount of decline in water level as a result of an increase in atmospheric pressure are calculated. The barometric efficiency can then be computed with Eq. (4.1). Drawdown data are adjusted for atmospheric-pressure changes occurring during an aquifer test by obtaining a record of atmospheric-pressure fluctuations and using Eq. (4.2).

$$\mathrm{BE} = \frac{\Delta W}{\Delta B_p} 100 \tag{4.1}$$

$$\Delta W = \frac{\mathrm{BE}(\Delta B_p)}{100} \tag{4.2}$$

where BE = barometric efficiency, in percent
ΔW = change in water level resulting from a change in atmospheric pressure, in feet
ΔB_p = change in atmospheric pressure, in feet of water

The tidal or river efficiency can be computed and drawdown data can be adjusted for any surface-water–stage changes occurring during an aquifer test by obtaining a record of surface-water–stage fluctuations during the test and using Eqs. (4.3) and (4.5) or (4.4) and (4.6).

$$\mathrm{TE} = \frac{\Delta W}{\Delta H} 100 \tag{4.3}$$

$$\mathrm{RE} = \frac{\Delta W}{\Delta H} 100 \tag{4.4}$$

$$\Delta W = \frac{\mathrm{TE}(\Delta H)}{100} \tag{4.5}$$

$$\Delta W = \frac{\mathrm{RE}(\Delta H)}{100} \tag{4.6}$$

where TE = tidal efficiency, in percent
RE = river efficiency, in percent
ΔW = change in water level resulting from a change in surface-water stage, in feet
ΔH = change in surface-water stage, in feet

The application or the removal of heavy loads in the vicinity of some artesian wells causes changes in their water levels. Fluctuations in water levels (see Roberts and Romine, 1947) sometimes occur when railroad trains or trucks pass aquifer test sites. Drawdown data must also be adjusted for these changes in loading before they are used to determine the hydrogeologic properties of aquifers and aquitards.

Figure 4.1 shows the effect of atmospheric-pressure fluctuations (Roberts and Romine, 1947) on the water level in a well which is located in Champaign County near the village of Savoy in east-central Illinois. The well was drilled in 1944 to

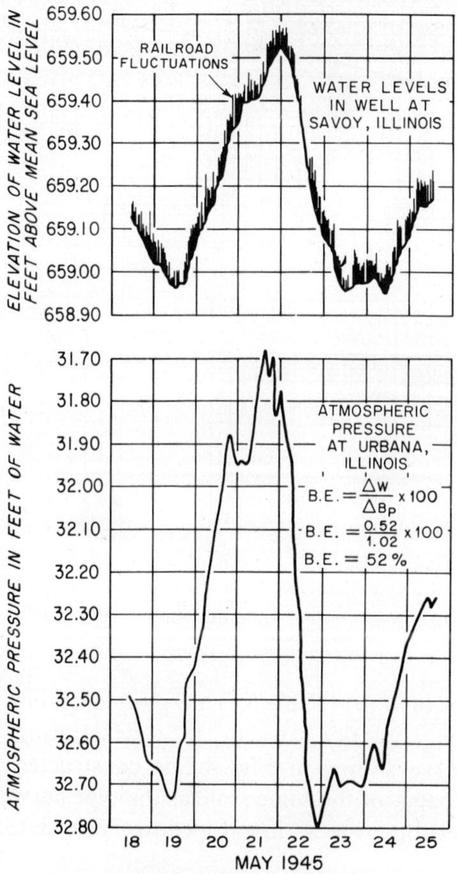

Fig. 4.1 Effect of atmospheric-pressure fluctuations on the water level in a well at Savoy, Illinois. (*After Roberts and Romine, 1947.*)

supply water for the construction of the University of Illinois airport, is 169 feet deep, and is cased with 8-inch-ID pipe. Thirteen feet of 7.5-inch-diameter, wirewound, 50-slot screen is exposed to the aquifer. The nonpumping level was about 70 feet below land surface in May 1945. The aquifer is 15 feet of glacial sand and gravel overlying shales of Pennsylvanian age. Till with a very low permeability and 154 feet thick overlies the aquifer. The barometric efficiency of the well was computed to be about 52 percent by substituting data on water-level and atmospheric-pressure changes in Eq. (4.1) as shown in Fig. 4.1.

4.2 Nonleaky Isotropic Artesian Aquifer with Fully Penetrating Wells and Constant-discharge Conditions

The coefficients of transmissibility and storage of nonleaky isotropic artesian aquifers with fully penetrating wells and constant-discharge conditions may be determined from aquifer test data by the following type-curve graphical method devised by

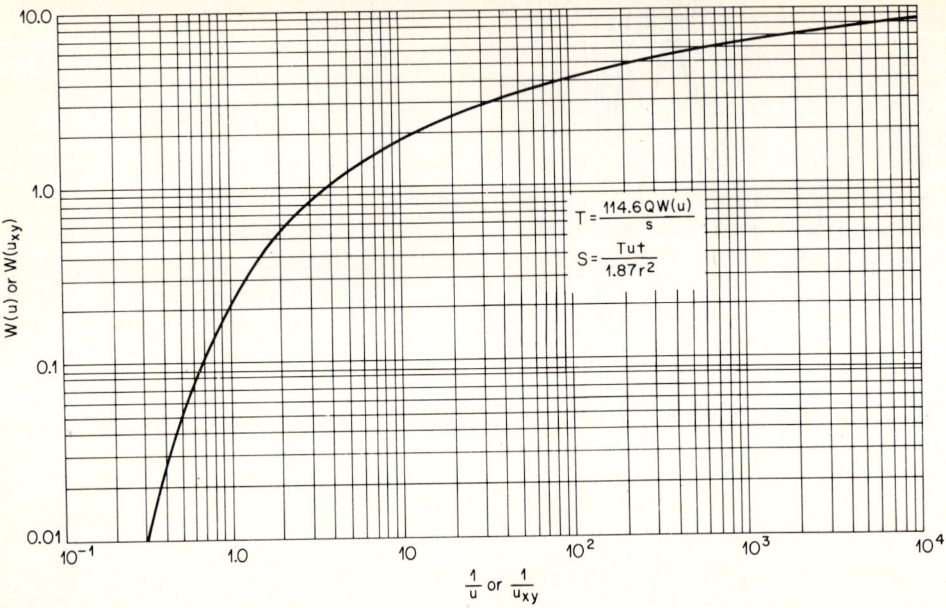

Fig. 4.2 Nonleaky artesian, fully penetrating, constant-discharge, time-drawdown type curve. (*Drawn by W. C. Walton.*)

Theis and described by Jacob (1940). Values of $W(u)$ (Table 3.1) are plotted against values of $1/u$ on logarithmic paper and the nonleaky artesian, fully penetrating, constant-discharge, time-drawdown type curve shown in Fig. 4.2 is constructed. Values of drawdown plotted on logarithmic paper of the same scale as the type curve against values of time after pumping started describe a time-drawdown field-data curve.

Equations (3.48) and (3.49) may be rewritten as

$$W(u) = \frac{T}{114.6Q} s \tag{4.7}$$

and

$$\frac{1}{u} = \frac{T}{1.87r^2S} t \tag{4.8}$$

or

$$\log W(u) = \log \left(\frac{T}{114.6Q}\right) + \log s \tag{4.9}$$

and

$$\log \frac{1}{u} = \log \left(\frac{T}{1.87r^2S}\right) + \log t \tag{4.10}$$

If Q is held constant, the parts of Eqs. (4.9) and (4.10) in parentheses are constant for a given aquifer test, and $W(u)$ is related to $1/u$ in the same manner that s is related to t. Thus, the time-drawdown type curve is analogous to the time-drawdown field-data curve.

The time-drawdown field-data curve is superposed on the type curve, keeping the $W(u)$ axis parallel with the s axis and the $1/u$ axis parallel with the t axis. In the matched position a point at the intersection of the major axes of the type curve is selected and marked on the time-drawdown field-data curve (the point also may be selected anywhere on the type curve). In the matched position the graph scales are displaced vertically by the quantity log $(114.6Q/T)$ and horizontally by the quantity log $(T/1.87r^2S)$. The coordinates of the match point $W(u)$, $1/u$, s, and t are substituted into Eqs. (3.48) and (3.49) to determine T and S. T is calculated using Eq. (3.48) with the $W(u)$ and s coordinates. S is determined using Eq. (3.49), the calculated value of T, and the $1/u$ and t coordinates of the match point.

The coefficients of transmissibility and storage can also be computed with distance-drawdown data. Values of $W(u)$ (Table 3.1) are plotted against values of u on logarithmic paper to obtain the nonleaky artesian, fully penetrating, constant-discharge, distance-drawdown type curve in Fig. 4.3. Values of s measured at the same time in several wells at various distances from the pumped well are plotted against the squares of the respective distances from the pumped well on logarithmic paper of the same scale as the type curve to obtain a distance-drawdown field-data curve. The distance-drawdown field-data curve is superposed on the distance-drawdown type curve, keeping the axes of the two graphs parallel. In the matched position a point at the intersection of the major axes of the type curve is selected and marked on the distance-drawdown field-data curve. Match-point coordinates $W(u)$, u, s, and r^2 are substituted into Eqs. (3.48) and (3.49) to determine T and S. Graphs of s versus r^2/t can also be used to compute hydrogeologic properties.

Values of drawdown also may be plotted against the logarithms of time after pumping started on semilogarithmic paper. The time-drawdown field data will yield a

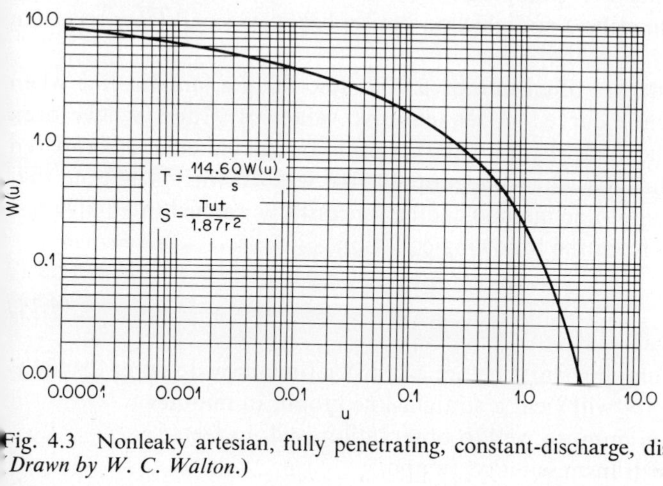

Fig. 4.3 Nonleaky artesian, fully penetrating, constant-discharge, distance-drawdown type curve. (*Drawn by W. C. Walton.*)

straight-line graph in the region where $u \leq 0.01$. The straight-line portion of the time-drawdown field-data graph is extrapolated to its intersection with the zero-drawdown axis. The slope of the straight line and Eq. (3.58) are used to determine the coefficient of transmissibility, and the zero-drawdown intercept and Eq. (3.61) are used to calculate the coefficient of storage. The coefficient of storage cannot be determined with any degree of accuracy from data for the pumped well because the effective radius of the pumped well is seldom known and drawdowns in the pumped well are often affected by well losses which cannot be determined precisely.

Drawdowns observed at the end of a particular pumping period in two or more observation wells at different distances from the pumped well can be plotted against the logarithms of the respective distances on semilogarithmic paper. The distance-drawdown field-data graph will yield a straight-line graph in the region where $u \leq 0.01$. The straight-line portion of the distance-drawdown graph is extrapolated to its intersection with the zero-drawdown axis. The slope of the straight line and Eq. (3.55) are used to determine the coefficient of transmissibility, and the zero-drawdown intercept and Eq. (3.62) are used to calculate the coefficient of storage.

The straight-line method is popular largely because of its simplicity of application and interpretation; however, as pointed out by Cooper and Jacob (1946), "the method is not applicable in some cases and it supplements, rather than supersedes, the type-curve method." They state further that "the method is designed especially for artesian conditions, but it may be applied successfully to tests of nonartesian aquifers under favorable circumstances."

The straight-line method is based on the fact that when u becomes small, a plot of drawdown against the logarithm of time after pumping started or distance from the pumped well describes a straight line. A semilogarithmic graph of values of $W(u)$ and u does not describe a straight line until $u \leq 0.01$. Deviation from a straight line becomes appreciable when u exceeds about 0.02. Therefore, a plot of drawdown against the logarithm of time after pumping started or the logarithm of distance from the pumped well cannot describe a straight line until u becomes small (say, less than 0.02).

Scattered drawdown data are often interpreted as describing a straight line when actually they plot as a gentle curve. After tentative values of T and S have been calculated, the region of the data where $u \leq 0.02$ should be calculated and compared with the region of data through which the straight line was drawn. The time that must elapse before the straight-line method can be successfully applied to aquifer test data is readily determined from the following equation

$$t_{s1} = \frac{1.35 \times 10^5 r^2 S}{T} \tag{4.11}$$

where t_{s1} = time after pumping starts before a semilog time-drawdown or distance-drawdown plot will yield a straight-line graph, in minutes

r = distance from pumped well to observation well, in feet

T = coefficient of transmissibility, in gpd/ft

S = coefficient of storage, fraction

Inserting the data for an artesian aquifer ($S = 0.0001$, $T = 17{,}000$ gpd/ft, and $r = 100$ ft) in Eq. (4.11) results in a t_{s1} of about 8 minutes. In contrast, t_{s1} is about two days for a water-table aquifer ($S = 0.2$, $T = 1{,}000{,}000$ gpd/ft, and $r = 100$ ft). Although these values of t_{s1} apply only to the particular assumptions made, they do point out that the straight-line method should be used with caution particularly in the water-table aquifer situation.

The determination of the storage coefficient by the straight-line method may involve appreciable error. The zero-drawdown intercept is poorly defined where the slope of the semilog plot is small. Intercepts often occur at points where the values of time are very small and minor deviations in extrapolating the straight line will result in large variations in computed values of the coefficient of storage.

4.3 Nonleaky Anisotropic Artesian Aquifer with Fully Penetrating Wells and Constant-discharge Conditions

The coefficients of transmissibility and storage of nonleaky anisotropic artesian aquifers with fully penetrating wells may be determined from aquifer test data by use of the following type-curve graphical method devised by Papadopulos (1965), provided three observation wells at different distances and directions from the pumped well are available.

A convenient rectangular coordinate system is chosen with the origin at the pumped well and the x and y coordinates of each observation well are recorded. Values of $W(u_{xy})$ (Table 3.1) are plotted against values of $1/u_{xy}$ on logarithmic paper and the nonleaky artesian, fully penetrating constant-discharge time-drawdown type curve in Fig. 4.2 is constructed. Values of drawdown for each of the three observation wells plotted on logarithmic paper of the same scale as the type curve against values of time after pumping started describe time-drawdown field-data curves.

The time-drawdown field-data curves are superposed on the type curve, keeping the $W(u_{xy})$ axis parallel with the s axis and the $1/u_{xy}$ axis parallel with the t axis. In the matched positions, points at the intersection of the major axes of the type curve are selected and marked on the time-drawdown field-data curves for the three observation wells. The coordinates of the match points $W(u_{xy})$ and t are substituted into Eq. (3.68) to determine $(T_{xx}T_{yy} - T_{xy}^2)$. All three match points should yield the same, or approximately the same, value for $(T_{xx}T_{yy} - T_{xy}^2)$. If they do not, judgment must be used to obtain an average value.

The coordinates of the match point $1/u_{xy}$ and t for each observation well, the coordinates (x,y) of the three observation wells, and the computed value of $(T_{xx}T_{yy} - T_{xy}^2)$ are substituted into Eq. (3.69), and three equations involving the products ST_{xx}, ST_{yy}, and ST_{xy} are obtained. The three equations are solved simultaneously for the products $T_{xx}S$, $T_{yy}S$, and $T_{xy}S$. The products ST_{xx}, ST_{yy}, and ST_{xy} are substituted in the equation

$$\frac{ST_{xx}ST_{yy} - S^2T_{xy}^2}{S^2} \tag{4.12}$$

and S is determined. The value of S is substituted into the products ST_{xx}, ST_{yy}, and ST_{xy}, and T_{xx}, T_{yy}, and T_{xy} are calculated.

After the components T_{xx}, T_{yy}, and T_{xy} of the transmissibility tensor are obtained by the type-curve method, the principal transmissibilities T_{xx} and T_{yy} and the orientation of the principal axes are determined by making use of tensor properties. The following equations (Papadopulos, 1965), obtained from the invariants and the rules of transformation of tensors, apply for all symmetric tensors of the second rank

$$T_{XX} = \tfrac{1}{2}\{T_{xx} + T_{yy} + [(T_{xx} - T_{yy})^2 + 4T_{xy}{}^2]^{\frac{1}{2}}\} \tag{4.13}$$

$$T_{YY} = \tfrac{1}{2}\{T_{xx} + T_{yy} - [(T_{xx} - T_{yy})^2 + 4T_{xy}{}^2]^{\frac{1}{2}}\} \tag{4.14}$$

$$\theta_A = \arctan \frac{T_{XX} - T_{xx}}{T_{xy}} \tag{4.15}$$

where θ_A is the angle between the x and the X axis, positive in a counterclockwise direction from the x axis, and restricted for convenience to the interval $0 \leq \theta_A < \pi$.

4.4 Nonleaky Isotropic Artesian Aquifer with Fully Penetrating Wells and Constant-drawdown Conditions

The coefficients of transmissibility and storage of nonleaky isotropic artesian aquifers with fully penetrating flowing wells may be determined from aquifer test data by use of the following type-curve graphical method devised by Jacob and Lohman (1952).

Values of $W(\lambda)$ (Table 3.2) are plotted against values of λ on logarithmic paper and the nonleaky artesian, fully penetrating constant-drawdown time-discharge type curve shown in Fig. 4.4 is constructed. Values of discharge plotted on logarithmic paper

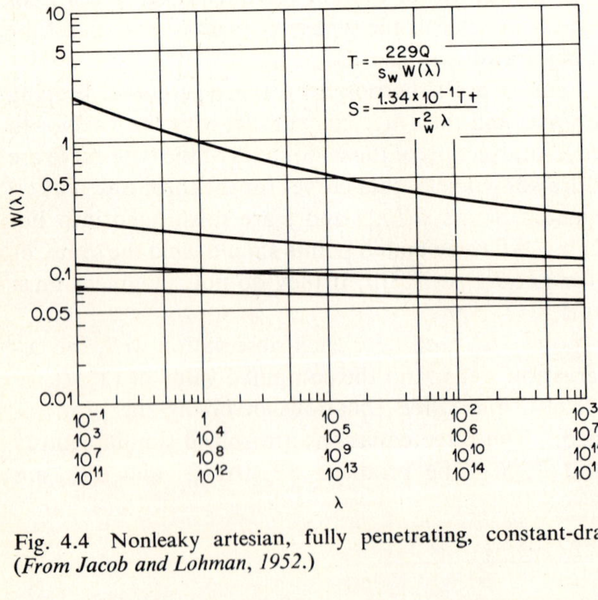

Fig. 4.4 Nonleaky artesian, fully penetrating, constant-drawdown, time-discharge type curve. (*From Jacob and Lohman, 1952.*)

of the same scale as the type curve against values of time after the well was uncapped describe a time-discharge field-data curve.

The time-discharge field-data curve is superposed on the type curve, keeping the $W(\lambda)$ axis parallel with the discharge axis and the λ axis parallel with the t axis. In the matched position a point at the intersection of the major axes of the type curve is selected and marked on the time-discharge field-data curve. The coordinates of the match point $W(\lambda)$, λ, Q, and t are substituted into Eqs. (3.71) and (3.72) to determine the hydrogeologic properties of the aquifer. T is calculated using Eq. (3.71) with the $W(\lambda)$ and Q coordinates. S is determined with Eq. (3.72), the calculated value of T, and the λ and t coordinates of the match point.

4.5 Nonleaky Isotropic Artesian Aquifer with Partially Penetrating Wells and Constant-discharge Conditions

The coefficients of transmissibility and storage of nonleaky isotropic artesian aquifers with partially penetrating wells and constant-discharge conditions may be determined from aquifer test data by the following type-curve graphical method.

Values of $W(u, r/m, \gamma)$ as in Table 3.3, are plotted against values of $1/u$ on logarithmic paper and families of nonleaky artesian, partially penetrating, constant-discharge, time-drawdown type curves for selected values of r/m and γ are constructed as shown in Figs. 4.5 to 4.7. Values of s plotted on logarithmic paper of the same

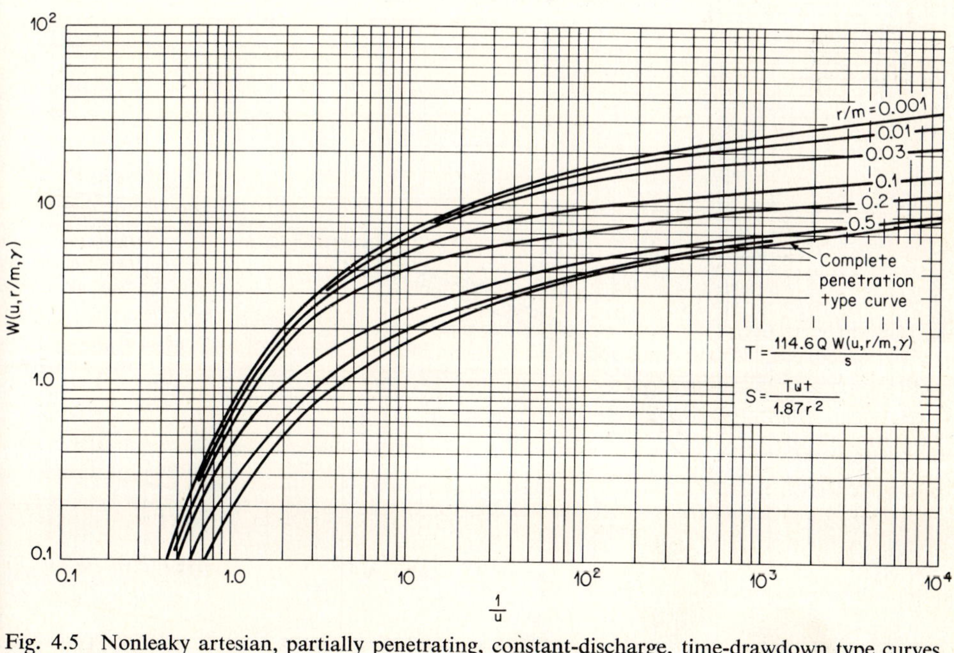

Fig. 4.5 Nonleaky artesian, partially penetrating, constant-discharge, time-drawdown type curves ($\gamma = 0.25$). (*Drawn by W. C. Walton.*)

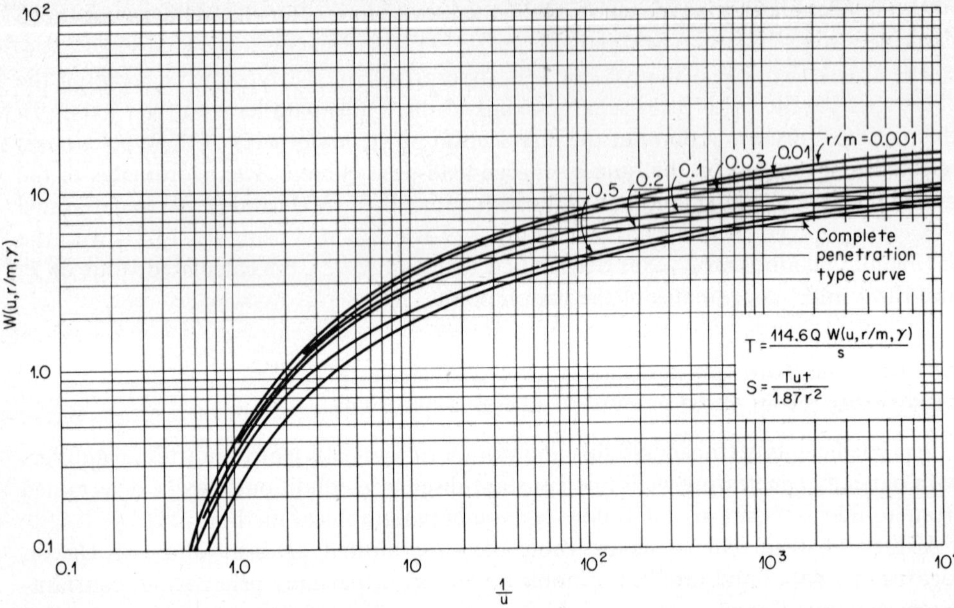

Fig. 4.6 Nonleaky artesian, partially penetrating, constant-discharge, time-drawdown type curves ($\gamma = 0.50$). (*Drawn by W. C. Walton.*)

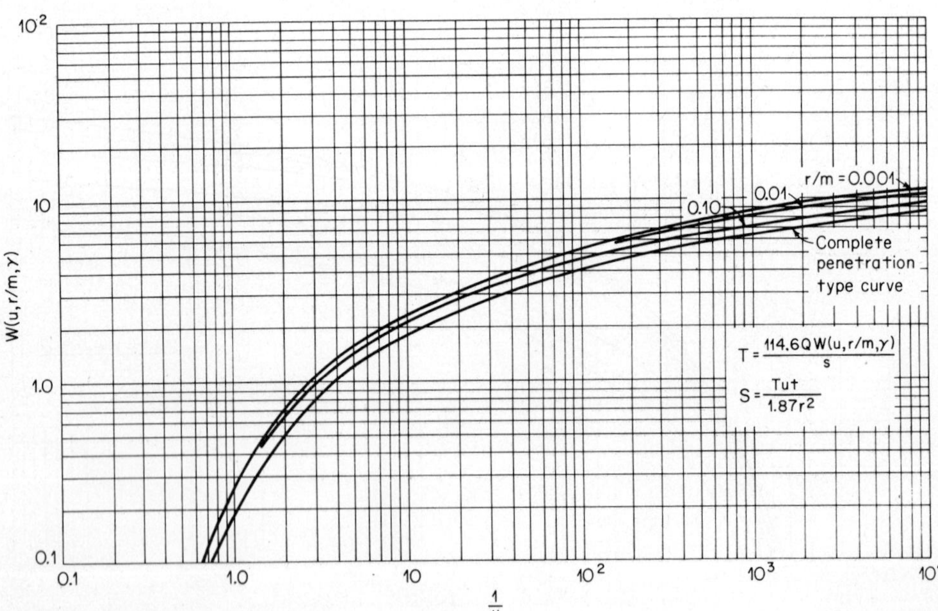

Fig. 4.7 Nonleaky artesian, partially penetrating, constant-discharge, time-drawdown type curves ($\gamma = 0.75$). (*Drawn by W. C. Walton.*)

scale as the type curves against values of t describe a time-drawdown field-data curve that is analogous to one of the families of type curves depending upon the values of r/m and γ.

The time-drawdown field-data curve is superposed on the appropriate r/m type curve, keeping the $W(u, r/m, \gamma)$ axis parallel with the s axis and the $1/u$ axis parallel with the t axis. In the matched position a point at the intersection of the major axes of the type curve is selected and marked on the time-drawdown field-data curve. The coordinates of the match point $W(u, r/m, \gamma)$, $1/u$, s, and t are substituted into Eqs. (3.76) and (3.77) to determine the hydrogeologic properties of the aquifer. T is calculated using Eq. (3.76) with the $W(u, r/m, \gamma)$ and s coordinates. S is determined with Eq. (3.77), the calculated value of T, and the $1/u$ and t coordinates of the match point.

4.6 Leaky Artesian Aquifer with Fully Penetrating Wells without Water Released from Storage in Aquitard and Constant-discharge Conditions

The coefficients of transmissibility and storage of a leaky artesian aquifer with fully penetrating wells without water released from storage in the aquitard and constant-discharge conditions and the coefficient of permeability of the overlying aquitard may be determined from aquifer test data by the following type-curve graphical method devised by Walton (1960).

Values of $W(u, r/B)$ (Table 3.5) are plotted against values of $1/u$ on logarithmic paper and a family of leaky artesian, fully penetrating, without water released from storage in aquitard, constant-discharge, time-drawdown type curves is constructed as shown in Fig. 4.8. Values of s plotted on logarithmic paper of the same scale as the type curves against values of t describe a time-drawdown field-data curve that is analogous to one of the family of type curves.

The time-drawdown field-data curve is superposed on the family of type curves, keeping the $W(u, r/B)$ axis parallel with the s axis and the $1/u$ axis parallel with the t axis. In the matched position a point at the intersection of the major axes of the type curve is selected and marked on the time-drawdown field-data curve. The coordinates of the match point $W(u, r/B)$, $1/u$, s, and t are substituted into Eqs. (3.86) and (3.87) to determine the hydrogeologic properties of the aquifer. T is calculated using Eq. (3.86) with the $W(u, r/B)$ and s match-point coordinates. S is determined with Eq. (3.87), the calculated value of T, and the $1/u$ and t coordinates of the match point. The value of r/B used to construct the particular type curve found to be analogous to the time-drawdown field-data curve is substituted in Eq. (3.88) to determine P'.

Interpretations of aquifer test data based solely on time-drawdown graphs are weak. Distance-drawdown data complement time-drawdown data, and the hydrogeologic properties of the aquifer and aquitard can be determined from distance-drawdown field-data curves (radial profiles of cones of depression). Values of

$W(u, r/B)$ (Table 3.5) are plotted against values of r/B on logarithmic paper and a family of leaky artesian, fully penetrating, without water released from storage in aquitard, constant-discharge, distance-drawdown type curves is constructed. Values of s plotted against values of r on logarithmic paper of the same scale as the type curves describe a distance-drawdown field-data curve that is analogous to one of the family of type curves. The distance-drawdown field-data curve is superposed on the family of type curves, keeping the $W(u, r/B)$ axis parallel with the s axis and the r/B axis parallel with the r axis. The distance-drawdown field-data curve is matched to one of the family of type curves. In the matched position a point at the intersection of the major axes of the type curve is selected and marked on the distance-drawdown field-data curve. Match-point coordinates $W(u, r/B)$, r/B, s, and r are substituted into Eqs. (3.86) and (3.88) to determine the coefficients of transmissibility and vertical permeability. The value of u/r^2 used to construct the particular type curve found to be analogous to the distance-drawdown field-data curve is substituted in Eq. (3.87) to compute S.

Under steady-state leaky artesian conditions, that is, when time-drawdown data fall on the flat portions of the family of type curves indicating that discharge is balanced by leakage, the cone of depression is described by Eq. (3.91). Jacob (1946) devised the following graphical method for determining values of T and P' under leaky artesian, fully penetrating, without water released from storage in aquitard, constant-discharge and steady-state conditions. A steady-state type curve is prepared

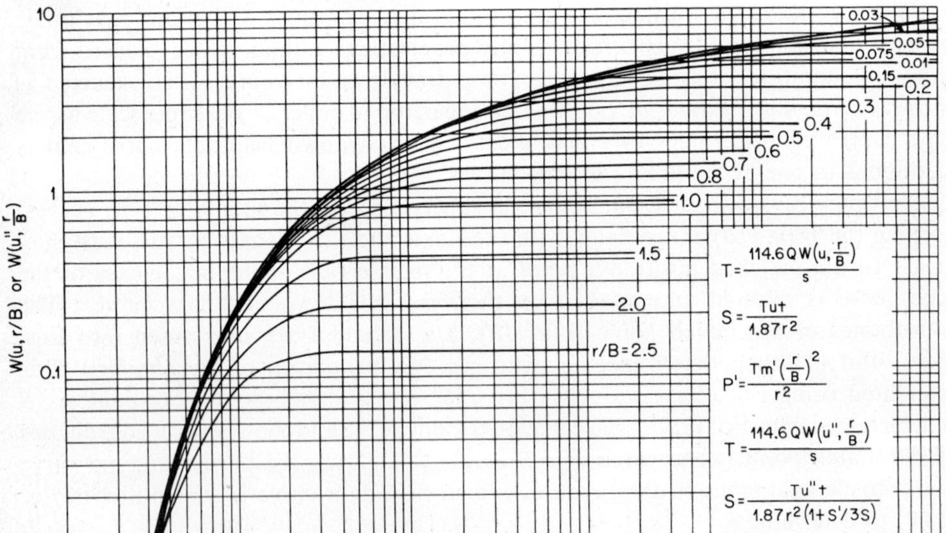

Fig. 4.8 Leaky artesian, fully penetrating, without water released from storage in aquitard, constant-discharge, time-drawdown type curves. (*After Walton, 1962a.*)

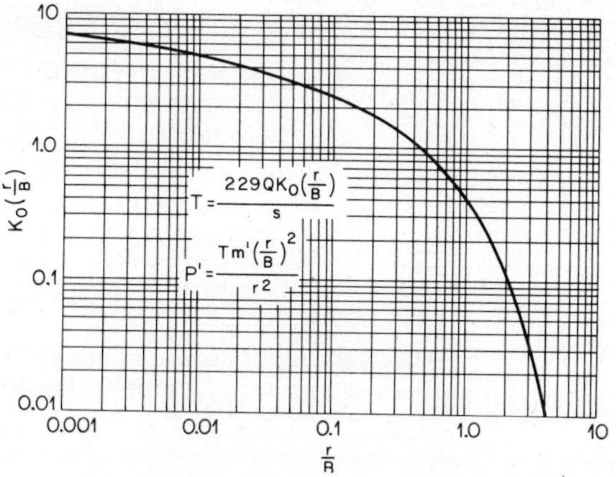

Fig. 4.9 Leaky artesian, fully penetrating, without water released from storage in aquitard, constant-discharge, steady-state, distance-drawdown type curve. (*From Walton, 1962a.*)

by plotting values of $K_0(r/B)$ (Table 3.6) against values of r/B on logarithmic paper as shown in Fig. 4.9. Aquifer test data collected under steady-state conditions are plotted on logarithmic paper of the same scale as the type curve with r as the abscissa and s as the ordinate to describe a distance-drawdown field-data curve. A match of the two curves is obtained by superposing the distance-drawdown field-data curve over the steady-state leaky artesian type curve, keeping the axes of the two graphs parallel. In the matched position a point at the intersection of the major axes of the steady-state leaky artesian type curve is selected and marked on the distance-drawdown field-data curve. Match-point coordinates $K_0(r/B)$, r/B, s, and r are substituted into Eqs. (3.91) and (3.88) to determine T and P'. The coefficient of storage cannot be computed by use of the steady-state type curve because, under such conditions of flow, the entire discharge of the well is derived from leakage sources only.

4.7 Leaky Artesian Aquifer with Fully Penetrating Wells without Water Release from Storage in Aquitard and Constant-drawdown Conditions

The coefficient of transmissibility and storage of a leaky artesian aquifer with fully penetrating flowing wells without water released from storage in the aquitard may be determined from aquifer test data by use of the following type-curve graphical method (see Walton, 1962a).

Values of $W(\lambda, r_w/B)$ (Table 3.7) are plotted against values of λ on logarithmic paper and a family of leaky artesian, fully penetrating wells, without water released from storage in aquitard and constant-drawdown, time-discharge type curves is constructed as shown in Fig. 4.10. Values of discharge plotted against values of time on logarithmic paper of the same scale as the type curves describe a time-discharge field-data curve that is analogous to one of the family of type curves.

The time-discharge field-data curve is superposed on the family of type curves, keeping the $W(\lambda, r_w/B)$ axis parallel with the Q axis and the λ axis parallel with the t

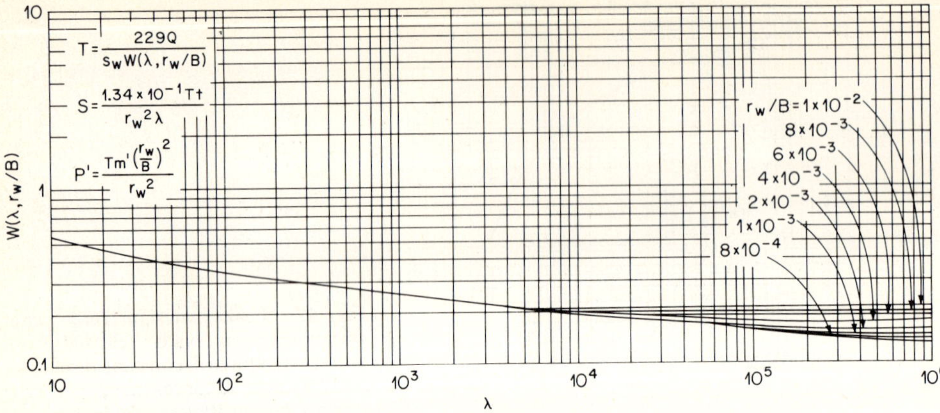

Fig. 4.10 Leaky artesian, fully penetrating, without water released from storage in aquitard, constant-drawdown, time-discharge type curves. (*Drawn by W. C. Walton.*)

axis. The time-discharge field-data curve is matched to one of the type curves and a point at the intersection of the major axes of the type curve is selected and marked on the time-discharge field-data curve. The coordinates of the match point $W(\lambda, r_w/B)$, λ, Q, and t are substituted in Eqs. (3.93) and (3.94) to determine T and S. T is calculated using Eq. (3.93) with the $W(\lambda, r_w/B)$ and Q match-point coordinates. S is determined using Eq. (3.94), the calculated value of T, and the λ and t coordinates of the match point. The value of r_w/B used to construct the particular type curve found to be analogous to the time-discharge field-data curve is substituted in Eq. (3.95) to compute P'.

4.8 Leaky Artesian Aquifer with Fully Penetrating Wells with Water Released from Storage in Aquitard and Constant-discharge Conditions

The coefficients of transmissibility and storage of a leaky artesian aquifer with fully penetrating wells with water released from storage in the aquitard and constant-discharge conditions may be determined from aquifer test data by use of the following type-curve graphical methods.

For the case where $t \leq m'S'/74.8P'$ and the aquitard is thick and relatively impermeable, values of $W(u,\psi)$ (Table 3.8) are plotted against values of u on logarithmic paper and a family of leaky artesian, fully penetrating, with water released from storage in aquitard, $t \leq m'S'/74.8P'$, constant-discharge, time-drawdown type curves is constructed as shown in Fig. 4.11. Values of drawdown plotted against values of time on logarithmic paper of the same scale as the type curves describe a time-drawdown field-data curve that is analogous to one of the family of type curves.

The time-drawdown field-data curve is superposed on the family of type curves, keeping the $W(u,\psi)$ axis parallel with the s axis and the $1/u$ axis parallel with the t

axis. The time-drawdown field-data curve is matched to one of the type curves and a point at the intersection of the major axes of the type curve is selected and marked on the time-drawdown field-data curve. The coordinates of the match point $W(u,\psi)$, $1/u$, s, and t are substituted in Eqs. (3.98) and (3.99) to determine T and S'. T is calculated using Eq. (3.98) with the $W(u,\psi)$ and s match-point coordinates. S' is determined with Eq. (3.99), the calculated value of T, and the $1/u$ and t coordinates of the match point. The value of ψ used to construct the particular type curve found to be analogous to the time-drawdown field-data curve is substituted in Eq. (3.100) to compute the ratio P'/S. If the coefficient of storage of the aquifer can be estimated with available hydrogeologic data, the value of P' can be determined from the ratio.

For the case where $t \leq 0.27m'S'/P'$ and the aquitard is thin and relatively permeable, the family of type curves in Fig. 4.8 may be used to determine the hydrogeologic properties of the aquifer and aquitard. Values of drawdown plotted against values of time on logarithmic paper of the same scale as the type curves describe a time-drawdown field-data curve that is analogous to one of the family of type curves.

The time-drawdown field-data curve is superposed on the family of type curves [in this case the vertical axis is $W(u'', r/B)$ and the horizontal axis is $1/u''$], keeping the $W(u'', r/B)$ axis parallel with the s axis and the $1/u''$ axis parallel with the t axis. The time-drawdown field-data curve is matched to one of the type curves and a point at the intersection of the major axes of the type curve is selected and marked on the time-drawdown field-data curve. The coordinates of the match point $W(u'', r/B)$, $1/u''$, s,

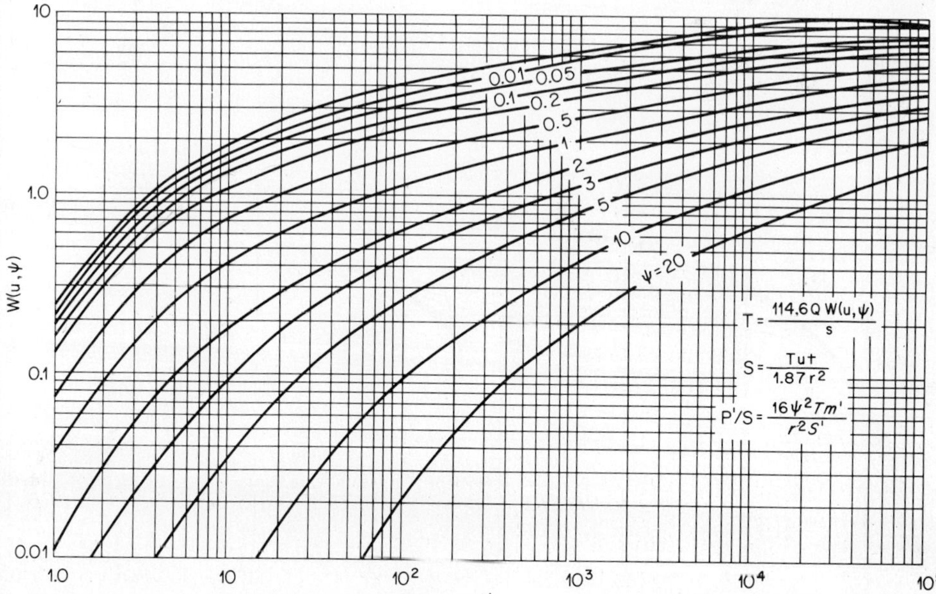

Fig. 4.11 Leaky artesian, fully penetrating, with water released from storage in aquitard, $t \leq m'S/74.8\,P'$, constant-discharge, time-drawdown type curves. (*Drawn by W. C. Walton.*)

and t are substituted in Eqs. (3.102) and (3.103) to determine T and the ratio $S(1 + S')/3S$. T is calculated using Eq. (3.102) with the $W(u'', r/B)$ and s match-point coordinates. The ratio $S(1 + S')/3S$ is determined with Eq. (3.103), the calculated value of T, and the $1/u''$ and t coordinates of the match point. If the coefficient of storage of the aquifer can be estimated with available hydrogeologic data, the value of S' can be determined from the ratio. The value of r/B used to construct the particular type curve found to be analogous to the time-drawdown field-data curve is substituted in Eq. (3.104) to determine P'.

4.9 Water-table Aquifer with Fully Penetrating Wells and Constant-discharge Conditions

The coefficient of transmissibility and storage of water-table aquifers with fully penetrating wells and constant-discharge conditions may be determined from aquifer test data by the following type-curve graphical method devised by Boulton (1963) and described by Prickett (1965). Values of $W(u_{ay}, r/D_t)$ (Table 3.9) are plotted against values of $1/u_a$ and $1/u_y$ on logarithmic paper and two families of water-table fully penetrating, constant-discharge, time-drawdown type curves are constructed as shown in Fig. 4.12. The type curves which lie to the left of the values of r/D_t are termed *type A curves* and are essentially the same as the set of leaky artesian type curves given by Walton (1962a). The type A curves which were obtained with the aid of Eq.

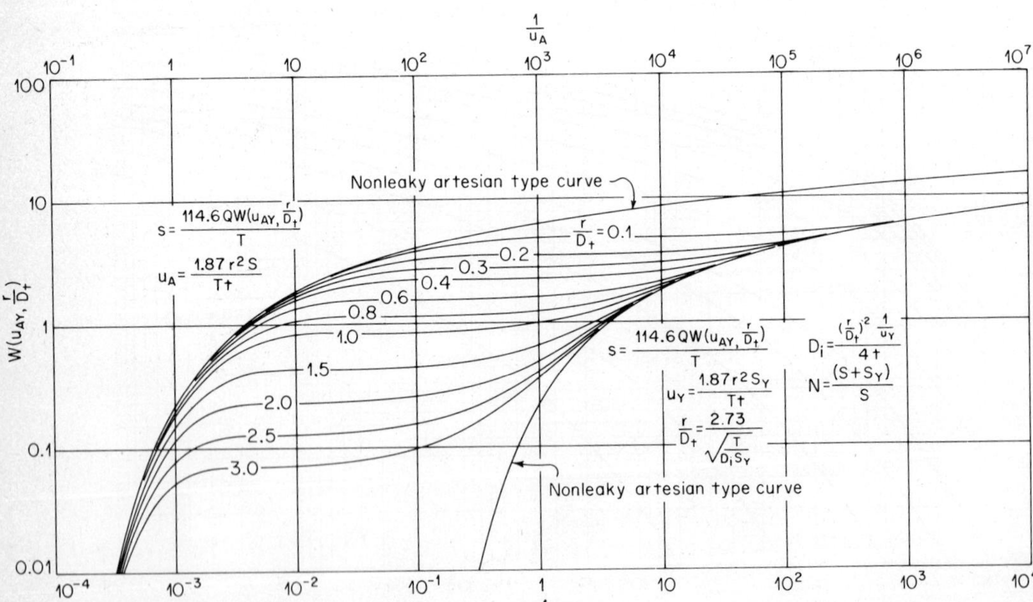

Fig. 4.12 Water-table, fully penetrating, constant-discharge, time-drawdown type curves. (*From Prickett, 1965.*)

(3.112) are used to analyze early time-drawdown data collected during an aquifer test. The type curves which are shown to the right of the values of r/D_t are termed *type Y curves* and were obtained with the aid of Eq. (3.111). The type Y curves are used to analyze late time-drawdown data.

The right-hand portion of each type A curve and the left-hand portion of each type Y curve in Fig. 4.12 approach a horizontal asymptote given by (Prickett, 1965)

$$W\left(u_{ay}, \frac{r}{D_t}\right) = 2K_0\left(\frac{r}{D_t}\right) \tag{4.16}$$

where $K_0(r/D_t)$ is the modified Bessel function of the second kind and zero order.

Equation (3.112) was derived on the basic assumption that N is infinite. Theoretically, the type A curves should then be plotted at an infinite distance to the left of the type Y curves in the logarithmic plot of $1/u_y$. However, for the practical case when N has a finite value, the type A curves, for small values of $1/u_y$, cannot be shown on a base of $1/u_{ay}$ before the value of N is known. Therefore, the early time values of $W(u_{ay}, r/D_t)$ are shown on a base of $1/u_a = (N - 1)(1/u_y)$.

A method of using the type curves for finding the values of T, S, S_y, N, D_i, and the delay index was described by Prickett (1965). Observed values of drawdown for an observation well at a given distance r from the pumped well are plotted on logarithmic paper against values of time t to describe a time-drawdown field-data curve. The time-drawdown field-data curve is analogous to two of the families of water-table type curves shown in Fig. 4.12. The time-drawdown field-data curve, plotted on logarithmic paper of the same scale as the type curves, is first superposed on the type A curves, keeping the $W(u_{ay}, r/D_t)$ axis parallel with the s axis and the $1/u_a$ axis parallel with the t axis. As much as possible of the early time-drawdown field-data is matched to a type A curve. The r/D_t value of the selected type A curve is noted. In the matched position a point at the intersection of the major axes of the type curves is selected and marked on the time-drawdown field-data curve. The coordinates of the match point s, $1/u_a$, $W(u_{ay}, r/D_t)$, and t are substituted in Eqs. (3.113) and (3.114) to determine the coefficients of transmissibility T and storage S. T is calculated using Eq. (3.113) with the $W(u_{ay}, r/D)$ and s coordinates. S is determined with Eq. (3.114), the calculated value of T, and the $1/u_a$ and t coordinates of the match point.

The time-drawdown field-data curve is moved horizontally and is superposed on the type Y curve. The selected type Y curve should have the same value of r/D_t as that used in matching the early time-drawdown field data. In the second matched position a point at the intersection of the major axes of the type curves is selected and marked on the time-drawdown field-data curve. The coordinates of the match points, $1/u_y$, $W(u_{ay}, r/D_t)$, and t are substituted in Eqs. (3.113) and (3.115) to determine the coefficients of transmissibility T and specific yield S_y. T is calculated using Eq. (3.113) with the $W(u_{ay}, r/D_t)$ and s coordinates and should have approximately the same value as the T calculated from early data. S_y is determined using Eq. (3.115), the calculated value of T, and the $1/u_y$ and t coordinates of the second match point. The reciprocal of the delay index, D_i, is determined by substituting the value r/D_t of the selected type curve in Eq. (3.117) with the second match-point coordinates $1/u_y$ and t. The value of N may be determined by substituting the calculated values of S and S_y into Eq. (3.119).

Type A and Y curves are joined by straight lines which are tangent to both curves. The field data which might not have been included when matching the early time-drawdown or late time-drawdown field data should fall along the straight lines. If the value of N is equal to or greater than about 100 (graphically determined), the slope of the tangential line joining the type A and Y curves will be essentially zero. It is evident that the type A and Y curves, which are constructed assuming N equals infinity, are still applicable when N has a finite value of 100 or larger, since they both have zero slope at their intersection. When N has a value between 100 and about 5, type curves for finite N are generally required in order to obtain a precise solution. However, a comparison between N finite type curves obtained from Boulton (1963) and finite N type curves obtained by the simple tangent method described above indicate that the difference between the two type curves is remarkably small even for values of N as small as 6.5 (Prickett, 1965).

Equations (3.113) to (3.115) were derived on the basis that the drawdown is very small in comparison to the total thickness of the aquifer. Gravity drainage of interstices decreases the saturated thickness and therefore the coefficient of transmissibility of the aquifer. Under water-table conditions, observed values of drawdown must be compensated for the decrease in saturated thickness by a method outlined by Jacob (1944) before the data can be used to determine the hydrogeologic properties of the aquifer. The following equation derived by Jacob (1944) is used to adjust drawdown data for decreases in the coefficient of transmissibility:

$$s_a = s_{wt} - \frac{{s_{wt}}^2}{2m} \tag{4.17}$$

where s_a = drawdown that would occur in an equivalent nonleaky artesian aquifer, in feet

s_{wt} = observed drawdown under water-table conditions, in feet

m = initial saturated thickness of aquifer, in feet

The coefficients of transmissibility T and storage S_y can be computed with distance-drawdown data after the effects of delayed gravity drainage have dissipated in observation wells. If an aquifer test under water-table conditions is of long enough duration, the effects of delayed gravity drainage cease to influence the drawdown in observation wells and the distance-drawdown field-data curves conform closely to the nonleaky artesian, fully penetrating, constant-discharge, distance-drawdown type curve (Fig. 4.3).

It should be emphasized that the effects of delayed gravity drainage must dissipate in all observation wells before distance-drawdown data can be used to compute the hydrogeologic properties of the aquifer. During the time when delayed gravity drainage is influencing drawdown in observation wells, the cone of depression is distorted and an analysis of distance-drawdown data during this time will lead to erroneous results.

4.10 Boundary Conditions

Aquifers are often delimited by two or more barrier boundaries, and time-drawdown data deviate more than once under the influence of two or more image wells. Suppose that the cone of depression intercepts two barrier boundaries and that drawdown data for an observation well are affected by two discharging image wells. The proper type curve is matched to early time-drawdown data for the observation well unaffected by the barrier boundaries, and the hydrogeologic properties of the aquifer and aquitard if present are computed. The type-curve trace is extended beyond early time-drawdown data. The type curve is then matched to later time-drawdown data affected by the closest barrier boundary. For a particular value of the appropriate well function, the s match-point coordinate for the second matched position of the type curve must be twice the s match-point coordinate for the first matched position of the type curve. The second type-curve trace is extended beyond later time-drawdown data. The type curve is again matched to the time-drawdown data, this time over the late time-drawdown data affected by both barrier boundaries. For a particular value of the appropriate well function, the s match-point coordinate for the third matched position of the type curve must be three times the s match-point coordinate for the first matched position of the type curve. The departure of type-curve traces 1 and 2, s_{i1}, at a convenient time t_{i1} is noted and the time t_{i2} at which the departure of type-curve traces 2 and 3 is a particular value, s_{i2}, is also noted. The times t_{p1} and t_{p2} at which the first type-curve trace intersects s values equal to s_{i1} and s_{i2}, respectively, are also noted. Values of t_{i1}, t_{i2}, t_{p1}, and t_{p2} are substituted into the following equations for computation of the distances from the observation well to both image wells

$$r_{i1} = r\sqrt{\frac{t_{i1}}{t_{p1}}} \tag{4.18}$$

$$r_{i2} = r\sqrt{\frac{t_{i2}}{t_{p2}}} \tag{4.19}$$

where r_{i1} = distance from observation well to closest image well, in feet
r_{i2} = distance from observation well to farthest image well, in feet
r = distance from observation to pumped well, in feet
t_{p1} or t_{p2} = time after pumping started for a particular value of s_{i1} or s_{i2} to be observed before either boundary becomes effective, in minutes
t_{i1} = time after pumping started when the departure of type-curve traces 1 and 2 is equal to s_{i1}, in days
t_{i2} = time after pumping started when the departure of type-curve traces 2 and 3 is equal to s_{i2}, in days

Multiple-boundary conditions can also be analyzed with time-departure curves. A time-departure curve is prepared by comparing the extension of the type-curve trace through early time-drawdown data and later time-drawdown data affected by the closest barrier boundary. Departures are plotted on logarithmic paper against values of time. The proper type curve is matched to the time-departure curves and r_{i1} is determined with the appropriate equation using values of the hydrogeologic properties computed from early data unaffected by the boundaries. The latter part of the time-departure data will again deviate from the second type-curve trace due to the effects of the farthest barrier boundary. As before, the departures are plotted on logarithmic paper against values of time to form a second time-departure curve. The proper type curve is matched to the second time-departure curve, and match-point coordinates and values of the hydrogeologic properties computed from early data unaffected by the boundaries are substituted into the appropriate equation for computation of r_{i2}. Any number of image wells can be located by repeating the processes outlined earlier.

4.11 Case Histories of Aquifer Test Analysis

Gridley, Illinois

Representatives of Layne-Western Company conducted an aquifer test (see Bruin and Hudson, 1955) on July 2, 1953, using a village well at Gridley, McLean County, Illinois. Village officials, L. A. Miller & Associates as consulting engineers, and the Illinois State Water Survey observed the test. A group of wells (Fig. 4.13) located

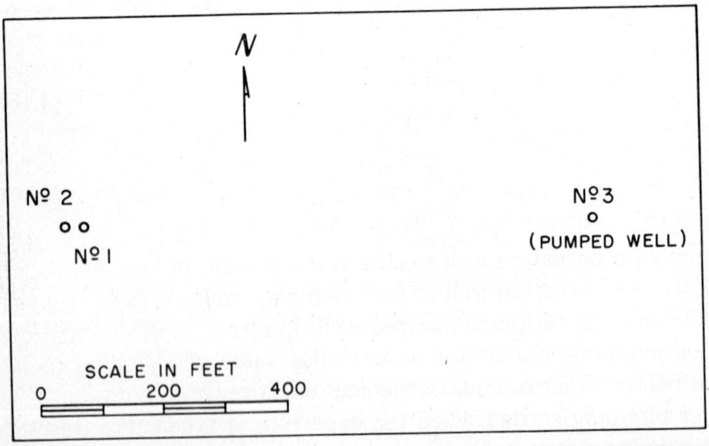

Fig. 4.13 Location of wells used in aquifer test at Gridley, Illinois. (*After Bruin and Hudson, 1955.*)

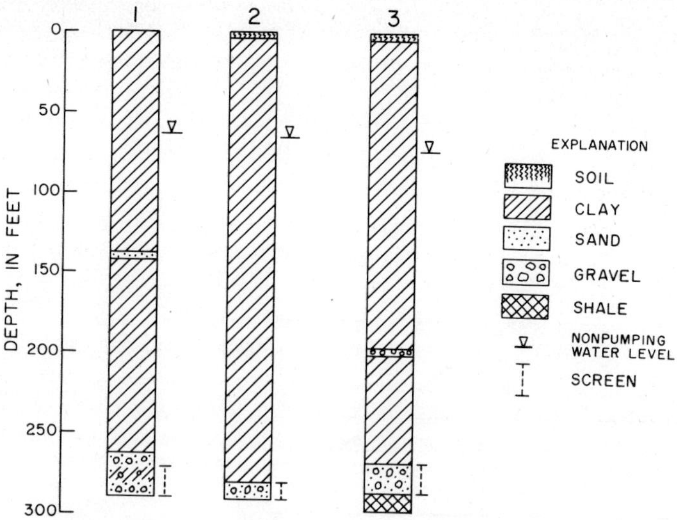

Fig. 4.14 Generalized graphic logs of wells used in aquifer test at Gridley, Illinois. (*From Walton, 1962a.*)

within the corporate limits of the village of Gridley in section 4, T26N, R3E was used. The generalized graphic logs of the wells are given in Fig. 4.14. The effects of pumping well 3 were measured in observation wells 1 and 2. Pumping was started at 9:45 A.M. on July 2 and was continued at a constant rate of 220 gpm for about 8 hours until 6:02 P.M.

Drawdowns in the pumped well and in the observation wells were plotted against time on semilogarithmic and logarithmic paper, respectively. The time-drawdown field-data graphs for the pumped well and for observation well 1 are given in Figs. 4.15 and 4.16, respectively. Time-drawdown data for the wells are given in Tables 4.1 and 4.2, respectively.

The time-drawdown field-data graph for well 1 was superposed on the nonleaky artesian, fully penetrating, constant-discharge, time-drawdown type curve (Fig. 4.2) because leakage was not measurable during the test. Match-point coordinates and Eqs. (3.48) and (3.49) were used to determine the hydrogeologic properties of the aquifer. Computations are given in Fig. 4.16.

A straight line was fitted to the time-drawdown field-data graph for the pumped well as shown in Fig. 4.15. The slope of the line was susbstituted into Eq. (3.58) to calculate the coefficient of transmissibility. The coefficient of storage cannot be determined because the effective radius of the pumped well is not known and the total drawdown in the pumped well is affected by well loss. The average values of T and S computed by using time-drawdown data are 11,000 gpd/ft and 2.2×10^{-5}, respectively.

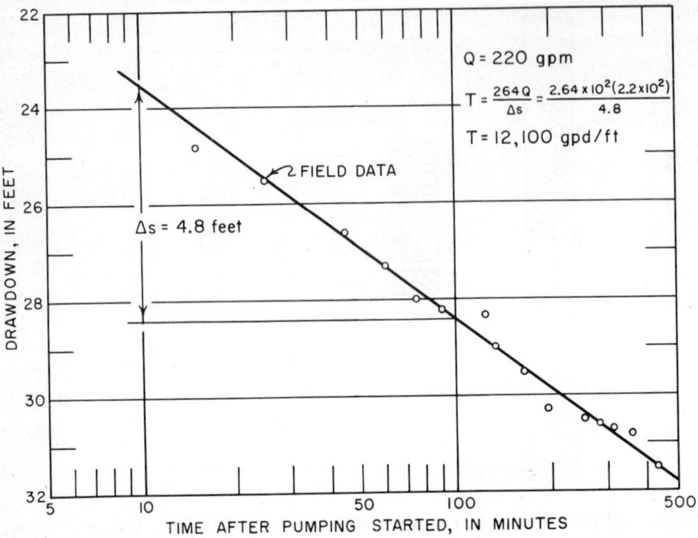

Fig. 4.15 Time-drawdown graph for well 3 at Gridley, Illinois. (*From Walton, 1962.*)

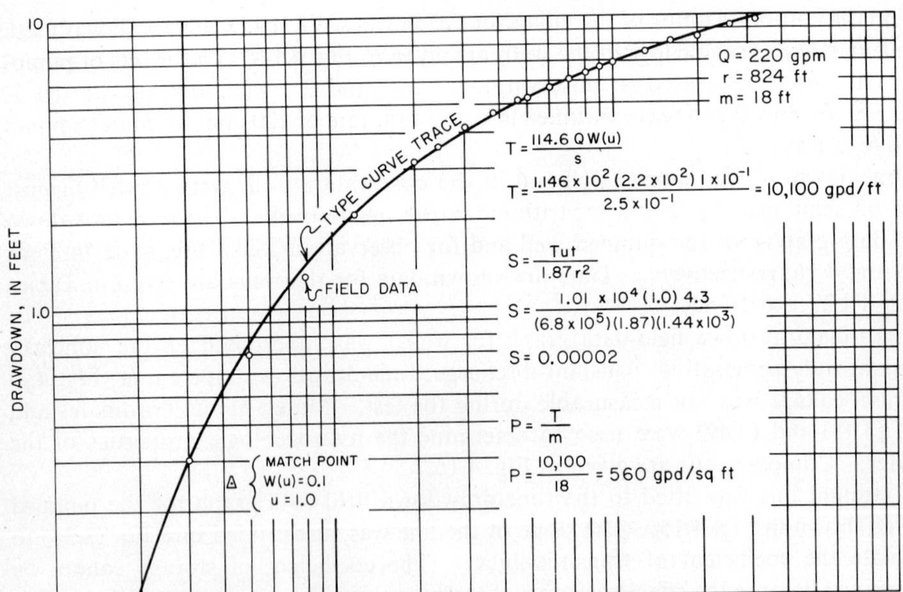

Fig. 4.16 Time-drawdown graph for well 1 at Gridley, Illinois. (*From Walton, 1962a.*)

TABLE 4.1 TIME-DRAWDOWN DATA FOR WELL 1 AT GRIDLEY, ILLINOIS. (*From Walton, 1962a.*)[a]

Time after pumping started, min	*Drawdown, ft*
3	0.3
5	0.7
8	1.3
12	2.1
20	3.2
24	3.6
30	4.1
38	4.7
47	5.1
50	5.3
60	5.7
70	6.1
80	6.3
90	6.7
100	7.0
130	7.5
160	8.3
200	8.5
260	9.2
320	9.7
380	10.2
500	10.9

[a] $Q = 220$ gpm and $r = 824$ ft.

TABLE 4.2 TIME-DRAWDOWN DATA FOR WELL 3 AT GRIDLEY, ILLINOIS. (*From Walton, 1962a.*)[a]

Time after pumping started, min	*Drawdown, ft*
15	24.8
25	25.5
45	26.6
60	27.3
76	28.0
90	28.2
132	29.0
166	29.5
195	30.3
256	30.5
282	30.6
314	30.7
360	30.8
430	31.5

[a] $Q = 220$ gpm.

Fairchild, Idaho

Time-recovery and distance-recovery data may be used to evaluate T and S. To determine recovery, the water-level trend during the pumping period is extrapolated through the recovery period, and differences between extrapolated states of the water level that would have been observed if the well had continued pumping and the water levels measured during the recovery period are computed. An aquifer test was conducted (see Walton, 1962b) on November 7, 1957, using a well (1S-14E-9dbl) in Camas County, Idaho. The well is within the corporate limits of the village of Fairfield, Idaho, in section 9, T1S, R14E. The log of the well is given in Table 4.3. The casing in the well is perforated between the depths of 372 and 495 feet; the top of the aquifer is at a depth of 371 feet and the base is at a depth of 497 feet.

The well was allowed to flow for a period of 4 hours from 1:00 P.M. until 5:00 P.M. The flow decreased continually during the test from about 61 to 50 gpm. However,

TABLE 4.3 LOG OF WELL (1S-14E-9dbl) IN CAMAS COUNTY, IDAHO. (*From Walton, 1962b.*)

Formation	*Thickness, ft*	*Depth, ft*
Topsoil	3	3
Clay, yellow and gravel	69	72
Clay, blue and gravel	27	99
Boulder, basalt	1	100
Sand, coarse	33	133
Clay, brown	5	138
Clay, blue	17	155
Clay, brown and blue	96	251
Sand, fine and gravel	10	261
Clay, yellow	3	264
Clay, yellow and sand	30	294
Clay, yellow	6	300
Sand, yellow, gravel and clay	30	330
Sand, brown and gravel	7	337
Clay	3	340
Sand and gravel	25	365
Clay, brown	6	371
Sand and gravel	21	392
Clay and sand	3	395
Clay, brown	5	400
Hardpan	3	403
Clay, soft, sand and gravel	12	415
Sand	7	422
Sand and gravel	39	461
Clay, brown and sand	6	467
Sand, gravel and clay	30	497
Granite, decomposed	38	535

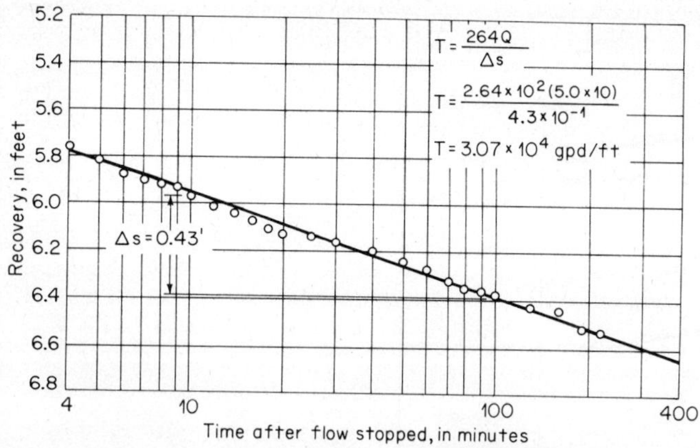

Fig. 4.17 Time-recovery graph for well 1S-14E-9dbl in Camas County, Idaho. (*From Walton, 1962b.*)

the rate of decrease in flow after about 30 minutes was very small, and for practical purposes the discharge of the well was constant during most of the flowing period. The well was capped at 5:00 P.M. and measurements of water-level recovery were made until 8:40 P.M.

Recovery of water levels was plotted against time on semilogarithmic paper as shown in Fig. 4.17. The slope of the straight line through the plotted points and Eq. (3.58) were used to determine the cofficient of transmissibility as indicated in Fig. 4.17. The computed T is about 30,000 gpd/ft. The coefficient of storage of the aquifer cannot be determined from the results of the test because well loss is appreciable and the effective radius of the well is unknown. However, based on available hydrogeologic data, groundwater in the vicinity of the well occurs under artesian conditions.

Grand Junction, Colorado

The coefficients of transmissibility and storage of several aquifers of the Grand Junction artesian basin in Mesa County, Colorado, were determined by Jacob and Lohman (1952) from aquifer tests on individual flowing wells. The Grand Junction artesian basin, as it was developed in 1946, comprised an area of 35 square miles that lies southwest of the city of Grand Junction, Colorado. The basin included about 25 flowing wells with artesian pressures ranging from a few feet to as much as 169 feet above the land surface. The yields of the wells ranged from less than 1 gpm to 20 or 30 gpm.

According to Jacob and Lohman (1952), the wells are supplied from one or more of three sandstones of Jurassic age. Most wells obtain water from the Entrada Sandstone, but a few wells tap a shallower sandstone (Salt Wash Sandstone Member) in the lower part of the Morrison Formation, and a few tap the deeper Wingate Sandstone. The Entrada is about 150 feet thick and consists of cemented fine-grained

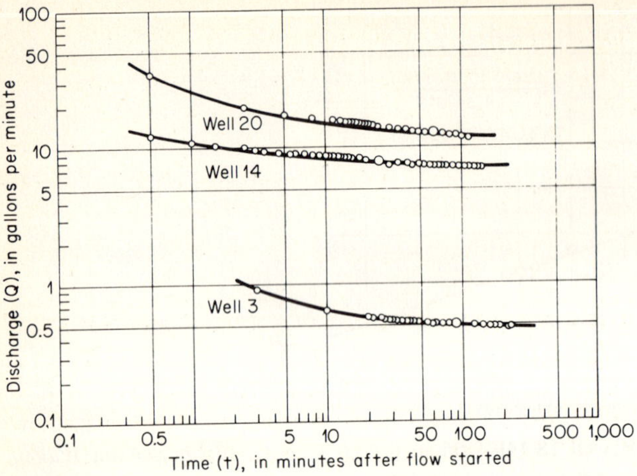

Fig. 4.18 Time-discharge graphs for wells at Grand Junction, Colorado. (*From Jacob and Lohman, 1952.*)

salmon or white sandstone. It is overlain by the Summerville Formation (mostly siltstone), which is succeeded above by the Morrison Formation. The Kayenta Formation, comprising 20 to 80 feet of sandstone and red shale, separates the Entrada from the underlying Wingate Sandstone. Although the Wingate is about 300 feet thick, its buff fine-grained sandstone generally is more firmly cemented and has a lower coefficient of transmissibility than the Entrada. The Wingate is underlain by about 100 feet of the Triassic Chinle Formation (mostly red siltstone), which in turn rests upon Precambrian granite, gneiss, and schist. The sandstone in the lower part of the Morrison is highly lenticular and only 20 to 30 feet thick and it yields less water than the others. The Morrison, which is about 600 feet thick and is mainly variegated shale, is overlain by the Cretaceous Burro Canyon Formation, Dakota Sandstone, and Mancos Shale.

Flow tests were made on 11 wells during the period 1946–1951 using an ink-well mercury gage (Lohman, 1947). Most wells were capped for 12 to 14 hours before testing. The wells were uncapped and left open for a period of about 3 hours.

TABLE 4.4 RESULTS OF CONSTANT-DRAWDOWN AQUIFER TESTS AT GRAND JUNCTION, COLORADO. (*After Jacob and Lohman, 1952.*)

Well no.	*Coefficient of transmissibility, gpd/ft*	*Coefficient of storage, fraction*
20	110	8×10^{-3}
14	120	8×10^{-5}
3	9.2	3×10^{-5}

Values of Q versus t from the tests on wells 20, 14, and 3 were plotted on logarithmic paper as shown in Fig. 4.18. Well 20 is in the Wingate Sandstone, well 14 is in the Entrada Sandstone, and well 3 is in the sandstone of the Morrison Formation. The time-discharge field-data graphs were superposed on the nonleaky artesian, fully penetrating, constant-drawdown, time-discharge type curve in Fig. 4.4. Match-point coordinates and Eqs. (3.71) and (3.72) were used to determine values of T and S. Computed values of T and S are given in Table 4.4.

Dieterich, Illinois

An aquifer test was made by G. E. Neher of the Illinois State Water Survey in cooperation with Marbry & Johnson, Inc., consulting engineers, and E. C. Baker & Sons, well contractor, on July 2 and 3, 1951 (see Walton, 1960). A group of wells (Fig. 4.19) located in Effingham County about 1 mile southwest of the corporate limits of the village of Dieterich, Illinois, in section 22, T7N, R7E was used. The generalized graphic logs of the wells are given in Fig. 4.20. The effects of pumping well 18 were measured in observation wells 15, 16, and 19. Pumping was started at 2:10 P.M. on July 2 and was continued for a period of about 20 hours at a constant rate of 25 gpm until 10:00 A.M. on July 3.

Drawdowns in the pumped well and observation wells were determined by comparing the extrapolated graphs of water levels measured before pumping started with graphs of water levels measured during the pumping period. Drawdowns were plotted against time on logarithmic paper. The time-drawdown field-data graph for observation well 19 is given, as an example, in Fig. 4.21. Time-drawdown data for observation well 19 are given in Table 4.5.

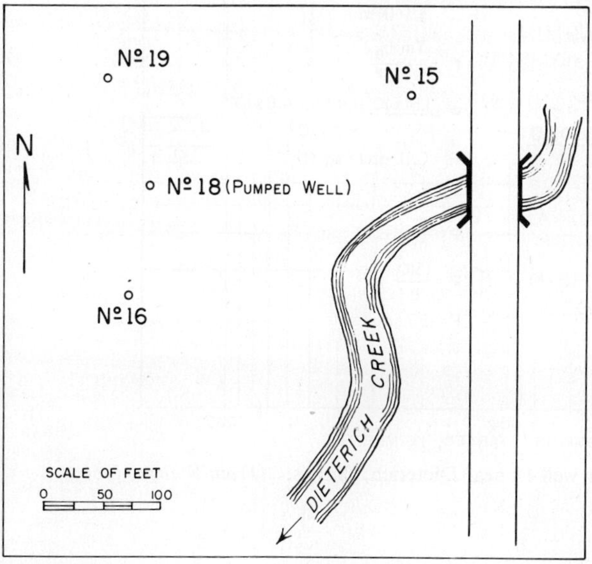

Fig. 4.19 Location of wells used in aquifer test near Dieterich, Illinois. (*From Walton, 1960.*)

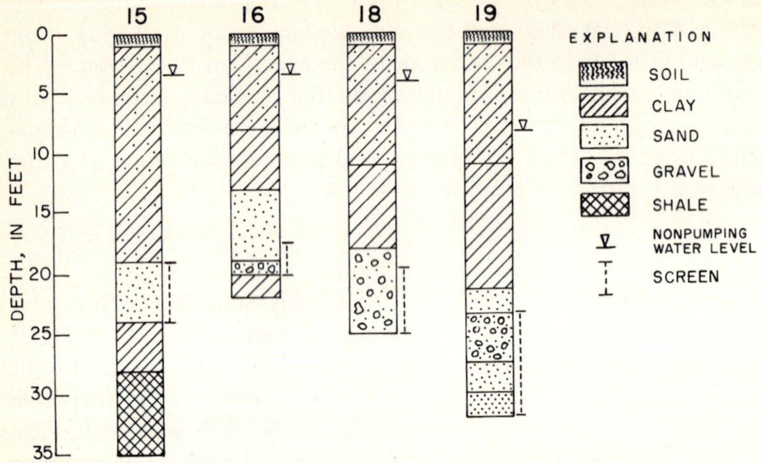

Fig. 4.20 Generalized graphic logs of wells used in aquifer test near Dieterich, Illinois. (*From Walton, 1960.*)

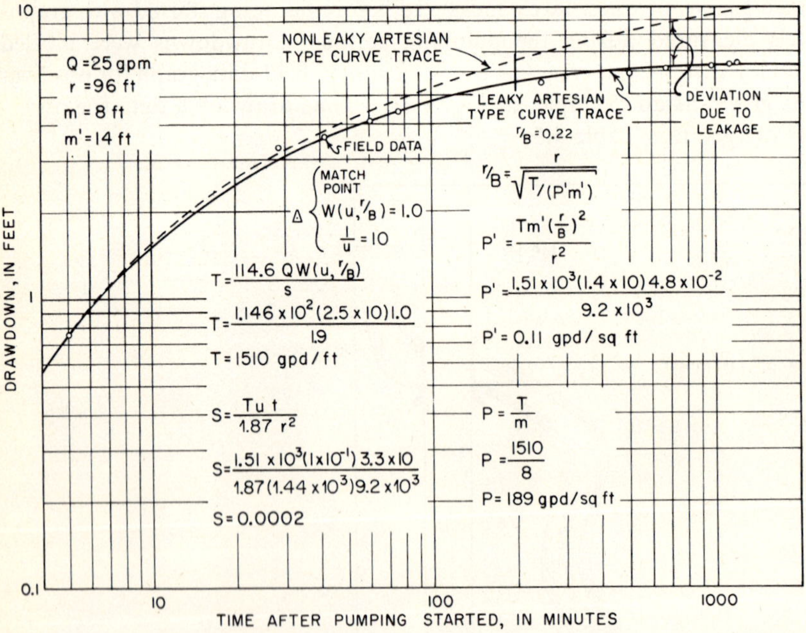

Fig. 4.21 Time-drawdown graph for well 19 near Dieterich, Illinois. (*From Walton, 1960.*)

TABLE 4.5 TIME-DRAWDOWN DATA FOR WELL 19 NEAR DIETERICH, ILLINOIS. (*From Walton, 1962a.*)[a]

Time after pumping started, min	*Drawdown, ft*
5	0.76
28	3.30
41	3.59
60	4.08
75	4.39
244	5.47
493	5.96
669	6.11
958	6.27
1,129	6.40
1,185	6.42

[a] $Q = 25$ gpm and $r = 96$ ft.

The time-drawdown field-data graph was superposed on the family of leaky artesian, fully penetrating, constant-discharge, time-drawdown type curves in Fig. 4.8. The time-drawdown field-data curve closely follows but falls slightly below the $r/B = 0.2$ type curve. By interpolation an $r/B = 0.22$ type curve was selected as analogous to the time-drawdown field-data curve. Match-point coordinates and an r/B value of 0.22 were substituted into Eqs. (3.86) to (3.88) to compute coefficients of transmissibility, permeability of the aquitard, and storage. Computations for well 19 are given in Fig. 4.21.

Drawdowns in observation wells 15, 16, and 19 (Table 4.6) at the end of the test when steady-state conditions prevailed were plotted on logarithmic paper against the distances, from the respective observation wells to the pumped well, to describe a distance-drawdown field-data curve. The leaky artesian, fully penetrating, without

TABLE 4.6 DISTANCE-DRAWDOWN DATA FOR TEST NEAR DIETERICH, ILLINOIS. (*From Walton, 1962a.*)[a]

Well no.	*Distance from pumped well, ft*	*Drawdown, ft*
15	234	3.25
16	92	6.70
19	94	6.42

[a] $Q = 25$ gpm and $t = 1{,}185$ min.

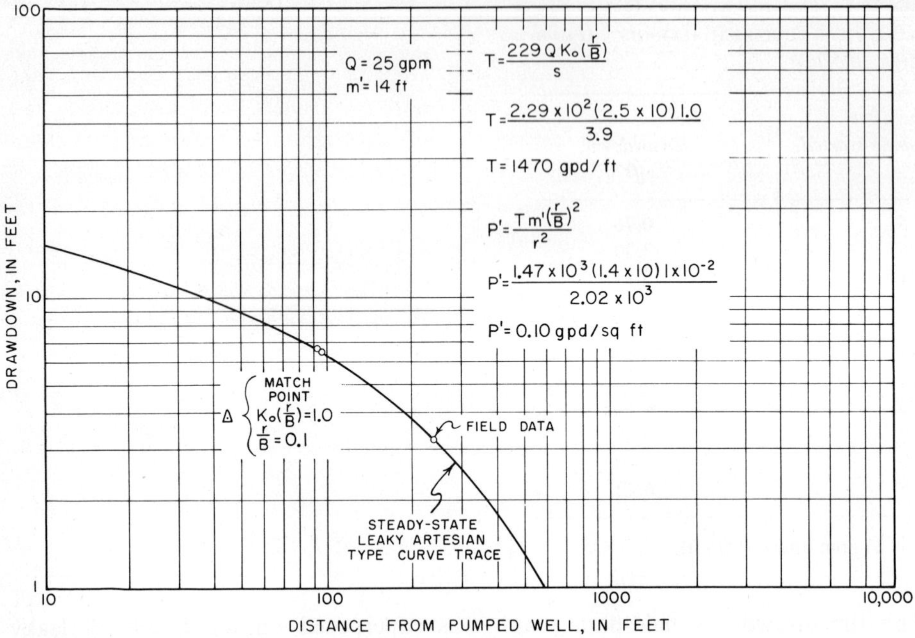

Fig. 4.22 Distance-drawdown graph for test near Dieterich, Illinois. (*From Walton, 1960.*)

water released from storage in aquitard, constant-discharge, steady-state, distance-drawdown type curve in Fig. 4.9 was matched to the distance-drawdown field-data curve, and match-point coordinates were substituted in Eqs. (3.91) and (3.88), for computation of coefficients of transmissibility and permeability of the aquitard as shown in Fig. 4.22.

The average values of T, S, and P' computed by using time-drawdown and distance-drawdown data are 1,500 gpd/ft, 0.0002, and 0.10 gpd/sq ft, respectively. As indicated by the distance-drawdown curve shown in Fig. 4.22, the 20-hour test sampled an area of the aquifer within a radius of roughly 2,000 feet. The coefficients computed from the results of the aquifer test represent the average hydrogeologic properties of the aquifer and aquitard within that cone of depression.

The test of Dieterich was chosen as an example to demonstrate analysis of data under leaky artesian conditions partly because, as shown in Fig. 4.19, the wells are near a possible source of recharge (Dieterich Creek). Two interpretations of the aquifer test data are therefore possible if the effects of partial penetration are excluded. The decrease in the time-rate of drawdown can be attributed either to the effects of leakage through the aquitard or to the effects of induced infiltration of surface water (recharge boundary). Available hydrogeologic data indicate that Dieterich Creek is not a recharge boundary. The streambed rests on clayey materials and has not cut into the aquifer. In addition, the streambed is silted and is relatively impermeable.

The streambed is only a few feet wide and streamflow (135 gpm at the time of the test) during the summer and fall months is low. The results of the aquifer test lend support to this interpretation. If Dieterich Creek were a recharge boundary, the cone of depression would be distorted and distance-drawdown and time-drawdown data would yield differing results. However, values of T and P' computed from both time-drawdown and distance-drawdown data agree, indicating that Dieterich Creek is not a recharge boundary.

Bracebridge, Ontario, Canada

An aquifer test was made in 1963 by International Water Supply Limited, London, Ontario, using test holes 1 to 3 (see Fig. 4.23) in the Peake Springs area near Bracebridge, Ontario, Canada (see Walton, 1964). The logs of wells in and near the test site are shown in Fig. 4.24. The effects of pumping test hole 2 were measured in test holes 1 and 3. Pumping was continued until water levels stabilized at a constant

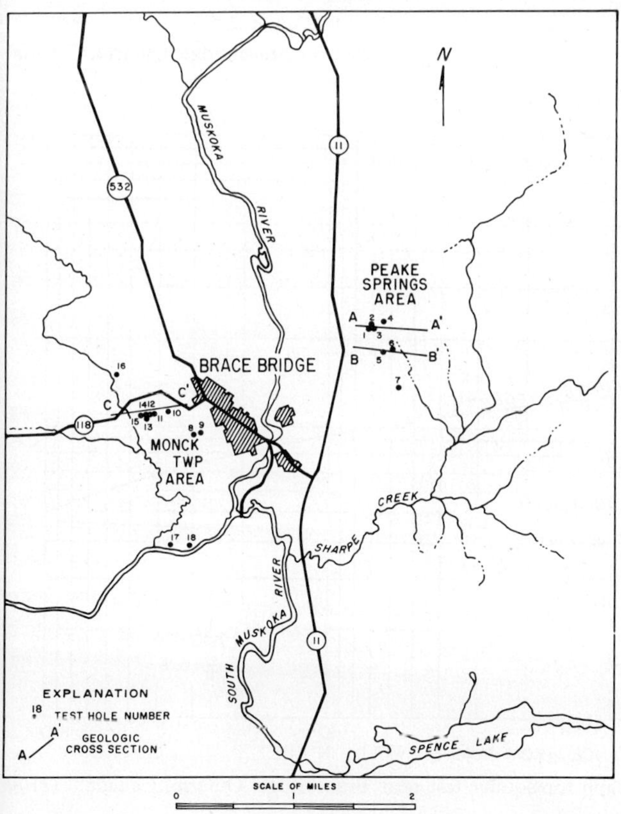

Fig. 4.23 Location of wells used in aquifer test near Bracebridge, Ontario, Canada. (*From Walton, 1964.*)

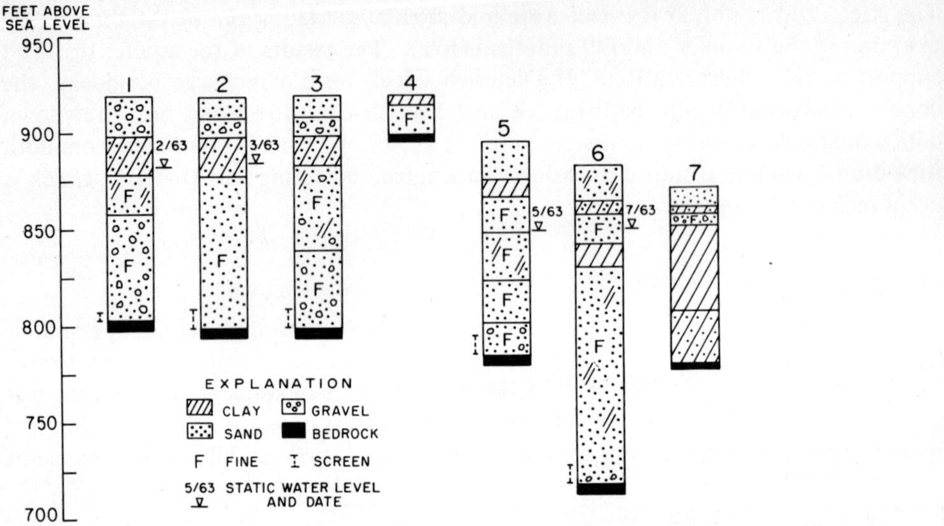

Fig. 4.24 Generalized graphic logs of wells used in aquifer test near Bracebridge, Ontario, Canada. (*From Walton, 1964.*)

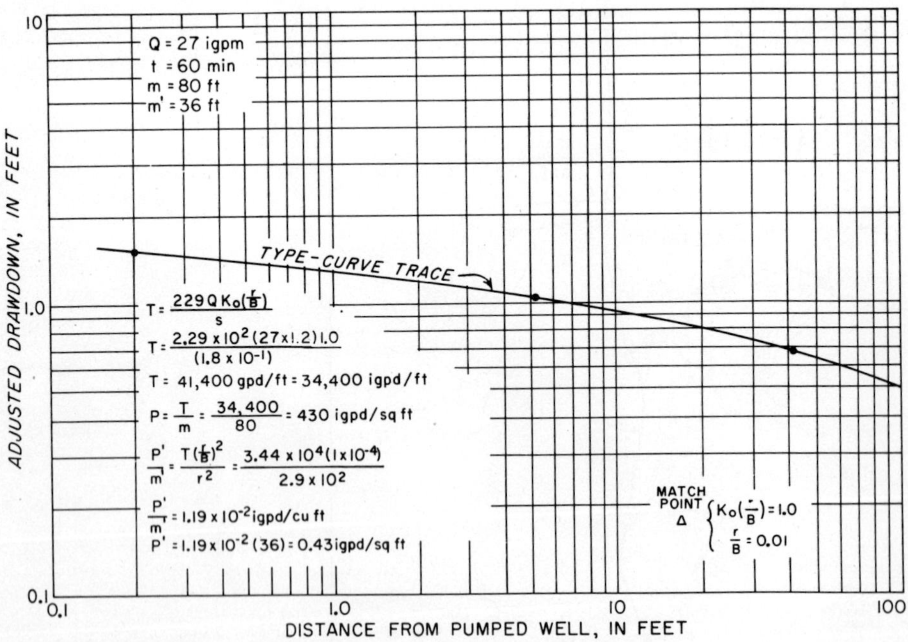

Fig. 4.25 Distance-drawdown graph for aquifer test near Bracebridge, Ontario, Canada. (*From Walton, 1964.*)

TABLE 4.7 COEFFICIENTS OF LEAKAGE AND PERMEABILITY FOR AQUITARDS IN ILLINOIS. (*From Walton, 1962a.*)

Location	*P′ gpd/sq ft*	*Δh ft*	*m′ ft*	*P′/m′ gpd/cu ft*	*Lithology*
Beecher City, Effingham County	0.25	. . .	12	2.1×10^{-2}	Drift, clay with considerable sand and gravel
Dieterich, Effingham County	0.10	. . .	14	7.1×10^{-3}	Drift, clay with considerable sand and gravel
Cowden, Shelby County	1.60	. . .	7	2.3×10^{-1}	Drift, sand and gravel with some clay
Assumption, Christian County	0.19	. . .	8	2.4×10^{-2}	Drift, clay with considerable sand and gravel
Mattoon, Coles County	0.63	. . .	12	5.2×10^{-2}	Drift, clay with considerable sand and gravel
Barry, Pike County	0.15	. . .	16	9.4×10^{-3}	Drift, clay with considerable sand and gravel
Winchester, Scott County	0.08	. . .	16	5.0×10^{-3}	Drift, clay with some sand and gravel
Arcola, Douglas County	0.04	. . .	70	5.7×10^{-4}	Drift, clay with some sand and gravel
Northeastern, Illinois	0.00005	300	200	2.5×10^{-7}	Dolomitic shale
West Chicago, DuPage County	0.0046	45	90	5.1×10^{-5}	Drift, clay with some sand and gravel and shaly dolomite
Downers Grove, DuPage County	0.0068	65	90	7.6×10^{-5}	Drift, clay with some sand and gravel and dolomite
Woodstock, McHenry County	0.012	30	80	1.5×10^{-4}	Drift, clay with some sand and gravel
Champaign-Urbana, Champaign County	0.01	50	120	8.3×10^{-5}	Drift, clay with some sand and gravel
Near Joliet, Will County	1.02	. . .	30	3.4×10^{-2}	Drift, sand and gravel with some clay
Champaign-Urbana, Champaign County	0.21	. . .	35	6.1×10^{-3}	Drift, clay with considerable sand and gravel
LaGrange, Cook County	0.008	30	40	2.0×10^{-4}	Drift, clay with some sand and gravel and dolomite
Libertyville, Lake County	0.009	40	200	4.5×10^{-5}	Drift, clay with some sand and gravel and dolomite
Chicago Heights, Cook County	0.011	25	35	3.2×10^{-4}	Drift, clay with some sand and gravel and dolomite

rate of 27 imperial gallons per minute (igpm). At the end of the test, water levels stabilized, indicating that leakage through the aquitard overlying the aquifer balanced discharge.

Drawdowns at the end of the test were adjusted for the effects of partial penetration of the pumped well and were plotted on logarithmic paper against distances from the pumped well. The distance-drawdown graph was superposed on the leaky artesian, fully penetrating, without water released from storage in aquitard, constant-discharge, steady-state, distance-drawdown type curve in Fig. 4.9, and the coefficients of transmissibility and permeability of the aquifer and permeability and leakage of the aquitard were computed as shown in Fig. 4.25.

The values of T, P, P'/m', and P' computed from the distance-drawdown graph are 34,400 imperial gallons per day (igpd) per foot, 430 igpd/sq ft, 1.19×10^{-2} igpd/cu ft, and 0.43 igpd/sq ft, respectively. The permeability of the aquifer is low and indicative of fine-grained sediments. The aquitard above the aquifer consists of brown clay and sand and packed fine sand and clay averaging 36 feet thick.

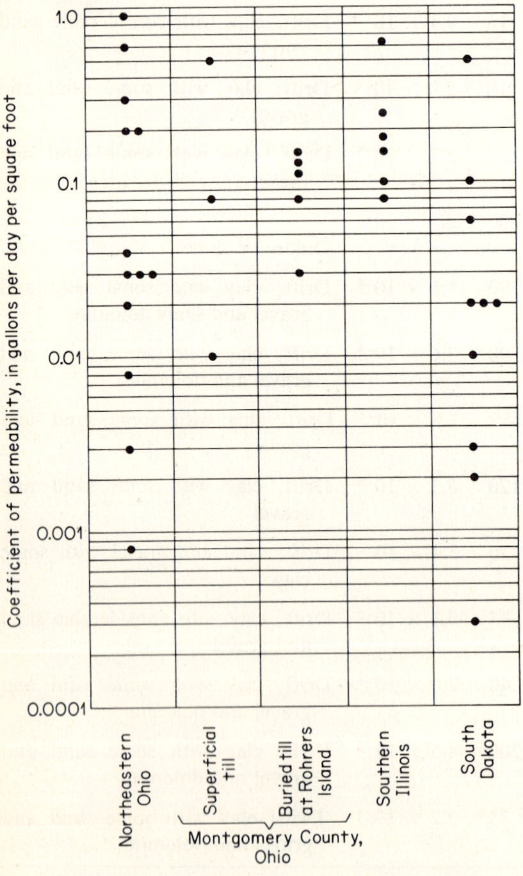

Fig. 4.26 Chart showing range of till permeability in Ohio, Illinois, and South Dakota. (*From Norris, 1962.*)

A summary of coefficients of leakage and permeability of aquitards computed from aquifer test data and other hydrogeologic investigations in Illinois is given in Table 4.7. A chart showing the range of till permeability in Ohio, Illinois, and South Dakota is given in Fig. 4.26. A coefficient of permeability of 0.0013 gpd/sq ft was computed by the author for the clay-with-sand-lenses aquitard of the Potomac Formation beneath the Chesapeake and Delaware Canal area in Delaware.

Lawrenceville, Illinois

An aquifer test (Prickett, 1965) was made by H. F. Smith and F. X. Bushman of the Illinois State Water Survey in cooperation with the Texas Company and the Heldt-Monroe Company, Water Well Contractors, on May 17 and 18, 1950. A group of wells (Fig. 4.27) located about 4 miles southeast of the corporate limits of the city of Lawrenceville, Illinois, in section 11, T3N, R11W was used for the test. The available drillers' logs of wells and test holes near the pumping test site are given in Table 4.8. The effects of pumping the 16-inch well were measured in observation wells 1 through 11. Pumping was started at 8:38 A.M. on May 17 and was continued for a period of about 24 hours at a constant rate of 1,000 gpm until 9:10 A.M. on May 18.

Drawdowns in the pumped well and observation wells were determined by comparing the extrapolated graphs of water levels measured before pumping started with the graphs of water levels measured during pumping. Drawdowns in observation wells were corrected for the effects of decrease in aquifer thickness and partial penetration losses with Eq. (4.17). Corrected drawdowns were then plotted against time on logarithmic paper. The time-drawdown field-data graph for observation well 5 is given, as an example, in Fig. 4.28.

The time-drawdown field-data graph was superposed on the family of water-table, fully penetrating, constant-discharge, time-drawdown type curves in Fig. 4.12. The time-drawdown field-data curve closely follows but falls slightly below the $r/D_t = 0.6$ type curve. By interpolation, a $r/D_t = 0.7$ type curve was selected as analogous to the time-drawdown field-data curve. Match-point coordinates and a r/D_t value of 0.7 were substituted into Eqs. (3.113) to (3.119) to compute the coefficients of transmissibility and storage of the aquifer, the delay index, and values of D_i and N. Computations for well 5 are given in Fig. 4.28.

Corrected drawdowns in observation wells 1 to 11 near the end of the test, when the effects of delayed gravity drainage ceased to influence the drawdowns, were plotted on logarithmic paper against the squares of the distances from the respective observation wells to the pumped well. The distance-drawdown type curve in Fig. 4.3 was matched to the distance-drawdown field data, and match-point coordinates were substituted into Eqs. (3.48) and (3.49) for computation of coefficients of transmissibility T and storage S_y as shown in Fig. 4.29.

The average values of T, S, S_y, D_i, N, and the delay index computed from all observation wells are 250,000 gpd/ft, 0.0023, 0.032, 1.1×10^{-2} minutes^{-2}, 14.9, and 91 minutes, respectively (Prickett, 1965). The coefficients computed from the results

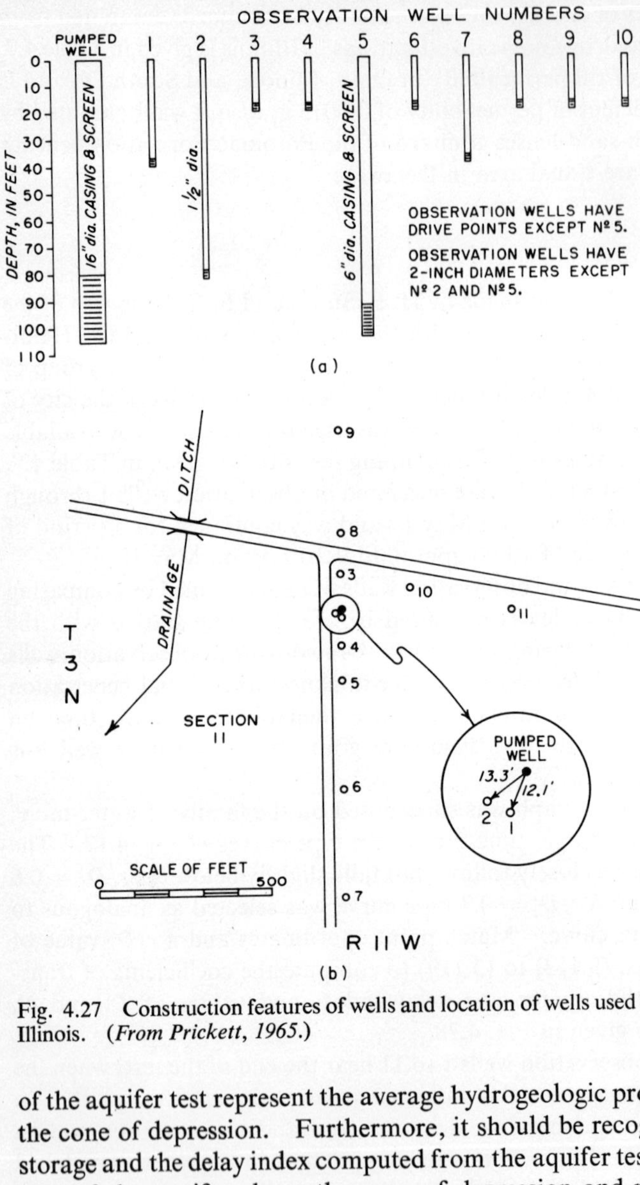

Fig. 4.27 Construction features of wells and location of wells used in aquifer test near Lawrenceville, Illinois. (*From Prickett, 1965.*)

of the aquifer test represent the average hydrogeologic properties of the aquifer within the cone of depression. Furthermore, it should be recognized that the coefficient of storage and the delay index computed from the aquifer test results apply largely to the part of the aquifer above the cone of depression and may not be representative of the aquifer as a whole.

It was previously stated that the effects of delayed gravity drainage become negligible after sufficient time has elapsed during the pumping test. Thus, the type Y curves shown in Fig. 4.12, for any particular value of r/D_t, eventually merge with the nonleaky artesian type curve shown as the lower curve of the family of type Y curves.

TABLE 4.8 LOGS OF WELLS AND TEST HOLES AT PUMPING TEST SITE NEAR LAWRENCEVILLE, ILLINOIS. (*From Prickett, 1965.*)

Formation	*Thickness, ft*	*Depth to bottom, ft*
Topsoil	5	5
Fine sand, medium gravel	5	10
Coarse sand, medium gravel	10	20
Coarse sand, large gravel	5	25
Fine gray sand	40	65
Medium sand, some gravel	10	75
Coarse sand, medium to large gravel	20	95
Fine sand	5	100
Medium sand, some gravel	5	105
Clay	5	5
Sand and gravel	10	15
Coarse sand	10	25
Fine sand	10	35
Coarse sand	5	40
Sand and gravel	5	45
Sand	20	65
Sand and gravel	18	83
Coarse gravel	12	95
Sand and rocks	4	99
Sand and gravel	6	105
Topsoil	5	5
Medium to fine sand and gravel	10	15
Medium to coarse sand, fine gravel	10	25
Fine gray sand	55	80
Medium sand, fine gravel	5	85
Coarse sand, medium gravel	5	90
Medium sand, medium gravel	5	95
Coarse sand, large gravel	5	100
Topsoil	1	1
Silty yellow gravel	4	5
Yellow sand and gravel	15	20
Sand	20	40
Sand and small gravel	10	50
Medium sand	23	73
Sand with clay streaks	10	83
Clay with sand streaks	14	97
Shale	3	97

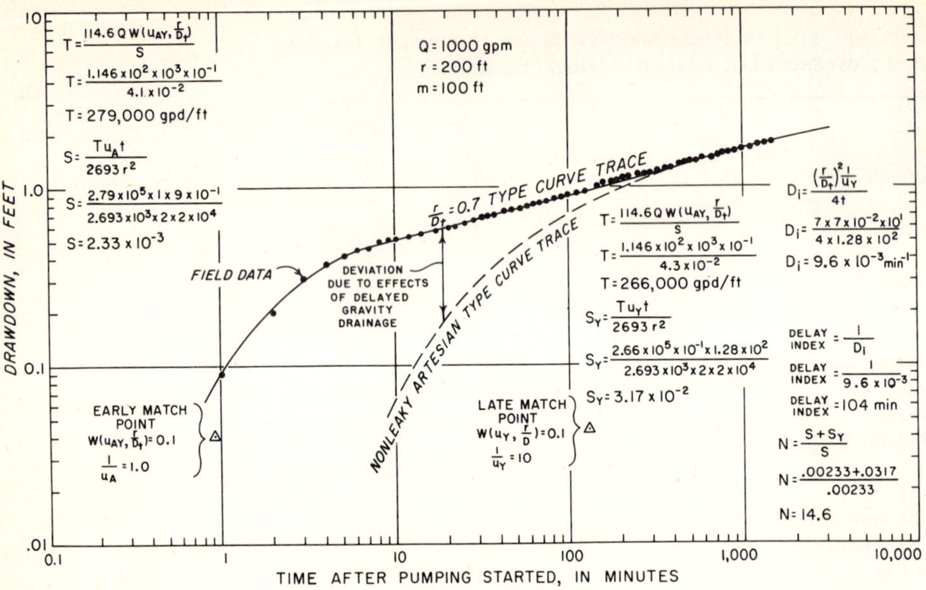

Fig. 4.28 Time-drawdown graph for observation well 5 near Lawrenceville, Illinois. (*From Prickett, 1965.*)

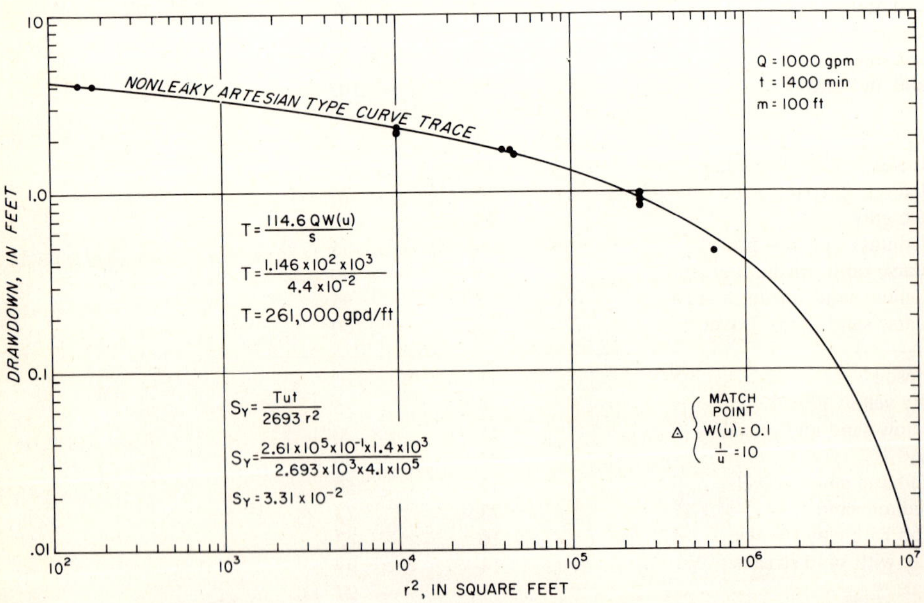

Fig. 4.29 Distance-drawdown graph for test near Lawrenceville, Illinois. (*From Prickett, 1965.*)

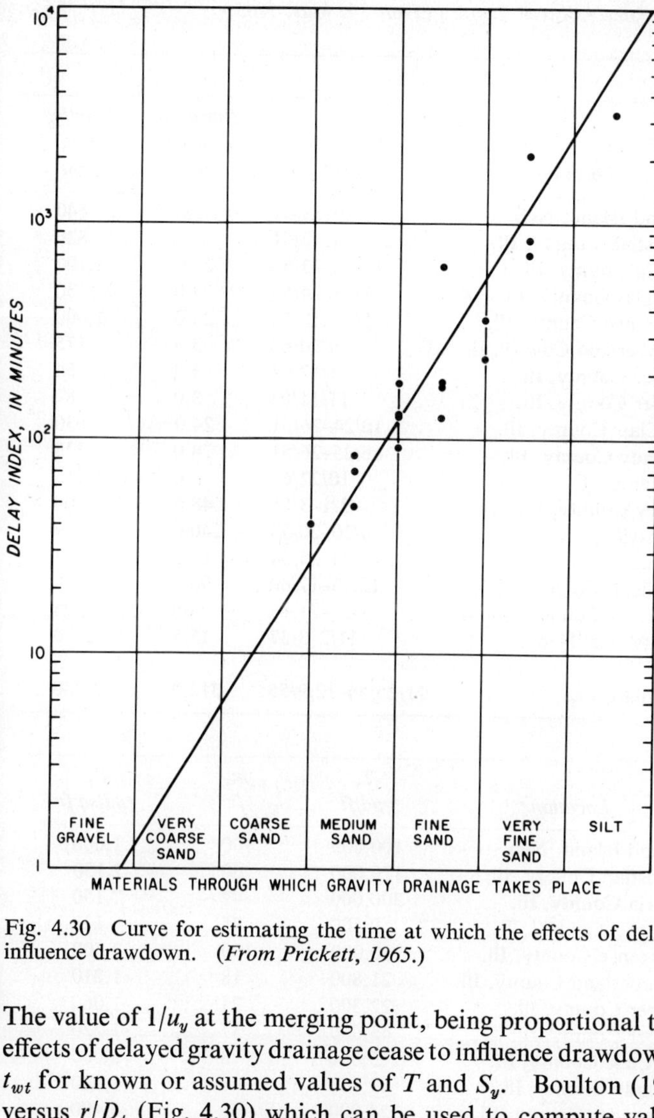

Fig. 4.30 Curve for estimating the time at which the effects of delayed gravity drainage cease to influence drawdown. (*From Prickett, 1965.*)

The value of $1/u_y$ at the merging point, being proportional to the time t_{wt} at which the effects of delayed gravity drainage cease to influence drawdown, can be used to compute t_{wt} for known or assumed values of T and S_y. Boulton (1963) gave a curve of $D_i t_{wt}$ versus r/D_t (Fig. 4.30) which can be used to compute values of t_{wt}. Consider the data for observation well 5: the data needed in order to compute t_{wt} are $r/D_t = 0.7$ and $D_i = 9.6 \times 10^{-3}$ minute^{-1}. Using a r/D_t value of 0.7 and the graph in Fig. 4.30, a value of $D_i t_{wt} = 3.74$ is obtained. Therefore, the time t_{wt} from the start of pumping at which the effects of delayed gravity drainage are negligible and at which true water-table conditions exist is

$$t_{wt} = \frac{3.74}{9.6 \times 10^{-3}} = 390 \text{ min}$$

TABLE 4.9 RESULTS OF WATER-TABLE AQUIFER TESTS. (*From Prickett, 1965.*)

Physical data

Owner	*Location*	*Date of test*	*Duration of test, hr*	*Pumping rate, gpm*
U.S. Geol. Surv.	Grand Island, Neb.	7/29–31/31	48.0	540
City of Taylorville	Christian County, Ill.	1/12/51	7.5	820
Caterpillar Tractor Co.	Peoria County, Ill.	12/22–23/50	24.0	1,100
Village of Hindsboro	Douglas County, Ill.	1/23–24/56	24.0	30
Texas Oil Co.	Lawrence County, Ill.	5/17–18/50	24.0	1,000
Village of Greenup	Cumberland County, Ill.	9/20/63	3.5	175
City of Newton	Jasper County, Ill.	11/12/63	3.0	64
City of Newton	Jasper County, Ill.	11/11/63	3.0	89
Mobil Oil Co.	St. Clair County, Ill.	10/24–26/61	24.0	630
E. I. DuPont Co.	Grundy County, Ill.	6/25–26/51	24.0	325
U.S. Forest Service	Alexander County, Ill.	10/22/64	6.0	1,515
City of Jerseyville	Jersey County, Ill.	3/1–3/55	48.0	300
Stafford Chemical Co.	Louisville, Ky.	8/10–20/53	240.0	609
Village of Yorkville	Kendall County, Ill.	5/24–25/54	16.0	203
Southern Ill. University	Madison County, Ill.	12/13–17/60	96.5	307
City of Wood River	Madison County, Ill.	1/20–21/62	24.0	495
Harry Pieper	Scottsbluff, Neb.	11/2–3/37	15.5	1,270
Louisiana Dept. Public Works	Tallulah, La.	11/26/56–12/9/56	312.0	3,200

Results

Owner	*Location*	*T, gpd/ft*	*m, ft*	*P, gpd/sq ft*
U.S. Geol. Surv.	Grand Island, Neb.	100,800	100	1,008
City of Taylorville	Christian County, Ill.	129,000	60	2,150
Caterpillar Tractor Co.	Peoria County, Ill.	300,000	42	7,150
Village of Hindsboro	Douglas County, Ill.	9,500	75	127
Texas Oil Co.	Lawrence County, Ill.	250,000	100	2,500
Village of Greenup	Cumberland County, Ill.	21,800	18	1,210
City of Newton	Jasper County, Ill.	22,300	21	1,060
City of Newton	Jasper County, Ill.	45,000	37	1,220
Mobil Oil Co.	St. Clair County, Ill.	225,000	73	3,080
E. I. DuPont Co.	Grundy County, Ill.	43,400	40	1,080
U.S. Forest Service	Alexander County, Ill.	324,000	60	5,400
City of Jerseyville	Jersey County, Ill.	157,000	36	4,490
Stafford Chemical Co.	Louisville, Ky.	84,000	53	1,580
Village of Yorkville	Kendall County, Ill.	7,050	40	176
Southern Ill. University	Madison County, Ill.	167,000	84	1,990
City of Wood River	Madison County, Ill.	130,000	60	2,170
Harry Pieper	Scottsbluff, Neb.	160,000	114	1,400
Louisiana Dept. Public Works	Tallulah, La.	250,000	140	1,780

Lithology: Aquifer (driller's log)	Lithology: Materials through which gravity drainage takes place (driller's log)	S	S_y	N	D_i, min^{-1}	Delay index $1/D_i$, min
Sand and gravel, fine to coarse	Sand, medium to coarse	...	0.200	...	2.5×10^{-2}	40
Sand and gravel, fine to coarse	Sand, medium	0.0410	0.161	4.9	2.1×10^{-2}	48
Sand and gravel, fine to coarse	Sand, medium	...	0.166	...	1.4×10^{-2}	71
Sand, fine to coarse, some silty	Sand, medium	0.0550	0.181	4.3	1.2×10^{-2}	83
Sand and gravel, fine to coarse	Sand, fine to medium	0.0023	0.032	14.9	1.1×10^{-2}	91
Sand, some silty, and gravel	Sand, medium, some silty	0.00079	0.051	65.6	8.1×10^{-3}	123
Sand, fine to coarse, some silty	Sand, fine to medium	0.00037	0.005	18.9	8.0×10^{-3}	125
Sand, fine to medium	Sand, fine to medium	0.00059	0.007	12.9	7.8×10^{-3}	128
Sand, fine to coarse, and gravel	Sand, fine	0.00105	0.109	104.8	5.8×10^{-3}	173
Sand and gravel, silty to fine to coarse	Sand, medium to coarse, silty	0.00630	0.250	40.5	5.6×10^{-3}	179
Sand, fine to very coarse	Sand, fine	0.00132	0.113	86.6	5.6×10^{-3}	179
Sand and gravel, fine to coarse, some silt and clay	Sand, silty to medium	0.00084	0.014	17.7	4.3×10^{-3}	233
Sand and gravel, fine to coarse, some silty	Sand, fine to medium, silty	...	0.192	...	2.8×10^{-3}	357
Sand and gravel, fine to coarse, some silty	Sand, fine to coarse, silty	0.00095	0.014	15.7	1.6×10^{-3}	625
Sand, fine to coarse, some clay streaks	Sand, fine with clay streaks	0.00053	0.021	40.6	1.4×10^{-3}	715
Sand, some fine and coarse gravel clay streaks	Sand, fine with clay streaks	0.00014	0.206	1470.0	1.2×10^{-3}	834
Sand and gravel	Sand, fine, and silt	0.00327	0.018	6.5	4.8×10^{-4}	2,080
Sand, gravel, coarse, to clay and silt on top	Clay, silt, and fine sand	0.00143	0.039	28.2	3.1×10^{-4}	3,230

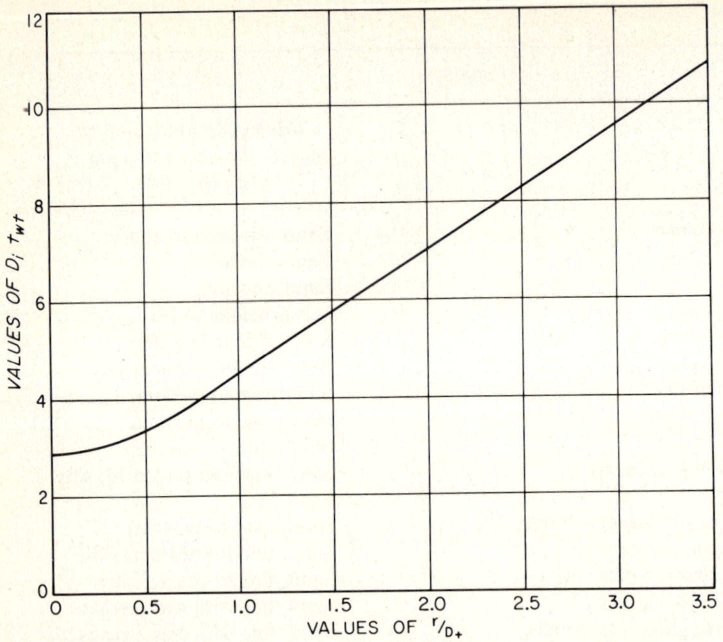

Fig. 4.31 Relationship of delay index to character of materials through which gravity drainage takes place. (*From Prickett, 1965.*)

Table 4.9 summarizes the water-table aquifer test data (Prickett, 1965) collected in Illinois, Kentucky, Louisiana, and Nebraska, and the coefficients of transmissibility, storage, delay index, permeability, D_i, and N computed from these data. The aquifer tests are listed in order of increasing delay index. Values of delay index given in Table 4.9 range between 40 and 3,230 minutes. The smallest delay index pertains to the medium to coarse sand materials through which gravity drainage took place during the aquifer test at Grand Island, Nebraska. The largest delay index relates to the clay, silt, and fine sand materials near Tallulah, Louisiana.

Values of delay index were plotted against the corresponding materials through which gravity drainage took place, as shown in Fig. 4.31 (Prickett, 1965). In general, the delay index increases as the coarseness of materials through which gravity drainage takes place decreases. The graph in Fig. 4.31 should aid in planning future aquifer tests. The values probably can be applied to geologically similar areas, at least in making rough quantitative investigations.

Rio Grande de San Miguel Basin, El Salvador

An aquifer test was made in 1962 by the United Nations Food and Agricultural Organization (FAO) (see Walton, 1965a). A production well and an observation well (D-6a) located in the lower basin of the Rio Grande de San Miguel, El Salvador,

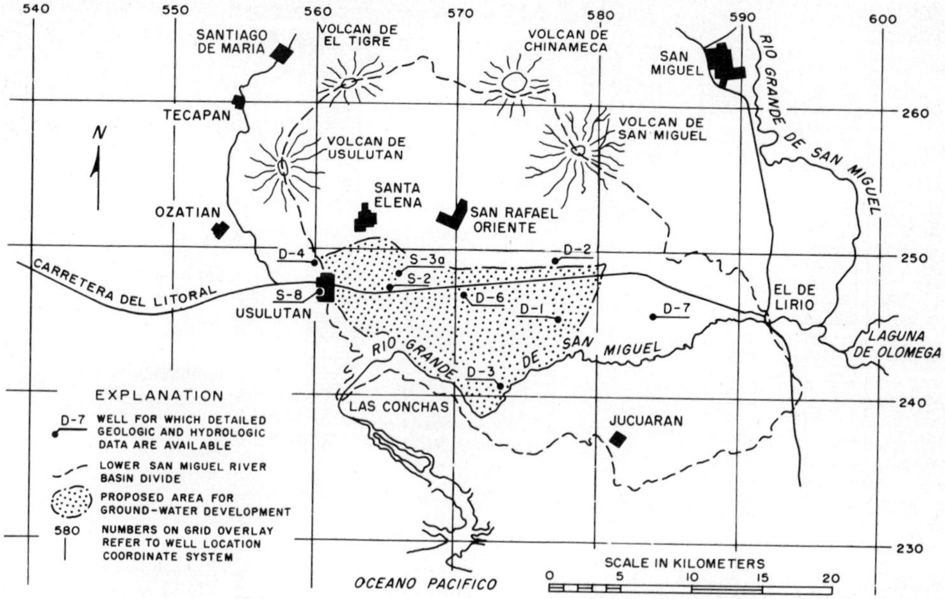

Fig. 4.32 Location of wells used in aquifer test near San Miguel, El Salvador. (*From Walton, 1965a.*)

as shown in Fig. 4.32 were used for the test. The logs of wells at and near the test site are given in Fig. 4.33. The aquifer tested consists primarily of volcanic ash and other pyroclastic materials. The effects of pumping well D-6 were measured in observation well D-6a. Pumping was continued for a period of about 24 hours at a constant rate of 1,000 gpm. Well D-6a is 50 feet from well D-6. The thickness of the aquifer tested is about 274 feet.

The time-drawdown field-data graph for well D-6a is given in Fig. 4.34. The time-drawdown field-data curve follows the $r/D_t = 0.4$ type-curve trace. Match-point coordinates and a r/D_t value of 0.4 were substituted into Eqs. (3.113) to (3.119) to compute coefficients of transmissibility and storage of the aquifer, the delay index, and values of D_i and N. Computations are given in Fig. 4.34. The values of T, P, S, S_y, D_i, N, and delay index are 280,000 gpd/ft, 1,020 gpd/sq ft, 7.3×10^{-3}, 1.44×10^{-1}, 1.21×10^{-2} minute^{-1}, 20.8, and 82.6 minutes, respectively.

St. David, Illinois

An aquifer test was made by W. H. Walker of the Illinois State Water Survey in cooperation with Klinger & Associates, consulting engineers, and Layne-Western Co., well contractor, June 2, 1960 (Walton, 1962a). A group of wells (Fig. 4.35) located about 1.5 miles north of the corporate limits of the village of St. David, Illinois, in section 10, T6N, R4E, in Fulton County was used. The generalized graphic logs of the wells are given in Fig. 4.36. The effects of pumping well 3-60

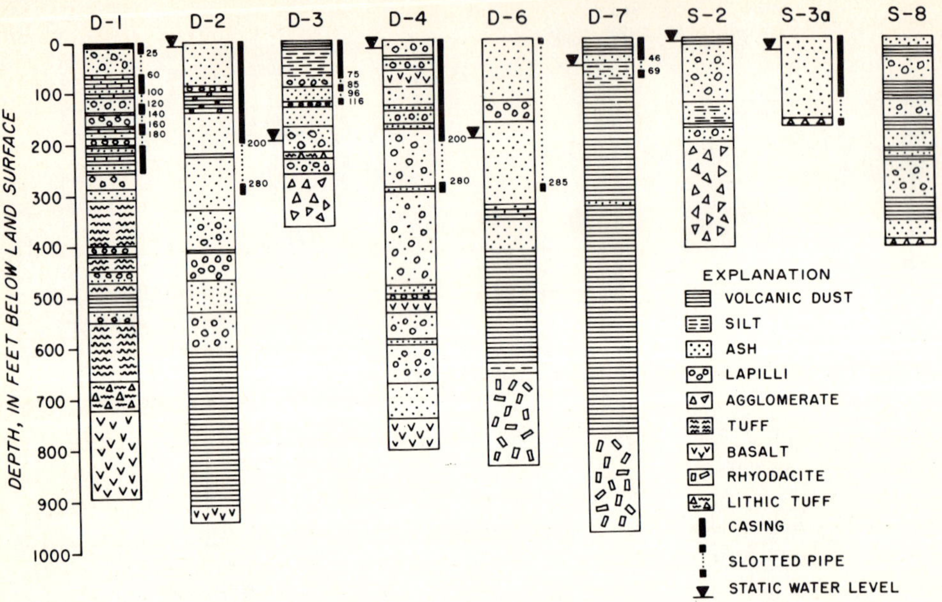

Fig. 4.33 Generalized graphic logs of wells used in aquifer test near San Miguel, El Salvador. (*From Walton, 1965a.*)

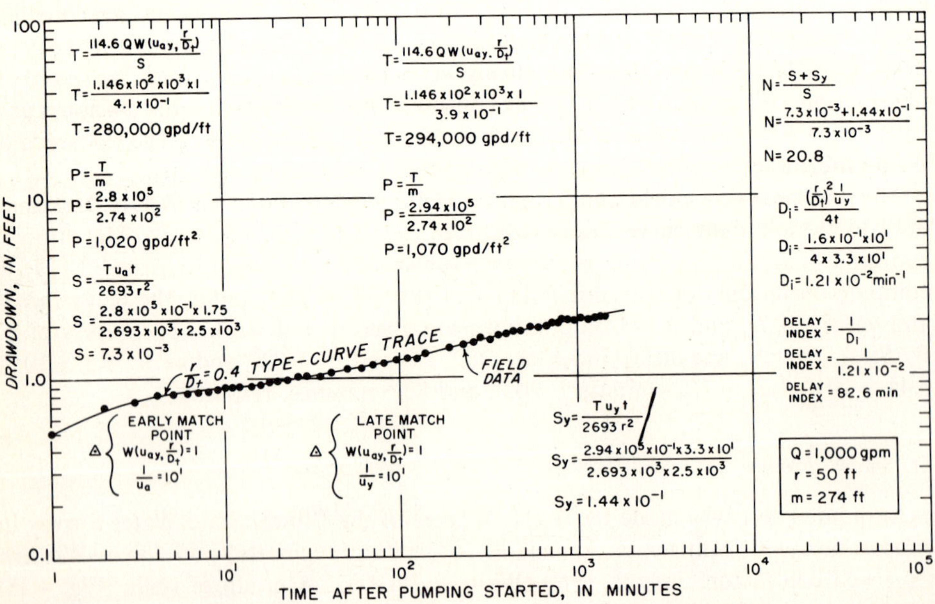

Fig. 4.34 Time-drawdown graph for aquifer test near San Miguel, El Salvador. (*From Walton, 1965a.*)

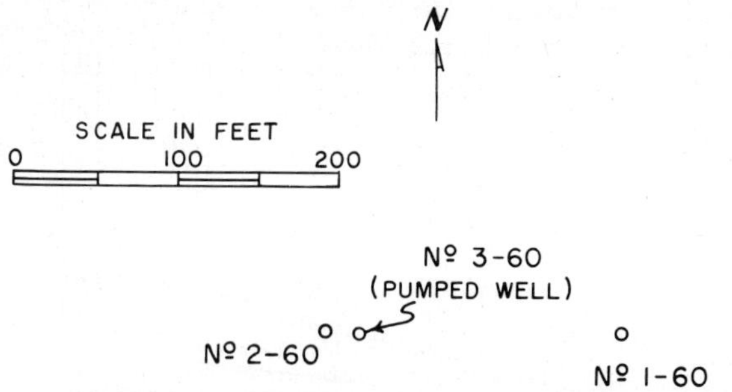

Fig. 4.35 Location of wells used in aquifer test near St. David, Illinois. (*From Walton, 1962a.*)

were measured in observation wells 1-60 and 2-60. Pumping was started at 10:00 A.M. and was continued for six hours at a constant rate of 62 gpm until 4:00 P.M.

Drawdowns in the observation wells were plotted against time on logarithmic paper (Figs. 4.37 and 4.38). The time-drawdown data for wells 1-60 and 2-60 are given in Table 4.10.

The nonleaky artesian, fully penetrating, constant-discharge, time-drawdown type curve in Fig. 4.2 was matched to early time-drawdown data, and values of T and S were calculated using Eqs. (3.48) and (3.49) as shown in Figs. 4.37 and 4.38. After about 12 minutes of pumping the time-rate of drawdown in the observation wells increased and field data deviated upward from type-curve traces, indicating the presence of a barrier boundary. The type curve was again matched to later time-drawdown data affected by the barrier boundary. The correctness of the new match position was judged by noting that, in the second match position, the s value opposite a $W(u)$ value of 1.0 is twice the s value opposite a $W(u)$ value of 1.0 in the first match position. The divergence of the two type-curve traces at a time t_{i1} was determined. The time t_{p1} at which the first type-curve trace intersects an s value equal to the divergence at t_{i1} was noted. The distances from the observation wells to the image wells associated with the barrier boundary were then computed with Eq. (4.18) as shown in Figs. 4.37 and 4.38.

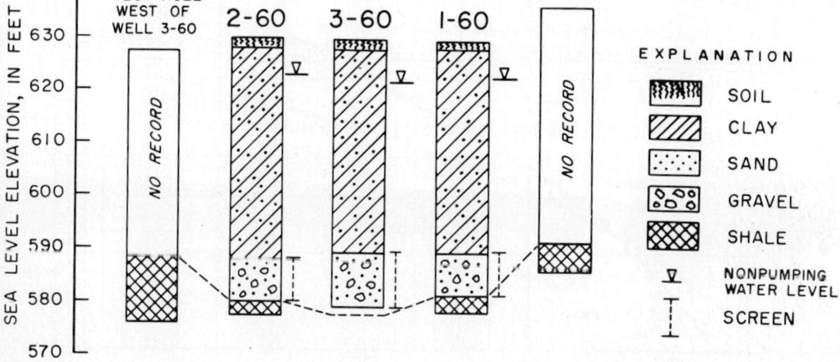

Fig. 4.36 Generalized graphic logs of wells used in aquifer test near St. David, Illinois. (*From Walton, 1962a.*)

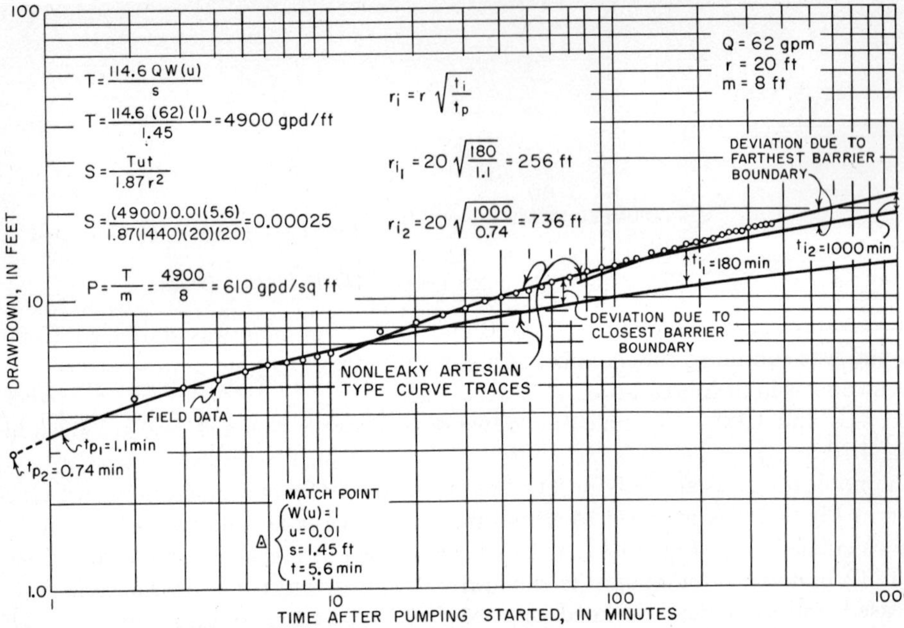

Fig. 4.37 Time-drawdown graph for well 1-60 near St. David, Illinois. (*From Walton, 1962a.*)

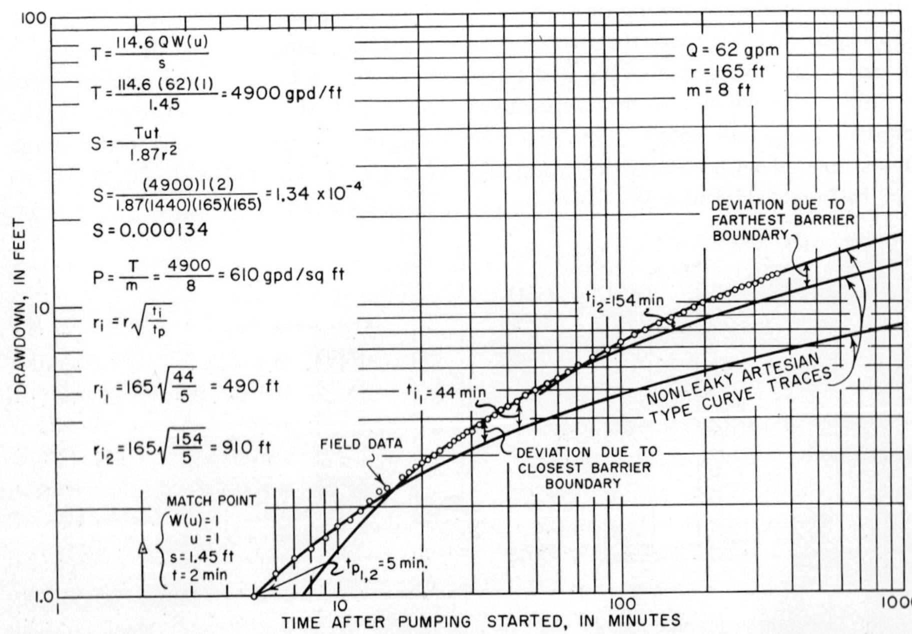

Fig. 4.38 Time-drawdown graph for well 2-60 near St. David, Illinois. (*From Walton, 1962a.*)

TABLE 4.10 TIME-DRAWDOWN DATA FOR TEST AT ST. DAVID, ILLINOIS. (*From Walton, 1962a.*)

Well 1-60[a]		Well 2-60[b]	
Time after pumping started, min	*Drawdown, ft*	*Time after pumping started, min*	*Drawdown, ft*
5	1.00	2	4.60
6	1.19	3	4.97
7	1.25	4	5.32
8	1.43	5	5.65
9	1.58	6	5.91
10	1.72	7	6.08
12	1.94	8	6.20
15	2.32	9	6.33
18	2.64	10	6.50
20	2.85	15	7.62
25	3.26	20	8.21
30	3.61	25	8.75
40	4.42	30	9.23
50	5.00	40	10.02
60	5.51	50	10.55
71	6.12	60	11.25
80	6.50	70	11.74
90	6.91	80	12.21
100	7.28	90	12.54
152	8.91	100	12.80
195	9.90	150	14.35
254	11.15	195	15.38
300	11.64	255	16.57
360	12.60	300	17.14
		360	17.78

[a] $Q = 62$ gpm and $r = 165$ ft.
[b] $Q = 62$ gpm and $r = 20$ ft.

After about 70 to 100 minutes of pumping, the time-rate of drawdown in the observation wells again increased, and field data deviated upward from the second type-curve trace, indicating the presence of a second barrier boundary. The type curve was matched to late time-drawdown data affected by both barrier boundaries. The correctness of the third match position was judged by noting that in the third match position the s value opposite a $W(u)$ value of 1.0 is three times the s value opposite a $W(u)$ value of 1.0 in the first match position. The divergence of the second and third type-curve traces at a time t_{i2} was determined. The time t_{p2} at which the first type-curve trace intersects an s value equal to the divergence at t_{i2} was noted. The distances from the observation wells to the image well associated with the second barrier boundary were then computed with Eq. (4.19).

The distances from each observation well to the image wells were scribed as arcs from the respective observation wells. Theoretically the arcs should intersect at common points, but the real aquifer is not a vertically bounded aquifer as assumed in the derivation of the image-well theory, and as a result the arcs and their intersections are dispersed. In addition, only two observation wells are available, so that the exact locations of the image wells, and therefore of the barrier boundaries, cannot be determined with aquifer test data alone.

As indicated in Fig. 4.36, the aquifer occurs as a fill in a buried valley in shale bedrock. Based on aquifer test data and geologic data, the boundaries were located at positions east and west of the pumped well. It should be pointed out that the boundaries determined with aquifer test data occur at distances at which the changes in thickness and hydrogeologic properties are great enough to constitute barier boundaries. The boundaries represent a rectangular section which is equivalent hydraulically to the real aquifer.

The average values of T and S computed from time-drawdown data are 4,700 gpd/ft and 0.0002, respectively. Based on an average saturated thickness of 8 feet, the coefficient of permeability is about 610 gpd/sq ft. Available data indicate that the aquifer is a thin strip of sand and gravel approximately 600 feet wide which trends northeast to southwest through the test-site area.

Estevan, Saskatchewan, Canada

An aquifer test was made by the Saskatchewan Research Council March 4 through March 12, 1965 (see Walton, 1965c). The group of wells shown in Fig. 4.39 located about 13 miles northwest of Estevan, Saskatchewan, Canada, was used. The generalized graphic logs and construction features of the wells are shown in Figs. 4.40 and 4.41. Johnson stainless steel continuous-slot screens were installed at selected depth intervals. The wells are all screened in approximately the same depth intervals and are open to the coarse sand and gravel materials near the base of the glacial deposits. The effects of pumping the production well were measured in observation wells 1 to 3.

Pumping was started at 2:00 P.M. on March 4 and was continued at a constant rate of 460 igpm for 11,520 minutes or eight days until 2:00 P.M. on March 12. The pumping rate was held constant by means of a gate valve installed in the discharge pipe. A circular orifice and a manometer tube installed in the end of the discharge pipe were used to measure the rate of pumping. The rate of pumping varied between 464 and 457 igpm. Water levels in the production well were frequently measured with a steel tape; water levels in the observation wells were continuously measured by means of recording gages. Atmospheric-pressure changes were measured with a recording Belfort microbarometer.

Drawdowns in the wells, determined by comparing water levels measured before pumping started with water levels measured during the pumping period, were

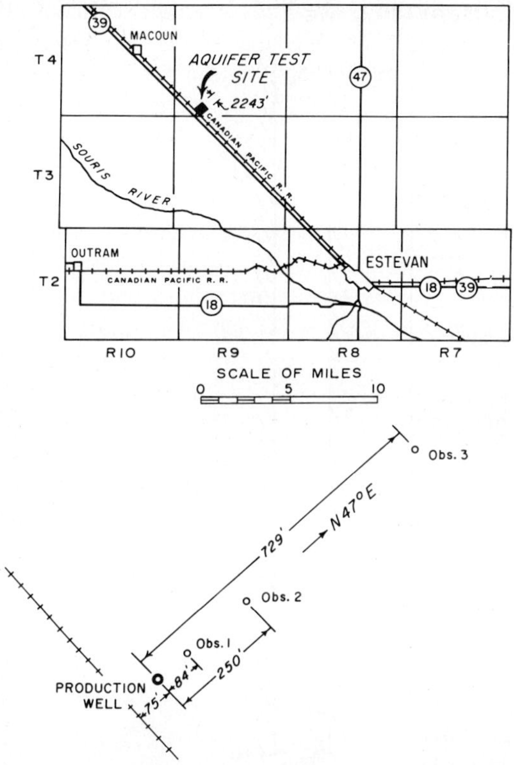

Fig. 4.39 Location of wells used in aquifer test near Estevan, Saskatchewan, Canada. (*From Walton, 1965c.*)

corrected for atmospheric-pressure changes and were plotted against time on logarithmic paper as shown in Fig. 4.42. The average barometric efficiency of the wells was determined to be 33 percent from records of water-level and atmospheric-pressure records collected February 12–16, 1965. Drawdowns in the wells at the end of the test are given in Table 4.11.

The time-drawdown field-data graphs for the observation wells were superposed on the nonleaky artesian, fully penetrating, constant-discharge, time-drawdown type curve in Fig. 4.2. Early time-drawdown data were matched to the type curve as illustrated in Fig. 4.42. Match-point coordinates $W(u)$, $1/u$, s, and t were substituted in Eqs. (3.48) and (3.49) to determine the T and S of the aquifer. The coefficient of transmissibility (188,000 igpd/ft) was divided by the average thickness (25 feet) of the aquifer affected by pumping to determine the coefficient of permeability (7,500 igpd/sq ft).

During the early portion of the aquifer test (prior to about 20 minutes after pumping started), leakage through the aquitard overlying the aquifer was not measurable. After about 3 minutes of pumping the time-rate of drawdown in the observation wells increased and field data deviated upward from the type-curve trace, indicating the presence of a barrier boundary. The type curve was again matched to drawdown

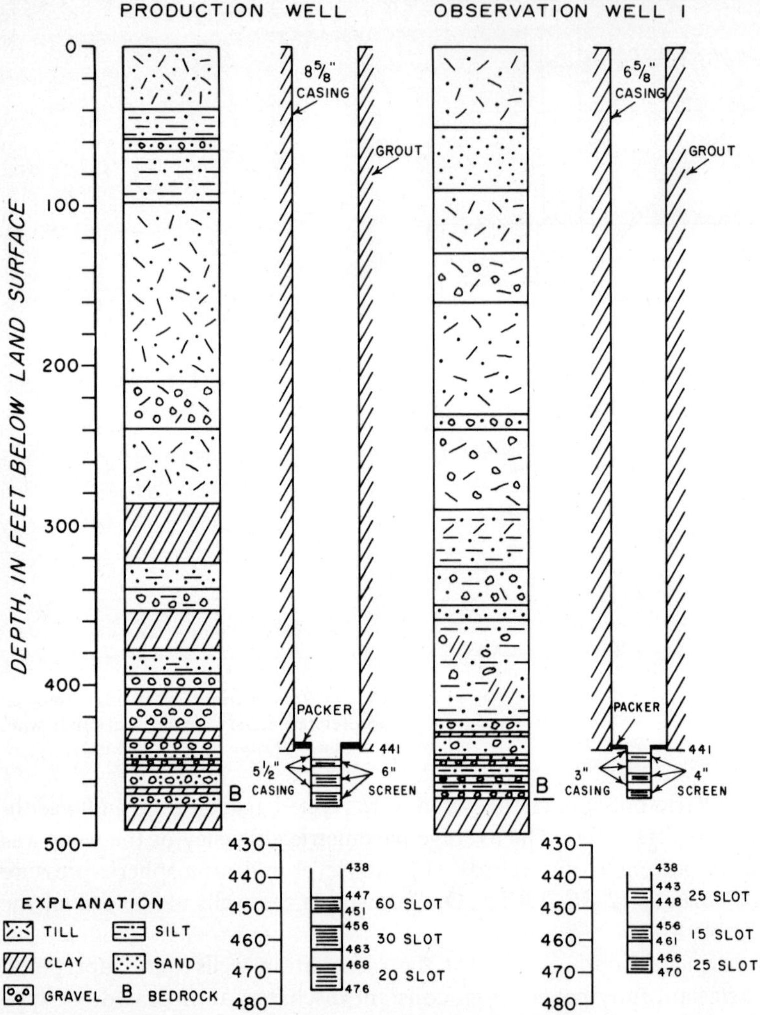

Fig. 4.40 Generalized graphic logs of production well and observation well 1 used in aquifer test near Estevan, Saskatchewan, Canada. (*From Walton, 1965c.*)

data for time values between 3.5 and 8.0 minutes. After about 8 minutes the time-rate of drawdown again increased, indicating the presence of a second barrier boundary. The type curve was again matched to drawdown data for time values between 8 and 15 minutes. The divergence of the three type-curve traces was determined and the distances from the observation wells to image wells associated with the two barrier boundaries were computed with Eqs. (4.18) and (4.19) as shown in Fig. 4.42.

After about 20 minutes of pumping, the time-rate of drawdown decreased as recharge in the form of leakage through the aquitard into the aquifer became appreciable and balanced in part the effects of the barrier boundaries. The time-rate of drawdown accelerated again after about 90 minutes of pumping, probably as water levels in the source bed above the aquitard started to decline appreciably. During the latter part of the test there was leakage through one or more aquitards into the source bed overlying the aquitard immediately above the aquifer. The vertical hydraulic gradient causing flow through the aquitard immediately overlying the aquifer was decreasing as leakage from higher and higher aquitards occurred and as water levels in the source bed above the aquitard declined.

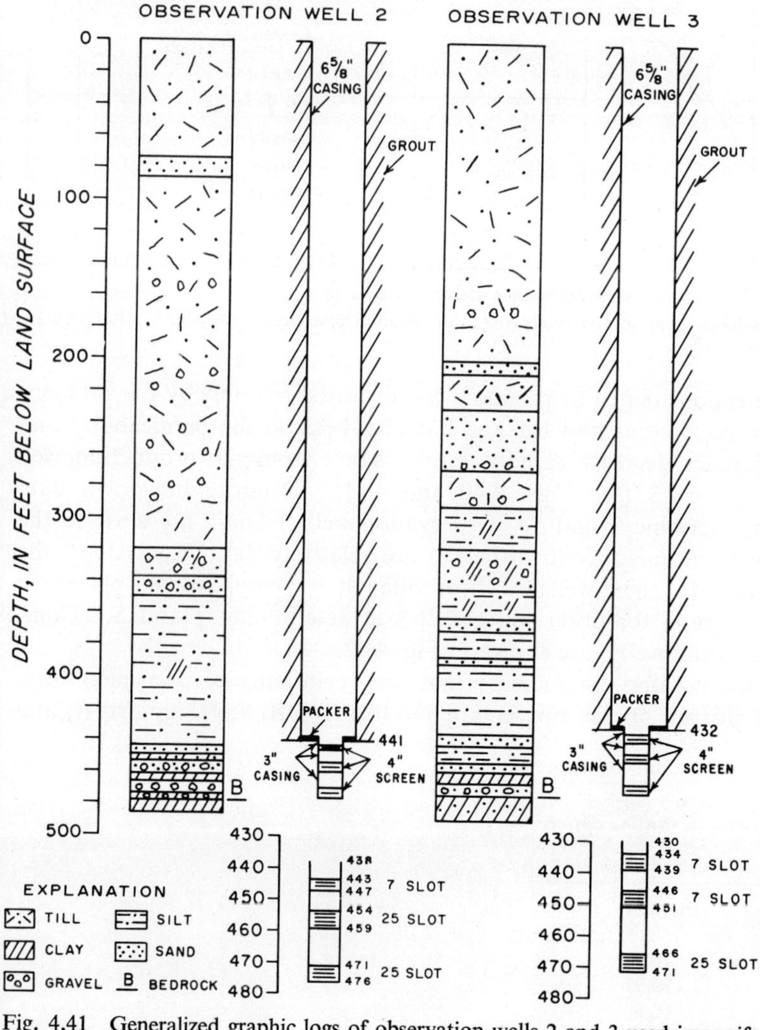

Fig. 4.41 Generalized graphic logs of observation wells 2 and 3 used in aquifer test near Estevan, Saskatchewan, Canada. (*From Walton, 1965c.*)

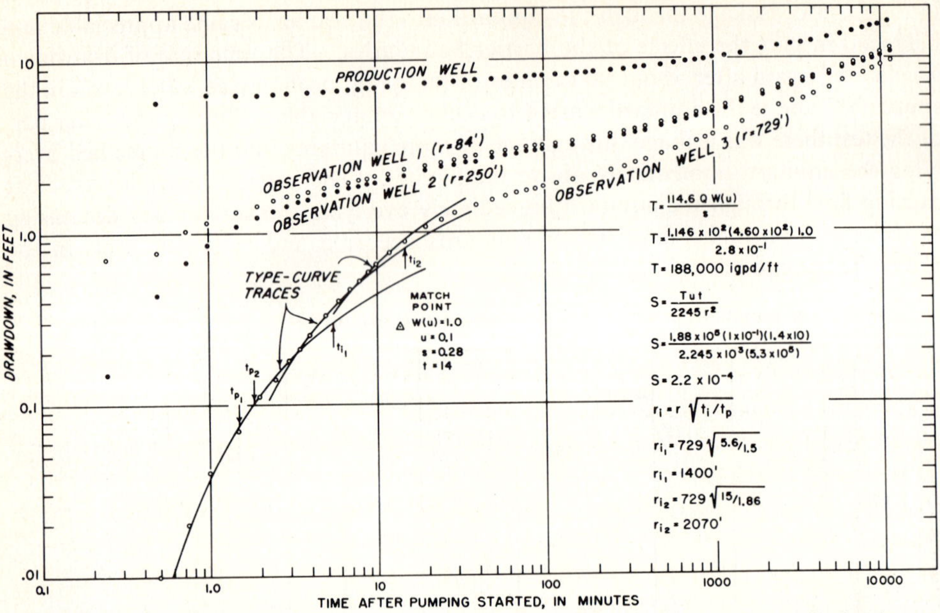

Fig. 4.42 Time-drawdown graphs for wells used in aquifer test near Estevan, Saskatchewan, Canada. (*From Walton, 1965c.*)

The cone of depression created by pumping was distorted not only by the presence of the two barrier boundaries and leakage but also because the permeability and thickness of the aquifer decrease rapidly with distance from the production well toward observation well 3 (see Figs. 4.40 and 4.41). Distance-drawdown data cannot be used to determine T and S. Observation wells 1 and 2 are close to the pumped well and their time-drawdown curves are relatively flat. Analysis of the time-drawdown curves for these wells was most difficult and emphasis was placed on the time-drawdown graph for observation well 3 in determining T and S. Computations for observation well 3 are shown in Fig. 4.42.

The results of the aquifer test indicate that the coefficients of transmissibility, permeability, and storage of the aquifer are 188,000 igpd/ft, 7,500 igpd/sq ft, and

TABLE 4.11 DRAWDOWNS IN OBSERVATION WELLS USED DURING AQUIFER TEST NEAR ESTEVAN, SASKATCHEWAN, CANADA. (*After Walton, 1965c.*)

Well	*Drawdown, ft*
Production	16.03
Observation 1	10.97
Observation 2	10.88
Observation 3	9.54

0.00022, respectively. The time-drawdown graphs and available geologic data suggest that the aquifer is a strip of sand and gravel approximately 1,700 feet wide which trends north-northwest through the production well. The coefficient of permeability is very high and the coefficient of storage is in the normal artesian range.

The coefficient of permeability of the aquitard overlying the aquifer cannot be determined with the leaky artesian, fully penetrating, without water released from storage in aquitard, constant-discharge, time-drawdown type curve because of the complicating effects of the barrier boundaries. In addition, the effective thickness of the aquitard is unknown due to the complex stratification of formations above the aquifer. The drawdown in observation well 3 for a pumping period of 30 minutes was studied to evaluate the leakage coefficient. Several values of P'/m' were assumed and the drawdown in observation well 3 for a pumping period of 30 minutes was computed taking into account the barrier boundaries and hydrogeologic properties determined from early test data. Computed drawdowns were computed with the observed drawdowns. The leakage coefficient used to compute a drawdown equal to the observed drawdown was assigned to the aquitard. A P'/m' of 4.5×10^{-3} igpd/cu ft was assigned to the leakage coefficient of the aquitard. The P'/m' of the aquitard is high (see Walton, 1960), suggesting that the hydraulic connection between the aquifer and overlying stratified sediments is good and that the coefficient of permeability of the aquitard is high.

A drawdown of 0.59 foot was measured in well GSC-3A penetrating the lower aquifer about 9 miles northeast of the production well as shown in Table 4.11, indicating that cones of depression under heavy pumping conditions will spread to great distances.

Zion, Illinois

An aquifer test (Mikels, 1952) was made by Ranney Method Water Supplies, Inc., on November 19–22, 1952. A group of wells (Fig. 4.43) located along the shore of Lake Michigan within the city limits of Zion, Illinois, in section 26, T46N, R12E in Lake County was used. The generalized graphic logs of the wells are given in Fig. 4.44. The effects of pumping well A-P were measured in observation wells A-S-1, A-S-2, A-N-1, A-E-1, A-E-2, and A-E-3. The lake stage was also measured throughout the test. Pumping was started at 11:12 A.M. on November 19 and was continued for about 72 hours at a rate of 99 gpm until 11:30 A.M. on November 22. A power failure resulted in a short shutdown from 10:30 P.M. on November 19 to 12:15 A.M. on November 20. The hydrographs of water levels in observation wells and changes in the lake stage are shown in Fig. 4.45. The maximum drawdown in the pumped well was 12.8 feet.

Drawdowns in the observation wells for a pumping period of 53 hours, before the large change in lake stage occurred, were determined by comparing the extrapolated graphs of water levels measured before pumping started with graphs of water levels after 53 hours of pumping. Drawdown data are given in Table 4.12. Figure 4.45

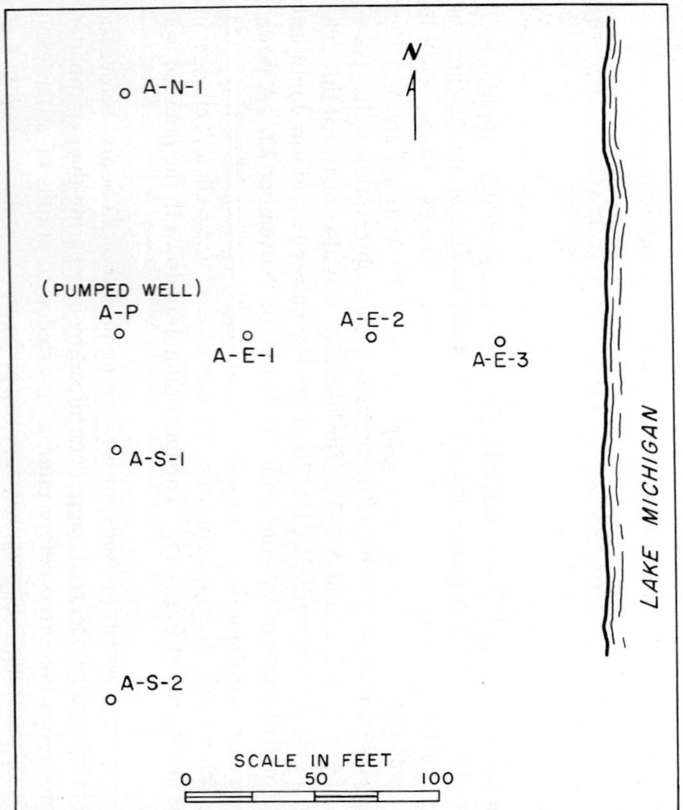

Fig. 4.43 Location of wells used in aquifer test at Zion, Illinois. (*From Mikels, 1952.*)

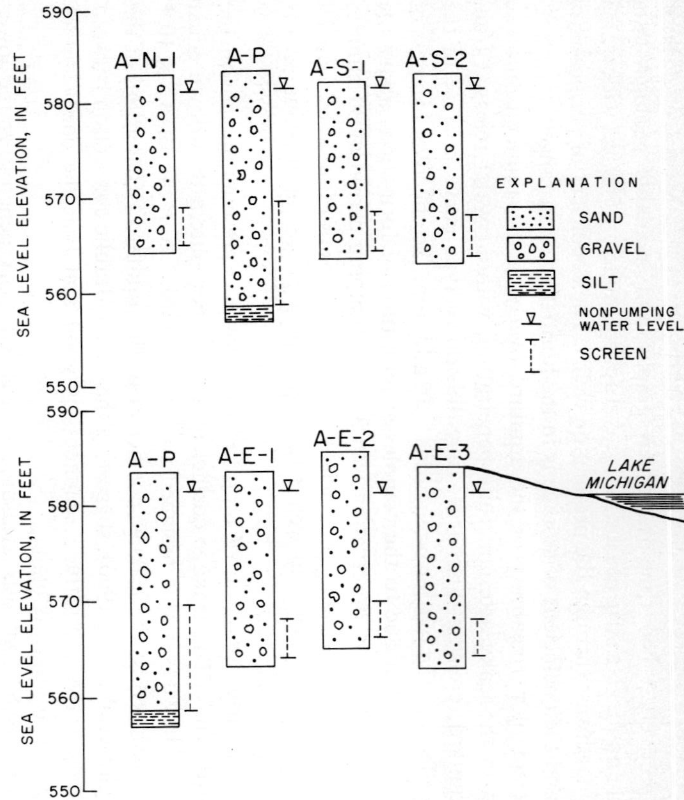

Fig. 4.44 Generalized graphic logs of wells used in aquifer test at Zion, Illinois. (*From Mikels, 1952.*)

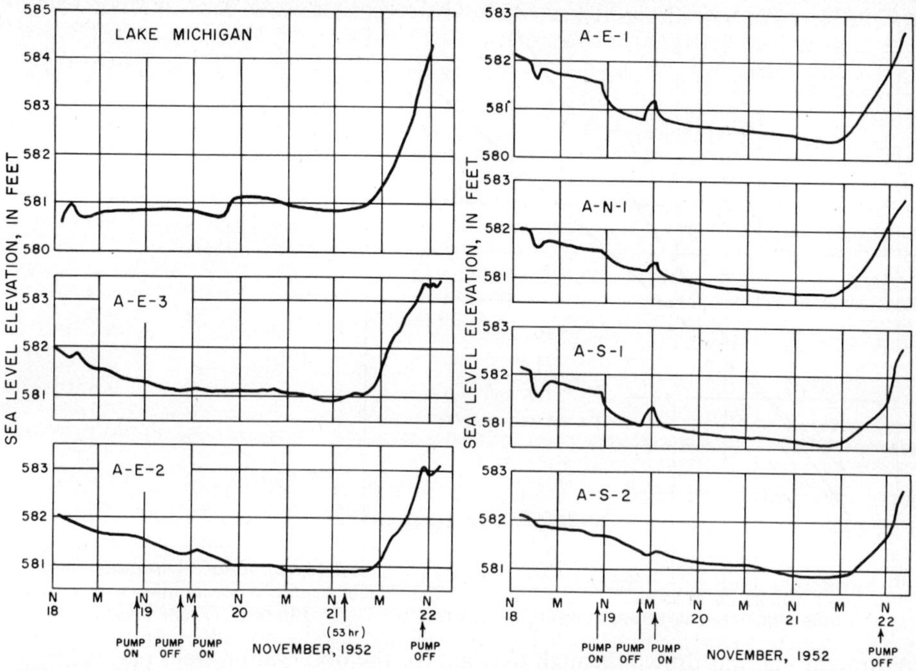

Fig. 4.45 Lake stage and water levels in wells during aquifer test at Zion, Illinois. (*From Walton, 1962a.*)

shows that the lake stage was about the same prior to the test as it was after 53 hours of pumping; therefore, water-level adjustments for changes in lake stage are unnecessary. Drawdowns in the observation wells adjusted for the effects of dewatering were plotted on semilogarithmic paper against the distances from the respective observation wells to the pumped well as shown in Fig. 4.46.

Table 4.12 Distance-drawdown Data for Aquifer Test at Zion, Illinois. (*From Walton, 1962a.*)[a]

Well no.	*Distance from pumped well, ft*	*Drawdown adjusted for dewatering, ft*
A-S-1	48	0.87
A-N-1	96	0.59
A-S-2	141	0.43
A-E-1	51	0.87
A-E-2	100	0.44
A-E-3	151	0.20

[a] $Q = 99$ gpm and $t = 3{,}180$ min.

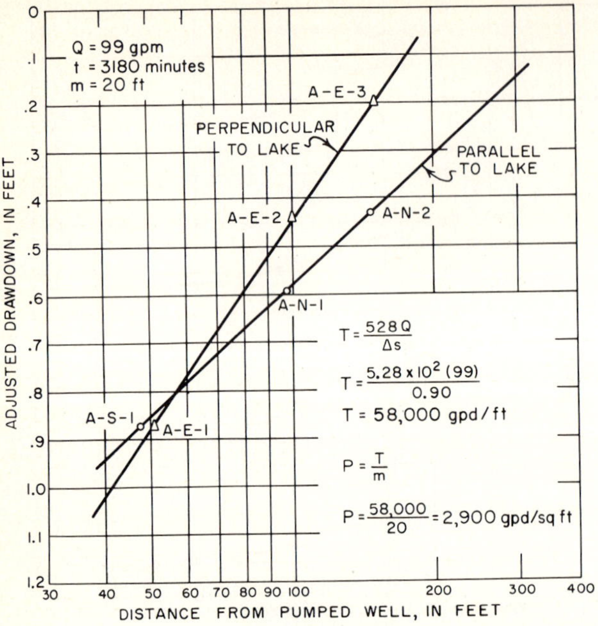

Fig. 4.46 Distance-drawdown graphs for aquifer test at Zion, Illinois. (*From Walton, 1962a.*)

A straight line was drawn through the data for the observation wells parallel to the lake, and the slope of the line was substituted into Eq. (3.55) for computation of *T*. The coefficient of transmissibility was found to be 58,000 gpd/ft and the coefficient of permeability is about 2,900 gpd/sq ft.

Evidence of a hydraulic interconnection between the aquifer and the lake is shown by the similarity between lake and observation-well hydrographs and by a comparison of slopes or hydraulic gradients of lines of wells perpendicular and parallel to the lake. As shown in Fig. 4.46, the slope of the line of wells perpendicular to the lake is much steeper than the slope of the line of wells parallel to the lake, indicating that the cone of depression was distorted by the presence of the lake, and that water from the lake was diverted into the cone of depression by induced infiltration. The time-rate of drawdown in the observation wells decreased a short time after pumping started; after a pumping period of 53 hours, water levels stabilized and near-equilibrium conditions prevailed.

Data in Table 4.12, the computed value of *T*, and Eqs. (3.130) and (3.131) were used to determine the distance to the effective line of recharge. The distance *a* was computed to be 206 feet. The image well associated with the recharge boundary was located on a line perpendicular to the lake shore and at a distance of 412 feet from the pumped well.

Distances from the observation wells to the image well were measured. Knowing *T*, *a*, and the distance from the observation wells to the image well, the coefficient of storage was computed with Eqs. (3.122) through (3.127). The average coefficient of storage is 0.01.

Springfield, Ohio

During 1955 and 1956 an intensive hydrogeologic study was made to determine the feasibility of developing a large water supply for the city of Springfield, Ohio, from a sand and gravel aquifer in an area along the Mad River, about 4 miles northwest of the city (Walton, 1963). The study included a test-hole drilling program and an aquifer test.

The unconsolidated deposits in the study area consist of recent silty alluvial materials and glacial outwash sand and gravel. These deposits are contained in a buried valley cut into the underlying bedrock of Ordovician and Silurian ages. The permeable outwash forming the aquifer averages about 100 feet thick and consists of permeable sand and gravel and occasional thin lenses of clayey materials. Logs of wells and test holes show that the aquifer is about 4,000 feet wide and trends southwest to northeast through the study area. The Mad River has been dredged into the aquifer.

The aquifer test was made by Schaefer and Walton, consulting groundwater hydrologists, and Black and Veatch, consulting engineers, during the period February 20 through 23, 1956. A group of six wells on a line approximately parallel to and 400 feet west of the Mad River as shown in Fig. 4.47 was used. Pumping was at a constant rate of 1,034 gpm.

Values of drawdown in wells N-1, N-2, N-3, and N-4 at a time 4,310 minutes after pumping started were plotted on semilogarithmic paper against values of distance from the pumped well, as shown in Fig. 4.47. A straight line was drawn through the points. The slope of the straight line per log cycle and the pumping rate were substituted into Eq. (3.55), and the coefficient of transmissibility was computed to be 547,000 pgd/ft.

The distance a from the pumped well to the recharge boundary was determined by substituting the computed value of the coefficient of transmissibility, the measured rate of pumping, and values of drawdowns in the observation wells into Eq. (3.130). The average distance a was found to be 686 feet.

The coefficient of storage was determined to be 0.01 by substituting the computed value of T and the drawdowns in the observation wells at the end of the test into Eqs. (3.122) through (3.127). The percentage of pumped water being diverted from the Mad River was estimated to be 93 with Eq. (3.133) and Fig. 3.9. Q_r is then equal to 960 gpm, based on Eq. (3.136).

Points within the streambed up and downstream from the pumped well where drawdown was negligible were located with Fig. 4.47, the hydrogeologic properties of the aquifer, and Eqs. (3.122) through (3.127). The reach of the streambed within the area of influence of pumping was computed to be about 8,000 feet and the average width of the streambed is 82 feet. The streambed area of infiltration is, therefore, about 656,000 square feet. These values of Q_r and A_r were inserted in Eq. (3.137), from which the infiltration rate I_a was computed to be about 9.2×10^4 gpd/acre. Values of h_r were computed with Eqs. (3.122) through (3.127), the hydrogeologic properties of the aquifer, and Fig. 4.47; the average computed value of h_r is 0.09 foot. From Eq. (3.138) the infiltration rate I_h was found to be about 1.0×10^6 gpd/acre/ft.

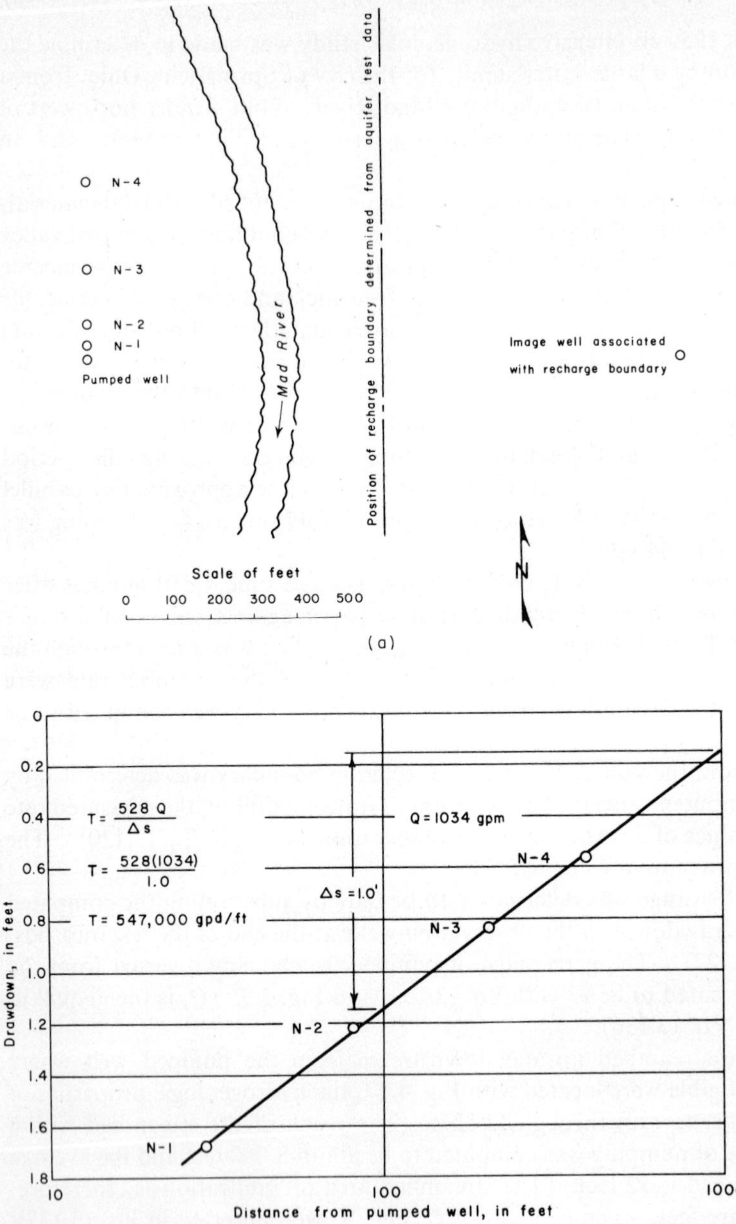

Fig. 4.47 Location of wells (*a*) and distance-drawdown data (*b*) for aquifer test near Springfield, Ohio. (*From Walton, 1963.*)

TABLE 4.13 SUMMARY OF INDUCED STREAMBED INFILTRATION AQUIFER TEST RESULTS. (*From Walton and Ackroyd, 1966.*)

Owner of wells	Location of test site	Date of test	Pumping period, days	Pumping rate, gpm	Hydrogeologic properties: T, gpd/ft	P, gpd/sq ft	S	I_a, gpd/acre	h_r, ft	I_h, gpd/acre ft	Temperature of surface water, °F
City of Springfield, Ohio	Along Mad River, about 4 miles northwest of Springfield, Ohio	2/20–23/56	3	1,034	547,000	5,470	0.01	92,000	0.09	1,000,000	39
The Southwestern Ohio Water Co., Cincinnati, Ohio	Along Miami River, 14 miles northwest of Cincinnati, Ohio	1954	5	1,175	345,000	2,760		224,000	1.33	168,000	82
City of Anderson, Indiana	Along White River, immediately upstream from the confluence of White River and Killbuck Creek at Anderson, Indiana	1953	1½	170	151,000	4,720	0.05	42,000	0.20	216,000	54
City of Canton, Ohio	Along Sandy Creek, 12 miles south of Canton, Ohio	6/20–27/55	7	1,020	204,000	1,700	0.01	331,000	0.46	720,000	69
City of Anderson, Indiana	Along White River, 1 mile west of Anderson, Indiana, ½ mile below sewage treatment plant	1/17–22/57	4½	355	169,000	4,000	0.07	52,100	1.31	39,800	35
Olin Mathison Chemical Corp.	Along Mississippi River, near East St. Louis, Illinois	2/13–17/59	4	7,000	100,000	1,100	0.10	418,000	1.37	305,000	33
Shell Oil Company	Along Mississippi River, near East St. Louis, Illinois	3/3–6/52	3	510	190,000	1,900	0.002	9,800	0.27	35,300	38
Monsanto Chemical Corp.	Along Mississippi River, near East St. Louis, Illinois	8/4–8/52	4	1,100	290,000	3,900	0.08	15,500	0.17	91,200	83

TABLE 4.14 SUMMARY OF INFILTRATION-RATE DATA FOR SLOW AND RAPID SAND FILTERS. (*From Walton, 1963.*)

	Slow sand filter	*Rapid sand filter*
Infiltration rate I_a	1 to 4 to 10 mgd	100 to 125 to 200 mgd
Size of bed	Large, $\frac{1}{2}$ acre	Small, $\frac{1}{100}$ to $\frac{1}{10}$ acre
Depth of bed	12 in. of gravel; 42 in. of sand, usually reduced to 24 in. by scraping	18 in. of gravel; 30 in. of sand or less
Size of sand	Effective size 0.25 to 0.3 to 0.35 mm; uniformity coefficient 2 to 2.5 to 3	0.45 mm and higher; uniformity coefficient 1.5 and lower
Loss of head h_r	0.2 ft initial to 4 ft final	1 ft initial to 9 ft final
Length of run between cleaning	20 to 30 to 60 days	12 to 24 to 40 hr
Infiltration rate I_h	5,000,000 to 20,000,000 gpd/acre/ft to 50,000,000 gpd/acre/ft initial; 250,000 to 1,000,000 to 2,500,000 gpd/acre/ft final	100,000,000 to 125,000,000 to 200,000,000 gpd/acre/ft initial; 11,000,000 to 14,000,000 to 22,000,000 gpd/acre/ft final

A summary of the induced streambed infiltration aquifer test data collected to date by the writer and infiltration rates computed therefrom are given in Table 4.13. Values of I_h range from 39,800 to 1,000,000 gpd/acre/ft. Fair and Geyer (1954) give infiltration rates of slow and rapid sand filters, summarized in Table 4.14, which may be compared with infiltration rates given in Table 4.13. All streambeds tested fall in the clogged slow sand filter category.

4.12 Case Histories of Flow-net Analysis

Joliet, Illinois

A flow-net analysis was made of the piezometric surface of the Cambrian-Ordovician Aquifer in the vicinity of the city of Joliet in northeastern Illinois to check the accuracy of the coefficients of transmissibility computed from the results of aquifer tests. The sequence, structure, and general characteristics of rocks in northeastern Illinois are shown in Fig. 4.48. The Cambrian-Ordovician Aquifer, consisting of the Galena-Platteville Dolomite, Glenwood–St. Peter Sandstone, Prairie du Chien Series, Trempealeau Dolomite, Franconia Sandstone, and Ironton-Galeville Sandstone, is underlain by shale beds of the Eau Claire Formation (aquiclude) which have a very low permeability, and overlain in large parts of northeastern Illinois by the Maquoketa Formation (aquitard) which confines the water in the aquifer under leaky artesian conditions.

The piezometric surface of the Cambrian-Ordovician Aquifer in 1958 is shown in Fig. 4.49. The area bounded by isopiestic lines having elevations of 200 and 150 feet in the vicinity of Joliet was selected for flow-net analysis. The quantity of

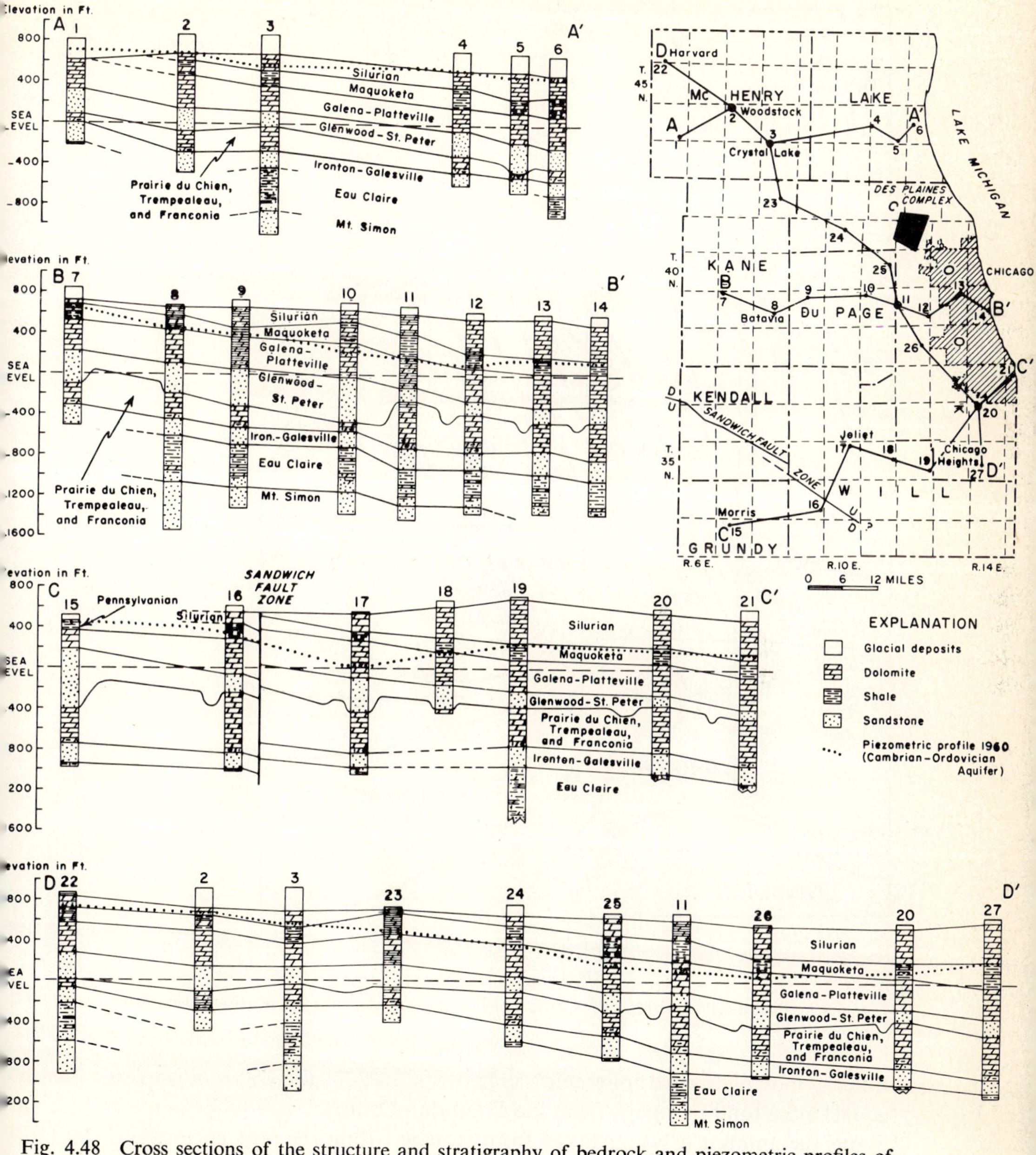

Fig. 4.48 Cross sections of the structure and stratigraphy of bedrock and piezometric profiles of Cambrian-Ordovician Aquifer in northwestern Illinois. (*After Suter et al., 1959.*)

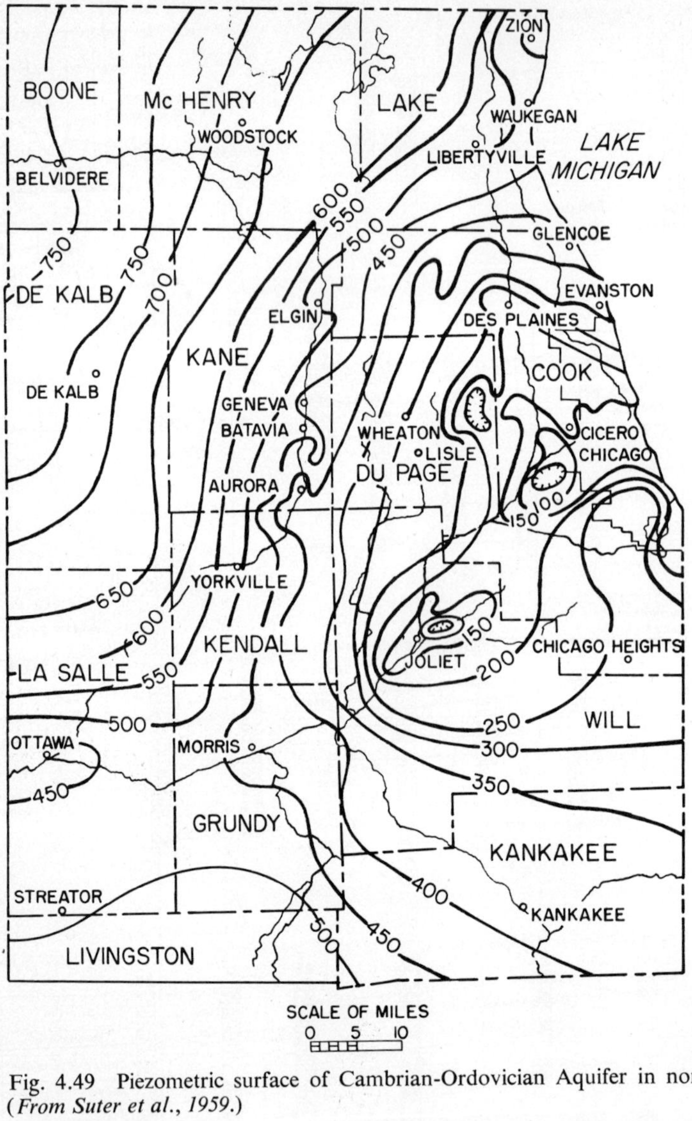

Fig. 4.49 Piezometric surface of Cambrian-Ordovician Aquifer in northeastern Illinois in 1958. (*From Suter et al., 1959.*)

water Q moving toward Joliet midway between the 200- and 150-foot isopiestic lines is equal to the total pumpage from the Cambrian-Ordovician Aquifer in the Joliet area minus the amount of water taken from storage within the area enclosed by the 150-foot isopiestic line. Because the coefficient of storage of the Cambrian-Ordovician Aquifer and the area enclosed by the 150-foot isopiestic line are very small, the amount of water taken from storage is negligible; therefore, Q is essentially equal to the total pumpage. Total pumpage from the Cambrian-Ordovician Aquifer in the Joliet area

TABLE 4.15 COEFFICIENTS OF TRANSMISSIBILITY COMPUTED FROM AQUIFER TEST DATA FOR THE JOLIET AREA, ILLINOIS. (*From Suter et al., 1959.*)

Well owner	*Depth of well, ft*	*Date of test*	*Pumping rate, gpm*	*Coefficient of transmissibility, gpd/ft*
Village of Elwood	1,645	1941	1,345	16,200
Kankakee Ordnance Works	1,649	1953	1,253	17,400
Kankakee Ordnance Works	1,569	1953	1,220	17,000
City of Joliet	1,620	1944	753	14,300
City of Joliet	1,544	1946	600	16,200
City of Joliet	1,608	1946	1,290	17,100
Illinois State Penitentiary 1	1,600	1948	650	19,100
Diagnostic Depot 3	1,600	1948	642	19,300
City of Lockport	1,572	1954	700	13,000
Village of Rockdale	1,586	1946	293	16,500
Village of Romeoville	1,537	1952	1,016	16,000

in 1958 was 11.6 mgd according to Suter et al. (1959). The hydraulic gradient I and the length of cross section L midway between the 300- and 150-foot isopiestic lines were estimated with Fig. 4.49 and Eq. (3.182). The average coefficient of transmissibility of the part of the Cambrian-Ordovician Aquifer within the Joliet cone of depression was computed, using the data mentioned above and Equation (3.182), and is 16.600 gpd/ft. The value of T from flow-net analysis compares favorably with the coefficients of transmissibility computed from the results of the aquifer tests in the Joliet area given in Table 4.15.

Northeastern Illinois

The Maquoketa Formation overlying the Cambrian-Ordovician Aquifer consists mostly of shale, dolomitic shale, and argillaceous dolomite and has a maximum thickness of about 250 feet in northeastern Illinois. The Cambrian-Ordovician Aquifer receives water from overlying glacial deposits largely in areas west of the border of the Maquoketa Formation shown in Fig. 4.50 where the Galena-Platteville Dolomite, the uppermost unit of the aquifer, is directly overlain by glacial deposits (Suter et al., 1959). Recharge of the glacial deposits in turn is derived from precipitation that falls locally. The piezometric-surface map for the Cambrian-Ordovician Aquifer in the year 1864 (see Fig. 4.51*a*) indicates that, under natural conditions, water entering or recharging the aquifer was discharged by vertical leakage upward through the Maquoketa Formation and by leakage into the Illinois River Valley, in areas to the east and south.

The changes in artesian pressure produced by pumping since the days of early settlement have been pronounced and widespread. Pumpage from deep wells

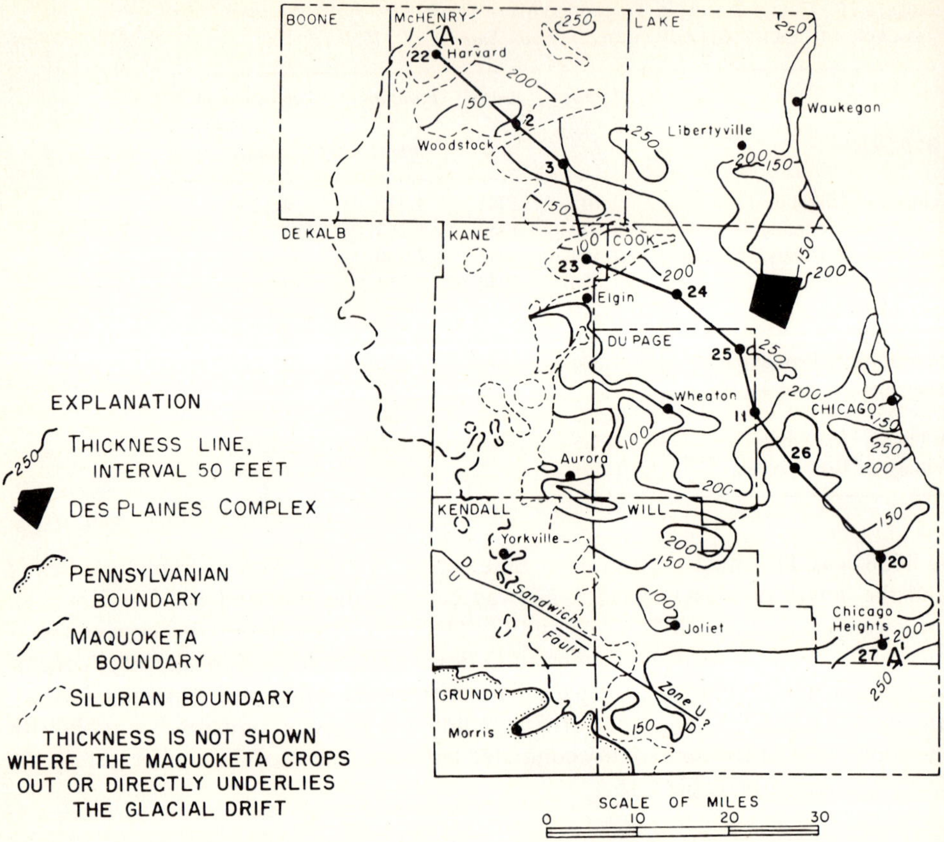

Fig. 4.50 Thickness of Maquoketa Formation in northeastern Illinois. (*From Suter et al., 1959.*)

increased from 200,000 gpd in 1864 to about 78 mgd in 1958. Figure 4.51*b* shows the decline of artesian pressure in the Cambrian-Ordovician Aquifer from 1864 to 1958 as the result of heavy pumping. The greatest declines, more than 600 feet, occurred in areas of heavy pumpage west of Chicago at Summit and Joliet. In 1958, the piezometric surface of the Cambrian-Ordovician Aquifer was several hundred feet below the water table in most of northeastern Illinois, and downward movement of water through the Maquoketa Formation was appreciable under the influence of large differentials in head between shallow deposits and the Cambrian-Ordovician Aquifer (Walton, 1960). The coefficient of permeability of the Maquoketa Formation was computed by flow-net analysis.

Flow lines were drawn from the groundwater divide in McHenry County toward the northern and southern boundaries of Cook County at right angles to the estimated piezometric-surface contours for 1864 as shown in Fig. 4.51*a*. The part of the aquifer (area 1) that is enclosed by the flow lines, the groundwater divide, and section B–B′ was

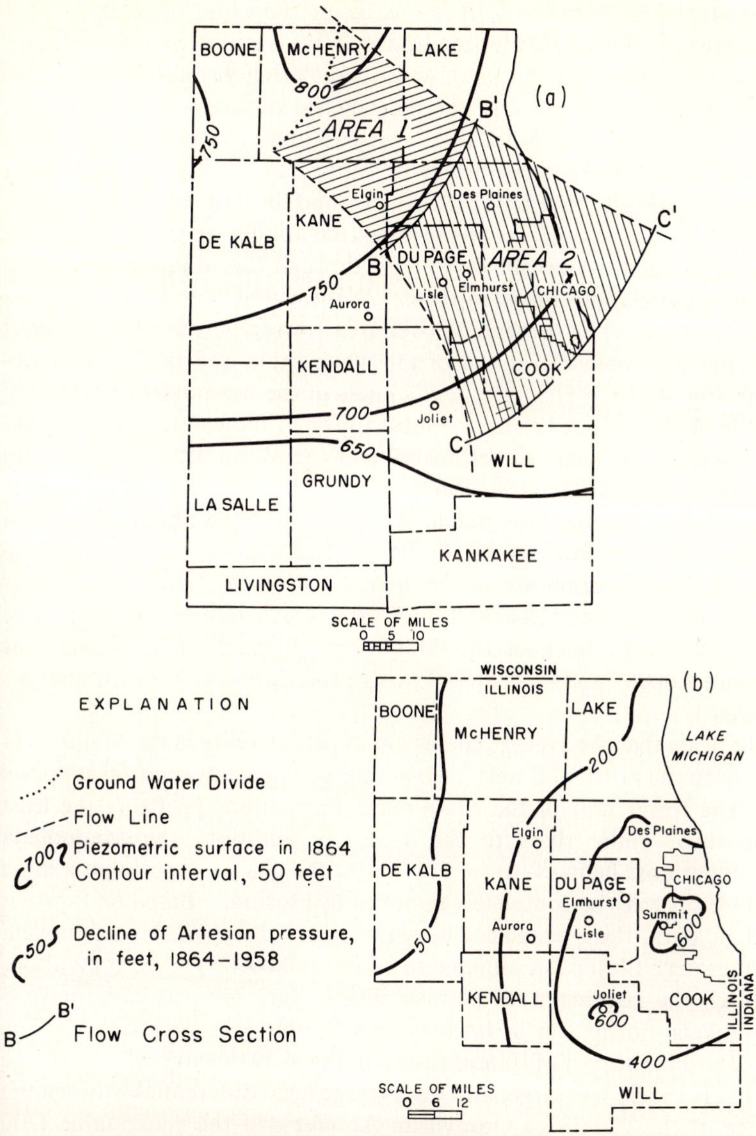

Fig. 4.51 Piezometric surface of Cambrian-Ordovician Aquifer in northeastern Illinois, about 1864 (*a*) and decline in artesian pressure in Cambrian-Ordovician Aquifer, 1864–1958 (*b*). (*From Walton, 1960.*)

considered. In 1864, the piezometric surface was below the water table and downward leakage through the Maquoketa Formation into the aquifer was occurring in area 1. Leakage was equal to the quantity of water percolating through section B–B′. At section B–B′ the hydraulic gradient of the piezometric surface was about 2 ft/mi and the distance between limiting flow lines was about 25 miles. Based on data given by Suter et al. (1959), the average coefficient of transmissibility of the aquifer at section B–B′ is about 19,000 gpd/ft. Using Eq. (3.182) and the data mentioned above, the quantity of water moving southeastward through the aquifer at section B–B′ was computed to be about 1 mgd. Leakage downward through the Maquoketa Formation in area 1 was therefore about 1 mgd in 1864. As measured from Fig. 4.51*a*, area 1 is about 750 square miles. The average Δh over area 1 was determined to be about 85 feet by comparing estimated elevations of the water table and the piezometric-surface contours in Fig. 4.51*a*. The average thickness of the Maquoketa Formation over area 1 from Fig. 4.50 is about 175 feet. Substitution of these data in Eq. (3.186) indicates that the average coefficient of permeability of the Maquoketa Formation in area 1 is about 0.0001 gpd/sq ft (Walton, 1960).

In 1864 the piezometric surface was above the water table southeast of section B–B′, and the quantity of water entering the aquifer in area 1 was discharged by leakage up through the Maquoketa Formation in the areas between the limiting flow lines southeast of section B–B′ in northeastern Illinois and northwestern Indiana. The average coefficient of permeability of the Maquoketa Formation in area 2 was estimated to be about 0.00007 gpd/sq ft by following procedures outlined for analysis of conditions in area 1.

Computations indicate that the average coefficient of permeability of the Maquoketa Formation increases to the north and west. Available geologic information supports this conclusion. The lower unit of the Maquoketa Formation, probably the least permeable of the three units, thins to the west. In addition, the Maquoketa Formation is the uppermost bedrock formation below the glacial deposits in a large part of area 1 and locally may be completely removed by erosion. Based on flow-net analysis and geologic data, the average coefficient of permeability of the Maquoketa Formation in northeastern Illinois is estimated to be about 0.00005 gpd/sq ft (Walton, 1960). Even though the coefficient of permeability is very low, leakage in 1958 through the Maquoketa Formation in northeastern Illinois was appreciable. The area of the aquitard within the part of Illinois shown in Fig. 4.50 through which leakage occurred (4,000 square miles) was large, and the average head differential between the piezometric surface of the Cambrian-Ordovician Aquifer and the water table (300 feet) was great. Computations made with Eq. (3.185) using the data given above, and assuming an average m' of 200 feet and a P' of 0.00005 gpd/sq ft, indicate that leakage was about 8.4 mgd or about 11 percent of the water pumped from deep sandstone wells in northeastern Illinois in 1958 (Walton, 1960).

The piezometric-surface map for 1958 in Fig. 4.49 was used to determine the rate of vertical leakage to the Cambrian-Ordovician Aquifer in areas where the Maquoketa Formation is missing (see Walton, 1965b). Flow lines were drawn from De Kalb

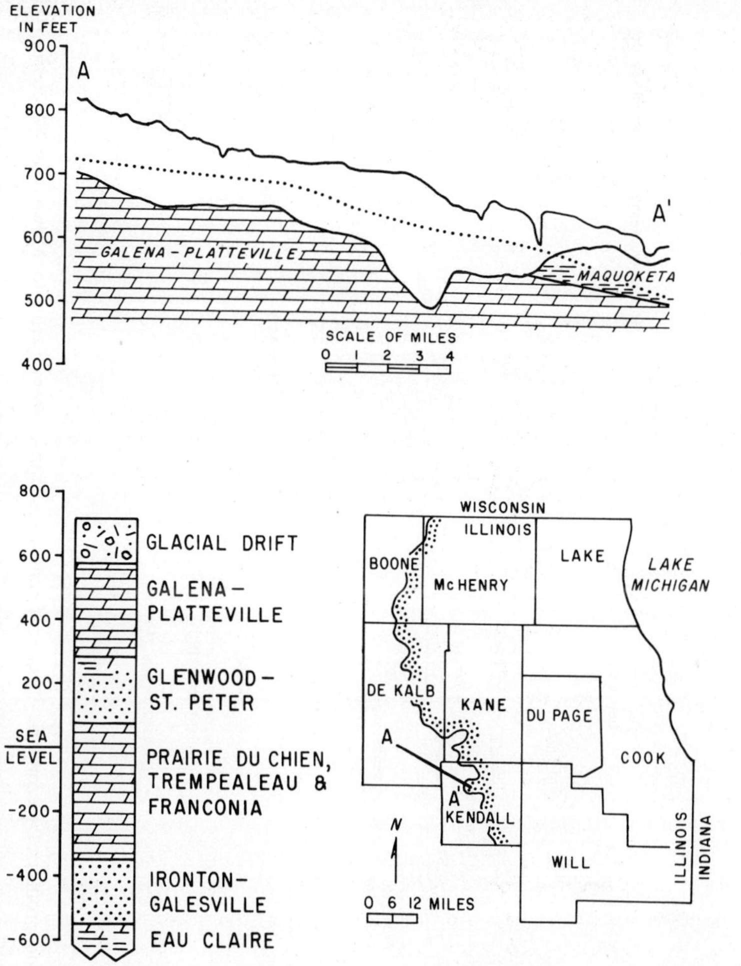

Fig. 4.52 Hydrogeologic cross section in DeKalb and Kendall Counties, Illinois. (*From Walton, 1965c.*)

County toward the cone of depression at Joliet to describe the flow channel shown in Fig. 4.49. The flow channel is north of the Sandwich Fault Zone (see Suter et al., 1959) and the sections A–A′ and B–B′ are both west of the border of the Maquoketa Formation where the Galena-Platteville Dolomite immediately underlies the glacial drift as shown in Fig. 4.52. In 1958 the piezometric surface was below the water table, and downward leakage through glacial deposits and underlying bedrock into the various units of the Cambrian-Ordovician Aquifer was occurring in the flow channel.

At sections A–A′ and B–B′ the hydraulic gradients of the piezometric surface were about 8.3 and 15.0 ft/mi, respectively. The distances between flow lines at sections

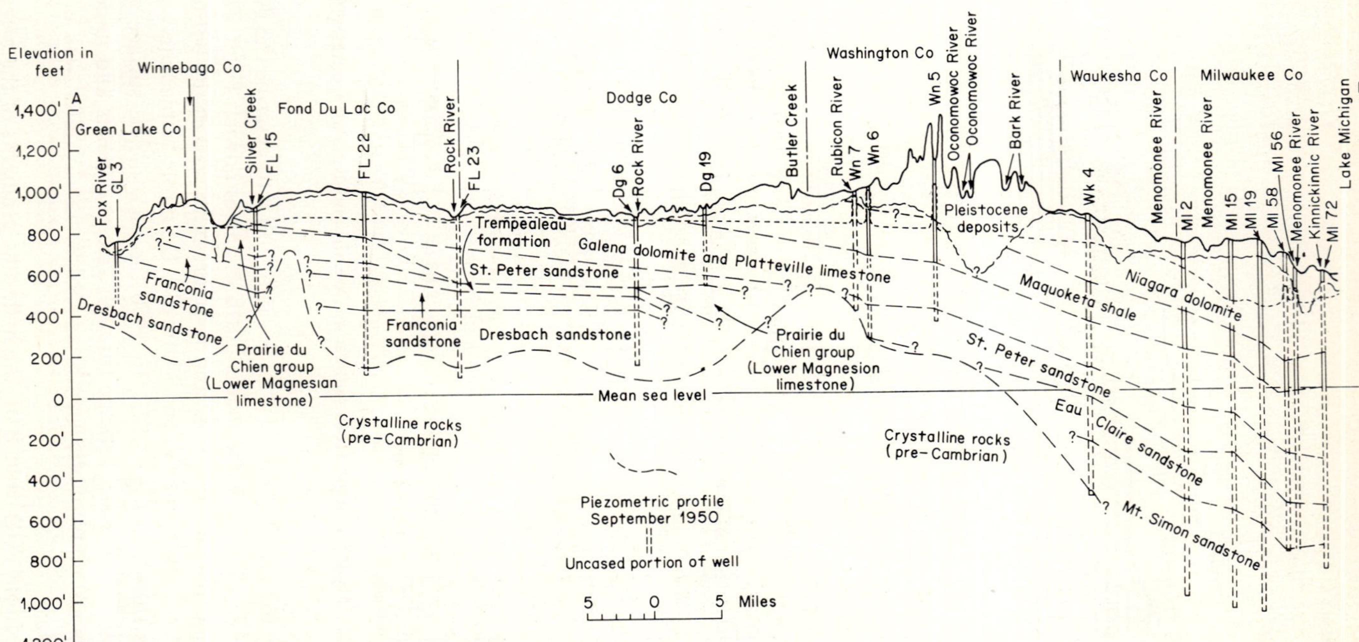

Fig. 4.53 Geologic cross sections and profiles of piezometric surface of sandstone aquifer from Berlin to Milwaukee, Wisconsin. (*From Foley et al., 1953.*)

A–A′ and B–B′ were 1.5 and 7 miles, respectively. The average coefficients of transmissibility and storage of the aquifer within the flow channel are 20,000 gpd/ft and 0.00034, respectively. The area within the flow channel is 100 square miles, and the average water-level decline in 1958 was 2 feet per year (fpy) or 0.0055 fpd. The recharge rate was computed to be 18,000 gpd/sq mi by substituting these data in Eqs. (3.182) and (3.190).

The leakage rate per unit area for the Cambrian-Ordovician Aquifer in 1958 in areas where the Galena-Platteville Dolomite is the uppermost bedrock is about 8.6 times the leakage rate per unit area for the aquifer in areas where the Maquoketa Formation overlies the aquifer, if vertical head losses are not considered. Taking into consideration vertical head losses, the leakage rate per unit area per foot of head loss in areas where the Maquoketa Formation is missing is about 64 times the leakage rate per unit area per foot of head loss in areas where the Maquoketa Formation overlies the aquifer.

Milwaukee-Waukesha area, Wisconsin

Three major aquifers underlie the Milwaukee-Waukesha area, Wisconsin: sandstones of Cambrian and Ordovician age, Niagara Dolomite of Silurian age, and sand and gravel deposits of Pleistocene (Foley et al., 1953). The sequence, structure, and general characteristics of these rocks are shown in Fig. 4.53. The Maquoketa Shale of Ordovician age acts as a more or less effective seal between the Pleistocene deposits and Niagara Dolomite above and the sandstone aquifer below. The sandstone aquifer consists of the Galena Dolomite, the Platteville Limestone, and the St. Peter Sandstone of Ordovician age, and the Eau Claire and Mount Simon Sandstones of Cambrian age.

A large number of deep wells in the sandstone aquifer are uncased in the Niagara Dolomite and allow transfer of water between the Niagara and the sandstone aquifers. The quantity of water recharged locally to the sandstone aquifer by leakage from the Niagara Dolomite through uncased deep wells was estimated by flow-net analysis of the piezometric-surface map for the sandstone aquifer (Foley et al., 1953).

There is very little leakage from the Niagara Dolomite in Waukesha County because there are only a few uncased deep wells. Most of the leakage occurs through downward movement in the many uncased deep wells in Milwaukee County. The leakage from the Niagara Dolomite into the sandstone aquifer was calculated by subtracting the amount of water moving through the sandstone aquifer into the Milwaukee County area, from the total pumpage from deep wells in the Milwaukee County area during a period when essentially no water was being withdrawn from or added to storage in the sandstone aquifer. Examination of hydrographs of wells in the Milwaukee-Waukesha area that are distant from the Milwaukee and Waukesha cones of depression showed no significant change in water levels during the month of May, 1950, indicating that essentially no water was being taken from or added to storage in the sandstone aquifer in that month.

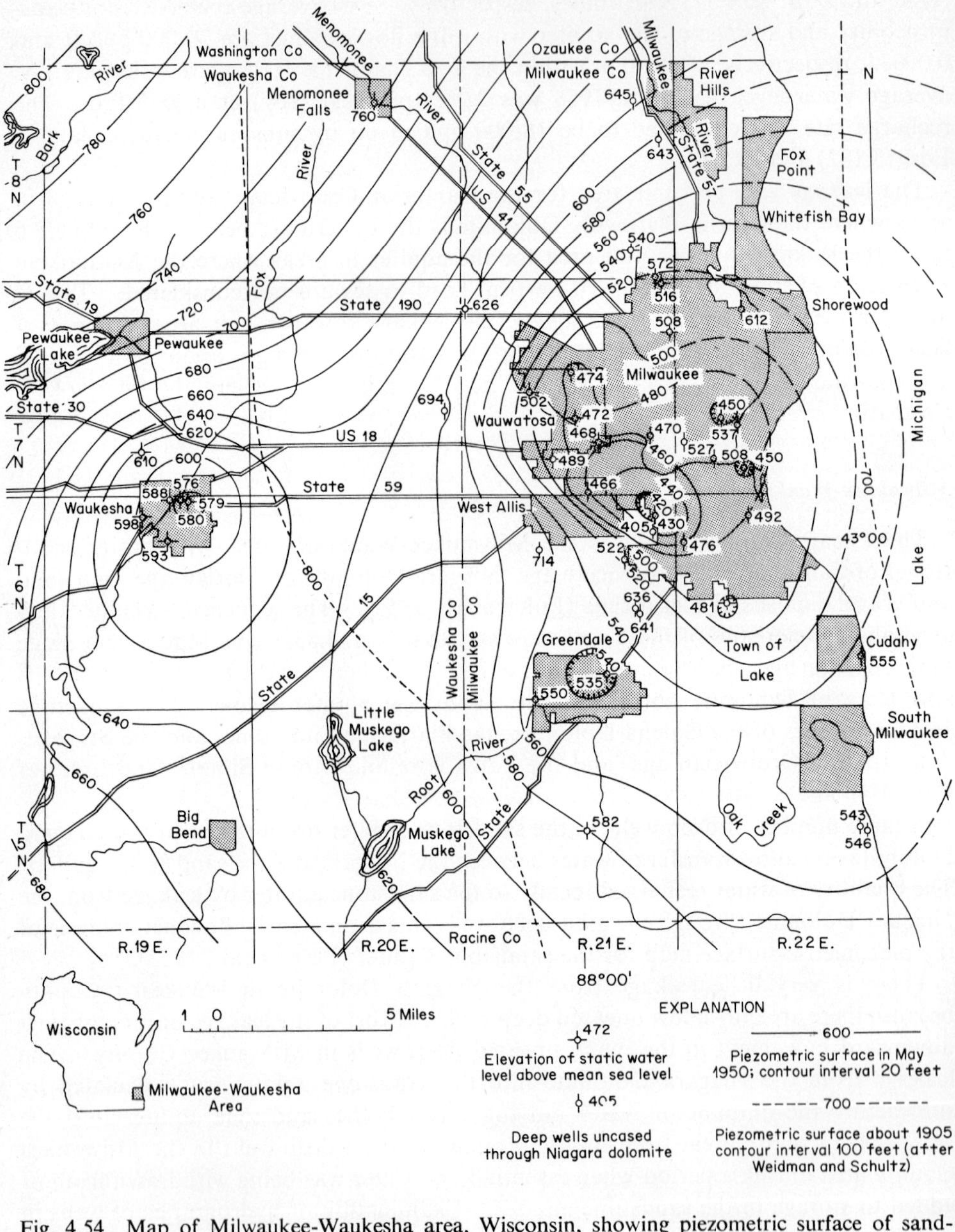

Fig. 4.54 Map of Milwaukee-Waukesha area, Wisconsin, showing piezometric surface of sandstone aquifer in May 1950. (*From Foley et al., 1953.*)

TABLE 4.16 RESULTS OF FLOW-NET ANALYSIS FOR MILWAUKEE-WAUKESHA AREA, WISCONSIN. (*After Foley et al., 1953.*)

Contour interval, ft	L *mi*	A' *sq mi*	W_A *mi*	T *gpd/ft*	I *ft/ft*	Q *mgd*
600–620	58.4	125.7	2.15	24,100	0.00175	13.0

The amount of water moving through the sandstone aquifer into Milwaukee County was computed from Eqs. (3.182) and (3.183). A piezometric-surface map of the sandstone aquifer was constructed from water-level data for May 1950. Figure 4.54 shows the piezometric-surface map. The length of the contours across which all the water arriving in Milwaukee County moves was obtained from the piezometric-surface map. The length L is the average length of the 620- and 600-foot contours between limiting flow lines. The hydraulic gradient I was calculated with Eq. (3.183), in which c_i is the contour interval (20 feet) and A' is the area between limiting flow lines and the piezometric contours. The results of aquifer tests made in Milwaukee County and at Waukesha indicate that the average coefficient of transmissibility of the sandstone aquifer between the 600- and 620-foot contours is 24,100 gpd/ft.

The amount of water flowing through the section of the aquifer midway between the 600- and 620-foot contours was calculated and the results are shown in Table 4.16. The amount of water moving into the Milwaukee County area in May 1950 was found to be about 13 mgd. The actual discharge from deep wells in Milwaukee County in May 1950 was about 18.5 mgd. Therefore, 5.5 mgd was recharged locally by leakage from the Niagara Dolomite through uncased deep wells into the sandstone aquifer. It is believed that most of this leakage occurs through the approximately 100 deep wells in Milwaukee County that are uncased in the Niagara Dolomite; leakage through the Maquoketa Shale was estimated to be very small.

Venice, Ohio

In 1956, the Southwestern Ohio Water Company was pumping groundwater at an average rate of about 13 mgd from a collector well field in the northwest corner of Colerain Township in northern Hamilton County, Ohio. The well field is located along the oxbow meander of the Miami River near Venice, Ohio, about 14 miles northwest of Cincinnati. The water-table map in the well-field area (Big Bend area) was studied to determine the amount of recharge from the river, the induced infiltration rate of the streambed, and the coefficient of transmissibility of the aquifer (Walton and Schaefer, 1956).

The Big Bend area is underlain by approximately 150 feet of unconsolidated deposits consisting mostly of sand and gravel. The average thickness of the sand and gravel deposits beneath the bed of the Miami River is 100 feet. These materials partly fill a deep buried valley now occupied in part by the Miami River. The thick deposits of sand and gravel were for the most part deposited during two periods of Pleistocene glaciation. The deep valley was incised in bedrock of sedimentary origin and of Ordovician age. The rocks that crop out along the sides of the valley are relatively impermeable shales and limestones. The buried valley is about $1\frac{3}{4}$ miles wide and is probably U-shaped, having a broad, relatively flat bottom and steep valley walls.

Controlled aquifer tests of the sand and gravel aquifer underlying the Big Bend area were made in 1950 and 1954. The data obtained during the tests, when analyzed, yielded the following results:

1. The average coefficients of transmissibility and permeability of the aquifer are 300,000 gpd/ft and 2,500 gpd/sq ft, respectively
2. The permeability of the aquifer increases to the north as does the saturated thickness of the aquifer
3. The average long-term coefficient of storage is 0.20

During the summer months of 1956, fourteen 2-inch-diameter observation wells were installed and the water levels in these wells together with those of 26 other observation wells were measured. Eight of the above-mentioned observation wells were driven in the riverbed. The configuration of the riverbed was determined in August 1956 by taking soundings at 11 points along the river.

Water levels in 40 observation wells in the Big Bend area were measured on August 31, 1956. The water-table map shown in Fig. 4.55 was then prepared by drawing contours through points having the same water-level elevation. A contour interval of 1 foot was used. From the water-table map the direction of movement of groundwater, the slope of the water table, areas where infiltration of surface water is occurring, and cones of depression caused by pumping water from the two collectors were determined.

The cones of depression around each collector were studied to check the coefficients of permeability computed from the results of the controlled aquifer tests. The contours having elevations of 512 and 513 feet above mean sea level were selected for analysis because they do not intercept the river and they are not in the zone of projected laterals of the collectors. Consider the approximate cylinders of the aquifer as outlined by the two contours with an elevation of 512 feet. All of the water pumped from collectors 1 and 2 moved through the walls of the two cylinders in accordance with Darcy's law (changes in storage were negligible). The coefficient of permeability of the aquifer in the vicinity of each collector was determined by substituting appropriate data in Eqs. (3.182) and (3.183). The coefficients of permeability of the aquifer in the vicinity of collectors 1 and 2 were computed to be 2,440 gpd/sq ft and 2,780 gpd/sq ft, respectively. These values agree closely with the coefficients of permeability determined from the results of the two aquifer tests and were used to study induced streambed infiltration.

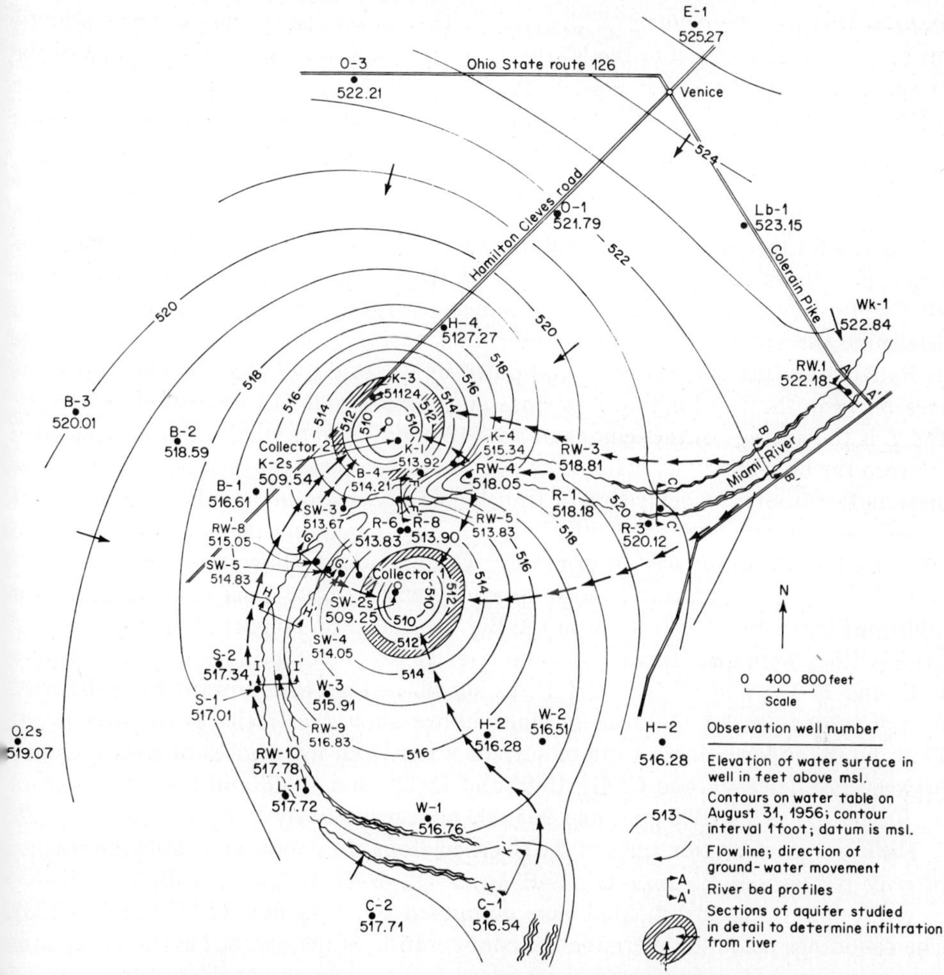

Fig. 4.55 Map showing water-table contours on August 31, 1956, in Big Bend area near Cincinnati, Ohio. (*From Walton and Schaefer, 1956.*)

The water-table contours in Fig. 4.55 bend in an upstream direction and in a downstream direction and roughly parallel the river in the oxbow area in a pattern which shows that a groundwater mound exists beneath the bed of the river. The slope of the water table between the collectors and the river is much greater than the slopes of the water table landward from the collectors. The water-table map is distorted because the aquifer was receiving large amounts of recharge from the river. Pumping from the collectors lowered groundwater levels below stream level which in turn induced surface water to flow downward through the riverbed and into the aquifer.

Contours of the water table north of the Colerain Pike Bridge and near the south end of the oxbow bend in an upstream direction, creating valleys in the water table corresponding to the channel of the river. This shows that groundwater is moving toward and discharging into the Miami River through seeps and springs north of the bridge and near the south end of the oxbow. Thus, recharge by the induced infiltration of surface water was confined to the reach of the river south of the Colerain Pike Bridge and north of the south end of the oxbow.

Flow lines were drawn from the river at the north and south ends of the oxbow toward the collectors as shown in Fig. 4.55. The four flow lines delimit the section of the aquifer through which the infiltrated water from the river moved toward the areas of withdrawal. The two sections of the aquifer included between the four limiting flow lines and the 512 and 513 water-table contours were studied in detail to determine the amount of induced infiltration occurring on August 31, 1956. The hydraulic gradients involved were computed by use of Eq. (3.183), in which A' is the area between the two limiting flow lines and the 512 and 513 water-table contours, and L is the average of the lengths of the 512 and 513 water-table contours included between the flow lines. A planimeter was used to measure the area between the flow lines and contours. Coefficients of transmissibility, hydraulic gradient, and aquifer cross-sectional data were then substituted in Eq. (3.182) and the amount of induced streambed infiltration was determined. Computations show that approximately 10,550,000 gpd, 62 percent of the amount of water pumped from the collectors, was infiltrated from the Miami River in the Big Bend area on August 31, 1956.

Flow lines were also drawn from the river at streambed profiles E–E′ and G–G′, E–E′ and D–D′, and G–G′ and I–I′, as shown in Fig. 4.55, toward the collectors. Computations similar to those outlined above show that recharge on August 31, 1956, by the induced infiltration of surface water from the reaches of river included between profiles E–E′ and G–G′, E–E′ and D–D′, and G–G′ and I–I′ amounted to 6,250,000 gpd, 1,280,000 gpd, and 965,000 gpd, respectively.

The average infiltration rates of the riverbed along the oxbow and along the reaches of river (sections E–E′ to G–G′, E–E′ to D–D′, G–G′ to I–I′, and B–B′ to K–K′) where recharge was investigated were computed by using Eqs. (3.137) and (3.138). The soundings made to determine the configuration of the riverbed at the 11 sections shown in Fig. 4.55 were plotted on standard graph paper and profiles of the riverbed were prepared (see Figs. 4.56 to 4.58). Groundwater levels beneath the riverbed, as indicated by the river wells, and the surface of the river, both computed to sea-level elevations, were drawn on the profiles. The average head loss (difference between the surface of the river and groundwater levels, or the bed of the river in the case where the water table is below the bed of the river) and the average width of the river at each profile were then determined. The lengths of river between sections were measured by means of a calibrated map-measuring instrument. Computations show that the average infiltration rate of the riverbed as determined from a study of infiltration between sections B–B′ and K–K′ is 168,000 gpd/acre/ft. The average infiltration rate of the riverbed varies from one location to another and decreases downstream from the collectors. The infiltration rates for the reaches of river included between sections E–E′ and G–G′, E–E′ and D–D′, and G–G′ and I–I′ were

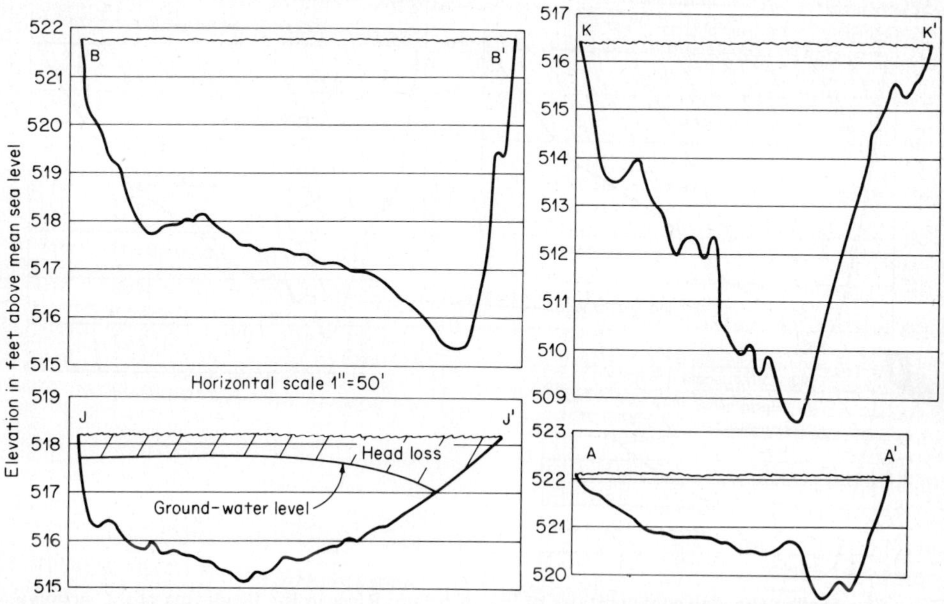

Fig. 4.56 Profiles showing configuration of bed of Miami River in Big Bend Area near Cincinnati, Ohio, at cross sections A-A′, B-B′, J-J′, and K-K′. (*From Walton and Schaefer, 1956.*)

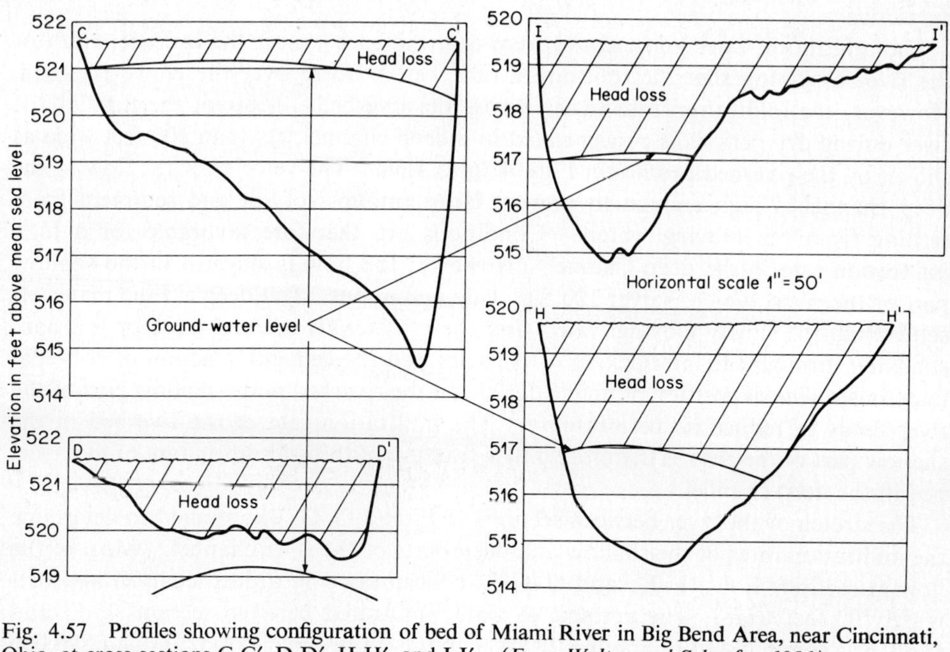

Fig. 4.57 Profiles showing configuration of bed of Miami River in Big Bend Area, near Cincinnati, Ohio, at cross sections C-C′, D-D′, H-H′, and I-I′. (*From Walton and Schaefer, 1956.*)

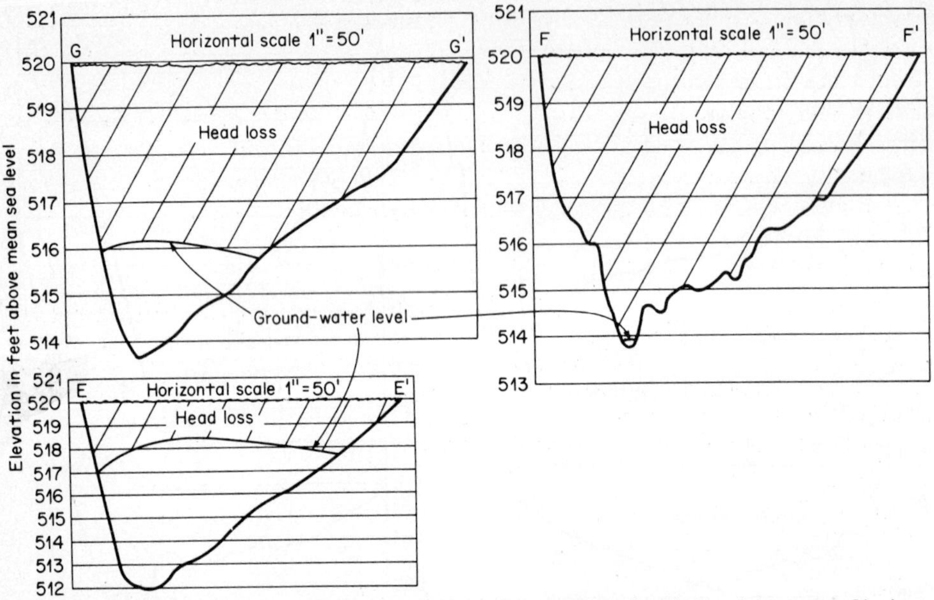

Fig. 4.58 Profiles showing configurations of bed of Miami River in Big Bend Area near Cincinnati, Ohio, at cross sections E-E′, F-F′, and G-G′. (*From Walton and Schaefer, 1956.*)

computed as 245,000 gpd/acre/ft, 229,000 gpd/acre/ft, and 63,500 gpd/acre/ft, respectively.

The infiltration rates given above were determined by using the average width of the river assuming that the infiltration rate was uniform over the entire riverbed. However, the infiltration rate varies across the riverbed. Most of the flow of the river during dry periods is concentrated in a deep channel less than 100 feet wide as shown by the riverbed profiles in Figs. 4.56 to 4.58. The velocity of the flow in the deep channel is high enough to prevent large amounts of silt and sediment from settling from the moving waters. Conditions are therefore favorable for a high infiltration rate in the deep channel. However, the flow is sluggish in the shallow part of the river, which is over 100 feet wide and about 2 feet deep. Fine materials settle from the slowly moving waters and these materials become more or less consolidated through the interlocking of particles and the cementing action of colloidal materials. The deposition of fine materials on the riverbed in the shallow part of the river tends to reduce its permeability. The infiltration rate of the riverbed in the shallow part of the river is therefore only a fraction of the infiltration rate of the riverbed in the deep channel.

The stretch of the river between sections E–E′ and G–G′ was studied to determine the infiltration rates of the shallow and deep parts of the river channel. Most of the flow at sections E–E′, F–F′, and G–G′ is concentrated in a deep channel approximately 95 feet wide. The amount of water infiltrated between sections E–E′ and G–G′ was computed to be about 6,250,000 gpd. It is estimated that about 5,890,000 gpd was infiltrated from the deep channel. The average head loss associated with the above-mentioned infiltration is 3.4 feet. Computations made using Eq. (3.138) show that the infiltration rate of the river in the deep channel is about 415,000 gpd/acre/ft. The infiltration rate of the riverbed in the shallow part is estimated to be about 40,000 gpd/acre/ft. The depth and width of the river in the Big Bend area during periods of low streamflow are controlled by two natural dams (rapids occur in the vicinity of sections D–D′ and I–I′). The relatively low average infiltration rate of the riverbed results from the ponded condition behind these rapids.

Problems

4.1 The saturated thickness of a nonleaky isotropic artesian aquifer infinite in areal extent is 100 feet. A production well fully penetrating the aquifer was continuously pumped at a constant rate of 1,500 gpm for a period of one day. The drawdowns given in Table P4.1 were observed at a distance of 300 feet from the production well in a fully penetrating observation well. Compute the coefficients of transmissibility, permeability, and storage of the aquifer.
Answer: $T = 358{,}000$ gpd/ft; $P = 3{,}580$ gpd/sq ft; $S = 0.00047$.

TABLE P4.1

Time after pumping started, min	*Drawdown, ft*
1	0.45
2	0.74
3	0.91
4	1.04
5	1.13
6	11.2
7	1.28
8	1.32
9	1.38
10	1.45
21	1.79
30	2.02
40	2.17
50	2.30
60	2.34
70	2.41
80	2.50
90	2.58
100	2.67
200	2.96
300	3.11
400	3.25
500	3.32
600	3.41
700	3.46
800	3.50
900	3.55
1,000	3.60
1,440	3.81

4.2 A production well fully penetrating a nonleaky isotropic artesian aquifer infinite in areal extent was continuously pumped at a constant rate of 500 gpm for a period of 200 days. The drawdowns given in Table P4.2 were observed at various distances from the production well in fully penetrating observation wells at the end of the 200-day pumping period. Compute the coefficients of transmissibility and storage of the aquifer.

TABLE P4.2

Distance from production well, ft	*Drawdown, ft*
100	8.40
1,000	5.65
10,000	2.84

Answer: $T = 93{,}000$ gpd/ft; $S = 0.0006$.

TABLE P4.3

Time after flow started, min	*Discharge, gpm*
1	1.04
2	0.90
3	0.86
4	0.81
5	0.79
6	0.76
7	0.73
8	0.72
9	0.71
10	0.70
20	0.63
30	0.61
40	0.59
50	0.57
60	0.56
70	0.55
80	0.54
90	0.53
100	0.52
200	0.48
300	0.47
400	0.45
500	0.44
600	0.43
700	0.42
800	0.42
900	0.41
1,000	0.41
1,440	0.40

4.3 A 1-foot-diameter production well in a nonleaky isotropic artesian aquifer infinite in areal extent has been flowing at a variable rate for a period of one day. The constant drawdown s_w in the well is 5 feet. The discharges given in Table P4.3 were observed in the fully penetrating production well. Compute the coefficients of transmissibility and storage of the aquifer.
Answer: $T = 106$ gpd/ft; $S = 0.0018$.

4.4 The saturated thickness of a nonleaky isotropic artesian aquifer infinite in areal extent is 600 feet. A partially penetrating production well is cased in the upper 450 feet of the aquifer and is screened in the lower 150 feet of the aquifer. The production well was continuously pumped at a constant rate of 1,000 gpm for a period of one day. The drawdowns given in Table P4.4 were observed at a distance of 300 feet from the production well in an observation well screened in the lower 150 feet of the aquifer. Compute the coefficients of transmissibility and storage of the aquifer.
Answer: $T = 239{,}000$ gpd/ft; $S = 0.00032$.

TABLE P4.4

Time after pumping started, min	*Drawdown, ft*
1	0.69
2	0.99
3	1.16
4	1.30
5	1.40
6	1.50
7	1.55
8	1.62
9	1.68
10	1.74
20	2.10
30	2.26
40	2.46
50	2.51
60	2.58
70	2.65
80	2.70
90	2.76
100	2.80
200	3.12
300	3.28
400	3.35
500	3.42
600	3.48
700	3.55
800	3.61
900	3.68
1,000	3.70
1,440	3.91

4.5 The saturated thickness of a leaky artesian aquifer infinite in areal extent without water released from storage in aquitard is 60 feet. The saturated thickness of the aquitard is 30 feet. A fully penetrating production well was continuously pumped at a constant rate of 1,018 gpm for a period of one day. The drawdowns given in Table P4.5 were observed at a distance of 2,100

TABLE P4.5

Time after pumping started, min	*Drawdown, ft*
50	0.02
60	0.03
70	0.04
80	0.06
90	0.07
100	0.09
200	0.23
300	0.36
400	0.43
500	0.50
600	0.56
700	0.60
800	0.65
900	0.68
1,000	0.70
1,440	0.75

feet from the production well in a fully penetrating observation well. Compute the coefficients of transmissibility, permeability, and storage of the aquifer and the permeability of the aquitard. *Answer:* $T = 182{,}000$ gpd/ft; $P = 3{,}030$ gpd/sq ft; $S = 0.002$; $P' = 0.87$ gpd/sq ft.

4.6 The saturated thickness of a leaky artesian aquifer infinite in areal extent without water released from storage in aquitard is 100 feet. The saturated thickness of the aquitard is 80 feet. A fully penetrating production well was continuously pumped at a constant rate of 695 gpm for a period sufficient for water levels to stabilize. The drawdowns given in Table P4.6 were observed

TABLE P4.6

Distance from production well, ft	*Drawdown, ft*
100	16.00
1,000	8.80
10,000	2.60

in fully penetrating observation wells at various distances from the production well at the end of the test. Compute the coefficient of transmissibility of the aquifer and the permeability of the aquitard.
Answer: $T = 57{,}000$ gpd/ft; $P' = 0.012$ gpd/sq ft.

4.7 The saturated thickness of a water-table aquifer infinite in areal extent is 75 feet. A production well fully penetrating the aquifer was continuously pumped at a constant rate of 1,100 gpm for a period of three days. The drawdowns given in Table P4.7 below were observed at a distance of 200 feet from the production well in a fully penetrating observation well. Compute the coefficients of transmissibility and storage, S_y, of the aquifer, the delay index, and t_{wt}.
Answer: $T = 173{,}000$ gpd/ft; $S_y = 0.09$; delay index $= 56$ min; t_{wt} 390 min.

TABLE P4.7

Time after pumping started, min	*Drawdown, ft*
10	0.21
20	0.24
30	0.28
40	0.31
50	0.34
60	0.37
70	0.40
80	0.43
90	0.46
100	0.50
200	0.77
300	0.99
400	1.12
500	1.27
600	1.38
700	1.45
800	1.56
900	1.62
1,000	1.70
2,000	2.15
3,000	2.45
4,320	2.76

4.8 A production well fully penetrating a nonleaky isotropic artesian aquifer delimited by two barrier boundaries was continuously pumped at a constant rate of 1,485 gpm for a period of 4 hours. The drawdowns given in the table below were observed at a distance of 300 feet in a fully penetrating observation well. Compute the coefficients of transmissibility and storage of the aquifer and the distances to the image wells associated with the barrier boundaries.
Answer: $T = 385{,}000$ gpd/ft; $S = 0.00034$; $r_{i1} = 2{,}450$ ft; $r_{i2} = 4{,}370$ ft.

TABLE P4.8

Time after pumping started, min	*Drawdown, ft*
2	0.80
3	0.92
4	1.06
5	1.17
6	1.23
7	1.32
8	1.37
9	1.43
10	1.48
20	1.88
30	2.11
40	2.34
50	2.52
60	2.70
70	2.83
80	3.00
90	3.17
100	3.30
200	4.21
300	4.43

4.9 The saturated thickness of a water-table aquifer hydraulically connected to a stream 1,500 feet wide is 100 feet. A fully penetrating production well 200 feet from the shore of the stream was continuously pumped at a constant rate of 510 gpm for a period of three days. The drawdowns given in the table below were observed at the end of the test in three fully penetrating observation wells at various distances from the production well on a line through the production well and parallel to the stream. Compute the coefficients of transmissibility and storage of the aquifer, the distance a to the effective line of recharge, and the infiltration rate of the streambed per foot of head loss, I_h.
Answer: $T = 210{,}000$ gpd/ft; $S = 0.002$; $I_h = 36{,}300$ gpd/acre/ft.

TABLE P4.9

Distance from production well, ft	*Drawdown, ft*
100	0.95
200	0.58
400	0.17

4.10 Given: $Q_c = 1$ mgd, $A_c = 750$ square miles, $\Delta h = 85$ feet, and $m' = 175$ feet. Compute P'.
Answer: 1×10^{-4} gpd/sq ft.

4.11 Given: A nonleaky isotropic artesian aquifer, $T = 17{,}000$ gpd/ft, $S = 0.0004$. A fully penetrating production well has been discharging 1 mgd at a continuous rate for a period of 1 year. Prepare a time-drawdown graph for a fully penetrating observation well located 1 mile from the production well. Prepare a distance-drawdown graph for a time-period of 1 year.

References

Boulton, N.S., 1963. Analysis of Data from Nonequilibrium Pumping Tests Allowing for Delayed Yield from Storage, *Proc. Inst. Civil Engrs.* (*London*), vol. 26, no. 6693.

Bruin, J., and H. E. Hudson, Jr., 1955. "Selected Methods for Pumping Test Analyses," Illinois State Water Surv., Rept. of Invest. 25.

Cooper, H. S., Jr., and C. E. Jacob, 1946. A Generalized Graphical Method for Evaluating Formation Constants and Summarizing Well-field History, *Trans. Am. Geophys. Union*, vol. 27, no. 4.

Fair, G. M., and J. C. Geyer, 1954. "Water Supply and Waste-water Disposal," John Wiley & Sons, Inc., New York.

Foley, F. C., W. C. Walton, and W. J. Drescher, 1953. "Ground-water Conditions in the Milwaukee-Waukesha Area, Wisconsin," U.S. Geol. Surv. Water Supply Paper 1229.

Jacob, C. E., 1940. On the Flow of Water in an Elastic Artesian Aquifer, *Trans. Am. Geophys. Union*, pt. 2.

———, 1944. "Notes on Determining Permeability by Pumping Tests under Water-table Conditions," U.S. Geol. Surv., mimeographed report.

———, 1946. Radial Flow in a Leaky Artesian Aquifer, *Trans. Am. Geophys. Union*, vol. 27, no. 2.

——— and S. W. Lohman, 1952. Nonsteady Flow to a Well of Constant Drawdown, in an Extensive Aquifer, *Trans. Am. Geophys. Union*, vol. 33., no. 4.

Lohman, S. W., 1947. "Ink-well Gage for Measuring Artesian Head," U.S. Geol. Surv., mimeographed report.

Mikels, F. C., 1952. "Report on Hydrogeological Survey for City of Zion, Illinois," Ranney Method Water Supplies, Inc., Columbus, Ohio.

Norris, S. E., 1962. "Permeability of Glacial Till," U.S. Geol. Surv., Research 1962.

Papadopulos, I. S., 1965. Nonsteady Flow to a Well in an Infinite Anisotropic Aquifer, Symp. Intern. Assoc. Sci. Hydrology, Dubrovinik.

Prickett, T. A., 1965. Type-curve Solution to Aquifer Tests under Water-table Conditions, *Ground Water*, vol. 3, no. 3.

Roberts, W. J., and H. E. Romine, 1947. Effect of Train Loading on the Water Level in a Deep Glacial-drift Well in Central Illinois, *Trans. Am. Geophys. Union*, vol. 28, no. 6.

Suter, Max, R. E. Bergstrom, H. F. Smith, G. H. Emrich, W. C. Walton, and T. E. Larson, 1959. "Preliminary Report on Ground-water Resources of the Chicago Region, Illinois," Illinois State Water Surv. Cooperative Ground-water Rept. 1.

Walton, W. C., 1960. "Leaky Artesian Aquifer Conditions in Illinois," Illinois State Water Surv., Rept. of Invest. 39.

——, 1962a. "Selected Analytical Methods for Well and Aquifer Evaluation," Illinois State Water Surv. Bull. 49.

———, 1962b. "Ground-water Resources of Camas Prairie, Camas and Elmore Counties, Idaho," U.S. Geol. Surv. Water Supply Paper 1609.

———, 1963. "Estimating the Infiltration Rate of a Streambed by Aquifer-test Analysis," Intern. Assoc. Sci. Hydrology, Gen. Assembly, Berkeley.

———, 1964. "Ground-water Conditions at Bracebridge, Ontario," typewritten report prepared for James F. MacLaren Limited, Toronto, Ontario, Canada.

———, 1965a. "Feasibility of Scheme of Ground-water Development in Lower Basin of Rio Grande de San Miguel, El Salvador," typewritten report prepared for Harza Engineering Co. International, Chicago.

———, 1965b. "Ground-Water Recharge and Runoff in Illinois," Illinois State Water Surv., Rept. of Invest. 48.

———, 1965c. "Potential Yield of a Sand and Gravel Aquifer in a Buried Valley near Estevan, Saskatchewan—Results of Aquifer and Well Production Tests and Evaluation of Ground-water Potential," typewritten report prepared for Saskatchewan Research Council, Saskatoon, Saskatchewan, Canada.

——— and E. J. Schaefer, 1956. "Report to the Southwestern Ohio Water Co. on Evaluation of the Potential Yield of the Aquifer in the Big Bend Area, Ohio," mimeographed report.

——— and E. A. Ackroyd, 1966. "Effects of Induced Streambed Infiltration on Water Levels in Wells during Aquifer Tests," Minn. Water Resources Research Center, Bull. 2.

CHAPTER 5

WELL-DESIGN CRITERIA, CONSTRUCTION, PRODUCTION TESTS, AND MAINTENANCE

Wells generally serve as devices for extracting groundwater from aquifers. The importance of wells, however, is not limited to water-supply problems. Wells serve such purposes as for subsurface exploration and observation, disposal of sewage or industrial wastes, draining agricultural lands, controlling salt-water encroachment, relieving pressures under dams or levees, and artificial recharge of aquifers.

Selection of the design features of a production well and a particular method of construction depends upon the quantity of water required, economic factors, the well's objective, and hydrogeologic conditions. Shallow low-capacity production wells are usually dug, bored, diven, or jetted, and high-capacity production wells are usually drilled by the cable-tool, hydraulic-rotary, or reverse-rotary method. Collector wells and infiltration galleries are sometimes constructed where special hydrogeologic conditions exist.

After a production well is drilled, it must be completed with casing and a screen or perforated casing in the case of a production well in unconsolidated deposits. In addition, the production well must be developed for optimum yield and tested before a pump is installed. Production wells must be properly sealed to prevent contamination from surface or subsurface sources. Most production wells need periodic maintenance. There are various methods for rehabilitating old production wells whose yields have decreased because of corrosion or incrustation.

The drilling and construction of many production wells can benefit from the professional assistance of groundwater hydrologists. Geologic and hydrologic information gives positive guidance concerning the proper location and completion of production wells and optimum pumping rates. In exploring for production-well sites, groundwater geologists, with the aid of aerial photographs, regional geologic maps, and rapid ground reconnaissance, can make preliminary conclusions pertaining to the occurrence and movement of groundwater at the rate of 100 square miles per day. Detailed groundwater exploration work can progress at the rate of 1 square mile per day (Davis and DeWiest, 1966).

Many problems of production-well location can profit from the application of geophysical exploration methods. Subsurface geophysical methods give positive guidance concerning the construction features of production wells. Geophysical techniques may eliminate the need for an extensive test-hole drilling program. If the geologic framework of the area is unfavorable for the use of geophysics and no inferences concerning geologic conditions can be gained from surface-exploration studies, the groundwater hydrologist must resort to test-hole drilling. The final designs of production wells are dependent to a large extent on the accurate well logs based on profession examination of drilling cutting samples. The proper collection and analysis of drilling cutting samples and the collection of records of water quality and water levels are important.

PRODUCTION-WELL-DESIGN CRITERIA AND SPECIFICATIONS

There are many criteria used in designing production wells and to date there are no standardized design criteria acceptable to all those concerned with water wells. The production-well-design criteria presented here are something of a compromise between the various criteria in the literature.

Production-well completion must provide for ready entrance of ground water into the production well. In consolidated formations, where the material of the well wall is stable, water enters directly into the uncased production well. Surface casing is required to prevent contamination of the production well and liners are often necessary to prevent caving zones from filling the production well. Holes and casings of deep production wells in consolidated formations are often telescoped in diameter size to allow drilling to great depths.

In unconsolidated deposits, a screen or perforated casing is required to hold back sediment and to allow water to flow into the production well without excessive head loss or the passage of fine materials during pumping. Openings may vary from the open bottom of production-well casings to all sorts of punched, perforated, and sawed slots to sophisticated screens.

The Mills Knife and hydraulic louver-type perforators are often used to make perforations in the field. The Mills Knife is a mechanically operated perforator which produces a vertical opening approximately 0.5 inch wide and 5 inches long. The louver-type perforator permits a more precise control of perforation size. The smallest practical openings are $\frac{1}{8}$ to $\frac{5}{32}$ inch (Ross, 1958). Machine-perforated casing is also available.

Commercial screens are available in various designs, diameters, slot sizes (see Fig. 5.1), and corrosion-resistant metals. Some screens are spiral-wound and slots are tapered with the widest space at the inner surface to prevent particles from accumulating within the openings. The lengths of screens, screen slot opening, and type of well are selected on the basis of a study of drilling cutting samples and hydrogeologic conditions. There are two types of screened production wells: (1) natural pack and (2) artificial pack. Materials surrounding the production well are developed in place in the case of the *natural-pack production well*; materials having a coarser uniform grain size than the natural formation are artificially placed around the production well in the case of the *artificial-pack production well*. In the natural-pack

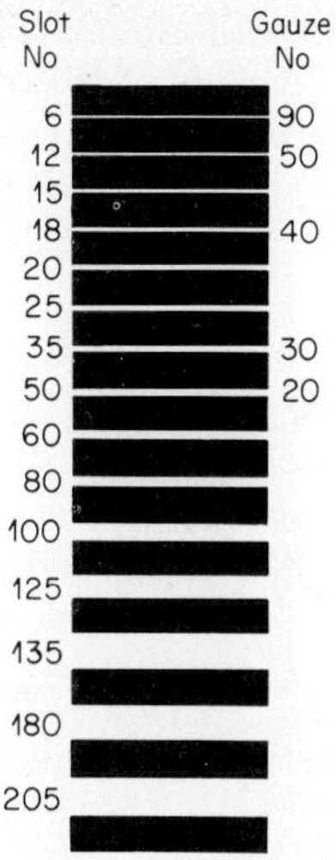

Fig. 5.1 Selected screen slot-opening sizes. Slot numbers designate widths of openings in thousandths of an inch; no. 10 is an opening of 10/1,000 inch, etc. Equivalent gauze numbers are shown on the right. (*Drawn by W. C. Walton.*)

case, development removes the finer material from the aquifer, so that only coarser material surrounds the screen. The materials around the production well are made more uniform in grain size and the sand and gravel are graded in such a way that fine deposits from the aquifer cannot clog the natural pack.

5.1 Grain-size Distribution and Size-frequency Curves

Production wells in unconsolidated deposits are designed based in part upon an analysis of the drilling cutting samples of the formation to be screened. Samples are dried and then reduced to a convenient working size by the quartering method. In the quartering method, a large sample is mixed and poured in a cone-shaped pile on a large flat surface. The pile is flattened and spread in a circle. The flat circular pile is divided in half. The two halves are divided in quarters. Two opposite quarters are removed and the remaining two quarters are mixed together. This is the reduced sample; if it is still too large, the quartering is repeated. The reduced sample will contain the same proportions of each size of particles as the original larger sample.

The reduced sample is sifted through a series of sieves. The following series of sieves are commonly used:

For fine sand

0.023 inch	28 mesh
0.016	35
0.012	48
0.008	65
0.006	100
Bottom pan	

For coarse sand

0.046 inch	14 mesh
0.033	20
0.016	28
0.012	48
0.008	65
Bottom pan	

Sand and gravel

0.131 inch	6 mesh
0.093	8
0.065	10
0.046	14
0.033	20
0.023	28
0.016	35
0.012	48
Bottom pan	

The sieves are nested with the finest one resting on a bottom pan and the coarsest at the top of the series. The nested sieves are then vibrated. When all the finer sand sifts through the top screen, it is removed and the sand retained is weighed on a scale. This process is repeated until all sizes are weighed, including the contents of the pan at the bottom which catches the material that passes through the finest sieve. Record is made of the quantities (by weight) of the sand retained on each of the screens and the quantity that passes through the finest openings.

A grain-size distribution graph is then prepared by plotting the cumulative percent retained by weight on the vertical scale against the logarithm of the sieve opening on the horizontal scale. The points of the graph are connected with a smooth curve.

Generally speaking, there are three elements essential to a complete description of a sample grading. One is the fineness, another is the slope of the grain-size distribution curve, and the third is the characteristic shape of the curve. The MIT grain-size classification (developed by the Massachusetts Institute of Technology) and the Bureau of Soils grain-size classification are useful in the description of fineness. Portions of the classification pertinent to production-well design are:

	MIT system	*Bureau of Soils system*
Fine gravel	0.080 to 0.375 inch	0.040 to 0.080 inch
Coarse sand	0.024 to 0.080	0.020 to 0.040
Medium sand	0.010 to 0.024	0.010 to 0.020
Fine sand	0.003 to 0.010	0.004 to 0.010
Silt and clay	Below 0.003	Below 0.002

A specific point, the effective grain size, on the grain-size distribution curve is also used as a general index of fineness and is often employed in designing production wells. The sieve size that retains 90 percent of the aquifer materials is the *effective size*.

The slope of the major portion of the grain-size distribution curve is also used in designing production wells. The term most frequently used to describe the slope is the *uniformity coefficient*. The uniformity coefficient C_u is the ratio of the sieve size that will retain 40 percent of the aquifer materials to the effective size. The uniformity coefficient is the average slope of the sand curve between the 90 percent and the 40 percent particle size. The lower its value, the more uniform is the grading of the sand between these limits. Larger values represent less uniform grading.

The *sorting coefficient* is another term that is used to indicate the relative slope of the grain-size distribution curve. The sorting coefficient is defined as the square root of the quotient of the 25 percent size divided by the 75 percent size.

The grain-size distribution curve for most granular materials that have been deposited in flowing water and by wave action is S-shaped. The S-shape of the curve becomes distorted when the material is a mixture of sand and gravel and the gravel content is at least 15 percent of the total. Gradings that give S-shaped curves usually have higher porosity than materials that exhibit a "tail" curve.

Characteristics of a given type of sand can often be identified by plotting a size-frequency curve. To develop this curve the weight of the grains held between each pair of the test sieves is expressed as a percentage of the total weight of the sample. Each of these values is then plotted against the average of the sieve openings which is taken as the average grain size of that fraction of the sample. The size-frequency curves for representative samples of formations sorted by flowing water show a tendency toward symmetrical shapes. For an alluvial sand the mode will be to the left of the halfway point between the ends of the curve. For a beach sand the mode will be almost exactly halfway between the ends.

5.2 Artificial- and Natural-pack Production Wells

Artificial-pack production wells in unconsolidated deposits are usually justified when the aquifer is homogeneous, has a uniformity coefficient less than 3.0, and has an effective grain size less than 0.01 inch (Ahrens, 1957). In addition, an artificial pack is sometimes needed to stabilize well-graded aquifers having a large percentage of fine materials in order to avoid excessive settlement of materials above the screen or to permit the use of larger screen slots. Artificial packs are often required to complete a production well in a poorly cemented sandstone.

Careful selection of the artificial pack is important to prevent the clogging of the pack with fine materials from the aquifer. The proper grain size for an artificial pack is selected on the basis of the mechanical analysis of the aquifer. A criterion that has been successfully used is that the ratio of the 50 percent sizes of the pack and the aquifer be 5 (Smith, 1954).

Artificial packs should range in thickness from 6 to 9 inches. To avoid segregation of bridging during placement, a uniform grain-size pack should be used. The screen slot opening should be designed so that at least 90 percent of the size fractions of the artificial pack are retained in the aquifer.

The design of a natural production well in unconsolidated deposits is somewhat more involved, as the screen governs in part the extent to which development is accomplished. One of the important factors in the design of natural-pack production-well screens is the width of the screen openings, referred to as *slot size*. The optimum slot size is selected principally on the basis of the grain-size distribution of the aquifer, but hydrogeologic conditions also play an important role. Slot sizes are selected which will allow a definite proportion of fine materials to pass into the production well. The fines are removed from the production well with a bailer, and such methods as surging are used to draw the finer material out of the aquifer around the production well and thereby provide improved permeability in the immediate vicinity of the screen.

With a uniformity coefficient greater than 6 (a heterogeneous aquifer) and in the case where the materials overlying the aquifer are fairly firm and will not easily cave, the sieve that retains 30 percent of the aquifer materials is generally selected as the slot size. With a uniformity coefficient greater than 6 and in the case where the materials overlying the aquifer are soft and will easily cave, the sieve size that retains 50 percent of the aquifer materials is selected as the slot size.

With a uniformity coefficient as low as 3 (a homogeneous aquifer) and in the case where the materials overlying the aquifer are fairly firm and will not easily cave, the sieve size that retains 40 percent of the aquifer materials is selected as the slot size. With a uniformity coefficient as low as 3 and in the case where the materials overlying the aquifer are soft and will easily cave, the sieve size that retains 60 percent of the aquifer materials is selected as the slot size.

A production well sometimes encounters several layers of sand and gravel having different grains sizes and gradations. If the 50 percent size of the materials in the coarsest aquifer is less than 4 times the 50 percent size of the materials in the finest aquifer, the slot size and pack, if needed, should be selected based on the mechanical analysis of the finest material (Ahrens, 1957). Otherwise, the slot size and pack should be tailored to individual layers.

5.3 Length of Screen

As long a screen as possible should be used to reduce entrance velocities and the effects of partial penetration of the aquifer by the production well. Under water-table conditions, optimum production-well specific capacity and yield are generally obtained by screening the lower 0.5 to 0.3 of the aquifer. The pumping level must be kept above the top of the screen; thus, the longer the screen, the less the available drawdown. The factors that affect the choice of screen length include: the open area per foot of screen, the character of the aquifer, the cost of the screen, the desired yield from the production well, and the desired service life of the production well.

Screen-length selection is often a compromise between the factors of cost and hydrogeology.

The screen length is based in part on the effective open area of a screen and an optimum screen entrance velocity (Walton, 1962). When a screen is placed in an aquifer, sediment will settle around it and partially block the slot openings. The amount of blocking depends largely upon the shape and type of slot and the shape, size, and sorting of the aquifer or pack material. On the average about one-half of the open area of the screen will be blocked by aquifer material. Thus, the effective open area averages about 50 percent of the actual open area of a screen.

To ensure that the production well has a long service life by avoiding migration of fine materials toward the screen and subsequent clogging of the well wall and screen openings, screen length is based on velocities between 2 and 12 fpm. In general, aquifers of low permeability are composed of finer-grained material than aquifers of high permeability. The possibility of clogging openings depends in large part upon the grain size of the finer aquifer materials, and there is a relationship between optimum screen entrance velocities and the coefficient of permeability of the aquifer. A study was made of several actual case histories of production-well failures due to the partial clogging of the well walls and screen openings, resulting in the values in Table 5.1 (see Walton, 1962).

The permeability of the aquifer is determined from aquifer tests. To prevent rapid clogging of the screen and formation, the length of the screen for a natural-pack production well is selected using Table 5.1 and the following equation (Walton, 1962):

$$S_L = \frac{Q}{7.48 A_0 V_c} \tag{5.1}$$

where S_L = optimum length of screen, in feet
Q = discharge of production well, in gpm
A_0 = effective open area per foot of screen, in square feet
V_c = optimum screen entrance velocity, in fpm

TABLE 5.1 OPTIMUM SCREEN ENTRANCE VELOCITIES. (*From Walton, 1962.*)

Coefficient of permeability, gpd/sq ft	*Optimum screen entrance velocities, fpm*
>6,000	12
6,000	11
5,000	10
4,000	9
3,000	8
2,500	7
2,000	6
1,500	5
1,000	4
500	3
<500	2

The same procedure is followed in selecting the optimum length of screen for an artificial-pack production well except that the average of the permeabilities of the aquifer and pack is used to determine the optimum entrance velocity.

5.4 Metal Requirements

The quality of groundwater and the screen-strength requirements must be considered in choosing the metal from which the screen is to be fabricated. If the groundwater is corrosive or encrusting, screens should be made of corrosion-resistant materials. The principal indicators of corrosive groundwater are (Anon, 1966)

TABLE 5.2 METALS USED IN FABRICATING SCREENS. (*From Anon, 1966.*)

Name of metal	*Analysis*[a]	*Cost factor*	*Best suited for*
Monel	70% nickel 30% copper	1.5	High sodium chloride water combined with dissolved oxygen as in sea water. Usually not needed for water suitable for drinking.
Supernickel	70% copper 30% nickel	1.2	Same as above, but not quite as corrosion resistant.
Everdur Silicon-bronze	96% copper 3% silicon 1% manganese	1.0	High total hardness, high sodium chloride water (not with dissolved oxygen present). High iron. Metal most often used for municipal and industrial production wells. Extremely resistant to acid treatment.
Stainless steel	74% steel 18% chromium 8% nickel	1.0	Hydrogen sulfide. Dissolved oxygen. Carbon dioxide. Iron bacteria. Strength second to Everdur. Used in municipal and industrial production wells.
Silicon red brass	83% copper 16% zinc 1% silicon	0.9	Used for same conditions as Everdur, but not quite as good. Not as strong as Everdur. Used in relatively inactive waters.
Armco iron	99.84% pure iron (double galvanized)	0.6	Not corrosion resistant, but functions satisfactorily in some areas. Used for irrigation wells in many areas where waters are relatively neutral.
Steel	99.38/99.72% iron 0.08/15% carbon 0.20/0.50% manganese (double galvanized)	0.5	Not corrosion resistant. Generally used only in temporary wells such as test wells or wells for dewatering. Has provided satisfactory service life in some areas of the southwestern U.S. where waters are noncorrosive and nonencrusting.

[a] The analyses shown are typical but are subject to some deviation from the percentages shown.

low pH, presence of dissolved oxygen, presence of hydrogen sulfide, total dissolved solids in excess of 1,000 parts per million (ppm), carbon dioxide in excess of 50 ppm, and chloride content greater than 500 ppm. The principal indicators of encrusting groundwater are total hardness greater than 330 ppm, total alkalinity greater than 300 ppm, iron content greater than 2 ppm, and pH greater than 8.0.

The removal of only a few thousandths of an inch of metal from screen openings by corrosion could permit the entrance of an excessive amount of fine deposits into a production well. Encrusting waters have the tendency to deposit minerals on the surface of the screen (plugging the openings) and in the pore spaces of the materials immediately adjacent to the screen. Mineral deposition is most often removed by acid treatment of a production well. Crenothrix or iron bacterial feed on dissolved iron and manganese and oxidize and precipitate them, causing minerals and bacteria to clog screen openings. Slime-producing bacteria are often removed with chlorine and the chlorine treatment is followed with acid treatment to redissolve the precipitated iron and manganese. If the screen is to be subjected to acid treatment, it must be made of corrosion-resistant metal.

Screens are often long and have to support a considerable pipe load. Also, screens are frequently subject to high collapse stresses (lateral pressure due to formation caving or sloughing). Extra-strength requirements can be met by using a metal having a high modulus of elasticity, increasing the screen-wall thickness, using larger or more vertical members and thicker horizontal members, or reducing the open area.

Well screens are usually made from one of the seven metals listed in Table 5.2.

5.5 Casing Diameter and Material

The diameter of the production-well casing should be two nominal sizes larger than the bowl size of the pump to prevent the pump shaft from binding, to reduce head losses, and to allow measurement of water levels in the well. The casing diameter may be reduced below the maximum anticipated pump setting depth. Suggested casing diameters for various pumping rates are given in Table 5.3.

TABLE 5.3 SUGGESTED CASING DIAMETER.

Pumping rate, gpm	*Diameter of well, in.*
<100	6
200	8
400	10
600	12
900	14
1,200	16
1,800	20
3,000	24
>3,000	30

The open area of a screen increases with the diameter of the screen; thus selection of the production-well diameter may depend upon the desired open area rather than the probable pump required.

In the selection of the type of casing it is necessary to consider the strain to which the pipe will be subjected during the installation and the corrosiveness of the water with which it comes in contact. Often the casing will be long and the strength requirements will be high. Wrought iron and steel pipes have both given satisfactory service in many locations.

5.6 Spacing of Production Wells

The problem of spacing production wells is frequently encountered. The farther apart wells are spaced, the less their mutual interference but the greater the cost of connecting pipeline and electrical equipment. The spacing of production wells is often dictated by practical considerations such as property boundaries and existing distribution of pipe networks. The following discussion is concerned only with the influence of aquifer characteristics and economics on the spacing of production wells.

Theis (1957) derived the following equation for determining the optimum production-well spacing in the simple case of two production wells pumping at the same rate from a thick and areally extensive aquifer

$$r_s = \frac{2.4 \times 10^8 C_p Q^2}{kT} \tag{5.2}$$

where r_s = optimum production-well spacing, in feet

C_p = cost to raise a gallon of water 1 foot, consisting largely of power charges, but also properly including some additional charges on the equipment, in dollars

k = capitalized cost for maintenance, depreciation, original cost of pipeline, etc., in dollars per year per foot of intervening distance

Q = pumping rate of each production well, in gpm

T = coefficient of transmissibility, in gpd/ft

For small values of T and Q, r_s from Eq. (5.2) is of no practical significance. Because of the effects of partial penetration which are appreciable within a distance of about twice the saturated thickness of the aquifer, $2m$, from a production well, it is generally advisable to space production wells at least a distance of $2m$ apart in aquifers 100 or less feet thick. Experience has shown that in the case of a multiple-well system consisting of more than two production wells the proper spacing between wells is at least 250 feet.

Production wells should be spaced parallel to and as far away from barrier boundaries and as near to the center of a buried valley as possible. Wells should be spaced on a line parallel to a recharge boundary and as close to the source of recharge as possible.

Theis (1941) derived the following equation to determine the permissible distance between production and disposal wells in an areally extensive isotropic aquifer

$$r_d = \frac{2Q}{TI} \tag{5.3}$$

where r_d = permissible distance between production and disposal wells to prevent recirculation of water, in feet
Q = pumping and disposal rate, in gpd
T = coefficient of transmissibility, in gpd/ft
I = natural hydraulic gradient of water table or piezometric surface, in ft/ft

5.7 Production-well Specifications

Strictly speaking, specifications for production-well construction work are contract documents related to the materials and equipment to be used, the methods of performing the work, and a description of the completed production well. However, it is a prevailing practice to include the notice to bidders, form of proposal and form of contract, general conditions, plans, and drawings, as well as technical conditions in the specifications. The general conditions include the following items: enumeration of contract documents; brief descriptions of work; definitions of owner, contractor, engineer, notice, and terms such as directed, acceptable, etc.; time of initiation and completion of work; liquidation damages for delay; extra work; contractor's insurance; maintenance; time and method of payment; and owner's right to make changes. The production well and its appurtenances such as the well pump, motor controls, pumphouse, and discharge piping may be covered by separate specifications or only as a part of a larger overall specification. The following discussion is concerned with the technical conditions of the specifications for a production well.

Specifications for a production well form the basis for inviting bids from qualified contractors. Properly prepared specifications will not be unduly restrictive and will encourage fair competition and bring reasonable bids. The National Institute of Municipal Law Officers recommends that the following factors be considered as a guide for determining the lowest responsible bidder: capacity and skill of the bidder to perform the contract, capacity to perform the work within the time specified, capacity of financial resources of the bidder, ability of the bidder to provide future maintenance and service, number and scope of conditions submitted with the bid, and a favorable price. With these factors in mind, the specification writer must describe the requirements in terms that will obtain a quality job at a favorable price. Specification writing requires the attention of personnel who know the technical aspects of production-well design and groundwater development and management.

The specification writer is generally not liable for construction defects if he has exercised ordinary skill and care in his work. However, some states have laws holding that, by implication, the specification writer warrants that the plans and specifications he prepares will accomplish the intended purpose. A contractor is not ordinarily liable for defects arising from faulty specifications. A contractor who has carried out his work in accordance with the plans and specifications has performed his contract. Defects that may appear as the result of defective plans and specifications are the fault of the party who prepares the plans and specifications.

While appearing to warrant the production of a certain quantity of water, many guaranteed-yield contracts for production wells actually do not, in practice, provide an absolute guarantee. Most such contracts have the effect of setting a price for the quantity said to be guaranteed which becomes the reference figure for negotiating a settlement at a lower price if the quantity is less than anticipated.

The technical conditions of specifications generally cover the following items: scope of work; permits, certificates, laws and ordinances; location; local conditions; boundaries of work; protection of site; general description of well; facilities or materials to be furnished by owner; competent workmen; casings—material, minimum weights and dimensions, drive shoes, joints; well screens—diameter, length, material, type, openings, strength, fittings; testing for yield and drawdown-time of test, test pump, auxiliary equipment, duration of test; grouting and sealing—grouting materials, placement of grout, grouting liners; testing for plumbness and alignment—requirements to test, description of test, requirements for plumbness and alignment; disinfection time of test, chlorine solution, requirements for disinfection of test pump; samples and records—samples of formations, record of casing pipe, liquidated damages, daily reports; protection of quality of water—precautions to be taken, corrective work, freedom from sand and turbidity; temporary capping; abandonment of well; and measurement and compensation.

Standard specifications for deep production wells have been published by the American Water Works Association (1958) and can be used as a guide in the preparation of specifications. These standard specifications must be supplemented and modified by the specification writer to fit the particular needs and conditions of his own case.

Some important factors to keep in mind when preparing specifications are discussed below (see *Johnson Drillers J.*, 1961). The term "standard-weight pipe" should only be used in reference to 6-inch diameter and smaller. Pipe in sizes larger than 6-inch is called *line pipe*. Eight-inch and larger sizes of line pipe are produced regularly in more than one wall thickness and weight. The heavier pipe should be specified where longer life under severe conditions is desired. If soil and water are only mildly corrosive, the lighter wall thickness may be chosen.

The American Standards Association has developed a system that classifies different wall thicknesses of a given size pipe by schedule numbers. Schedules 10, 20, 30, 40, and 120 designate the most common weights. Six-inch and smaller standard-weight pipe is all schedule 40. Extra-heavy pipe in these sizes is schedule 80 and double-extra-heavy is schedule 120. Table 5.4 gives wall thicknesses, weights, and schedule numbers for 8-inch and larger sizes of line pipe.

Where a corrosion-resistant material such as stainless steel is needed for well casing or liner, the wall thickness can be considerably less. Schedule 10 stainless steel pipe

TABLE 5.4 WALL THICKNESSES AND WEIGHTS OF API STANDARD LINE PIPE. (*From Johnson Drillers J., 1961.*)

Nominal size, in.	*Wall thickness, in.*	*Plain end, wt. per ft-lb*	*ASA schedule no.*
8	0.277	24.70	30
8	0.322	28.55	40
10	0.279	31.20	...[a]
10	0.307	34.24	30
10	0.365	40.48	40
12	0.330	43.77	30
12	0.375	49.56	...[a]
14 OD	0.375	54.57	30
16 OD	0.375	62.58	30
18 OD	0.375	70.59	...[a]
20 OD	0.375	78.60	20

[a] No ASA schedule number corresponding to the wall thickness shown for pipe of this size.

should be adequate for diameters up to 12 inches. This assumes that welded joints would be employed since screwed joints for stainless steel pipe present several problems.

Production-well screens are fabricated in two series of sizes. One series, with sizes referred to as *nominal diameters*, provides screens designed to be installed by telescoping through pipe of the same size designations. Thus, an 8-inch nominal-diameter screen is designed to telescope through 8-inch-ID pipe. The second regular series of sizes provides pipe-size screens which are the same inside diameter as pipes of the corresponding designations. An 8-inch pipe-size screen is made with nearly the same inside diameter and outside diameter as 8-inch steel pipe.

Not all responsibility for estimating subsurface conditions should be placed on the contractor's shoulders. The joint form of contract worked out by a Committee of the American Society of Civil Engineers and the Associated General Contractors includes the following paragraphs under the headings "Familiarity with Work" and "Changed Conditions."

Section 7. The owner shall make known to all prospective bidders all information that he may have on subsoil conditions in the vicinity of the work, topographic maps, or other information that might assist the bidder in properly evaluating the amount and character of the work that might be required. Such information is given, however, as being the best factual information available without the assumption of responsibility as to its accuracy or for any conclusions that the Contractor might draw therefrom.

Section 8. The Contractor shall promptly, and before such conditions are disturbed, notify the Owner in writing of: (1) subsurface or latent physical conditions at the site differing materially from those indicated in this Contract; or (2) unknown physical conditions at the site, of an unusual nature, differing materially from those ordinarily encountered and from those generally recognized as inherent in the work of the character provided for in this Contract. The Contracting Officer shall promptly investigate the conditions, and if he finds that such conditions do so materially differ and cause an increase or decrease in the cost of or the time required for the performance of this Contract, an equitable adjustment shall be made and the contract modified accordingly.

Adoption of these paragraphs as part of the general conditions in construction contracts would help to reduce cost of construction, encourage more responsible bidding, and decrease litigation.

PRODUCTION-WELL CONSTRUCTION METHODS

5.8 Vertical Wells

Production wells of high capacity for industrial, irrigation, or municipal use are generally constructed by drilling. *Well drilling* involves intensive mechanical abrasion of the bottom of the hole and accompanying rapid movement of fluid, outward and upward from the face of the cutting tool to the "bit." The drilling fluid suspends the cuttings and is the medium for their removal from the hole. Two drilling methods are most commonly used in the construction of production wells (see Bowman, 1911; War Dept., 1943; Gordon, 1958; and Anon, 1966). One employs cable tools and the other hydraulic-rotary or reverse-rotary tools. As a general rule, cable tools are employed in areas underlain by consolidated rock (sandstone, limestone, shale, etc.), and both cable and hydraulic-rotary tools are used in areas underlain by unconsolidated rock (sand, gravel, clay, etc.), the latter to an increasing degree.

Cable-tool drilling is accomplished by crushing the rock with hammerlike blows delivered by a heavy chisel-shaped bit, suspended on a cable. The bit rises and falls freely in the hole, the springlike action of the cable causing it to deliver a sharp, quick blow; and the lay of the cable gives a twist on each fall that turns the bit a few degrees. When water-yielding zones are penetrated, they must be cased off or sealed with grout if the hole is to be deepened. As drilling continues, the action of the bit is impeded by the accumulation of cuttings. The bit is periodically removed from the hole and a "bailer" (a cylindrical device, generally with a flap valve in the bottom) is run in the hole. The bailer is allowed to fall to the bottom of the hole where it strikes the water, causing a rapid surge of water and cuttings upward within it. The bailer is surged (rapidly raised and lowered) a few times and withdrawn from the hole with cuttings. A string of cable tools consists of a rope socket, a set of jars (to aid in loosening tools stuck in a hole), a drill stem (for weight and length), and a drilling bit. The tools together may weigh several thousand pounds. Tools are lifted by a mast and a multiline hoist and engine, forming a cable-tool drilling rig, which is usually truck-mounted.

Hydraulic-rotary drilling is accomplished by the combined effects of a rotating bit on the lower end of a hollow forged-steel drill pipe and high-velocity jet streams of drilling fluid directed downward and outward through holes in the face of the bit. The annular space between the drill collar (a length of very heavy drill pipe immediately above the bit) and the borehole is so narrow that there is a violent upward flow of drilling fluid laden with cuttings. The rotation of the bit further increases the lateral component of mud-fluid energy, tending to cause enlargement of the hole diameter. Above the drill collar the annular space between the drill pipe and hole wall generally is larger, and accordingly the drilling fluid rises more slowly to the land surface where the drill cuttings settle out in a pit.

Its load decreased, the mud fluid is then recirculated. The drill rod is turned by a rotating table which permits the rod to slide downward as the hole deepens.

The *reverse-rotary drilling* method uses water instead of mud and operates as a suction dredging method in which cuttings are removed by a rotating suction pipe. Hydrostatic pressure, acting against fine-grained material deposits, supports the walls of the hole.

In order to install and maintain pumping equipment, the production well should be straight and plumb. The alignment of a production well should be kept within practical limits. The standard specifications for deep wells adopted by the American Water Works Association (1958) include the following requirements:

Plumbness and alignment shall be tested by lowering into the well to a depth at least as great as the lowest anticipated pump setting a section of pipe 40 feet long or a dummy of the same length. The outer diameter of the pump shall not be more than 0.5 inch smaller than the diameter of that part of the casing or hole being tested. If a dummy is used it shall consist of a rigid spindle with 3 rings, each ring being 12 inches wide. The rings shall be truly cylindrical and shall be spaced one at each end of the dummy and one ring in the center thereof. The central member of the dummy shall be rigid so that it will maintain the alignment of the axis of the rings.

Should the dummy fail to move freely throughout the depth to be tested or should the well vary from the vertical in excess of 0.67 of the smallest inside diameter or beyond the limitations of this test, the plumbness and alignment of the well should be corrected by the contractor at his own expense.

The engineer may waive the requirements for plumbness if, in his judgment, (a) the contractor has exercised all possible care in constructing the well and the defect is due to circumstances beyond his control; (b) the utility of the completed well will not be materially affected; and (c) the cost of necessary remedial measures will be excessive.

There are several methods of installing screens in production wells (*Johnson Drillers J.*, 1954). In the case of the natural-pack production well, the casing may be run to the bottom of the hole, the inside of the casing cleaned out, the screen lowered to the bottom, and then the casing may be pulled back to expose the screen (pull-back method). In another method, the hole is drilled to the full depth of the production well and is held open by the drilling fluid. The screen, fitted with a closed bail bottom or an open bottom and short extension pipe, is washed down through any drilling cuttings that may have settled in the hole. The open bottom is plugged with lead wood or cement grout and the lead packer at the top is expanded by use of a swedge block. In the bail-down method, the hole is drilled to the proper depth to set the casing and the casing is sealed or grouted in place. The plug in the bottom of the hole is drilled out. The screen, fitted with a bail-down shoe and a string of pipe, is sunk into the formation below the casing by operating a bailer and other tools inside the string of tools. The wash-down method is similar to the bail-down method, except the screen is fitted with a self-enclosing bottom into which is connected a string of wash-line pipe. Lightweight drilling mud is pumped through the wash line to produce a high-velocity jet of fluid through the self-enclosing bottom. The jetting out allows the screen to sink; sand comes up inside the casing.

In the artificial-pack production well, the outer casing is set to the required depth and suspended in position with clamps or slips at the ground surface. The inner casing is then run to the bottom of the hole. Suitable guides are welded to the inner casing and in the open hole below. The screen is telescoped through the inner casing. The properly selected gravel-pack material is fed into the annular space around the inner casing. The inner casing is then lifted in stages as the gravel-pack material is placed, keeping the lower 3 or 4 feet of the pipe in the gravel at all times. When the full length of the screen surface is exposed, the inner casing is held in position and enough gravel is added to fill in several feet above the screen.

A production well must be developed to remove drilling cuttings and drilling mud which has migrated into the well wall and the surrounding formation, and to remove the fine silt and sand around the screen to produce a coarser and more uniform sand and gravel envelope around the production well (see Anon, 1966). In addition to this, development grades the material in the water-bearing formation around the screen in such a way that it is in a "stabilized" condition. Development may be accomplished by a variety of procedures including pumping, surging, injection of compressed air, and backwashing.

Surge plungers are one of the most effective devices for developing a production well. Surge plungers consist of plungers, pistons, or blocks which are operated in the casing of a production well for the purpose of surging the water in a sand or gravel formation to produce an action which will loosen the finer sand or gravel grains and carry them, or aid in carrying them, through the screen openings and into the production well, from which they may be removed by any convenient means. Surge plungers may be classified as either of two general types: (1) solid surge plungers or (2) valved surge plungers.

The solid surge plunger may be operated on the bottom of a string of cable-tool drilling tools. The plunger is worked up and down in the casing, surging the water back and forth through the screen. After the surging has continued for a reasonable period, the plunger is hoisted and the bailer is run into the production well to clean it out. It is the purpose of a valve-type surge plunger to produce a surging or in-and-out movement of the water through the screen and the formation around the screen, with a greater inrush of water than outrush. When a valve-type plunger is operated in a production well, the water is pulled forcibly into the production well on the upstroke of the plunger. There being no check valve in the production well, the water recedes as the plunger is driven on the downstroke, thus backwashing the production well. The friction of the rapidly moving plunger with its relatively small valve ports will exert a downward pressure on the receding water also, and will give more force to the backwashing action. The more rapidly the plunger is operated, the more downward the force will be exerted. Sometimes a combination surge plunger and bailer is used in the development process.

It is often helpful to pull the casing and expose the screen in stages, developing the production well by surging and backwashing each time the casing is lifted. Adding a small proportion of one of the several polyphosphates to the water that is circulated to displace the drilling fluid often helps considerably in mud removal during development. The phosphates disperse the clay in the drilling mud and reduce its gel strength. The phosphates that work effectively in helping the removal are tetra sodium pyrophosphate, sodium tripolyphosphate, sodium hexametaphosphate, and sodium septaphosphate.

Production wells furnishing drinking water should be properly sealed to prevent contamination from surface or subsurface sources. To keep water from entering a production well, the annular space outside of the casing should be filled with cement grout. The top of the well should contain a watertight seal. Surface covers around the production well should be made of concrete and should slope away from the production well.

A necessary final step in production-well completion is thorough disinfection of the production well and its appurtenances to kill any bacteria that may be present. A chlorine solution is the simplest and most effective agent for disinfecting or sterilizing a production well, pump, storage tank, or piping system. Highly chlorinated water for this purpose may be prepared by dissolving calcium hypochlorite, sodium hypochlorite, or gaseous chlorine in water.

Grouting is the process of forcing a suitable slurry into inaccessible places such as the annular space around a production-well casing and the crevices, cracks, or other voids in subsurface rocks. The pressure employed to force the grout into place may be either gravity or pump pressure. In production-well construction, grouting is used primarily to produce an impervious seal around casings. The slurry most commonly used for grouting consists of portland cement and water. A little clay is often added to the cement-water mixture to improve its properties. For some applications, the grout may be a clay-water mixture such as heavy drilling mud.

5.9 Collector Wells

A *collector well* is constructed by lowering a cylindrical concrete caisson into unconsolidated deposits, sealing the bottom with a concrete plug, and jacking perforated pipes horizontally into the aquifer. The construction features of collector wells installed in the United States were described by Spiridonoff (1964). The caisson is a poured-in-place, reinforced-concrete, vertical shaft, 13 feet in diameter. The wall thickness of the caisson is 18 or 24 inches, depending upon whether the caisson is shallow or deep. The bottom of the caisson is a heavy reinforced-concrete slab.

One or more tiers of horizontal perforated steel pipe (laterals) are connected to valved ports in the caisson wall about 4 feet from the bottom of the caisson. The ports are usually no less than 22.5° apart. The laterals usually have an $8\frac{5}{8}$-inch outside diameter and a $\frac{3}{8}$-inch wall thickness. Special copper-alloy steel is frequently used. The slots in the pipe, cut with a saw, are $\frac{3}{8}$ inch wide and $1\frac{1}{2}$ inches long; the perforated area equals about 20 percent of the total surface area. The total lengths of individual laterals range from about 100 to 250 feet. In general, 8-foot lengths of pipe are used. The laterals may fan out in a semicircle or in a full circle like the spokes in a wheel.

Laterals are driven into the aquifer with two hydraulic jacks of 100-ton capacity each. These are placed against the opposite points on the wall of the caisson. As the laterals are jacked into position, the fine materials in the aquifer are removed, thus developing a natural pack around the perforated pipe. Vertical-turbine pumping units are used to pump the water which flows into the caisson through the laterals.

Since the development of the first collector wells in the early 1930s, more than 200 collector wells have been constructed (see Spiridonoff, 1964, and Gridley, 1952). It is not uncommon for the collector wells to yield several million gallons per day. Collector wells are particularly effective in shallow sand and gravel deposits near streams. The yields of collector wells near streams are generally substantially increased during high stages of the streams.

The problem of estimating the sustained yields of collector wells is frequently encountered. The sustained yield of the collector is often empirically estimated by simulating the collector with a vertical production well having the same specific capacity as the collector well. The effective radius of a vertical production well that simulates a collector well in the case of a radial lateral pattern covering the entire circumference and having equal-length laterals is equal to 75 and 85 percent of the average length of individual laterals. (Mikels and Klaer, 1956.) Several effective radii of vertical production wells that simulate collector wells having lateral patterns covering only part of the circumference and laterals of unequal length are given in Table 5.5.

TABLE 5.5 EFFECTIVE RADII OF COLLECTOR WELLS. (*From Mikels and Klaer, 1956.*)

Vertical production-well effective radius, ft	*Collector-well construction features*
72	Five laterals projected in a fan-shaped pattern ranging in length from 60 to 120 feet and totaling 496 feet around 180° of the circumference.
60	Four laterals projected to lengths of about 100 feet each around 90° of the circumference.
74	Seven laterals 136 to 176 feet in length around 130° of the circumference.

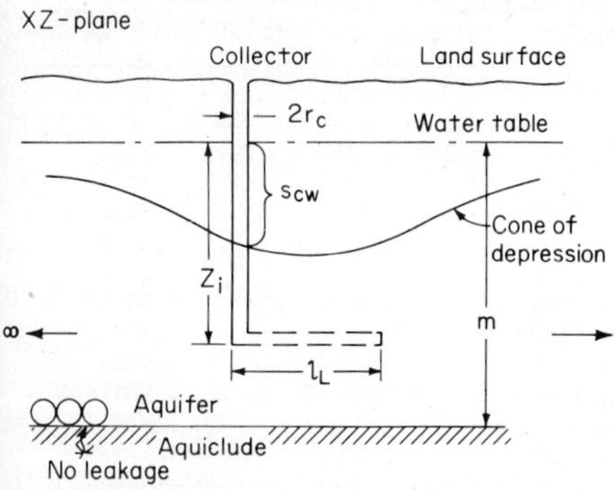

Fig. 5.2 Water-table aquifer with collector well. (*After Hantush, 1964.*)

Analytical solutions for problems of flow toward collector wells, based on the assumptions that collector wells are partially penetrating horizontal wells and the laterals are line sinks, were obtained by Hantush and Papadopulos (1962). Figure 5.2 shows a collector well penetrating a homogeneous, isotropic water-table aquifer infinite in areal extent and of the same thickness throughout. The radial laterals lie in the same horizontal plane, are equal in length, and are so oriented that each will drain an equal section of the aquifer. The discharge of all of the laterals is the same. The collector well is pumped at a constant rate Q.

Assuming the maximum drawdown of the water table is small, relative to the original saturated thickness of the aquifer (say, less than 25 percent), the radius of the caisson is small compared with the lengths of the laterals, $t > 2.5m^2/(Pm/S_y)$ and $> 5(r_c^2 + l_L^2)/(Pm/S_y)$, $l_L > 0.5m$, and $r_L < m/2\pi$, the drawdown s_{cw} in a collector well of four or more symmetrically located laterals can be approximated by the following equation (Hantush, 1964)

$$s_{cw} = \frac{Q}{4\pi PmN_L}\left\{W\left(\frac{l_L^2}{4Pmt/S_y}\right) + \frac{N_L - 1}{l_L}\left[l'W\left(\frac{l'^2}{4Pmt/S_y}\right)\right.\right.$$
$$\left. - r_cW\left(\frac{r_c^2}{4Pmt/S_y}\right)\right] + 2N_L + \frac{m}{2l_L}\ln\left\{\frac{(m/\pi r_L)^2}{2[1 - \cos\pi(2Z_i + r_L)]/m}\right\}$$
$$\left. + \frac{4m(N_L - 1)}{\pi l_L}\sum_{n=1}^{M'}\frac{1}{n}\left[\frac{\pi}{2} - W\left(\frac{n\pi r_c}{m}, 0\right)\right]\left[\cos\frac{n\pi Z_i}{m}\cos\frac{n\pi(Z_i + r_L)}{m}\right]\right\} \qquad (5.4)$$

in which M' is an integer large enough so that $M' > m/2r_c$. The functions $W(u)$ and $W(u,0)$ are defined:

$$W(u) = -0.5772 - \ln u + u - \frac{u^2}{2.2!} + \frac{u^3}{3.3!} - \frac{u^4}{4.4!} + \cdots$$

$$W(u,0) = \int_0^u K_0(y)\,dy$$

Tabular values of $W(u)$ are given in Table 3.1 and tabular values of $W(u,0)$ are given in Table 5.6.

TABLE 5.6 VALUES OF $W(u,0)$. (*From Hantush, 1964.*)

u	$W(u,0) = 0$	0.1	0.2	0.3	0.4	0.5	0.6	0.7	0.8	0.9	1.0
0	0.0	0.3422	0.5471	0.7020	0.8256	0.9271	1.0119	1.0836	1.1447	1.1972	1.2425
1	1.2425	1.2818	1.3159	1.3457	1.3718	1.3946	1.4146	1.4323	1.4478	1.4616	1.4737
2	1.4737	1.4844	1.4939	1.5023	1.5098	1.5164	1.5223		1.5321		1.5400
3	1.5400		1.5462		1.5511		1.5550		1.5582		1.5606
4	1.5606		1.5626		1.5642		1.5655		1.5666		1.5674
5	1.5674		1.5681		1.5686		1.5690		1.5694		1.5696
6	1.5696		1.5699		1.5700		1.5702		1.5703		1.5704
7	1.5704		1.5705		1.5705		1.5706		1.5706		1.5707
8	1.57066		1.57068		1.57071		1.57072		1.57074		1.57075
9	1.57075		1.57076		1.57076		1.57077		1.57078		1.57078
10	1.57078		1.57078		1.57078		1.57078		1.57079		1.57079

Equation (5.4) may be rewritten in the gallon-day-foot system of units in abbreviated form as

$$s_{cw} = \frac{114.6Q}{TN_L}\left\{W(u) + \frac{N_L - 1}{l_L}\left[l' W(u_b) - r_c W(u_c) + 2N_L\right.\right.$$

$$\left. + \frac{m}{2l_L}\ln\left\{\frac{(m/\pi r l_L)^2}{2[1 - \cos\pi(2Z_i + r_L)]/m}\right\} + \frac{4m(N_L - 1)}{\pi L_L}\sum_{n=1}^{M'}\frac{1}{n}\left[\frac{\pi}{2} - W\left(\frac{n\pi r_c}{m}, 0\right)\right]\right.$$

$$\left. \times \cos\frac{n\pi}{m} z_i \cos\frac{n\pi}{m}(Z_i + r_L)/m\right\} \tag{5.5}$$

where
$$u = \frac{1.87 l_L^2}{Pmt/S_y} \tag{5.6}$$

$$u_b = \frac{1.87 l'^2}{Pmt/S_y} \tag{5.7}$$

$$u_c = \frac{1.87 r_c^2}{Pmt/S_y} \tag{5.8}$$

l_L = length of each lateral, in feet
$l' = l_L + r_c$, in feet
m = initial saturated thickness of aquifer, in feet
N_L = number of laterals
P = coefficient of permeability, in gpd/sq ft
Q = discharge of collector well, in gpm
r_c = radius of caisson, in feet
r_L = radius of each lateral, in feet
S_y = specific yield of aquifer, fraction
s_{cw} = drawdown in collector caisson, in feet
$T = Pm$ = coefficient of transmissibility, in gpd/ft
t = time after pumping started, in days
Z_i = vertical position of laterals, in feet (see Fig. 5.2)

Values of

$$\sum_{n=1}^{M'} \frac{1}{n}\left[\frac{\pi}{2} - W\left(\frac{n\pi r_c}{m}, 0\right)\right] \cos\frac{n\pi}{m} Z_i \cos\frac{n\pi}{m}(Z_i + r_L)$$

for practical ranges of m, Z_i, and r_L assuming $r_c = 6.5$ feet are given in Table 5.7.

The drawdown in a collector well under artesian conditions can be computed with Eqs. (5.5) through (5.8) by substituting S for S_y, where S is the coefficient of storage. The drawdown in a collector well near a stream can often be estimated with Eqs. (5.5) through (5.8) and the image-well theory. The image well associated with the recharge boundary is assumed to be a fully penetrating vertical recharging well.

For $t > 5l_L^2/(Pm/S_y)$, $l_L > 0.5m$, $r_L < m/2\pi$, the drawdown in the collector well is (Hantush, 1964)

$$s_{cw} > \frac{Q}{2\pi Pm} \ln \frac{R_c}{l_L} \tag{5.9}$$

where

$$R_c = 4.08\sqrt{\frac{Pmt}{S_y}} \left\{\frac{(m/\pi r_L)^2}{2[1 - \cos \pi(2Z_i + r_L)]/m}\right\}^{\frac{m}{4l_L}} \tag{5.10}$$

WELL-PRODUCTION TESTS

After a production well is completed, it should be tested to determine its productivity. Constant-rate and/or variable-rate well-production tests are made to measure the productivity of a production well. In the *constant-rate well-production test*, the production well is pumped at a constant rate for a period of at least 8 hours and water levels in the production well are measured at frequent intervals. In the *variable-rate well-production test*, the production well is operated during successive periods usually about 1 hour in duration at constant fractions of full capacity, and water levels in the production well are measured at frequent intervals. Sometimes the drawdown in nearby observation wells is measured during the well-production tests. The productivity of a production well is generally expressed by the term *specific capacity*, which is defined as the ratio of the pumping rate and the drawdown, or Q/s_T.

The total drawdown s_T in a production well has all or some of the following components upon hydrogeologic and production-well conditions: the drawdown s (*aquifer loss*) due to laminar flow of water through the aquifer toward the well, plus the drawdown s_{wL} (*well loss*) due to the turbulent flow of water through the screen or well face and inside the casing to the pump intake, plus the drawdown s_p due to the partial penetration of the pumped well, plus the drawdown s_d due to dewatering a portion of an aquifer, plus the drawdown s_b due to barrier boundaries of the aquifer, minus the buildup s_r due to recharge boundaries of the aquifer. Stated as an equation,

$$s_T = s + s_{wL} + s_p + s_d + s_b - s_r \tag{5.11}$$

TABLE 5.7 VALUES OF $\sum_{1}^{M'} \frac{1}{n}\left[\frac{\pi}{2} - W\left(\frac{n\pi r_c}{m}, 0\right)\right] \cos\frac{n\pi Z_i}{m} \cos\frac{n\pi}{m}(Z_i + r_L)$

(Prepared by Adrian Visocky, Illinois State Water Survey.)

$r_L = 4$ in.

m	$Z_i/m = 0.75$	0.80	0.85	0.90
20	0.1681	0.2231	0.2603	0.3266
30	0.2778	0.3478	0.4451	0.5417
40	0.3744	0.4589	0.5735	0.7119
50	0.4494	0.5396	0.6723	0.8462
60	0.5340	0.6378	0.7669	0.9582
70	0.5954	0.7209	0.8598	1.0549
80	0.6582	0.7813	0.9378	1.1338
90	0.7024	0.8361	1.0111	1.2081
100	0.7715	0.8824	1.0638	1.2684

$r_L = 8$ in.

m	$Z_i/m = 0.75$	0.80	0.85	0.90
20	0.1758	0.2314	0.2866	0.3320
30	0.2793	0.3589	0.4566	0.5503
40	0.3797	0.4662	0.5824	0.7224
50	0.4536	0.5462	0.6820	0.8572
60	0.5386	0.6531	0.7744	0.9695
70	0.5994	0.7259	0.8651	1.0654
80	0.6612	0.7869	0.9438	1.1451
90	0.7051	0.8399	1.0162	1.2098
100	0.7741	0.8877	1.0683	1.2767

$r_L = 12$ in.

m	$Z_i/m = 0.75$	0.80	0.85	0.90
20	0.1831	0.2392	0.2934	0.3371
30	0.2851	0.3672	0.4647	0.5579
40	0.3867	0.4729	0.5925	0.7315
50	0.4574	0.5518	0.6917	0.8680
60	0.5427	0.6478	0.7822	0.9804
70	0.6029	0.7310	0.8715	1.0753
80	0.6635	0.7914	0.9486	1.1525
90	0.7075	0.8442	1.0209	1.2176
100	0.7766	0.8899	1.0735	1.2831

5.10 Well Loss

The components of drawdown and buildup except well loss can be computed with equations given in previous chapters. Well loss may be represented approximately by the following equation (Jacob, 1946)

$$s_{wL} = CQ^2 \tag{5.12}$$

where s_{wL} = well loss, in feet
C = well-loss constant, in sec²/ft⁵
Q = discharge, in cubic feet per second (cfs)

Rorabaugh (1953) derived a similar equation for determining well loss and presents a more exact method for evaluation of well loss when a large range of pumping rates is encountered. The relation between the well-loss constant and drawdown due to well loss for several selected pumping rates is shown in Fig. 5.3.

If the aquifer is fairly permeable and the production well can be pumped at rates exceeding 25 gpm, the value of C in Eq. (5.12) may often be computed from the data collected during a variable-rate well-production test, by using the following equation (Jacob, 1946)

$$C = \frac{\Delta s^i/\Delta Q_i - \Delta s^{i-1}/\Delta Q_{i-1}}{\Delta Q_{i-1} + \Delta Q_i} \tag{5.13}$$

The Δs terms in Eq. (5.13) represent increments of drawdown produced by each increase ΔQ in the rate of pumping. The dimensions of Δs and ΔQ are feet and cfs, respectively. Increments of drawdown are determined by taking in each case the difference between the observed water level and the extension of the preceding water-level curve. Steps of any length of time may be used provided the Δs values chosen

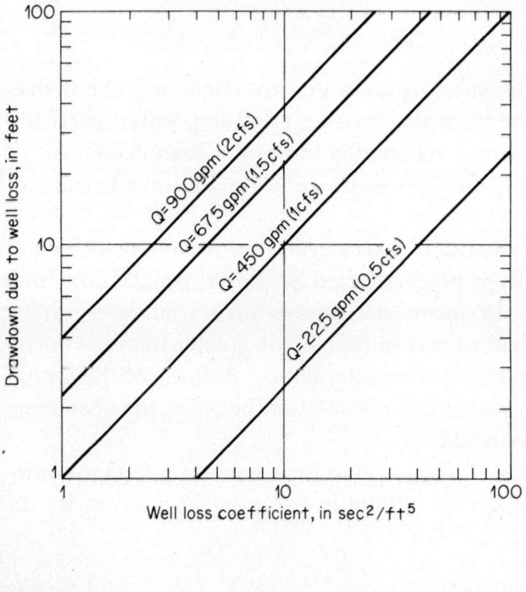

Fig. 5.3 Relation between well-loss constant and drawdown due to well loss for selected pumping rates. (*Drawn by W. C. Walton.*)

are for the same length of time in each step. For steps 1 and 2 Eq. (5.13) may be rewritten as

$$C_{1 \text{ and } 2} = \frac{\Delta s_2/\Delta Q_2 - \Delta s_1/\Delta Q_1}{\Delta Q_1 + \Delta Q_2} \tag{5.14}$$

For steps 2 and 3, Eq. (5.13) may be written as

$$C_{1 \text{ and } 2} = \frac{\Delta s_3/\Delta Q_3 - \Delta s_2/\Delta Q_2}{\Delta Q_2 + \Delta Q_3} \tag{5.15}$$

Equation (5.13) assumes that the well is stable and that C does not change during the variable-rate well-production test. However, newly completed production wells and old wells are sometimes unstable and the value of C is affected by changes in pumping rates. The value of C computed for steps 1 and 2 may be greater or less than that computed for 2 and 3 if the production well is unstable. Sand and gravel often shift outside a production well during step 3 under the influence of high rates of pumping, resulting in either the development or clogging of the pores of the aquifer immediately behind the production-well screen. If the value of C for steps 2 and 3 is considerably less than that for steps 1 and 2, it is probable that development has occurred during the variable-rate well-production test. A large increase in C with higher pumping rates indicates clogging of the production-well screen or wall. If development during the pumping period is large, $\Delta s_2/\Delta Q_2$ will be greater than $\Delta s_3/\Delta Q_3$ and solution of Eq. (5.15) will be impossible. Thus, it is often possible to appraise the stability of a well with variable-rate well-production test data and Eqs. (5.14) and (5.15).

5.11 Specific-capacity Data

In many cases, especially in reconnaissance groundwater investigations, the hydrogeologic properties of an aquifer must be estimated based on well-log, water-level, and specific-capacity data. High specific capacities usually indicate a high coefficient of transmissibility, and low specific capacities generally indicate low coefficients of transmissibility.

The specific capacity of a well cannot be an exact criterion of the coefficient of transmissibility because specific capacity is often affected by partial penetration, well loss, and hydrogeologic boundaries. In most cases these factors adversely affect specific capacity and the actual coefficient of transmissibility is greater than the coefficient of transmissibility computed from specific-capacity data. Because of the usefulness of rough estimates of T, an examination of the relation between the coefficient of transmissibility and specific capacity is useful.

The theoretical specific capacity of a well discharging at a constant rate in a homogeneous, isotropic, nonleaky artesian aquifer infinite in areal extent is given by the following equation

$$\frac{Q}{s} = \frac{T}{264 \log \left(\frac{Tt}{2{,}693 r_w^2 S}\right) - 65.5} \tag{5.16}$$

where $\frac{Q}{s}$ = specific capacity, in gpm/ft

Q = discharge, in gpm
s = drawdown, in feet
T = coefficient of transmissibility, in gpd/ft
S = coefficient of storage, fraction
r_w = nominal radius of well, in feet
t = time after pumping started, in minutes

The equation assumes that: (1) the production well penetrates and is uncased through the total saturated thickness of the aquifer, (2) well loss is negligible, and (3) the effective radius of the production well has not been affected by the drilling and development of the production well and is equal to the nominal radius of the production well.

The coefficient of storage of an aquifer can usually be estimated with the well-log and water-level data. Because specific capacity varies with the logarithm of $1/S$, large errors in estimated coefficients of storage result in comparatively small errors in coefficients of transmissibility estimated with specific-capacity data.

The relationships between the specific capacity and the coefficient of transmissibility for artesian and water-table conditions are shown in Figs. 5.4 to 5.9. Pumping periods of 2 minutes, 10 minutes, 60 minutes, 8 hours, 24 hours, and 180 days; a radius of 6 inches; and storage coefficients of 0.0001 and 0.02 were assumed in constructing the graphs. These graphs may be used to obtain rough estimates of the coefficients of transmissibility from specific-capacity data. The coefficient of storage is estimated from well-log and water-level data and a line based on the estimated S is drawn parallel to the lines on one of the Figs. 5.4 to 5.9, depending upon the pumping period. The coefficient of transmissibility is selected from the point of intersection of the S line and the known specific capacity.

As shown by Eq. (5.16), the specific capacity varies with the logarithm of $1/r_w^2$. Large increases in the radius of a well result in comparatively small changes in Q/s. The relationship between specific capacity and the radius of a well assuming $T =$ 17,000 gpd/ft, $S = 0.0004$, and $t = 1$ day is shown in Fig. 5.10*a*. A 30-inch-diameter well has a specific capacity of about 8 percent more than that of a 16-inch-diameter well; a 16-inch-diameter well has a specific capacity of about 3 percent more than that of a 12-inch-diameter well.

Specific capacity decreases with the period of pumping as shown in Fig. 5.10*b*, because the drawdown continually increases with time as the cone of influence of the well expands. For this reason, it is important to state the duration of the pumping period for which a particular value of specific capacity is computed. The graph of

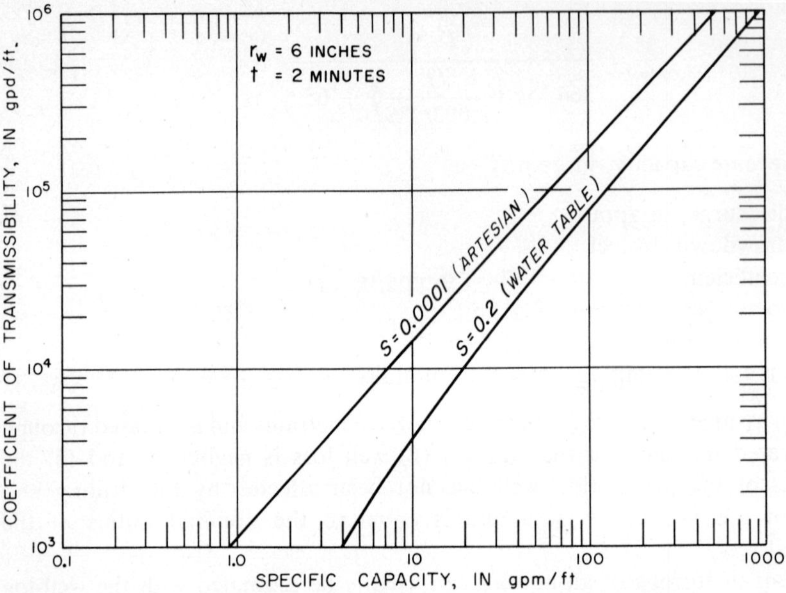

Fig. 5.4 Graphs of specific capacity versus coefficient of transmissibility for a pumping period of 2 minutes. (*From Walton, 1962.*)

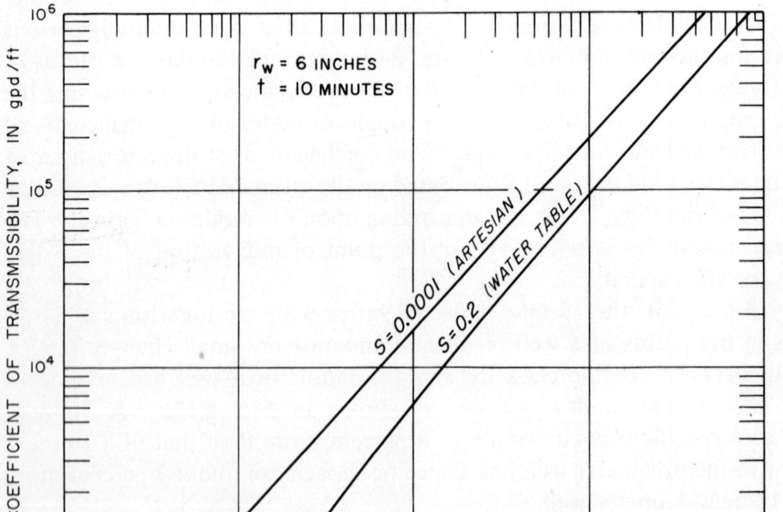

Fig. 5.5 Graphs of specific capacity versus coefficient of transmissibility for a pumping period of 10 minutes. (*From Walton, 1962.*)

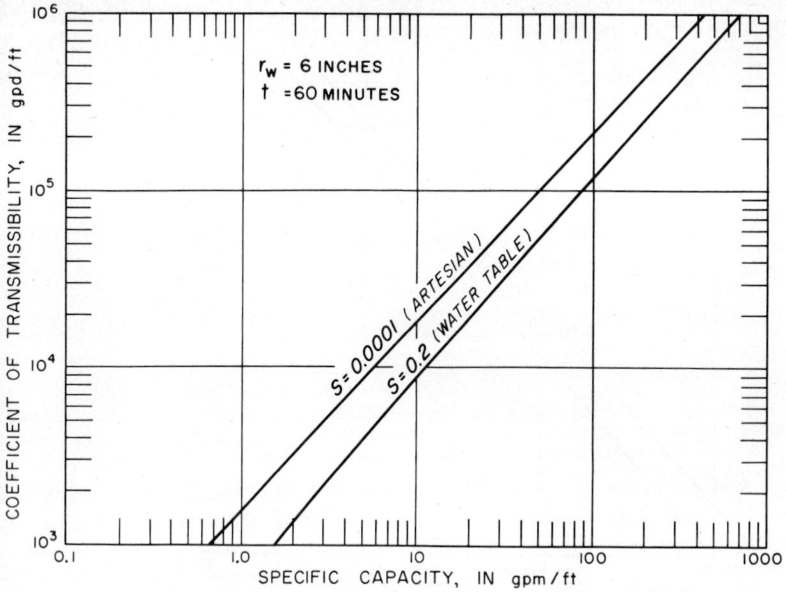

Fig. 5.6 Graphs of specific capacity versus coefficient of transmissibility for a pumping period of 60 minutes. (*From Walton, 1962.*)

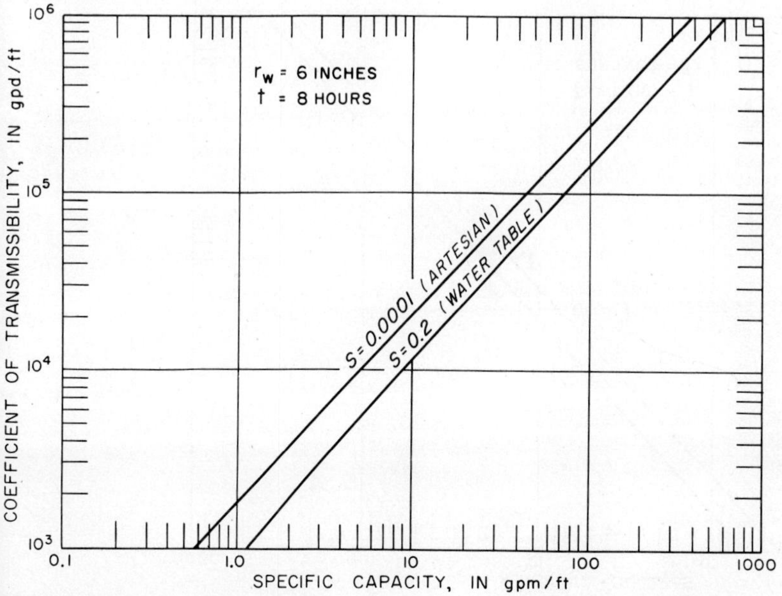

Fig. 5.7 Graphs of specific capacity versus coefficient of transmissibility for a pumping period of 8 hours. (*From Walton, 1962.*)

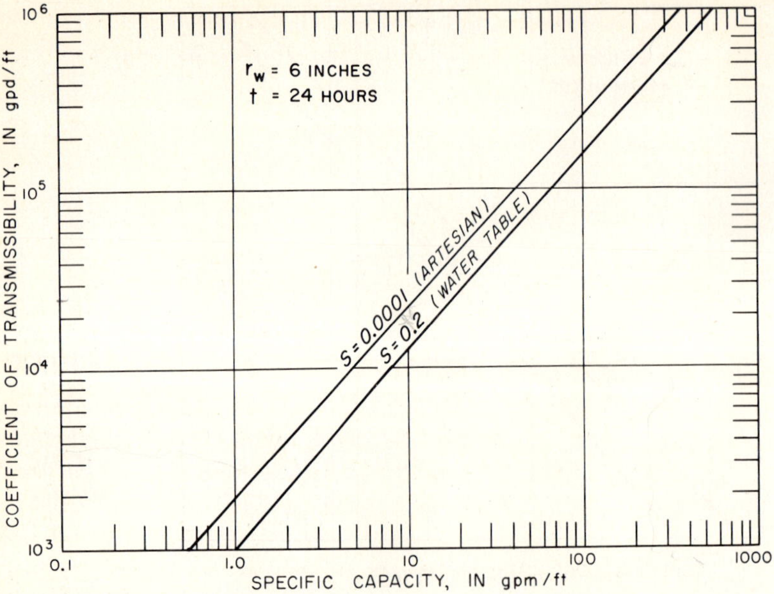

Fig. 5.8 Graphs of specific capacity versus coefficient of transmissibility for a pumping period of 24 hours. (*From Walton, 1962.*)

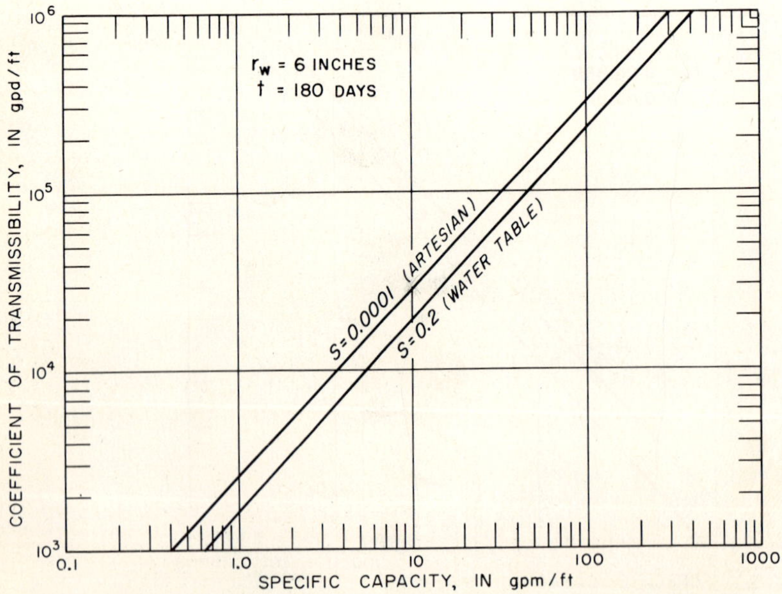

Fig. 5.9 Graphs of specific capacity versus coefficient of transmissibility for a pumping period of 180 days. (*From Walton, 1962.*)

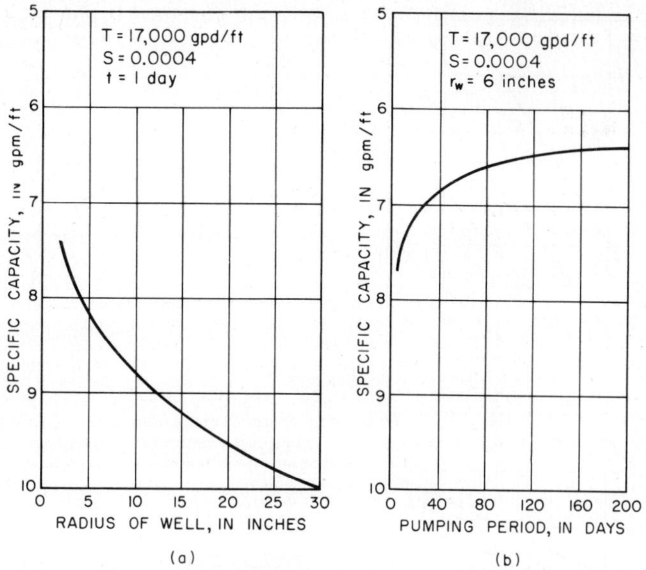

Fig. 5.10 Graphs of specific capacity versus well radius (*a*) and pumping period (*b*). (*From Walton, 1962.*)

specific capacity versus pumping period in Fig. 5.10*b* was constructed assuming $T = 17{,}000$ gpd/ft, $S = 0.0004$, and $r_w = 6$ inches.

The specific capacity of a well that is screened opposite only part of an aquifer (Q/s_p) can be expressed in terms of the specific capacity of a well that is screened opposite the full thickness of the same aquifer as (Turcan, 1963)

$$\frac{Q}{s_p} = \frac{Q}{s}\left[K_p\left(1 + 7\sqrt{\frac{r_w}{2K_p m}}\cos\frac{K_p\pi}{2}\right)\right] \tag{5.17}$$

where K_p = ratio of length of screen to saturated thickness of the aquifer, fraction
r_w = radius of the pumped well, in feet
m = thickness of the aquifer, in feet

Equation (5.17) was derived for steady-state conditions; it is not applicable unless the rate of drawdown has become so slow that for all practical purposes steady-state conditions can be assumed to exist. Even if the well is screened opposite the middle part of the aquifer, or is screened opposite two or more different intervals, Eq. (5.17) provides results that are satisfactory for most general engineering and design purposes. The quantity in brackets in Eq. (5.17) is commonly referred to as the *productivity ratio* F_p. Equation (5.17) may be rewritten in abbreviated form as (Turcan, 1963)

$$\frac{Q}{s_p} = \frac{Q}{s}F_p \tag{5.18}$$

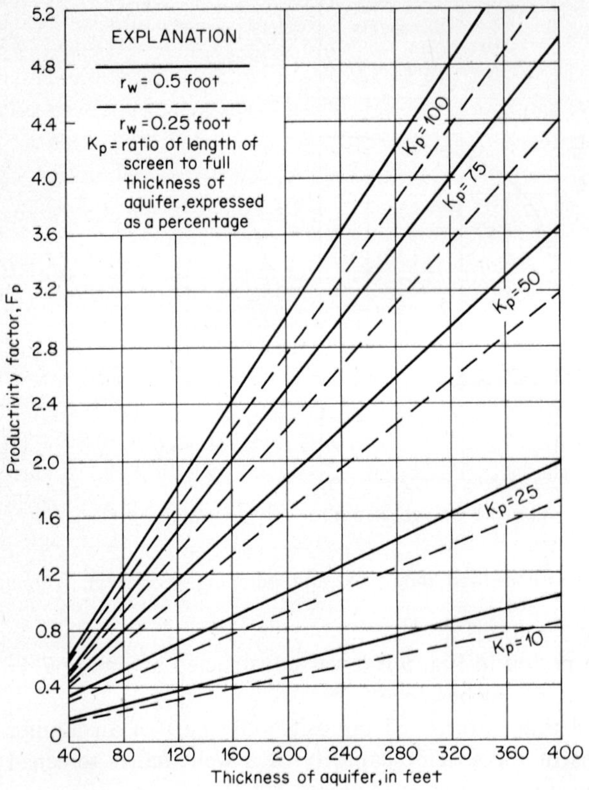

Fig. 5.11 Relation of aquifer thickness and productivity factor to well radius and percent of aquifer screened. (*From Turcan, 1963.*)

A graph presented by Muskat (1937) was modified (Turcan, 1963) by extension of the curves and changes in nomenclature to provide an easy means for determining the productivity ratio for wells screened opposite different percentages of the full thickness of the aquifer. The productivity factor is a dimensionless term which expresses the relative yield of wells that are screened opposite different percentages of aquifer thickness. The curves in Fig. 5.11 can be both extrapolated and interpolated with considerable accuracy because, for aquifers more than 50 feet thick, the relation of the productivity factor to aquifer thickness is for practical purposes linear (Muskat, 1937).

The methods of statistical analysis and specific-capacity data can be of great help in appraising the role of individual units of multiunit aquifers as contributors of water (Walton, 1962). The productivity of some bedrock aquifers, expecially dolomite aquifers, is inconsistent, and it is impossible to predict with a high degree of accuracy the specific capacity of a production well before drilling at any location. However, the probable range of specific capacities of a well can often be estimated on frequency graphs.

Suppose that specific-capacity data are available for wells penetrating one or several units of a multiunit aquifer and it is required to estimate the range in productivity and relative consistency in productivity of the three units. Specific capacities are divided by the total depths of penetration to obtain specific capacities per foot of penetration. Wells are segregated into categories depending upon the units penetrated by wells. Specific capacities per foot of penetration for wells in each category are tabulated in order of magnitude, and frequencies are computed with the following equation derived by Kimball (1946)

$$F_s = \frac{m_0}{n_w + 1} 100 \tag{5.19}$$

where m_0 is the order number, n_w is the total number of wells, and F_s is the percentage of wells whose specific capacities are equal to, or greater than, the specific capacity of order number m_0.

Values of specific capacity per foot of penetration are then plotted against the percentage of wells on logarithmic probability paper. Straight lines are fitted to the data. If specific capacities per foot of penetration decrease as the depth of wells and number of units penetrated increase, the upper units are judged to be more productive than the lower units.

Unit-frequency graphs can be constructed from the category-frequency graphs by the process of subtraction, taking into consideration uneven distribution of wells in the categories. The slope of a unit-frequency graph varies with the inconsistency of production, a steeper line indicating a greater range in productivity.

MAINTENANCE OF PRODUCTION WELLS

The life of a production well will be limited unless it is constructed to permit both a high level of initial efficiency and ready redevelopment, and unless it is pumped at the proper design rate. Some production wells will with continuous heavy pumping eventually become partially clogged; thus, periodic maintenance is a requirement. With the use of modern materials and construction techniques and appropriate maintenance, most production wells could last for 40 years or more. Many production wells that were initially efficient have been observed to decline in efficiency with time but to recover almost initial efficiency after cleaning.

There are four basic causes of declines in well production: (1) lowering of water levels; (2) inefficient pump operation caused by worn, corroded, or plugged parts; (3) deposits of scale, corrosion products, and microorganism growths in the production-well screen and on the well face; and (4) clogging of the production-well screen by mud, sand, and silt. Rehabilitation of the production well and pump can help in recovering lost production if the decline is due to any of the last three causes.

Continuous data collections should be a routine matter for all high-capacity production wells. Everyone agrees that it is important to determine the capacity, drawdown, and water level in a new production well. It is also necessary to check these factors frequently enough to compile an operating history of the production well. By comparison of such data, collected over a period of time, it is possible to detect a loss of production efficiency and many times to determine the cause of such loss. With this forewarning, repair and maintenance work can be accomplished at opportune times and complete breakdowns avoided.

5.12 Corrosion and Incrustation of Casing and Screen

Corrosion of metals is the chemical action by the environment to which they are exposed, resulting in a deterioration or eating away of the material (Briggs, 1949). Corrosion of production-well screens is not as prevalent as incrustation, but is more harmful. Through corrosion, metal is substantially weakened and screen openings are enlarged. There are several different forms of corrosion including the following: direct chemical corrosion—simple solution of the metal as a result of the chemical nature of the water (Bennison, 1953), dezincification of brass and graphitization of iron, intergranular corrosion, a localized chemical reaction at the boundaries of metallic crystals; galvanic or bimetallic corrosion—a localized attack resulting when two dissimilar metals are connected in an electrolyte; and concentration cell corrosion—the form of electric cell formed when one metal is placed in two different concentrations of the same electrolyte.

Most groundwaters are only mildly corrosive, if at all. Corrosion may be offset by the use of protective coatings or of highly corrosion-resistant metals in casings and screens. Certain types of bronze-copper silicon alloys are excellent materials for production-well screens from the standpoint of corrosion resistance, cost, strength, and weldability.

Incrustation is an accumulation of extraneous material in and about the openings of the screen. Sometimes silt and clay are carried toward the screen in suspension. Scalelike carbonates often cement the sand grains together around a screen. Incrustation is generally made up largely of the bicarbonates and sulfates of calcium, magnesium, sodium, and iron. Iron and manganese compounds are the most difficult to break down and remove. Silt and clay may add to the problem as do iron bacteria and slime-forming organisms such as Crenothrix.

Pumping a production well reduces the static head in the vicinity of the well. This reduction in pressure, plus an increase in water velocity, releases some of the carbon dioxide in groundwater as a gas and, thus, reduces the ability of the water to carry the dissolved minerals in solution. Minerals such as iron hydroxide, calcium carbonate, and magnesium carbonate are then deposited in the pump, in the production-well screen openings, and in the well walls. Mineral deposition on the upper portion of the screen will occur if the top is exposed by overpumping and lowering of the water level below the top of the screen.

Iron bacteria and other slime-forming organisms can live in groundwater by feeding on ammonia, methane, and carbon dioxide. The production of slime is a result of the life cycle of the organisms, and iron is changed to insoluble mineral salts, thus augmenting incrustation.

Incrustation and corrosion can occur together under certain conditions. The products of corrosion of iron can cause incrustation. There are no wholly effective safeguards against corrosion or incrustation. The action can be retarded by periodic cleaning of production wells, providing a screen with maximum possible inlet area to reduce velocity, and reducing pumping rates.

5.13 Rehabilitation of Production Wells

Rehabilitation is generally defined as the treatment of a production well by mechanical, chemical, or other means for the purpose of recovering the lost production of the well due to incrustation or other clogging of screens of the formation adjacent to the well bore. No single treatment is suitable for all production wells. There are two successful methods for rehabilitating screened production wells: (1) pulling the encrusted screens, treating them for removal of the encrusting material, and resetting them; and (2) treating the screens and the water-yielding formation directly around them with acid or other chemicals without pulling the screens.

An outline of the application of rehabilitation methods is given in Table 5.8.

TABLE 5.8 APPLICATIONS OF VARIOUS WELL-REHABILITATION METHODS. (*From Erikson, 1961.*)

Method	*Unconsolidated aquifers*	*Consolidated sandstone*	*Consolidated limestone*
Muriatic acid[a] followed by chlorine	Removes iron, sulfur, and carbonate deposits	Not usually effective	Sometimes beneficial; best results obtained by pressure acidizing
Polyphosphate followed by chlorine	Removes fine silt, clay, colloids, disseminated shale, and soft iron deposits	Not usually effective	Not usually effective
Dynamiting	Not recommended	Effective for all types of well-screen deposits	Effective when very large charges are used
Compressed air	Removes plugging deposits of silt and fine sand in areas adjacent to screens	Not used	Not used
Dry ice	Same as compressed air	Used only rarely, to remove cuttings from the face of a new production well	Not usually effective
Surging	Same as compressed air	Rarely used	Rarely used
Chlorine[b]	Removes iron and slime-forming bacteria	Removes iron and slime-forming bacteria	Removes iron and slime-forming bacteria
Caustic soda	Removes oil scum left by oil-lubricating pumps	Removes oil scum left by oil-lubricating pumps	Removes oil scum left by oil-lubricating pumps

[a] Not to be used with concrete screens.
[b] Usually used in a concentration of 500 ppm.

The process of surging involves the movement of water back and forth through the well openings in an effort to remove the fine particles of material clogging the screen and/or the well wall. The use of a surge block is the most effective method of surging. Other methods of obtaining surging action utilize compressed air and dry ice. Surging is often used in conjunction with chemicals to increase their effectiveness. Some of the chemicals used are phosphates, chlorine, acids, and wetting agents.

Shooting wells has increased yields. Shots of 50 to 200 pounds of 80 percent high-velocity gelatin have been used. Vibratory explosion is a form of dynamiting in which the explosive is divided into many small charges and arranged to fire in rapid sequence, thus producing a vibrating effect. Gas released from the explosion creates a bubble which expands and contracts, gradually decreasing in size until it is finally damped out. Surging action is produced by the bubbles.

Muriatic acid, a commercial grade of hydrochloric acid which contains a small percentage of impurities, is normally used in acid treatment. The concentration usually injected into the production well is 15 percent. An inhibiting agent is added to the acid before use to delay attack on the metals in the production well. Antifoam additives are available to increase the effectiveness of the acid and prevent overflowing due to foaming. Stabilizing chemicals are frequently added to inhibited muriatic acid in order to prevent secondary precipitation of dissolved iron deposits. Surging of the acid solution is accomplished by means of the pump. Muriatic acid is never used on a production well with a concrete casing or screen; citric acid is often used in this case. Pressure acidizing consists of injecting large quantities of acid solution into the production well at high pressures, thereby producing deep radial penetration.

When production wells are operated at high rates of pumping, the pressure of the water in dolomite and limestone aquifers is greatly reduced in the immediate vicinity of the well bore and in fractures extending some distance from the well bore. Because of the decline in artesian pressure, carbon dioxide is liberated and the water is unable to hold in solution its load of mineral salts. Consequently calcium carbonate is precipitated in the openings of the well face and well wall, and the permeability of the well face and well wall is greatly reduced. This clogging is especially noticeable in those production wells with pumping levels below the top of the dolomite or limestone aquifer. The yields of clogged production wells have often been restored to their original values by acid treatment.

According to Muskat (1937), acid treatment will be relatively ineffective unless the dolomite or limestone aquifer has extended fractures or the openings are partially clogged. Increases up to 50 percent for production wells of initially moderate or high capacity may be explained on the assumption that the width of the water-yielding openings of a small radial zone about the well bore has been increased and/or that materials partially clogging the well wall have been removed. Moderate increases, 50 to 500 percent, may be explained on the assumption that there are extended fractures in the dolomite or limestone aquifer which are penetrated and widened by the action of the acid or that mild clogging is the principal factor in determining the initial yield of the production well. Increases larger than 500 percent can only be explained on the assumption that there are extended fractures in the dolomite or limestone aquifer which are penetrated and widened by the acid or that there was initially a condition of almost complete clogging of the fractures.

The use of polyphosphates for production-well rehabilitation is generally confined to wells in unconsolidated deposits. They can successfully be used for cleaning screened sand and gravel production wells which are plugged with fine sand, silt, clays, colloids, and disseminated shales. The polyphosphates react with and hasten the removal of these materials, whether they are naturally found in the aquifer or placed there during drilling operations. Polyphosphates can also be used to dissolve certain forms of soft iron deposits, and where these are present a considerable saving can be effected over the use of acid treatment. Polyphosphates range from very acidic to very basic.

Chlorine is sometimes used alone as a complete production-well treatment, but it is more frequently used as a supplement following an acid or polyphosphate treatment. It is almost always used as a final chemical treatment before putting a production well back in service. A concentration of 50 ppm of available chlorine in the well provides effective disinfection, and a concentration of 200 ppm destroys certain types of iron bacteria. The destruction of sulfur and slime bacteria can be accomplished with chlorine concentrations of 500 ppm.

When the internal parts of a deep-well pump become corroded or covered with deposits, loss of efficiency and pumping capacity results. If it is apparent from data that the pump requires attention and the unit has been in continuous service for a period of 3 to 4 years or more, it would probably be advisable to pull and dismantle the pump for cleaning and repair. If, however, the pump has been installed for a shorter period and the nature of the deposit is known from past experience, it is often advantageous to clean the pump in place. This is usually done with a 15 percent concentration of inhibited muriatic acid (with a stabilizer added to hold the material in solution) or with a polyphosphate compound. Occasionally, both acid and polyphosphate treatments are used, particularly if the exact nature of the deposits is not known. The material is surged with the pump and discharged, followed by a 500-ppm chlorine solution treatment.

The median costs for the various rehabilitation methods are (Koenig, 1960a):

Surging	$380
Vibratory explosion	$660
Pressure acidizing	$950
All methods	$1,050
Fracturing	$1,250
Shooting	$1,800

Fracturing and pressure acidizing often show a unit cost as high as that of the original production well. Surging, however, shows a unit cost greater than that of the original production well for only about 23 percent of the treatments. For all treatments combined, the unit cost of rehabilitation is the greater for about 15 percent of the cases.

When considered by type of formation, data show that rehabilitation success has been definitely greater in consolidated formations than in unconsolidated formations. Consolidated formations show a median of 141 percent improvement, whereas unconsolidated formations show only 45 percent improvement (Koenig, 1960b). When analyzed by treatment type, data show a quite consistent pattern by all three criteria, the methods (arranged in ascending order of effectiveness) being surging, shooting, vibratory explosion, pressure acidizing, and fracturing (Koenig, 1960b).

Production-well rehabilitation has definitely been technically successful; some of the more advanced techniques that have already been adopted as conventional practice in crude-oil–production technology have shown higher performance records than the conventional techniques in the water-well industry. Judging by the wide geographic coverage of successful rehabilitation treatments, rehabilitation could probably be much more widely applied to maintain groundwater production.

CASE HISTORIES OF PRODUCTION WELLS

Practical applications of many of the equations and methods described in previous sections of this chapter are shown in the following case histories involving actual field data. These case histories illustrate several production-well-design principles, the analysis of well-production test data, and the appraisal of the success of production-well rehabilitation.

5.14 Design of Sand and Gravel Production Wells

The results of studies on the mechanical analyses of drilling cutting samples of two aquifers in Illinois will demonstrate some of the principles involved in the design of sand and gravel production wells. Suppose that it is desired to estimate the optimum yield and to design an 8-inch-diameter production well with a continuous-slot screen 30 feet long. The coefficients of transmissibility, permeability, and storage of the aquifer are 15,000 gpd/ft, 500 gpd/sq ft, and 0.0003, respectively, and there are 125 feet of available drawdown. The grain-size distribution curves for samples of aquifer in which the production well is to be completed are given in Fig. 5.12. The mechanical analyses are for actual samples taken from a test well penetrating a deeply buried sand and gravel aquifer and owned by the city of Woodstock, Illinois.

The 50 percent size of the coarsest sample is less than 4 times the 50 percent size of the finest sample; therefore, the slot size and artificial pack are not tailored to individual samples, but are based on the finest sample. The curves in Fig. 5.12 show that the effective grain size of the materials of the finest sample is 0.006 inch and the uniformity coefficient is about 2. An artificial-pack production well is indicated because the effective grain size is less than 0.01 inch and the uniformity coefficient is less than

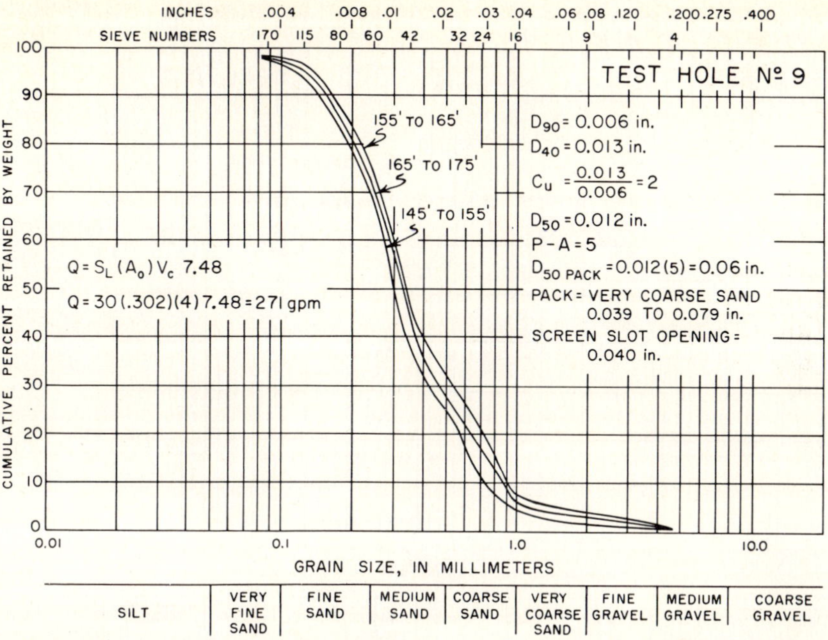

Fig. 5.12 Mechanical analyses of drilling cutting samples from an aquifer at Woodstock, Illinois. (*From Walton, 1962.*)

3. The 50 percent size of the materials of the finest sample is 0.012 inch; thus, with a pack-to-aquifer ratio of 5, a very coarse sand pack with particles ranging in diameter from about 0.04 to 0.08 inch is indicated. To retain 90 percent of the size fractions of the pack, a slot size of 0.04 inch (40 slot) is required. An artificial-pack thickness of 6 inches is adequate.

The permeability of the artificial pack is estimated to be about 1,500 gpd/sq ft, and from Table 5.1 optimum screen entrance velocities of 5 and 3 fpm are indicated for the pack and aquifer, respectively. The average of the velocities for the pack and aquifer, 4 fpm, is selected for use in determining the optimum yield of the production well. The actual open area of an 8-inch-diameter, 40-slot, continuous-slot screen is 0.604 square foot for each foot of screen. The effective open area is estimated to be about 50 percent of the actual open area, or 0.302 square foot for each foot of screen. Substitution of data on the length of screen, effective open area, and optimum screen entrance velocity into Eq. (5.1) results in the conclusion that the optimum yield of the production well is about 271 gpm. Computations are shown in Fig. 5.12. Computations made with the nonleaky artesian equation, the given hydrogeologic properties, and available drawdown indicate that the aquifer would yield more than 600 gpm to an 8-inch production well. However, pumping the production well at continuous rates exceeding 271 gpm would probably result in the production well having a short service life.

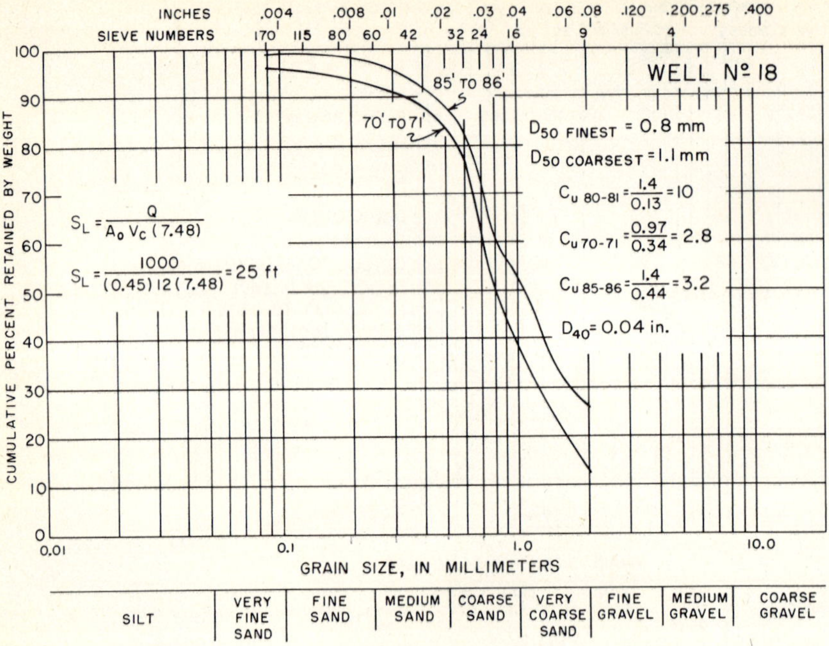

Fig. 5.13 Mechanical analyses of drilling cutting samples from an aquifer near Mossville, Illinois. (*From Walton, 1962.*)

For a demonstration of the design of a natural-pack production well, consider the grain-size distribution curves in Fig. 5.13. The mechanical analyses are for samples taken from a test well owned by the Caterpillar Tractor Company located near Mossville, Illinois. The coefficients of transmissibility, permeability, and storage of the sand and gravel aquifer are 340,000 gpd/ft, 8,100 gpd/sq ft, and 0.09, respectively. The saturated thickness of the aquifer is 42 feet.

The 50 percent size of the materials in the finest sample is less than 4 times the 50 percent size of the materials in the coarsest sample; therefore, the slot size does not have to be tailored to individual samples, but is based on the mechanical analysis of the finest sample. The effective grain sizes of both samples are greater than 0.01 inch and the average uniformity coefficient is 3. A natural-pack production well is therefore indicated. The materials overlying the aquifer will not easily cave, and so the sieve size (0.04 inch) that retains 40 percent of the aquifer materials is selected as the proper slot size.

Suppose a pumping rate of 1,000 gpm is desired. The proper length of the screen can be estimated with Eq. (5.1). The optimum screen entrance velocity, from Table 5.1, is equal to 12 fpm, and the proper diameter of the screen is estimated to be 12 inches. The effective open area of a continuous-slot, 40-slot, 12-inch-diameter screen is about 0.45 square foot for each foot of screen. Substitution of the above data in Eq. (5.1) indicates that the screen should be about 25 feet long. Computations are given in Fig. 5.13. Alternate designs are possible if the well diameter is not limited by pump requirements. Computations indicate that a 24-inch-diameter, 40-slot, continuous-slot screen that is 15 feet long will also yield 1,000 gpm with a long service life. A shorter 8-inch screen can be used if an artificial-pack production well is constructed instead of a natural pack. Analysis of Fig. 5.13 indicates that a 100-slot, continuous-slot screen that is 22 feet long will also yield 1,000 gpm.

5.15 Production Wells in Bedrock Aquifers

The water-yielding properties of crystalline rocks are dependent on the occurrence of joints and faults and the extent of weathering. Interstitial openings caused by weathering are mostly at depths of less than 100 feet. Joints are less abundant and openings along joint planes are smaller as depth increases. Openings along fault surfaces also tend to close with depth. These geological observations, which indicate a decrease in rock permeability (see Fig. 5.14), were verified by a study of production-well yields and water-injection tests (Davis and Turk, 1964). Mean yields per foot of well production are 0.23 to 0.30 gpm at 100 feet but only 0.013 to 0.04 gpm at 1,000 feet. Mean injection rates per foot of drill hole under 100 psi pressure are 0.11 to 0.4 gpm at 100 feet but only 0.014 to 0.038 gpm at 1,000 feet. Median values of both production-well yields and injection rates are from one-half to one-third of the mean value.

Geologic structure is more important than rock type in determining yields and injection rates. The decrease in rock permeability with depth indicates an increase in the unit cost of water with depth. The optimum depth of a production well is, therefore, determined largely by economic factors. Unless geologic factors are favorable, production wells in crystalline rocks should be less than 600 feet deep. In general, domestic wells should be less than 150 to 200 feet deep.

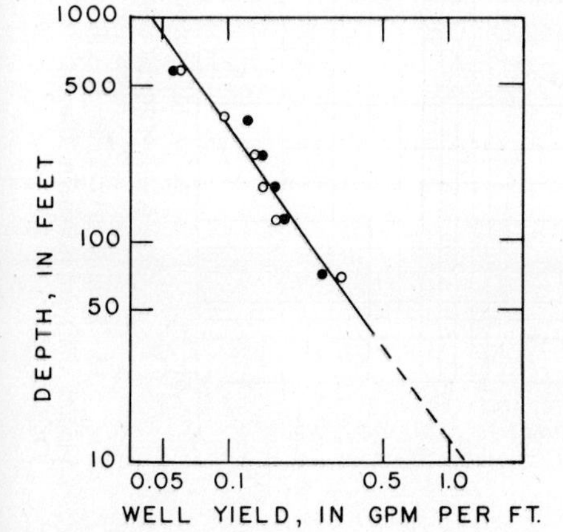

Fig. 5.14 Yields of wells in crystalline rocks of eastern United States. (*From Davis and Turk, 1964.*)

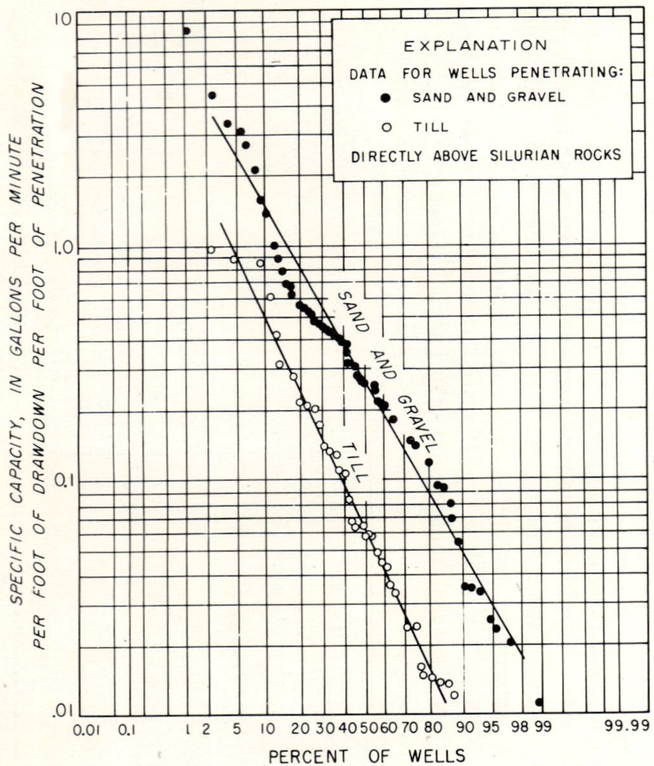

Fig. 5.15 Relation between specific capacity and character of glacial drift in northern Illinois. (*From Csallany and Walton, 1963.*)

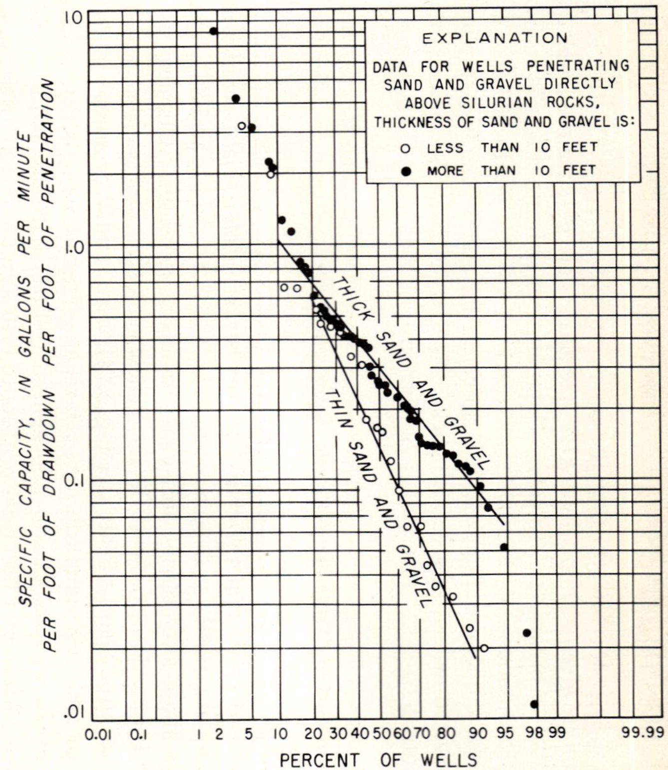

Fig. 5.16 Relation between specific capacity and thickness of sand and gravel directly above Silurian rocks in northern Illinois. (*From Csallany and Walton, 1963.*)

Northern Illinois—Specific capacities of dolomite wells

In northern Illinois large quantities of groundwater are withdrawn from production wells in shallow dolomite aquifers of Silurian and Ordovician age (Csallany and Walton, 1963). Silurian and Ordovician rocks are encountered at depths ranging from a few to several hundred feet. Silurian rocks range in thickness from a few to more than 450 feet; the maximum known thickness of the Galena-Platterville Dolomite of Ordovician age exceeds 350 feet. The Silurian Dolomite and the Galena-Platteville Dolomite yield moderate to large quantities of groundwater to production wells. Dolomite beds of the Maquoketa Formation of Ordovician age between the Galena-Platteville Dolomite and the Silurian Dolomite yield small quantities of water to production wells. The average depth of shallow dolomite production wells is about 140 feet, and most of the production wells of recent design are finished 12 to 16 inches in diameter.

About 1,000 well-production tests were made during the period 1921 to 1961 by the Illinois State Water Survey on more than 800 shallow dolomite wells in northern Illinois. Statistical analysis of specific-capacity data provided a basis for determining whether or not significant relationships exist between the yields of production wells and hydrogeologic controls. Probable ranges in yields of shallow dolomite production wells in undeveloped areas were estimated from specific-capacity frequency graphs, aquifer thickness and areal geology maps, and water-level data.

The specific capacities based on the well-production tests were adjusted for well losses and to a common radius and pumping period. Adjusted specific capacities were then divided by the total depths of penetration below the tops of shallow dolomite aquifers to determine adjusted specific capacities per foot of penetration.

The specific-capacity frequency graphs in Figs. 5.15 to 5.18 were prepared with Eq. (5.19) to determine the relations between production-well yields and the following geologic controls: (1) the glacial deposits penetrated by the production well immediately above bedrock are predominantly till; (2) the glacial deposits penetrated by the production well immediately above bedrock are predominantly sand and gravel; (3) sand and gravel deposits immediately above bedrock are less than 10 feet thick; (4) sand and gravel deposits immediately above bedrock exceed 10 feet in thickness; (5) the production well is in a bedrock valley; and (6) the production well is in a bedrock upland. Figure 5.15 indicates that the productivity of the Silurian rocks is greater in areas where sand and gravel overlie the dolomite than it is in areas where relatively impermeable till directly overlies the dolomite. Figure 5.16 indicates that the productivity of the Silurian rocks increases as the thickness of sand and gravel overlying the dolomite increases. Graphs in Fig. 5.17 indicate that the productivity of the Silurian rocks is greater in bedrock uplands than it is in bedrock valleys. Average specific-capacity frequency graphs for cases (1) where Silurian drift rocks are overlain by glacial deposits and (2) where Silurian rocks are overlain by bedrock are shown in Fig. 5.18. The graphs indicate that in northern Illinois, the Silurian rocks yield much more water to production wells in areas where the Silurian rocks are the uppermost bedrock than they do in areas where Silurian rocks are overlain by bedrock. Specific-capacity frequency graphs show that in areas where the Galena-Platteville Dolomite is overlain by the bedrock, the productivity of the aquifer is about the same as the productivities of the Maquoketa Formation and Silurian rocks in areas where these rocks are also overlain by bedrock.

The productivity of the shallow dolomite aquifers is very inconsistent, and it is impossible to predict with a high degree of accuracy the yield of a production well before drilling at any location in undeveloped areas. However, the probable yields of production wells can be estimated, based on specific-capacity frequency graphs, aquifer thickness and areal geology maps, and water-level data. Probable ranges of specific capacities of shallow dolomite production wells in northern Illinois were estimated as the products of the specific capacities per foot of penetration measured in 50 percent of existing production wells and aquifer thickness (Csallany and Walton, 1963). It was assumed that wells completely penetrate the Silurian rocks or the Galena-Platteville Dolomite. Aquifer thicknesses were estimated from geologic maps

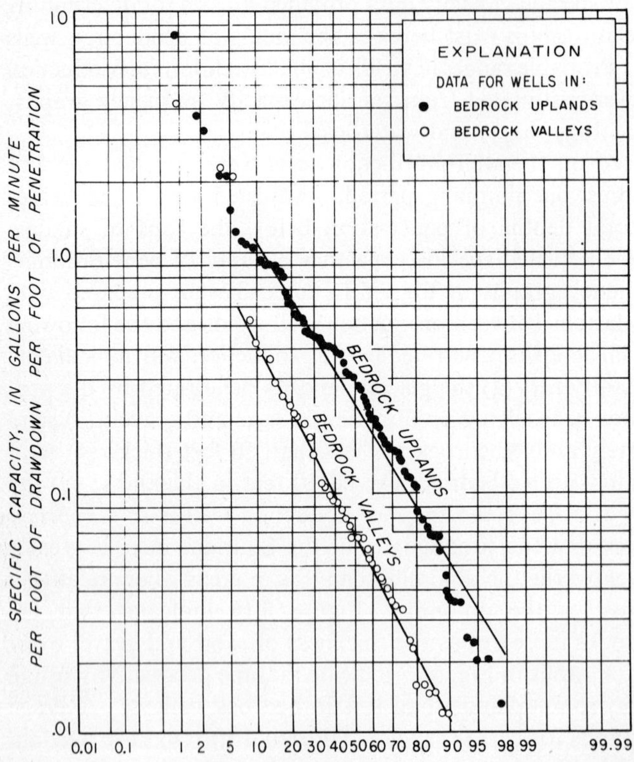

Fig. 5.17 Relation between specific capacity and bedrock topography in northern Illinois. (*From Csallany and Walton, 1963.*)

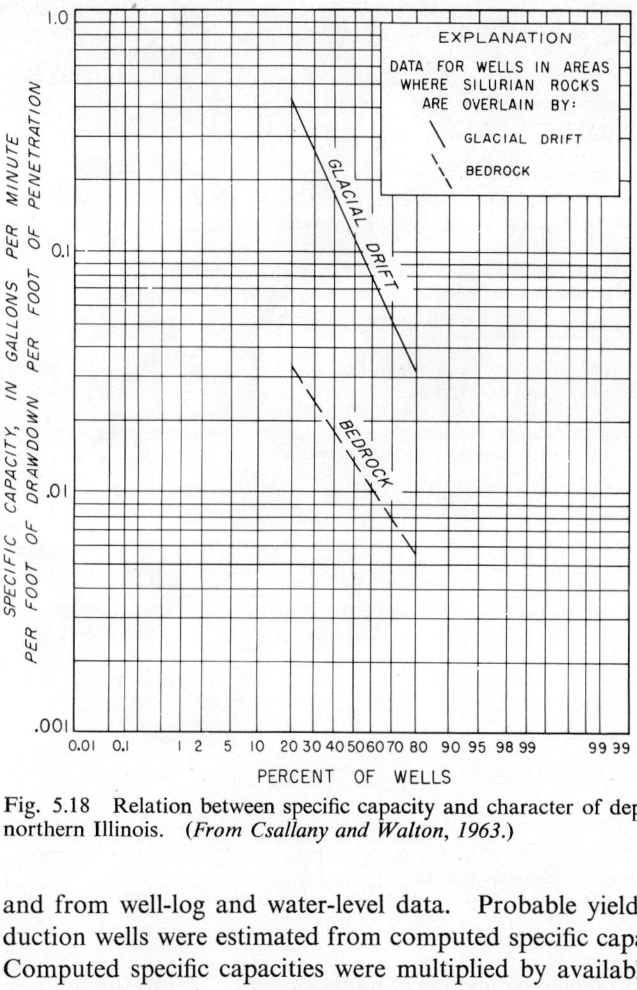

Fig. 5.18 Relation between specific capacity and character of deposits overlying Silurian rocks in northern Illinois. (*From Csallany and Walton, 1963.*)

and from well-log and water-level data. Probable yields of shallow dolomite production wells were estimated from computed specific capacities and water-level data. Computed specific capacities were multiplied by available drawdowns to determine probable yields of production wells. Pumping levels were limited to depths below the top of the shallow dolomite aquifer equal to about one-third of the thickness of the aquifer. Under such conditions, the most productive water-yielding openings in the shallow dolomite aquifer will be discharging freely into the production well, and the maximum use of upper openings is attained. Additional discharge with pumping levels much below the upper one-third of the shallow dolomite aquifer will be accompanied by rapidly declining specific capacities and yields of production wells.

The probable range of yields of shallow dolomite production wells in northern Illinois is shown in Fig. 5.19. It is possible to drill what is essentially a dry hole at any location; however, from data for 50 percent of existing production wells, the chances of obtaining a well with a production of 250 gpm or more are good in all areas except

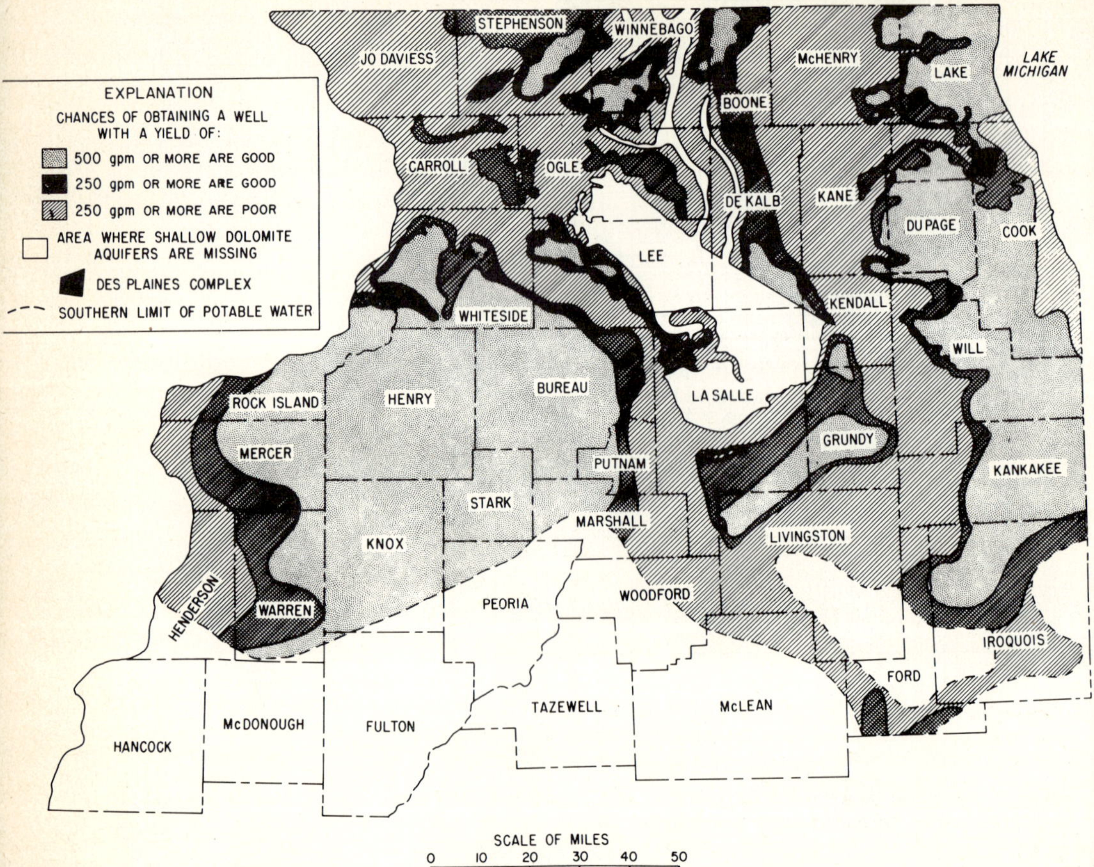

Fig. 5.19 Estimated yields of shallow dolomite production wells in northern Illinois. (*From Csallany and Walton, 1963.*)

where the Silurian rocks and Galena-Platteville Dolomite are thin or where the Maquoketa Formation is the uppermost bedrock. The chances of obtaining a well with a production of 500 gpm or more are good in large portions of northeastern Illinois. It is probable that the yield of the shallow dolomite aquifers is high enough to support heavy industrial or municipal development in large parts of northern Illinois.

DuPage County, Illinois—Specific capacities of dolomite wells

In DuPage County in northeastern Illinois, rocks of Silurian age immediately underlie the glacial drift and in many areas yield large quantities of water to production wells. In addition, dolomite beds in the Maquoketa Formation of Ordovician age encountered beneath rocks of Silurian age yield small to moderate quantities of groundwater to production wells. Specific capacities of production wells range from 0.6 to 530 gpm/ft and average 42 gpm/ft. Rocks of Silurian age are the Alexandrian Series overlain by the Niagaran Series (see Zeizel et al., 1962). The thickness of the Silurian rocks increases from less than 50 feet in the western part of the county to more than 200 feet in portions of the southeastern part. Many production wells extend through the Silurian strata and penetrate dolomite beds of the Maquoketa Formation. The thickness of the Maquoketa Formation ranges from 85 to 227 feet and averages about 175 feet in DuPage County. The yields of production wells vary greatly from place to place.

Many production wells penetrate only rocks of the Niagaran (N) Series, some wells penetrate rocks of both the Niagaran and Alexandrian (A) Series, and other wells penetrate rocks of the Niagaran and Alexandrian Series and the Maquoketa Formation (M). The total depth of penetration of the production wells and the depth of penetration of the production wells into each unit were determined from well logs and sample studies of drilling cuttings. In general, the specific capacity of a

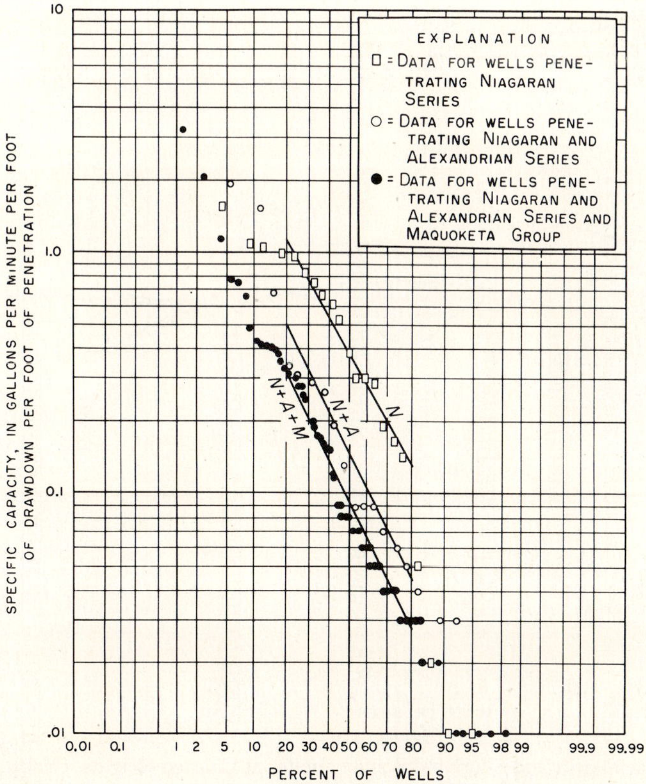

Fig. 5.20 Specific-capacity frequency graphs for dolomite production wells in DuPage County, Illinois. (*From Zeizel et al., 1962.*)

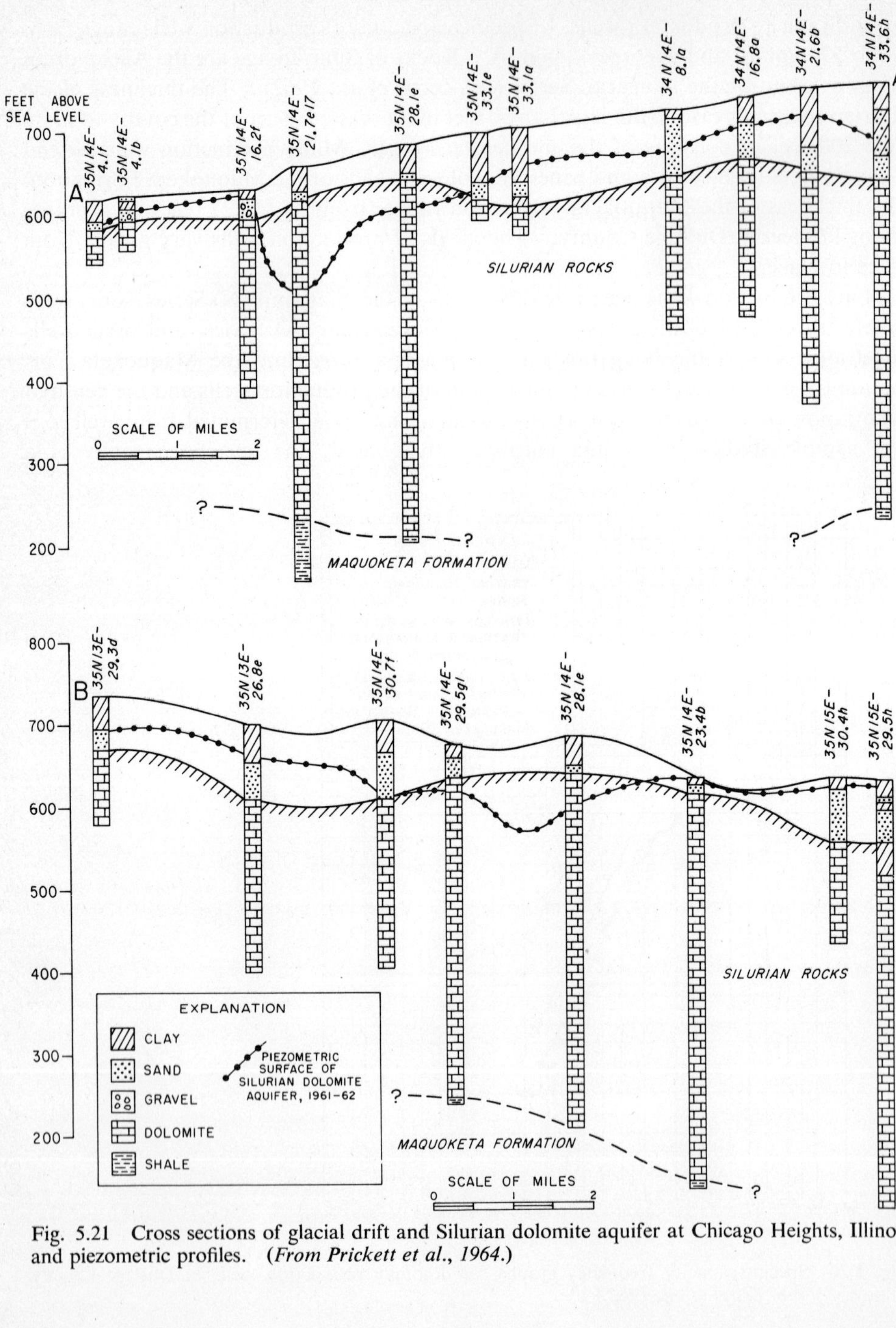

Fig. 5.21 Cross sections of glacial drift and Silurian dolomite aquifer at Chicago Heights, Illinois, and piezometric profiles. (*From Prickett et al., 1964.*)

dolomite production well increases with the depth of penetration and number of units penetrated; however, the upper part of the Silurian rocks is usually the most productive rock. Specific capacities were divided by the total depth of penetration of production wells to obtain the specific capacities per foot of penetration.

Production wells were segregated into three categories, N, $N + A$, and $N + A + M$, depending on the units penetrated by the production wells. Specific capacities per foot of penetration for production wells in each of the three categories were tabulated in order of magnitude, and frequencies were computed with Eq. (5.19). Values of specific capacity per foot of penetration were then plotted against the percentage of production wells on logarithmic probability paper as shown in Fig. 5.20. Specific capacities per foot of penetration decrease as the depth of penetration of production wells and number of units penetrated increase, indicating that both the Niagaran and Alexandrian Series are more productive than the Maquoketa Formation, and the Niagaran Series is more productive than the Alexandrian Series (Walton and Neill, 1963).

The dolomite unit-frequency graphs in Fig. 5.20 were constructed by the process of subtraction taking into consideration uneven distribution of production wells in the three categories. Based on the slopes of the dolomite unit-frequency graphs, the Maquoketa Formation is much less consistent in production than both the Niagaran and Alexandrian Series, and the productivity of the Alexandrian Series is more consistent than the productivity of the other dolomite units.

Chicago Heights, Illinois

Because of close spacing of production wells and well fields and the heavy pumpage at Chicago Heights in northeastern Illinois, extensive dewatering of upper portions of the Silurian dolomite aquifer has taken place (see Fig. 5.21). Thicknesses of dewatered dolomite were determined by comparing nonpumping water levels in wells and the bedrock surface map (Prickett et al., 1964). A map showing thickness of dewatered dolomite was prepared with data for individual wells, the piezometric-surface map, and the bedrock topography map. As shown in Fig. 5.22, the area within which dewatering has taken place is about 9 square miles, the greatest thickness of dewatered dolomite is located around a group of production wells owned by the Inland Steel Company and Chicago Heights, and it exceeds 170 feet.

It has been shown by Walton and Neill (1963) and Zeizel et al. (1962) that the specific capacity per foot of penetration of dolomite production wells decreases as the depths of production wells increase, indicating that the upper part of the Silurian dolomite aquifer is much more productive than the lower part. To determine the magnitude of decreases in specific capacities of dolomite production wells in Chicago Heights as the result of dewatering of the uppermost productive portion of the Silurian dolomite aquifer, specific-capacity data for production wells within and outside the area of dewatering were compared.

The total depths of penetration of production wells into the dolomite were determined from well logs. Adjusted specific capacities were divided by the total depth of penetration to obtain adjusted specific capacities per foot of penetration. Production wells were segregated into two categories, depending upon whether they were located within or outside the area of dewatering shown in Fig. 5.22. Adjusted specific capacities per foot of penetration for production wells in the two categories were tabulated in order of magnitude, and frequencies were computed with Eq. (5.19). Values of specific capacity per foot of penetration were then plotted against percent of wells on logarithmic probability paper as shown in Fig. 5.23.

The frequency graphs indicate that the specific capacities of production wells within the area of dewatering are much less than the specific capacities of production wells outside the area of dewatering. Based on the specific capacities with a frequency of 50 percent, it is estimated that the specific capacities of production wells within the area of dewatering have decreased about 65 percent as the result of dewatering. Specific-capacity data indicate that the coefficient of transmissibility of the Silurian dolomite aquifer averages about 65,000 gpd/ft in areas where no dewatering of the dolomite has taken place; the coefficient of transmissibility in areas where extensive dewatering of the dolomite has taken place averages about 22,000 gpd/ft.

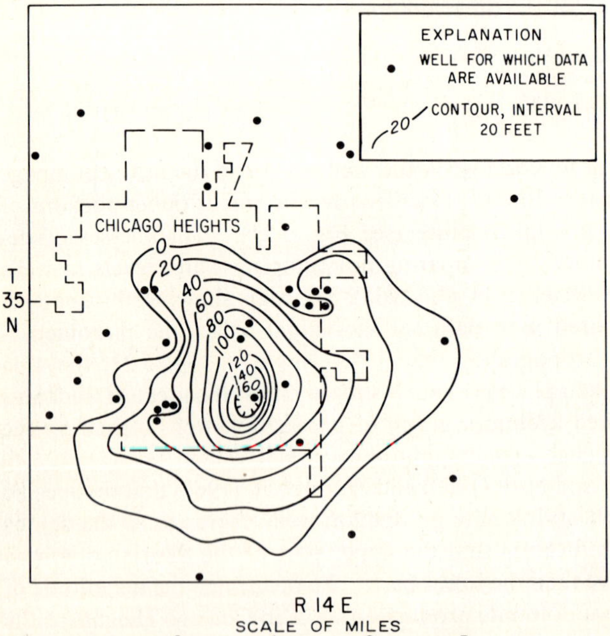

Fig. 5.22 Thickness of dewatered Silurian dolomite aquifer at Chicago Heights, Illinois. (*From Prickett et al., 1964.*)

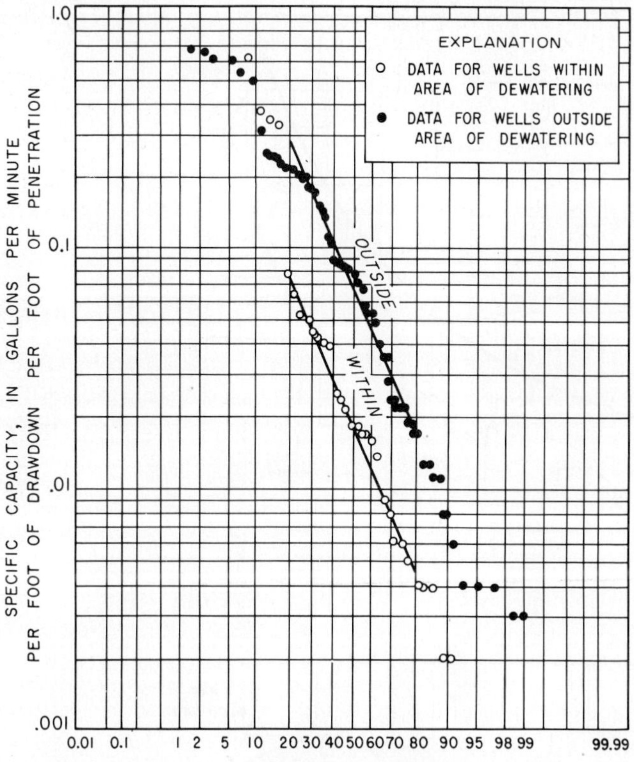

Fig. 5.23 Specific-capacity frequency graphs for shallow dolomite production wells in Chicago Heights, Illinois. (*From Prickett et al., 1964.*)

Northern Illinois—Deep sandstone wells

Most deep sandstone production wells in northern Illinois tap several bedrock units and are multiunit wells. The Galena-Platteville Dolomite, Glenwood–St. Peter Sandstone, and Prairie du Chien Series of Ordovician age and the Trempealeau Dolomite, Franconia Formation, Ironton-Galesville Sandstone, and Mt. Simon Sandstone of Cambrian age yield appreciable quantities of water to production wells. During the period 1906 to 1960 well-production tests were made by the Illinois State Water Survey on more than 500 deep sandstone production wells in northern Illinois. Specific-capacity data were studied by Walton and Csallany (1962) to determine the role of the individual bedrock units encased in deep sandstone production wells as contributors of water.

Production wells were grouped into categories according to the units uncased in the production wells. The yields of the individual units were ascertained, taking into consideration that the specific capacity of a multiunit well is the numerical sum of the specific capacities of the individual production wells.

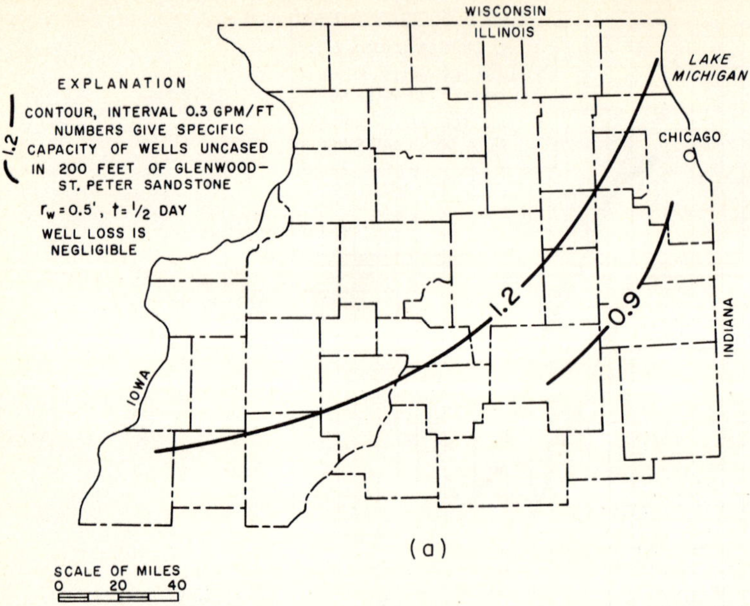

(a)

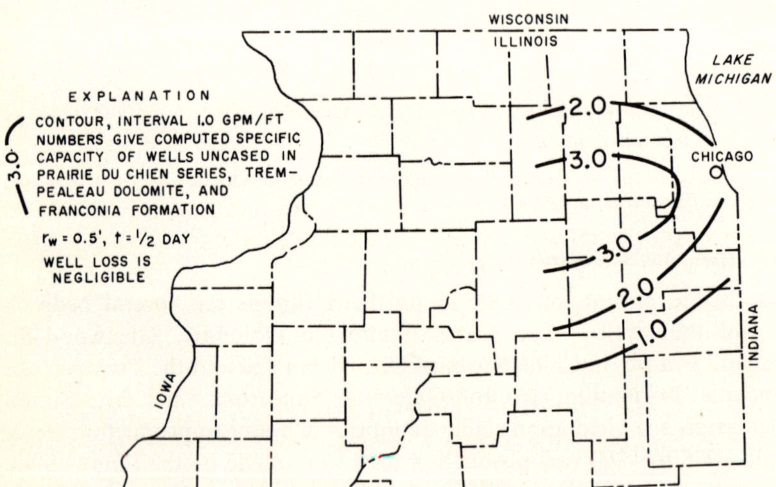

(b)

Fig. 5.24 Specific capacities of production wells uncased in Glenwood–St. Peter Sandstone (*a*) and Prairie du Chien Series, Trempeauleau Dolomite, and Franconia Formation (*b*) in northeastern Illinois. (*From Walton and Csallany, 1962.*)

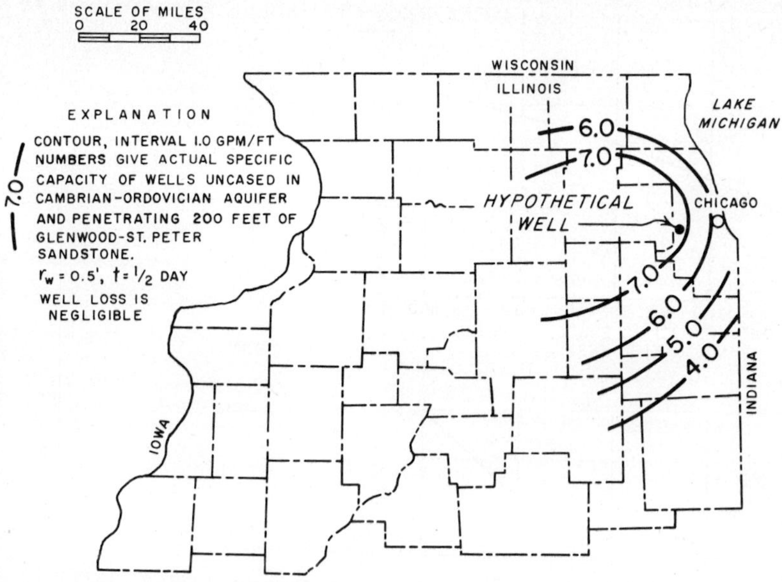

Fig. 5.25 Specific capacities of production wells uncased in Cambrian-Ordovician Aquifer. (*From Walton and Csallany, 1962.*)

Regional variations in yields are shown for the Glenwood–St. Peter Sandstone, the Prairie du Chien Series, Trempealeau Dolomite, and Franconia Formation in Fig. 5.24, and for the Cambrian-Ordovician Aquifer in Fig. 5.25. The specific capacity of a production well increases with the thickness of the Glenwood–St. Peter Sandstone but is not directly proportional to thickness as shown in Fig. 5.26*a*. The yields of production wells per foot of penetration in the Mt. Simon Aquifer are fairly constant with depths throughout northern Illinois as shown in Fig. 5.26*b*.

Granite City, Illinois

A variable-rate well-production test (Bruin and Hudson, 1955) was conducted by E. G. Jones and Jack Bruin of the Illinois State Water Survey on an irrigation well owned by J. R. Thompson. The well was constructed by the Thorpe Concrete Well Company and was completed on November 4, 1953. The well is located approximately 1,300 feet east and 1,450 feet north of the southwest corner of section 29, T4N, R9W, near Granite City, Illinois. The driller's log of the well is given in Table 5.9. The bottom 60 feet of the well is screened with porous concrete screen and the upper 46 feet is cased with concrete casing. The casing and screen have an inside diameter of 30 inches and an outside diameter of 40 inches.

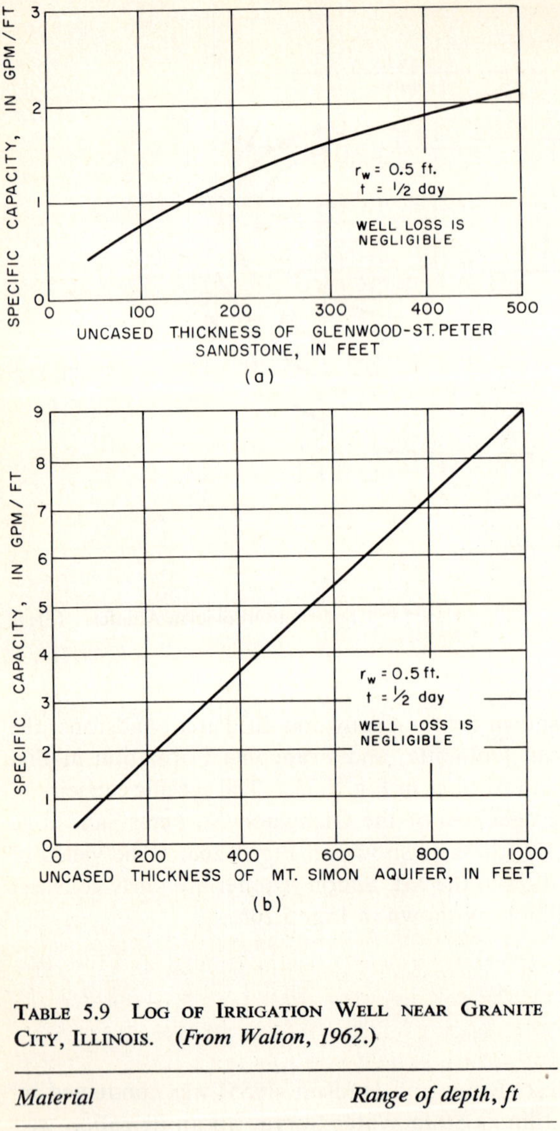

Fig. 5.26 Relation between specific capacity of a production well and uncased thickness of Glenwood–St. Peter Sandstone (*a*) and uncased thickness of Mt. Simon Aquifer (*b*) in northern Illinois. (*From Walton and Csallany, 1962.*)

TABLE 5.9 LOG OF IRRIGATION WELL NEAR GRANITE CITY, ILLINOIS. (*From Walton, 1962.*)

Material	*Range of depth, ft*
Sandy clay	0–23
Yellow medium-coarse sand	23–28
Fine gray sand	28–44
Medium-fine gray sand	44–52
Sand	52–60
Coarse, clean sand	60–84
Medium-coarse sand	84–100
Coarse sand and boulders	100–106

TABLE 5.10 DATA FOR WELL-PRODUCTION TEST NEAR GRANITE CITY, ILLINOIS. (*From Bruin and Hudson, 1955.*)

Time	*Feet to water*	*Pumping rate, gpm*
9:13 A.M.	23.95	0
9:23	23.95	
9:45	Started pumping	
9:50	29.13	1,000
9:55	29.23	1,000
10:00	29.30	1,000
10:10	29.30	1,000
10:15	29.32	1,000
10:20	29.37	1,000
10:30	29.35	1,000
10:35	29.36	1,000
10:52	29.40	1,000
11:00	29.39	1,000
11:25	29.43	1,000
11:30	30.97	1,280
11:50	31.01	1,280
12:00	31.02	1,280
12:15 P.M.	31.04	1,280
12:30	31.08	1,280
12:45	31.08	1,280
12:50	31.90	1,400
1:00	31.95	1,400
1:15	31.90	1,400
1:30	31.90	1,400
1:40	31.93	1,400

The well-production test was started at 9:45 A.M. on April 27, 1954, and was continued for about 4 hours until 1:40 P.M. The well was pumped at three rates, 1,000, 1,280, and 1,400 gpm. Well-production test data are given in Table 5.10.

The data were plotted on semilogarithmic paper with the pumping levels on the arithmetic axis and the time after pumping started on the logarithmic axis as shown in Fig. 5.27. It can be noted that the time-drawdown curve at 1,000 gpm has a slope of 0.21 foot per log cycle. The slope of the semilogarithmic time-drawdown curve is directly proportional to Q. The slope at 1,280 gpm was estimated to be about 0.269 foot per log cycle by multiplying the slope at 1,000 gpm by the ratio of 1,280:1,000. By similar procedure the slope at 1,400 gpm was computed to be 0.294 foot per log cycle. These slopes were used to extrapolate water-level trends beyond periods of pumping as shown by the dashed lines in Fig. 5.27. These extrapolations were used to obtain increments of drawdown produced by each increase in the rate of pumping. Increments of drawdown were determined for a pumping period of 1 hour and are given in Table 5.11.

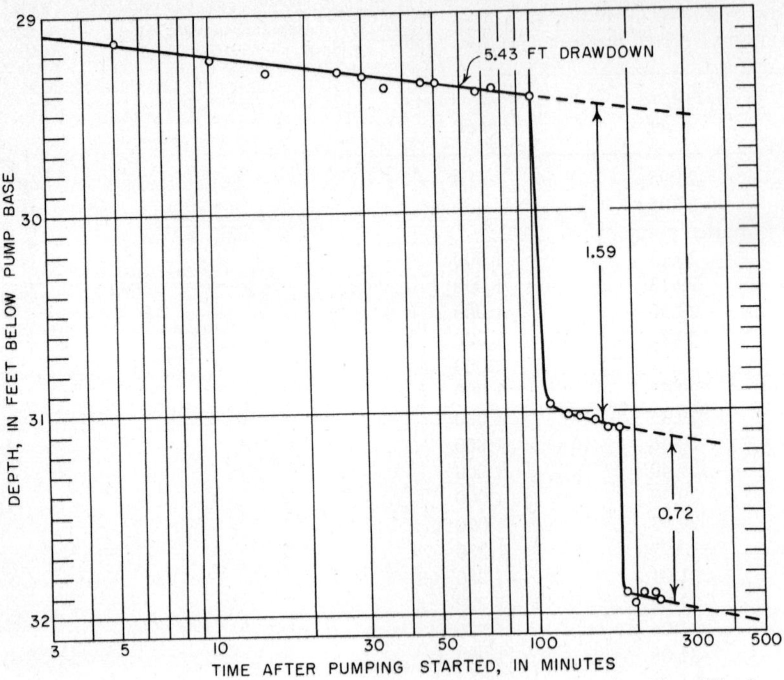

Fig. 5.27 Time-drawdown graph for irrigation well near Granite City, Illinois. (*From Bruin and Hudson, 1955.*)

Data in Table 5.11 were substituted into Eqs. (5.14) and (5.15) for computations of C as shown below

For steps 1 and 2

$$C_{1 \text{ and } 2} = \frac{1.59/0.62 - 5.43/2.22}{2.22 + 0.62}$$

$$C_{1 \text{ and } 2} = 0.04 \text{ sec}^2/\text{ft}^5$$

For steps 2 and 3

$$C_{2 \text{ and } 3} = \frac{0.77/0.27 - 1.59/0.62}{0.62 + 0.27}$$

$$C_{2 \text{ and } 3} = 0.28 \text{ sec}^2/\text{ft}^5$$

TABLE 5.11 DRAWDOWNS AND PUMPING RATES FOR WELL 1 NEAR GRANITE CITY, ILLINOIS

Step	*Q, gpm*	*ΔQ, gpm*	*ΔQ, cfs*	*Δs, ft*
1	1,000	1,000	2.22	5.4
2	1,280	280	0.62	1.59
3	1,400	120	0.27	0.72

The value of C for steps 2 and 3 is greater than the value of C for steps 1 and 2. However, both values of C are very low and it is probable that the values of C differ in this case not because the well is unstable but because of slight inaccuracies in measured drawdowns and pumping rates. The average value of C is about 0.16 sec^2/ft^5. The well loss at the 1,400-gpm pumping rate was computed to be about 1.0 foot with Eq. (5.12) as shown below:

$$s_{wL} = 0.16\ (1{,}440/449)^2 = 1.0 \text{ ft}$$

The total drawdown in the well at 1,400 gpm was 7.74 feet. Thus, well loss amounted to about 13 percent of the total drawdown in the irrigation well for a pumping rate of 1,400 gpm.

Iuka, Illinois

A variable-rate well-production test was made by Sandor Csallany of the Illinois State Water Survey in cooperation with Marbry and Johnson, Inc., consulting engineers, and E. C. Baker and Sons, well contractors, on a sandstone production well owned by the village of Iuka, Illinois (see Walton, 1962). The production well is located about 600 feet west and 1,300 feet south of the northeast corner of section 18, T2N, R4E, in Marion County. The log and construction features of the production well and three observation wells are shown in Fig. 5.28. The locations of the wells are shown in Fig. 5.29. The two major water-yielding zones were encountered in wells 1, 2, and 3 between the depths of 20 and 31 feet and 40 and 85 feet. The well-production test was started at 11:20 A.M. on May 29, 1961, and was continued for about 4 hours until 3:30 P.M. Well 3 was pumped at three rates, 5.4, 9.3, and

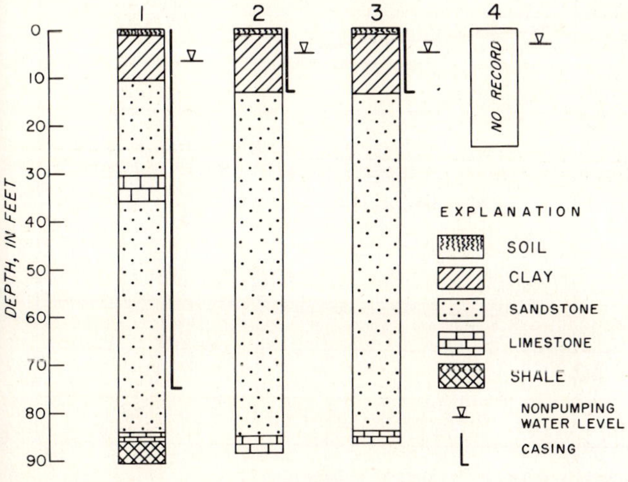

Fig. 5.28 Generalized graphic logs of wells used in well-production test at Iuka, Illinois. (*From Walton, 1963.*)

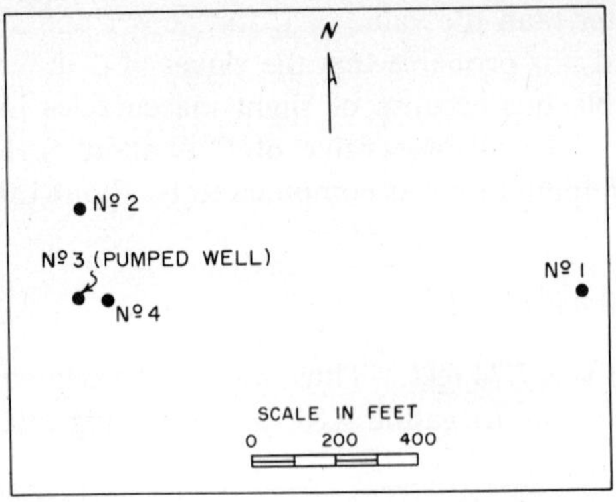

Fig. 5.29 Map showing location of wells used in well-production test at Iuka, Illinois. (*From Walton, 1962.*)

12.7 gpm. Drawdowns were measured in the production well and in the three observation wells.

Drawdowns in wells 2, 3, and 4 were plotted against time on semilogarithmic paper. The time-drawdown graph for well 3 is given in Fig. 5.30 and time-drawdown graphs for wells 2 and 4 are given in Fig. 5.31. Time-drawdown data for the wells are tabulated in Table 5.12. There was no significant drawdown in well 1.

During the first step with a pumping rate of 5.4 gpm the pumping level in well 3 remained above the top of the upper water-yielding zone. The specific capacity of the production well for a pumping rate of 5.4 gpm and a pumping period of 1 hour is 0.35 gpm/ft. Water levels in the shallow observation well (well 4), which penetrates

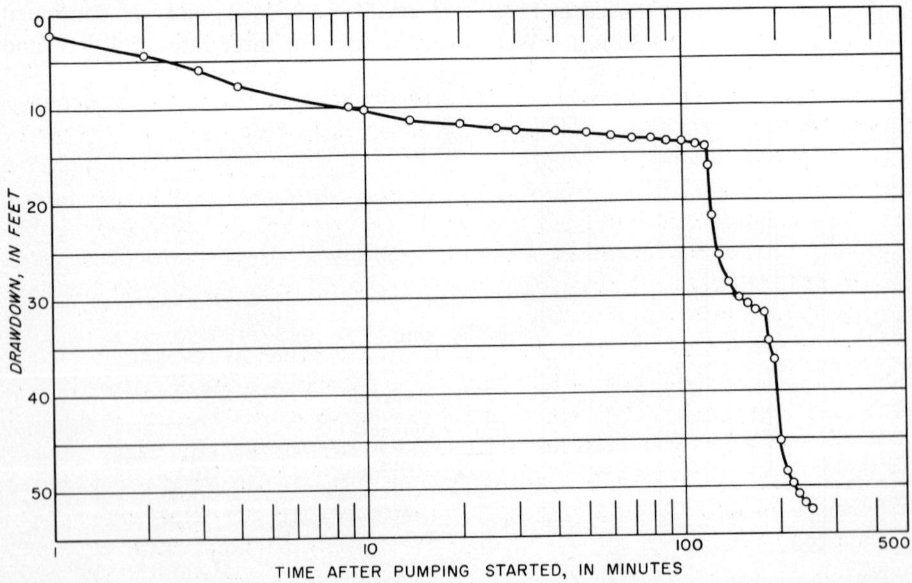

Fig. 5.30 Time drawdown for well 3 at Iuka, Illinois. (*From Walton, 1962.*)

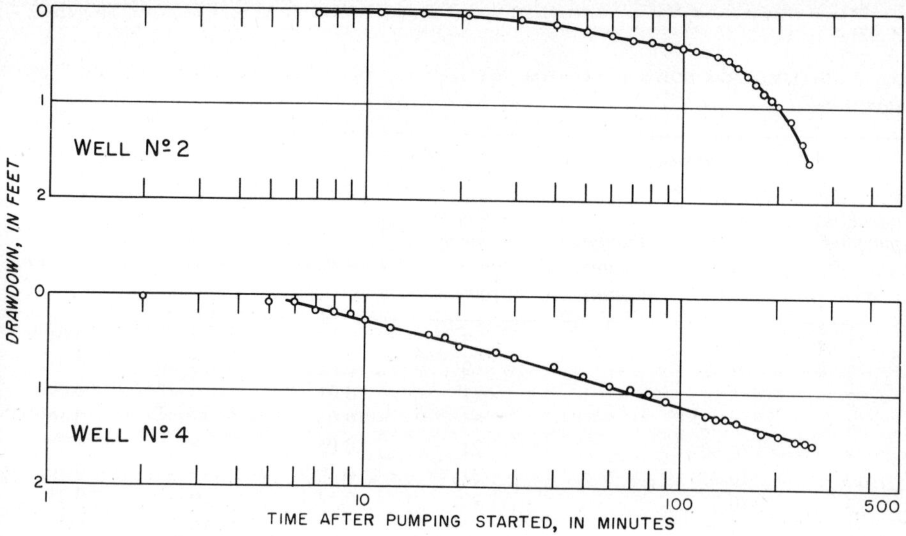

Fig. 5.31 Time drawdown for wells 2 and 4 at Iuka, Illinois. (*From Walton, 1962.*)

only the upper water-yielding zone, and in the deep observation well (well 2), which penetrates both the upper and lower water-yielding zones, were similarly affected during the first step.

The pumping level in well 3 declined below the upper water-bearing zone during the second step with a pumping rate of 9.3 gpm. The specific capacity of the production well for a pumping rate of 9.3 gpm and a pumping period of 1 hour is 0.28 gpm/ft and is much less than the specific capacity for a pumping rate of 5.4 gpm. Water levels in the shallow observation well were not affected by the increase in pumping rate, whereas the time-rate of drawdown in the deep observation well increased about in proportion to the pumping rate. When the pumping level declined below the upper water-yielding zone there was free flow from the upper water-yielding openings. Thus, the maximum contribution for the upper zone was attained during the second step. As shown by the time-drawdown graph for well 4, discharge from the upper zone was not appreciably increased, indicating that most of the increase in discharge in step 2 was obtained from the lower zone.

The pumping level in well 3 declined below the top of the lower zone during the third step with a pumping rate of 12.7 gpm. The specific capacity of the production well for a pumping rate of 12.7 gpm and a pumping period of 1 hour is 0.17 gpm/ft and is less than the specific capacities for pumping rates of 5.4 and 9.3 gpm. Water levels in the shallow observation well were not affected by the increase in pumping rate, whereas the time-rate of drawdown in the deep observation well increased about in proportion to the pumping rate. The specific capacity during the third step is less than the specific capacity during the second step because there was free flow from some of the openings in the lower zone and the openings in the basal part of the lower zone

TABLE 5.12 TIME-DRAWDOWN DATA FOR WELL-PRODUCTION TEST AT IUKA, ILLINOIS. (*From Walton, 1962.*)

Well 3 (pumped well)			*Well 2*		*Well 4*	
Time after pumping started, min	*Drawdown, ft*	*Pumping rate, gpm*	*Time after pumping started, min*	*Drawdown, ft*	*Time after pumping started, min*	*Drawdown, ft*
1	2.10	5.5	7	0.01	2	0.01
2	4.42		9	0.01	5	0.08
3	5.84		11	0.01	6	0.09
4	7.63		15	0.02	7	0.16
9	10.00		21	0.05	8	0.18
10	10.23	5.4	31	0.10	9	0.20
13	10.84		40	0.14	10	0.25
15	11.09		50	0.21	16	0.40
17	11.37	5.4	60	0.25	20	0.46
20	11.82	5.4	70	0.29	30	0.62
23	11.88		80	0.32	40	0.72
26	12.09	5.4	90	0.35	50	0.81
30	12.33		100	0.38	60	0.92
40	12.68	5.4	120	0.42	70	0.95
45	12.79	5.4	130	0.45	80	1.00
50	12.89	5.4	140	0.50	90	1.09
55	13.04	5.4	150	0.59	100	1.11
60	13.15	5.4	160	0.68	120	1.22
70	13.34	5.4	170	0.75	130	1.25
80	13.54	5.4	180	0.84	140	1.27
90	13.79	5.4	190	0.90	150	1.30
100	13.93	5.4	200	0.95	160	1.32
121	16.27	9.3	210	1.03	170	1.37
130	25.84	9.3	220	1.14	180	1.42
140	28.99	9.3	230	1.27	194	1.42
150	30.13		240	1.39	203	1.43
160	30.85	9.3	250	1.59	220	1.46
170	31.48				230	1.48
180	31.84	9.3			240	1.50
193	36.82	12.7			250	1.52
203	45.89	12.7	*Well 1*			
220	49.82					
230	51.00	12.7	23	0.03		
240	51.92		65	0.03		
250	52.28		105	0.04		

and the openings in the basal part of the lower zone were called upon for much of the increase in pumpage.

From the above discussion, it is obvious that erroneously optimistic predicted yields of the production well under higher rates of pumping would occur if the specific capacities for step 1 or 2 were used in the computations. Experience has shown that the constant-rate method of production-well testing when applied to production wells tapping bedrock aquifers, especially dolomite aquifers, has sometimes resulted in erroneously optimistic predicted yields of production wells under higher rates of pumping. The variable-rate well-production test provides data that can be analyzed to obtain more accurate predictions of production-well yields under various pumping-rate conditions. Drawdown at the end of each pumping period is plotted against the corresponding pumping rate and a curve is drawn through the points. The drawdown in the production well caused by a planned rate of discharge may be read directly from the curve or approximated for higher pumping rates by projecting the curve.

DuPage County, Illinois—Well-production tests

Well-production tests were made on several Silurian dolomite wells in DuPage County, Illinois. Analysis of available data (Zeizel et al., 1962) indicates that the well-loss constant is a function of (1) the specific capacity and therefore the hydrogeologic properties of the Silurian dolomite aquifer, and (2) the position of the pumping level in relation to the top of the Silurian dolomite aquifer. High values of C were computed for production wells having low specific capacities and low values of C were computed for production wells having high specific capacities. Apparently turbulence and therefore well loss increase as the coefficient of transmissibility of the aquifer decreases. The size and/or number of openings in the dolomite aquifer decrease with the coefficient of transmissibility.

The well-loss constant increases greatly when water levels are lowered below the top of the Silurian dolomite aquifer. As the pumping level declines below the top of the aquifer, maximum contribution from openings in the upper part of the dolomite above the pumping level is attained and future increases in pumping are obtained from the openings below the pumping level. A greater burden is placed upon lower openings and well loss is greatly increased.

Graphs in Figs. 5.32 and 5.33 were prepared showing the relation between specific capacity and C for the two cases, when the pumping level is above the top of the aquifer and when the pumping level is below the top of the aquifer.

5.16 Acid Treatment of Production Wells to Increase Yield

Acid treatment has been used successfully to rehabilitate old shallow dolomite production wells and to develop newly constructed production wells in northern

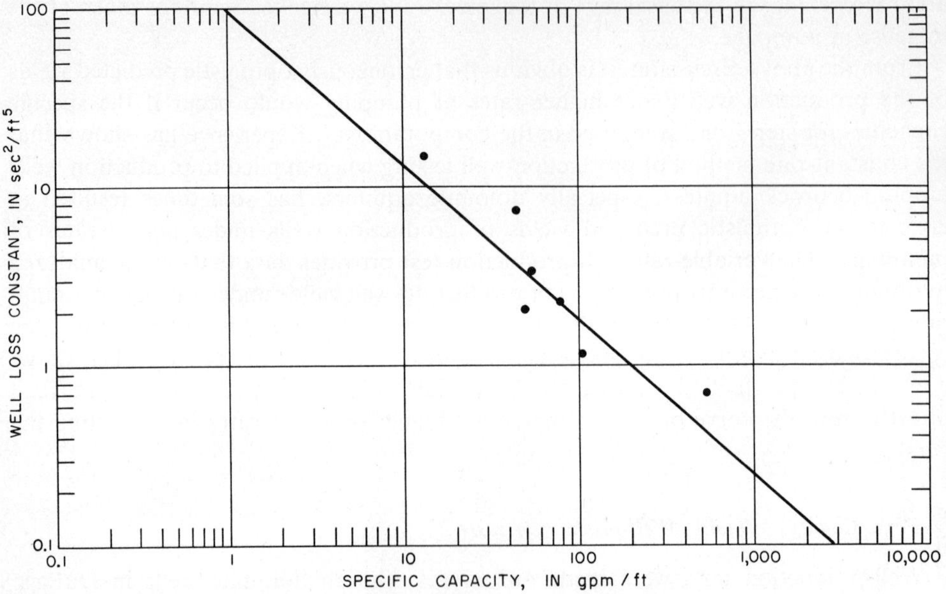

Fig. 5.32 Well-loss constant versus specific capacity. Pumping levels are above top of Silurian dolomite aquifer in DuPage County, Illinois. (*From Zeizel et al., 1962.*)

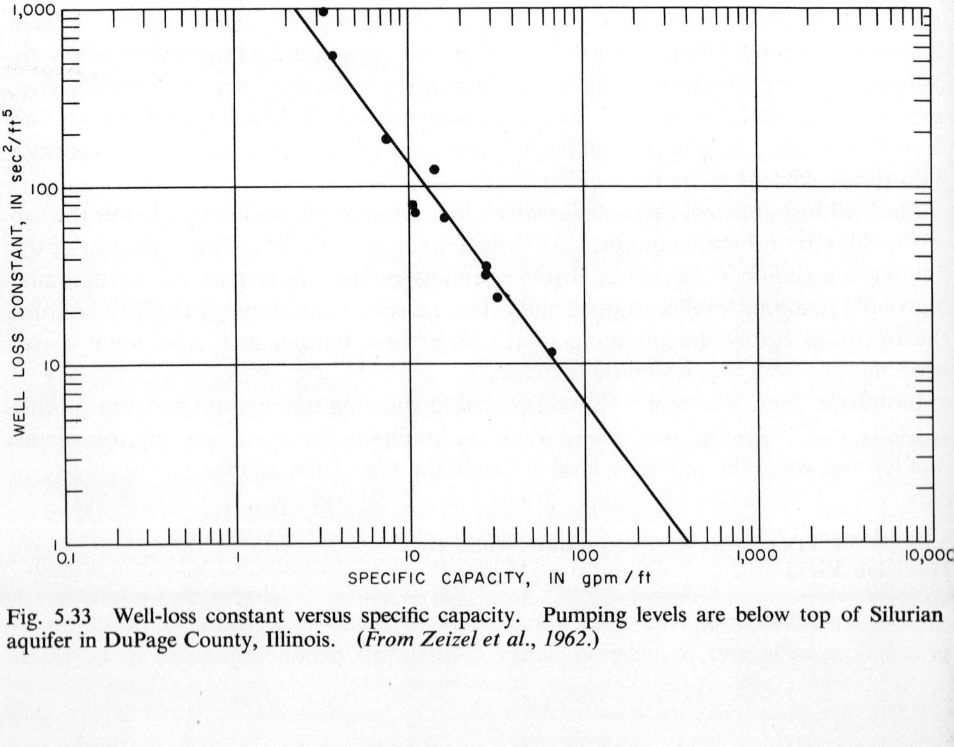

Fig. 5.33 Well-loss constant versus specific capacity. Pumping levels are below top of Silurian aquifer in DuPage County, Illinois. (*From Zeizel et al., 1962.*)

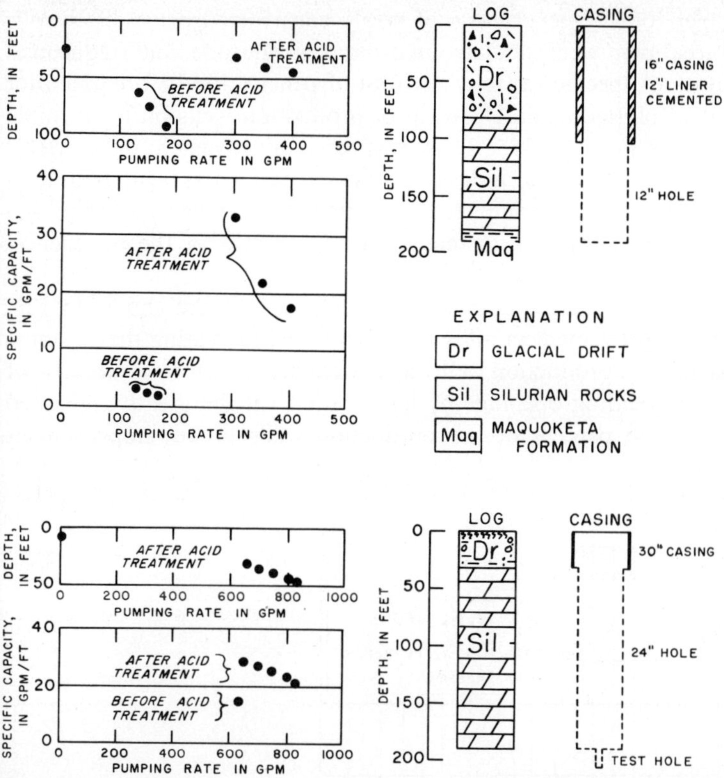

Fig. 5.34 Well-production test data showing results of acid treatment of dolomite production wells in DuPage County, Illinois. (*From Csallany and Walton, 1963.*)

Illinois (Csallany and Walton, 1963). Many production wells have been treated with inhibited 15 percent hydrochloric acid in quantities ranging from 100 to 400 gallons. The pump and the discharge column are often removed from the production well during the treatment period. Acid is introduced through a temporary line extending to a position near the bottom of the production well. The solution is allowed to stand under pressure in the production well for periods ranging from one-half hour to four days and averaging about one day. The pump is then reinstalled, and the spent acid is removed from the production well during pumping periods ranging from 1 to 8 hours.

Well-production tests were made on many production wells before and after acid treatment. There is an extremely large range (0 to 1,190 percent) in improvement. Most improvements over 100 percent were recorded for rehabilitated production wells; improvements generally less than 100 percent were recorded for newly constructed production wells. In several cases where little or no improvement was observed, the acid was allowed to stand in the well for short periods (1 hour or less). The results of acid treatment of two production wells are shown graphically in Fig. 5.34.

Specific capacities per foot of penetration of newly constructed production wells before and after acid treatment were tabulated in order of magnitude, and frequencies were computed. Values of specific capacity per foot of penetration before and after acid treatment were then plotted against percent of production wells on logarithmic probability paper as shown in Fig. 5.35. There was some improvement in the yields of 80 percent of the production wells, indicating that in most cases acid treatment will increase the initial yields of newly completed dolomite production wells. Improvement increases as the specific capacity increases. In 50 percent of the production wells tested, improvement averaged over 100 percent.

Frequency graphs for rehabilitated dolomite production wells are shown in Fig. 5.36. There was some improvement in all production wells, indicating that in most cases the specific capacity of a production well whose yield has deteriorated because of partial clogging of water-yielding openings by incrustation can be greatly increased with acid treatment. In 50 percent of the production wells tested, improvement averaged over 150 percent.

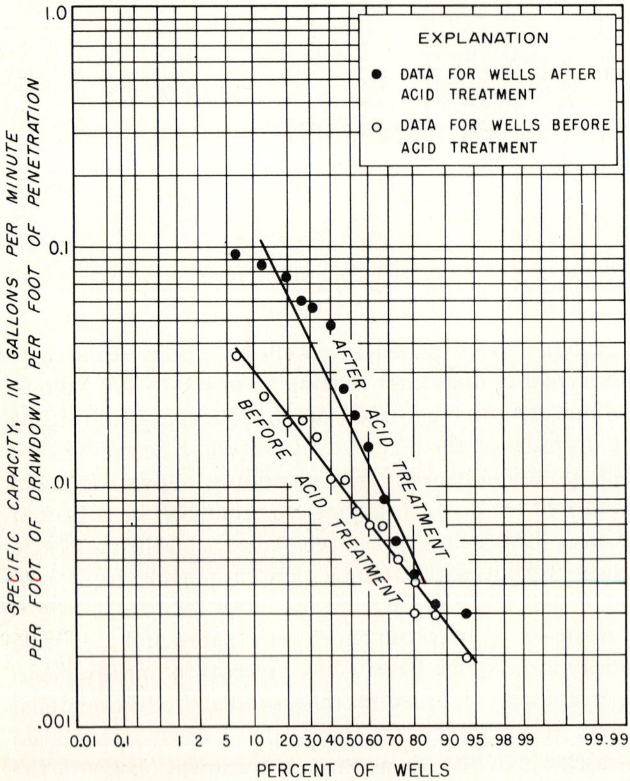

Fig. 5.35 Effects of acid treatment on specific capacities of newly constructed dolomite production wells in northern Illinois. (*From Csallany and Walton, 1963.*)

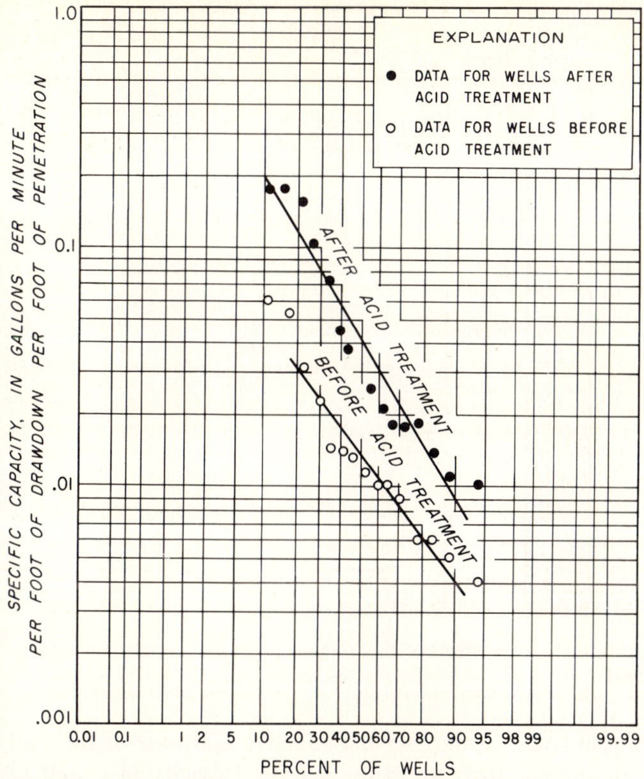

Fig. 5.36 Effects of acid treatment on specific capacities of old dolomite production wells in northern Illinois. (*From Csallany and Walton, 1963.*)

5.17 Shooting Production Wells to Increase Yield

Explosives have been successfully used to develop newly constructed deep sandstone production wells in northern Illinois (Walton and Csallany, 1962). Many production wells are shot with nitroglycerine opposite several areas in the well bore. Shooting is normally accomplished with liquid or solidified nitroglycerine. Shots of approximately 100 to 600 pounds of 80 to 100 percent nitroglycerine are usually exploded opposite the most permeable zones of a formation and are commonly spaced vertically 20 feet apart. Shots are often exploded opposite the lower 80 feet of the Ironton-Galesville Sandstone and occasionally opposite the middle 80 feet of the St. Peter Sandstone. The explosions loosen quantities of rock, varying from a few cubic feet to several hundred yards, that have to be bailed out of the production wells. The effects of shooting are illustrated in Fig. 5.37.

Careful study (Walton and Csallany, 1962) of the effects of shooting suggests that in most cases the yields of deep sandstone production wells are increased because: (1) the hole is enlarged, and (2) fine materials deposited during drilling on the well

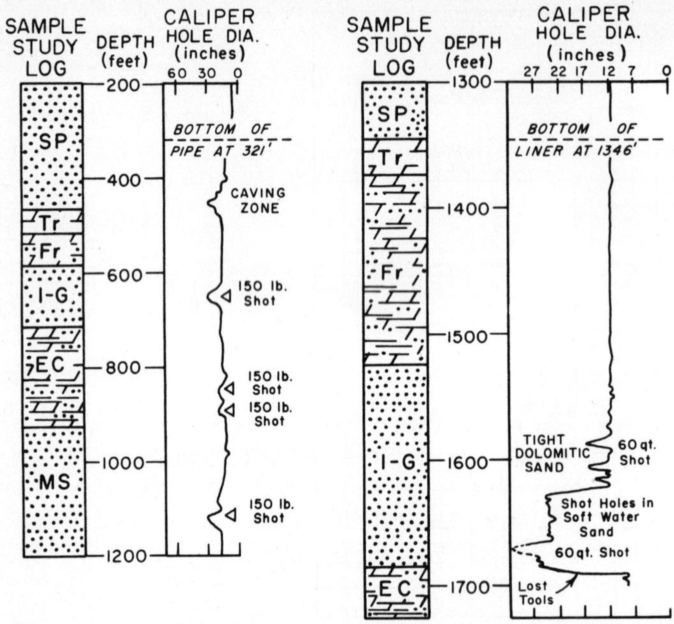

Fig. 5.37 Caliper logs of selected wells showing effects of shooting deep sandstone wells in northern Illinois. (*After Bays and Folk, 1944.*)

face and in the well wall are removed. Enlarging the effective diameter of the well bore by shooting results in an average increase in the specific capacity of a well of about 10 percent. The yield of a newly completed production well is on the average increased about 20 percent by removing fine materials from the well face and well wall. Average increases in the yields of newly completed production wells uncased in the various units or aquifers are listed in Table 5.13.

The city of Lansing, Michigan, obtains its water supply principally from production wells drilled into sandstone of Pennsylvanian age. The sandstone is encountered at depths of 50 to 150 feet and extends to a depth of 400 to 450 feet below the land surface. Production wells commonly have original specific capacities of 12 gpm/ft. The yields of the production wells decrease rapidly with continuous heavy pumping partly because of the partial clogging of well walls (Erickson and Wright, 1957) with fine particles of sandstone and shale and the precipitation of iron and calcium carbonate. Shooting has been used to keep the production wells operating at original capacity.

By 1956, 71 production wells has been shot; 48 of these have been shot once, 21 twice, and 2 three times. Sandstone not closer than 10 feet from shale formations, or 50 feet from the bottom of the production-well casing, is shot. Shots in the sandstone were made at 5-foot intervals. Fifty percent nitroglycerine dynamite was used and 10-pound shots were made at each 5-foot interval. The 10-pound charges

TABLE 5.13 RESULTS OF SHOOTING DEEP SANDSTONE WELLS IN NORTHERN ILLINOIS. (*From Walton and Csallany, 1962.*)

Units or aquifers uncased in well	*Average increase in specific capacity due to shooting, %*
Glenwood–St. Peter Sandstone	38
Cambrian-Ordovician Aquifer	22
Ironton-Galesville Sandstone	30
Cambrian-Ordovician and Mt. Simon Aquifers	25

were set off separately. After the shooting was completed, a well-drilling rig was moved in and the sandstone which had been loosened was bailed out. As much as 50 yards of material was commonly removed. Specific-capacity increases up to 137 percent were observed. Usually the diameter of the production well was increased from 14 to 24 inches on the first shooting, with further increases on the second and third shootings. Data concerning the cost of shooting and cleaning 10 production wells are:

Labor	$10,385
Material	2,325
Equipment rental	5,014
Miscellaneous	147
Total	$17,871
Average per production well	$1,787
Cost per mgd increase yield	$8,510

The Southern California Water Company has successfully used a vibratory-explosive method for production-well rehabilitation. The casings used by the company are thin and of large diameter. These casings, which are perforated at selected depths, consist of continuous telescoping 4-foot sections with the inner and other joints staggered. Buried jacks pull them together and force them down the hole as drilling advances. The casing may consist of two sections of pipe ranging from 14 to 8 gage and from 10 to 30 inches in diameter. Because the strength of this type of casing leaves much to be desired, heavy shooting would seriously damage the casings.

A series of light shots of selected explosives with as low as one-ninth the strength (320 calories per gram) of 1 foot of pentaerythrite tetranitrate are spaced on various leads and fired at specified intervals of time throughout the length of the perforated area. The explosives produce a continuing series of shock waves safe to use in thin-walled casings, yet powerful enough to dislodge material from perforations and the interstices of the surrounding formation (Ebaugh, 1950). Water is forced out through the perforations and back into the hole 10 times for each shot. This method is patented for water- and oil-well redevelopment. Generally the pump does not have to be removed and the amount of material removed is surprisingly small.

Problems

5.1 A drilling cutting sample taken from a test hole in a sand and gravel aquifer was subjected to a mechanical analysis with the following results:

Size opening, mm		*Weight retained, grams*
6.68		0
4.70		4.7
3.33		4.7
1.65		16.5
0.417		51.9
0.417		24.8
0.074		8.3
	Total	118.0

Plot a grain-size distribution curve for the sample. What is the effective grain size and the uniformity coefficient of the sample?
Answer: $D_{90} = 0.0005$; $C_u = 9$.

5.2 Assume that the materials overlying the aquifer in Problem 5.1 are fairly firm and will not easily cave. Determine the proper slot size for a natural-pack production well based on the grain-size distribution curve.
Answer: Slot size, 0.05 in.

5.3 The grain-size distribution curve for a drilling cutting sample from a test hole in a sand aquifer has the following properties: $D_{90} = 0.007$ in., $C_u = 2$, $D_{50} = 0.025$ in. Determine the proper pack and slot size for an artificial-pack production well.
Answer: Pack size $= 0.125$ in.; slot size, 0.10 in.

5.4 The grain-size distribution curve for a drilling cutting sample from a test hole in a sand and gravel aquifer has the following properties: $D_{40} = 0.060$, $D_{90} = 0.014$ in., $C_u = 4.70$. The coefficient of permeability of the aquifer is 3,000 gpd/sq ft. The materials overlying the aquifer will not easily cave. Suppose that a production well with a capacity of 1,000 gpm and a 16-inch-diameter continuous-slot screen with a slot opening of 0.060 inch is to be constructed. The effective open area of the screen is 0.640 sq ft per foot of screen. How long should the screen be, in order to ensure a long service life for the production well?
Answer: 26 ft.

5.5 Given: An infinite nonleaky artesian aquifer, $T = 17{,}000$ gpd/ft, $S = 0.0004$, $t =$ one day, $r_w = 2$, 4, 8, and 12 inches. Construct a graph showing the relation between specific capacity and radius of the production well.

5.6 Given: An infinite nonleaky artesian aquifer, $T = 17{,}000$ gpd/ft, $S = 0.0004$, $r_w = 6$ inches, $t = 1$ hour, 30 days, and 200 days. Construct a graph showing the relation between specific capacity and time.

5.7 Given: Infinite nonleaky artesian and water-table aquifers, $T = 17{,}000$ gpd/ft, $t =$ one day, $r_w = 6$ inches, $S = 0.0002$, 0.002, 0.02, and 0.20. Construct a graph showing the relation between specific capacity and coefficient of storage of the aquifer.

5.8 Given: An infinite nonleaky artesian aquifer, $S = 0.0004$, $t =$ one day, $r_w = 6$ inches, $T = 100$, 1,000, 10,000, and 100,000 gpd/ft. Construct a graph showing the relation between specific capacity and coefficient of transmissibility.

5.9 Given: An infinite water-table aquifer, $S = 0.020$, $t =$ one day, $r_w = 6$ inches, $T = 100$, 1,000, 10,000, and 100,000 gpd/ft. Construct a graph showing the relation between specific capacity and coefficient of transmissibility.

5.10 A variable-rate well-production test was conducted with the following results:

Step	*Pumping rate, gpm*	Δs, *ft*
1	100	3.25
2	151	2.47
3	199	2.81

Determine the well-loss coefficient and the well loss for a pumping rate of 152 gpm.
Answer: $C = 21.6$ sec²/ft⁵, $s_{wL} = 2.46$ ft.

5.11 A variable-rate well-production test was conducted with the following results:

Step	*Pumping rate, gpm*	Δs, *ft*
1	106	2.16
2	173	1.80
3	252	1.92

Is the well stable? Compute the well-loss constant for steps 1 and 2 and for 2 and 3.
Answer: The well is unstable; slight development probably occurred during the third step. $C_{1\text{ and }2} = 7.55\ \text{sec}^2/\text{ft}^5$; $C_{1\text{ and }2+3} = 3.93\ \text{sec}^2/\text{ft}^5$.

5.12 Well-production tests were made on 14 wells in sand and gravel aquifers within a drainage basin. The following coefficients of permeability were estimated for well test sites: 1,850, 1,450, 1,200, 1,100, 1,070, 1,030, 920, 820, 750, 750, 650, 500, 420, and 370 gpd/sq ft. Construct a coefficient of permeability frequency graph. Determine the coefficient of permeability equaled or exceeded at 50 percent of well test sites.
Answer: $P = 820$ gpd/sq ft.

5.13 Two production wells are spaced 250 feet apart in an infinite water-table aquifer with an original saturated thickness of 50 feet. The coefficients of transmissibility and storage of the aquifer are 100,000 gpd/ft and 0.20, respectively. The production well has a radius of 6 inches. Compute the yield of the two-well system assuming that partial penetration effects are negligible and assuming a pumping period of 180 days and an available drawdown of 25 feet.
Answer: 1,100 gpm.

5.14 Given: Electric power = $1\frac{1}{3}$ cents per kwh, pipe line and electric wiring = $10 per foot, capitalized at 5 percent; $Q = 500$ gpm, $T = 17{,}000$ gpd/ft; pump is 100 percent efficient; infinite nonleaky artesian aquifer. Determine the optimum spacing of a two-well system.
Answer: 295 ft.

References

Ahrens, T. P., 1957. Well Design Criteria, *Water Well J.*, September and November.

American Water Works Association, 1958. "AWWA Standard for Deep Wells," New York.

Anon, 1966. "Ground Water and Wells," UOP-Johnson Div., 315 North Pierce St., St. Paul, Minn.

Bays, C. A., and S. H. Folk, 1944. "Geophysical Logging of Water Wells in Northeastern Illinois," Illinois State Geol. Surv. Circ 113.

Bennison, E. W., 1953. Fundamentals of Water Well Operation and Maintenance, *J. Am. Water Works Assoc.*, March.

Bowman, I., 1911. "Well-drilling Methods," U.S. Geol. Surv. Water Supply Paper 257.

Briggs, G. F., 1949. Corrosion and Incrustation of Well Screens, *J. Am. Water Works Assoc.*, January.

Bruin, J., and H. E. Hudson, Jr., 1955. "Selected Methods for Pumping Test Analysis," Illinois State Water Surv., Rept. of Invest. 25.

Csallany, S., and W. C. Walton, 1963. "Yields of Shallow Dolomite Wells in Northern Illinois," Illinois State Water Surv., Rept. of Invest. 46.

Davis, S. N., and R. J. M. DeWiest, 1966. "Hydrogeology," John Wiley & Sons, Inc., New York.

——— and L. J. Turk, 1964. Optimum Depth of Wells in Crystalline Rock, *Ground Water*, vol. 2, no. 2.

Ebaugh, R. M., 1950. Water Well Redevelopment by Explosives, *J. Am. Water Works Assoc.*, February.

Erickson, C. R., 1961. Cleaning Methods for Deep Wells and Pumps, *J. Am. Water Works Assoc.*, February.

——— and R. C. Wright, 1957. Maintenance of Rock Wells, *J. Am. Water Works Assoc.*, June.

Gordon, R.W., 1958. "Water Well Drilling with Cable Tools," Bucyrus-Erie Co., South Milwaukee.

Gridley, H. K., 1952. Installation and Performance of Radial Wells in Ohio River Gravel, *J. Am. Water Works Assoc.*, December.

——— and I. S. Papadopulos, 1962. Flow of Ground Water to Collector Wells, *Proc. Am. Soc. Civil Engrs.*, vol. 88, no. HY5.

Hantush, M. S., 1964. Hydraulics of Wells, in Ven te Chow (ed.), "Advances in Hydroscience," vol. 1, Academic Press Inc., New York.

Jacob, C. E., 1946. Drawdown Test to Determine Effective radius of Artesian Well, *Proc. Am. Soc. Civil Engrs.*, vol. 79, no. 5.

Johnson Drillers J., 1954. Edward E. Johnson, St. Paul, July–August.

———, 1961. Edward E. Johnson, St. Paul, July–August, September–October.

Kimball, B. F., 1946. Assignment of Frequencies to a Completely Ordered Set of Sample Data, *Trans. Am. Geophys. Union*, vol. 27.

Koenig, L., 1960a. Economic Aspects of Water Well Stimulation, *J. Am. Water Works Assoc.*, vol. 52, no. 5.

———, 1960b. Survey and Analysis of Well Stimulation Performance, *J. Am. Water Works Assoc.*, vol. 2, no. 2.

Mikels, F. C., and F. H. Klaer, Jr., 1956. "Application of Groundwater Hydraulics to the Development of Water Supplies by Induced Infiltration," Intern. Assoc. Sci. Hydrology, Symp. Darcy, Dijon, Publ. 41.

Muskat, M., 1937. "The Flow of Homogeneous Fluids through Porous Media," McGraw-Hill Book Company, New York.

Prickett, T. A., L. R. Hoover, W. H. Baker, and R. T. Sasman, 1964. "Groundwater Development in Several Areas in Northeastern Illinois," Illinois State Water Surv., Rept. of Invest. 47.

Rorabaugh, M. I., 1953. Graphical and Theoretical Analysis of Step Drawdown Test of Artesian Well, *Proc. Am. Soc. Civil Engrs.*, vol. 79, separate no. 362.

Ross, R., Jr., 1958. Water Well Construction on Formations Characteristic of the Southwest, *J. Am. Water Works Assoc.*, June.

Smith, H. F., 1954. "Gravel Packing Water Wells," Illinois State Water Surv. Circ. 44.

Spiridonoff, S. V., 1964. Design and Use of Radial Collector Wells, *J. Am. Water Works Assoc.*, June.

Suter, M., R. E. Bergstrom, H. F., Smith, G. H. Emrich, W. C. Walton, and T. E. Larson, 1959. "Preliminary Report on Ground-water Resources of the Chicago Region, Illinois," Illinois State Water Surv. and Geol. Surv. Cooperative Ground-water Rept. 1.

Theis, C. V., 1957. "The Spacing of Pumped Wells," U.S. Geol. Surv. Ground Water Branch, Ground Water Notes, Hydraulics, no. 31, open-file report.

Turcan, A. N., Jr., 1963. "Estimating the Specific Capacity of a Well," U.S. Geol. Surv. Profess, Paper 450-E.

Walton, W. C., 1962. "Selected Analytical Methods for Well and Aquifer Evaluation," Illinois State Water Surv. Bull. 49.

——— and S. Csallany, 1962. "Yields of Deep Sandstone Wells in Northern Illinois," Illinois State Water Surv., Rept. of Invest. 43.

——— and J. C. Neill, 1963. Statistical Analysis of Specific Capacity Data for a Dolomite Aquifer, *J. Geophys. Res.*, vol. 68, no. 8.

War Dept., 1943. "Well Drilling," Tech. Manual 5-297, Washington, D.C.

Zeizel, A. J., W. C. Walton, R. T. Sasman, and T. A. Prickett, 1962. "Ground-water Resources of DuPage County, Illinois," Illinois State Water Surv. Cooperative Ground-water Rept. 2.

CHAPTER 6

GROUNDWATER RECHARGE AND RUNOFF

Recharge rates for aquifers must be estimated before groundwater resources can be evaluated and before the consequences of the utilization of aquifers can be forecast. In many places, the major sources of recharge to aquifers are direct precipitation on intake areas and/or downward percolation of stream runoff. Recharge from precipitation on intake areas is irregularly distributed in time and place. In humid areas, most recharge from precipitation commonly occurs during spring months when evapotranspiration is small and soil moisture is maintained at or above field capacity by frequent rains. During summer and early fall months evapotranspiration and soil moisture requirements are so great that little precipitation percolates to the water table except during periods of excessive rainfall. Recharge during winter months whenever the ground is frozen is negligible.

Only a small fraction of the annual precipitation percolates downward to the water table. A large proportion of precipitation runs overland to streams or is discharged by the process of evapotranspiration before it reaches aquifers. The amount of precipitation that reaches the zone of saturation depends upon several factors. Among these are the character and thickness of the soil and other deposits above and below the water table; the topography; vegetal cover; land use; soil moisture content; the depth to the water table; the intensity, duration, and seasonal distribution of rainfall; the occurrence of precipitation as rain or snow; and the air temperature and other meteorological factors (humidity, wind, etc.).

Recharge direct from precipitation and by infiltration of surface water involves the vertical downward movement of groundwater under the influence of vertical head differentials. Thus, recharge involves vertical leakage of water through deposits. The quantity of vertical leakage varies from place to place and is controlled by the permeability and thickness of the deposits through which leakage occurs, the head differential between sources of water and the aquifer, and the area through which leakage occurs.

At places, deeply buried aquifers are recharged in part by the vertical leakage of water through thick unconsolidated deposits, including clayey materials. The water level in shallow, 10- to 30-feet-deep, dug or bored wells fluctuates through a wide range in response to periods of above- or below-normal precipitation; in drought years many shallow dug wells go dry. However, water stored in thick unconsolidated deposits is available to deeply buried aquifers, so that drought periods often have little influence on water levels in these aquifers. Groundwater storage in deposits above aquifers and in aquitards permits pumping for limited periods of time at rates greater than recharge. However, many aquifers are greatly limited in areal extent and thickness, and pumping at rates much above recharge rates for extended periods results in the rapid depletion of these aquifers.

Artificial recharging techniques have been employed to add water to aquifers. A variety of methods have been developed including water spreading, recharging through pits, excavations, wells and shafts, and pumping to induce recharge from surface-water bodies. Recharge to aquifers by infiltration of surface waters occurs when the water table is below the water surface of a stream, and the streambed is permeable. The rate of infiltration depends upon several factors: the surface-water temperature, the permeability of the streambed and the aquifer, the thickness of the streambed, the position of the water table, and the depth of water in the stream.

Streamflow consists of surface runoff and groundwater runoff. *Surface runoff* is here defined as precipitation that finds its way into the stream channel without infiltrating into the soil. *Groundwater runoff* is precipitation that infiltrates into the soil or to the water table and then percolates into the stream channel. Groundwater runoff includes bank storage and interflow. Data on groundwater runoff can be useful in estimating recharge to aquifers; however, studies indicate that no simple relation exists between groundwater runoff and the potential recharge or practical sustained yields of aquifers. In many areas, not all groundwater runoff can be diverted into cones of depression. However, with sound professional judgement, rough estimates of recharge to aquifers, which will be useful in making preliminary evaluations of groundwater resources, can be made from groundwater runoff and hydrogeologic data. If recharge by induced infiltration of surface water is excluded, recharge to most deeply buried aquifers will be less than groundwater runoff under natural conditions. If decreases in evapotranspiration due to lowering of water levels under heavy pumping conditions and recharge by induced infiltration of surface water are excluded, maps of groundwater runoff can be useful in setting upper limits on the potential yields of groundwater reservoirs. The relations between groundwater runoffs during years of near-, below, and above-normal precipitation and basin characteristics such as geologic environment, topography, and land use may be determined by statistical analysis.

6.1 Recharge by Vertical Leakage through Deposits

The rate of recharge by vertical leakage, through deposits under pumping conditions, may be expressed mathematically by the following form of Darcy's law

(Walton, 1965)

$$\frac{Q_c}{A_c} = 2.8 \times 10^7 \frac{P'}{m'} \Delta h \tag{6.1}$$

where $\frac{Q_c}{A_c}$ = recharge rate, in gpd/sq mi

Q_c = vertical leakage through deposits, in gpd

A_c = recharge area within which vertical leakage is being diverted to a pumping center, in square miles

P' = coefficient of permeability of deposits, in gpd/sq ft

m' = saturated thickness of deposits, in feet

Δh = difference between the head in the aquifer and in the source bed, above deposits through which leakage occurs, in feet

Equation (6.1) assumes that recharge balances pumpage and that no water is being taken from storage. The recharge rate varies with the vertical head loss, associated with leakage of water through deposits. The recharge rate per unit area, Q_c/A_c, being dependent upon vertical head loss, is not constant but varies in space and time. The recharge rate is generally greatest in the deepest parts of cones of depression and decreases with distance from a pumping center. The recharge rate increases as the piezometric surface declines and the vertical head loss increases. The recharge rate per unit area is at a maximum when the piezometric surface of the aquifer is at the base of the deposits through which leakage occurs, provided the head in the source bed above the deposits remains fairly constant.

The recharge rate per unit area is valid for one and only one average vertical head loss. On the other hand, the recharge rate per unit area per foot of head loss remains constant, as long as the saturated thickness and permeability of the deposits through which leakage occurs does not change and the piezometric surface does not decline below the base of the deposits. Thus, the recharge rate per unit area per foot of head loss is much more meaningful than the recharge rate per unit area. The recharge rate per unit area per foot of head loss P'/m' is the leakage coefficient (Hantush, 1956) and can be expressed by the following equation

$$\frac{P'}{m'} = \frac{Q_c}{A_c\, \Delta h(2.8 \times 10^7)} \tag{6.2}$$

Differences in recharge rates per unit area per foot of head loss from place to place cannot be attributed only to differences in the permeability of deposits. Recharge rates per unit area per foot can vary from place to place because of variations in the saturated thickness of deposits, as well as variations in permeabilities. The permeabilities of deposits through which leakage occurs are expressed as coefficients of permeability (P') and can be computed by multiplying leakage coefficients by the saturated thickness of deposits.

Data concerning recharge rates for aquifers in Illinois (Walton, 1965) are summarized in Table 6.1. Recharge rates per unit area vary from 1,330 to 500,000 gpd/sq

TABLE 6.1 SUMMARY OF RECHARGE RATES FOR AQUIFERS IN ILLINOIS

Location	*Recharge rate $Q_c/A_c/\Delta h$, gpd/sq mi/ft*	*Recharge rate Q_c/A_c, gpd/sq mi*	*Lithology of deposits above aquifer or permeable zones*	*Aquifer*
Northeastern Illinois	15.7	1,330	Maquoketa Formation, largely shale	Cambrian-Ordovician
DeKalb and Kendall Counties	300	18,000	Glacial drift and units of Cambrian-Ordovician Aquifer	Cambrian-Ordovician
DuPage County	1,160	64,000	Glacial drift, largely till and shaly dolomite	Dolomite of Silurian age
	1,840	138,000	Glacial drift, largely till and dolomite	
	1,600	136,000	Glacial drift, largely till and dolomite	
	1,760	158,000	Glacial drift, largely till and dolomite	
LaGrange, Cook County	5,300	161,000	Glacial drift, largely till and dolomite	Dolomite of Silurian age
Chicago Heights, Cook County	9,000	225,000	Glacial drift, largely till and dolomite	Dolomite of Silurian age
Libertyville, Lake County	1,300	52,000	Glacial drift, largely till and shaly dolomite	Glacial sand and gravel
Woodstock, McHenry County	4,200	127,000	Glacial drift, largely till	Glacial sand and gravel
Near Joliet Will County		200,000	Glacial drift, largely silt and sand	Glacial sand and gravel
Champaign-Urbana, Champaign County	2,300	115,000	Glacial drift, largely till	Glacial sand and gravel
Havana region, Mason and Tazewell Counties		258,000	Glacial drift, largely till	Glacial sand and gravel
		279,000	Glacial drift, largely till	
		500,000	Glacial drift, largely sand gravel	
		486,000	Glacial drift, largely sand and gravel	
East St. Louis, Madison and St. Clair Counties		347,000	Glacial drift and alluvium, largely sand and gravel	Glacial sand and gravel
		343,000	Glacial drift and alluvium, largely sand and gravel	
		299,000	Glacial drift and alluvium, largely sand and gravel	
		370,000	Glacial drift and alluvium, largely sand and gravel	
Panther Creek basin, Woodford, Livingston, and McLean Counties		380,000	Glacial drift, largely till	Glacial drift

mi; recharge rates per unit area per foot of head loss vary from 15.7 to 9,000 gpd/sq mi/ft.

Harder and Drescher (1954) report that in Langlade County, Wisconsin, in a sandy area the average annual recharge over an area of 8.56 square miles was 0.45 foot per year. This is over 18 percent of the total annual precipitation. The annual recharge in Goshen County, Wyoming, was estimated to be 5 percent of the annual precipitation of 14 inches, or 0.058 foot per year (Rapp et al., 1957). Babcock and Bjorklund (1956) referred to an estimate by Morgan (1946) of 6 percent of the annual precipitation, or 0.07 foot per year, for the recharge in the Cheyenne area of Wyoming. Keech and Dreeszen (1959) computed an average recharge of 0.13 foot per year, or about 6.5 percent of the annual precipitation, for a small area in Clay County Nebraska, and stated that this was typical for the county. In another part of Nebraska Newport (1959) estimated that the recharge varies from 0.42 foot per year in an area of sand dunes to less than 1 inch per year for areas underlain by relatively impermeable rock. Near Olds, Alberta, the annual recharge has been calculated to be about 0.04 foot, or 3 percent of the annual precipitation. The annual rate of recharge to the groundwater reservoir over a small basin near Devon, Alberta, is 0.031 foot per year, or about 2 percent of the total annual precipitation (Farvolden et al., 1963). This rate may be representative of the rate of recharge over a much larger area of the Alberta "parkland" region.

Subsurface percolation of water, from basin-margin mountains, directly into aquifers in the valley basins is appreciable and should be considered as recharge. Studies of precipitation, runoff, and evapotranspiration along the western front of the Wasatch Mountains indicate that about 30,000 acre feet of water per year is available for subsurface percolation along a 30-mile study area (see Feth, 1964).

6.2 Artificial Recharge

Artificial recharge may be defined as the practice of increasing, by artificial means, the amount of water that enters a groundwater reservoir. Artificial recharge has application in waste disposal, secondary oil recovery, and land-subsidence problems as well as water-supply problems. The need for artificial recharge has been brought about by an increasing demand for groundwater as a source of fresh water. The specific purposes for which artificial recharge is practiced are to:

1. Conserve and dispose of runoff and flood waters
2. Supplement the quantity of groundwater available
3. Reduce or eliminate the decline in the water level of groundwater reservoirs
4. Reduce, prevent, or correct salt-water intrusion
5. Store water to reduce costs of pumping and piping
6. Store clear, cool water in winter, for use during the summer
7. Allow heat exchange by diffusion through the ground
8. Obtain suspended-solids removal by filtration through the ground

An artificial recharge installation may serve more than one purpose. In certain areas, for example, artificial recharge not only adds water to the available groundwater supply but also is a means of disposing of storm-water runoff. In another instance, artificial recharge to bar salt-water intrusion is also increasing the available supply of fresh water and alleviating a ground-subsidence condition that has been in progress for years.

Artificial recharge for water conservation and flood control has been important in Southern California since about 1895 (Todd, 1957). In the Santa Clara River valley in California, flood waters are stored behind dams and released slowly, so that the water is absorbed in streambeds and off-channel ponds. Also, water is spread over receptive orchard land. Thus, water that would otherwise be largely wasted in this semiarid region is conserved by storage in the ground and is available to meet water requirements. In addition, a degree of control of flood waters is obtained. Similar installations are in operation at other locations in California.

The Duhernal Water Co., situated along the South River near Old Bridge, New Jersey, supplemented the available groundwater from its well field by impounding water behind a dam constructed in 1930 across a stream. Thus, by spreading water and increasing the head of water over the outcropping sands, the natural recharge of groundwater is supplemented. Later, low-level dams constructed across small feeder streams further increased the area over which water was spread.

The most significant artificial recharge operations in the eastern United States are being carried out at numerous locations on Long Island, New York, particularly in Nassau County. Because of its rapidly increasing population and industrial growth, Long Island recognized the need to protect and conserve its sources of fresh-water supplies at an early date. At first, definite steps were taken to prevent pollution and contamination of the streams (Welsh, 1957). Because of the increasing transition from an agricultural to an urban use of land, the runoff from storm waters greatly increased. Instead of percolating into the ground, vast quantities of runoff water were wasted to tidewaters through storm sewers. The increasing demand for water and the decreasing quantity of water percolating into the ground were causing declining groundwater levels that, unabated, would result in serious salt-water intrusion, which was occurring in the populous Brooklyn area at the western end of Long Island. Since 1933, the New York State Water Power and Control Commission (now Water Resources Commission) has required that water pumped from new production wells and used for cooling and air conditioning be returned to the aquifer from which it was withdrawn.

In 1936, Nassau County developed, and since that time has actively pursued, a long-range water-conservation program. Under this program, the use of recharge basins has resulted in the return to the ground of large quantities of storm water which otherwise would have been wasted to tidewater. The application of artificial recharge methods on Long Island has made a major contribution to the maintenance of the groundwater level.

As a result of groundwater overdevelopment, extensive damage, caused by sea-water intrusion, has already occurred in numerous groundwater basins adjacent to the coast of California (Anon, 1958). A comprehensive investigation program was initiated in 1951 to determine the criteria for the control and prevention of sea-water intrusion. This program included a large-scale investigation of the feasibility, with regard to hydraulics, of creating a pressure ridge in confined aquifers by means of injection wells, of using fresh water, and the effectiveness of such a ridge in preventing sea-water intrusion. One conclusion was that direct recharge or maintenance of a fresh-water ridge above sea level may be a method to prevent or abate sea-water intrusion, depending on geologic and hydrologic conditions and on the cost and quality of available supplemental water supplies for recharge.

At Amarillo, Texas, large water-storage facilities close to the city are needed to supply peak demands. An investigation indicated that it is feasible to recharge a production well field near the city limits from a production well field several miles away (Anon, 1957). Underground storage space is available in the aquifer because of a decline in the level of the groundwater table. Pumping the distant wells throughout the year at a fairly constant but lower than maximum rate would supply the city's winter demands and provide a surplus to be stored in the production well field near the city. Summer demands in excess of the pipeline capacity from the distant fields would be met by pumping stored water from the nearby field. Thus, the pipeline to the new, distant well fields would be smaller than would be necessary if the lines were to meet peak seasonal water demands. Other advantages of groundwater storage as compared to surface storage are: (1) no losses by evaporation, (2) virtually no construction cost in preparing the reservoir for storage, and (3) utilization of equipment at the distant production well fields, which otherwise would stand idle during the winter months.

Large quantities of groundwater are used by the distilleries in Louisville, Kentucky. Increased industrial growth in the vicinity of the distilleries during World War II placed an additional demand on groundwater, resulting in a serious lowering of the water table. In order to provide adequate cooling water for summer use, the distilleries recharged the aquifer through their production wells with clarified river water, purchased from the city during the winter months, when the river water was cold. Then they withdrew water from the production wells when the temperature of the river water was too warm.

Recharge from induced filtration, by pumping of production wells located near surface sources, is one example of suspended-solids removal by filtration through the ground. An armament plant located near Charleston, Indiana, on the Ohio River, needed large quantities of water, of which a sizable amount had to be clear water. Investigations showed that the glacial outwash sands and gravels underlying the land terrace along the river were hydraulically connected with the river, from which large quantities of water could be drawn by pumping water from the glacial materials. Consequently, several collector wells were installed. Not only was this an economical method to obtain the needed water, but also the water came clear, because the turbidity in the river water was removed by the sands in the bottom of the river. In addition, the temperature of the water did not reach as high or as low a level as that of the river, and the concentration of dissolved solids in the water was more constant than that of the river. Also, the bacteriologic quality of the water was improved, so that only a nominal dosage of chlorine was necessary to ensure its use for drinking purposes.

Although artificial recharge is a potential means of solving some water-supply problems, each application must be evaluated to determine if it is physically and economically feasible. The geologic and hydrologic conditions that may affect the recharge must be evaluated in each location where artificial recharge is to be used. Also, the recharge water must be analyzed to determine its adequacy—that is, to determine whether it is chemically compatible with the groundwater and whether it requires pretreatment to avoid the clogging of the soil. In addition, the most suitable method of recharge for the application must be selected, and its cost determined.

Spreading methods may be classified as flooding, basin ditch or furrow, natural channel, and irrigation. In relatively flat topography, water may be diverted to spread evenly over a large area. Tests indicate that highest infiltration rates occur on areas with undisturbed vegetation and soil covering. Compared with other spreading methods, flood spreading costs the least for land preparation (Todd, 1959).

Water may be recharged by releasing it into basins, which are formed by excavation, or by construction of dikes or small dams. The most common system consists of individual basins fed by pumped water from nearby surface-water sources. Most basins require periodic cleaning, by scraping of the bottom surface when dry. Series of basins have been successfully built and operated in abandoned stream channels. In alluvial plains, basins may parallel existing channels with water being led into the upper basin by canal.

In the ditch or furrow method, water is distributed to a series of ditches, or furrows, which are shallow, flat-bottomed, and closely spaced, to obtain maximum water-contact area. Gradients of major feeder ditches should be sufficient to carry suspended material through the system. Deposition of fine-grained material clogs soil surface openings.

Water spreading in a natural stream channel may use any of the three methods described. Whatever method is adopted, the primary purpose is to extend the time and area over which water is recharged from a naturally influent channel.

The economy of water spreading hinges upon maintenance of a high infiltration rate. Typical rate curves, however, show a pronounced tendency to decrease with time. The initial decrease is attributed to dispersion and swelling of soil particles after wetting; the subsequent increase accompanies elimination of entrapped air by solution in passing water, while the final gradual decrease results from microbial growths clogging the soil pores. Generally, recharge rates decrease as the mean particle size of soil on a spreading area decreases. Efforts to maintain soil pores free for water passage have led to additions of organic matter and chemicals to the soil as well as to growing vegetation on the spreading area. Alternating wet and dry periods on a basin generally furnish a greater total recharge than does continuous spreading, in spite of the fact that water is in contact with the soil for as little as one-half of the total time.

Studies in small ponds have confirmed that infiltration rates are directly proportional to the head of water. Where less pervious strata lie below the surface stratum, the recharge rate depends on the rate of subsurface lateral flow.

Sewage constitutes a valuable source of water for artificially recharging groundwater. In the present day when conservation, reclamation, and reuse of water are receiving increasing emphasis, sewage recharge is being practiced and has been practiced in a variety of ways throughout the world. Because of higher concentrations of suspended matter and bacteria, spreading rates for sewage effluents are lower than for fresh water. Recharging waste waters from a vegetable quick-freezing plant near Seabrook, New Jersey, has been accomplished by spraying the effluent from large rotating nozzles over a wooded area.

A recharge well may be defined as a well which admits water from the surface to aquifers. Its flow is the reverse of a production well, but its construction may or may not be the same. Most problems of recharge through wells involve clogging of the well and aquifer. Principal causes of clogging are air entrainment, suspended particles in the recharge water, and microorganisms. As water is pumped from a production well, fine material present in the aquifer is carried through the coarser particles surrounding the well and into the well. Any silt carried by water into a recharge well is filtered out and tends to clog the aquifer surrounding the well. Similarly, recharge water may carry large amounts of dissolved air, tending to reduce the permeability of the aquifer by air binding. Recharge water may contain bacteria which can form growths on the well screen and the surrounding formation, thereby reducing the effective flow area. Chemical constituents of the recharge water may differ sufficiently from the normal groundwater to cause undesired chemical reactions, as for example, a base exchange in aquifers containing sizable fractions of silt and clay. These factors all act to reduce recharge rates; as a result, well recharging has been limited to a few areas where local conditions and experience have shown the practicality of the method.

Although artificial recharge is widely practiced, little information has been assembled regarding its costs. Perhaps the reason lies in the fact that recharge projects vary greatly in size, purpose, and method of operation. Land and water costs are other significant variables. Yet, if engineers concerned with water resources management are to evaluate the economic feasibility of recharging, cost data must be collected and analyzed. Todd (1957) has summarized benefits of and alternatives to artificial recharge projects and costs of selected artificial recharge projects.

6.3 Groundwater Mounds in Response to Artificial Recharge from Rectangular or Circular Areas

Artificial recharge by spreading waters on rectangular or circular areas is not uncommon in practice. Solutions describing the rise and fall of the water table, in response to deep percolation from such areas, have been developed by Hantush (1967). Figure 6.1 is a diagrammatic representation of a water-table aquifer artificially recharged from a rectangular or circular area. The aquifer is assumed to be infinite in areal extent, homogeneous, and isotropic. The hydrogeologic properties of the aquifer are constant in time and space, and the water table is assumed to remain below the bottom of the recharge area.

The rise of the water table in response to a vertically downward uniform rate of recharge that is supplied from a rectangular area can be estimated with the following equation (Hantush, 1967)

$$h_m^2 - h_i^2 = \frac{W_m \bar{m} t}{15 S_y} \left\{ W^*\left[1.37(b_m + x)\sqrt{\frac{S_y}{Tt}}, 1.37(a_m + y)\sqrt{\frac{S_y}{Tt}}\right] + W^*\left[1.37(b_m + x)\sqrt{\frac{S_y}{Tt}}, 1.37(a_m - y)\sqrt{\frac{S_y}{Tt}}\right] + W^*\left[1.37(b_m - x)\sqrt{\frac{S_y}{Tt}}, 1.37(a_m + y)\sqrt{\frac{S_y}{Tt}}\right] + W^*\left[1.37(b_m - x)\sqrt{\frac{S_y}{Tt}}, 1.37(a_m - y)\sqrt{\frac{S_y}{Tt}}\right]\right\} \tag{6.3}$$

where h_i = initial height of water table above aquiclude, in feet
h_m = height of water table above aquiclude with recharge, in feet
W_m = recharge rate, in gpd per unit area
$\bar{m}$ = $0.5(h_i + h_m)$, in feet
t = time after recharge starts, in days
S_y = specific yield of aquifer, fraction
b_m = one-half width of recharge area, in feet
x, y = coordinates of observation point in relation to center of recharge area, in feet
T = coefficient of transmissibility, in gpd/ft
a_m = one-half length of recharge area, in feet

The function of W^* is defined by

$$W^*(\alpha_m, \beta_m) = \int_0^1 \operatorname{erf}\left(\frac{\alpha_m}{\sqrt{\tau_m}}\right) \operatorname{erf}\left(\frac{\beta_m}{\sqrt{\tau_m}}\right) d\tau_m$$

Values of $W^*(\alpha_m, \beta_m)$ for the practical range of α_m and β_m are given in Table 6.2. The decay of the water table is given by (Hantush, 1967)

$$h_m^2 - h_i^2 = Z(x,y,t) - Z(x, y, t - t_0) \tag{6.4}$$

where $Z(x,y,t)$, $Z(x, y, t - t_0)$ is the right-hand member of Eq. (6.3) and t_0 is the time after recharge ceases, in days.

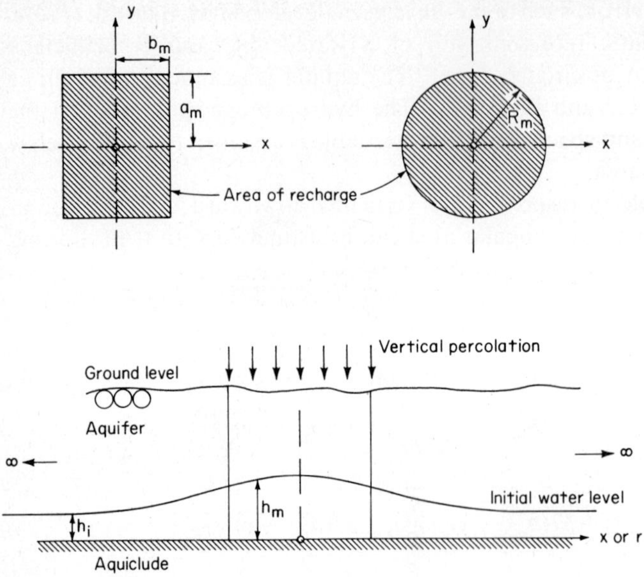

Fig. 6.1 Diagrammatic representation of a water-table aquifer artificially recharged from a rectangular or circular area. (*From Hantush, 1967.*)

The rise of the water table in response to a vertically downward uniform rate of recharge that is supplied from a circular area can be estimated with the following equation (Hantush, 1967)

$$h_m^2 - h_i^2 = \frac{2W_m R_m^2}{P} \int_0^\infty (1 - e^{-m_c \beta_m^2}) J_1(\beta_m) J_0(m_m \beta_m) \frac{\partial \beta_m}{\beta_m^2} \tag{6.5}$$

where $m_c = \dfrac{P\bar{m}t}{7.48 S_y R_m^2}$ (6.6)

$$m_m = \frac{r}{R_m} \tag{6.7}$$

P = coefficient of permeability, in gpd/sq ft
R_m = radius of recharge area, in feet
r = distance from center of recharge area to observation point, in feet
J_0 = the zero-order Bessel function of the first kind
J_1 = first-order Bessel function of the first kind and other symbols have been defined

when $t \geqq \dfrac{3.74r^2}{P\bar{m}/S_y}$ and $r \leqq R_m$

$$(h_m^2 - h_i^2) = \frac{W_m R_m^2}{2P}\left[W(u_0) - \left(\frac{r}{R_m}\right)^2 e^{-u_0} + \frac{1}{u_0}(1 - e^{-u_0})\right] \tag{6.8}$$

Table 6.2 Values of the Function $W^*(\alpha,\beta) = \int_0^1 \operatorname{erf}\left(\frac{\alpha_m}{\sqrt{\tau_m}}\right) \operatorname{erf}\left(\frac{\beta_m}{\sqrt{\tau_m}}\right) d\tau_m$. *(From Hantush, 1967.)*

α_m	$\beta_m = 0.02$	0.04	0.06	0.08	0.10	0.14	0.18	0.22	0.26	0.30	0.34	0.38	0.42	0.46	0.50	0.54	0.58	0.62
0.02	0.0041	0.0073	0.0101	0.0125	0.0146	0.0184	0.0216	0.0243	0.0267	0.0288	0.0306	0.0322	0.0337	0.0349	0.0361	0.0371	0.0380	0.0387
0.04	0.0073	0.0135	0.0188	0.0236	0.0278	0.0353	0.0416	0.0470	0.0518	0.0559	0.0596	0.0628	0.0657	0.0683	0.0705	0.0725	0.0743	0.0759
0.06	0.0101	0.0188	0.0266	0.0335	0.0398	0.0509	0.0602	0.0684	0.0754	0.0817	0.0871	0.0920	0.0963	0.1001	0.1035	0.1065	0.1091	0.1115
0.08	0.0125	0.0236	0.0335	0.0425	0.0508	0.0652	0.0776	0.0884	0.0978	0.1060	0.1133	0.1197	0.1254	0.1305	0.1350	0.1389	0.1425	0.1456
0.10	0.0146	0.0278	0.0398	0.0508	0.0608	0.0786	0.0939	0.1072	0.1188	0.1290	0.1381	0.1461	0.1532	0.1595	0.1650	0.1700	0.1744	0.1783
0.14	0.0184	0.0353	0.0509	0.0652	0.0786	0.1025	0.1232	0.1414	0.1573	0.1714	0.1839	0.1949	0.2048	0.2135	0.2212	0.2281	0.2343	0.2397
0.18	0.0216	0.0416	0.0602	0.0776	0.0939	0.1232	0.1490	0.1716	0.1916	0.2094	0.2251	0.2391	0.2515	0.2626	0.2724	0.2812	0.2890	0.2959
0.22	0.0243	0.0470	0.0684	0.0884	0.1072	0.1414	0.1716	0.1984	0.2222	0.2433	0.2621	0.2789	0.2938	0.3071	0.3189	0.3295	0.3389	0.3472
0.26	0.0267	0.0518	0.0754	0.0978	0.1188	0.1573	0.1916	0.2222	0.2494	0.2737	0.2954	0.3147	0.3320	0.3474	0.3612	0.3735	0.3844	0.3941
0.30	0.0288	0.0559	0.0817	0.1060	0.1290	0.1714	0.2094	0.2433	0.2737	0.3009	0.3252	0.3470	0.3665	0.3839	0.3995	0.4134	0.4257	0.4368
0.34	0.0306	0.0596	0.0871	0.1133	0.1381	0.1839	0.2251	0.2621	0.2954	0.3252	0.3520	0.3761	0.3976	0.4169	0.4341	0.4495	0.4633	0.4756
0.38	0.0322	0.0628	0.0920	0.1197	0.1461	0.1949	0.2391	0.2789	0.3147	0.3470	0.3761	0.4022	0.4256	0.4466	0.4654	0.4823	0.4973	0.5108
0.42	0.0337	0.0657	0.0963	0.1254	0.1532	0.2048	0.2515	0.2938	0.3320	0.3665	0.3976	0.4256	0.4508	0.4734	0.4937	0.5119	0.5281	0.5427
0.46	0.0349	0.0683	0.1001	0.1305	0.1595	0.2135	0.2626	0.3071	0.3474	0.3839	0.4169	0.4466	0.4734	0.4975	0.5191	0.5385	0.5559	0.5715
0.50	0.0361	0.0705	0.1035	0.1350	0.1650	0.2212	0.2724	0.3189	0.3612	0.3995	0.4341	0.4654	0.4937	0.5191	0.5420	0.5626	0.5810	0.5975
0.54	0.0371	0.0725	0.1065	0.1389	0.1700	0.2281	0.2812	0.3295	0.3735	0.4134	0.4495	0.4823	0.5119	0.5385	0.5626	0.5842	0.6036	0.6209
0.58	0.0380	0.0743	0.1091	0.1425	0.1744	0.2343	0.2890	0.3389	0.3844	0.4257	0.4633	0.4973	0.5281	0.5559	0.5810	0.6036	0.6238	0.6420
0.62	0.0387	0.0759	0.1115	0.1456	0.1783	0.2397	0.2959	0.3472	0.3941	0.4368	0.4756	0.5108	0.5427	0.5715	0.5975	0.6209	0.6420	0.6609
0.66	0.0394	0.0773	0.1136	0.1484	0.1818	0.2445	0.3020	0.3547	0.4027	0.4466	0.4865	0.5227	0.5556	0.5854	0.6122	0.6364	0.6582	0.6778
0.70	0.0401	0.0785	0.1154	0.1509	0.1849	0.2488	0.3075	0.3612	0.4104	0.4553	0.4962	0.5334	0.5672	0.5977	0.6254	0.6503	0.6728	0.6929
0.74	0.0406	0.0796	0.1171	0.1531	0.1876	0.2526	0.3123	0.3671	0.4172	0.4630	0.5048	0.5429	0.5774	0.6087	0.6371	0.6627	0.6857	0.7064
0.78	0.0411	0.0806	0.1185	0.1550	0.1900	0.2559	0.3166	0.3722	0.4232	0.4699	0.5125	0.5513	0.5865	0.6185	0.6475	0.6736	0.6972	0.7184
0.82	0.0415	0.0814	0.1198	0.1567	0.1921	0.2589	0.3203	0.3768	0.4286	0.4760	0.5192	0.5587	0.5946	0.6272	0.6567	0.6834	0.7074	0.7291
0.86	0.0419	0.0822	0.1209	0.1582	0.1940	0.2615	0.3237	0.3808	0.4333	0.4813	0.5252	0.5653	0.6017	0.6348	0.6648	0.6920	0.7165	0.7386
0.90	0.0422	0.0828	0.1219	0.1595	0.1957	0.2638	0.3266	0.3844	0.4374	0.4860	0.5305	0.5711	0.6080	0.6416	0.6721	0.6996	0.7245	0.7469
0.94	0.0425	0.0834	0.1228	0.1607	0.1971	0.2658	0.3292	0.3875	0.4411	0.4902	0.5351	0.5762	0.6136	0.6476	0.6784	0.7063	0.7316	0.7543
0.98	0.0428	0.0839	0.1236	0.1617	0.1984	0.2676	0.3314	0.3902	0.4442	0.4938	0.5392	0.5807	0.6184	0.6528	0.6840	0.7123	0.7378	0.7608
1.00	0.0429	0.0842	0.1239	0.1622	0.1990	0.2684	0.3324	0.3914	0.4457	0.4955	0.5410	0.5827	0.6206	0.6552	0.6865	0.7150	0.7406	0.7638
1.20	0.0437	0.0858	0.1263	0.1654	0.2030	0.2740	0.3396	0.4001	0.4558	0.5070	0.5540	0.5969	0.6362	0.6719	0.7044	0.7339	0.7605	0.7846
1.40	0.0441	0.0866	0.1275	0.1669	0.2049	0.2767	0.3431	0.4043	0.4608	0.5127	0.5603	0.6039	0.6438	0.6801	0.7132	0.7432	0.7704	0.7949
1.80	0.0444	0.0871	0.1283	0.1680	0.2062	0.2785	0.3454	0.4071	0.4641	0.5165	0.5645	0.6086	0.6489	0.6856	0.7190	0.7494	0.7769	0.8018
2.00	0.0444	0.0871	0.1284	0.1681	0.2064	0.2787	0.3457	0.4075	0.4645	0.5169	0.5651	0.6092	0.6495	0.6863	0.7198	0.7502	0.7778	0.8027
2.20	0.0444	0.0872	0.1284	0.1682	0.2065	0.2788	0.3458	0.4076	0.4646	0.5171	0.5653	0.6094	0.6497	0.6865	0.7200	0.7505	0.7781	0.8030
2.50	0.0444	0.0872	0.1284	0.1682	0.2065	0.2788	0.3458	0.4077	0.4647	0.5172	0.5653	0.6095	0.6498	0.6867	0.7202	0.7506	0.7782	0.8032
3.00	0.0444	0.0872	0.1284	0.1682	0.2065	0.2789	0.3458	0.4077	0.4647	0.5172	0.5654	0.6095	0.6499	0.6867	0.7202	0.7506	0.7782	0.8032

TABLE 6.2 (*Continued*)

α_m	$\beta_m = 0.62$	0.66	0.70	0.74	0.78	0.82	0.86	0.90	0.94	0.98	1.00	1.20	1.40	1.80	2.00	2.20	2.50	3.00
0.02	0.0387	0.0394	0.0401	0.0406	0.0411	0.0415	0.0419	0.0422	0.0425	0.0428	0.0429	0.0437	0.0441	0.0444	0.0444	0.0444	0.0444	0.0444
0.04	0.0759	0.0773	0.0785	0.0796	0.0806	0.0814	0.0822	0.0828	0.0834	0.0839	0.0842	0.0858	0.0866	0.0871	0.0871	0.0872	0.0872	0.0872
0.06	0.1115	0.1136	0.1154	0.1171	0.1185	0.1198	0.1209	0.1219	0.1228	0.1236	0.1239	0.1263	0.1275	0.1283	0.1284	0.1284	0.1284	0.1284
0.08	0.1456	0.1484	0.1509	0.1531	0.1550	0.1567	0.1582	0.1595	0.1607	0.1617	0.1622	0.1654	0.1669	0.1680	0.1681	0.1682	0.1682	0.1682
0.10	0.1783	0.1818	0.1849	0.1876	0.1900	0.1921	0.1940	0.1957	0.1971	0.1984	0.1990	0.2030	0.2049	0.2062	0.2064	0.2065	0.2065	0.2065
0.14	0.2397	0.2445	0.2488	0.2526	0.2559	0.2589	0.2615	0.2638	0.2658	0.2676	0.2684	0.2740	0.2767	0.2785	0.2787	0.2788	0.2788	0.2789
0.18	0.2959	0.3020	0.3075	0.3123	0.3166	0.3203	0.3237	0.3266	0.3292	0.3314	0.3324	0.3396	0.3431	0.3454	0.3457	0.3458	0.3458	0.3458
0.22	0.3472	0.3547	0.3612	0.3671	0.3722	0.3768	0.3808	0.3844	0.3875	0.3902	0.3914	0.4001	0.4043	0.4071	0.4075	0.4076	0.4077	0.4077
0.26	0.3941	0.4027	0.4104	0.4172	0.4232	0.4286	0.4333	0.4374	0.4411	0.4442	0.4457	0.4558	0.4608	0.4641	0.4645	0.4646	0.4647	0.4647
0.30	0.4368	0.4466	0.4553	0.4630	0.4699	0.4760	0.4813	0.4860	0.4902	0.4938	0.4955	0.5070	0.5127	0.5165	0.5169	0.5171	0.5172	0.5172
0.34	0.4756	0.4865	0.4962	0.5048	0.5125	0.5192	0.5252	0.5305	0.5351	0.5392	0.5410	0.5540	0.5603	0.5645	0.5651	0.5653	0.5653	0.5654
0.38	0.5108	0.5227	0.5334	0.5429	0.5513	0.5587	0.5653	0.5711	0.5762	0.5807	0.5827	0.5969	0.6039	0.6086	0.6092	0.6094	0.6095	0.6095
0.42	0.5427	0.5556	0.5672	0.5774	0.5865	0.5946	0.6017	0.6080	0.6136	0.6184	0.6206	0.6362	0.6438	0.6489	0.6495	0.6497	0.6498	0.6499
0.46	0.5715	0.5854	0.5977	0.6087	0.6185	0.6272	0.6348	0.6416	0.6476	0.6528	0.6552	0.6719	0.6801	0.6856	0.6863	0.6865	0.6867	0.6867
0.50	0.5975	0.6122	0.6254	0.6371	0.6475	0.6567	0.6648	0.6721	0.6784	0.6840	0.6865	0.7044	0.7132	0.7190	0.7198	0.7200	0.7202	0.7202
0.54	0.6209	0.6364	0.6503	0.6627	0.6736	0.6834	0.6920	0.6996	0.7063	0.7123	0.7150	0.7339	0.7432	0.7494	0.7502	0.7505	0.7506	0.7506
0.58	0.6420	0.6482	0.6728	0.6857	0.6972	0.7074	0.7165	0.7245	0.7316	0.7378	0.7406	0.7605	0.7704	0.7769	0.7778	0.7781	0.7782	0.7782
0.62	0.6609	0.6778	0.6929	0.7064	0.7184	0.7291	0.7386	0.7469	0.7543	0.7608	0.7638	0.7846	0.7949	0.8018	0.8027	0.8030	0.8032	0.8032
0.66	0.6778	0.6953	0.7110	0.7250	0.7375	0.7486	0.7584	0.7671	0.7748	0.7816	0.7846	0.8064	0.8171	0.8243	0.8252	0.8255	0.8257	0.8257
0.70	0.6929	0.7110	0.7272	0.7417	0.7546	0.7660	0.7762	0.7852	0.7932	0.8002	0.8034	0.8259	0.8370	0.8445	0.8454	0.8458	0.8460	0.8460
0.74	0.7064	0.7250	0.7417	0.7566	0.7698	0.7816	0.7921	0.8014	0.8096	0.8168	0.8201	0.8434	0.8549	0.8627	0.8636	0.8640	0.8642	0.8642
0.78	0.7184	0.7375	0.7546	0.7698	0.7834	0.7956	0.8063	0.8159	0.8243	0.8317	0.8351	0.8591	0.8710	0.8789	0.8799	0.8803	0.8805	0.8805
0.82	0.7291	0.7486	0.7660	0.7816	0.7956	0.8080	0.8190	0.8288	0.8374	0.8450	0.8485	0.8731	0.8853	0.8935	0.8945	0.8949	0.8951	0.8951
0.86	0.7386	0.7584	0.7762	0.7921	0.8063	0.8190	0.8302	0.8402	0.8491	0.8569	0.8604	0.8855	0.8980	0.9065	0.9075	0.9079	0.9081	0.9081
0.90	0.7469	0.7671	0.7852	0.8014	0.8159	0.8288	0.8402	0.8504	0.8594	0.8674	0.8710	0.8966	0.9094	0.9180	0.9191	0.9195	0.9197	0.9197
0.94	0.7543	0.7748	0.7932	0.8096	0.8243	0.8374	0.8491	0.8594	0.8686	0.8767	0.8803	0.9064	0.9195	0.9282	0.9294	0.9298	0.9300	0.9300
0.98	0.7608	0.7816	0.8002	0.8168	0.8317	0.8450	0.8569	0.8674	0.8767	0.8849	0.8886	0.9151	0.9284	0.9373	0.9384	0.9389	0.9391	0.9391
1.00	0.7638	0.7846	0.8034	0.8201	0.8351	0.8485	0.8604	0.8710	0.8803	0.8886	0.8924	0.9191	0.9324	0.9414	0.9426	0.9430	0.9432	0.9433
1.20	0.7846	0.8064	0.8259	0.8434	0.8591	0.8731	0.8855	0.8966	0.9064	0.9151	0.9191	0.9472	0.9614	0.9709	0.9722	0.9726	0.9728	0.9729
1.40	0.7949	0.8171	0.8370	0.8549	0.8710	0.8853	0.8980	0.9094	0.9195	0.9284	0.9324	0.9614	0.9759	0.9858	0.9871	0.9875	0.9878	0.9878
1.80	0.8018	0.8243	0.8445	0.8627	0.8789	0.8935	0.9065	0.9180	0.9282	0.9373	0.9414	0.9709	0.9858	0.9959	0.9972	0.9972	0.9979	0.9980
2.00	0.8027	0.8252	0.8454	0.8636	0.8799	0.8945	0.9075	0.9191	0.9294	0.9384	0.9426	0.9722	0.9871	0.9972	0.9985	0.9990	0.9992	0.9934
2.20	0.8030	0.8255	0.8458	0.8640	0.8803	0.8949	0.9079	0.9195	0.9298	0.9389	0.9430	0.9726	0.9875	0.9977	0.9990	0.9995	0.9997	0.9998
2.50	0.8032	0.8257	0.8460	0.8642	0.8805	0.8951	0.9081	0.9197	0.9300	0.9391	0.9432	0.9728	0.9878	0.9979	0.9992	0.9997	1.0000	1.0000
3.00	0.8032	0.8257	0.8460	0.8642	0.8805	0.8951	0.9081	0.9197	0.9300	0.9391	0.9433	0.9729	0.9878	0.9980	0.9933	0.9998	1.0000	1.0000

and when $t \geqq \dfrac{3.74R_m^2}{P\bar{m}/S_y}$ and $r \geqq R_m$

$$(h_m^2 - h_i^2) = \frac{W_m R_m^2}{2P}[W(u) + 0.5u_0e^{-u}] \quad (6.9)$$

where
$$u_0 = \frac{1.87R_m^2S_y}{P\bar{m}t} \quad (6.10)$$

$$u = \frac{1.87r^2S_y}{P\bar{m}t} \quad (6.11)$$

The decay of the water table is given by (Hantush, 1967)

$$h_m^2 - h_i^2 = Z(r,t) - Z(r, t - t_0) \quad (6.12)$$

where $Z(r,t)$, $Z(r, t - t_0)$ is the right-hand member of the pertinent equation for the water-table rise and other symbols have been defined.

6.4 Groundwater Runoff

Principles of separating the streamflow hydrograph into its two major components, surface runoff and groundwater runoff, are not well developed. Groundwater runoff, several days after precipitation ceases, is readily determined; however, groundwater runoff under flood hydrographs is the subject of much discussion. Wisler and Brater (1959) described groundwater runoff under flood hydrographs.

As the water surface in the stream rises higher than the adjacent water table, groundwater inflow into the stream channel ceases temporarily and the direction of flow reverses, creating bank storage. Bank storage continues to increase as long as the water level in the stream is higher than the water table. As soon as the stream stage starts to fall, the direction of flow again reverses, and for a time, because of the accumulated bank storage, the groundwater contribution to the stream is considerably increased. As soon as the bank storage is drained out, the groundwater flow decreases.

Even though groundwater runoff into the stream channel ceases temporarily during periods of flood, groundwater continues to percolate toward the stream, creating groundwater storage in the lowlands adjacent to the stream channel. As soon as the stream stage starts to fall, groundwater runoff is considerably increased, not only because of the accumulated bank storage, but also because of the accumulated groundwater storage in the floodplain. When bank and groundwater storage are drained out, groundwater runoff will generally be greater than before precipitation occurred, because, during most flood periods, precipitation infiltrating into the groundwater reservoir causes the water table to rise and the hydraulic gradient toward the stream to increase.

Groundwater runoff under flood hydrographs may be estimated in the following manner (see Linsley et al., 1958). A straight line is drawn from the point of rise to the hydrograph N_b days after the peak; N, defined as the time after the peak of the streamflow hydrograph at which surface runoff terminates, is approximated by $N_b = A_b^{0.2}$, where A_b is the drainage-basin area in square miles. Another line is constructed by projecting the recession of the streamflow after the storm back under the hydrograph to a point under the inflection point of the falling limb; an arbitrary rising limb is sketched from the point of rise of the hydrograph to connect with the projected streamflow recession. Groundwater runoff is estimated as the average of these two lines taking into consideration frozen ground and climatic conditions. On the basis of studies made by Schicht and Walton (1961), it is believed that the procedures described give reasonably accurate estimates of groundwater runoff under flood hydrographs, even though the lines do not describe the time sequence of events occurring during periods of flood.

Flow-duration curves can be used in making comparisons of the flow characteristics of streams (Cross and Hedges, 1959). The shape of the flow-duration curve is governed in part by the water-yielding properties and areal of the unconsolidated and consolidated deposits within a drainage basin. The more nearly horizontal the curve, the greater are the values of the water-yielding properties and/or the areal extent of deposits. Thus, the shape of the flow-duration curve is an index of the effects of the geology of a basin on streamflow.

Grain-size frequency-distribution curves (see Dapples, 1959) are somewhat analogous to flow-duration curves, in that their shapes are indicative of the water-yielding properties of deposits (personal communication, J. E. Hackett, Illinois State Geological Survey). A measure of the degree to which all the grains approach one size, and therefore the slope of the grain-size frequency-distribution curve, is the sorting. One parameter of sorting is obtained by the ratio (Pettijohn, 1949) $(D_{25}/D_{75})^{\frac{1}{2}}$, where D_{25} is the size that has 25 percent larger and 75 percent smaller grains in the distribution, and D_{75} is the size that has 75 percent larger and 25 percent smaller grains in the distribution. Because geology, and therefore grain-size frequency distribution, affects streamflow to a great degree, the parameter selected to describe the slope of the flow-duration curve is the ratio $(Q_{25}/Q_{75})^{\frac{1}{2}}$, where Q_{25} is the streamflow equaled or exceeded 25 percent of the time, and Q_{75} is the streamflow equaled or exceeded 75 percent of the time. Streamflow data (for most basins) over the range Q_{25} to Q_{75} describe straight lines.

The baseflow-duration curve is a useful technique for the study of groundwater runoff (see Kunkle, 1962). The baseflow-duration curve is a cumulative frequency curve that shows the percentage of time during which specified daily groundwater runoffs were equaled or exceeded in a given time period. Specified groundwater runoffs are selected to give a uniform vertical spacing of points throughout the range of groundwater runoff as plotted on logarithmic paper. The days are counted when groundwater runoff was equal to or greater than one specified runoff, but less than the next higher specified runoff. This gives the sum of the days when groundwater runoff fell within the specified runoff ranges. The sums are cumulated, beginning with the highest discharge interval. The cumulated values are divided by the total number of days to give the percentage of total time that each specified runoff was equaled or exceeded. The values are then plotted on logarithmic probability paper and a curve is fitted to the points. For comparative work, the groundwater runoff values are divided by the area of the drainage basin, in order to eliminate this variable. The area beneath the baseflow-duration curve represents the groundwater runoff per square mile during the total observation period.

6.5 Hydrologic Budget

A *hydrologic budget* is a quantitative statement of the balance between the total water gains and losses of a basin, for a period of time. The budget considers all waters, surface and subsurface, entering and leaving or stored within a basin. Water entering a basin is equated to water leaving a basin, plus or minus changes in basin storage.

Many drainage basins are contiguous to headwater reaches of streams, and, except in the vicinity of stream gaging stations, the boundaries of the basins are reasonably congruous with groundwater and topographic divides. There is no surface or subsurface flow into or out of the basins except subsurface underflow from the basins in the vicinity of stream gaging stations. Water stored on the surface of the basins, in ponds, is very small, and discharge from wells is mostly for domestic and livestock use and is not significant. Thus, for many basins, several items of the general hydrologic budget can be eliminated because they do not measurably affect the balance between water gains and losses.

Under these specific conditions, precipitation, including rain and snow, is the source of water entering many basins and is the only water gain considered in the hydrologic budget. Water leaving basins includes streamflow, evapotranspiration, and subsurface underflow. Water is stored beneath the surface in soils and in the groundwater reservoir. Changes in storage of water in the soil are reflected in changes in soil moisture, and changes in water levels in wells indicate changes in storage of water in the groundwater reservoir.

Under these specified conditions, for a given period of time, water gain is balanced by water loss, plus or minus changes in basin storage. Stated as an equation, the hydrologic budget is (Schicht and Walton, 1961)

$$P_a = R_u + ET + U \pm \Delta S_s \pm \Delta S_g \tag{6.13}$$

where P_a is precipitation, R_u is streamflow, ET is evapotranspiration, U is subsurface underflow, ΔS_s is change in soil moisture, ΔS_g is change in groundwater storage, and all quantities are volumes during the period for which the balance is made.

Streamflow R_u consists of a surface runoff R_s and groundwater runoff R_g. Surface runoff reaches streams more rapidly and is discharged from the basins within a few days. Groundwater percolates slowly toward and reaches streams gradually. A few (three to five) days in many basins, after precipitation ceases, there is no surface runoff and streamflow is derived entirely from groundwater runoff.

Evapotranspiration may be subdivided into two parts, according to the source of the water discharged into the atmosphere, as follows: (1) surface and soil evapotranspiration ET_s, and (2) groundwater evapotranspiration ET_g. The part of evapotranspiration derived from soil moisture and by evaporation from the surfaces of water, vegetation, buildings, and other objects is surface and soil evapotranspiration; the part derived from the water table is groundwater evapotranspiration. In many areas, most of the precipitation falling during summer months is evaporated or transpired before it is able to infiltrate to the water table. Thus, a large proportion of the water discharged annually by evapotranspiration processes is not derived from the groundwater reservoir. However, in areas where the water table is shallow and the capillary fringe extends to land surface, large quantities of groundwater are discharged into the atmosphere by vegetation, through the process of transpiration.

Evapotranspiration depends primarily upon meteorological factors, available soil moisture and groundwater, and type of soil and vegetation. At many places, evapotranspiration is very small during the winter, increases rapidly in the spring, reaches a maximum generally in July, and decreases rapidly as the growing season closes in the autumn. This variation with the time of the year is approximately the same from year to year. Evapotranspiration is often greater than precipitation during the summer months and much less than precipitation during winter and early spring months.

Evapotranspiration can be determined by balancing Eq. (6.13). Soil moisture, one of the hydrologic factors, is generally not measured during investigations; therefore, daily, weekly, and monthly evapotranspiration cannot be appraised. However, soil moisture is near field capacity during January of most years, and annual change in soil moisture is very small. Equation (6.13) can be rewritten for an annual inventory period as

$$ET = P_a - R_u - U \pm \Delta S_g \tag{6.14}$$

Application of Eq. (6.14) requires that the annual change in soil moisture not be significant. Subsurface underflow out of basins occurs in the vicinity of stream gaging stations where water-table contours are not everywhere at right angles to the streams. Underflow can be estimated from Darcy's equation. Precipitation is measured in the field.

The change in mean groundwater during an inventory period, ΔH_s, multiplied by the gravity yield Y_g, of the deposits within the zone of groundwater fluctuation is equal to the change in groundwater storage, ΔS_g. Stated as an equation

$$\Delta S_g = \Delta H_s(Y_g) \tag{6.15}$$

The gravity drainage of clayey materials, such as glacial till, may be very slow. The gravity yield of the deposits beneath basins can be determined from the hydrologic budget. Equation (6.13) contains two factors, evapotranspiration and change in soil moisture, which are not generally measured during studies. However, at many places during the winter and early spring months (December, January, February, and early March) evapotranspiration and soil moisture change are very small (Thornwaite et al., 1958). A reasonable estimate of evapotranspiration for periods during winter and early spring months can be made. Soil moisture change can be eliminated and evapotranspiration estimated without introducing serious error in the hydrologic budget. Equation (6.13) may be rewritten for inventory periods during winter and early spring months when the water table is rising as

$$Y_g = \frac{P_a - R_u - ET - U}{\Delta H_s} \tag{6.16}$$

Equation (6.16) can be used to estimate the gravity yield of deposits for periods when the soil moisture change is not significant.

6.6 Groundwater Budget

A part of precipitation on basins percolates down through the soil and, upon reaching the water table, becomes groundwater. Some groundwater is discharged to streams as groundwater runoff and some is discharged into the atmosphere by the processes of evapotranspiration. For a given period of time, precipitation reaching the water table (groundwater recharge) is balanced by groundwater runoff, underflow, and evapotranspiration, plus or minus changes in groundwater storage. This balance expresses a *groundwater budget* and may be stated as (Schicht and Walton, 1961)

$$P_g = R_g + ET_g + U \pm \Delta S_g \tag{6.17}$$

where P_g is groundwater recharge, R_g is groundwater runoff, ET_g is groundwater evapotranspiration, U is subsurface underflow, and ΔS_g is change in groundwater storage.

Net groundwater runoff depends in part upon the position of the water table with respect to stream stage, because, associated with a particular mean groundwater stage, there is a related hydraulic gradient and a consequent discharge of groundwater into a stream. The relationship between mean groundwater stage and groundwater runoff can often be determined by plotting mean groundwater stages against streamflow on corresponding dates, when streamflow consists entirely of groundwater runoff. Periods are selected assuming that surface runoff is complete within N_b days after rainfall and that in the following protracted period of fair weather, streamflow is all groundwater runoff.

In summer months evapotranspiration is very effective in reducing groundwater runoff. With the same mean groundwater stage, groundwater runoff is much less in August than in February. Separate rating curves of mean groundwater stage versus groundwater runoff must be prepared for dates April through October, when groundwater evapotranspiration is great, and for dates November through March, when groundwater evapotranspiration is very small.

Groundwater runoff under flood hydrographs and during rainless periods can be obtained from the rating curves and plotted beneath the streamflow hydrograph to describe a hydrograph of groundwater runoff. The rating curves give reasonably accurate estimates of groundwater runoff under flood hydrographs because accumulated groundwater storage, drained out as the stream stage starts to fall, approximately balances the negative groundwater outflow at flood stages.

At many places, groundwater runoff generally is at a maximum during spring and early summer months and is least in late summer and fall months. Groundwater runoff often increases in the fall, even though the mean groundwater stage declines, because groundwater evapotranspiration decreases rapidly during that period. Annual groundwater runoff depends upon antecedent soil moisture and groundwater stage conditions, as well as the amount and distribution of annual precipitation.

Groundwater continuously percolates toward streams; however, the roots of plants and soil capillaries intercept and discharge into the atmosphere some of the water which otherwise would become groundwater runoff. Evapotranspiration from the groundwater reservoir is largely a function of the season and the mean groundwater stage. At many places, groundwater evapotranspiration is very small from November to April, increases rapidly in the late spring, reaches a maximum generally in July, and decreases rapidly as the growing season closes in the autumn. The opportunity for evapotranspiration is large at places where the water table is within a few feet of the surface and it generally decreases as the mean groundwater stage declines.

Groundwater evapotranspiration often can be estimated from rating curves of mean groundwater stage versus groundwater runoff. The difference in groundwater runoff between the curve for the period April through November and the curve for the period November through March is approximate groundwater evapotranspiration.

Evapotranspiration and soil moisture requirements have first priority on precipitation. Rainfall percolates to the water table to recharge the groundwater reservoir during periods when precipitation is in excess of evapotranspiration and soil moisture requirements and the ground is not frozen. At many places, groundwater recharge is greatest in spring months of heavy rainfall and least in winter, summer, and fall months. In the summer little precipitation reaches the water table except during periods of excessive rainfall. Precipitation cannot infiltrate through frozen soil to recharge the groundwater reservoir. Groundwater recharge is indicated when the mean groundwater stage rises or declines less than is necessary to balance groundwater runoff and evapotranspiration.

The amount of precipitation which reaches the water table can be estimated from Eq. (6.17). Groundwater runoff and evapotranspiration are determined from the mean groundwater stage runoff rating curves. To compute changes in groundwater storage, appropriate values of gravity yield must be substituted in Eq. (6.15), based on the average period of groundwater decline preceding the inventory period.

Groundwater recharge also can be estimated from data on significant rises in the mean groundwater stage, using a method described by Rasmussen and Andreasen (1959). Recession limbs of the hydrograph of mean groundwater stage are projected to dates on which the following peaks of water-level rise occur, thus taking into consideration groundwater runoff. The amount of recharge occurring during individual rises in the water table is equal to the product of the individual rise and the gravity yield.

6.7 Case Histories of Studies to Determine Groundwater Recharge Rates

DuPage County, Illinois

In DuPage County, about 6 miles west of the corporate limits of Chicago, permeable zones, within a Silurian dolomite aquifer, encountered at an average depth of 100 feet, are recharged by the vertical leakage of water through glacial drift and less permeable zones of the dolomite aquifer. Except in small areas in the north-central, western, and southwestern parts of the county, the bedrock surface beneath the glacial drift is formed by Silurian rocks of the Niagaran Series. The Alexandrian Series occurs immediately below the glacial drift, in narrow bands associated with buried-bedrock valleys in north-central parts of the county, and in fairly extensive areas in the southwest corner of the county.

The Niagaran Series is composed chiefly of dolomite; however, shaly dolomite beds occur at the base, and reefs and associated strata are present in upper formations. The maximum known thickness is 175 feet, and thicknesses of more than 50 feet are common in the eastern two-thirds of the county. Basal shaly dolomite beds have a fairly uniform total thickness of about 30 feet, except in the southwestern part of the county where the beds are missing. The Alexandrian Series is composed chiefly of dolomite; shale and argillaceous dolomite beds occur near the base, and relatively pure dolomite is present in upper formations. The thickness of the rocks commonly exceeds 40 feet and reaches 90 feet at some places.

Mean annual precipitation of 34.2 inches occurs in DuPage County. Mean annual temperature is 49.6°F.

Recharge rates for the dolomite aquifer in DuPage County were estimated with a piezometric-surface map and past records of pumpage and water levels (Zeizel et al., 1962). A comparison of hydrographs for dolomite production wells and pumpage graphs indicates that water-level declines are directly proportional to pumping rates, and water levels stabilize within a short time after each increase in pumpage. Thus, in the past, recharge has balanced discharge. In order to determine areas of recharge, a piezometric-surface map was prepared with water-level data collected during August 1960. Total withdrawal from the dolomite aquifer in August 1960 was about 20 mgd. Recharge areas A_c for production wells in the West Chicago area (1), Wheaton–Glen Ellyn–Lombard area (2), Downers Grove–Westmont–Clarendon Hills–Hinsdale area (3), and Argonne National Laboratory area (4), shown in Fig. 6.2, were delineated by flow-net analysis of the piezometric-surface map. The boundaries of recharge areas enclose areas within which vertical leakage is being diverted to pumping centers.

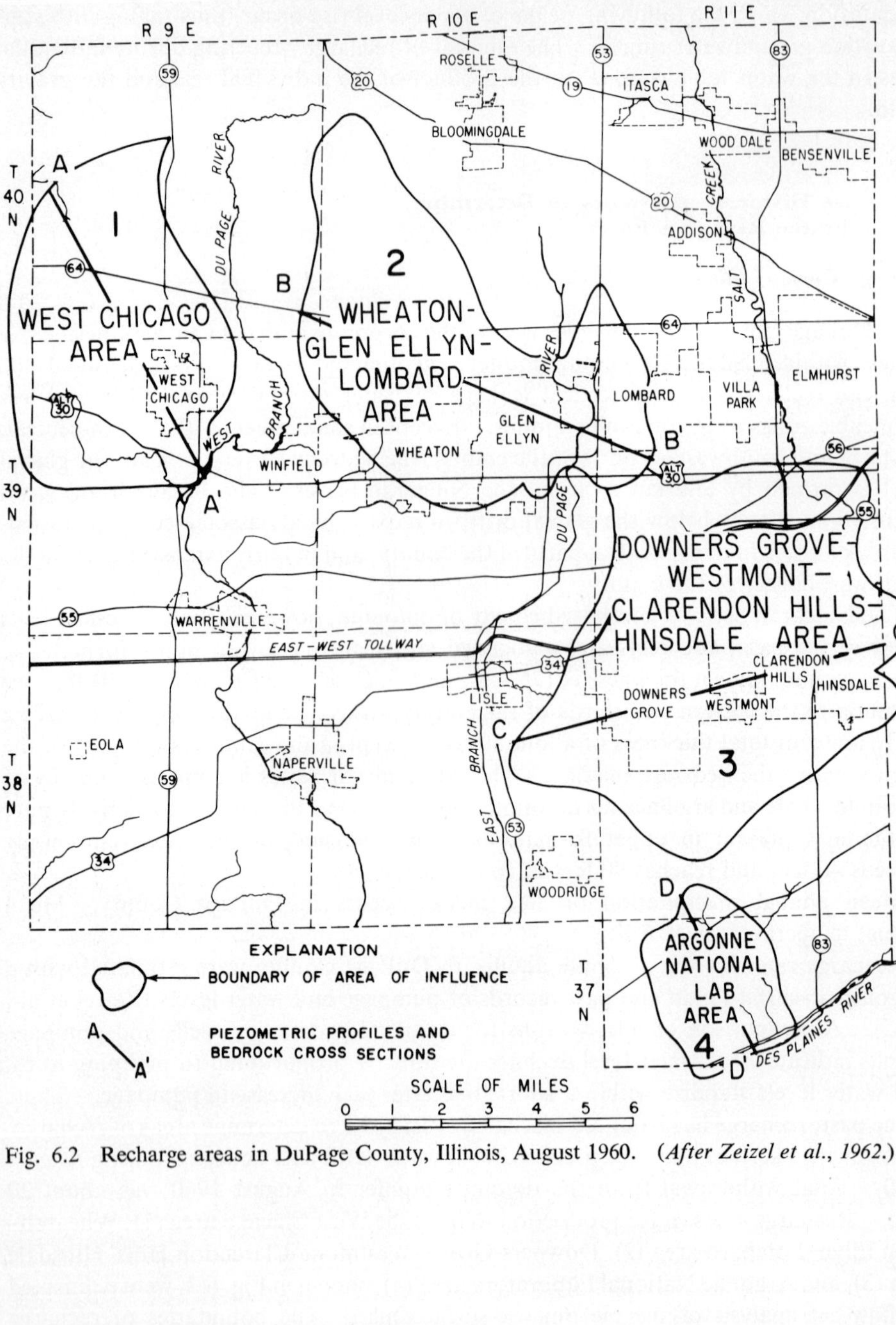

Fig. 6.2 Recharge areas in DuPage County, Illinois, August 1960. (*After Zeizel et al., 1962.*)

TABLE 6.3 RATES OF RECHARGE FOR DOLOMITE AQUIFER IN DUPAGE COUNTY. (*From Zeizel et al., 1962.*)

Area	*Pumpage in 1960, mgd*	*Recharge area, sq mi*	*Recharge rate, gpd/sq mi*
1	1.8	28.0	64,000
2	4.5	32.5	138,000
3	6.3	46.2	136,000
4	1.2	7.6	158,000

Measured recharge areas, pumpage data, and recharge rates per unit area computed as the quotient of pumpage and area (Q_c/A_c) are given in Table 6.3. Except for area 1, recharge rates per unit area are approximately the same. On a gross basis, the glacial drift deposits are similar in character and thickness in the four areas and slightly more permeable in area 1 than in the other three areas. The profiles in Fig. 6.3 show that the piezometric surface is more than 50 feet below the ground surface in most of the areas. Average vertical hydraulic gradients do not differ appreciably from area to area. Thus, the low recharge rate in area 1 cannot be explained by differences in the character and thickness of the glacial drift or average vertical hydraulic gradients. In study areas 2, 3, and 4, most of the water pumped is obtained from the thick rocks of the Niagaran Series, above the shaly dolomite unit. The rocks of the Niagaran Series are thin in area 1, and much of the water withdrawn from production wells is obtained from rocks of the Alexandrian Series, below the shaly dolomite unit. The shaly dolomite unit greatly retards the vertical movement of groundwater and is responsible for the low recharge rate in area 1.

Areas in DuPage County, where recharge will be low and about the same as in the West Chicago area under heavy pumping conditions, were delineated by assuming that recharge will be limited east of the Niagaran-Alexandrian contact, where rocks of the Niagaran Series overlying the shaly dolomite unit are less than 25 feet thick. In addition, the recharge rate will be low in areas where water is obtained from the Maquoketa Formation because permeable dolomite beds of the formation are interbedded with, and often overlain by, beds of dolomite shale with a very low permeability. Areas where recharge will be about the same as in study areas 2, 3, and 4 were delineated by assuming that recharge will be high west of the Niagaran-Alexandrian contact, where rocks of the Alexandrian Series are more than 25 feet thick, and east of the Niagaran-Alexandrian contact where rocks of the Niagaran Series overlying the shaly dolomite unit are more than 25 feet thick. A map showing estimated recharge rates under heavy pumping conditions for the dolomite aquifer is shown in Fig. 6.4. It is probable that recharge to the dolomite aquifer under heavy pumping conditions will average about 140,000 gpd/sq mi in parts of the eastern one-third of the county, and about 60,000 gpd/sq mi in large areas of the western two-thirds of the county.

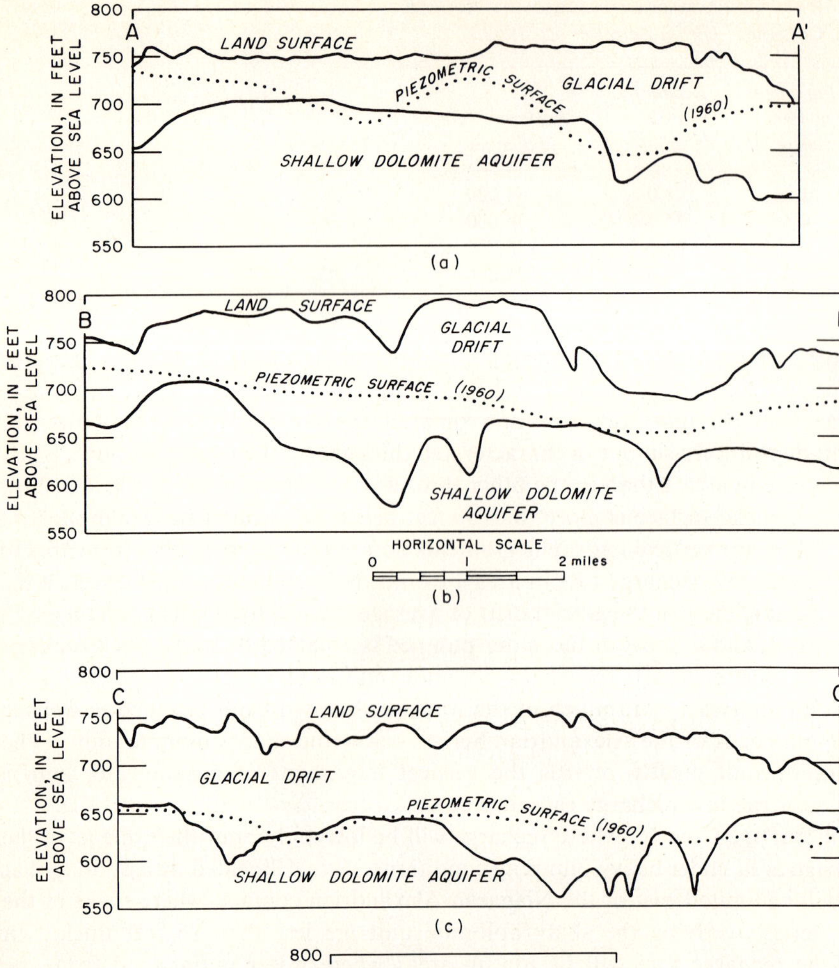

Fig. 6.3 Hydrogeologic cross sections in West Chicago area (*a*), Wheaton–Glen Ellyn–Lombard area (*b*), Downers Grove–Hinsdale area (*c*), and Argonne area (*d*), Illinois. (*From Zeizel et al., 1962.*)

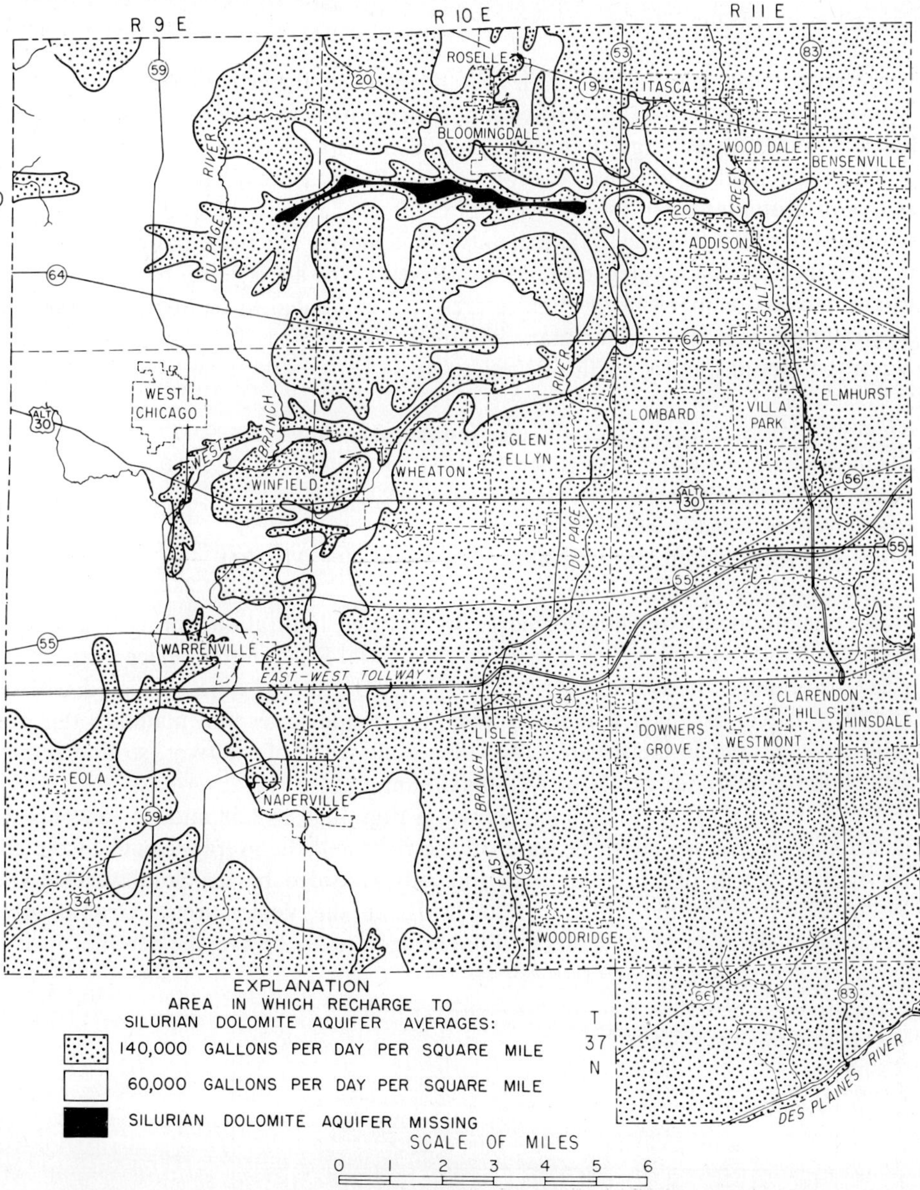

Fig. 6.4 Estimated recharge rates for dolomite aquifer in DuPage County, Illinois. (*From Zeizel et al., 1962.*)

Venice, Ohio

Hydrogeologic data for the valley-fill deposits in the Miami River valley near Venice, Ohio, about 14 miles northwest of Cincinnati were studied to gain insight into the magnitude of recharge by induced infiltration under complex stream-stage and groundwater level conditions (Walton et al., 1967). As explained in Chap. 4, the source of groundwater in the Venice area is the glacial outwash and sand and gravel in the Miami River valley. These outwash deposits are principally recharged by induced infiltration of water in the Miami River. The average infiltration rate of the Miami River streambed in the Venice area was computed to be about 168,000 gpd/acre/ft (see Chap. 4). A summary of streamflow records for the Miami River at Hamilton, about 8 miles north of Venice, is:

Maximum daily discharge: 73,900 cfs, January 22, 1959
Mean discharge: 3,214 cfs, October 30, 1931 to September 30, 1960
Minimum daily discharge: 155 cfs, September 27, 1941
Maximum recorded discharge: 108,000 cfs, January 21, 1959
Minimum recorded discharge: 100 cfs, September 26, 27, 1941

The magnitude of potential recharge by induced infiltration during a dry period 1952 through 1954 from a 2-mile reach of the Miami River in the Venice area, and the effects of changes in stream stage on induced infiltration, were estimated with Eq. (3.139). It was assumed that heavy pumping would lower and maintain the water table to stages below the streambed of the 2-mile reach of the river, so that induced infiltration at all times would be at a maximum.

The profiles of the river channel, shown in Figs. 4.56 to 4.58 and 6.5, were used to determine the average depth of water in the river and the average width of the river for various stages of the river as periodically recorded by the Southwestern Ohio Water Company at the Colerain Pike bridge in the Venice area. Graphs of the elevation of the river at the Colerain Pike bridge versus the average depth of water in

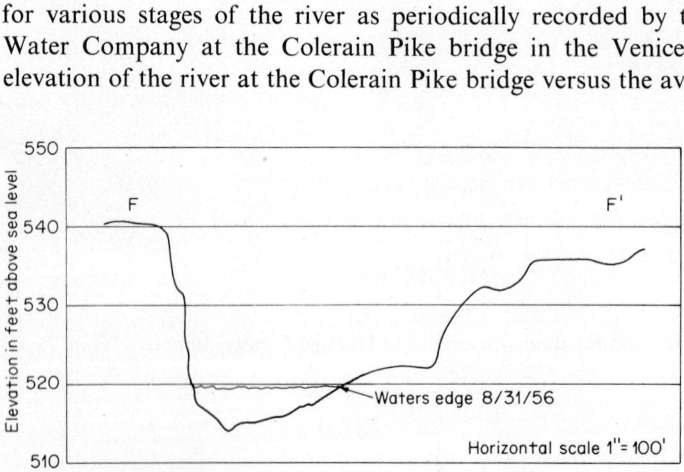

Fig. 6.5 Profile of Miami River channel in Venice area, Ohio. (*From Walton et al., 1967.*)

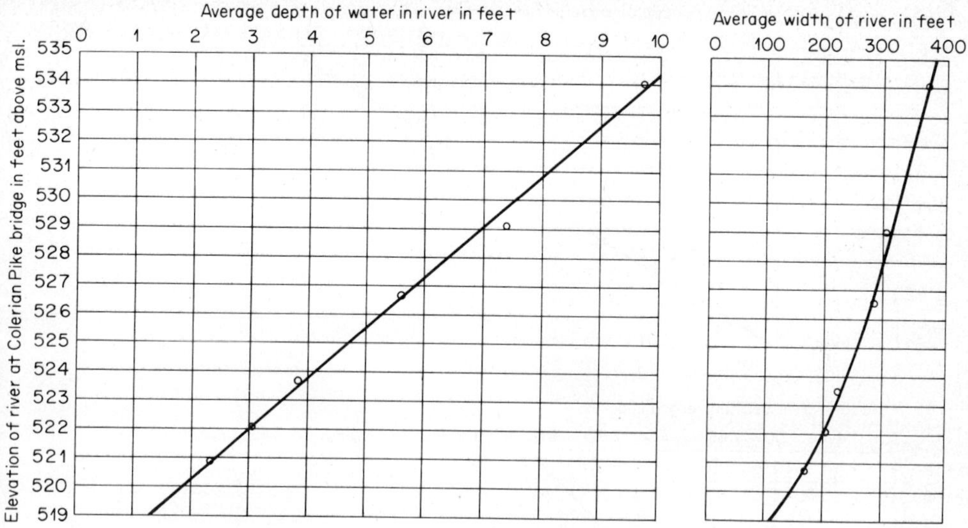

Fig. 6.6 Graphs of elevation of Miami River at Colerain Pike bridge versus average depth of water in river and average width of river in Venice area, Ohio. (*From Walton et al., 1967.*)

the river and the average width of the river were prepared (Fig. 6.6). The relation between gage height readings at Hamilton, Ohio, collected by the U.S. Geological Survey and the elevation of the river at the Colerain Pike bridge was established. The graph of the gage readings at Hamilton versus the average depth of the river in the Venice area was then prepared (Fig. 6.7) so that daily streamflow records collected at Hamilton could be used to compute potential induced streambed infiltration. Considering the length of river south of the Colerain Pike bridge available for infiltration, 11,000 feet, a graph of average depth of water in the river and acres of riverbed submerged versus the elevation of the river at Colerain Pike bridge was prepared as shown in Fig. 6.7.

The average infiltration rate of the Miami River was computed with water-table data for August 1956, when the average temperature of the river water was about 82°F. During the winter months the temperature of the river water declines to about 39°F according to records collected by the Champion Paper and Fibre Company at Hamilton, Ohio (see Rose, 1951). Because infiltration varies with the temperature of the surface water, with a particular stream stage the infiltration rate during the winter months will be only about 52 percent of the infiltration rate during the summer months. Values of I_t for use in estimating potential recharge were therefore determined, taking into consideration surface-water temperature data.

Streamflow records, the graphs in Figs. 4.56 to 4.58, 6.5, and 6.7, the average infiltration rate, and surface-water temperature data were substituted in Eq. (3.139) to compute the maximum potential recharge by the induced infiltration of surface water from the 2-mile reach of the Miami River south of the Colerain Pike bridge for each day of the years 1952 through 1954. Estimated daily potential recharge

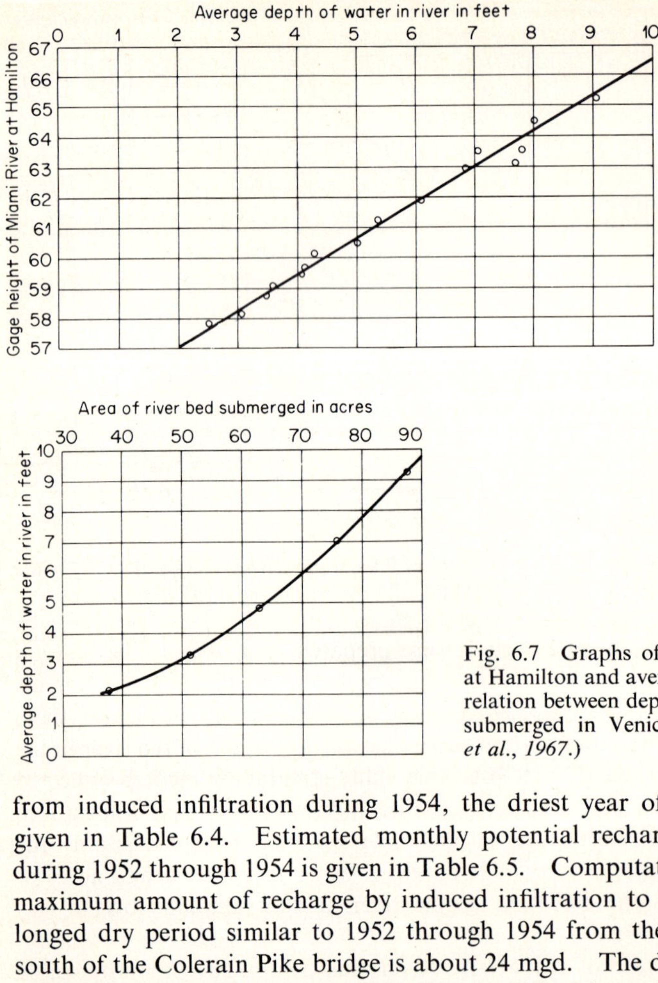

Fig. 6.7 Graphs of gage heights of Miami River at Hamilton and average depth of water in river and relation between depth of river and area of riverbed submerged in Venice area, Ohio. (*From Walton et al., 1967.*)

from induced infiltration during 1954, the driest year of the three-year period, is given in Table 6.4. Estimated monthly potential recharge by induced infiltration during 1952 through 1954 is given in Table 6.5. Computations show that the average maximum amount of recharge by induced infiltration to be expected during a prolonged dry period similar to 1952 through 1954 from the 2-mile reach of the river south of the Colerain Pike bridge is about 24 mgd. The data illustrate the profound influence of changes in stream stage on potential recharge by induced infiltration. It is apparent that quantitative estimates of the potential yields of aquifers, recharged principally by induced infiltration of surface water, must take into consideration changes in stream stage.

6.8 Case Histories of Hydrologic and Groundwater Budget Studies

Central Illinois

To determine how groundwater recharge varied from year to year, month to month, and basin to basin, hydrologic and groundwater budgets were prepared for three small watersheds in central Illinois (Schicht and Walton, 1961). During study

TABLE 6.4 ESTIMATED DAILY POTENTIAL RECHARGE FROM INDUCED INFILTRATION DURING 1954 IN VENICE AREA, OHIO. (*From Walton, et al., 1967.*)

Day	January			February			March			April			May			June		
	h_r, *ft*	A_r, *acres*	R_I, *mgd*	h_r, *ft*	A_r, *acres*	R_I, *mgd*	h_r, *ft*	A_r, *acres*	R_I, *mgd*	h_r, *ft*	A_r, *acres*	R_I, *mgd*	h_r, *ft*	A_r, *acres*	R_I, *mgd*	h_r, *ft*	A_r, *acres*	R_I, *mgd*
1	2.3	41	8.3	2.8	47	11.5	2.8	47	13.9	5.3	66	44.0	3.1	50	21.6	2.9	48	22.2
2	2.3	41	8.3	2.8	47	11.5	2.9	48	14.7	4.5	60	34.0	3.3	52	23.9	3.2	51	26.0
3	2.3	41	8.3	2.7	46	10.8	3.2	51	17.3	4.0	57	28.8	3.7	55	28.4	2.9	48	22.2
4	2.3	41	8.3	2.7	46	10.8	3.1	50	16.4	3.7	55	25.6	3.7	55	28.4	2.8	47	21.0
5	2.3	41	8.3	2.7	46	10.8	3.1	50	16.4	4.2	58	30.6	3.6	54	27.2	2.8	47	21.0
6	2.3	41	8.3	2.6	45	10.2	3.0	49	15.5	4.3	59	32.0	3.5	54	26.3	2.7	46	19.8
7	2.3	41	8.3	2.5	43	9.4	2.8	47	13.9	4.5	60	34.0	3.3	52	23.9	2.7	46	19.8
8	2.3	41	8.3	2.5	43	9.4	2.8	47	13.9	4.5	60	34.0	3.1	50	21.6	2.9	48	22.2
9	2.3	41	8.3	2.5	43	9.4	2.8	47	13.9	4.5	60	34.0	3.0	49	20.5	3.3	52	27.4
10	2.2	40	7.7	2.5	43	9.4	2.8	47	13.9	4.3	59	32.0	3.0	49	20.5	3.6	54	31.1
11	2.3	41	8.3	2.5	43	9.4	2.8	47	13.9	3.9	56	27.5	3.0	49	20.5	3.6	54	31.1
12	2.3	41	8.3	2.5	43	9.4	2.8	47	13.9	3.7	55	25.6	2.9	48	19.3	3.5	54	30.2
13	2.3	41	8.3	2.5	43	9.4	2.8	47	13.9	3.6	54	24.5	2.9	48	19.3	3.3	52	27.4
14	2.3	41	8.3	2.5	43	9.4	2.7	46	13.1	3.6	54	24.5	2.9	48	19.3	3.1	50	24.7
15	2.3	41	8.3	2.5	43	9.4	2.7	46	13.1	3.6	54	24.5	2.9	48	19.3	3.1	50	24.7
16	2.3	41	8.3	2.5	43	9.4	2.7	46	13.1	3.8	56	26.9	2.8	47	18.4	2.8	47	21.0
17	2.3	41	8.3	2.7	46	10.8	2.7	46	13.1	4.8	63	38.0	2.8	47	18.4	3.7	55	32.5
18	2.4	42	8.8	2.7	46	10.8	2.7	46	13.1	5.5	67	46.5	2.8	47	18.4	5.3	66	56.0
19	2.3	41	8.3	2.7	46	10.8	2.8	47	13.9	4.7	62	36.8	2.8	47	18.4	4.2	58	38.9
20	2.4	42	8.8	2.7	46	10.8	3.1	50	16.4	4.2	58	30.6	2.8	47	18.4	3.6	54	31.1
21	2.7	46	10.8	2.7	46	10.8	2.9	48	14.7	3.9	56	27.5	2.7	46	17.3	3.2	51	26.0
22	2.6	45	10.2	2.7	46	10.8	2.8	47	13.9	3.6	54	24.5	2.7	46	17.3	3.0	49	23.4
23	2.5	43	9.4	2.7	46	10.8	2.9	48	14.7	3.8	56	26.9	2.5	43	15.0	2.9	48	22.2
24	2.4	42	8.8	2.7	46	10.8	2.8	47	13.9	3.8	56	26.9	2.6	45	16.4	2.8	47	21.0
25	2.4	42	8.8	2.7	46	10.8	2.8	47	13.9	3.5	54	23.8	2.6	45	16.4	2.7	46	19.8
26	2.6	45	10.2	2.7	46	10.8	2.8	47	13.9	3.4	53	22.7	2.6	45	16.4	2.6	45	18.7
27	2.8	47	11.5	2.6	45	10.2	2.8	47	13.9	3.3	52	21.6	2.6	45	16.4	2.6	45	18.7
28	3.7	55	17.8	2.7	46	10.8	2.8	47	13.9	3.3	52	21.6	2.6	45	16.4	2.6	45	18.7
29	3.7	55	17.8				2.8	47	13.9	3.2	51	20.6	2.6	45	16.4	2.5	43	17.2
30	3.2	51	14.3				5.4	67	38.2	3.2	51	20.6	2.6	45	16.4	2.5	43	17.2
31	2.9	48	12.4				6.3	72	48.0				2.6	45	16.4			

TABLE 6.4 (*Continued*)

Day	July			August			September			October			November			December		
1	2.5	43	18.1	2.2	40	14.8	2.5	43	16.7	2.3	41	13.2	2.6	45	12.4	2.8	47	11.5
2	2.4	42	16.9	2.4	42	16.9	2.4	42	15.6	2.2	40	12.3	2.6	45	12.4	2.8	47	11.5
3	2.5	43	18.1	2.4	42	16.9	2.4	42	15.6	2.2	40	12.3	2.6	45	12.4	2.8	47	11.5
4	2.5	43	18.1	2.4	42	16.9	2.3	41	14.6	2.2	40	12.3	2.6	45	12.4	2.7	46	10.8
5	2.5	43	18.1	2.5	43	18.1	2.3	41	14.6	2.3	41	13.2	2.6	45	12.4	2.7	46	10.8
6	2.5	43	18.1	3.7	55	34.2	2.3	41	14.6	2.4	42	14.1	2.6	45	12.4	2.5	43	4.9
7	2.3	41	15.8	3.7	55	34.2	2.3	41	14.6	2.5	43	15.0	2.5	43	11.4	2.6	45	10.2
8	2.4	42	16.9	3.4	53	30.3	2.3	41	14.6	2.4	42	14.1	2.6	45	12.4	2.6	45	10.2
9	2.5	43	18.1	3.1	50	26.1	2.3	41	14.6	2.4	42	14.1	2.6	45	12.4	2.5	43	9.4
10	2.5	43	18.1	2.9	48	23.4	2.3	41	14.6	2.3	41	13.2	2.6	45	12.4	2.5	43	9.4
11	2.4	42	16.9	2.7	46	20.8	2.2	40	13.6	2.4	42	14.1	2.6	45	12.4	2.5	43	9.4
12	2.4	42	16.9	2.5	43	18.1	2.2	40	13.6	2.8	47	18.4	2.6	45	12.4	2.4	42	8.8
13	2.4	42	16.9	2.5	43	18.1	2.2	40	13.6	2.6	45	16.4	2.5	43	11.4	2.5	43	9.4
14	2.4	42	16.9	2.5	43	18.1	2.3	41	14.6	2.8	47	18.4	2.5	43	11.4	2.5	43	9.4
15	2.3	41	15.8	2.4	42	16.9	2.4	42	15.6	3.2	51	22.8	2.4	42	10.7	2.5	43	9.4
16	2.3	41	15.8	2.5	43	18.1	2.4	42	15.6	3.2	51	22.8	2.5	43	11.4	2.5	43	9.4
17	2.2	40	14.8	2.8	47	22.1	2.3	41	14.6	3.5	54	26.3	2.5	43	11.4	2.5	43	9.4
18	2.2	40	14.8	3.3	52	28.8	2.3	41	14.6	3.2	51	22.8	2.5	43	11.4	2.5	43	9.4
19	2.3	41	15.8	2.8	47	22.1	2.2	40	13.6	3.1	50	21.6	2.5	43	11.4	2.5	43	9.4
20	2.4	42	16.9	2.7	46	20.8	2.3	41	14.6	2.9	48	19.3	2.5	43	11.4	2.5	43	9.4
21	4.5	60	45.3	2.6	45	19.7	2.4	42	15.6	2.8	47	18.4	2.4	42	10.7	2.5	43	9.4
22	2.8	47	22.1	2.4	42	16.9	2.3	41	14.6	2.8	47	18.4	2.4	42	10.7	2.5	43	9.4
23	2.5	43	18.1	2.4	42	16.9	2.3	41	14.6	2.8	47	18.4	2.4	42	10.7	2.5	43	9.4
24	2.4	42	16.9	2.4	42	16.9	2.3	41	14.6	2.9	48	19.3	2.4	42	10.7	2.5	43	9.4
25	2.3	41	15.8	2.4	42	16.9	2.2	40	13.6	2.8	47	18.4	2.4	42	10.7	2.4	42	8.8
26	2.3	41	15.8	2.5	43	18.1	2.2	40	13.6	2.8	47	18.4	2.4	42	10.7	2.4	42	8.8
27	2.3	41	15.8	3.3	52	28.8	2.2	40	13.6	2.8	47	18.4	2.4	42	10.7	2.4	42	8.8
28	2.3	41	15.8	3.1	50	26.1	2.3	41	14.6	2.7	46	17.3	2.5	43	11.4	3.3	52	15.0
29	2.3	41	15.8	2.8	47	22.1	2.3	41	14.6	2.6	45	16.4	2.6	45	12.4	4.2	58	21.3
30	2.3	41	15.8	2.7	46	20.8	2.2	40	13.6	2.6	45	16.4	2.7	46	13.1	5.3	66	30.3
31	2.2	40	14.8	2.6	45	19.7				2.6	45	16.4				4.9	64	28.4

TABLE 6.5 ESTIMATED MONTHLY POTENTIAL RECHARGE BY INDUCED INFILTRATION DURING 1952, 1953, AND 1954 IN VENICE AREA, OHIO. (*From Walton et al., 1967.*)

1952		*1953*		*1954*	
Month	R_I, *mgd*	*Month*	R_I, *mgd*	*Month*	R_I, *mgd*
Jan.	56.8	Jan.	24.9	Jan.	9.6
Feb.	39.8	Feb.	20.7	Feb.	10.3
Mar.	50.8	Mar.	31.1	Mar.	16.2
Apr.	54.2	Apr.	25.3	Apr.	29.1
May	35.7	May	48.4	May	19.8
June	31.9	June	28.4	June	25.1
July	25.0	July	22.3	July	17.7
Aug.	23.6	Aug.	20.0	Aug.	21.2
Sept.	20.8	Sept.	15.0	Sept.	14.5
Oct.	16.7	Oct.	13.0	Oct.	16.8
Nov.	12.9	Nov.	10.1	Nov.	11.7
Dec.	14.5	Dec.	8.6	Dec.	12.0
Annual	31.8	Annual	22.4	Annual	17.0

periods ranging from 2.5 to 8 years, the Illinois State Water Survey and cooperating state and federal agencies measured precipitation on, stream discharge from, and groundwater levels in parts of the drainage basins of Panther, Hadley, and Goose Creeks. The study areas are in north-central, west-southwestern, and east-central Illinois, respectively. Information pertaining to Panther Creek basin is presented in detail to illustrate the influence of climatic conditions on groundwater recharge. Data for the other two basins are summarized and compared with data for Panther Creek basin to show the influence of basin characteristics on groundwater recharge.

The Panther Creek basin is about 30 miles east of Peoria and about 20 miles north of Bloomington. The part of Panther Creek drainage basin considered, hereafter referred to as "the basin," is approximately between 88°52′ and 89°07′ west longitude, and between 40°44′ and 40°54′ north latitude. The basin covers 95 square miles, mostly in Woodford County, although small parts are in Livingston and McLean Counties, and is in T26N to T28N and R1E and R3E. The basin is above a stream gaging station, about 4 miles northwest of the city of El Paso.

The basin lies in the Till Plains section of the Central Lowland physiographic province. The topography consists mostly of gently undulating uplands. Rolling topography is found in belts on moraines along the west, northeast, and east edges of the basin. The uplands are eroded in the immediate vicinity of Panther Creek in the extreme southwest corner of the area, where the topography is more diversified. Except in the southwest corner adjacent to Panther Creek, where the elevation of the land surface declines about 50 feet in a distance of 0.25 mile, the relief seldom exceeds 20 ft/mi.

Panther Creek is the principal stream and flows in a generally southwestward course. A small tributary, East Branch Panther Creek, drains the southern quarter of the basin and flows westward to a confluence with Panther Creek, 4 miles northwest of El Paso. The average gradients of Panther Creek and East Branch are 4.7 and 5.0 ft/mi, respectively. The water table was very near the surface, and shallow ponds, swamps, and poorly drained areas were widespread prior to settlement. Extensive surface and subsurface drainage was necessary to permit agricultural development.

The population of the basin is chiefly rural and, according to the U.S. Bureau of the Census, had a density of about 37 persons per square mile in 1950. At the time of the study about 80 percent of the basin was cleared and cultivated; the remainder was pasture, woodland, and farm lots. The cleared land was devoted to three major crops, field corn, oats, and soybeans for beans, and to other crops such as alfalfa, clover and timothy hay, winter wheat, rye, and sweet corn.

The basin lies in the north temperate zone. Its climate is characterized by warm summers and moderately cold winters. The mean length of the growing season is about 170 days. Based on records collected by the U.S. Weather Bureau, the mean annual temperature is 51°F. June, July, and August are the hottest months with mean temperatures of 71, 76, and 73°F, respectively; January is the coldest month, with a mean temperature of 25°F. Mean monthly temperatures during December, January, and February are below 32°F. Normal annual precipitation is 33.6 inches, based on 1900-to-1944 U.S. Weather Bureau records. The months of greatest precipitation are April, May, June, August, and September, each having an average of less than 2 inches. Monthly and annual precipitation, 1950 through 1958, is given in Table 6.6. Precipitation was above, near, and below normal in 1951, 1952, and 1956,

TABLE 6.6 MONTHLY AND ANNUAL PRECIPITATION IN INCHES, 1950 TO 1958, PANTHER CREEK BASIN, ILLINOIS. (*From Schicht and Walton, 1961.*)

Month	*1950*	*1951*	*1952*	*1953*	*1954*	*1955*	*1956*	*1957*	*1958*
Jan.	4.90	1.41	1.01	1.36	1.23	1.92	0.14	1.51	1.02
Feb.	2.71	2.88	1.19	1.19	2.11	1.50	1.45	1.16	0.45
Mar.	1.13	3.58	2.73	4.38	3.95	1.55	0.73	1.64	0.33
Apr.	5.99	4.20	4.66	1.94	4.46	4.28	2.39	7.47	2.56
May	1.07	2.93	3.36	2.06	4.58	3.53	3.24	4.42	2.57
June	6.91	7.16	7.07	3.52	2.58	2.81	0.89	4.64	5.07
July	6.42	8.40	2.18	6.29	4.42	3.12	3.22	2.28	6.05
Aug.	0.62	4.11	4.47	1.22	5.18	4.33	3.23	1.96	4.24
Sept.	3.83	2.34	1.43	2.32	0.81	1.86	1.08	1.31	1.82
Oct.	0.90	2.99	0.64	0.71	3.42	3.71	0.40	5.14	0.64
Nov.	1.81	2.70	2.31	0.72	1.75	0.83	1.54	2.08	2.52
Dec.	0.78	1.54	1.57	2.53	1.61	0.35	1.18	2.75	0.49
Annual	37.07	44.24	32.62	28.24	36.10	29.79	19.49	36.36	28.40

respectively. The annual maximum precipitation amounts, occurring on an average of once in 5 and once in 50 years, are 39 and 50 inches, respectively; annual minimum amounts expected for the same intervals are 30 and 24 inches, respectively. The mean annual snowfall is 24 inches. On the average more than 28 days have 1 inch or more, and more than 13 days have 3 inches or more, of ground snow cover in a year. The average depth of maximum frost penetration is 26 inches.

The soils of the basin were divided into four groups by Smith et al. (1927): upland prairie, upland timber, swamp and bottomland, and terrace soils. Except for small areas adjacent to Panther Creek and East Branch, upland prairie soils predominate, and these are largely very dark gray to dark brown silt loams, formed under prairie vegetation from thin loess. The surface layer is a very dark gray to dark brown silt loam, 6 to 8 inches thick, which is medium in organic matter, and slightly to medium acid. The subsurface is a light silty clay loam, very dark grayish brown, and 6 to 8 inches thick. The subsoil, beginning at a depth of 12 to 16 inches, is a brown to dark grayish yellow silty clay. In a small area in the north-central part of the basin, the surface layer is a brown to dark brown heavy silt loam, 8 to 10 inches thick, or a granular black clay loam to silty clay loam, 8 to 10 inches thick. These materials are high in organic matter and nitrogen and slightly acid to neutral. The subsurface layer is a brown or pale yellowish-brown silt loam, a very dark gray or grayish-black clay loam, or silt clay loam, ranging from yellowish-brown to dark gray.

The upland prairie soils occur on 1 to 6 percent slopes. Surface drainage is moderate, and artificial drainage is often required for agricultural development. The permeability is moderately slow, but underdrainage by tiles is satisfactory under proper farm management. The materials beneath the subsoils to depths of 40 to 60 inches are compact calcareous glacial tills, except in a small area in the north-central part of the basin where there is stratified silt and sand, or stratified clay, silt, and sand. The permeability of the materials beneath the subsoils is moderate to slow.

Thick deposits of glacial drift, chiefly of Wisconsinan age, cover the bedrock and constitute the main features of the present land surface. The deposits are composed predominantly of unstratified clayey materials called glacial till, but include some stratified beds of silt sand and gravel, as shown by logs of wells in Fig. 6.8. The average thickness of the glacial drift on the bedrock uplands is about 100 feet. Along the eastern edge of the basin, in a bedrock valley, it may reach a thickness of more than 290 feet (see log of well MCL 26N3E-4.5c in Fig. 6.8). Sand and gravel, ranging in thickness from a few inches to more than 40 feet, occur as irregular lenses or layers in the till. These deposits are discontinuous and are limited greatly in areal extent. In general, because of the complex glacial history, the character of the drift varies greatly, both vertically and horizontally. However, for the basin as a whole, the character of the drift in relation to the occurrence and movement of groundwater is fairly uniform.

The surficial glacial deposits are immediately underlain by bedrock formations of Pennsylvanian age, consisting predominantly of shale with alternating thin beds of limestone, sandstone, siltstone, fire clay, and coal. These formations are situated

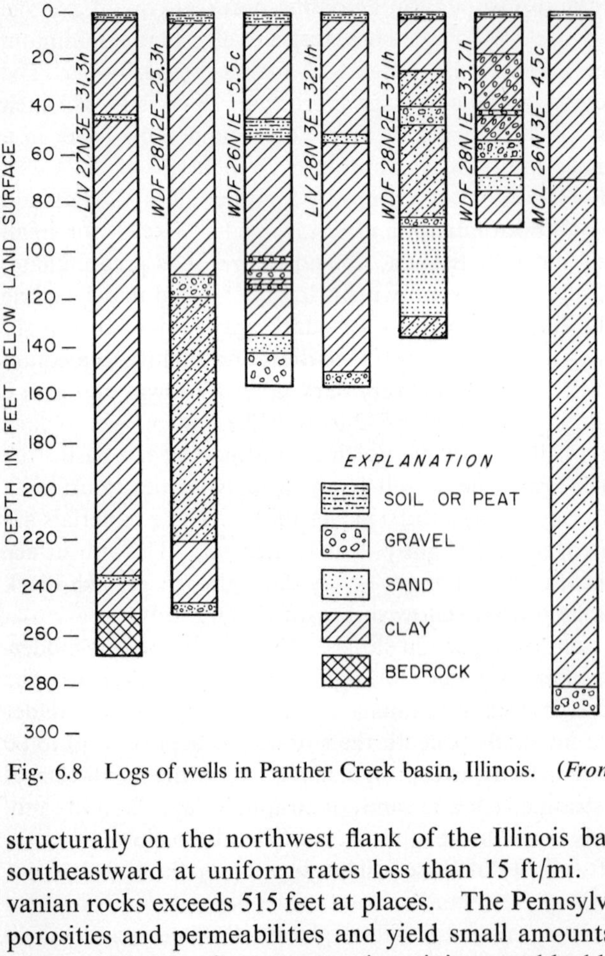

Fig. 6.8 Logs of wells in Panther Creek basin, Illinois. (*From Schicht and Walton, 1961.*)

structurally on the northwest flank of the Illinois basin, and dip regionally south-southeastward at uniform rates less than 15 ft/mi. The thickness of the Pennsylvanian rocks exceeds 515 feet at places. The Pennsylvanian rocks generally have low porosities and permeabilities and yield small amounts of water to wells from interconnected cracks, fractures, crevices, joints, and bedding planes. Practically speaking, the rocks are important because they act as a barrier to deep percolation. The bedrock-surface topography is relatively flat except for a bedrock valley, according to Horberg (1957). The bedrock surface slopes eastward along the eastern edge toward the bedrock valley, and its elevation declines from about 600 to 450 feet, in a distance of 4 miles.

Daily mean streamflow at the gaging station northwest of El Paso was plotted for 1951, 1952, and 1956. In general, streamflow is high in the winter and spring and low in the summer and fall. In the winter, melting of accumulated snow often produces disproportionately high streamflow for short periods of time. Daily mean streamflow exceeded 6,000 cfs in July 1951, and was less than 0.1 cfs during parts of August and September 1956. Monthly and annual streamflow during 1951, 1952, and 1956, expressed in inches of water over the basin, is given in Table 6.7. Streamflow was

TABLE 6.7 MONTHLY AND ANNUAL STREAMFLOW IN INCHES, 1951, 1952, AND 1956, PANTHER CREEK BASIN, ILLINOIS. (*From Schicht and Walton, 1961.*)

	1951			*1952*			*1956*		
Month	R_s	R_g	R_u	R_s	R_g	R_u	R_s	R_g	R_u
Jan.	0.61	0.16	0.77	0.39	0.77	1.16	Neg	0.01	0.01
Feb.	2.85	0.15	3.00	0.08	0.57	0.65	0.14	0.08	0.22
Mar.	0.97	0.30	1.27	0.43	1.57	2.00	0.01	0.04	0.05
Apr.	1.08	1.44	2.52	0.65	1.94	2.59	0.03	0.03	0.06
May	0.12	0.82	0.94	0.06	0.82	0.88	0.34	0.08	0.42
June	1.80	0.56	2.36	1.03	1.10	2.13	0.04	0.07	0.11
July	3.63	1.13	4.76	Neg	0.27	0.27	0.04	0.03	0.07
Aug.	0.16	0.22	0.38	0.01	0.04	0.05	0.01	0.01	0.02
Sept.	0.03	0.10	0.13	Neg	0.02	0.02	Neg	Neg	Neg
Oct.	0.07	0.22	0.29	Neg	0.01	0.01	Neg	Neg	Neg
Nov.	0.97	0.55	1.52	Neg	0.02	0.02	Neg	0.01	0.01
Dec.	0.05	0.35	0.40	0.01	0.03	0.04	Neg	0.01	0.01
Annual	12.34	6.00	18.34	2.66	7.16	9.82	0.61	0.37	0.98

greatest in 1951 largely as a result of above-normal precipitation during that year, and was least in 1956 when precipitation was much below normal. Several conditions were responsible for the low streamflow in 1956. Precipitation was below normal during most of 1955; consequently, the mean groundwater stage was low at the beginning of 1956. Precipitation during 1956 was only slightly in excess of evapotranspiration and soil moisture requirements. Very little precipitation reached the water table and the mean groundwater stage and groundwater runoff were abnormally low throughout the year.

Rating curves were prepared to determine the relationship between mean groundwater stage and groundwater runoff. Fluctuations of the water table in the basin were plotted on hydrographs. Daily averages of groundwater levels in wells in the basin were computed for selected dates when streamflow consisted entirely of groundwater runoff. Three to five days after precipitation ceases, there is no surface runoff, and streamflow is derived entirely from groundwater runoff. Mean groundwater stages were then plotted against groundwater runoff on corresponding dates as shown in Fig. 6.9. Closed circles represent sets of data for dates November to April, when evapotranspiration is at a minimum; open circles represent sets of data for dates April through October, when evapotranspiration is great.

Daily mean groundwater stages during 1951, 1952, and 1956 were plotted as yearly hydrographs. Groundwater runoff corresponding to each mean groundwater stage was read directly from the rating curves in Fig. 6.9. During protracted rainless periods actual streamflow is groundwater runoff. Monthly and annual groundwater and surface runoff, expressed in inches of water over the basin, were computed by Schicht and Walton (1961) and are given in Table 6.7. These data indicate that groundwater

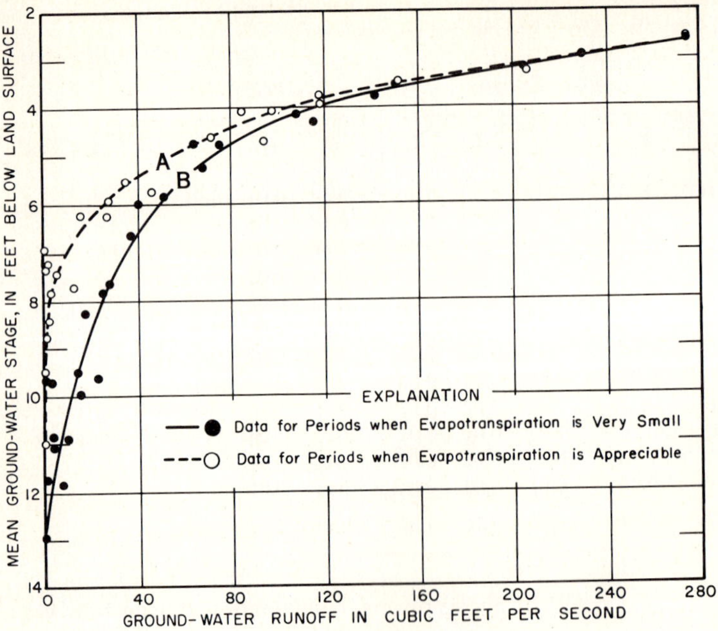

Fig. 6.9 Rating curves of mean groundwater stage versus groundwater runoff for gaging station in Panther Creek basin, Illinois. (*From Schicht and Walton, 1961.*)

runoff is at a maximum during spring and early summer months and is least in late summer and fall months. More than half of annual groundwater runoff occurs during the first six months of the year. Annual groundwater runoff depends upon antecedent groundwater stage conditions as well as the amount and distribution of annual precipitation. Groundwater runoff was less in 1951 than in 1952, although precipitation was much greater in 1951 than in 1952. Groundwater stages, and consequently groundwater runoff, during the first six months of 1952 were higher than those for the same period in 1951 because of excessive precipitation during the summer months of 1951 and near-normal precipitation in 1950. During extended dry periods groundwater runoff is reduced greatly. Groundwater runoff was very small during 1956 because precipitation was much below normal in 1955 and 1956. Groundwater runoff amounted to 33, 73, and 38 percent of streamflow in 1951, 1952, and 1956, respectively.

The basin is contiguous to headwater reaches of Panther Creek, and except in the vicinity of the stream gaging station northwest of El Paso, the boundaries of the basin are reasonably congruous with groundwater and topographic divides. There is no surface or subsurface flow into or out of the basin, except subsurface underflow from the basin in the vicinity of the stream gaging station. Water stored on the surface of the basin in ponds is very small, and withdrawals from wells is mostly for domestic and livestock use and is not significant. Thus, for this basin, several items of the general hydrologic budget can be eliminated because they do not measureably affect the balance between water gains and losses.

Precipitation, including rain and snow, is the source of water entering the basin and is the only water gain considered in the hydrologic budget. Water leaving the basin included streamflow, evapotranspiration, and subsurface underflow. Water is stored beneath the surface in soils and in the groundwater reservoir. Changes in storage of water in the soil are reflected in changes in soil moisture, and changes in water levels in wells indicate changes in storage of water in the groundwater reservoir.

Soil moisture, one of the hydrologic factors, was not measured during these investigations; therefore, daily, weekly, and monthly evapotranspiration cannot be appraised. However, soil moisture is near field capacity during January of most years, and annual change in soil moisture is very small. Annual values of evapotranspiration in 1951, 1952, and 1956 estimated by balancing Eq. (6.14), assuming annual change in soil moisture is negligible, are given in Table 6.8. Methods used to determine underflow and change in groundwater storage will be described later. The range in annual evapotranspiration is much less than the range in annual precipitation. Evapotranspiration was less in 1952, a year of near-normal precipitation, than in 1951, a year of above-normal precipitation, even though the average temperature during the growing season of 1951 was below normal and the average temperature during the growing season of 1952 was above normal. The ratio of evapotranspiration and precipitation was 56 percent in 1951, 73 percent in 1952, and 96 percent in 1956.

Estimates of daily groundwater evapotranspiration during 1951, 1952, and 1956 were computed from the mean groundwater stages and the groundwater stage runoff

TABLE 6.8 MONTHLY AND ANNUAL EVAPOTRANSPIRATION IN INCHES IN 1951, 1952, AND 1956, PANTHER CREEK BASIN, ILLINOIS. (*From Schicht and Walton, 1961.*)

	1951			*1952*			*1956*		
Month	ET_s	ET_g	ET	ET_s	ET_g	ET	ET_s	ET_g	ET
Jan.		Neg			Neg			Neg	
Feb.		Neg			Neg			Neg	
Mar.		Neg			Neg			Neg	
Apr.		0.08			0.13			0.06	
May		0.27			0.43			0.11	
June		0.18			0.18			0.12	
July		0.05			0.47			0.13	
Aug.		0.34			0.33			0.14	
Sept.		0.23			0.28			0.12	
Oct.		0.04			0.19			0.06	
Nov.		Neg			Neg			Neg	
Dec.		Neg			Neg			Neg	
Annual	23.52	1.19	24.71	21.93	2.01	23.94	18.01	0.74	18.75

rating curves in Fig. 6.9. Monthly and annual groundwater evapotranspiration is given Table 6.8. These data indicate that monthly groundwater evapotranspiration is greatest generally during July and August, and that annual groundwater evapotranspiration is least during dry years. The ratio of groundwater evapotranspiration to total evapotranspiration was 5, 8, and 4 percent in 1951, 1952, and 1956, respectively.

Subsurface underflow out of the basin occurs in the vicinity of the stream gaging station northwest of El Paso. Underflow can be estimated from Darcy's equation. The width of the lowlands adjacent to Panther Creek through which underflow occurs is about 500 feet. Based on the bedrock surface and topographic maps, the thickness of the glacial drift is estimated to be less than 25 feet, and the hydraulic gradient of the water table in the vicinity of the stream gaging station is estimated to be less than 50 ft/mi. The coefficient of transmissibility of the deposits through which underflow occurs is low and probably in the magnitude of 500 gpd/ft. By substituting these data in Darcy's equation, underflow was computed to be about 0.01 cfs and is so small that it was omitted from budget computations.

A reasonable estimate of evapotranspiration for periods during winter and early spring months is 0.3 inch per month. Soil moisture change can be eliminated and evapotranspiration estimated to average 0.3 inch per month without introducing serious error in the hydrologic budget. Computations of gravity yield were made using Eq. (6.16) and data for nine inventory periods during winter and early spring months, 1951 through 1958. Values of Y_g were plotted against the average time of drainage preceding the inventory periods. These data indicate that the average gravity yield of the glacial deposits increases at a diminishing rate from about 1 percent for a drainage period of 10 days to about 8 percent for a drainage period of 140 days. Extrapolation of the data suggests that the average specific yield of glacial deposits beneath the basin is about 12 percent. Monthly increases or decreases in groundwater storage during 1951, 1952, and 1956 were estimated by multiplying mean groundwater stage changes by appropriate values of Y_g. The data on changes in groundwater storage appear in Table 6.9.

With a few possible exceptions the water table rose or declined less than was necessary to balance groundwater runoff and evapotranspiration during portions of every month of 1951, 1952, and 1956. There was, therefore, some groundwater recharge in most months of these years. Monthly and annual groundwater recharge was estimated by balancing Eq. (6.17) and is given in Table 6.9.

Groundwater recharge during the three years ranged from 8.38 inches (400,000 gpd/sq mi) in 1951 to 0.87 inch (41,000 gpd/sq mi) in 1956, and was 8.03 inches (380,000 gpd/sq mi) in 1952. Groundwater recharge was 19 percent of precipitation during a year of above-normal precipitation, 4.5 percent of precipitation during a year of below-normal precipitation, and 25 percent of precipitation during a year of near-normal precipitation.

Data in Table 6.9 show the pronounced adverse effects of extended dry periods on groundwater recharge. Monthly groundwater recharge is largest in spring months of

TABLE 6.9 MONTHLY AND ANNUAL GROUNDWATER RECHARGE IN INCHES IN 1951, 1952, AND 1956, PANTHER CREEK BASIN, ILLINOIS. (*From Schicht and Walton, 1961.*)

	1951		*1952*		*1956*	
Month	P_G	ΔS_g	P_G	ΔS_g	P_G	ΔS_g
Jan.	0.44	+0.28	0.69	−0.08	Neg	−0.01
Feb.	0.20	+0.05	0.57	Neg	0.29	+0.21
Mar.	1.16	+0.86	1.71	+0.14	Neg	−0.04
Apr.	2.20	+0.68	1.92	−0.15	0.11	+0.02
May	0.89	−0.20	1.11	−0.14	0.20	+0.01
June	0.79	+0.05	1.36	+0.08	0.09	−0.10
July	1.03	−0.15	0.15	−0.59	0.06	−0.10
Aug.	0.41	−0.15	0.18	−0.19	0.06	−0.09
Sept.	0.12	−0.21	0.03	−0.27	0.02	−0.10
Oct.	0.03	−0.23	0.02	−0.18	0.02	−0.04
Nov.	0.88	+0.33	0.02	Neg	0.01	Neg
Dec.	0.23	−0.12	0.27	+0.24	0.01	Neg
Annual	8.38	+1.19	8.03	−1.14	0.87	−0.24

heavy rainfall and least in summer and fall months. Most ordinary summer rains have little or no effect on the water table. However, occasionally there was appreciable groundwater recharge when summer rainfall was in excess of evapotranspiration and soil moisture requirements. In February, March, and the early part of April 1951, precipitation was above normal; however, groundwater recharge was only moderate. Temperatures during part of November and December of 1950 and March and the early part of April of 1951 were below normal. As a result, there was a snow cover over frozen ground much of February and March which impeded the infiltration of precipitation to the water table. Thus, most precipitation in February and March was discharged from the basin by surface runoff.

Data in Figs. 6.10 and 6.11 indicate that groundwater recharge generally is at a maximum during April and most recharge occurs prior to July. In many years there

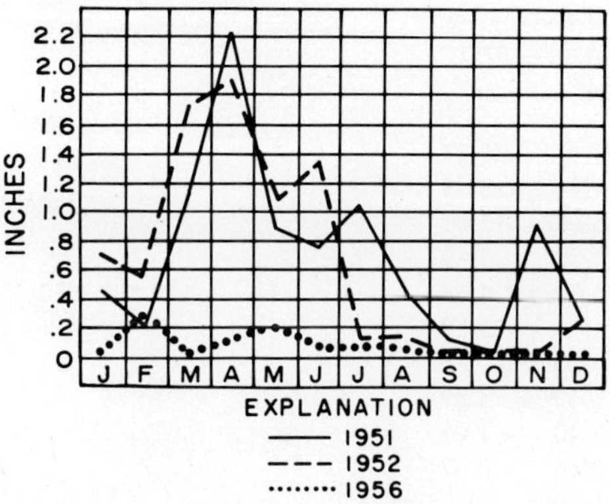

Fig. 6.10 Monthly groundwater recharge, Panther Creek basin, Illinois. (*From Walton, 1965.*)

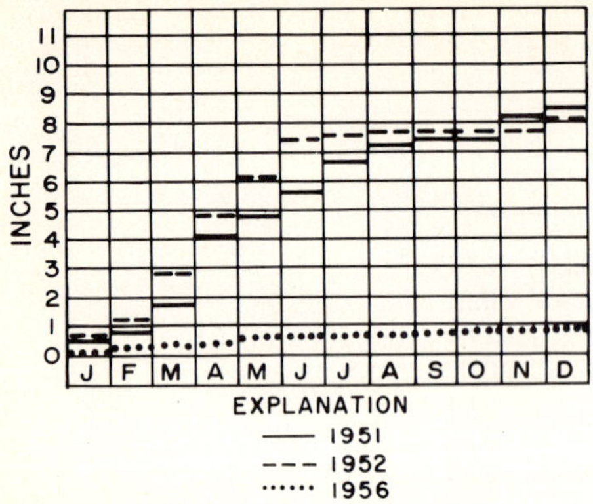

Fig. 6.11 Cumulative monthly groundwater recharge, Panther Creek basin, Illinois. (*From Walton, 1965.*)

is very little recharge during the five-month period July through November. Thus, water must be taken from storage within shallow aquifers for periods of at least five months to balance discharge.

Comparative results of annual hydrologic and groundwater budget factors for Panther, Hadley, and Goose Creek basins are given in Table 6.10. Data for years during which precipitation was near normal are presented. A comparison of the characteristics of the basins is given in Table 6.11. Groundwater recharge is much greater in Panther and Goose Creek basins than in Hadley Creek basin. The lower groundwater recharge in Hadley Creek basin is probably due to rugged upland topography and thin unconsolidated deposits.

Data in Figs. 6.12 and 6.13 indicate that groundwater runoff generally is at a maximum during spring and early summer months, and is least in late summer and fall months. More than half of annual groundwater runoff occurs during the first six

TABLE 6.10 COMPARISON OF BUDGET FACTORS FOR BASINS IN CENTRAL ILLINOIS. (*From Schicht and Walton, 1961.*)

Budget factor	*Panther Creek 1952, in.*	*Goose Creek 1957, in.*	*Hadley Creek 1957, in.*
Precipitation	32.62	37.18	39.73
Streamflow	9.82	9.48	13.93
Surface runoff	2.66	5.68	12.04
Groundwater runoff	7.16	3.80	1.89
Evapotranspiration	23.94	24.30	24.68
Surface and soil evapotranspiration	21.93	21.10	23.80
Groundwater evapotranspiration	2.01	3.20	0.88
Groundwater recharge	8.03	10.40	3.89
Change in groundwater storage	−1.14	+3.40	+1.05
Underflow	Neg	Neg	0.07

TABLE 6.11 COMPARISON OF CHARACTERISTICS OF BASINS IN CENTRAL ILLINOIS. (*From Schicht and Walton, 1961.*)

Characteristics	*Panther Creek*	*Goose Creek*	*Hadley Creek*
Topography	Gently undulating uplands	Level uplands	Rugged uplands
Average stream gradient	4.7 ft/mi	3.9 ft/mi	16.6 ft/mi
Vegetal cover	80% corn, oats, and soybeans; 20% pasture, woodland, and farm lots	86% corn, oats, soybeans, alfalfa, hay, wheat, rye; 14% pasture, woodland, and farm lots	40% row crops, small grain, and hay; 60% pasture, woodland, and farm lots
Soil	Upland prairie silt loams	Drummer silty clay loam and Flanagan silt loam	Upland prairie and timber silt loams
Unconsolidated deposits	100 feet of glacial till	175 feet of glacial till	25 feet of loess and 50 feet of glacial till
Bedrock formations	Shale of Pennsylvanian age	Shale of Pennsylvanian age	Shale of Mississippian and Pennsylvanian age
Average depth to water table	7 feet (below land surface)	8 feet (below land surface)	20 feet (below land surface)
North latitude	40°44′–40°54′	40°05′–40°13′	39°41′–39°50′
Mean annual temperature	51°F	53°F	55°F
Mean annual precipitation	33.6 in.	37.0 in.	36.0 in.

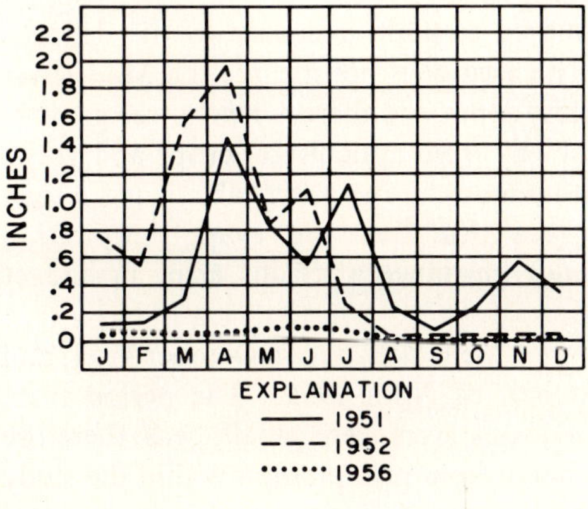

Fig. 6.12 Monthly groundwater runoff, Panther Creek basin, Illinois. (*From Walton, 1965.*)

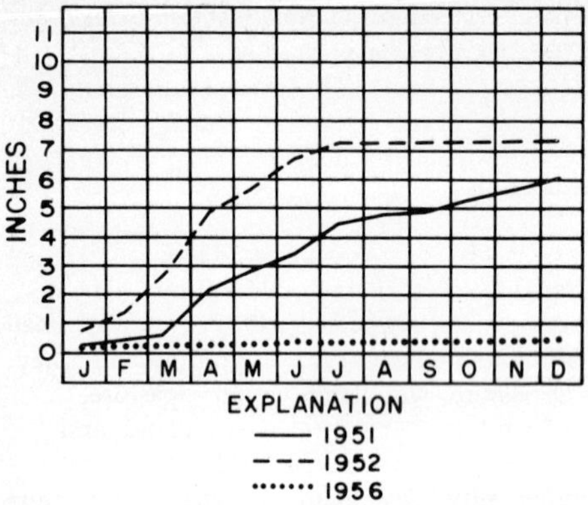

Fig. 6.13 Cumulative monthly groundwater runoff, Panther Creek basin, Illinois. (*From Walton, 1965.*)

months of the year. Annual groundwater runoff depends upon antecedent groundwater stage conditions as well as the amount and distribution of annual precipitation. Groundwater runoff is greatly reduced during extended dry periods.

Fairborn, Ohio

A groundwater budget study (Walton and Scudder, 1960) was made during the period October 1954 to May 1955 in the Fairborn area, Ohio, about 7 miles northeast of Dayton. The study area is underlain by thick permeable outwash sand and gravel deposits. Water entering the sand and gravel deposits includes recharge from precipitation on the deposits, recharge by induced infiltration of surface water, and subsurface flow from limestone of Silurian age and glacial deposits bordering the study area. Water leaving the sand and gravel deposits includes subsurface outflow (underflow) down the Mad River valley, water that is discharged from production wells, water that is discharged into surface streams (groundwater runoff), and water that is discharged into the atmosphere by evapotranspiration.

The inventory area (14 square miles) used to compute recharge to the deposits from precipitation is shown in Fig. 6.14. The area does not include the Mad River bottom lands, where groundwater is at shallow depth and there is heavy tree growth. The loss from the area by transpiration is chiefly from agricultural crops and grasslands and is smaller than the loss in the bottom lands. The water table is more than 10 feet below the land surface in much of the area; therefore, evapotranspiration is small enough so that its omission does not introduce a serious error in budget computations.

The lowest seasonal water levels in observation wells occurred in October 1954, and peak water levels occurred during the last half of April 1955. The period from October to May, when evapotranspiration losses were very small, was therefore selected as the time unit for determining recharge from precipitation within the study area.

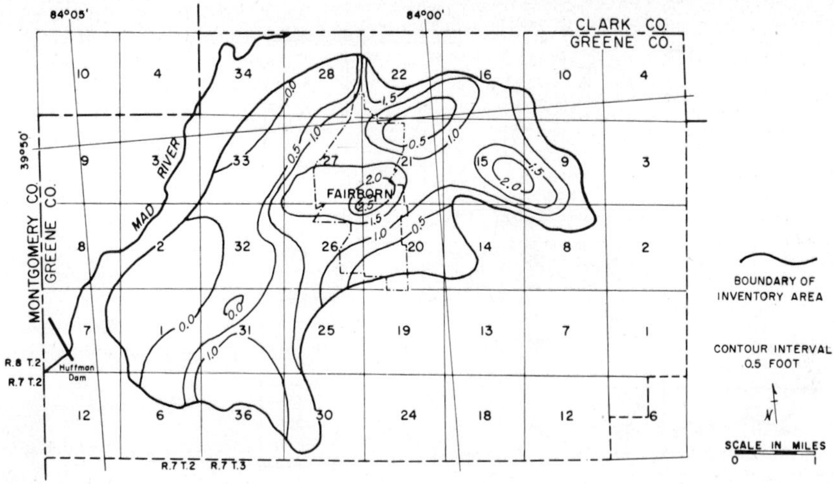

Fig. 6.14 Map of Fairborn area, Ohio, showing boundary of inventory area. (*From Walton and Scudder, 1960.*)

The total discharge of groundwater from the study area in the inventory period was about 10.4 mgd. The average discharge of ditches and small streams within the study area was approximately 4 mgd. The water pumped from production wells was discharged outside the basin to the Mad River through sewage systems. Discharge of groundwater into the small natural streams within the inventory area was negligible.

Groundwater entering the inventory area includes recharge from precipitation within the area, recharge by the induced infiltration from small streams, and subsurface flow from the valley walls. Hebble Creek is the only small watercourse traversing the area that has a significant sustained flow. Pumping from production wells along Hebble Creek at an Air Force base and from production wells at cement companies in the area has lowered the water table below the surface of the stream. Recharge by induced infiltration of water in Hebble Creek during the inventory period was computed to be about 1.9 mgd, based on streamflow records. Several small drainage ditches discharge water into a gravel pit about 1.5 miles east of the Huffman Dam. The water impounded in a gravel pit during periods of heavy rainfall enters the groundwater reservoir. Recharge from the gravel pit during the inventory period was estimated to be about 20,000 gpd. The subsurface flow of water into the inventory area from about 11 miles of valley wall was computed to be about 1.75 mgd, or about 160,000 gpd per mile of valley wall, by studying the movement of water through the valley-train deposits near the foot of the valley wall. Calculations were based on Darcy's law, the hydraulic gradients shown on the water-table map, permeabilities estimated from drillers' logs and aquifer tests, and data on the saturated thickness of aquifer.

During the inventory period recharge exceeded discharge and the water table rose (see Fig. 6.15). The saturated volume of the aquifer within the study area was

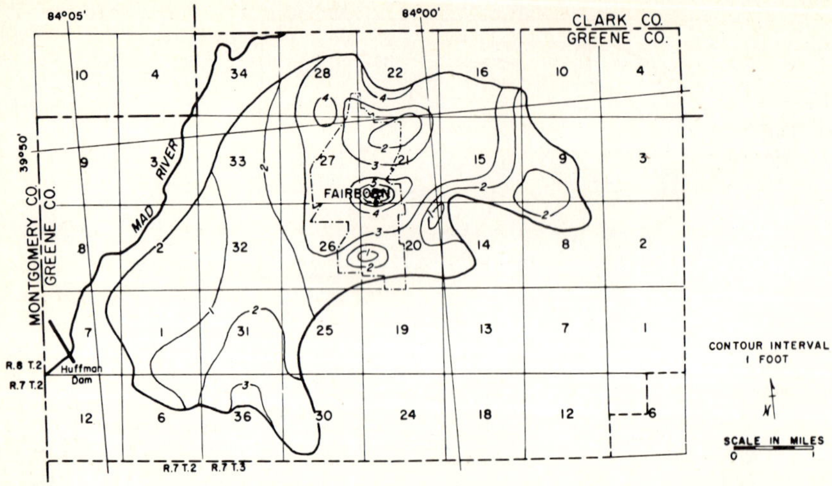

Fig. 6.15 Map of Fairborn area, Ohio, showing rise in water level, October 1954 to May 1955. (*From Walton and Scudder, 1960.*)

increased by approximately 20,000 acre feet, equivalent to 4,000 acre feet of water, on the basis of a specific yield of 20 percent.

For the inventory period, the following equation may be written to express the balance of recharge to, discharge from, and increase in storage in the aquifer

$$\Delta S = R_a + R_I + R_s - Q - R_g \tag{6.18}$$

where ΔS = volumetric increase of water stored in aquifer
R_a = recharge from precipitation within the area
R_I = recharge by the induced infiltration of surface water
R_s = subsurface flow from valley walls
Q = discharge from wells
R_g = discharge of groundwater into streams

Recharge from precipitation within the inventory area was calculated to be about 8,300 acre feet, or about 11 inches of precipitation, by substituting data discussed in preceding paragraphs in Eq. (6.18). Precipitation during the inventory period was 18.58 inches, 0.18 inch below normal. Recharge therefore amounted to about 60 percent of the total precipitation during the inventory period.

Records of water levels indicate that most of the annual recharge from precipitation occurs during the months covered by the inventory period. Very little rainfall reaches the water table in the late spring and summer months because during the growing season the evapotranspiration often exceeds precipitation. Annual recharge to the study area is therefore estimated to be about 12 inches of precipitation. A recharge of 12 inches is equivalent to about a third of the annual precipitation. This applies only to the sand and gravel deposits in the Miami River. From a study of the hydrology of the Mad River valley made by Houk (1921), annual recharge to the till deposits on the uplands is estimated to be about 8 inches of precipitation.

6.9 Case Histories of Studies to Determine Relation between Groundwater Runoff and Basin Characteristics

Streamflow data in Water Supply Papers published by the U.S. Geological Survey were used to determine annual groundwater runoff from 109 drainage basins in Illinois and 38 drainage basins in Minnesota and to determine the extent that geology and other basin characteristics affect streamflow (Walton, 1965, and Ackroyd et al., 1967). Streamflow data for years of near-normal, below-normal, and normal precipitation in each basin shown in Figs. 6.16 and 6.17 were investigated.

Daily mean streamflow at the stream gaging stations in Figs. 6.16 and 6.17 were plotted on semilogarithmic hydrograph paper. Hydrographs were divided into two components, surface runoff and groundwater runoff, with streamflow hydrograph separation methods outlined earlier in this chapter. Flow-duration curves for the stream gaging stations also were used in investigations. The parameter selected to describe the slope of the flow-duration curves was the ratio $(Q_{25}/Q_{75})^{\frac{1}{2}}$.

Annual groundwater runoff estimates during years of near-, below-, and above-normal precipitation and flow-duration curve ratios are listed in Tables 6.12 and 6.13. Distributions of groundwater runoff during years of near-normal precipitation are shown in Figs. 6.18 and 6.19.

The general characteristics of the drainage basins in Illinois and Minnesota were determined from published reports and maps. As shown in Fig. 6.20, the most common bedrocks in Illinois are those of the Paleozoic age. Exposures of bedrock are limited chiefly to areas in the extreme western and southern parts of the state. Only small scattered supplies of groundwater are available from Pennsylvanian, Mississippian, and Devonian rocks. These bedrock formations were classified as relatively impermeable and cover all but the northern third of Illinois. Tertiary, Cretaceous Silurian, Ordovician, and Cambrian bedrock formations are favorable aquifers and yield large quantities of water to wells. These rocks were classified as permeable and cover the northern third and extreme southern tip of the state.

Precambrian rocks directly underlie the glacial drift in much of the northern and central parts of Minnesota and in scattered areas in the southwest, as shown in Fig 6.21. Generally southwestward-dipping Paleozoic strata overlie the Precambrian rocks in the southeastern part, and westward-dipping Cretaceous rocks overlie Precambrian and Paleozoic rocks in a belt one to five counties wide, along the west edge of the state. Exposures of bedrock are found chiefly in northeastern and southeastern parts of the state, in the Minnesota River valley, and in the lower reaches of the Mississippi and St. Croix River valleys. The Paleozoic and Late Precambrian rocks include nearly all the highly productive aquifers in the bedrock. These rocks and the Cretaceous Dakota Sandstone of Mesozoic age were classified as permeable bedrock.

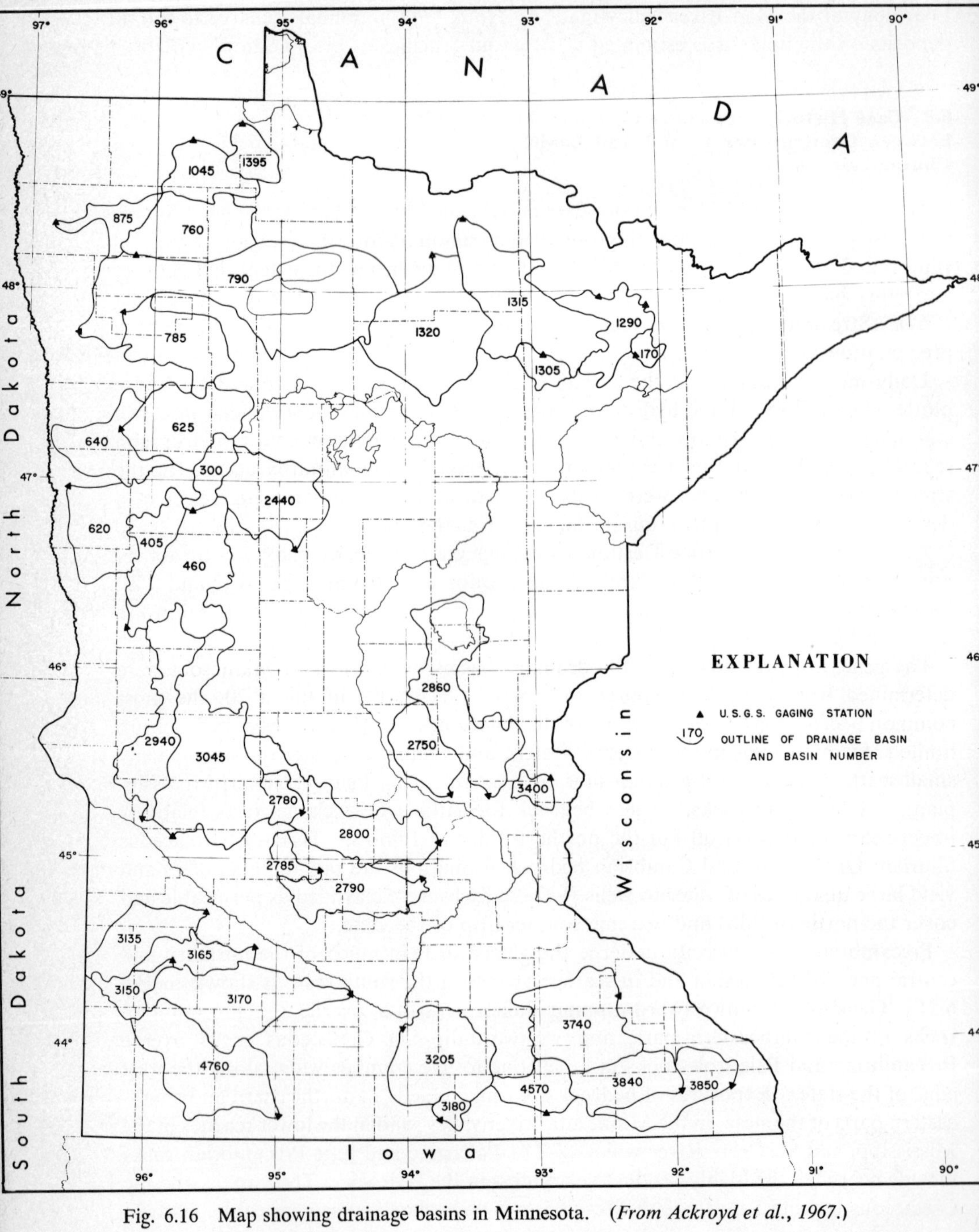

Fig. 6.16 Map showing drainage basins in Minnesota. (*From Ackroyd et al., 1967.*)

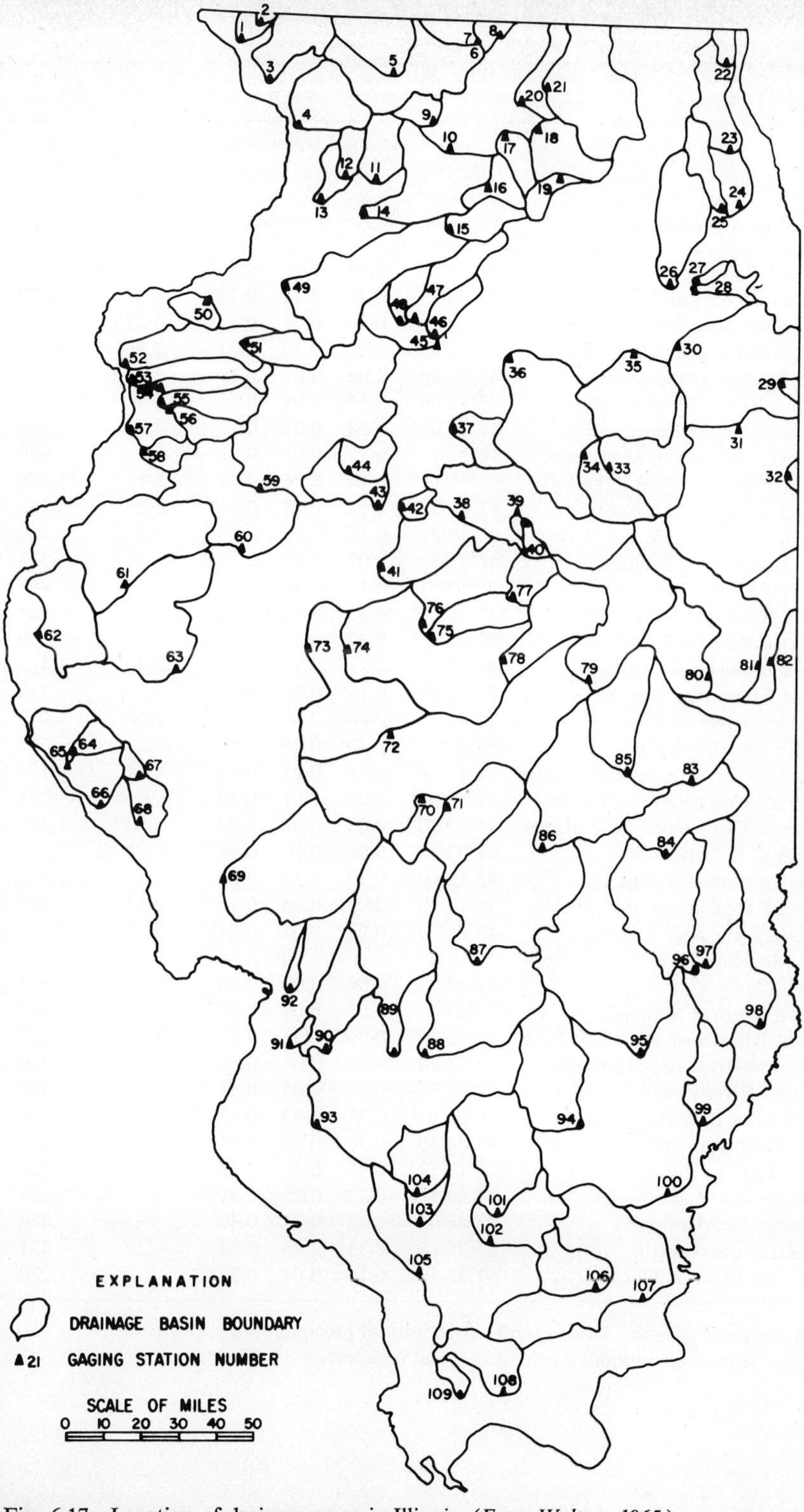

Fig. 6.17 Location of drainage areas in Illinois. (*From Walton, 1965.*)

TABLE 6.12 ANNUAL GROUNDWATER RUNOFF FROM 38 DRAINAGE BASINS IN MINNESOTA. (*From Ackroyd et al., 1967.*)

Basin no.	*Location of gaging station*	*Degrees north latitude*	*Annual groundwater runoff, cfs/sq mi*			*Ratio*[b] $(Q_{25}/Q_{75})^{\frac{1}{2}}$	*Basin area, sq mi*
			Near[a]	*Below*[a]	*Above*[a]		
170	Embarrass River at Embarrass	47°39′24″	0.42	0.55	0.55	...	94
300	Otter Tail River near Detroit Lakes	46°50′	0.21	0.12	0.23	2.37	270
405	Pelican River near Fergus Falls	46°20′10″	0.15	0.07	0.15	2.22	482
460	Otter Tail River near Fergus Falls	46°12′35″	0.20	0.10	0.21	2.36	1,830
620	Buffalo River near Dilworth	46°57′40″	0.06	0.02	0.10	2.58	1,040
625	Wild Rice River at Twin Valley	47°16′00″	0.11	0.05	0.26	2.46	888
640	Wild Rice River at Hendrum	47°16′05″	0.08	0.08	0.12	2.78	1,600
760	Thief River near Thief River Falls	48°11′	0.17	0.02	0.28	...	959
785	Clearwater River at Red Lake Falls	47°53′15″	0.14	0.26	0.27	2.10	1,370
790	Red Lake River at Crookston	47°46′32″	0.10	0.04	0.19	2.26	5,280
875	Middle River at Argyle	48°20′27″	0.16	0.03	0.24	2.08	265
1,045	Roseau River near Malung	48°37′30″	0.07	0.07	0.49	3.70	573
1,290	Vermillion River near Tower	47°57′31″	0.43	0.43	0.49	...	483
1,305	Sturgeon River near Chisholm	47°40′25″	0.47	0.44	0.52	2.16	187
1,315	Little Fork River at Little Fork	48°24′	0.37	0.45	0.60	...	1,730
1,320	Big Fork River at Big Falls	48°12′	0.27	0.34	0.55	2.40	1,460
1,395	Warroad River near Warroad	48°52′00″	0.15	0.06	0.43	...	110
2,440	Crow Wing River at Nimrod	46°39′	0.40	0.38	0.45	1.40	1,010
2,750	Elk River near Big Lake	45°20′	0.30	0.25	0.49	...	615
2,780	Middle Fork Crow River near Spicer	45°15′45″	0.18	0.05	0.52	4.74	179
2,785	South Fork Crow River at Cosmos	44°56′05″	0.06	0.01	0.32	...	221
2,790	South Fork Crow River near Mayer	44°54′20″	0.03	0.07	0.29	5.80	1,170
2,800	Crow River at Rockford	45°05′15″	0.05	0.10	0.34	2.68	2,520
2,860	Rum River near St. Francis	45°19′40″	0.35	0.38	0.56	...	1,360
2,940	Pomme de Terre River at Appleton	45°12′10″	0.05	0.03	0.17	2.27	905
3,045	Chippewa River near Milan	45°06′39″	0.08	0.02	0.21	3.13	1,870
3,135	Yellow Medicine River near Granite Falls	44°43′	0.04	0.02	0.30	3.33	653
3,150	Redwood River at Marshall	44°27′	0.03	0.09	0.21	3.16	307
3,165	Redwood River near Redwood Falls	44°31′25″	0.06	0.03	0.31	3.74	697
3,170	Cottonwood River near New Ulm	44°17′40″	0.09	0.04	0.37	...	1,280
3,180	Blue Earth River near Bricelyn	43°37′50″	0.16	0.04	0.37	...	132
3,205	LeSueur River near Rapidan	44°06′40″	0.31	0.09	0.59	3.57	1,100
3,400	Sunrise River near Stacey	45°24′30″	0.21	0.26	0.51	...	167
3,740	Zumbro River at Zumbro Falls	44°17′12″	0.46	0.31	0.67	...	1,130
3,840	Root River near Lanesboro	43°44′58″	0.22	0.28	0.51	...	615
3,850	Root River near Houston	43°46′05″	0.42	0.35	0.49	...	1,270
4,570	Cedar River near Austin	43°38′10″	0.33	0.26	0.82	...	425
4,760	Des Moines River at Jackson	43°37′10″	0.14	0.04	0.34	...	1,220

[a] Indicates data for years of near-, below-, and above-normal precipitation.
[b] Ratio $(Q_{25}/Q_{75})^{\frac{1}{2}}$ is used as an index to the flow-duration curve.

TABLE 6.13 ANNUAL GROUNDWATER RUNOFF FROM 109 DRAINAGE BASINS IN ILLINOIS. (*From Walton, 1965.*)

Basin no.	*Gaging station location*	*Annual groundwater runoff, cfs/sq mi*			*Ratio* $\left(\frac{Q_{25}}{Q_{75}}\right)^{\frac{1}{2}}$	*Basin*	
		Near[a]	*Below*[a]	*Above*[a]		*Area, sq mi*	*North latitude, degrees*
1	Galena River at Galena	0.36	0.27	0.51	1.52	192	42°24′50″
2	East Fork, Galena River at Council Hill	0.30	0.18	0.40	1.67	20.1	42°28′05″
3	Apple River near Hanover	0.30	0.19	0.50	1.85	244	42°15′05″
4	Plum River below Carroll Creek	0.26	0.15	0.49	2.04	231	42°06′50″
5	Pecatonica River at Freeport	0.41	0.28	0.58	1.53	1,330	42°18′13″
6	Pecatonica River at Shirland	0.38	0.29	0.59	1.33	2,540	42°26′10″
7	Sugar River at Shirland	0.40	0.32	0.53	1.37	757	42°26′10″
8	Rock River at Rockton	0.39	0.28	0.59	1.70	6,290	42°27′05″
9	Leaf River at Leaf	0.31	0.20	0.47	1.70	102	42°07′35″
10	Rock River at Oregon	0.36	0.27	0.56	1.64	8,120	42°01′00″
11	Elkhorn Creek near Penrose	0.29	0.20	0.41	1.61	153	41°54′10″
12	Rock Creek near Coleta	0.27	0.17	0.39	1.77	81.6	41°55′00″
13	Rock Creek near Morrison	0.25	0.18	0.36	1.61	143	41°49′50″
14	Rock River at Como	0.39	0.28	0.58	1.64	8,700	41°47′00″
15	Green River at Amboy	0.20	0.10	0.35	2·24	199	41°42′35″
16	Kyte River near Flag Center	0.13	0.08	0.31	2.48	125	41°56′00″
17	Killbuck Creek near Monroe Center	0.16	0.08	0.30	2.31	114	42°05′55″
18	South Branch, Kishwaukee River near Fairdale	0.19	0.07	0.40	2.98	386	42°06′40″
19	South Branch, Kishwaukee River at DeKalb	0.17	0.03	0.50	6.29	70	41°55′50″
20	Kishwaukee River at Perryville	0.24	0.13	0.50	2.24	1,090	42°11′45″
21	Kiswaukee River at Belvidere	0.24	0.16	0.43	1.97	525	42°15′20″
22	Des Plaines River near Gurnee	0.04	0.01	0.38	8.84	215	42°20′40″
23	Des Plaines River near Des Plaines	0.14	0.05	0.60	4.90	374	42°04′55″
24	Des Plaines River at Riverside	0.12	0.05	0.48	3.95	635	41°49′20″
25	Salt Creek at Western Springs	0.22	0.09	0.62	3.16	122	41°49′35″
26	DuPage River at Troy	0.32	0.19	0.61	2.30	325	41°31′20″
27	Spring Creek at Joliet	0.29	0.18	0.48	1.91	19.7	41°31′45″
28	Hickory Creek at Joliet	0.18	0.11	0.38	2.30	107	41°31′10″
29	Kankakee River at Momence	0.55	0.40	0.72	1.60	2,340	41°09′36″
30	Kankakee River near Wilmington	0.37	0.24	0.58	1.91	5,250	41°20′48″
31	Iroquois River near Chebanse	0.24	0.10	0.52	3.65	2,120	41°00′29″
32	Iroquois River at Iroquois	0.22	0.13	0.50	2.91	682	40°49′25″
33	North Fork, Vermilion River	0.09	0.02	0.28	7.55	184	40°50′08″
34	Vermilion River at Pontiac	0.10	0.03	0.27	5.40	568	40°52′40″
35	Mazon River near Coal City	0.07	0.01	0.29	30.6	470	41°17′10″
36	Vermilion River at Lowell	0.15	0.05	0.37	4.46	1,230	41°15′18″
37	Crow Creek near Washburn	0.20	0.06	0.40	7.75	123	40°57′15″
38	Mackinaw River near Congerville	0.24	0.05	0.46	5.79	764	40°37′25″
39	Hickory Creek above Lake Bloomington	0.30	0.01	0.59		10.1	40°38′15″
40	Money Creek above Lake Bloomington	0.28	0.03	0.56	9.50	51.9	40°37′13″
41	Mackinaw River near Green Valley	0.26	0.10	0.45	3.26	1,100	40°26′43″
42	Farm Creek at East Peoria	0.11	0.06	0.26	3.10	60.9	40°40′05″
43	Kickapoo Creek near Peoria	0.16	0.07	0.36	3.10	296	40°40′55″
44	Kickapoo Creek near Kickapoo	0.18	0.07	0.36	3.70	120	40°48′00″
45	Bureau Creek at Bureau	0.25	0.17	0.45	2.16	481	41°16′40″
46	East Bureau Creek near Bureau	0.12	0.04	0.32	5.20	101	41°20′06″
47	Bureau Creek at Princeton	0.20	0.08	0.48	4.19	186	41°21′55″
48	West Bureau Creek at Wyanet	0.16	0.07	0.39	4.70	83.3	41°21′54″
49	Green River near Geneseo	0.28	0.16	0.43	2.03	958	41°29′20″
50	Mill Creek at Milan	0.16	0.07	0.34	3.22	62.5	41°26′35″
51	Edwards River near Orion	0.19	0.09	0.38	2.83	163	41°16′20″
52	Edwards River near New Boston	0.23	0.14	0.42	2.57	434	41°11′15″
53	Pope Creek near Keithsburg	0.20	0.10	0.37	2.73	171	41°07′45″

TABLE 6.13 (*Continued*)

Basin no.	Gaging station location	Annual groundwater runoff, cfs/sq mi			Ratio $\left(\frac{Q_{25}}{Q_{75}}\right)^{\frac{1}{2}}$	Basin	
		Near[a]	Below[a]	Above[a]		Area, sq mi	North latitude, degrees
54	North Henderson Creek near Seaton	0.18	0.08	0.40	3.21	66.4	41°05′25″
55	Henderson Creek near Little York	0.17	0.08	0.34	3.16	151	41°02′35″
56	Cedar Creek at Little York	0.25	0.13	0.46	2.82	128	41°00′50″
57	Henderson Creek near Oquawka	0.19	0.12	0.38	2.40	428	41°00′05″
58	South Henderson Creek at Biggsville	0.32	0.13	0.62	3.65	81.4	40°51′25″
59	Spoon River at London Mills	0.22	0.08	0.40	3.29	1,070	40°42′51″
60	Spoon River at Seville	0.28	0.14	0.45	3.16	1,600	40°29′10″
61	LaMoine River at Colmar	0.29	0.14	0.46	3.05	655	40°19′45″
62	Bear Creek near Marcelline	0.10	0.05	0.15	3.60	348	40°08′34″
63	LaMoine River at Ripley	0.22	0.09	0.38	3.46	1,310	40°01′31″
64	Hadley Creek at Kinderhook	0.22	0.08	0.32	3.00	72.7	39°41′35″
65	Hadley Creek near Shinn	0.24	0.13	0.35	2.73	73.6	39°39′55″
66	The Sny River at Atlas	0.32	0.17	0.52	2.58	451	39°30′20″
67	Bay Creek at Pittsfield	0.15	0.05	0.22	4.12	39.6	39°37′30″
68	Bay Creek at Nebo	0.26	0.12	0.38	3.84	162	39°26′35″
69	Macoupin Creek near Kane	0.14	0.07	0.19	3.46	875	39°14′00″
70	South Fork, Sangamon River at Kincaid	0.30	0.12	0.58	4.25	510	39°34′40″
71	South Fork, Sangamon River at Taylorville	0.36	0.17	0.60	3.00	427	39°30′25″
72	Sangamon River at Riverton	0.30	0.13	0.44	3.94	2,560	39°50′34″
73	Sangamon River near Oakford	0.34	0.15	0.55	3.16	5,120	40°07′25″
74	Salt Creek near Greenview	0.38	0.17	0.59	2.83	1,800	40°08′01″
75	Kickapoo Creek near Lincoln	0.28	0.10	0.46	3.86	306	40°11′30″
76	Sugar Creek near Hartsburg	0.32	0.08	0.50	3.75	335	40°13′10″
77	Kickapoo Creek near Heyworth	0.19	0.08	0.35	4.32	71.4	40°21′00″
78	Salt Creek near Rowell	0.36	0.13	0.58	3.34	334	40°07′00″
79	Sangamon River at Monticello	0.33	0.14	0.58	3.84	550	40°01′40″
80	Salt Fork, Vermilion River near Homer	0.39	0.14	0.58	3.06	344	40°03′20″
81	Vermilion River near Catlin	0.32	0.14	0.56	3.58	959	40°06′09″
82	Vermilion River near Danville	0.35	0.14	0.60	4.08	1,280	40°05′53″
83	Embarrass River near Oakland	0.28	0.10	0.48	4.32	535	39°40′50″
84	Embarrass River near Diona	0.34	0.15	0.60	4.25	903	39°20′40″
85	Kaskaskia River near Arcola	0.36	0.12	0.52	4.95	390	39°40′50″
86	Kaskaskia River at Shelbyville	0.33	0.09	0.58	5.50	1,030	39°24′20″
87	Kaskaskia River at Vandalia	0.31	0.18	0.48	3.46	1,980	38°57′35″
88	Kaskaskia River at Carlyle	0.32	0.18	0.55	4.25	2,680	38°36′42″
89	Shoal Creek near Breese	0.17	0.10	0.28	3.16	760	38°36′35″
90	Silver Creek near Lebanon	0.13	0.08	0.18	3.56	335	38°35′40″
91	Canteen Creek at Caseyville	0.28	0.18	0.40	2.92	22.5	38°38′35″
92	Indian Creek at Wanda	0.09	0.04	0.15	5.30	37.0	38°50′30″
93	Kaskaskia River at New Athens	0.28	0.17	0.44	3.38	5,220	38°19′22″
94	Skillet Fork at Wayne City	0.08	0.04	0.18	5.65	475	38°21′25″
95	Little Wabash River at Wilcox	0.17	0.08	0.23	4.07	1,130	38°38′05″
96	Embarrass River at Ste Marie	0.30	0.18	0.42	3.26	1,540	38°56′10″
97	North Fork, Embarrass River near Oblong	0.16	0.09	0.22	3.60	304	39°00′35″
98	Embarrass River at Lawrenceville	0.40	0.28	0.62	3.33	2,260	38°43′25″
99	Bonpas Creek at Browns	0.06	0.02	0.12	18.6	235	38°22′50″
100	Little Wabash River at Carmi	0.19	0.10	0.39	5.76	3,090	38°03′40″
101	Big Muddy River near Benton	0.10	0.05	0.20	5.89	498	37°59′40″
102	Big Muddy River at Plumfield	0.15	0.06	0.28	7.06	753	37°54′05″
103	Beacuoup Creek near Matthews	0.14	0.08	0.18	3.87	291	37°58′00″
104	Beaucoup Creek near Pinckneyville	0.07	0.03	0.18	6.11	227	38°03′40″
105	Big Muddy River at Murphysboro	0.25	0.12	0.50	7.06	2,170	37°44′55″
106	Middle Fork, Saline River near Harrisburg	0.08	0.05	0.14	5.82	198	37°44′25″
107	Saline River near Junction	0.17	0.10	0.30	7.75	1,040	37°41′52″
108	Cache River at Forman	0.35	0.18	0.58	7.41	242	37°21′00″
109	Big Creek near Wetang	0.22	0.17	0.35	3.32	32.2	37°19′00″

[a] Words indicate that data are for years of near-, below-, and above-normal precipitation.

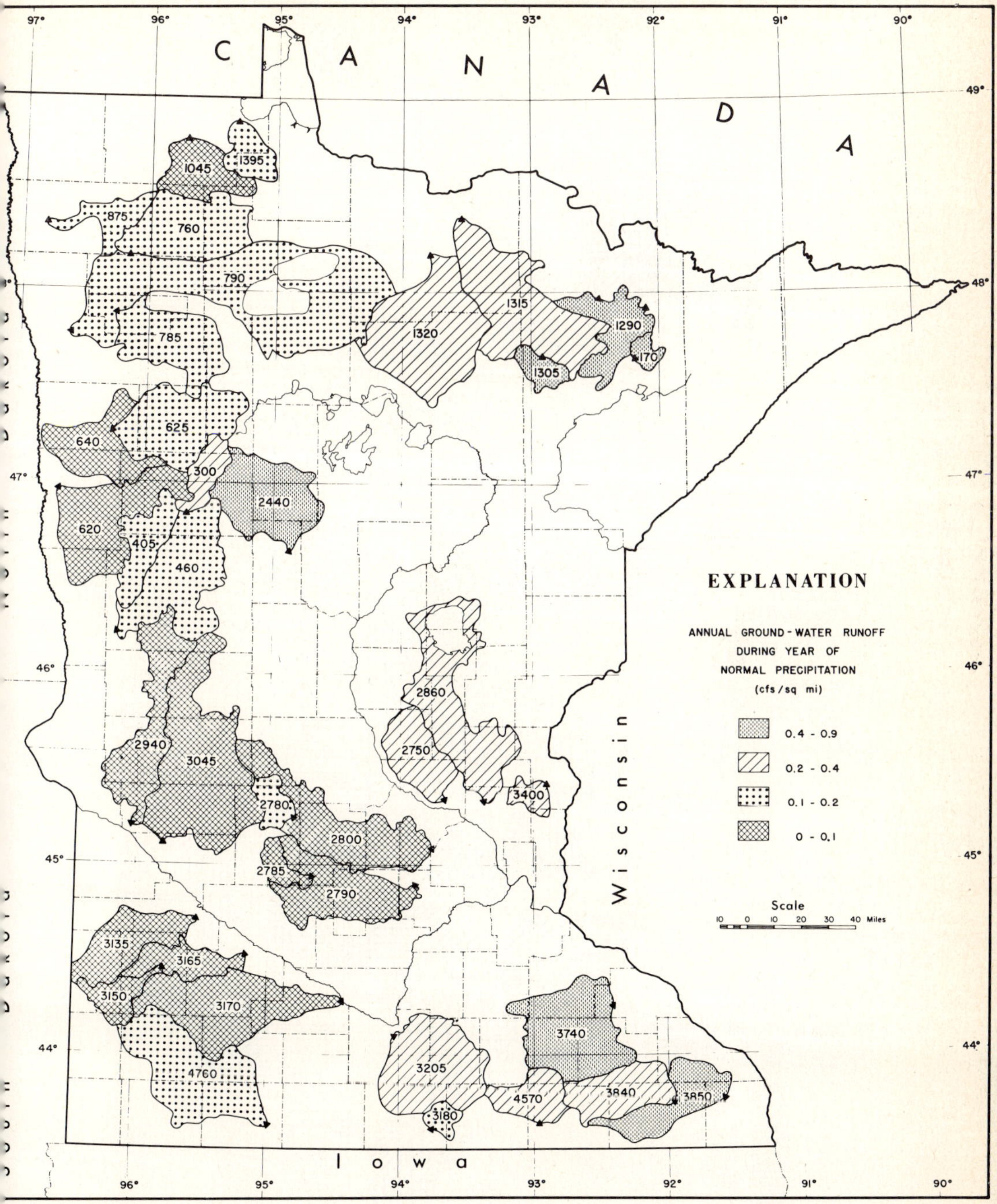

Fig. 6.18 Map showing distribution of annual groundwater runoff in Minnesota during a year of normal precipitation. (*From Ackroyd et al., 1967.*)

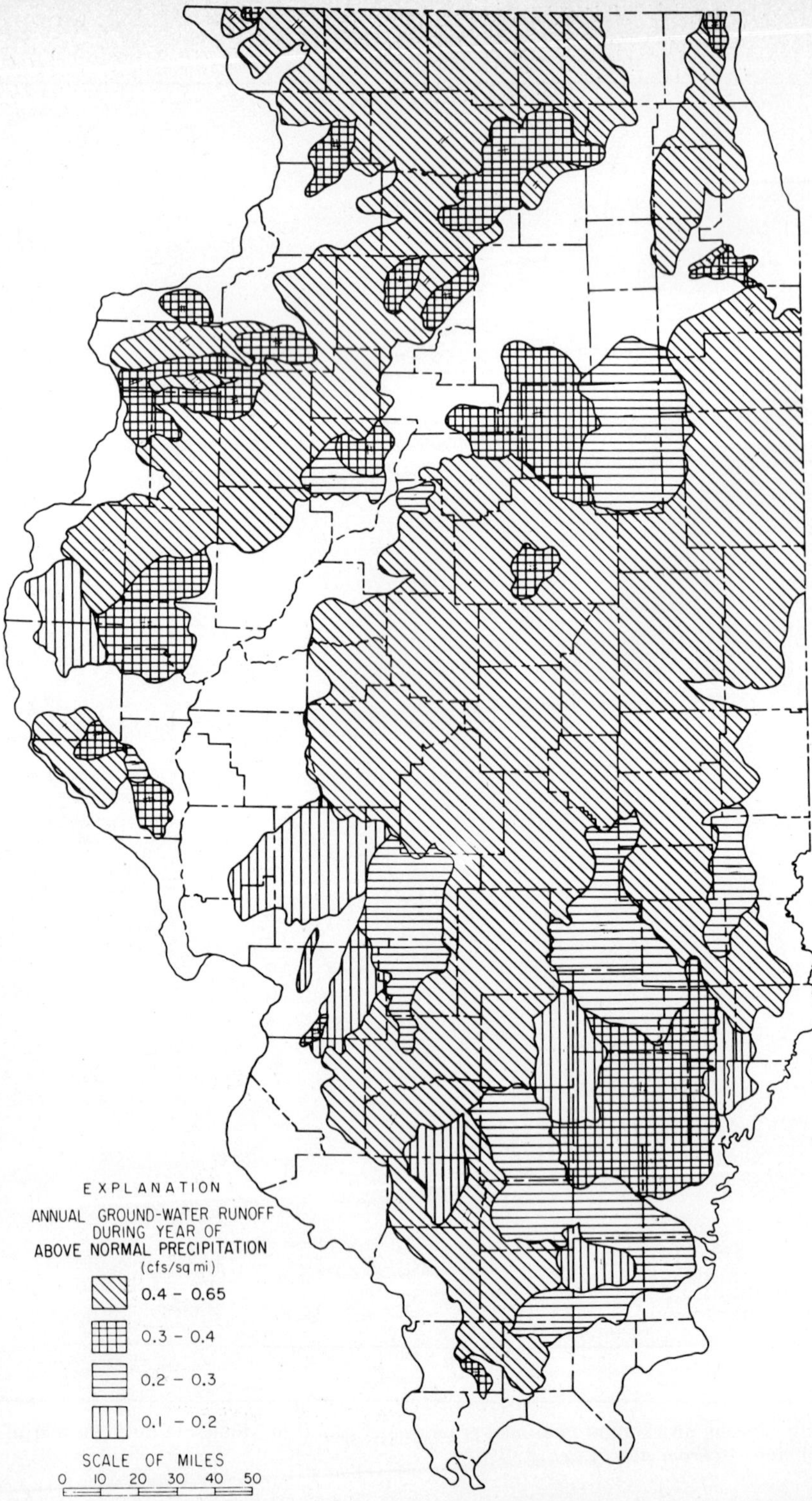

Fig. 6.19 Distribution of annual groundwater runoff in Illinois during a year of normal precipitation. (*From Walton, 1965.*)

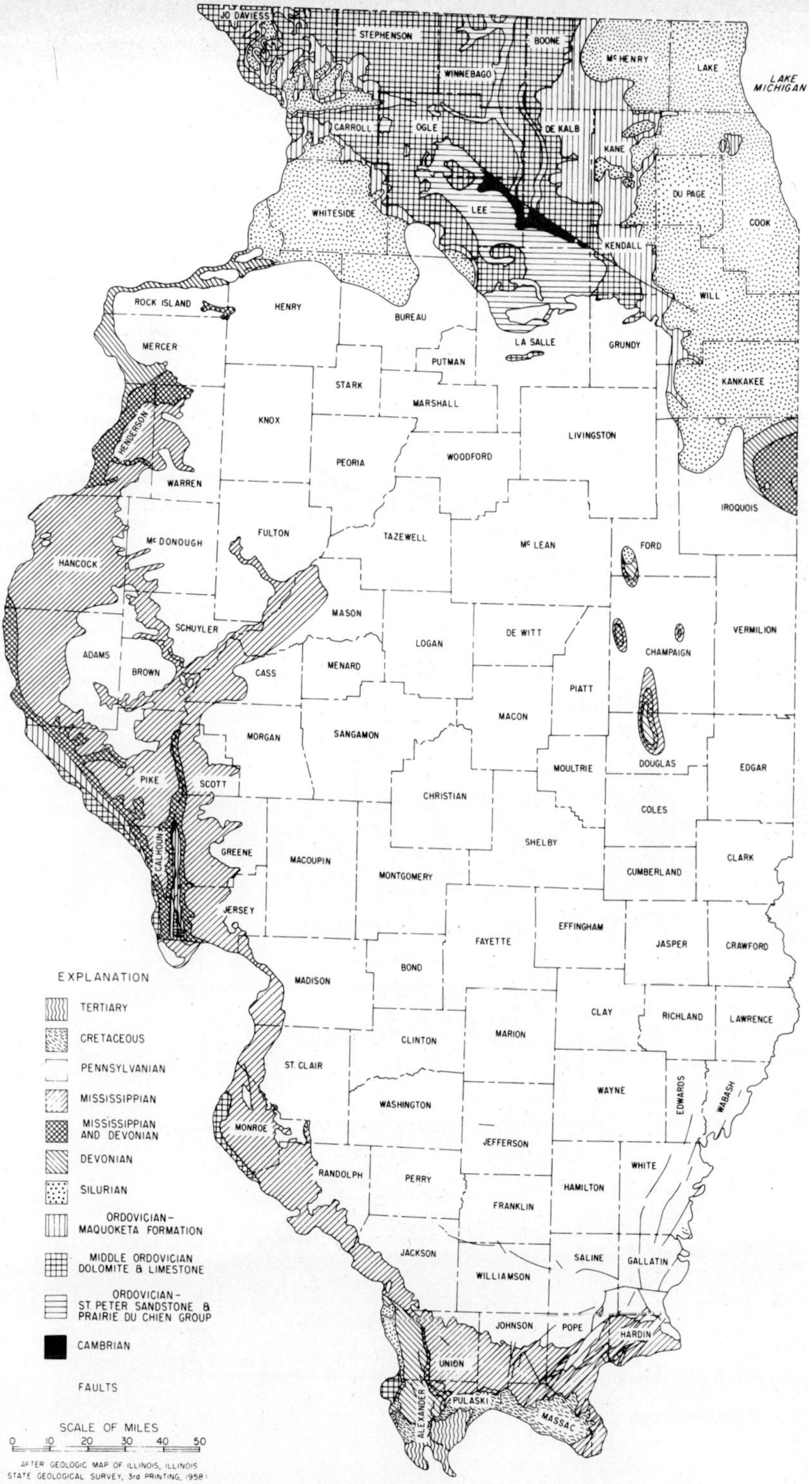

Fig. 6.20 Generalized bedrock geology of Illinois. (*From Walton, 1965.*)

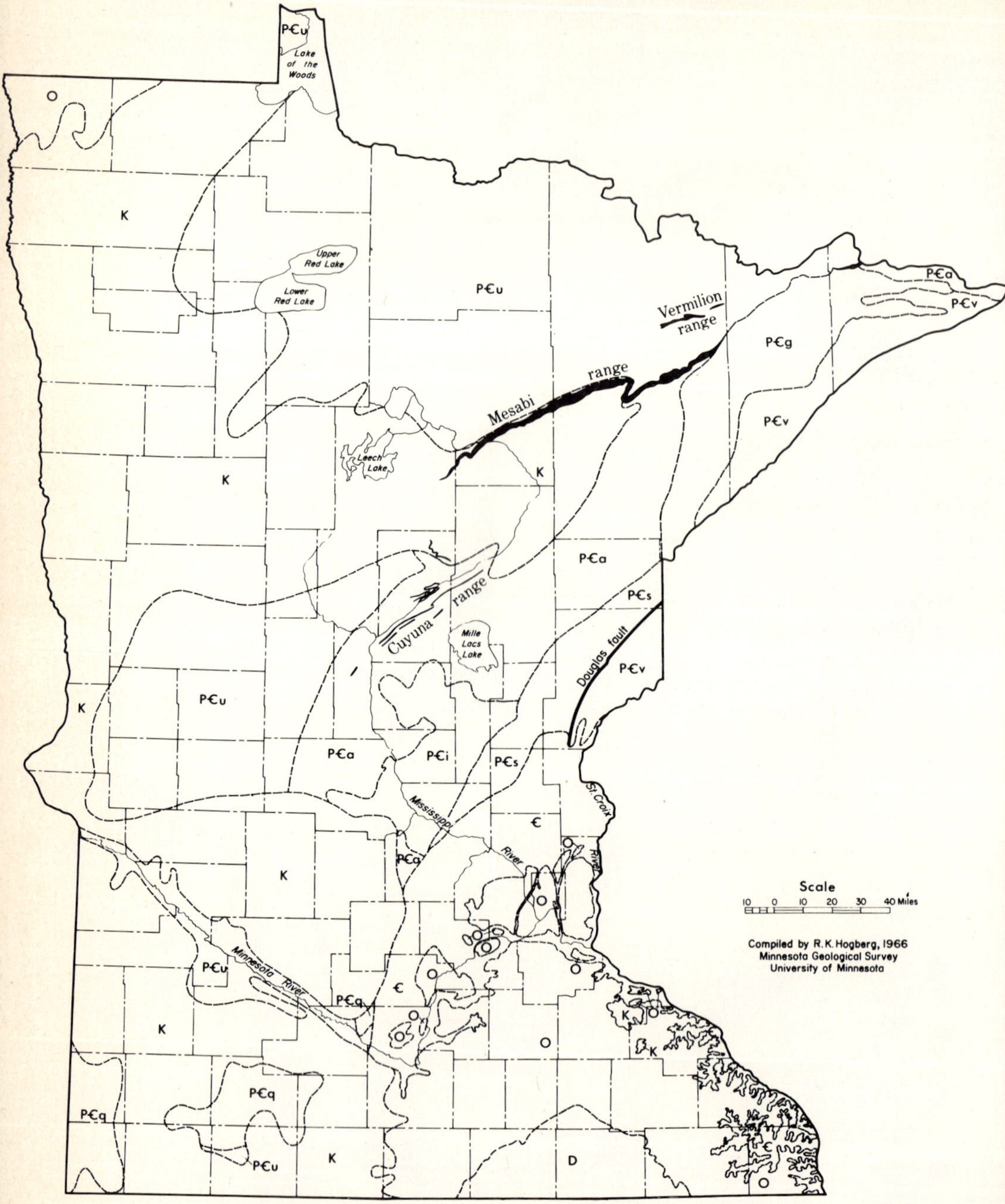

Fig. 6.21 Generalized bedrock geology of Minnesota. (*From Ackroyd et al., 1967.*)

EXPLANATION

MESOZOIC

Cretaceous

K

Sandstone and shale

(Coleraine Formation northern Minnesota; Windrow Formation southern Minnesota)

PALEOZOIC

Devonian

D

Cedar Valley Formation

(Dolomite and limestone)

Ordovician

O

Dolomite, sandstone and shale

(Prairie du Chien Group, Platteville Formation, Decorah Shale, Galena Formation, Dubuque Formation, Maquoketa Formation-oldest to youngest)

Cambrian

€

Sandstone and minor shale

(Dresbach Formation, Franconia Formation, St. Lawrence Formation and Jordan Sandstone-oldest to youngest)

PRECAMBRIAN

Keweenawan

P€s

Sandstone and shale

(Fond du Lac Formation and Hinckley Sandstone)

P€g

Duluth Gabbro Complex

(Multiple intrusions of gabbro, anorthosite and granophyre; age is 1. 1 billion years)

P€v

North Shore Volcanic Group

(Dominantly basaltic lava flows)

PRECAMBRIAN

age uncertain

P€q

Sioux Quartzite

(Dominantly quartzite with thin beds of argillite or pipestone)

P€i

Penokean intrusive rocks

(Quartz diorite, quartz monzonite, granodiorite, and granite; age is 1. 7 billion years)

Animikie

P€a

Argillite and graywacke

(Virginia Argillite, Thomson Slate and Rove Argillite)

Iron-formation

(Biwabik Iron-formation of Mesabi range and Trommald Formation of Cuyuna district)

Lower Precambrian

P€u

Undifferentiated metamorphic and intrusive rocks

(Includes Ely Greenstone, Knife Lake Group, Giants Range Granite and Vermilion Granite of northern Minnesota; Soudan Iron-formation is interbedded in the Ely Greenstone; and gneisses and intrusive igneous rocks of the Minnesota River Valley)

Fig. 6.22 Generalized bedrock topography of Illinois. (*After Horberg. 1957*)

The undifferentiated iron formations, metamorphic and igneous rocks, and Cretaceous shales yield only small scattered supplies of groundwater and were classified as relatively impermeable bedrock.

The bedrock in Illinois is scored by numerous stream valleys and their tributaries as shown in Figs. 6.22 and 6.23. The bedrock surface in Minnesota has not been mapped in detail. Only the general features of bedrock topography are known.

Since most streams in Illinois and Minnesota are cut into materials laid down during the glacial epoch, the character of glacial deposits is an important characteristic of basins. Most parts of Illinois and Minnesota are mantled by unconsolidated deposits left by the glaciers. Only small areas in the extreme northwest corner, southwest border, and southern tip of Illinois lie entirely outside the glacial boundary as shown in Fig. 6.24. Large areas in western, south-central, and southern Illinois are covered by glacial drift of Illinoian age. The drift cover is relatively thin and seldom exceeds 75 feet in thickness. In the area of the Wisconsinan glacial drift in the east-central and northern parts of Illinois, drift is thicker. The glacial drift is several hundred feet thick in deep buried bedrock valleys and contains permeable sand and gravel deposits. Ground moraine is the predominating surface deposit; glacial till is the principal constituent of the deposit. Morainic ridges are prominent features of surface deposits in northeastern Illinois. Interbedded sand and gravel are more common in morainic ridges than in ground moraine. Outwash plain deposits of sand and gravel are generally associated with moraines and lie in front of them. Closely associated with the outwash plains are deposits of water-sorted material in major stream valleys. The most important and extensive areas of lakebed sediments are in northeastern and southern Illinois. Deposits of loess cover much of the western half of Illinois to a depth of 6 feet or more; much of the eastern half of the state is covered with loess varying from 2 to 4 feet in thickness. Possibilities for occurrence of sand and gravel within the glacial drift are shown in Fig. 6.25. Best possibilities are confined mostly to major bedrock valley areas. Poor possibilities cover large areas in southern and western Illinois.

Only small areas in the extreme southeast corner of Minnesota lie entirely outside of the glacial boundary. Wisconsinan deposits cover all but the southwest and southeast corners of the state as shown in Fig. 6.26. Pre-Wisconsinan deposits are limited to small areas in the southwest and southeast corners of the state where widespread deposits of loess occur. The bedrock surface is buried in a large part of the state to a depth of more than 200 feet as shown in Fig. 6.27. The drift thickness is great in the southwest and scanty in the southeast. Ground moraine covers much of the state. Moraines are prominent features of surface deposits in central and south-central Minnesota. The Patrician red drift in northeastern and east-central parts of the state is sandy and contains little clay and limestone pebbles (see Fig. 6.28). This is in contrast to the gray drift covering much of the state, which is generally silty or clayey, rather than sandy, and highly calcareous because of the abundant fragments of limestone. The major areas of lakebed sediments are shown in Fig. 6.29. The

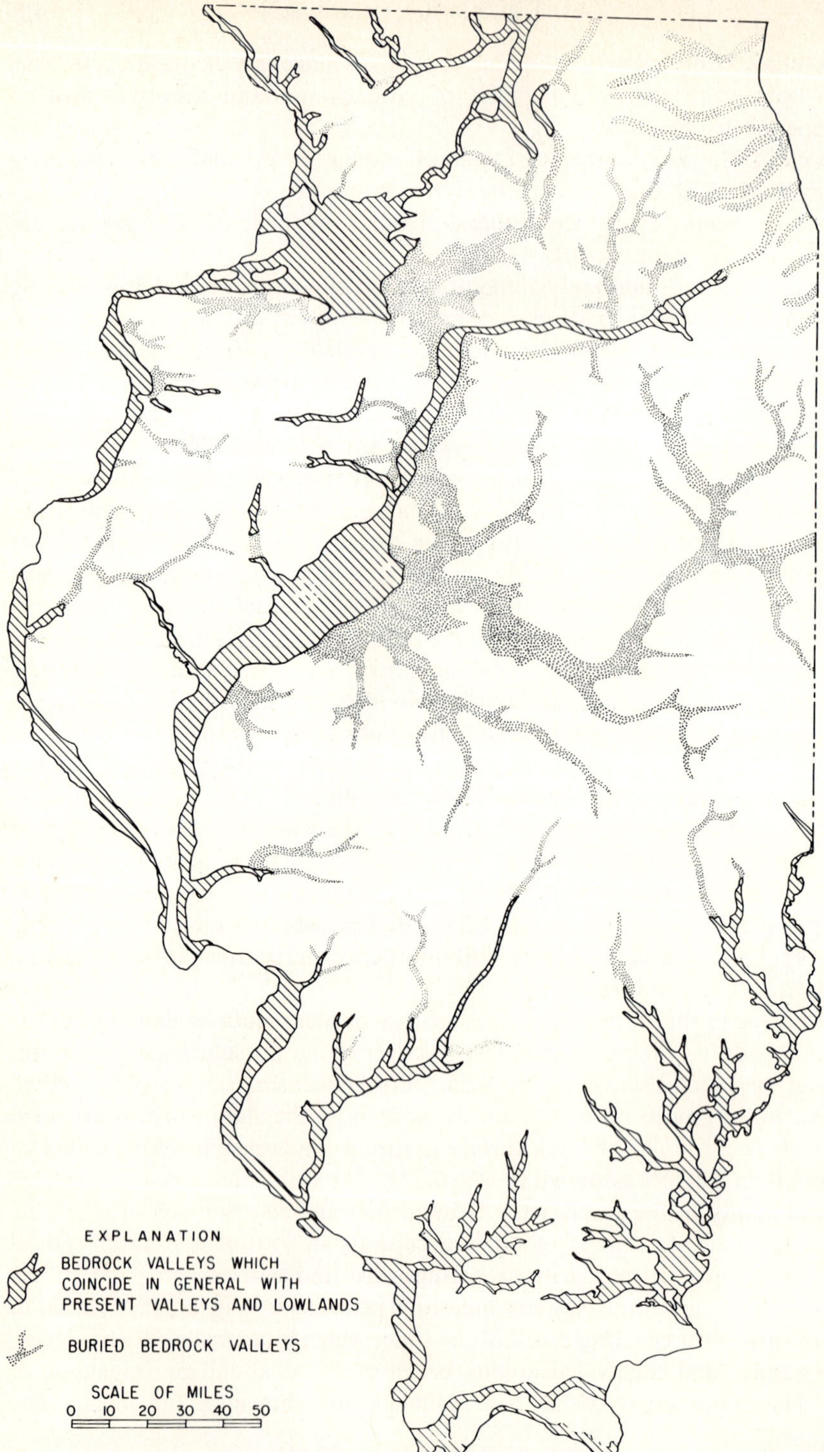

Fig. 6.23 Major bedrock valleys in Illinois. (*After Horberg, 1957.*)

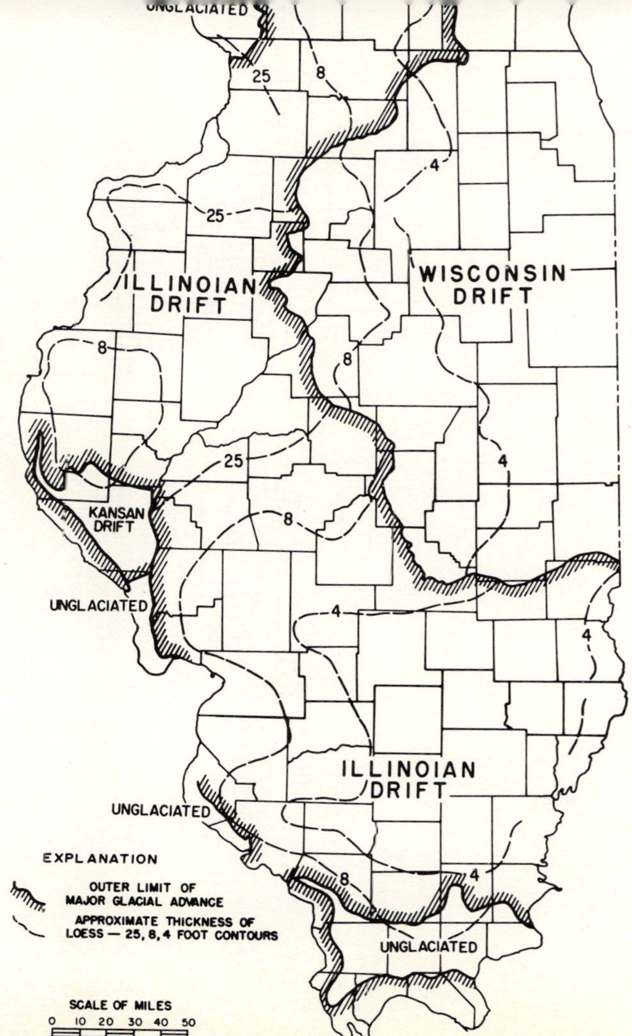

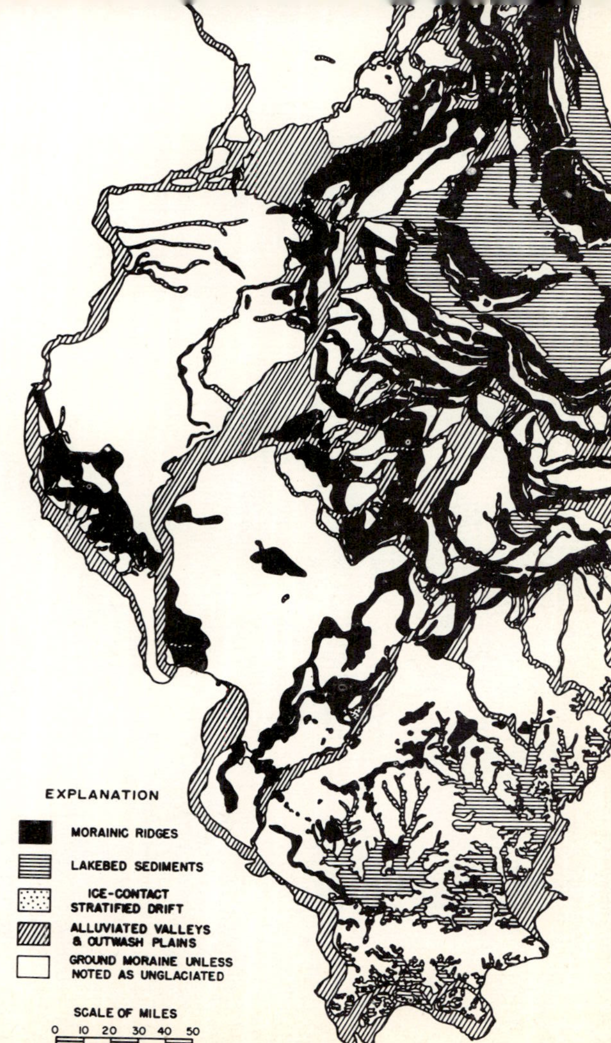

Fig. 6.24 Outer limits of major advances and thickness of loess (*a*) and surficial deposits (*b*) in Illinois. (*After Thornburn, 1960.*)

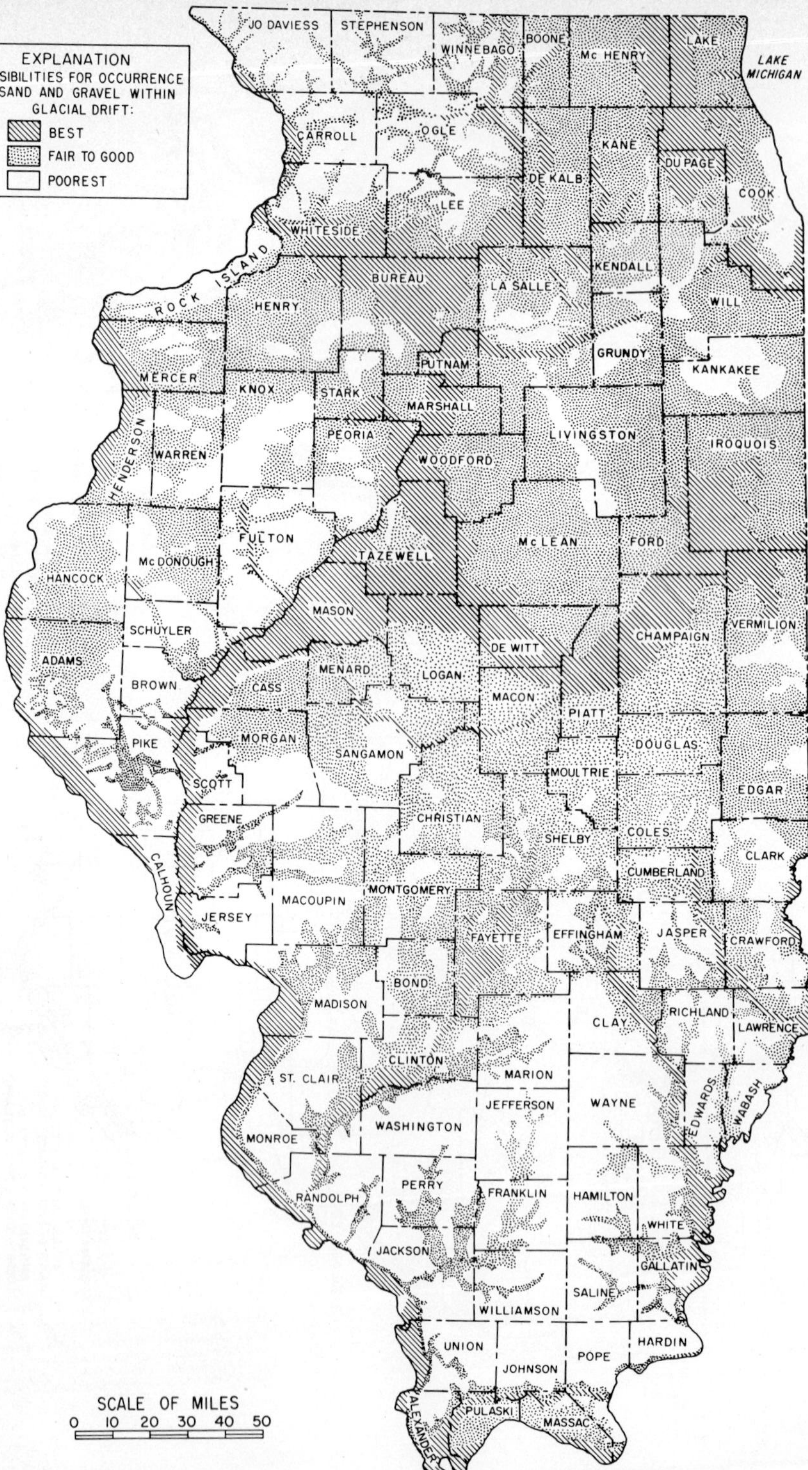

Fig. 6.25 Possibilities for occurrence of sand and gravel within glacial drift in Illinois. (*From Walton, 1965.*)

Fig. 6.26 Map showing moraines of Wisconsin stage of glaciation in Minnesota. (*From Leverett, 1932.*)

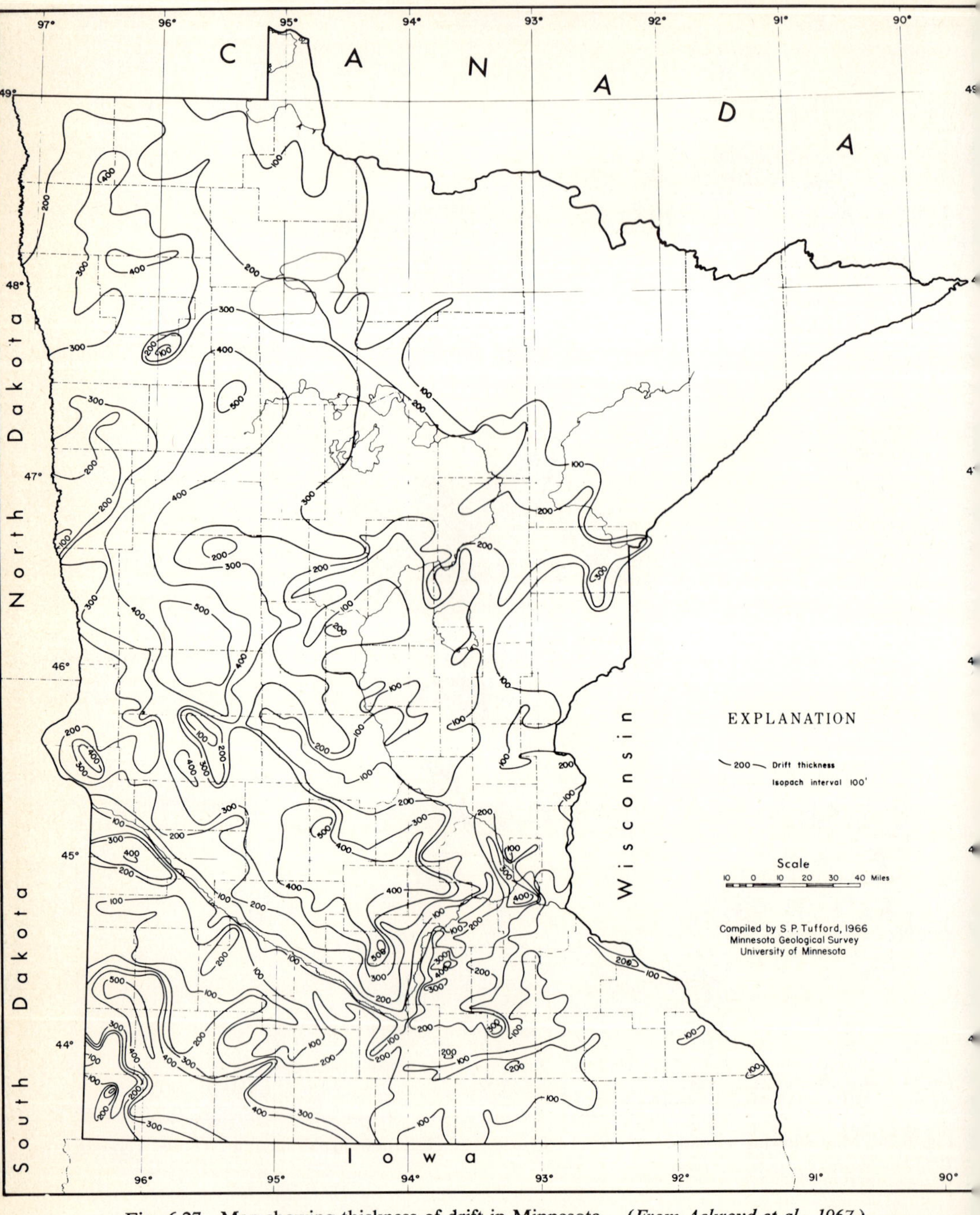

Fig. 6.27 Map showing thickness of drift in Minnesota. (*From Ackroyd et al., 1967.*)

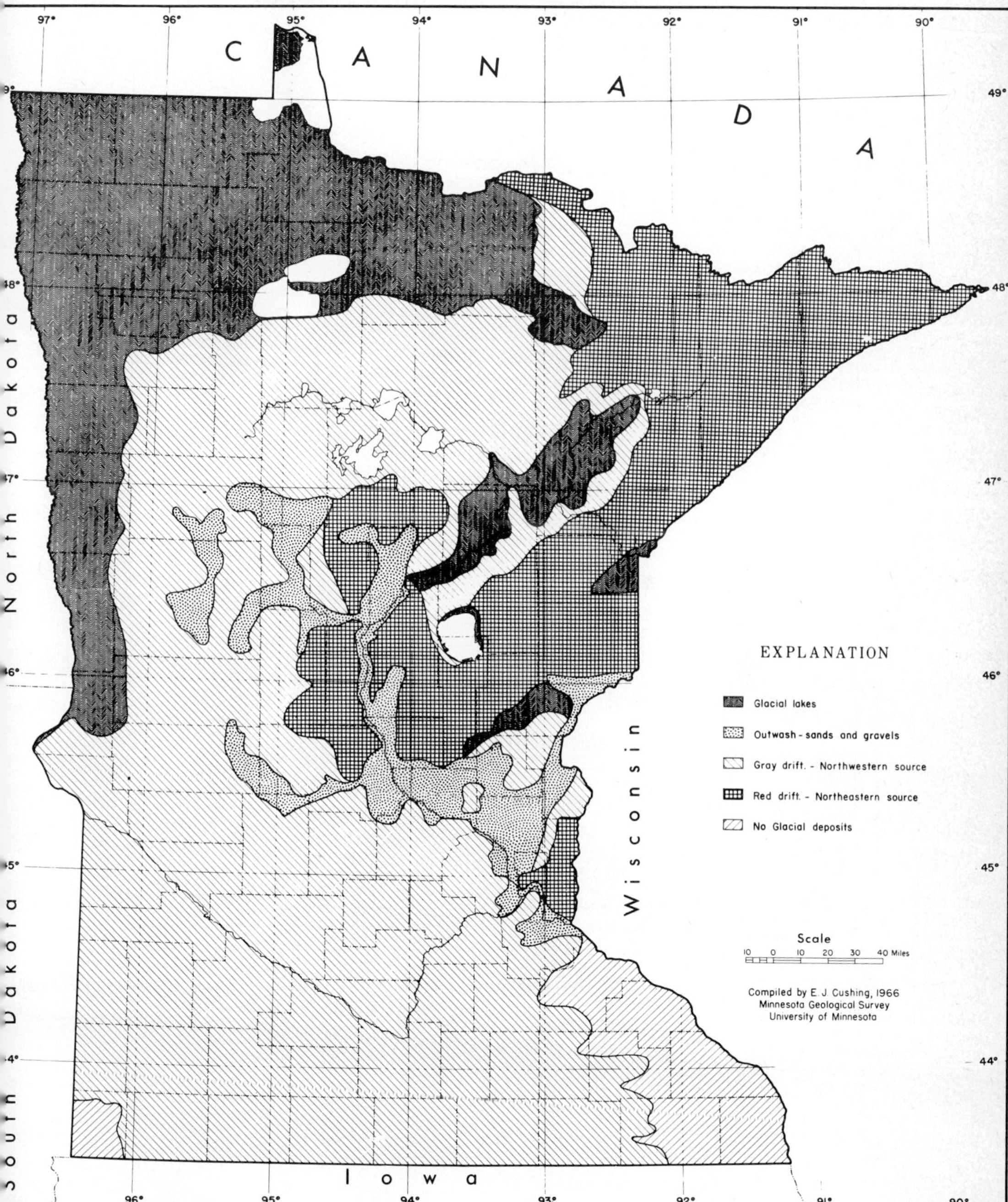

Fig. 6.28 Map showing major surficial deposits in Minnesota. (*From Ackroyd et al., 1967.*)

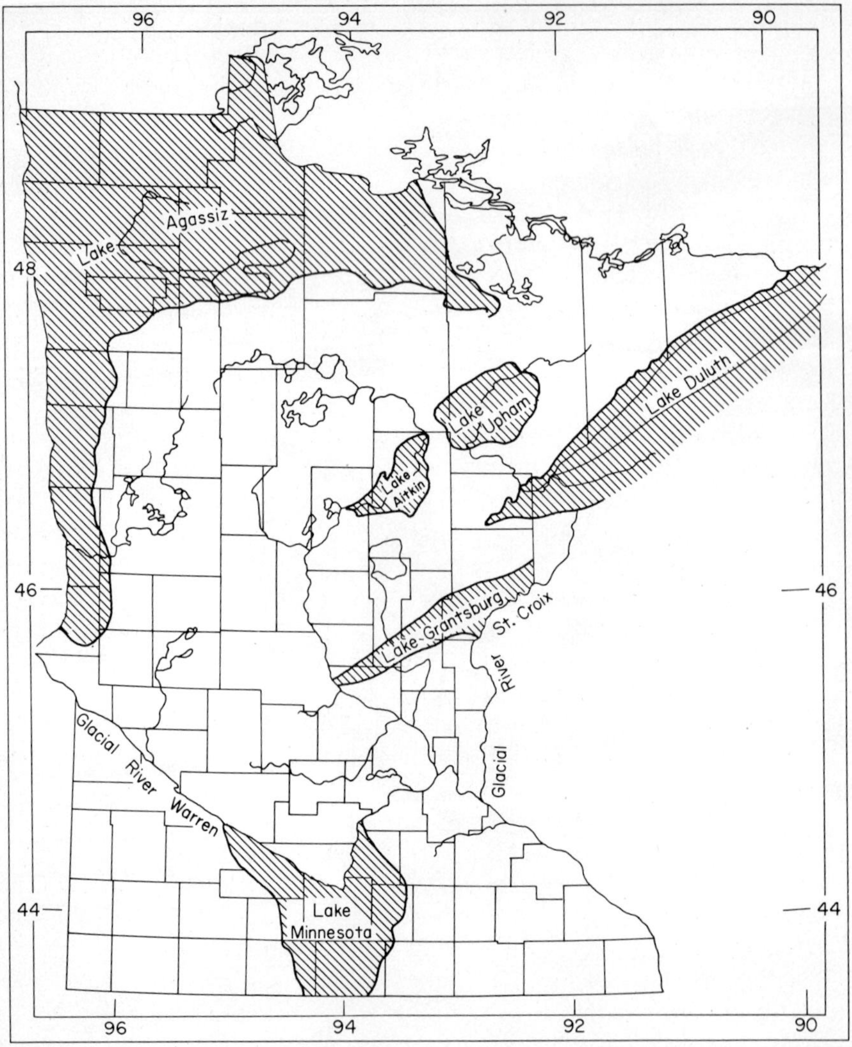

Fig. 6.29 Map showing major glacial lakes and rivers in Minnesota. (*From Schwartz and Thiel, 1963.*)

largest lakebed area is associated with Glacial Lake Agassiz in northwestern Minnesota. Areas of extensive surface sands and gravels are shown in Fig. 6.28.

Illinois is essentially a prairie plain. The relief over most of the state is moderate to slight; large-scale relief features are generally absent. Most of Minnesota lies between 1,000 and 1,500 feet above sea level and is level to rolling prairie or forest land, interspersed with irregular areas of glacial moraines whose hills rise from 50 to 300 feet above the surrounding land.

Woodland areas generally constitute but a few percent of acreage in Illinois except in the extreme southern and southwestern parts of the state where woodlands commonly occupy more than 20 percent of many areas. At the time the white man's settlement reached Minnesota, approximately 31 million acres or 61 percent of the state's land area was forest covered. By 1953 this acreage had been reduced to 19 million acres (38 percent of the state's land area). Approximately 80 percent of Minnesota's forest land is located in its 16 northeastern counties as shown in Fig. 6.30.

Average annual precipitation in Illinois increases from 32 inches in the north to 46 inches in the south as shown in Fig. 6.31. The growing season ranges from about 165 days in the north to more than 175 days in the south. Average annual precipitation in Minnesota is about 19 inches at the west-central edge, 20 inches at the northwest corner, 25 inches in the southwest, and 32 inches in the southeast as shown in Fig. 6.32. A short growing season of 90 to 110 days prevails in the northeast and along the Canadian border, but elsewhere ranges from 130 to 160 days.

Minnesota embraces an area of 84,068 square miles, of which about 4,059 square miles are covered by lakes. The lakes are not evenly distributed throughout the state; they show a sparse distribution in southeastern and northwestern counties and in other parts they are numerous. The state is reputed to have a larger area of swamps, bogs, and other marshes (wetlands) than any other state. They cover several thousands of square miles mostly in the northern part of the state.

The characteristics of the drainage basins in Illinois and Minnesota were determined by superposing maps showing the location of basins over such maps as those in Figs. 6.20 to 6.30. The relations between groundwater runoff and flow-duration curve ratios and basin characteristics were determined by statistical analysis. Basins were segregated into categories depending upon the gross characteristics of the basins. The range (10 and 90 percent frequencies) and median (50 percent frequency) of groundwater runoff and ratios were determined for each category from frequency graphs.

Groundwater runoff and ratios for basin characteristic categories in Illinois are shown in Table 6.14. These data and Figs. 6.33 and 6.34 suggest that groundwater runoff is greatest from glaciated and unglaciated basins having considerable surface and gravel and underlain by permeable bedrock. Groundwater runoff is least from glaciated basins with surface lakebed sediments and underlain by impermeable bedrock. Morainic ridges may yield slightly more water to streams than ground moraine. The least ratios are for basins underlain by permeable bedrock having large areal extent. The presence of surface sand and gravel deposits tends to reduce ratios;

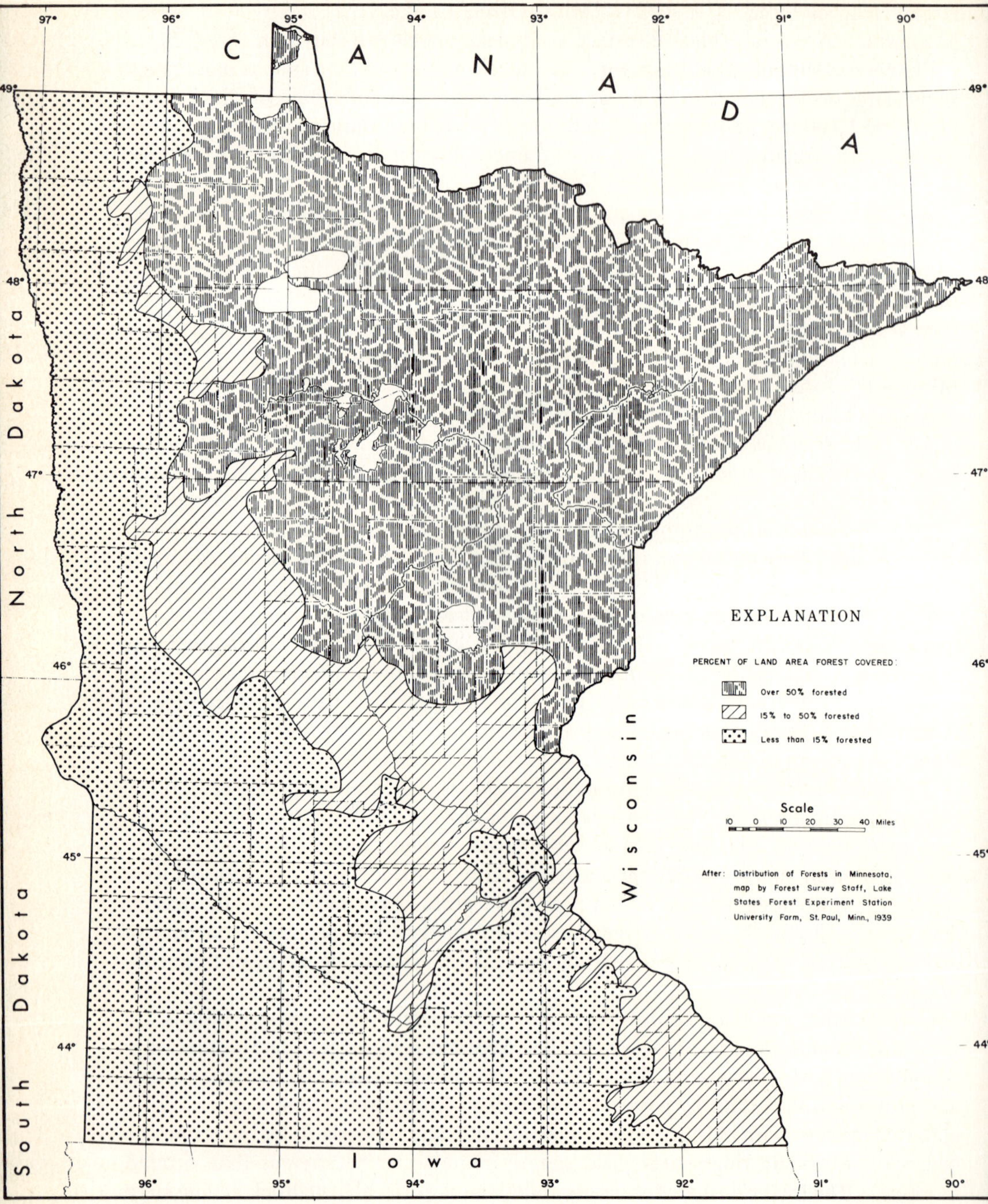

Fig. 6.30 Map showing present forest cover in Minnesota. (*From Ackroyd et al., 1967.*)

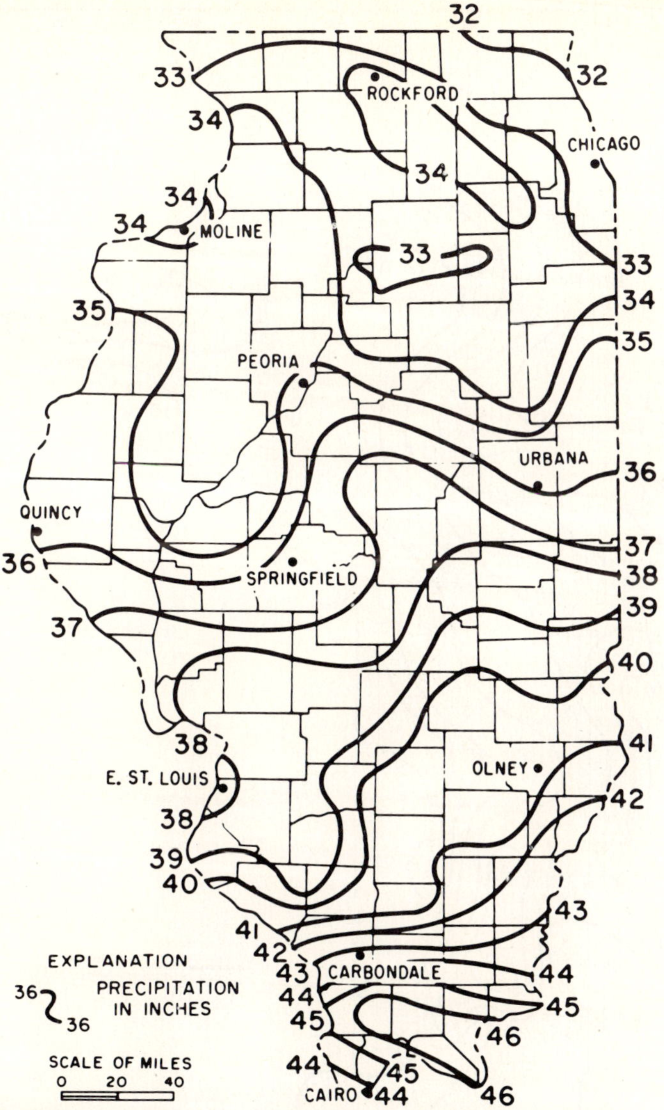

Fig. 6.31 Average annual precipitation in Illinois. *(From Walton, 1965.)*

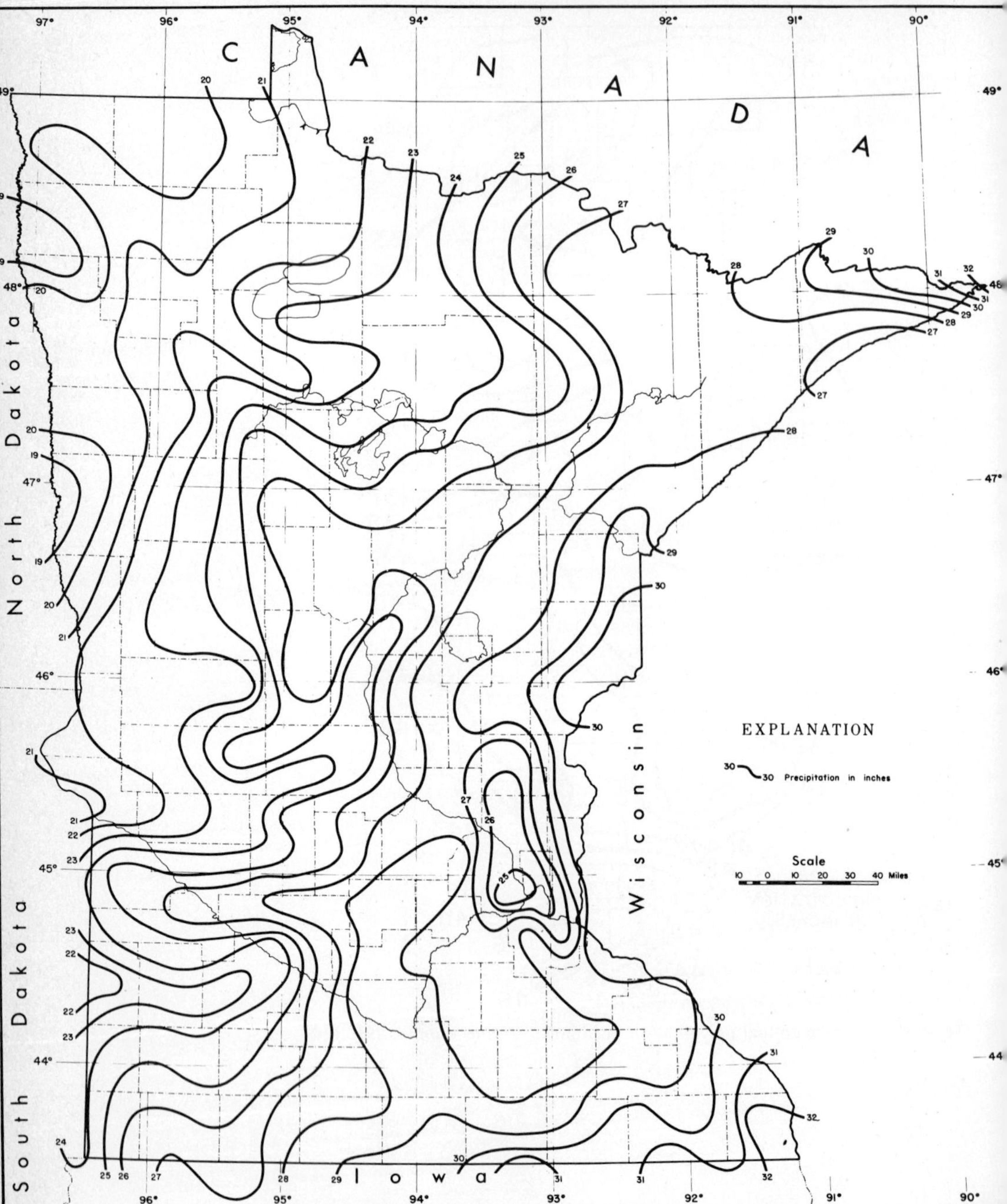

Fig. 6.32 Map showing average annual precipitation in Minnesota. (*From Ackroyd et al., 1967.*)

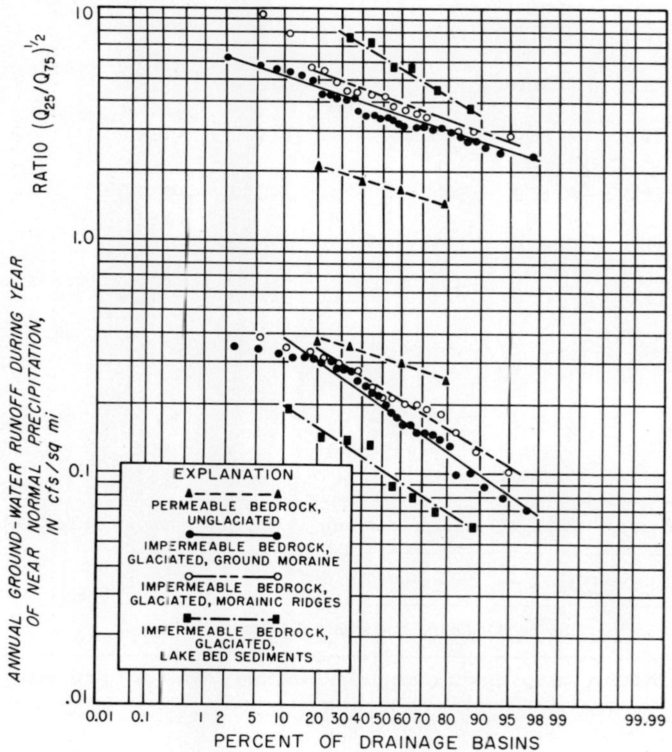

Fig. 6.33 Relation between annual groundwater runoff and ratios and character of surface deposits in Illinois. (*From Walton, 1965.*)

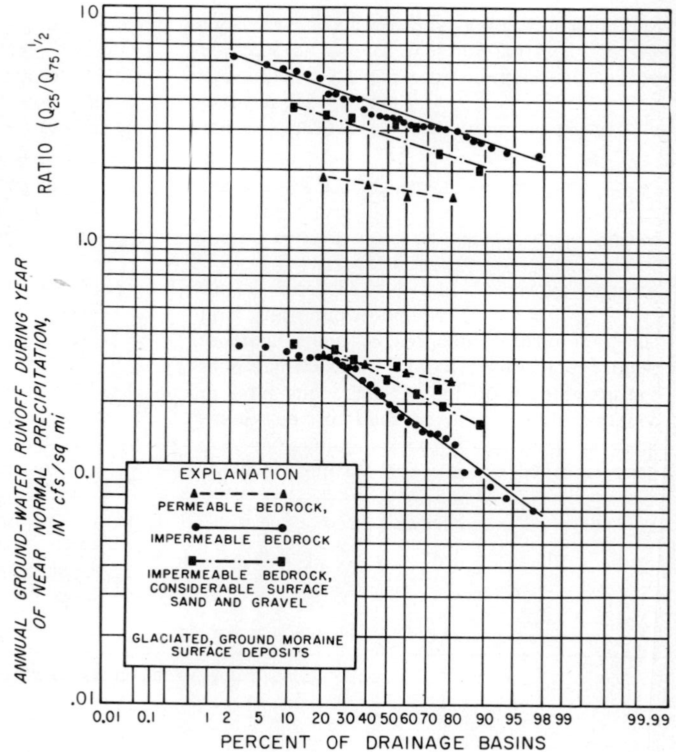

Fig. 6.34 Relation between annual groundwater runoff and ratios and character of bedrock in Illinois. (*From Walton, 1965.*)

TABLE 6.14 ANNUAL GROUNDWATER RUNOFF AND BASIN CHARACTERISTICS IN ILLINOIS. (*From Walton, 1965.*)

	Annual groundwater runoff, cfs/sq mi						*Ratio* $(Q_{25}/Q_{75})^{\frac{1}{2}}$		
	Near[a]		*Below*[a]		*Above*[a]				
Category	*Range*	*Median*	*Range*	*Median*	*Range*	*Median*	*Range*	*Median*	*Basin characteristics*
1	1.13 to 0.36	0.22	0.05 to 0.18	0.09	0.19 to 0.57	0.33	2.7 to 5.2	3.7	Glaciated, relatively impermeable bedrock, thick drift (commonly exceeding 50 feet), bedrock valleys; fair or poor possibility for occurrence of sand and gravel within drift, some surface sand and gravel of limited areal extent, ground moraine, slight to moderate stream gradient; little forest and woodland
2	0.24 to 0.38	0.28	0.17 to 0.22	0.19	0.32 to 0.59	0.40	1.5 to 1.9	1.7	Same as category 1 except permeable bedrock
3	0.20 to 0.38	0.29	0.09 to 0.18	0.14	0.35 to 0.56	0.44	2.8 to 3.8	3.2	Same as category 1 except considerable surface sand and gravel
4	0.15 to 0.38	0.23	0.04 to 0.17	0.08	0.29 to 0.62	0.43	2.7 to 6.2	4.0	Same as category 1 except morainic ridges
5	0.17 to 0.39	0.25	0.05 to 0.18	0.10	0.38 to 0.70	0.52	1.7 to 4.6	2.8	Same as category 1 except morainic ridges and permeable bedrock
6	0.17 to 0.43	0.27	0.07 to 0.22	0.14	0.38 to 0.72	0.52	2.0 to 5.0	3.1	Same as category 1 except morainic ridges and considerable surface sand and gravel
7	0.06 to 0.20	0.11	0.01 to 0.12	0.04	0.12 to 0.49	0.24	3.2 to 13	6.5	Same as category 1 except lakebed sediments

8	0.25 to 0.43	0.32	0.13 to 0.31	0.19	0.41 to 0.58	0.48	1.4 to 2.4	1.7	Unglaciated, permeable bedrock, thin unconsolidated deposits, bedrock valleys, some surface sand and gravel of limited areal extent, rugged topography, steep stream gradient, considerable forest and woodland
9	0.14 to 0.42	0.27	0.11 to 0.22	0.17	0.20 to 0.54	0.42	1.5 to 6.0	4.7	Same as category 8 except impermeable bedrock
10		0.30		0.10		0.49		3.9	Same as category 1 or 4 except major buried-bedrock valleys
11		0.29		0.07		0.48		4.2	Same as category 1 or 4 except no major buried-bedrock valleys
12		0.24		0.14		0.41		2.8	Same as category 1 or 4 except drift of Illinoian age
13		0.23		0.11		0.35		3.2	Same as category 1 or 4 except drift of Kansan age, rugged topography, steep stream gradient, and considerable forest and woodland
14		0.23		0.09		0.46		3.7	Same as category 1 or 4 except drift of Wisconsinan age
15		0.20		0.08		0.41		3.2	Same as category 1 or 4 except normal precipitation less than 38 inches and growing season less than 175 days
16		0.30		0.13		0.45		3.4	Same as category 1 or 4 except normal precipitation greater than 38 inches and growing season greater than 175 days

[a] Words indicate that data are for years of near-, below-, and above-normal precipitation.

however, the limited areal extent and thickness of many surface sand and gravel deposits result in small reductions.

The presence of major buried-bedrock valleys with accompanying greater drift thicknesses tends to significantly increase groundwater runoff during years of below-normal precipitation. Thick layers of till of low permeability in buried-bedrock-valley areas greatly retard the upward vertical movement of water toward streams which only partially penetrate the upper part of the drift. Upward vertical movement of water toward streams is greatest during dry periods when the head difference between the water table and the piezometric surface of deeply buried deposits is greatest. Groundwater runoff increases with increases in precipitation despite the fact that evapotranspiration also increases from north to south.

Groundwater runoffs and ratios for basin characteristic categories in Minnesota are shown in Table 6.15. Groundwater runoff is greatest from basins in categories 1, 5, 7, and 8; it is least from basins in categories 2, 3, 4, and 6. In general, groundwater runoff is increased by the presence of lakes and wetlands, outwash deposits, Patrician red drift deposits, loess deposits, and permeable bedrock; groundwater runoff from basins in categories 1 and 2 suggests that, at least in the northern part of the state, annual groundwater runoff during years of near- and below-normal precipitation is greatly affected by the degree of lake and wetland cover. Basins having lake and wetland covers in excess of 5 percent of their areas have more than twice as much annual groundwater runoff as basins with less lake and wetland covers. However, during years of above-normal precipitation lake and wetland cover has a relatively small effect on groundwater runoff. The ratios $(Q_{25}/Q_{75})^{\frac{1}{2}}$ for basins with small and large lake and wetland cover do not differ appreciably, lending support to the conclusion that the degree of lake and wetland cover and not the water-yielding properties and/or the areal extent of deposits is responsible for the great difference in groundwater runoffs from basins in categories 1 and 2. Thus, more than half of the apparent groundwater runoff basins in category 1 during years of near- and below-normal precipitation is not groundwater runoff as defined earlier, but instead is water slowly released from storage within lakes and wetlands. In general, lakes and wetlands seem to act as natural reservoirs which sustain and regulate streamflow during rainless periods.

Examination of categories 2 and 3 shows that groundwater runoff from basins having average annual precipitation rates less than 25 inches is significantly greater during years of below-normal precipitation than it is from similar basins having average annual precipitation rates greater than 25 inches. The data for categories 1, 2, and 4 suggest that recharge to groundwater reservoirs increases consistently with increases in precipitation, but in the southern parts of the state, higher temperatures result in large evapotranspiration rates and decreased groundwater runoffs. Thus, the annual rate of precipitation and the temperatures have profound and counteracting influences on groundwater runoff. Groundwater runoff data and ratios $(Q_{25}/Q_{75})^{\frac{1}{2}}$ for basins in categories 1, 2, and 4 suggest that the presence of lakes and wetlands increases streamflow in about the same proportion as evapotranspiration decreases streamflow.

Streamflow from northern parts of the state during extended rainless periods of years of near- and below-normal precipitation would be less than half as much as it is if lakes and wetlands were not widespread, or if the average temperature in the northern part of the state was comparable to that in the southern part of the state.

Groundwater runoff data from basins in categories 1, 3, and 5 indicate that groundwater runoff increases appreciably as the amount of surficial sand and gravel increases. The presence of deposits of surficial sand and gravel also reduces the ratio $Q_{25}/Q_{75})^{\frac{1}{2}}$. Groundwater runoff increases directly with an increase in the amount of sand and gravel, and the difference in runoff between wet and dry periods is decreased as the quantity of sand and gravel increases. The data for categories 2 and 6 and 3, 5, and 7 suggest that gray drift deposits do not yield appreciably more water to streams than do lakebed sediments, and groundwater runoff from Patrician red drift is about the same as from gray drift that contains outwash deposits and much greater than from lakebed sediments and gray drift. Patrician red drift is more permeable than gray drift, which may have a slightly higher average permeability than lakebed sediments. The lakebed sediments probably include substantial amounts of surficial sand and sandy silt. Highest groundwater runoffs are from basins in southeastern parts of the state that have a thick loess cover and are directly underlain by permeable bedrock, as shown by the data for category 8.

Groundwater runoffs and ratios $(Q_{25}/Q_{75})^{\frac{1}{2}}$ from basins in Minnesota were compared with groundwater runoffs and ratios of basins having similar characteristics in Illinois. Groundwater runoff from basins containing ground-moraine deposits (see category 1) is much less in Minnesota than in Illinois, even though the ratios $(Q_{25}/Q_{75})^{\frac{1}{2}}$ are about the same for both states. Probably, groundwater runoff from ground moraine is less in Minnesota than in Illinois because the precipitation is much less. Groundwater runoff apparently increases from north to south in the state because of increased precipitation, despite the fact that evapotranspiration also increases from north to south. It is also possible that on a gross basis the ground moraine composed of gray drift in Minnesota is less permeable than the ground moraine in Illinois. Groundwater runoff from basins having lakebed sediments (see category 2) is about the same in both states, despite the fact that precipitation is much less in Minnesota. However, the ratios $(Q_{25}/Q_{75})^{\frac{1}{2}}$ for basins in Minnesota are much less than those for basins in Illinois. The surficial lakebed sediments have much higher contents of sand and sandy silt (for example, extensive beach ridges) in Minnesota than in Illinois. Groundwater runoff from lakebed sediments in Minnesota probably would be much less, perhaps 50 percent less, if sand and sandy silts were not so widely spread.

Probably other important relations between groundwater runoff and basin characteristics exist in addition to those determined during the study. For example, it was not possible to determine from available data the relationship between the degree of forest cover and groundwater runoff. The profound influences of geology, lake and wetland cover, and precipitation on groundwater runoff, however, are apparent.

Estimates of annual groundwater runoff from ungaged basins may be made with reasonable accuracy from the data given in Tables 6.14 and 6.15 and information on

TABLE 6.15 ANNUAL GROUNDWATER RUNOFF AND BASIN CHARACTERISTICS IN MINNESOTA. *(From Ackroyd et al., 1967.)*

Basin category	*Annual groundwater runoff, cfs/sq mi*						Ratio $(Q_{25}/Q_{75})^{\frac{1}{2}}$		Major basin characteristics
	Near[a]		*Below*[a]		*Above*[a]				
	Range	*Average*	*Range*	*Average*	*Range*	*Average*	*Range*	*Average*	
1	0.15 to 0.27	0.21	0.07 to 0.34	0.18	0.15 to 0.55	0.31	2.22 to 2.40	2.33	Glaciated, ground moraine, gray drift, relatively impermeable bedrock, thick drift (commonly exceeding 50 feet), moderate topography, lakes and wetlands cover more than 5 percent of basin, forests cover more than 15 percent of basin, annual precipitation less than 25 inches, growing season less than 130 days
2	0.03 to 0.11	0.07	0.05 to 0.09	0.07	0.21 to 0.26	0.24	2.46 to 3.16	2.86	Same as category 1 except lakes and wetlands cover less than 5 percent of basin
3	0.04 to 0.16	0.10	0.02 to 0.04	0.03	0.30 to 0.37	0.34	3.33 to 3.74	3.54	Same as category 1 except lakes and wetlands cover less than 5 percent of basin, forests cover less than 15 percent of basin, annual precipitation greater than 25 inches, growing season greater than 130 days

4	0.05 to 0.08	0.07	0.02 to 0.03	0.03	0.17 to 0.21	0.19	2.27 to 3.13	2.70	Same as category 1 except forests cover less than 15 percent of basin, and growing season greater than 130 days
5	0.21 to 0.40	0.32	0.25 to 0.38	0.29	0.45 to 0.56	0.50	1.40	1.40	Same as category 1 except has some outwash deposits, and annual precipitation greater than 25 inches
6	0.06 to 0.17	0.12	0.02 to 0.08	0.04	0.10 to 0.43	0.27	2.08 to 3.7	2.43	Same as category 1 except has lakebed sediments, lakes, and wetlands cover less than 5 percent of basin, and topography is subdued
7	0.30 to 0.43	0.36	0.25 to 0.43	0.35	0.49 to 0.56	0.51			Same as category 1 except has red drift, lakes and wetlands cover less than 5 percent of basin, and annual precipitation is greater than 25 inches
8	0.31 to 0.59	0.42	0.09 to 0.35	0.26	0.29 to 0.82	0.57	3.57	3.57	Thick loess cover, permeable bedrock, moderate topography, lakes and wetlands cover less than 5 percent of basin, forests cover more than 15 percent of basin, precipitation more than 25 inches, and growing season greater than 130 days

[a] Indicate data for years of near-, below-, and above-normal precipitation.

the characteristics of the basins. To make the estimates, list the general basin characteristics in the manner of the tables, then assign the basin to one of the categories given in the tables. Next, note the range and average groundwater runoff for basins of the category assigned to the basin, and then estimate the annual groundwater runoff from the tables.

As large areas in Illinois and Minnesota are covered by glacial drift more than 50 feet thick, many sand and gravel aquifers and the bedrock aquifers are deeply buried, and recharge to the aquifers is derived from vertical leakage through layers of till having a low permeability. Vertical leakage with maximum vertical hydraulic gradients commonly is much less than recharge to surface deposits. As a result of the aquifers being deeply buried, not all groundwater runoff can be diverted into cones of depression. Even under heavy pumping conditions, there is some lateral as well as vertical movement of water in the surface deposits. Also, some groundwater runoff is bank storage and interflow which cannot be entirely diverted into cones of depression at some distance from streams. Precipitation and, therefore, groundwater runoff and recharge are unevenly distributed throughout the year. There are periods of time during the wet spring months when the rate of recharge to surface deposits greatly exceeds, at least temporarily, maximum rates of vertical leakage of water to cones of depression in deeply buried aquifers. If recharge by induced infiltration of surface water is excluded, recharge to most deeply buried aquifers will be less than groundwater runoff under natural conditions.

Cursory consideration may suggest that a map showing the distribution of groundwater runoff could be used directly to evaluate the potential yields of groundwater reservoirs. However, areas of high or low groundwater runoffs are not necessarily areas of high or low potential yield. Groundwater runoff may be high largely because of the presence of surficial sand and gravel deposits. But, because of their limited areal extent and thickness, the surficial sand and gravel deposits may be capable of yielding only small amounts of water to production wells and well fields. Also, groundwater runoff can be low because of the lack of surficial sand and gravel and the presence of lakebed sediments. However, deeply buried sand and gravel deposits or bedrock aquifers, extending beyond basin boundaries, may be capable of yielding moderate amounts of water to production wells and well fields. Thus, a map showing the distribution of groundwater runoff can be misleading if used directly to delineate favorable areas for development of groundwater resources. It is apparent that no simple relation exists between the amount of groundwater runoff and the potential yields of aquifers.

Problems

6.1 Review available U.S. Geological Survey Water Supply Papers and state groundwater reports and catalog data concerning groundwater recharge rates. Prepare a paper summarizing information concerning recharge rates and their relation to hydrogeologic conditions.

6.2 In the following illustration, assume a pumping center exists at point A. Production wells have been discharging at an average rate of 1.86 mgd from a deeply buried artesian aquifer. Recharge

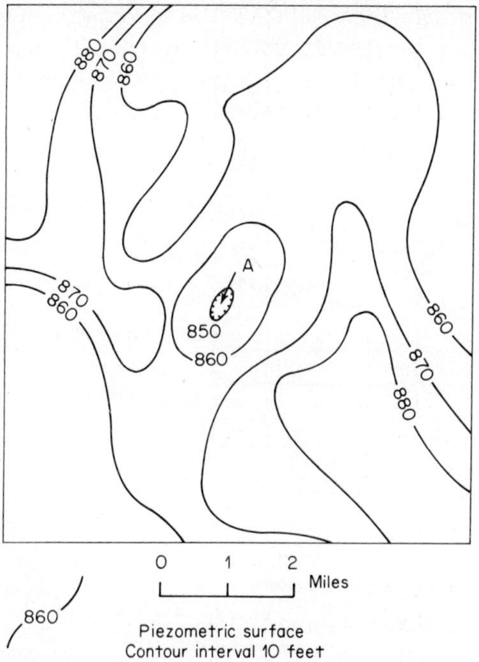

Fig. P6.2

by the vertical leakage of water through an overlying aquitard balances discharge. The average difference in head between the water table in the source bed above the aquitard and the piezometric surface of the aquifer is 30 feet. $P = 0.012$ gpd/sq ft and $m' = 80$ feet. Compute the average recharge rate.
Answer: 4,200 gpd/sq mi/ft.

6.3 Given the following mean groundwater stage graph and streamflow hydrograph (these data are for drainage basin covering 72.7 square miles) estimate monthly and annual groundwater runoff.
Answer:

Month	R_g, *in.*
Jan.	0.04
Feb.	0.08
Mar.	0.13
Apr.	0.56
May	0.56
June	0.36
July	0.06
Aug.	0.03
Sept.	Neg
Oct.	Neg
Nov.	0.01
Dec.	0.06
Annual	1.89

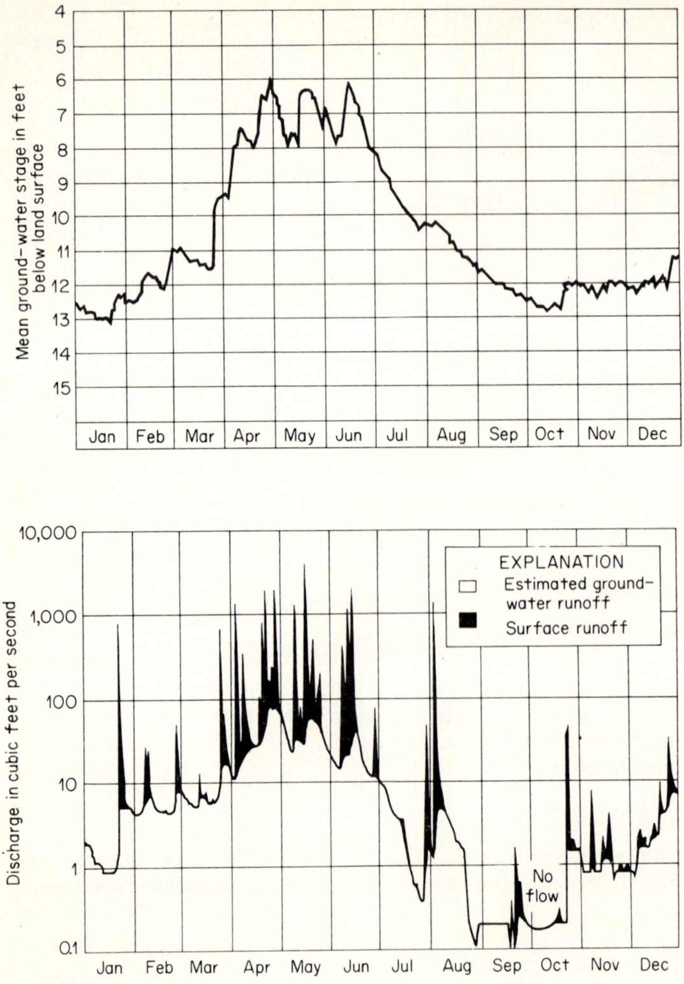

Fig. P6.3

6.4 Using information available in the files of state and federal agencies, study hydrogeologic, water-level, and pumpage data for a pumping center in your state and estimate the average recharge rate for the pumping center. Describe the factors governing the average recharge rate.

References

Ackroyd, E. A., W. C. Walton, and D. L. Hills, 1967. "Groundwater Contribution to Streamflow and Its Relation to Basin Characteristics in Minnesota," Minn. Geol. Surv., Rept. of Invest. 6.

Anon, 1957. "Artificial Recharge Experiments at McDonald Well Field, Amarillo, Texas," State Bd. of Water Engrs. Bull. 570, 1957.

Anon, 1958. "Sea Water Intrusion in California," State Dept. of Water Resources, Div. of Resources Planning, Bull. 63, 1958.

Cross, W. P., and R. E. Hedges, 1959. "Flow Duration of Ohio Streams," Ohio Div. of Water Bull. 31.

Economics of Ground Water Recharge, *J. Hydraulics Div., Am. Soc. Civil Engrs.*, July, 1965.

Fair, G. M., and J. C. Geyer, 1954. "Water Supply and Waste-water Disposal," John Wiley & Sons, Inc., New York.

Farvolden, R. N., W. A. Meneley, E. G. Le Breton, D. H. Lennox, and P. Meyboom, 1963. "Early Contributions to the Groundwater Hydrology of Alberta," Research Council of Alberta Bull. 2.

Feth, J. H., 1964. Hidden Recharge, *Ground Water*, vol. 2, no. 4.

Hantush, M. S., 1956. Analysis of Data from Pumping Tests in Leaky Aquifers, *Trans. Am. Geophys. Union*, vol. 37, no. 6.

———, 1967. Growth and Decay of Groundwater Mounds in Response to Uniform Percolation, *Water Resources Res.*, vol. 3., no. 1.

Harder, A. H., and W. J. Drescher, 1954. "Groundwater Conditions in Southwestern Langlade County, Wisconsin," U.S. Geol. Surv. Water Supply Paper 1294.

Horberg, L., 1950. "Bedrock Topography of Illinois," Illinois State Geol. Surv. Bull. 73.

———, 1957. "Bedrock Surface of Illinois," Illinois State Geol. Surv., map.

Houk, I. E., 1921. "Rainfall and Runoff in the Miami River Valley," Miami Conserv. Dist. Tech. Repts., pt. 8.

Keech, C. F., and V. H. Dreeszen, 1959. "Geology and Groundwater Resources of Clay County, Nebraska," U.S. Geol. Surv. Water Supply Paper 1468.

Kunkle, G. R., 1962. The Base Flow-duration Curve, a Technique for the Study of Groundwater Discharge from a Drainage Basin, *J. Geophys. Res.*, vol. 67, no. 4.

Leverett, F., 1932. "Quaternary Geology of Minnesota and Parts of Adjacent States," U.S. Geol. Surv. Profess. Paper 161.

Linsley, R. K., Jr., M. A. Kohler, and J. L. Paulhus, 1958. "Hydrology for Engineers," McGraw-Hill Book Company, New York.

Morgan, A. M., 1946. "Progress Report on Geology and Groundwater Resources of the Cheyenne Area, Wycoming," U.S. Geol. Surv., open-file rept., Washington, D.C., and Cheyenne, Wyo.

Newport, T. G., 1959. "Groundwater Resources of the Lower Niobrara River and Ponca Creek Basins, Nebraska and South Dakota," U.S. Geol. Surv., Water Supply Paper 1460-G.

Pettijohn, F. J., 1949. "Sedimentary Rocks," Harper & Row, Publishers, Incorporated, New York.

Rapp. J. R., F. N. Visher, and R. T. Lettleton, 1957. "Geology and Groundwater Resources of Goshen County, Wycoming," U.S. Geol. Surv. Water Supply Paper 1377.

Rasmussen, W. C., and G. E. Andreasen, 1959. "Hydrologic Budget of the Beaverdam Creek Basin, Maryland," U.S. Geol. Surv. Water Supply Paper 1472.

Rose, N. A., 1951. "Results of Investigation for Groundwater Development at Big Bend Well Field, Hamilton County, Ohio," Southwestern Ohio Water Co.

Schicht, R. J., and W. C. Walton, 1961. "Hydrologic Budgets for Three Small Watersheds in Illinois," Illinois State Water Surv., Rept. of Invest. 40.

Schwartz, G. M., and G. A. Thiel, 1963. "Minnesota's Rocks and Waters," Minnesota Geol. Surv. Bull. 37, rev. ed.

Smith, R. S., E. E. DeTurk, F. C. Bauer, and L. H. Smith, 1927. "Woodford County Soils," Univ. Ill. Agr. Exp. Sta. Soil Rept. 36.

Thornburn, T. H., 1960. "Surface Deposits of Illinois," Univ. Ill. Civil Eng. Studies, Soil Mechanics Ser. 3.

Thornwaite, C. W., J. R. Mather, and D. F. Carter, 1958. Three water balance maps of eastern North America, Resources for the Future, Inc., Washington, D.C.

Todd, D. K., 1957. "Artificial Recharge of Groundwater, an Annotated Bibliography through 1954," open-file report, U.S. Geol. Surv., Washington, D.C.

———, 1959. "Groundwater Hydrology," John Wiley & Sons, Inc., New York.

Walton, W. C., and G. D. Scudder, 1960. "Groundwater Resources of the Valley-train Deposits in the Fairborn Area, Ohio," Ohio Div. of Water Tech. Rept. 3.

Walton W. C., 1965. "Groundwater Recharge and Runoff in Illinois," Illinois State Water Surv., Rept. of Invest. 48.

———, D. L. Hills, and G. M. Grundeen, 1967. "Recharge from Induced Streambed Infiltration under Varying Groundwater-level and Stream-stage Conditions," Minn. Water Resources Res. Center Bull. 6.

Welsh, W. F., 1957. "Water Supply Problems in Nassau County, Long Island," presented at Albany, N.Y., at Ann. Convocation of New York Dist. Personnel of Water Resources Div., U.S. Geol. Surv.

Wisler, C. O., and E. F. Brater, 1959. "Hydrology," John Wiley & Sons, Inc., New York.

Zeizel, A. J., W. C. Walton, R. T. Sasman, and T. A. Prickett, 1962. "Groundwater Resources of DuPage County, Illinois," Illinois State Water Surv. Cooperative Groundwater Rept. 2.

CHAPTER 7

QUALITY OF GROUNDWATER

In groundwater resource evaluation, the quality of groundwater is of nearly equal importance to quantity. The chemical, physical, and bacterial characteristics of groundwater determine its usefulness for municipal, commercial, industrial, agricultural, and domestic water supplies. Development provides opportunities for pollution of groundwater and consideration must be given to the protection of quality. The study of groundwater quality involves a description of the occurrence of the various constituents in groundwater and the relation of these constituents to water use. In addition, groundwater quality data give important clues to the geologic history of rocks and indications of groundwater recharge, discharge, movement, and storage.

Chemical analysis of groundwater includes the determination of the concentrations of inorganic constituents. The analysis also includes measurement of pH and specific electrical conductance. Temperature, color, turbidity, odor, and taste are evaluated in a physical analysis. Bacterial analysis generally consists of tests to detect the presence of coliform organisms. Dissolved gases and radionuclides in groundwater are reported in some analyses.

Examination of water in the field to determine its approximate chemical character is often useful. Commercial equipment for determining specific electrical conductance in the field is available. Portable laboratory kits containing all necessary glassware and solutions for simple chemical tests such as for chloride, hardness, and alkalinity also are available. The principal value of field testing lies in getting approximate data that can be used to plan a sampling program. Field testing is inherently less accurate than work done in a laboratory. Standard laboratory methods of water analysis are specified by the American Public Health Association and others (1955).

CHEMICAL ANALYSIS

Substances commonly determined in chemical analyses and expressed as ions comprise the *cations* (positively charged ions), including calcium, magnesium, sodium, and potassium, and the *anions* (negatively charged ions), including sulfate, chloride, fluoride, nitrate, and those contributing to alkalinity which are usually expressed in terms of an equivalent amount of carbonate and bicarbonate. Dissolved gases such as carbon dioxide and hydrogen sulfide, when determined, are usually reported in undissociated form. The acidity of a water is generally expressed as concentration of hydrogen ions.

7.1 Expression of Analyses

Total dissolved constituents generally are only a small fraction of one percent of the total weight of a groundwater sample. Concentrations of ions are commonly reported by weight in ppm. One ppm means one part by weight of dissolved constituent in a million parts by weight of solution. A similar measure is milligrams per liter (mg/liter); ppm and mg/liter are numerically almost the same if the concentration of dissolved solids is low and the specific gravity of the water is nearly 1.0. Grains per gallon is another measure; 1 grain/gal = 17.1 ppm.

For convenience, concentrations are often expressed in equivalents per million (epm) calculated by dividing ppm by the equivalent weight of the ion under consideration. The equivalent weight is equal to the atomic weight divided by the valence. The sum of the cations and the sum of the anions expressed in epm will each equal half of the total concentration if the analysis is correct and complete. Conversion factors for converting ppm to epm are given in Table 7.1.

The ability of groundwater to conduct an electrical current is called its electrical conductance or electrical conductivity. Conductance is the reciprocal of resistance and is measured in mhos. The conductance of groundwater is a function of temperature, type of ions present, and concentration of various ions. Specific conductance (the ability of a cube 1 centimeter on a side to conduct electrical current) readings

TABLE 7.1 CONVERSION FACTORS: PARTS PER MILLION TO EQUIVALENTS PER MILLION. (*From Hem, 1959.*)

Ion	*Multiply by*	*Ion*	*Multiply by*
Aluminum (Al^{3+})	0.11119	Iron (Fe^{3+})	0.05372
Barium (Ba^{++})	0.01456	Lead (Pb^{++})	0.00965
Bicarbonate (HCO_3^-)	0.01639	Lithium (Li^+)	0.14409
Bromide (Br^-)	0.01251	Magnesium (Mg^{++})	0.08224
Calcium (Ca^{++})	0.04990	Manganese (Mn^{++})	0.03640
Carbonate (CO_3^{-})	0.03333	Manganese (Mn^{4+})	0.07281
Chloride (Cl^-)	0.02820	Nitrate (NO_3^-)	0.01613
Chromium (Cr^{6+})	0.11536	Phosphate (PO_4^{3-})	0.03159
Copper (Cu^{++})	0.03148	Potassium (K^+)	0.02558
Fluoride (F^-)	0.05263	Sodium (Na^+)	0.04350
Hydrogen (H^+)	0.99206	Strontium (Sr^{++})	0.02282
Hydroxide (OH^-)	0.05880	Sulfate (SO_4^{--})	0.02082
Iodide (I^-)	0.00788	Sulfide (S^{--})	0.06237
Iron (Fe^{++})	0.03581	Zinc (Zn^{++})	0.03059

are usually adjusted to 25°C so that variations in conductance are a function only of the concentration and type of dissolved constituents present. Pure water has a conductance of 0.055 micromho at 25°C. Distilled water commonly has a conductance of 0.5 to 5.0 micromhos. Rainwater will usually range from 5.0 to 30 micromhos, ocean water from 45,000 to 55,000 micromhos, and normal groundwater from 30 to 2,000 micromhos (Davis and DeWiest, 1966). Calcium bicarbonate and calcium sulfate water generally have the lowest conductance and sodium chloride water the highest conductance for a given total concentration of dissolved solids. For an approximation of total dissolved solids in ppm in fresh water, the specific conductance of the water in micromhos should be multiplied by 17.

The hydrogen-ion concentration is expressed as the pH, which is the reciprocal of the logarithm (base 10) of the hydrogen-ion concentration in moles per liter. A neutral water has a pH of 7.0, which indicates an equal number of H^+ and OH^- ions; an alkaline water has a pH of more than 7.0, which indicates a preponderance of OH^- ions; and an acidic water has a pH of less than 7.0, which indicates a preponderance of H^+ ions. The pH of pure water at 25°C is 7.0.

When material goes into solution in water the pH is commonly changed. Most groundwater has pH values between 5.0 and 8.0. Very high pH values above 8.5 are usually associated with sodium carbonate-bicarbonate waters; moderately high pH values are commonly associated with waters high in bicarbonate. Very low pH values below 4.0 are associated with waters containing free acids. Moderately low pH values may be associated with small amounts of mineral acids from sulfide sources or with organic acids.

7.2 Dissolved Constituents and Gases

Most of the major, secondary, minor, and trace constituents dissolved in groundwater and information concerning ranges of concentration are given in Table 7.2.

Total dissolved solids include all solid material in solution, whether ionized or not. It does not include suspended sediments, colloids, or dissolved gases. A rapid determination of total dissolved solids can be made by measuring the electrical conductance (EC) of a groundwater sample. For most natural groundwaters in the range of 100 to 5,000 micromhos/cm at 25°C, 1 ppm $= 1.56(EC) \times 10^6$.

The water that falls upon the earth as rain or snow contains only small quantities of dissolved mineral matter. As soon as it reaches the earth, however, it begins to react with the minerals of the soil and rocks with which it comes into contact. The amount and character of the mineral matter dissolved by precipitation depend upon the chemical composition and physical structure of the rocks with which they have been in contact, the temperature, the pressure, the duration of the contact, the materials already in solution, hydrogen- and hydroxyl-ion concentrations (pH), and redox potential (Eh). The solvent action of the water is assisted by the presence in solution of carbon dioxide, derived from the atmosphere as the water fell as precipitation, or from the soil through which it passes, where it is formed by organic processes.

TABLE 7.2 DISSOLVED CONSTITUENTS IN GROUNDWATER. (*After Davis and DeWiest, 1966.*)

Major constituents (range of concentration 1.0 to 1,000 ppm)	
Sodium	Bicarbonate
Calcium	Sulfate
Magnesium	Chloride
Silica	
Secondary constituents (range of concentration 0.01 to 10.0 ppm)	
Iron	Carbonate
Strontium	Nitrate
Potassium	Fluoride
Boron	
Minor constituents (range of concentration 0.00001 to 0.1 ppm)	
Antimony	Lithium
Aluminum	Manganese
Arsenic	Molybdenum
Barium	Nickel
Bromide	Phosphate
Cadmium	Rubidium
Chromium	Selenium
Cobalt	Titanium
Copper	Uranium
Germanium	Vanadium
Iodide	Zinc
Lead	
Trace constituents (range of concentration generally less than 0.001 ppm)	
Beryllium	Ruthenium
Bismuth	Scandium
Cerium	Silver
Cesium	Thallium
Gallium	Thorium
Gold	Tin
Indium	Tungsten
Lanthanum	Ytterbium
Niobium	Yttrium
Platinum	Zirconium
Radium	

The effect of soluble impurities occurring in rocks on water quality can be far out of proportion to the importance of such salts in the mineral composition of the rocks. The composition of the rock though of major importance is only one of many factors affecting water quality. There are no clear-cut relationships between water quality and rock types in all areas (Hem, 1959).

Analyses illustrating the water quality in certain rock types at selected places are shown in Figs. 7.1 to 7.5 to give the reader some indication of the composition of waters in various aquifers. Analyses from igneous rocks are illustrated by bar

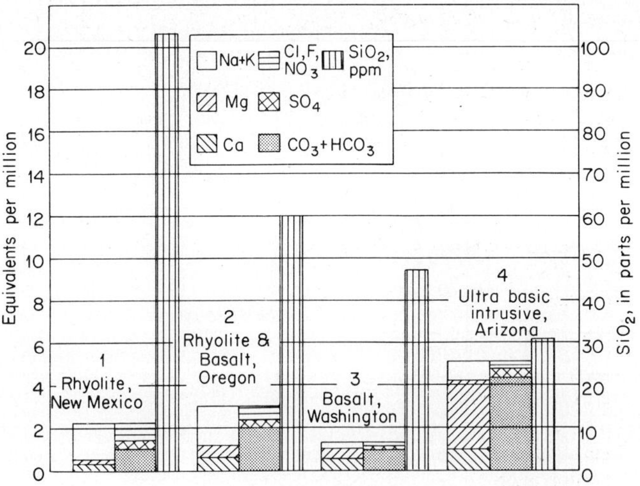

Fig. 7.1 Chemical analyses of waters from selected igneous rocks. (*After Hem, 1959.*)

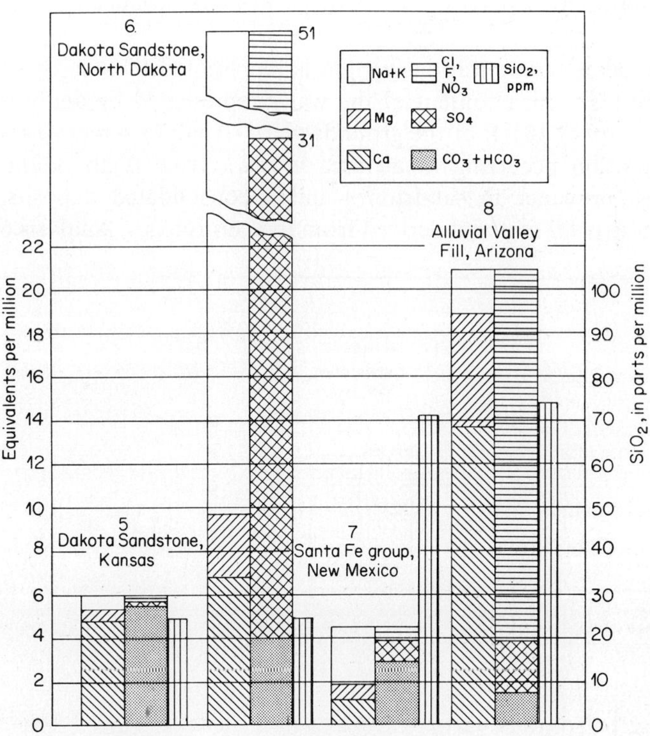

Fig. 7.2 Chemical analyses of waters from selected sandstone and unconsolidated deposits. (*After Hem, 1959.*)

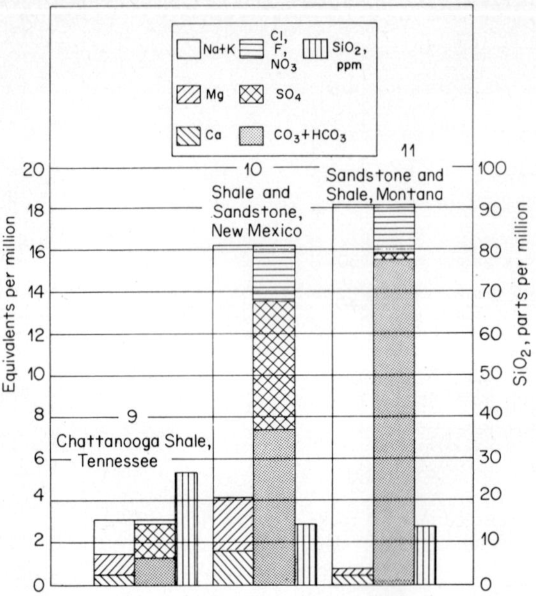

Fig. 7.3 Chemical analyses of waters from selected shales and shaly sandstones. (*After Hem, 1959.*)

diagrams in Fig. 7.1. The silica concentration in ppm is represented by a separate bar for each analysis. The high silica content of the water represented by analysis 1 is due to the high temperature of 101°F of the groundwater. Analysis 4 represents a water of unusual composition occurring in an area where olivine is abundant. Figure 7.2 shows analyses for water in sandstones and unconsolidated deposits. Analysis 7 is for water from detrital material derived from igneous rocks. Analysis 6

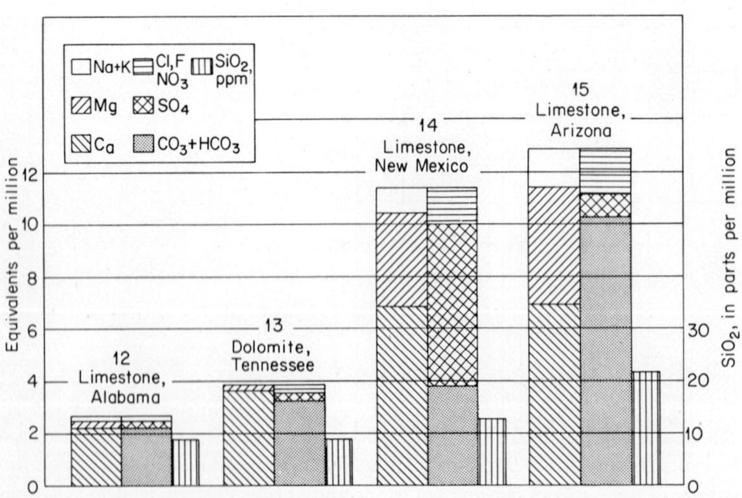

Fig. 7.4 Chemical analyses of waters from selected limestones and dolomites. (*After Hem, 1959.*)

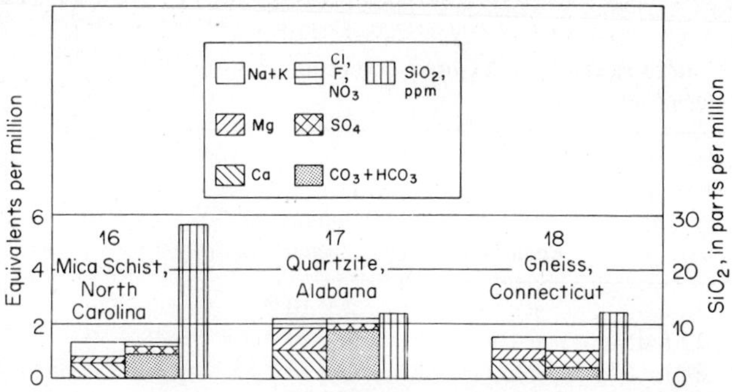

Fig. 7.5 Chemical analyses of waters from selected metamorphic rocks. (*After Hem, 1959.*)

represents water from a sandstone in which the flow of groundwater is sluggish and connate or saline residues in the formation have been taken into solution in the water. The character of the water in analysis 8 has been strongly influenced by irrigation return flow. Analyses in Fig. 7.3 represent waters from shales. The character of the water in analysis 9 is probably influenced by types of rock other than shale. Analysis 10 represents a more typical rather highly mineralized water from shale. Analyses from limestones and dolomites are illustrated in Fig. 7.4. Analysis 12 represents water from relatively pure limestones. Analysis 13 is for water from a dolomite. Water from impure limestone containing dolomite and gypsum is represented by analysis 14. Figure 7.5 shows analyses for waters from metamorphic rocks.

Groundwater in contact with sedimentary rocks of marine origin derive most of their calcium from the solution of calcite, aragonite, dolomite, anhydrite, and gypsum (Davis and DeWiest, 1966). In igneous and metamorphic rocks, weathering also releases calcium from such minerals as apatite, wollastonite, fluorite, and various members of the feldspar, amphibole, and pyroxene groups. The common sources of magnesium in groundwater are dolomite in sedimentary rocks; olivine, biotite, hornblende, and augite in igneous rocks; and serpentine, talc, diopside, and tremolite in metamorphic rocks.

The primary sources of most sodium in groundwater is from the release of soluble products during the weathering of plagioclase feldspars (Davis and DeWiest, 1966). In areas of evaporite deposits, the solution of halite is also important. Clay minerals may release large quantities of exchangeable sodium. Less important sources of sodium are the minerals nepheline, sodalite, stilbite, natrolite, jadeite, arfriedsonite, glaucophane, and aegirite. These minerals are locally abundant in some igneous and metamorphic rocks, but are quantitatively of minor importance in comparison with the feldspars. Common sources of potassium are the products formed by the weathering of orthoclase, microline, biotite, leucite, and nepheline in igneous and metamorphic rocks. Waters percolating through evaporite deposits may contain very large quantities derived from the dissolution of sylvite and niter.

TABLE 7.3 MAXIMUM PERMISSIBLE CONCENTRATION OF RADIONUCLIDES IN GROUNDWATER. (*From Federal Register, November 17, 1960.*)

Element	*Mass no. of radioisotope*	*Half-life years, y; days, d; hours, h*	*Radiation*	*MPC above natural background μc/ml in solution in water*
Barium	131	13 d	γ	2×10^{-4}
	140	12.8 d	β^-, γ	3×10^{-5}
Bromine	82	36 h	β^-, γ	3×10^{-4}
Calcium	45	153 d	β^-	9×10^{-6}
Carbon	14	5,600 y	β^-	8×10^{-4}
Cerium	144	290 d	β^-, γ	1×10^{-5}
Cesium	135	2.9×10^6 y	β^-	1×10^{-4}
	137	33 y	β^-, γ	2×10^{-5}
Chlorine	36	4×10^5 y	β^-	8×10^{-5}
Chromium	51	27.8 d	γ	2×10^{-3}
Cobalt	57	270 d	β^+, γ	5×10^{-4}
	60	5.3 y	β^-, γ	5×10^{-5}
Hydrogen	3	12.4 y	β^-	3×10^{-3}
Iodine	129	1.72×10^7 y	β^-, γ	4×10^{-7}
	131	8.04 d	β^-, γ	2×10^{-6}
Phosphorus	32	14.3 d	β^-	2×10^{-5}
Plutonium	238	92 y	α, γ	5×10^{-6}
	239	2.4×10^4 y	α, γ	5×10^{-6}
	240	6,580 y	α	5×10^{-6}
	242	5×10^5 y	α	5×10^{-6}
Radium	226[a]	1,620 y	α, γ	1×10^{-8}
	228	6.7 y	β^-	3×10^{-8}
Radon	222	3.83 d	α	A gas
Rubidium	86	18.7 d	β^-, γ	7×10^{-5}
	87	6×10^{10} y	β^-	1×10^{-4}
Ruthenium	103	40 d	β^-, γ	8×10^{-8}
	106	1 y	β^-	1×10^{-5}
Sodium	22	2.6 y	β^+, γ	4×10^{-5}
Strontium	89	51 d	β^-	1×10^{-5}
	90[a]	29 y	β^-	1×10^{-7}
Sulfur	35	88 d	β^-	6×10^{-5}
Uranium	235	7.1×10^8 y	α	3×10^{-5}
	238	4.5×10^9 y	α, γ	4×10^{-5}
Zinc	65	245 d	β^+, γ	1×10^{-4}

[a] The U.S. Public Health Service has adopted the following limits in its recommendations of drinking water standards: radium-226, 3×10^{-9} μc/ml; strontium-90, 1×10^{-9} μc/ml.

Some of the important minerals and mineral groups which may contain large amounts of iron are pyroxenes, amphiboles, magnetite, and pyrite.

Most chloride in groundwater comes from ancient sea water entrapped in sediments, solution of halite and related minerals in evaporite deposits, concentration by evaporation of chloride contributed by precipitation, and solution of dry fallout from the atmosphere. Most nitrate in groundwater is derived from organic sources or from industrial and agricultural chemicals.

Dissolved gases occur in concentrations from 1 to 100 ppm in most groundwater (Davis and DeWiest, 1966). Some production wells produce large amounts of methane, hydrogen sulfide, and carbon dioxide. Common dissolved gases include oxygen, nitrogen, carbon dioxide, hydrogen sulfide, methane, oxides of nitrogen, sulfur dioxide, and ammonia. These gases are commonly derived from the atmosphere and from decaying organic matter. In general, the solubility of a gas in water varies inversely with temperature and directly with pressure. A strong odor is imported by less than 1 ppm of H_2S in groundwater.

RADIONUCLIDES IN GROUNDWATER

Radioactive material provides the means of dating and tracing groundwater (Davis and DeWiest, 1966). Unstable nuclides that tend to change spontaneously to other species of nuclides through various decay reactions are called radioactive nuclides, or radionuclides. Nuclides with the same number of protons but with varying masses are isotopes of a single element. Many of the elements have more than 10 isotopes. Most of these are radioactive and have been produced artifically. Nuclear-decay reactions produce alpha (α), beta (β), and gamma (γ) radiation.

Small amounts of radiation which are received by the body as the result of the ingestion of radionuclides are of importance in hydrogeologic studies. Standards giving the maximum permissible concentration (MPC) of radionuclides in groundwater have been established. Table 7.3 gives the MPC of various important radionuclides.

The quantity of radionuclides present is measured indirectly by measuring the number of decay reactions during a given length of time. The basic unit of measure used is the curie, which is defined as the number of decay reactions per unit of time of one gram of radium, 3.7×10^{10} disintegrations per second. Microcuries (μc) are commonly used.

The largest number of natural radionuclides in groundwater are called primordial radionuclides which have exceptionally long half-lives (Davis and DeWiest, 1966). Some of these radionuclides decay to daughter products that are also radioactive. The most abundant primordial radionuclides are K^{40}, Rb^{87}, Th^{232}, U^{235}, and U^{238}. The daughter products of importance are Rn^{222} and Ra^{226}, which come from U^{238}. Another important group of natural radionuclides originates from cosmic-ray activation of the stable nuclides N^{14}, O^{16}, and Ar^{40} and have half-lives of less than 1 million years.

Uranium is present in groundwater generally in amounts ranging from 0.05 to 10 parts per billion (ppb); the median is about 1.5 ppb (Davis and DeWiest, 1966). The most abundant isotope of radium, Ra^{226}, is one of the decay products of U^{238} with a half-life of more than a few days. Most water from sedimentary aquifers has a radium concentration of less than 1×10^{-9} μc/ml; the median is probably about 3×10^{-6} μc/ml.

Radon, Rn^{222}, has a half-life of 3.8 days and is produced continuously in aquifers by the disintegration of the parent nuclide, Ra^{226}. Normal groundwater contains from less than 1×10^{-7} μc/ml to about 3×10^{-5} μc/ml radon; the median is about 2×10^{-6} μc/ml. Normal groundwater contains about 2×10^{-9} μc/ml of the radioisotope K^{40} with a half-life of 12.4 years.

Before 1952 the heavy isotope of hydrogen, H^3, or tritium content of most precipitation ranged from about 1 to 10 tritium units [1 tritium unit (TU) produces roughly 3.2×10^{-3} $\mu\mu$c/ml activity]. Testing of thermonuclear devices since 1952 has greatly increased the H^3 content of precipitation and surface water. Water entering the ground since 1954 should have had initially 10 TU and in many areas more than 20 TU. Within 40 to 60 years the tritrium content of pre-1954 water will be below most methods of detection. It is possible to identify water as water that has originated from precipitation which had resided in or out of contact with the atmosphere for more than a decade (Davis and DeWiest, 1966).

The dating of organic material by its C^{14} content is a well-established procedure. The C^{14} content might be used to date groundwater by removing all the carbon, most of which is in the carbonate ion, and counting the activity of the carbon. Unfortunately, the carbonate in most groundwater comes from various sources, of which the atmosphere may, under some circumstances, be only of secondary importance (Davis and DeWiest, 1966). Dates of groundwater based on the radiocarbon method will indicate ages that are in general older than the water. The total amount of C^{14} in the atmosphere has increased only a few percent above the pre-1954 level as a result of the testing of nuclear devices.

The environment is likely to be contaminated through the refinement and processing of uranium ore, initial production of nuclear fuels and explosives, reprocessing used reactor elements, discharge of cooling water which has been exposed to nuclear activation, escape of volatile material from evaporation and burning, dispersion of products of nuclear explosions, and the release of radionuclides used in science and medicine. The safe disposal of wastes from reactor operations and fuel reprocessing is one of the major problems in the widespread utilization of nuclear power. Disposal practices depend on the radioactivity of the waste, the general chemical character of the waste, and the physical environment in the area of disposal. The radioactivity of liquid waste is broadly referred to as low level if it has fractions of a microcurie per gallon, intermediate level if it has less than a few curies per gallon but greater than a microcurie per gallon, and high level if it has more than a few curies per gallon. Low-level wastes have been disposed into the subsurface at the National Reactor Testing Station in Idaho, Savanna River Plant in South Carolina, and at Hanford,

Washington. Sorption on soil particles plus decay of the radionuclides with short half-lives has prevented undesirable movement of contamination. A number of schemes for the eventual disposal of high-level wastes are under study.

ADSORPTION AND SULFATE REDUCTION

The surface of a solid which is in contact with a gas or a liquid is covered with a thin layer of molecules or ions. The layer is held to the solid surface by physical or chemical forces. The phenomenon is called adsorption. The skeletons of aquifers may have a large surface area per unit volume and hence a considerable capacity for physical adsorption.

One type of adsorption exists in which the forces of chemical bonding between ions is involved. This type of adsorption is termed *chemisorption* and replacement of adsorbed ions by ions in solution is termed *ion exchange* (Hem, 1959). The capacity for chemisorption is mostly a result of unsaturation of the chemical bonding in the crystal lattice of the solid having this capacity. In most of the important natural exchange media there is an excess of negative bonds in certain areas of the crystal lattice, and positively charged ions are attracted to these areas. Divalent cations are usually but not always held more tightly than nonvalent ones, but the exchange process is essentially a reversible chemical reaction which obeys the law of mass action. Most of the ion exchange reactions involve cations and are often referred to as base exchange.

Although many natural minerals have some exchange capacity, certain of the clay minerals are especially important, both because of high exchange capacity and wide distribution in sediments. The clays of the montmorillonite group are probably the most important. Waters softened by ion exchange have been recognized in many areas and may contain practically no calcium or magnesium. They have a wide range of dissolved solids concentration and among the anions bicarbonate, sulfate, or chloride are present in many different proportions. The sodium bicarbonate type of water is probably the most common (Hem, 1959). Normally, calcium and magnesium ions from solution replace adsorbed sodium on the exchange material. The release of calcium and magnesium from exchange position can be observed in some aquifers that become contaminated with sea water. The zone of contact of fresh and sea water often contains water that is higher in calcium and magnesium than could be explained by assuming a simple admixture of sea water and fresh water.

The chemical reduction of sulfur from the fully oxidized S^{6+} state in the form of sulfate ions to the S^{--} or sulfide state commonly occurs in groundwaters. The reactions are said to occur in the presence of certain types of bacteria (Hem, 1959). Water is probably required as a medium for the reactions, and the solid products formed would have to be carried away to permit the reactions to go to completion.

Water which has been subjected to sulfate reduction can be expected to contain some hydrogen sulfide unless conditions have existed which allowed the gas to escape or react with metals to form metallic sulfides. Many of these waters have a high content of bicarbonate and often of carbon dioxide which can be gained at the expense of sulfate.

PHYSICAL ANALYSIS

The determination of color in groundwater is based on a comparison of a column of the water sample with a column of equal height of an arbitrary standard whose color is rated at 500 (Hem, 1959). This standard is made up from 1 gram of cobalt chloride and 1.245 grams of potassium chloroplatinate to which is added 100 milliliters of concentrated hydrochloric acid and the whole diluted to 1 liter with distilled water. Color is expressed in units between 0 and 500 or more based on the standard. Color determination may indicate the presence of organic materials in aquifers and is significant in the evaluation of drinking-water supplies where color is not desirable.

Turbidity is a measure of the suspended and colloidal matter in groundwater, such as clay, silt, organic matter, and microscopic organisms. Turbidity limits are often included in production-well specifications. Small particles of ferric oxide may come from rusty casings and pump columns. Concentrations of suspended sediment in excess of 5 ppm may cause trouble. Tastes and odors may be derived from bacteria, dissolved gases, mineral matter, or phenols.

The temperature of groundwater at depths between 30 and 60 feet normally exceeds the mean annual air temperature by 2 or 3°F (see Collins, 1925). A shallow groundwater temperature map is given in Fig. 7.6. Below shallow depths, the temperature increases in accordance with the geothermal gradient approximately 1°C for each

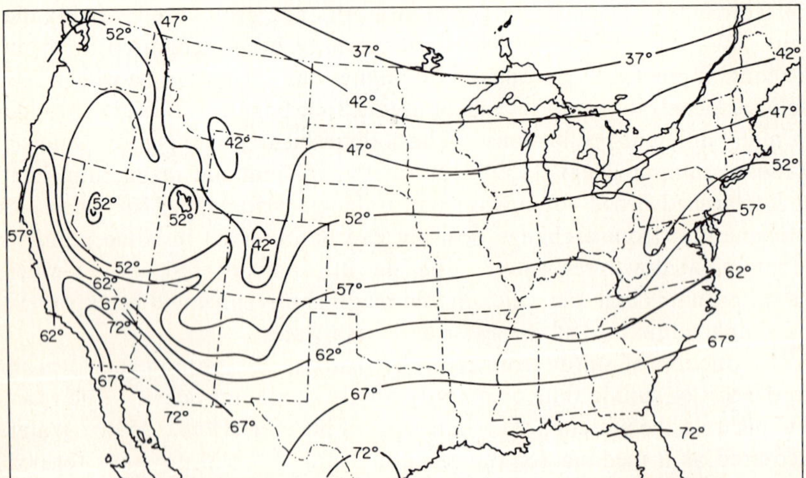

Fig. 7.6 Approximate temperature of shallow groundwater in conterminous United States. (*After Collins, 1925.*)

100 feet of additional depth. Annual temperature variations under ordinary circumstances are less than 1°F.

BACTERIAL ANALYSIS

Pathogenic organisms, fortunately, are rarely found in groundwater. Most cases of contamination are caused by poor well construction or are associated with bedrock aquifers in which large openings afford direct connection between the surface and groundwater. Several types of organic matter may occur in groundwater but rarely in amounts greater than 15 ppm.

Standard tests to determine the presence or absence of bacteria of the coliform group are made as an indication of the safety of groundwater for drinking purposes.

INTERPRETATION OF CHEMICAL ANALYSES

In the interpretation of groundwater chemical data, the analyses are correlated with one another and with related information. Studies commonly involve statistical analysis of the data and preparation of maps, diagrams, and graphs. A first step in data interpretation is the evaluation of the available data, with respect to their completeness, accuracy, and reliability. Water analyses then are generally presented in tabular form. A number of computer programs in Fortran IV have been developed using an IBM-7040 computer to store, retrieve, and manipulate water-quality data (Morgan et al., 1966).

Expression of the relationship of one ion to another or to the total dissolved solids in terms of a ratio may make resemblances or differences among groundwaters apparent. When a large number of analyses for a single source are available, a simple arithmetical average and extremes may shed light on the significance of groundwater quality data.

Salinity profiles for aquifers have been used as an effective means of showing groundwater quality changes over distances. Maps showing the areal distribution of the concentration of dissolved solids or of individual constituents often are helpful.

The first grouping of groundwater quality data for study purposes should be according to aquifers. A study of the analyses for waters from each aquifer will then show how uniform the composition of water in the aquifer may be. After the correlation of analyses with the stratigraphy of the region, a consideration of the areal variation in water quality in each aquifer and the relation of water quality in aquifers is usually desirable. This often is best accomplished through preparation of maps and profiles. A study of maps will often give some indication as to hydrogeologic or mineralogic conditions in the aquifers.

In many studies, the relationship of surface waters to groundwaters is important. The comparison of chemical analyses of waters from both surface and aquifers will often aid in establishing the existence and extent of such relationships.

A graph showing the changes with respect to time of some property of groundwater is generally termed a *hydrograph*. Hydrographs of chemical quality may be useful to illustrate the characteristics of quality fluctuations with time.

7.3 Graphical Treatment of Chemical Analyses

The system of graphical presentation of chemical analyses proposed by Collins (1923) is widely used. The chemical analysis is represented by a vertical bar graph whose total height is proportional to the total concentration of the cations and anions expressed in equivalents per million. The bar is divided into six subdivisions to show the concentration of individual cations and anions as illustrated in Fig. 7.1. Cations are plotted on the left half of the vertical bar and anions on the right.

The Stiff (1951) graphical method using four parallel horizontal axes and one vertical axis may be useful in making comparisons of waters, especially highly mineralized ones. Four cations are plotted along each axis to the left of the zero point and four anions on the right. Concentrations are expressed in equivalents per million. Connecting points representing anions and cations give a closed figure or "pattern" whose shape is more or less characteristic of a given kind of water as shown in Fig. 7.7.

Most natural groundwaters can be represented as solutions of three cationic constituents, calcium, magnesium, and the alkali metals; and of three anionic constituents, sulfate, chloride, and those contributing to alkalinity. The composition of groundwater may be represented conveniently by trilinear plotting (see Hem, 1959). Trilinear diagrams are very useful in bringing out the chemical relationships among groundwaters in more definite terms than is possible with other plotting methods.

Piper (1953) developed a form of the trilinear diagram which is an effective tool in segregating analysis data for critical study with respect to sources of the dissolved constituents in groundwaters, modifications in the character of a water as it passes through an area, and related geochemical problems. For the Piper trilinear diagram, a groundwater is treated substantially as though it contained three cation constituents (Mg, Na, and Ca) and three anion constituents (Cl, SO_4, and HCO_3). Less abundant

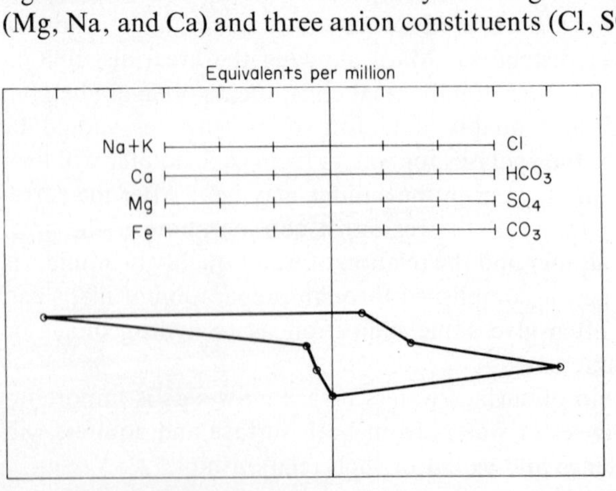

Fig. 7.7 Chemical analyses represented by patterns. (*After Hem, 1959.*)

TABLE 7.4 INFORMATION CONCERNING COMMON AND MINOR CONSTITUENTS OF GROUNDWATER FOR USE IN PIPER TRILINEAR DIAGRAM. *(From Piper, 1953.)*

Cations	*Reciprocal of combining weight*	*Anions*	*Reciprocal of combining weight*
Alkaline earths		Weak acids	
Calcium (Ca^{++})	0.04990	Bicarbonate (HCO_3^-)	0.01639
Barium (Ba^{++})	0.01456	Carbonate (CO_3^{--})	0.03333
Strontium (Sr^{++})	0.02282	Tetraborate ($B_4O_7^{--}$)	0.01288
Magnesium (Mg^{++})	0.08224	Orthophosphate (PO_4^{3-})	0.03157
Alkalies		Strong acids	
Sodium (Na^+)	0.04348	Sulfate (SO_4^{--})	0.02082
Potassium (K^+)	0.02558	Chloride (Cl^-)	0.02820
Rubidium (Rb^+)	0.01170	Bromide (Br^-)	0.01251
Lithium (Li^+)	0.14409	Fluoride (F^-)	0.05263
Ammonium (NH_4^+)	0.05543	Nitrate (NO_3^-)	0.01613

constituents than these major cations and anions are summed with the major constituents to which they are related in chemical properties as indicated in Table 7.4.

The Piper trilinear diagram (Fig. 7.8) combines three distinct fields for plotting, two triangular fields at the lower left and lower right, respectively, and an intervening diamond-shaped field. All three fields have scales reading in 100 parts. In the triangular field at the lower left, the percentage reacting values of the three cation groups (Ca, Mg, Na) are plotted as a single point according to conventional trilinear coordinates. The three anion groups (HCO_3, SO_4, and Cl) are plotted likewise in the triangular field at the lower right. Thus, two points on the diagram, one in each of the two triangular fields, indicate the relative concentrations of the several dissolved constituents of a groundwater. The subtotal of all cation equivalents per million is taken as the 100 percent base for computing percentage reacting values of the several cation variables; likewise for the several anion variables. The central diamond-shaped field is used to show the overall chemical character of the groundwater by a third single-point plotting, which is at the intersection of rays projected from the plottings of cations and anions. The position of this plotting indicates the relative composition of a groundwater in terms of the cation-anion pairs that correspond to the four vertices of the field. The three trilinear plottings will show the essential chemical character of a groundwater according to the relative concentration of its constituents, but not according to the absolute concentrations. Because the absolute concentrations commonly are decisive in many problems of interpretation, it is convenient to indicate the plotting in the central field by a circle whose area is proportional to the absolute concentration of the water. The chemical character of waters containing free acid in substantial quantities (hydrogen is present as a cation) cannot be fully represented on the diagram (Piper, 1953).

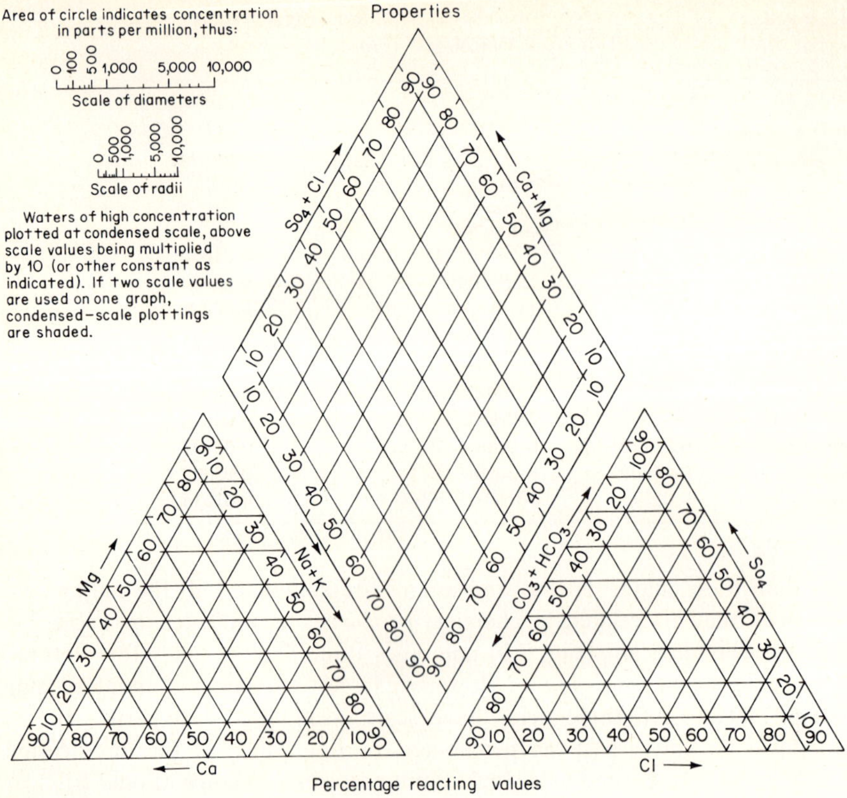

Fig. 7.8 Piper trilinear diagram. (*After Piper, 1953.*)

Distinct groundwater quality types can be quickly discriminated by their plottings in certain subareas of the diamond-shaped field, as indicated by Fig. 7.9 and the following explanation: area 1, alkaline earths exceed alkalies; area 2, alkalies exceed alkaline earths; area 3, weak acids exceed strong acids; area 4, strong acids exceed weak acids; area 5, carbonate hardness ("secondary alkalinity") exceeds 50 percent—that is, chemical properties of the groundwater are dominated by alkaline earths and weak

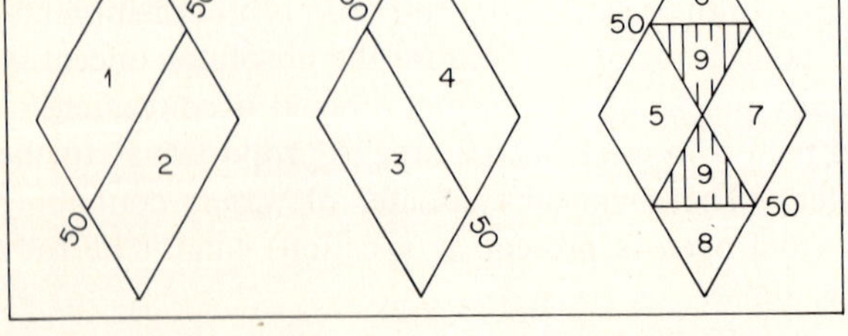

Fig. 7.9 Subdivisions of diamond-shaped field of Piper trilinear diagram. (*After Piper, 1953.*)

acids; area 6, noncarbonate hardness ("secondary salinity") exceeds 50 percent; area 7, noncarbonate alkali ("primary salinity") exceeds 50 percent—that is, chemical properties are dominated by alkalies and strong acids; ocean water and many brines plot in this area, near its right-hand vertex; area 8, carbonate alkali ("primary alkalinity") exceeds 50 percent—here plot the groundwaters which are inordinately soft in proportion to their content of dissolved solids; area 9, no one cation-anion pair exceeds 50 percent (Piper, 1953).

The solution to many hydrogeologic problems involving apparent mixtures of groundwaters is facilitated by use of the trilinear diagram. Mixtures of two groundwaters in all proportions, if all products remain in solution, plot in the three fields on the respective straight lines that joint the points representing the respective chemical characters of the two groundwaters mixed (Piper, 1953).

To demonstrate conclusively that a certain groundwater is a quantitative mixture of two other groundwaters, neither diluted, concentrated, nor chemically modified after the mixing, one graphic criterion and one graphic-algebraic criterion must be satisfied (Piper, 1953). By the graphic criterion, in all three fields of the diagram the apparent mixture must plot on straight lines between the plottings of its two inferred components. Also, the area-concentration plottings in the central field must conform to the principle that the concentration of a mixture is necessarily greater than the least, but less than the greatest, of the several concentrations of its components.

By the graphic-algebraic criterion, decisive proof of a quantitative mixture is accomplished when, for the groundwater of intermediate composition, agreement is shown between analytical data and corresponding values computed from Eq. (7.1) for total concentration and from Eqs. (7.2) and (7.3) for concentration of individual chemical constituents (Piper, 1953).

$$E_m = \frac{E_a E_b (a_i + b_i)}{a_i E_a + b_i E_b} \tag{7.1}$$

$$V_a = \frac{b_i E_b}{a_i E_a + b_i E_b} \quad \text{and} \quad V_b = \frac{a_i E_a}{a_i E_a + b_i E_b} \tag{7.2}$$

$$C_m = C_a V_a + C_b V_b \tag{7.3}$$

where C_a = concentration in component water A, in epm or ppm as desired
C_b = concentration in component water B, in epm or ppm as desired
C_m = concentration in the mixture, in epm or ppm as desired
V_a = proportionate volume in mixture M of water having composition A
V_b = proportionate volume of water having composition B
E_a = concentration of water A, in epm
E_b = concentration of water B, in epm
E_m = concentration of the mixture, in epm
a_i = intercept between the plottings of A and M, measured in any of the three fields of the diagram and at any convenient scale
b_i = intercept between the plottings of B and M

The chemical character of a mixture of groundwaters will plot at the center of gravity of the plottings of the respective components, each having been weighted according to its concentration and its proportionate volume in the mixture. Based on this principle, graphic-algebraic criteria for decisive proof of quantitative mixtures are perhaps feasible for a three-component system but become involved for systems with more than three components (Piper, 1953).

Changes in the chemical character of a natural groundwater by solution of progressively increasing amounts of some particular mineral must plot on a set of straight-line vectors directed in each of the three fields toward the point representing the chemical composition of the mineral.

If two groundwaters (or a groundwater and a mineral) react chemically (for example, sulfate reduction and base exchange) when brought together and some product or products of the reaction form in solid phase, the chemical character of the products remaining in solution will not plot on the straight line joining the points that represent the two reacting groundwaters. Rather, it will plot on the extension of the straight line drawn from (1) the point that represents the composition of the solid phase to (2) the point that indicates the proportionate volumes and compositions of the two reacting groundwaters (Piper, 1953). The chemical composition of a natural groundwater may undergo complex changes by an interplay of several or numerous processes. The trilinear diagram can often shed light on these processes.

All the trilinear systems omit silica from consideration. Because of the importance of silicates in rocks, consideration of silica in water is often desirable where correlations between water analyses and kinds of source rocks are to be made (Hem, 1959).

RELATIONSHIP OF GROUNDWATER QUALITY TO USE

Most interpretations of groundwater quality data are made to determine if the water is satisfactory in quality for a proposed use. Whether a groundwater of a given quality is suitable for a particular purpose depends on the criteria or standards of acceptable quality for that use. Quality limits of water supplies for drinking water and for industrial and agricultural purposes, will now be described.

7.4 Drinking Water

Most drinking-water supplies in the conterminous United States conform to standards established by the U.S. Public Health Service (1962). The following chemical substances should not be present in a water supply in excess of the listed concentrations where, in the judgment of the reporting agency and the certifying authority, other more suitable water supplies are or can be made available.

Alkyl benzene sulfonate (ABS)	0.5 mg/liter
Arsenic (As)	0.01
Chloride (Cl)	250
Copper (Cu)	1.0
Carbon chloroform extract (CCE)	0.2
Cyanide (CN)	0.01
Iron (Fe)	0.3
Manganese (Mn)	0.05
Nitrate (NO_3)[a]	45
Phenols	0.001
Sulfate (SO_4)	250
Total dissolved solids	500
Zinc (Zn)	5

[a] In areas in which the nitrate content of water is known to be in excess of the listed concentration, the public should be warned of the potential dangers of using the water for infant feeding.

The presence of the following substances in excess of the concentrations listed shall constitute grounds for rejection of the water supply:

Arsenic (As)	0.5 mg/liter
Barium (Ba)	1.0
Cadmium (Cd)	0.01
Chromium (hexavalent) (Cr^{6+})	0.05
Cyanide (CN)	0.2
Lead (Pb)	0.05
Selenium (Se)	0.01
Silver (Ag)	0.05

When fluoride is naturally present in drinking water, the concentration should not average more than the appropriate upper limit given in the list that follows. Presence of fluoride in average concentrations greater than two times the optimum values shall constitute grounds for rejection of the water supply. The recommended control limits of fluoride concentrations for various annual averages of maximum daily air temperatures are as follows:

	Lower	*Optimum*	*Upper*
50.0–53.7°F	0.9 mg/liter	1.2 mg/liter	1.7 mg/liter
53.8–58.3°	0.8	1.1	1.5
58.4–63.8°	0.8	1.0	1.3
63.9–70.6°	0.7	0.9	1.2
70.7–79.2°	0.7	0.8	1.0
79.3–90.5°	0.6	0.7	0.8

Water supplies shall be approved without further consideration of other sources of radioactivity intake of radium-226 and strontium-90 when the water contains these substances in amounts not exceeding 3 and 10 $\mu\mu c$/liter, respectively. When these concentrations are exceeded, a water supply shall be approved by the certifying authority if surveillance of total intakes of radioactivity from all sources indicates that such intakes are within the limits recommended by the Federal Radiation Council for control action. In the known absence of strontium-90 and alpha emitters, the water supply is acceptable when the gross beta concentrations do not exceed 1,000 $\mu\mu c$/liter. Gross beta concentrations in excess of 1,000 $\mu\mu c$/liter shall be grounds for rejection of the water supply except when more complete analyses indicate that concentrations of nuclides are not likely to cause exposures greater than the Radiation Protection Guides as approved by the President on recommendation of the Federal Radiation Council.

The presence of organisms of the coliform group as indicated by samples examined shall not exceed the following limits:

When 10-ml standard portions are examined, not more than 10 percent in any month shall show the presence of the coliform group. When 100-ml standard portions are examined, not more than 60 percent in any month shall show the presence of the coliform group. When the membrane filter technique is used, the arithmetic mean coliform density of all standard samples examined per month shall not exceed one per 100 ml.

Drinking water should contain no impurity which would cause offense to the sense of sight, taste, or smell. Under general use, the following limits should not be exceeded.

Turbidity	5 units
Color	15 units
Threshold odor number	3

7.5 Industrial Use

It should be apparent that the quality requirements of waters used in different industrial processes vary widely. Thus, makeup water for high-pressure boiler must meet extremely exacting criteria (see Table 7.5) whereas water of as low a quality as sea water can be satisfactorily employed for cooling of condensers. Even within each industry, criteria cannot be established; instead, only recommended limiting values or ranges can be stated. From an extensive study of industrial water quality, a committee of the New England Water Works Association (Anon, 1940) suggested limits for selected industries given in Table 7.6.

Of almost equal importance for industrial purposes as quality of a water supply is the relative constancy of the various constituents. In some areas, groundwaters have been used extensively for industrial purposes because of their low and relatively constant temperature.

TABLE 7.5 SUGGESTED WATER-QUALITY TOLERANCE FOR BOILER FEED WATER (ALLOWABLE LIMITS IN ppm). *(From Progress Report of the Committee on Quality Tolerances of Water for Industrial Uses, 1940.)*

Pressure, psi	*0–150*	*150–250*	*250–400*	*Over 400*
Turbidity	20	10	5	1
Color	20	40	5	2
Oxygen consumed	15	10	4	3
Dissolved oxygen[a]	1.4	0.14	0	0
Hydrogen sulfide (H_2S)	5[b]	3[b]	0	0
Total hardness as $CaCo_3$	80	40	10	2
Sulfate-carbonate ratio (ASNE)				
($Na_2SO_4:Na_2CO_3$)	1:1	2:1	3:1	3:1
Aluminum oxide (Al_2O_3)	5	0.5	0.05	0.01
Silica (SiO_2)	40	20	5	1
Bicarbonate (NCO_3)[a]	50	30	5	0
Carbonate (CO_3)	200	100	40	20
Hydroxide (OH)	50	40	30	15
Total solids[c]	3,000–500	2,500–500	1,500–100	50
pH value (minimum)	8.0	8.4	9.0	9.6

[a] Limits applicable only to feed water entering boiler, not to original water supply.
[b] Except when odor in live steam would be objectionable.
[c] Depends on design of boiler.

It is technically possible to treat any water to give it a desired composition. However, if the water requires extensive treatment, it may not be economically feasible to utilize the supply.

7.6 Agricultural Use

Range cattle in the western United States seem to be able to use water containing 5,000 ppm of dissolved solids, but a high proportion of sodium or magnesium and sulfate is very undesirable for stock use. A concentration of more than 0.5 ppm selenium in water could be very poisonous to animals.

Factors to be considered in evaluating the usefulness of groundwaters for irrigation are: the total concentration of dissolved solids, the concentration of individual constituents, the relative proportions of some of the constituents, the nature and composition of the soil and subsoil, the topography of the land, the position of the water table, the amounts of groundwater used and the method of applying it, the kinds of crops grown, and the climate of the area. Most of the soluable matter in irrigation groundwater remains behind in the soil. In order to maintain the productivity of irrigated soil on a permanent basis, excess soluble matter left in the soil from irrigation must be removed by leaching the topsoil and allowing the resulting solution to percolate to the water table. If the water table rises excessively because of irrigation, this process of drainage disposal of salts may not be effective. Lack of proper drainage to remove excess salts and groundwater will cause the abandonment of land.

TABLE 7.6 SUGGESTED WATER-QUALITY TOLERANCES FOR INDUSTRIAL USES (ALLOWABLE LIMITS IN ppm). (*From Anon, 1940.*)

Industry or use	*Turbidity*	*Color*	*Odor and taste*	*Iron as* Fe	*Manganese as* Mn	*Total solids*	*Hardness as* $CaCO_3$	*Alkalinity as* $CaCO_3$	*Hydrogen sulfide*	*Health*	*pH*	*Other requirements*
Air conditioning	···	···	Low	0.5	0.5	···	···	···	1.0			No corrosiveness or slime formation
Baking	10	10	Low	0.2	0.2	···	···	···	0.2	Potable	···	
Boiler feed												No corrosiveness or scale formation
Pressure (psi)												
0–150	20	80	···	···	···	3,000–5,000	80	···	5	···	8.0	
150–250	10	40	···	···	···	2,500–500	40	···	3	···	8.4	
250–400	5	5	···	···	···	1,500–100	10	···	0	···	9.0	
> 400	1	2	···	···	···	50	2	···	0	···	9.6	
Brewing and distilling												NaCl 275
Light beer, gin	10	···	Low	0.1	0.1	500	···	75	0.2	Potable	6.5–7.0	
Dark beer, whiskey	10	···	Low	0.1	0.1	1,000	···	150	0.2	Potable	7.0	
Canning												
Legumes	10	···	Low	0.2	0.2	···	25–75	···	1.0	Potable	···	
General	10	···	Low	0.2	0.2	···	···		1.0	Potable	···	
Carbonated beverages	2	10	Low	0.2	0.2	850	250	50–100	0.2	Potable	···	Organic matter infinitesimal, oxygen consumed 1.5
Confectionery	···	···	Low	0.2	0.2	100	···	···	0.2	Potable	7.0	
Cooling	50	···	···	0.5	0.5	···	50	···	5	···	··· ···	No corrosiveness or slime formation
Food, general	10	···	Low	0.2	0.2	···	···	···	···	Potable	···	
Ice	5	5	Low	0.2	0.2	1,300	···	···	···	Potable	···	SiO_2 10
Laundering	···	···	···	0.2	0.2	···	50	···	···	···		
Plastics, clear	2	2	···	0.02	0.02	200	···	···	···	···		
Paper and pulp												No grit or corrosiveness
Ground wood	50	20	···	1.0	0.5	···	180	···	···	···	···	
Kraft pulp	25	15	···	0.2	0.1	300	100	···	···	···	···	
Soda and sulfide pulp	15	10	···	0.1	0.05	200	100	···	···	···	···	
High-grade light papers	5	5	···	0.0	0.05	200	50	···		···	···	No slime formation
Rayon (Viscose)												
Pulp production	5	5	···	0.05	0.03	100	8	50	···	···	···	OH 8, Al_2O_3 8, SiO_2 25, Cu 5
Manufacture	0.3	···	···	0.0	0.0	···	55	···	···	···	7.8–8.3	
Steel manufacture	···	···	···	···	···	···	50	···	···	···	6.8–7.0	Temp. 75°F, Cl 175, suspended matter 25, minimum organic content and corrosiveness
Sugar manufacture	···	···	···	0.1	···	···	···	···	···	···	···	Ca 20, Mg 10, SO_4 20, Cl 20, HCO_3 100
Synthetic rubber	···	···	···	···	···	···	50	···	···	···	···	Oxygen consumed 3.0, minimum organic content and corrosiveness
Tanning	20	10–100	···	0.2	0.2	···	50–135	135	···	···	···	OH 8
Textiles												
General	5	20	···	0.25	0.25	···	···	···	···	···	···	Constant composition; residual alumina < 0.5
Dyeing	5	5–20	···	0.25	0.25	200	···	···	···	···	···	

TABLE 7.7 RATING OF IRRIGATION WATER FOR VARIOUS CROPS ON THE BASIS OF BORON CONCENTRATION IN WATER. (*After Wilcox, 1955.*)

Classes of water		*Sensitive crops,*	*Semitolerant crops,*	*Tolerant crops,*
Rating	*Grade*	*ppm*	*ppm*	*ppm*
1	Excellent	< 0.33	< 0.67	< 1.00
2	Good	0.33–0.67	0.67–1.33	1.00–2.00
3	Permissible	0.67–1.00	1.33–2.00	2.00–3.00
4	Doubtful	1.00–1.25	2.00–2.50	3.00–3.75
5	Unsuitable	> 1.25	> 2.50	> 3.75

In many large irrigated areas, open ditches, buried drain tile, or drainage wells are used to provide drainage and the necessary leaching of soluble matter from the soil. Essentially all the salts added in the irrigation groundwater are recovered in the drainage water. A favorable salt balance must be maintained in an irrigated area if irrigation is to continue permanently.

Boron is essential to proper plant nutrition; however, a small excess over the needed amount is toxic to some plants (Hem, 1959). Table 7.7 shows permissible boron concentrations, and relative tolerances of some plants to boron are given in Table 7.8.

TABLE 7.8 RELATIVE TOLERANCE OF PLANTS TO BORON.[a] (*After U.S. Dept. of Agr. Handbook 60, 1954, and Hem, 1959.*)

Sensitive	*Tolerant*	*Semitolerant*
Lemon	Carrot	Lima bean
Grapefruit	Lettuce	Sweet potato
Avocado	Cabbage	Bell pepper
Orange	Turnip	Pumpkin
Thornless blackberry	Onion	Zinnia
Apricot	Broadbean	Oat
Peach	Gladiolus	Milo
Cherry	Alfalfa	Corn
Persimmon	Garden beet	Wheat
Kadota fig	Mangel	Barley
Grape	Sugar beet	Olive
Apple	Date palm	Ragged robinrose
Pear	Palm	Field pea
Plum	Asparagus	Radish
American elm	Athel	Sweetpea
Navy bean		Tomato
Jerusalem artichoke		Cotton
English walnut		Potato
Black walnut		Sunflower
Pecan		

[a] In each group, the plants first named are considered to be more sensitive and the last named more tolerant.

The U.S. Salinity Laboratory staff (1954) has proposed the use of the sodium adsorption ratio (SAR) for studying the suitability of groundwaters for irrigation purposes. It is defined by

$$\mathrm{SAR} = \frac{\mathrm{Na}^+}{\sqrt{(\mathrm{Ca} + \mathrm{Mg})/2}} \tag{7.4}$$

where all concentrations are expressed in epm.

A soil high in exchangeable sodium is very undesirable for agriculture because it can become deflocculated and tends to have a relatively impermeable crust. This condition is promoted by waters of high SAR and is reversed by waters containing a high proportion of calcium and magnesium (Hem, 1959). Soil amendments such as gypsum or lime may correct the situation.

A diagram for evaluation of irrigation waters on the basis of SAR suggested by the U.S. Salinity Laboratory Staff (1954) is given in Fig. 7.10. Table 7.9, based on the same sources, shows the relative tolerance of crops to the salinity of the water. Within each group, the crops are listed in order of increasing salt tolerance.

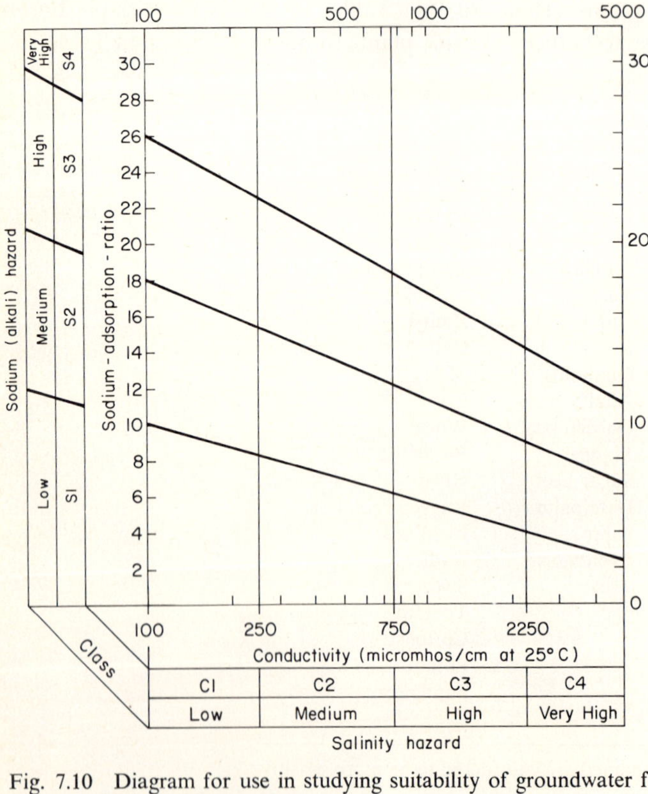

Fig. 7.10 Diagram for use in studying suitability of groundwater for irrigation purposes. (*After U.S. Dept. of Agr. Handbook 60, 1954.*)

TABLE 7.9 RELATIVE TOLERANCES OF CROPS TO SALT CONCENTRATIONS. (*After U.S. Dept. of Agr. Handbook 60, 1954.*)

Crop division	*Low salt tolerance*	*Medium salt tolerance*	*High salt tolerance*
Fruit crops	Avocado Lemon Strawberry Peach Apricot Almond Plum Prune Grapefruit Orange Apple Pear	Cantaloupe Date Olive Fig Pomegranate	Date palm
Vegetable crops[a]	$EC \times 10^3 = 3$ Green beans Celery Radish $EC \times 10^3 = 4$	$EC \times 10^3 = 4$ Cucumber Squash Peas Onion Carrot Potatoes Sweet corn Lettuce Cauliflower Bell pepper Cabbage Broccoli Tomato $EC \times 10^3 = 10$	$EC \times 10^3 = 10$ Spinach Asparagus Kale Garden beet $EC \times 10^3 = 12$
Forage crops[a]	$EC \times 10^3 = 2$ Burnet Ladino clover Red clover Alsike clover Meadow foxtail White dutch clover $EC \times 10^3 = 4$	$EC \times 10^3 = 4$ Sickle milkvetch Sour clover Cicer milkvetch Tall meadow oatgrass Smooth brome Big trefoil Reed canary Meadow fescue Blue grama Orchard grass Oats (hay) Wheat (hay) Rye (hay) Tall fescue Alfalfa Hubam clover Sudan grass Dallis grass	$EC \times 10^3 = 12$ Bird's-foot trefoil Barley (hay) Western wheat grass Canada wild rye Rescue grass Rhodes grass Bermuda grass Nuttall alkali grass Salt grass Alkali sacaton $EC \times 10^3 = 18$

TABLE 7.9 (*Continued*)

Crop division	*Low salt tolerance*	*Medium salt tolerance*	*High salt tolerance*
Forage crops[a]		Strawberry clover Mountain brome Perennial rye grass Yellow sweet clover White sweet clover $EC \times 10^3 = 12$	
Field crops[a]	$EC \times 10^3 = 4$ Field beans	$EC \times 10^3 = 6$ Castorbeans Sunflower Flax Corn (field) Sorghum (grain) Rice Oats (grain) Wheat (grain) Rye (grain) $EC \times 10^3 = 10$	$EC \times 10^3 = 10$ Cotton Rape Sugar beet Barley (grain) $EC \times 10^3 = 16$

[a] Electrical conductivity (EC) values represent salinity levels at which a 50 percent decrease in yield may be expected as compared to yield on nonsaline soil under comparable growing conditions. Concentrations refer to soil water.

CHANGES IN QUALITY

Sewage and industrial wastes entering an aquifer can pollute groundwater. Other causes of deterioration of groundwater quality are listed in Table 7.10. Groundwaters are subject to saline contamination in those areas where salt water is in contact with fresh water or where opportunity exists for such a contact. The effects of man's activities affects the relation between sea water and fresh water. Possible sources of pollution are connected with large-scale artificial recharge operations (including recharge with reclaimed sewage) and waste disposal operations. The problem of salt-water intrusion into highly developed aquifers is serious in many coastal areas and can occur elsewhere if there is a system into which salt water may enter upon reduction of hydrostatic head of fresh water. The mixing of waters of different sources by their simultaneous motion in the aquifer is a tool for quality control.

The temperature of the water withdrawn from a production well field recharged by induced infiltration of river water follows a cycle governed by the seasonal fluctuations of the river-water temperature (Burdick, 1942; Jeffords, 1945; Kazmann, 1947; Rorabaugh, 1951). The temperature range is less than that of the river and the highs and lows occur later than the highs and lows of the river. The temperature range of the discharged water and the time lag depend on a large number of variables (Rorabaugh, 1951): temperature of river water, distance of the production well field from river,

TABLE 7.10 CAUSES OF DETERIORATION OF GROUNDWATER QUALITY. (*From California State Water Resources Bd., 1955.*)

Contamination and pollution

- Domestic and municipal sewage
- Industrial wastes
  - Organic wastes
    - Food processing
    - Lumber processing
  - Mineral wastes
    - Metal-processing industries
    - Mining and ore-extraction industries
    - Oil industries
    - Chemical industries
    - Miscellaneous
  - Cooling water
- Solid and semisolid refuse

Degradation

- Effects of development, use, and reuse of water
  - Irrigation return water
    - Surface drainage
    - Percolation
  - Interchange between aquifers due to improperly constructed, defective abandoned wells
  - Interchange between aquifers due to differentials in pressure levels remote from excessive withdrawal
  - Overdraft conditions
    - Sea water intrusion
    - Salt balance
    - Upward or lateral diffusion of connate brines and/or juvenile water due to overpumping
  - Contamination from the surface due to improperly constructed wells
- Natural causes
  - Inflow and/or percolation of juvenile water from highly mineralized springs and streams
- Other causes
  - Accelerated erosion
  - Mineralization resulting from plant transpiration and/or evaporation

spacing of wells, pumping rate, volume of aquifer, porosity of aquifer amount and temperature of groundwater flow from the land side, and the specific heat of the material in the aquifer. The groundwater being pumped at a given time flows from the river to the production well field by various paths and at rates determined by the hydraulic gradients and properties of the aquifer; thus water entering the aquifer near the production well field where gradients are steepest travels to production wells a very short time compared to water entering the aquifer some distance upstream or downstream. In other words, the groundwater pumped at a given time is made up of water that entered the aquifer at different points, at different times, and at different

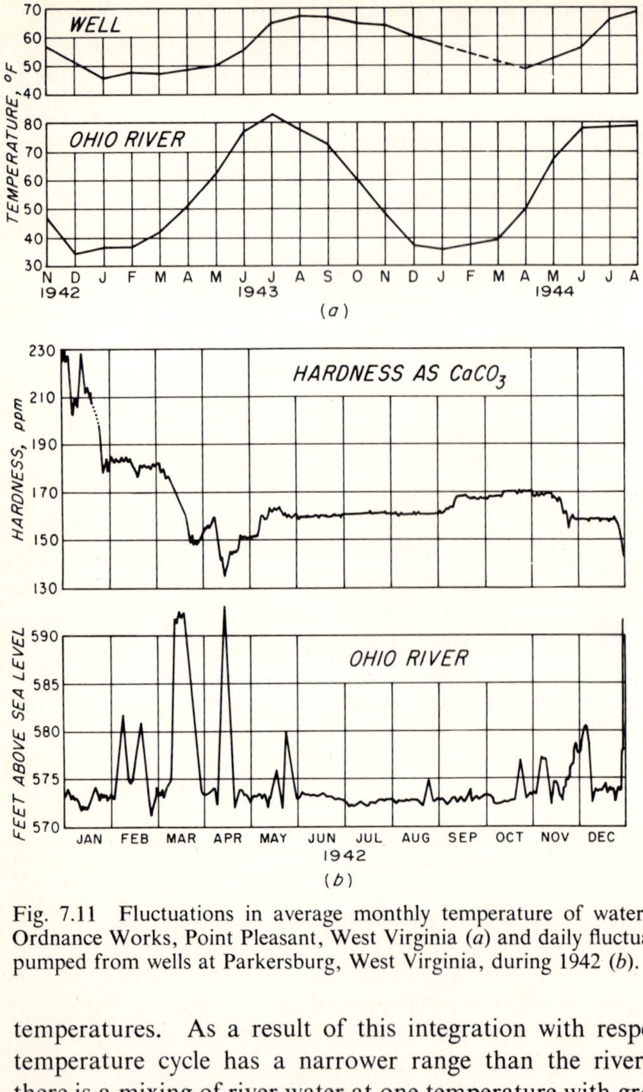

Fig. 7.11 Fluctuations in average monthly temperature of water pumped from a well at West Ordnance Works, Point Pleasant, West Virginia (*a*) and daily fluctuations in total hardness of water pumped from wells at Parkersburg, West Virginia, during 1942 (*b*). (*After Jeffords, 1945.*)

temperatures. As a result of this integration with respect to time, the discharge-temperature cycle has a narrower range than the river-temperature cycle. Also, there is a mixing of river water at one temperature with groundwater at a more nearly uniform temperature, which further reduces the range. There is an exchange of heat between the surface water and the sand as water at one temperature flows into sand at a different temperature. In addition, there may be movement of heat by radiation within the aquifer as well as to or from the bedrock or overlying material. Further complications are introduced by the changes in viscosity associated with changes in temperature; changes in viscosity alter the rate of flow, thereby retarding or accelerating all the other effects in portions of the aquifer.

Figure 7.11*a* illustrates the variations in the temperature of the groundwater pumped from a production well at the West Virginia Ordnance Works along the Ohio River at Point Pleasant, West Virginia. The temperature of groundwater from the production well near the river fluctuated between 47 and 68°F whereas the temperature of water in the Ohio River fluctuated between 36 and 79°F.

A water supply developed through induced infiltration of river water is more nearly uniform in chemical quality than a supply directly from the stream. Because of the mixing of waters that left the stream at different times, sudden changes in the quality of the stream are not reflected in the groundwater.

Normal groundwater along a river varies slightly in chemical character locally; however, the concentrations of the ions in the river water vary greatly. Although the quality of the river water varies in accordance with the season and flow, its hardness and alkalinity are generally below that for normal groundwater at least during most of the year. The hardness and alkalinity of the groundwater from aquifers that obtain a substantial portion of recharge from the river increase gradually during long periods of low river stage, generally during the summer and fall, and decrease abruptly during periods of flood or high river stage, such as occur typically in the early spring. Figure 7.11*b* illustrates the relationship of hardness for a production well field that receives important quantities of water from the Ohio River at Parkersburg, West Virginia.

The performances of many production well fields under a variety of groundwater conditions have shown that under normal conditions, the water produced by induced infiltration is free from turbidity, pathogenic bacteria and organic matter, and tastes and odors. The number of exceptions to this statement are very few and for the most part obtain water by induced infiltration from streams carrying exceptionally heavy organic loads (Klaer, 1963).

CASE HISTORIES OF QUALITY STUDIES

7.7 Clyde, Ohio

An investigation of groundwater quality conditions in the vicinity of Service Plaza 5 on the Ohio Turnpike in northern Ohio was made in 1956 to find the closest location to the service plaza at which a reasonably soft groundwater supply could be obtained (Schaefer and Walton, 1956). The water supply in 1956, obtained from two bedrock production wells on the service plaza property, was unsatisfactory because of the excessively high content of dissolved solids in the water, which could not be removed by the existing treatment plant. In the dishwashing process used at the service plaza the high content of dissolved solids resulted in streaking of dishes, glassware, and silverware, and streaking was noticeable on cars washed at the service plaza.

Service Plaza 5 is located near the intersection of State Routes 510 and 12, about 4 miles north of Clyde, in Sandusky County, Ohio. The investigation covered a 12-square-mile area surrounding the service plaza as shown in Fig. 7.12. The general topography is that of a flat plain with the land surface sloping gently to the northwest. From a topographic standpoint, therefore, it would be advantageous if a suitable site

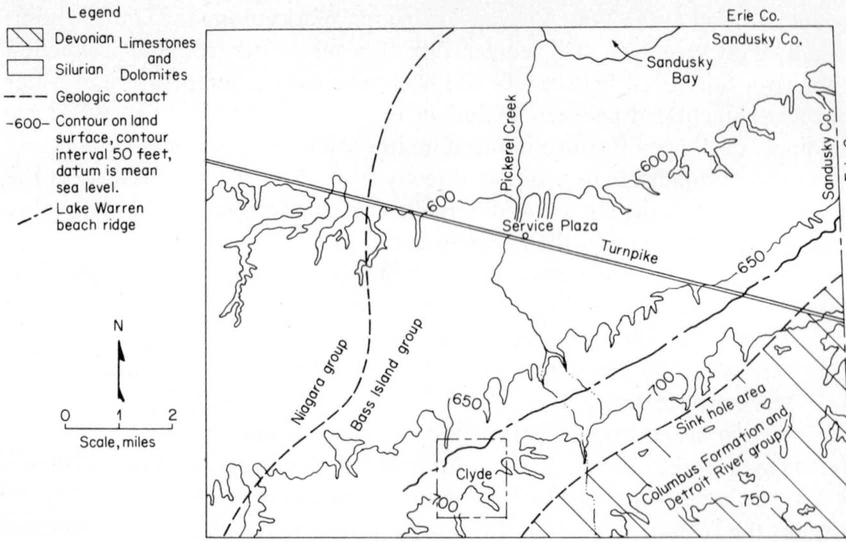

Fig. 7.12 Location of study area in vicinity of Clyde, Ohio. (*From Schaefer and Walton, 1956.*)

for a new production well field could be found to the south, southeast, or east of the service plaza to obtain a downslope of the pipeline. The glacial lake shorelines and beaches in the southeastern part of the area where the surface is rolling are the only conspicuous relief features. Sink holes are numerous and underground drainage takes place through cavernous limestone formations in the higher, southeastern part of the area.

The area is underlain by limestones and dolomites of Upper Silurian and Middle Devonian age. Structurally the rocks lie on the eastern flank of the Cincinnati arch near the axis where the formations dip to the southeast at a rate of about 17 feet per mile. Most of the surface of the bedrock is overlain by clay and fine materials left by the lakes formed during the retreat of the Wisconsinan glacier of Pleistocene age. Sandy beach materials cover the limestones and dolomites in the southeastern part of the area.

Rocks of Devonian age occur only in the southeastern part of the area where limestones and dolomites of the Columbus Formation and the Detroit River lie near the surface (see Fig. 7.12). The Columbus Formation consists of light gray to light brown limestone which is medium to massive in bedding and fine and dense in texture. The Columbus Formation changes from a low magnesian limestone in the upper portion to a limy dolomite in the lower part. An extensive system of joints and other fractures has developed in the Columbus Formation and these joints are connected to fissures that lead to sink holes and underground channels. Springs are numerous in the southeastern part of Sandusky County.

The Columbus Formation is underlain by dolomites of the Detroit River group. The upper formation, the Lucas, of the Detroit River group is a somewhat calcareous dolomite, light blue to drab in color, with beds ranging from 3 to 12 inches in thickness. The lower formation, the Amherstburg, is a true dolomite in composition, massive in bedding and slight drab to brownish gray in color. The textural, bedding, and joint features of the Detroit River group are like the Columbus Formation, favorable for the development of sink holes and underground openings leading to spring zones.

Rocks of Upper Silurian age underlie most of the area. The Silurian-Devonian contact cuts diagonally across the southeast part of the area passing about 1 mile east of Clyde.

The service plaza lies on the Put-in-Bay Dolomite of the Base Island group. The Put-in-Bay is a solomite of fair purity, gray to drab to light brown in color, crystalline and open in texture, and thin to massive in bedding. The rock is broken by joints, bedding planes, and solution cavities and is a fair source of groundwater. Formations of the Bass Island group contain pyrite and organic matter and vary in texture from hard and dense to open; hence the groundwaters are commonly charged with hydrogen sulfide and various soluble salts. The anhydrite and gypsum deposits of northern Ohio belong to the Bass Island group of rocks.

Dolomites of the Niagara group underlie the Bass Island group. The Bass Island–Niagara contact cuts across the area, as shown in Fig. 7.12, about $2\frac{1}{2}$ miles northwest of the service plaza. The Guelph Formation of the Niagara group is a highly crystalline dolomite, light gray or bluish gray in color, and open and porous in texture. The formation is massive in structure and contains 1 to 4 percent celestite or strontium sulfate. The deepest well at the service plaza in 1956 probably terminated in the Guelph Formation.

The most striking features of the bedrock-surface map, shown in Fig. 7.13, are the conspicuous buried valley in the western part of the area and the relatively abrupt cliff trending northeast to southwest in the eastern part of the area.

The area was glaciated by both the Illinoian and Wisconsian ice sheets. The glacial drift averages 50 feet in thickness and, with the exception of the deposits in the southeast corner of the area, consists largely of fine-grained lacustrine clays. In many places the clays are interbedded with irregular lenses or layers of sand and gravel. Deposits of water-washed sand and gravel often occur at the base of the glacial deposits. The dense clays act as aquitards over the underlying rocks and are largely responsible for the flowing artesian wells in the service plaza area. A geologic cross section and profile of the piezometric surface along the turnpike is given in Fig. 7.14.

The bedrock east of the Clyde-Castalia road in the southeast part of Sandusky County is covered with beach materials consisting of sand and sandy silt with some gravel. These permeable deposits are rather loose in texture and transmit water readily to the underlying consolidated rocks. Recharge to the dolomite aquifer underlying the service plaza is derived from precipitation falling on the sandy area. This fact led to the tentative conclusion, which was later verified, that the softest water would probably be found in the rocks beneath or near the recharge area.

The groundwater in the dolomite aquifer underlying the service plaza moves in a general northwesterly direction from the recharge areas in southeastern Sandusky

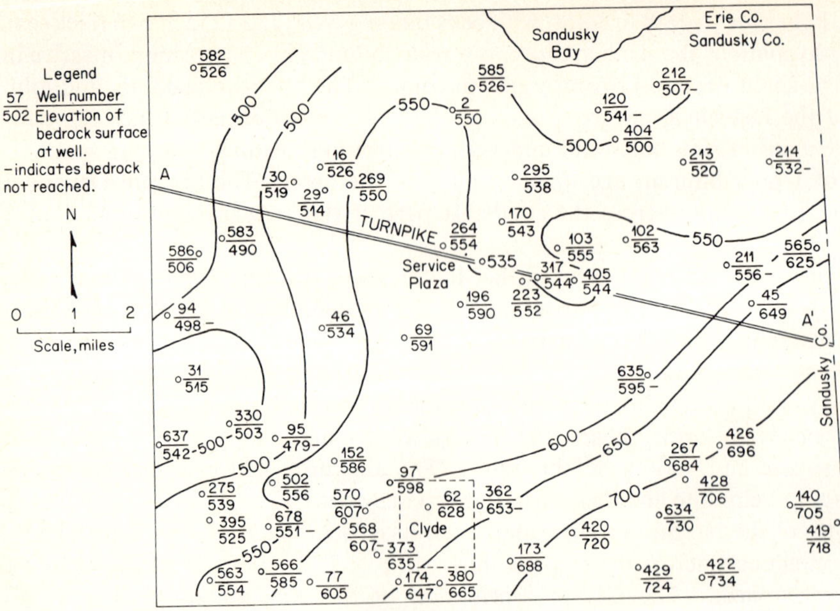

Fig. 7.13 Bedrock topography in vicinity of Clyde, Ohio. (*From Schaefer and Walton, 1956.*)

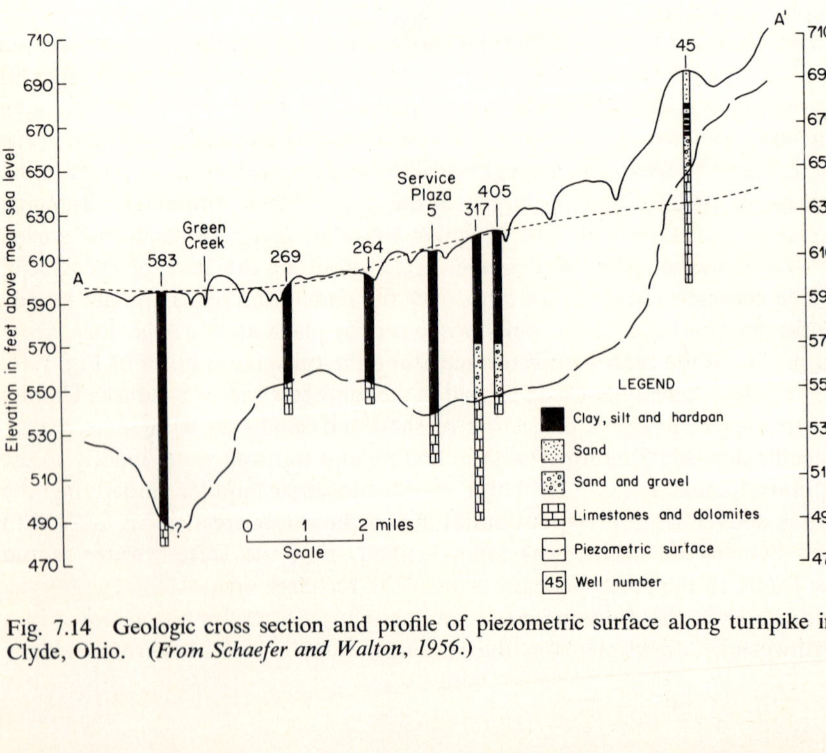

Fig. 7.14 Geologic cross section and profile of piezometric surface along turnpike in vicinity of Clyde, Ohio. (*From Schaefer and Walton, 1956.*)

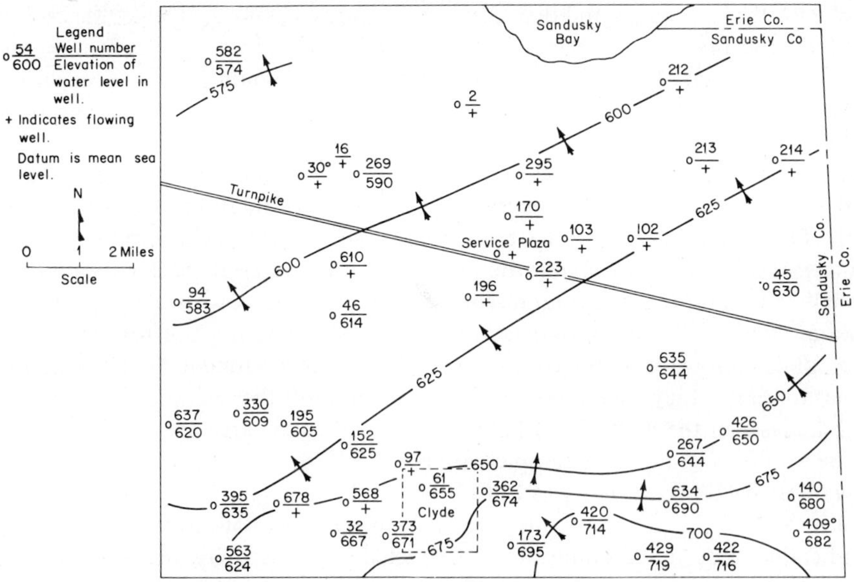

Fig. 7.15 Piezometric surface of bedrock aquifer in vicinity of Clyde, Ohio. (*From Schaefer and Walton, 1956.*)

County to discharge areas in Sandusky Bay and is confined by the relatively impermeable clays that cover the aquifer. When the production wells at the service plaza were first drilled in bedrock they flowed with a head about 14 feet above land surface. Flowing wells with pressure heads of from 5 to 15 feet above land surface are common along the turnpike from Townsend west to the Sandusky River and locally are called fountains. Figure 7.15 is a contour map showing the hydrostatic pressure for the water in the aquifer underlying the service plaza. The sand and gravel deposits overlying the consolidated rocks are hydraulically connected with the limestones and dolomites and water moves into the permeable glacial materials from the underlying rocks. As a result the chemical quality of the water in the glacial deposits is no better than in the rocks below.

The principal soluble material in limestones and dolomites are carbonates, chiefly of calcium. The solution of calcium carbonate, with some magnesium carbonate, is therefore the primary action when waters containing carbon dioxide attack the mineral matter in the limestones and dolomites. The amount of carbonates that can be dissolved largely depends upon the carbon dioxide content of the groundwater and the calcium and magnesium carbonate content of the rocks. As groundwater percolates downwater and laterally it continues to dissolve calcium and magnesium carbonate until the carbon dioxide content of the groundwater is exhausted. Thereafter the carbonate hardness of the groundwater tends to remain constant.

The sulfate minerals anhydrite, $CaSO_4$, gypsum, $CaSO_4 \cdot 2H_2O$, and celestite, $SrSO_4$, are found with the limestones and dolomites in Sandusky County. Unlike the carbonate minerals, the amount of sulfates dissolved in groundwater is not dependent on the carbon dioxide content of the water but on dissolving action only. Hence the period of contact of the water with sulfate minerals in the aquifer is practically the only governing factor with respect to the dissolved sulfate content of the water. The sulfate content of the water increases continuously as the water flows from the intake area to the discharge area, and forms the largest portion of the total dissolved solids despite the fact that more carbonates than sulfates are present in the aquifer.

The analyses of water showed that the production-well water at the service plaza is exceptionally hard and that the total dissolved solids are so excessive as to render the groundwater troublesome and unsatisfactory for most purposes. Unfortunately, the dissolved solids cannot be removed by the 1956 zoolite treatment facilities at the service plaza, nor can they be removed economically by any other method of treatment.

Partial analyses of the water from 15 wells scattered in the area are given in Table 7.11. The locations of the wells sampled are given in Fig. 7.16. Inspection of Table 7.11 shows that the total alkalinity as $CaCO_3$ and carbonate hardness as $CaCO_3$ of the 13 samples vary only a small amount from well to well and both average about 250 ppm, indicating that the groundwaters exhausted their capacity to take calcium carbonate into solution after a relatively short distance of travel underground from the recharge area. However, the noncarbonate hardness as $CaCO_3$ increases from 50 ppm in well 1 to 3,376 ppm in well 11 and the total solid increases from 340 ppm in well 1 to 8,124 ppm in well 11.

The data on the total hardness of the groundwaters from the 15 wells were plotted on the map shown in Fig. 7.16 and contours were drawn through points of equal total hardness. A study of the map shows that the total hardness of the groundwater increases in a northwesterly direction at a rate of about 270 ppm per mile from less than 300 ppm in the sink-hole area to 3,598 ppm near Sandusky Bay. The contours roughly parallel the contours on the piezometric surface in the aquifer, demonstrating that the total hardness increases in the direction of groundwater flow as the duration of the contact of the water with the rocks increases.

TABLE 7.11 CHEMICAL QUALITY OF THE GROUNDWATERS IN THE VICINITY OF CLYDE, OHIO (ppm). (*From Schaefer and Walton, 1956.*)

Sample no.	*1*	*2*	*3*	*5*	*6*	*7*	*8*	*9*	*10*	*11*	*13*	*14*	*15*
Total alkalinity as $CaCO_3$	230	254	292	266	198	228	226	254	258	222	206	260	350
Total hardness as $CaCO_3$	280	372	382	1,584	414	1,734	1,768	360	1,098	3,598	582	798	600
Carbonate hardness as $CaCO_3$	230	254	292	266	198	228	226	254	258	222	206	260	350
Noncarbonate hardness as $CaCO_3$	50	118	90	1,318	216	1,506	1,542	106	840	3,376	376	538	250
Total solids	340	610	482	2,332	710	2,518	2,622	462	1, 546	8,124	866	1,138	820

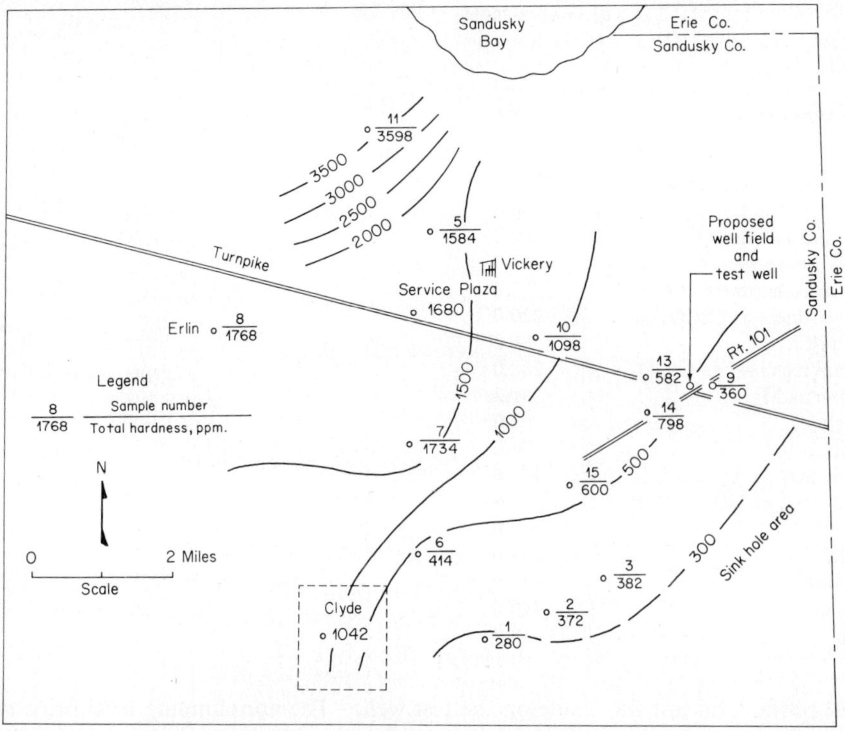

Fig. 7.16 Hardness of groundwater in vicinity of Clyde, Ohio. (*From Schaefer and Walton, 1956.*)

The map in Fig. 7.16 shows that the closest source of groundwater of satisfactory quality (414 ppm total hardness) is about $3\frac{1}{2}$ miles south of the service plaza. Ground water of satisfactory quality (500 ppm) could also be obtained about 4 miles southeast of the service plaza along the turnpike right-of-way. The closest site for a new production well field is about $3\frac{1}{2}$ miles due south of the service plaza. However, such a development would require purchase of right-of-way for a pipeline and a parcel of land on which production wells could be drilled. Wells at a second site, 4 miles southeast of the service plaza, could be drilled on the turnpike property and the pipeline run within the turnpike right-of-way. It was, therefore, recommended that a production well field be developed approximately 4 miles southeast of the service plaza along the turnpike right-of-way at the location shown in Fig. 7.16. Two production wells 8 inches in diameter, 150 feet deep, and spaced 500 feet apart would be satisfactory.

A test well was drilled at the recommended site to a depth of 122 feet. Gray impermeable clay was encountered from the surface to a depth of 20 feet, and dark gray limestone from that depth to the bottom of the well at 122 feet. As finally constructed the well has 25 feet of blank 8-inch-diameter casing with a 30-foot length of 8-inch-diameter perforated pipe screen attached to the bottom of the casing.

TABLE 7.12 CHEMICAL ANALYSIS OF WATER FROM A TEST WELL NEAR CLYDE, OHIO. (*From Schaefer and Walton, 1956.*)

Chemical constituents	*ppm*
Calculated carbon dioxide as CO_2	5.0
Phenol alkalinity as $CaCO_3$	0
Total alkalinity as $CaCO_3$	236.0
Total hardness as $CaCO_3$	502.0
Carbonate hardness as $CaCO_3$	236.0
Noncarbonate hardness as $CaCO_3$	286.0
Calcium hardness as $CaCO_3$	220.0
Calcium as Ca	88.0
Magnesium hardness as $CaCO_3$	282.0
Magnesium as Mg	81.2
Chlorides as Cl	7.0
Total iron as Fe	Trace
Sulfates as SO_4	183.6
Dissolved silica as SiO_2	7.0
Corrosion factor	+4.0
Total solids	574.0
Votile solids	134.0
Fixed solids	440.0
pH value	8.2

A well-production test was made on the test well. The nonpumping level prior to pumping was 24.35 feet below the surface. With a pumping rate of 100 gpm, at the end of the test the water level was about 89.2 feet below the surface. Computations showed that a total supply of 200 gpm, 100 gpm from each production well, could be obtained without causing excessive drawdown in each production well.

A water sample was taken during the well-production test from the test well. The results of the analysis of the water sample are given in Table 7.12.

The hardness of about 502 ppm agrees closely with that which would be predicted from the hardness contour map in Fig. 7.16.

7.8 East-central Iowa

Kunkle (1965) determined groundwater runoff from a small stream in Iowa by measurement of streamflow and quality of water in the stream. Specific electrical conductance was used as a descriptive measure of chemical quality.

The study area consists of 19.5 square miles above a gaging station on Four Mile Creek near Traer in east-central Iowa. Upland areas are underlain by loess that rests on till, and lowlands that are underlain by alluvium contribute groundwater to the creek.

A recording conductance meter at the gaging station shows that when streamflow is composed entirely of groundwater runoff the specific conductance of the creek water is uniformly 520 micromhos, ± 1 percent. The conductance of water from wells in the loess and alluvium ranges from 418 to 635 micromhos and averages 545 micromhos.

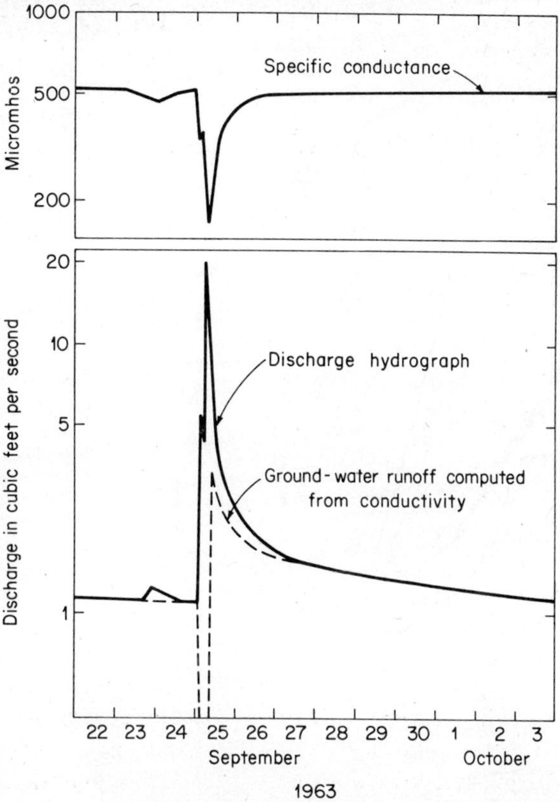

Fig. 7.17 Hydrographs showing water discharge, specific conductance, and computed groundwater runoff in Four Mile Creek near Traer, Iowa. (*From Kunkle, 1965.*)

The relation between magnitude of a flood discharge resulting from a 1.10-inch rainstorm and that of the specific conductance of water in the creek is shown in Fig. 7.17. A representative conductance value for surface runoff of 160 micromhos was determined from conductance data for peak runoffs, surface detention, and drainage from grassed ditches. Groundwater runoff shown in Fig. 7.17 was estimated by simultaneous solution of the following equations (Kunkle, 1965):

$$520R_g + 160R_s = C_C R_u \tag{7.5}$$

$$R_g + R_s = R_u \tag{7.6}$$

where R_g is groundwater runoff, R_s is surface water runoff, C_C is specific conductance of streamflow water, and R_u is total streamflow. The dashed line on Fig. 7.17 shows groundwater runoff. The computed discharge values include appreciable amounts of bank storage.

7.9 Atlantic Coastal Plain, New Jersey

Seaber (1962) used data on the chemical character of water in the Englishtown Formation of the Atlantic Coastal plain of New Jersey to determine the relation between

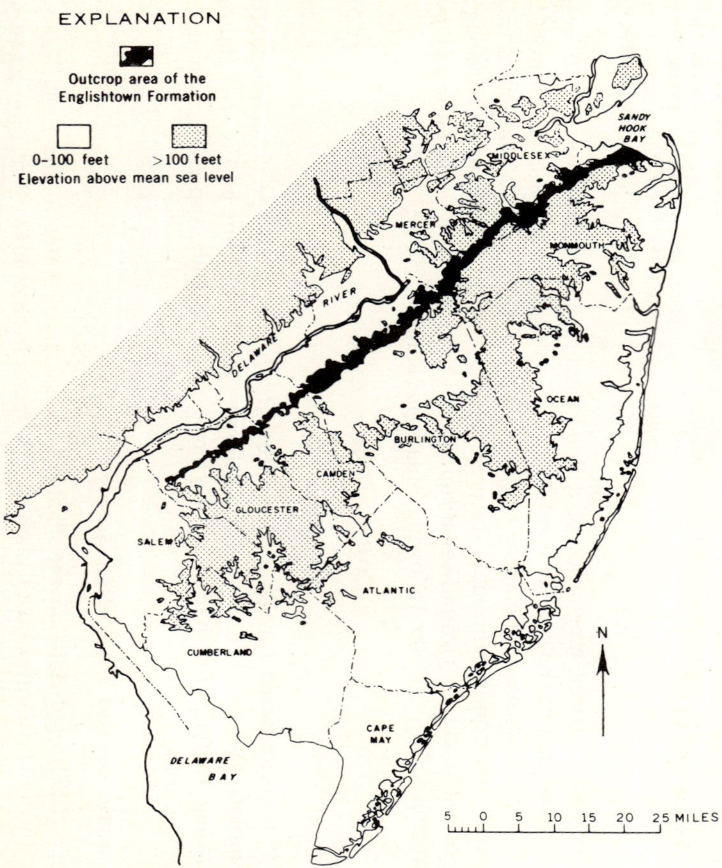
EXPLANATION
Outcrop area of the Englishtown Formation
0-100 feet
>100 feet
Elevation above mean sea level
SANDY HOOK BAY
MIDDLESEX
MERCER
MONMOUTH
RIVER
DELAWARE
OCEAN
BURLINGTON
CAMDEN
GLOUCESTER
SALEM
ATLANTIC
CUMBERLAND
CAPE MAY
DELAWARE BAY
N
5 0 5 10 15 20 25 MILES

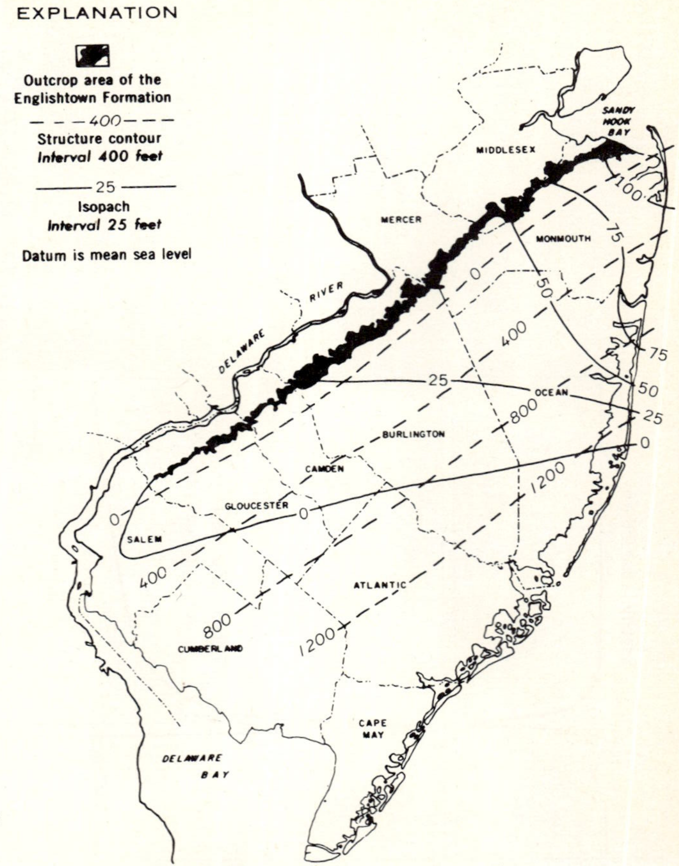
EXPLANATION
Outcrop area of the Englishtown Formation
400
Structure contour
Interval 400 feet
25
Isopach
Interval 25 feet
Datum is mean sea level
SANDY HOOK BAY
MIDDLESEX
MERCER
MONMOUTH
RIVER
DELAWARE
OCEAN
BURLINGTON
CAMDEN
GLOUCESTER
SALEM
ATLANTIC
CUMBERLAND
CAPE MAY
DELAWARE BAY

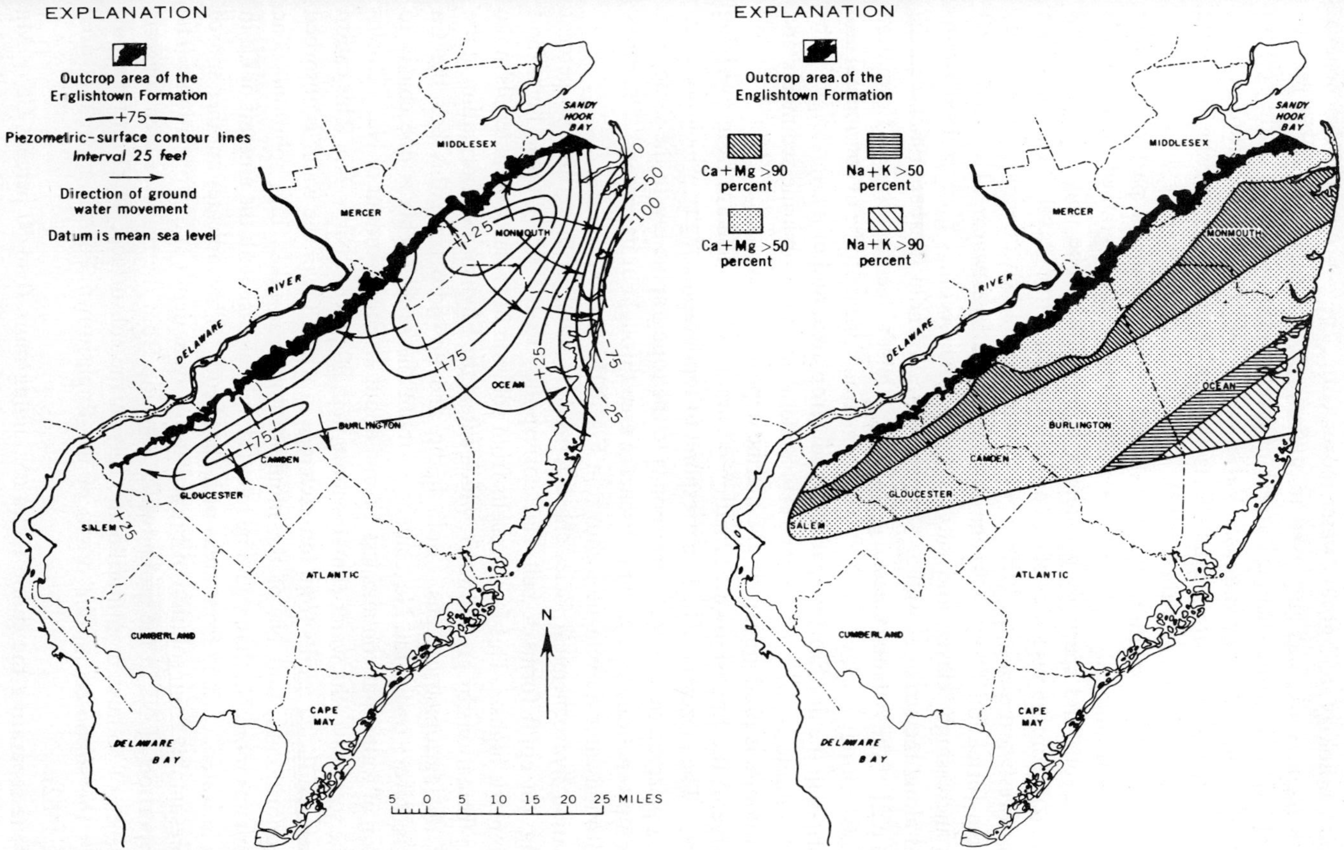

Fig. 7.18 Maps showing relation between topography, geology, groundwater movement, and cation hydrochemical facies in Englishtown Formation of Atlantic Coastal Plain in New Jersey. (*From Seaber, 1962.*)

topography, lithology, and groundwater movement. The concept of hydrochemical facies (see Back, 1961) was employed in studies. A hydrochemical facies may be thought of as a diagnostic chemical aspect of groundwater indicative of the relative proportions of the anions or cations. Variations in the relative proportions of Ca, Mg, Na, and K reflect environmental effects and can be used as diagnostic indicators of hydrochemical relations (Seaber, 1962).

The Englishtown Formation, where exposed in the northeast part of the area (Fig. 7.18), is dominantly light-colored well-sorted medium-grained sand that is slightly micaceous and sparingly glauconitic and lignitic. The unit contains a few laminae and discontinuous lenses of clay and silt and some fine gravel. The formation thins southwestward along the strike, but appears to thicken downdip. The sandy part of the formation forms a wedge that thins southwestward. The sandy unit becomes increasingly clayey and silty and the sand becomes finer grained southwestward along the strike as well as southeastward downdip (Seaber, 1962). In the southern part of the Atlantic Coastal plain of New Jersey, the Englishtown Formation is composed entirely of clay and silt. The formation appears to be of continental and deltaic origin in the northeastern part of the outcrop area, and of marine origin in the south. Kaolinite is the predominant clay mineral in the nonmarine sediments and montmorillonite is predominant in the marine deposits of the coastal plain.

Recharge of the Englishtown Formation occurs mostly by leakage from overlying sediments. The piezometric highs correspond to topographic highs, which are separated by a piezometric low that corresponds to the topographic low in the same area. The outcrop functions as a natural-discharge area throughout its extent. The direction of groundwater movement is shown in Fig. 7.18.

The cation hydrochemical facies characterized by water of more than 90 percent Ca + Mg (in epm) forms a belt that corresponds to the topographic highs and the piezometric highs. This facies results from solution of calcareous materials in the overlying formations by groundwater, which enters the Englishtown Formation in the areas of the piezometric highs. Both updip and downdip from this belt the Ca content is lower; updip the Na content is about the same. This is due chiefly to infiltration of water that contains less Ca and about an equal amount of Na. Downdip to the southeast, the water contains more than 90 percent Na + K. This cation hydrochemical facies results from ion exchange in which Ca of the water is absorbed by montmorillonite, and Na of the montmorillonite is taken into solution. The water contains more Na than could be released by exchange with the amount of Ca in solution in waters coming from areas of higher head. This indicates another source of Ca, presumably from solution of calcareous fossils at depth in the formation. The high proportion of Na in the groundwater downdip from the recharge area suggests the presence of sodium montmorillonitic clay in the subsurface, which in turn substantiates the supposed marine origin of the Englishtown Formation in that area (Seaber, 1962).

The zone separating the cation facies containing more than 50 percent Ca + Mg from that containing more than 50 percent Na + K is considered to be an "ion-exchange front." This front is presumably moving slowly east-southeastward at an indefinite rate in the direction of the movement of the water within the formation. The present alignment and position of the front is governed more by the piezometric surface that existed prior to pumping from the Englishtown Formation than by the present piezometric surface (Seaber, 1962).

7.10 Estevan, Saskatchewan, Canada

During 1963 and 1964, the Saskatchewan Research Council (see Walton, 1964) conducted a groundwater quality study in conjunction with a groundwater level study in the Estevan area, Saskatchewan (see case histories of groundwater exploration in Chap. 2 of this book). Data on groundwater quality assisted in the design of a test-hole drilling program aimed at delineating the sand and gravel aquifers in the buried Missouri River system in the vicinity of Estevan.

Geologic and hydrologic data suggested that permeable glacial sand and gravel deposits of large areal extent occur between the depths of 250 and 450 feet in the buried valley which cuts into the Ravenscrag Formation. The Ravenscrag Formation consists of poorly consolidated interbedded sandstones and shales with coal beds.

Water-level data indicated that the horizontal flow of groundwater in large parts of the Estevan area at depths between 250 and 450 feet was toward the buried valley. There is also vertical flow of groundwater from the surface downward through glacial drift to depths between 250 and 450 feet. It was reasoned that if there was appreciable horizontal and vertical flow of groundwater into the deeply buried sand and gravel deposits, as suggested by water-level data, the water in the sand and gravel deposits would be a mixture of waters from the Ravenscrag Formation and overlying glacial drift. Data on groundwater quality were collected and studied to evaluate the validity of conclusions reached with water-level data.

Samples of water from 46 existing wells in the Estevan area were collected and analyzed. Wells from which water samples were collected were segregated into the following four categories: (1) wells terminating within the upper 50 feet of glacial drift, (2) wells terminating at depths between 150 and 250 feet in glacial drift, (3) wells terminating at depths between 300 and 450 feet in glacial drift (in the deeply buried sand and gravel deposits), and (4) wells terminating in bedrock (Ravenscrag Formation).

The contrast in chemical content of waters from wells in the four categories is demonstrated by the water-analysis diagrams in Fig. 7.19. Waters from wells in categories 1 and 2 are somewhat similar in chemical content but much different in chemical content than waters from wells in category 4. Waters from wells in category 3 are intermediate in chemical content of waters from wells in categories 1 and 2 and 4. The water from wells in category 3 has many of the characteristics of the chemical content of waters from wells in category 4. Thus, water in the sand and gravel deposits between the depths of 250 and 450 feet in the buried valley appears

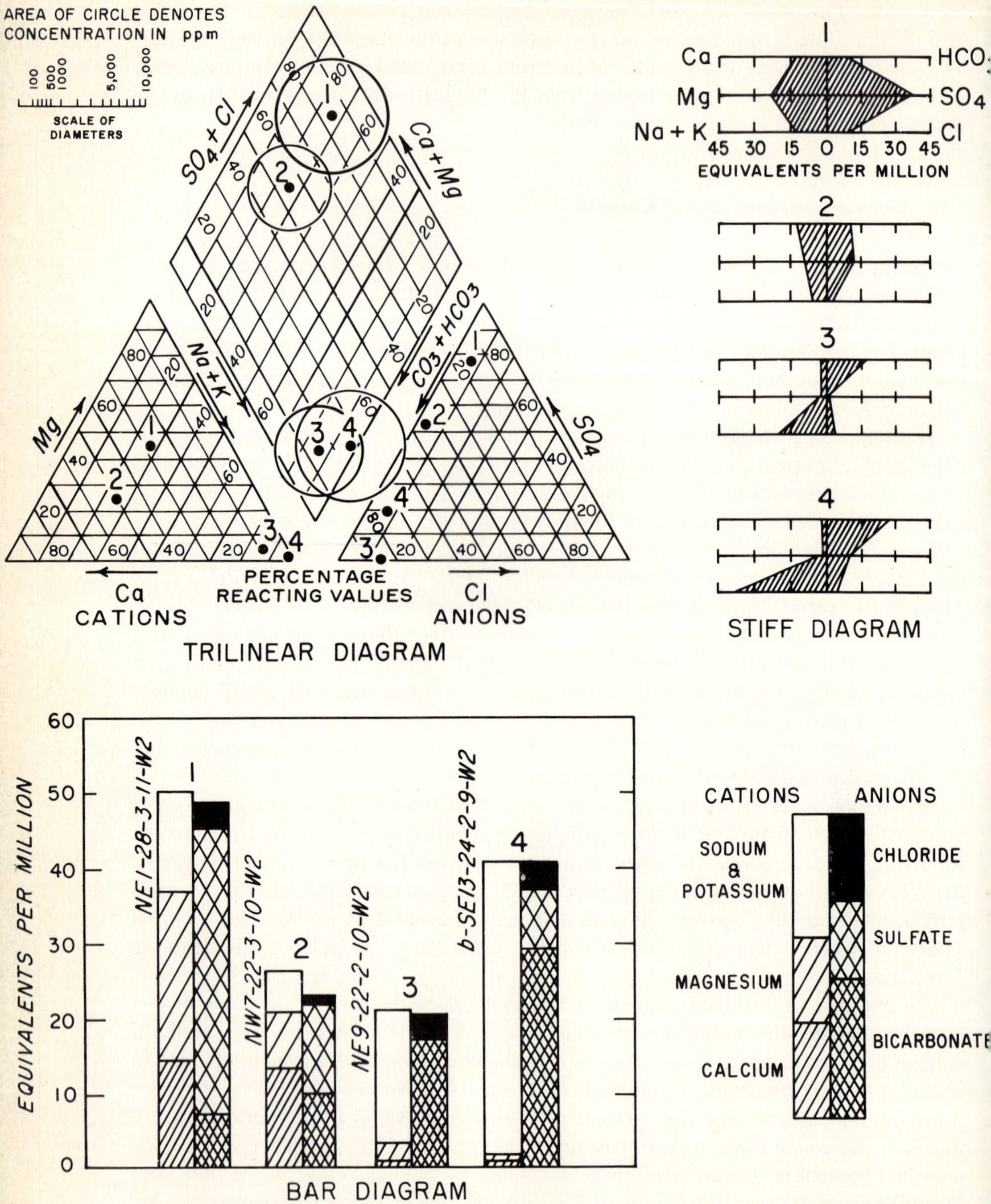

Fig. 7.19 Water analysis diagrams for Estevan area, Saskatchewan, Canada. (*From Walton, 1964.*)

to be a mixture of waters from the Ravenscrag Formation and the overlying glacial drift.

A study of the trilinear diagram indicates that water in the deeply buried sand and gravel deposits is not a simple mixture of waters from the Ravenscrag Formation and the glacial drift with all products remaining in solution. The composition of waters has come about by chemical changes as the result of an interplay of several processes including probably sulfate reduction and base exchange.

Data on the quality of groundwaters confirmed conclusions reached with water-level data that extensive and permeable sand and gravel deposits occur at depths between 250 and 450 feet in the buried valley. Based on quality of water, water-level, and geologic data, it was possible to design a much more efficient test-hole drilling program than would have been possible without these studies. Thus, it is possible to use groundwater quality data as a geophysical tool in exploring for groundwater.

Problems

7.1 The chemical analyses of groundwater from several wells are given in Table P7.1. Plot bar and Stiff diagrams for each well.

TABLE P7.1

	Chemical content, ppm		
Constituent	*Well 1*	*Well 2*	*Well 3*
SiO_2	135.00	36.0	14.0
Al	...	0.1	...
Fe	0.06	...	9.0
Mn	...	...	...
Ca	5.20	173.0	696.0
Mg	1.00	32.0	204.0
Na	28.00	40.0	462.0
K	3.60	2.6	9.1
HCO_3	50.00	218.0	72.0
SO_4	7.80	268.0	2,130.0
Cl	3.50	135.0	850.0
F	12.00	0.1	1.5
CO_3	0	0	0
NO_3	0.20	7.5	...
Dissolved solids	309.00	820.0	4,400.0

7.2 The chemical analyses of groundwaters from several wells are given in Table P7.2. Using Fig. 7.10, classify waters with respect to salinity and alkali hazards and their suitability for irrigation use.

TABLE P7.2

Chemical content, ppm; reacting values in italics given in epm

Well no.	Silica (SiO_2)	Iron (Fe)	Calcium (Ca)	Magnesium (Mg)	Sodium (Na)	Potassium (K)	Bicarbonate (HCO_3)	Carbonate (CO_3)	Sulfate (SO_4)	Chloride (Cl)	Fluoride (F)	Nitrate (NO_3)	Boron (B)	Dissolved solids	Hardness as $CaCO_3$ (all carbonate hardness)	Percent sodium	Specific conductance in micromhos per cm^3 at 25°C	pH
5	10	0.57	3.8 *0.19*	1.9 *0.16*	280 *12.17*	5.0 *0.13*	609 *9.98*	35 *1.17*	53 *1.10*	13 *0.37*	0.6 *0.03*	0 *0*	…	703	18	96	1,100	8.3
11	13	0.07	2.7 *0.13*	1.0 *0.08*	129 *5.61*	1.0 *0.03*	258 *4.23*	22 *0.73*	45 *0.94*	6.0 *0.17*	0.4 *0.02*	0 *0*	0.26	347	10	96	564	8.8
12	15	0.49	8.2 *0.41*	3.4 *0.28*	98 *4.26*	1.9 *0.05*	250 *4.10*	…	39 *0.81*	4.6 *0.13*	0.3 *0.02*	*0.1* *0*	0.11	294	34	85	468	8.2
15	16	1.1	13 *0.65*	5.0 *0.41*	87 *3.78*	2.5 *0.06*	246 *4.03*	…	38 *0.79*	4.6 *0.13*	0.3 *0.02*	0.2 *0*	0.11	289	53	77	457	8.0
16	20	0.36	17 *0.85*	5.8 *0.48*	91 *3.95*		257 *4.21*	…	43 *0.90*	5.2 *0.15*	0.3 *0.02*	0.1 *0*	0.06	309	66	75	487	…
19	14	0.42	14 *0.70*	6.7 *0.55*	173 *7.52*	1.6 *0.04*	276 *4.52*	12 *0.40*	162 *3.37*	15 *0.42*	0.5 *0.03*	0.1 *0*	0.16	535	62	85	836	8.4
	18	0.17	8.8 *0.44*	3.4 *0.28*	232 *10.09*	2.6 *0.07*	323 *5.29*	4 *0.13*	229 *4·77*	20 *0.56*	0.8 *0.04*	0.0 *0*	…	678	36	93	1,060	8.3
20	16	0.28	5.6 *0.28*	3.0 *0.25*	109 *4.74*	1.4 *0.04*	262 *4.29*	…	38 *0.79*	6.4 *0.18*	0.4 *0.02*	0.2 *0*	0.07	309	26	89	481	8.3
21	17	0.46	20 *1.00*	6.7 *0.55*	80 *3.48*		248 *4.06*	…	40 *0.83*	4.2 *0.12*	0.3 *0.02*	0.1 *0*	0.06	291	78	69	464	…
22	16	0.24	26 *1.30*	8.4 *0.69*	72 *3.15*	6.8 *0.17*	261 *4.28*	…	42 *0.87*	5 *0.14*	0.3 *0.02*	0 *0*	…	305	100	59	467	8.2
23	16	0.24	29 *1.45*	8.1 *0.67*	74 *3.20*	7.8 *0.20*	269 *4.41*	…	44 *0.92*	6 *0.17*	0.4 *0.02*	0 *0*	0.04	318	106	58	488	8.0
25	14	0.05	3.8 *0.19*	1.8 *0.15*	121 *5.24*		238 *3.90*	18 *0.60*	43 *0.90*	5.5 *0.16*	0.3 *0.02*	0.1 *0*	0.11	325	17	94	515	…
26	16	0.05	1.8 *0.09*	0.7 *0.06*	201 *8.74*		363 *5.95*	40 *1.21*	58 *1.21*	13 *0.37*	0.5 *0.03*	0 *0*	0.37	510	8	98	818	…
27	14	0.08	1.6 *0.08*	0.7 *0.06*	145 *6.31*	1.0 *0.03*	212 *3.47*	53 *1.77*	48 *1.00*	6.0 *0.17*	0.3 *0.02*	0.1 *0*	0.40	374	7	97	588	8.5
28	11	0.05	3.8 *0.19*	2.5 *0.21*	159 *6.92*	3.6 *0.09*	258 *4.23*	48 *1.60*	57 *1.19*	13 *0.37*	0.3 *0.02*	0 *0*	…	426	20	93	684	8.4
29	18	0.11	4.6 *0.23*	3.1 *0.25*	113 *4.91*		239 *3.92*	41 *0.47*	41 *0.85*	4.5 *0.13*	0.3 *0.02*	0.1 *0*	0.11	316	24	91	495	…
30	17	0.10	7.0 *0.35*	3.8 *0.31*	108 *4.71*		238 *3.90*	12 *0.40*	44 *0.92*	4.5 *0.13*	0.3 *0.02*	0.1 *0*	0.09	314	33	88	495	…
31	19	0.10	16 *0.80*	8.5 *0.70*	92 *4.00*		267 *4.38*	…	45 *0.94*	5.8 *0.16*	0.3 *0.02*	0.1 *0*	0.18	318	75	73	505	…
	19	0.19	15 *0.75*	8.0 *0.66*	110 *4.78*	3.3 *0.08*	289 *4,74*	…	55 *1.15*	9.2 *0.26*	0.6 *0.03*	0.1 *0*	…	362	70	76	579	7.9
32	17	0.11	8.6 *0.43*	4.8 *0.39*	102 *4.44*	2.2 *0.06*	254 *4.16*	…	39 *0.81*	6.2 *0.17*	0.3 *0.02*	0.2 *0*	0.04	306	41	83	466	8.5
33	23	2.7	30 *1.80*	8.2 *0.67*	78 *3.40*		278 *4.56*	…	51 *1.06*	8.0 *0.23*	200.	0.1 *0*	0.04	342	124	58	543	…
45	13	0.05	1.2 *0.06*	1.0 *0.08*	194 *8.43*		335 *5.49*	44 *1.47*	57 *1.19*	14 *0.39*	0.5 *0.03*	0.1 *0*	0.46	490	7	98	798	…

7.3 Review the reference given below and prepare a short paper discussing the use of trilinear diagrams by the authors.

Piper, A. M., et al., 1953. "Native and Contaminated Waters in the Long Beach–Santa Ana Area, California," U.S. Geol. Surv. Water Supply Paper 1136.

7.4 Review the references given below and prepare a short paper describing the techniques of mapping water-quality characteristics used by the authors.

Drescher, W. J., 1953. "Ground-water Conditions in Artesian Aquifers in Brown County, Wis.," U.S. Geol. Surv. Water Supply Paper 1190.

Gatewood, J. S., T. W. Robinson, B. R. Colby, J. D. Hem, and L. C. Halpenny, 1950. "Use of Water by Bottomland Vegetation in Lower Safford Valley, Ariz.," U.S. Geol. Surv. Water Supply Paper 1103.

Hill, R. A., 1940. Geochemical Patterns in Coachella Valley, Calif., *Trans. Am. Geophys. Union*, vol. 21.

7.5 Using information in the files of state and federal water resources agencies, study the quality of groundwater in an area within your state. Prepare a short report relating the quality of water to possible uses.

7.6 Review the reference given below and prepare a short report describing the techniques used by the author in studying radionuclides in groundwater.

Reichert, S. O., 1962. Radionuclides in Groundwater at the Savannah River Plant Waste Disposal Facilities, *J. Geophys. Res.*, vol. 67.

References

Am. Public Health Assoc., Am. Water Works Assoc., and Federation of Sewage and Ind. Wastes Assocs., 1955. "Standard Methods for the Examination of Water, Sewage and Industrial Wastes," 10th ed., Am. Public Health Assoc., New York.

Anon, 1940. Progress Report of the Committee on Quality Tolerances of Water for Industrial Uses, *J. New Engl. Water Works Assoc.*, vol. 54, 1940.

Back, W., 1961. "Techniques for Mapping Hydrochemical Facies," U.S. Geol. Surv. Profess. Paper 424-D.

Burdick, C. B., 1942. Ground Water—A Vital National Resource, Midwest Problems, *J. Am. Water Works Assoc.*, vol. 37.

California State Water Resources Bd., 1955. "Water Utilization and Requirements of California," Bull. 2, vol. 2.

Collins, W. D., 1923. Graphical Representation of Water Analyses, *Ind. Eng. Chem.*, vol. 15.

———, 1925. "Temperature of Water Available for Industrial Use in the United States," U.S. Geol. Surv. Water Supply Paper 520-F.

Davis, S. N., and R. J. M. DeWiest, 1966. "Hydrogeology," John Wiley & Sons, Inc., New York.

Hem, J. D., 1959. "Study and Interpretation of the Chemical Characteristics of Natural Water," U.S. Geol. Surv. Water Supply Paper 1473.

Jeffords, R. M., 1945. Recharge to Water-bearing Formations along the Ohio Valley, *J. Am. Water Works Assoc.*, vol. 37.

Kazmann, R. G., 1947. Discussion, Apparent Changes in Water Storage during Floods at Peoria, Ill., *Trans. Am. Geophys. Union*, vol. 28.

Klaer, F. H., Jr., 1963. Bacteriological and Chemical Factors in Induced Infiltration, *Ground Water*, vol. 1, no. 1.

Kunkle, G. R., 1965. "Computation of Groundwater Discharge to Streams during Floods, or to Individual Reaches during Base Flow, by Use of Specific Conductance," U.S. Geol. Surv. Profess. Paper 525-D.

Morgan, C. O., R. J. Dingman, and J. M. McNellis, 1966. Digital Computer Methods for Water Quality Data, *Ground Water*, vol. 4, no. 3.

Piper, A. M., 1953. "A Graphic Procedure in the Geochemical Interpretation of Water Analyses," U.S. Geol. Surv. Ground Water Note 12.

Rorabaugh, M. I., 1951. "Streambed Percolation in Development of Water Supplies," U.S. Geol. Surv. Ground Water Note 25.

Schaefer, E., and W. C. Walton, 1956. "Report on Groundwater Conditions on Ohio Turnpike in the Vicinity of Clyde, Ohio," typewritten report prepared for Ohio Turnpike Commission.

Seaber, P. R., 1962. "Cation Hydrochemical Facies of Ground Water in the Englishtown Formation, New Jersey," U.S. Geol. Surv. Profess. Paper.

Stiff, H. A., Jr., 1951. The Interpretation of Chemical Water Analysis by means of Patterns, *J. Petrol.*, tech. pt. 15.

Todd, D. K., 1959. "Ground Water Hydrology," John Wiley & Sons, Inc., New York.

U.S. Dept. of Health, Education, and Welfare, 1962. "Public Health Service Drinking Water Standards," U.S. Public Health Service Publ. 956.

U.S. Salinity Lab. staff, 1954. "Diagnosis and Improvement of Saline and Alkali Soils," U.S. Dept. Agr. Handbook 60.

Walton, W. C., 1964. "Potential Yield of a Sand and Gravel Aquifer in a Buried Valley near Estevan, Saskatchewan," typewritten report prepared for Saskatchewan Res. Council.

Wilcox, L. V., 1955. "Classification and Use of Irrigation Waters," U.S. Dept. of Agr. Circ. 969.

CHAPTER 8

HYDROGEOLOGIC SYSTEMS ANALYSIS

The groundwater hydrologist is concerned chiefly with the quantitative description of the hydrogeologic properties and boundaries of aquifers and aquitards, and with the response of aquifers to development and management practices. It is his responsibility to evaluate groundwater resources and to forecast the consequences of the utilization of aquifers. Proper planning for the development of groundwater resources requires testing of all possible schemes of development and appraising of the relative merits of various alternates. Groundwater resource development and management personnel are concerned with the practical sustained yields of wells and aquifers and the interrelation between surface water and groundwater.

In dealing with problems pertaining to the use of groundwater resources, the groundwater hydrologist must relate pumpage to water-level change with reference to time and space. He must determine the decline in water levels due to the withdrawal of water from aquifers. The two factors to consider are the cause and effect: pumpage and decline in water levels. The hydrogeologic characteristics and dimensions of the aquifer and existing aquitards and the boundaries of the aquifer are of the utmost importance in relating cause and effect. Computations of the practical sustained yield of aquifers require study of the cause-and-effect relationship. The *practical sustained yield* is here defined as the rate at which groundwater can be continuously withdrawn without lowering water levels to critical stages, exceeding recharge, or causing undesirable changes in water quality. *Critical water levels* may occur when pumping water levels are below tops of well screens or when they reach other undesirable stages.

The factors to be considered in calculating aquifer response are the coefficients of permeability and/or transmissibility and storage of the aquifer and the coefficients of permeability and storage of existing aquitards. Declines in water levels can be evaluated only if the widths, lengths, and thicknesses of the aquifer and existing aquitards are known. Boundaries, such as folds, faults, or relatively impervious layers of shale or clay, and beds of lakes, rivers, and other bodies of surface water influence the response of an aquifer to pumping. Cause cannot be related to effect without a complete hydrogeologic systems analysis and until hydrogeologic maps are available, which describe the following factors: (1) coefficient of permeability and/or transmissibility of the aquifer; (2) coefficient of storage of the aquifer; (3) area of the aquifer; (4) saturated thickness of the aquifer; (5) coefficients of permeability and storage of existing aquitards; (6) saturated thicknesses of existing aquitards; and (7) location, extent, and nature of the aquifer and aquitard boundaries. These prescribed maps must encompass all nonhomogeneous and irregular hydrogeologic conditions.

The response of an aquifer to pumping cannot be predicted with great precision unless the basic definition of the hydrogeologic conditions of the aquifer is precise. However, the task of defining the hydrogeologic conditions is difficult, because available basic data are seldom sufficient to permit rigorous descriptions of aquifers, and economic limitations often prohibit the collection of extensive and detailed information concerning the complexities of these aquifers. Insufficient basic data require much interpretation, extrapolation, and application of hydrogeologic principles together with intuition in the preparation of the requisite quantitative hydrogeologic maps.

The incompleteness of basic data need not necessarily dictate that quantitative evaluations of cause-and-effect relations be rarely made. Approximate solutions based on existing data, when properly qualified according to the quality and quantity of data, can be of great importance. Refinement and periodic reevaluation of solutions after the collection of additional data will be necessary and should be foreseen. Actually, the study of cause-and-effect relations based on existing data illuminates the deficiencies of information and the benefit-cost ratio for the collection of additional data.

Analytical and computer techniques have been devised that allow the study of cause-and-effect relationships. Analytical methods involving model aquifers and mathematical models and computer methods involving electric analog models or digital computers are described in the following sections.

ANALYTICAL APPROACH

Analytical methods have been applied chiefly to relatively uniform aquifers with simple geometry and have been used primarily for problems involving only parts of aquifers or aquifers of small areal extent. Electric analog computer methods are more realistic and adaptable than analytical methods. However, the realism and versatility of analog computers will depend largely upon the availability of abundant and accurate basic data. In some cases, basic data are not sufficient to warrant a rigorous description of complex aquifer conditions and analytical methods may be as useful as electric analog computer methods.

Recognized departures from ideal conditions do not necessarily dictate that analytical methods are rarely used. Such departures emphasize the need for sound professional judgment in the application of analytical methods to existing hydrogeologic conditions and in properly qualifying results according to the extent of departures. With appropriate recognition of hydrogeologic controls, there are many practical ways of circumventing analytical difficulties posed by complicated field conditions. Many aquifers can be highly idealized with little sacrifice in accuracy of analysis (Walton and Neill, 1960).

8.1 Model Aquifers and Mathematical Models

In applying analytical methods to field problems (see Walton and Walker, 1961), the hydrogeologic boundaries of the aquifer evident from areal studies must be idealized to fit comparatively elementary geometric forms such as wedges and infinite or semi-infinite rectilinear strips. Boundaries are assumed to be straight-line demarcations. The gross hydrogeologic properties of the aquifer and aquitards, if present, are considered in evaluating the effects of boundaries, and the detailed hydrogeologic property variations are considered in estimating the interference between wells and well fields.

Actual groundwater conditions are simulated with *model aquifers* which have straight-line boundaries and an effective width, length, and thickness. The aquifer is sometimes overlain by an aquitard which has an effective thickness. The areal extent of an aquifer often varies laterally or longitudinally and the construction of model aquifers requires that the complex aquifer area be converted to an equivalent uniform area.

Mathematical models are based on the hydrogeologic properties of model aquifers, the image-well theory, and groundwater equations. Problems associated with hydrogeologic boundaries are simplified to the consideration of an infinite aquifer in which real and image wells operate simultaneously. The effects of real and image wells are computed with appropriate groundwater formulas. It should be recognized that idealized mathematical models describe the drawdown least accurately in the immediate vicinity of boundaries. The greater the distance to the boundary from the observation point, the smaller will be the error involved with the approximation. Analytical methods involving the heat-conduction equation with or without distributed sources or sinks are sometimes used with model aquifers to solve problems.

Records of past pumpage and water levels may be used to establish whether assumed mathematical models satisfy the hydrogeologic limits of an aquifer. If computed and actual water-level declines agree, the mathematical model provides a means of evaluating the future performance of wells and aquifers. Equations commonly used with mathematical models and helpful conversion factors are given in Table 8.1.

For a demonstration of the application of analytical methods to groundwater resource evaluation, the solution to a simple hypothetical groundwater development problem will be described in detail. The problem is to estimate the practical sustained yield of two identical production wells in an aquifer, assuming no water-quality problems. The aquifer is a strip of sand and gravel averaging 50 feet in thickness and 0.25 mile wide. The aquifer extends beyond the expected influence of pumping. The average hydrogeologic properties of the aquifer, the coefficients of transmissibility and storage, are 50,000 gpd/ft and 0.0005, respectively. An aquitard composed of

TABLE 8.1 COMMONLY USED MATHEMATICAL-MODEL EQUATIONS AND SELECTED CONVERSION FACTORS

Mathematical-model equations

Isotropic nonleaky artesian aquifer with fully penetrating wells and constant-discharge conditions

$$s = \frac{114.6Q}{T} W(u) \qquad u = \frac{1.87r^2S}{Tt}$$

Isotropic nonleaky artesian aquifer with partially penetrating wells and constant-discharge conditions

$$s = \frac{114.6Q}{T} W\left(u, \frac{r}{m}, \gamma\right) \qquad u = \frac{1.87r^2S}{Tt}$$

$$\gamma = \frac{m - m_a}{m}$$

Isotropic leaky artesian aquifer with fully penetrating wells and constant-discharge conditions without water released from storage in aquitard

$$s = \frac{114.6Q}{T} W\left(u, \frac{r}{B}\right) \qquad u = \frac{1.87r^2S}{Tt}$$

$$\frac{r}{B} = \frac{r}{\sqrt{T/(P'/m')}} \qquad s = \frac{229Q}{T} K_0\left(\frac{r}{B}\right)$$

Isotropic water-table aquifer with fully penetrating wells and constant-discharge conditions

$$s = \frac{114.6Q}{T} W\left(u_{ay}, \frac{r}{D_t}\right) \qquad u_a = \frac{1.87r^2S}{Tt}$$

$$u_y = \frac{1.87r^2S_y}{Tt} \qquad \frac{r}{D_t} = \frac{2.73r}{\sqrt{T/D_iS_y}}$$

$$D_i = \frac{(r/D_t)^2(1/u_y)}{4t}$$

where s = drawdown, in feet
Q = discharge, in gpm
T = coefficient of transmissibility of aquifer, in gpd/ft
S = coefficient of storage of aquifer, fraction
r = distance from production well to observation point, in feet
t = time after pumping started, in days
m = saturated thickness of aquifer, in feet
m_a = distance from top of aquifer to top of screen, in feet
P' = coefficient of permeability of aquitard, in gpd/sq ft
m' = saturated thickness of aquitard, in feet
S_y = specific yield of aquifer, in feet
$W(u)$—see Table 3.1
$W\left(u, \frac{r}{m}, \gamma\right)$—see Table 3.3
$W\left(u, \frac{r}{B}\right)$—see Table 3.5
$K_0\left(\frac{r}{B}\right)$—see Table 3.6
$W\left(u_{ay}, \frac{r}{D_t}\right)$—see Table 3.9

TABLE 8.1 (*Continued*)

Selected conversion factors and constants
1 inch of mercury = 1.13 feet of water
1 cfs = 449 gpm
1 $sec^2/ft^5 = 2.02 \times 10^5\ ft/gpm^2$
1 gpd = 6.95×10^{-4} gpm
1 mgd = 695 gpm
1 mgd = 1.54 cfs
1 cfs = 6.46×10^5 gpd
1 sq mi = 2.79×10^7 sq ft
1 inch of water over 1 sq mi = 1.74×10^7 gal
1 inch of water over 1 sq mi per year = 4.77×10^4 gpd
1 acre = 43,560 sq ft
1 sq mi = 640 acres
1 cu ft = 7.48 gal
1 gal = 1.34×10^{-1} cu ft
1 foot of water = 0.4335 lb/sq in.
1 day = 1.44×10^3 minutes
Specific weight of water = 62.4 lb/cu ft
Reciprocal of bulk modulus of elasticity of water = 3.3×10^{-6} sq in./lb
1 ft/mi = 1.89×10^{-4} ft/ft
1 gal lift 1 foot = 3.15×10^{-6} kwh

clayey materials having an average thickness of 20 feet and a coefficient of permeability of 0.10 gpd/sq ft overlies the aquifer. It is assumed that no water will be released from storage within the aquitard during pumping. The aquitard is in turn overlain by a source bed; the water table in the source bed is assumed to remain constant during pumping and recharge from local precipitation is assumed to balance withdrawals from the source bed by leakage of water through the aquitard into the aquifer. The fully penetrating production wells, 12 inches in diameter, are spaced 500 feet apart and are located as shown in Fig. 8.1*a*. The available drawdown is 40 feet. Well loss is assumed to be negligible. Under these conditions, how much water can be continuously withdrawn from the two production wells without exceeding the specified allowable drawdown?

The model aquifer in Fig. 8.1*b* is constructed. From leaky artesian equations it is estimated that, when water levels stabilize at permanent stages and discharge is balanced by leakage through the aquitard, a 12-inch-diameter fully penetrating well in an infinite aquifer ($T = 50{,}000$ gpd/ft, $p' = 0.10$ gpd/sq ft, and $m' = 20$ feet) would have a specific capacity of about 28 gpm/ft. With an available drawdown of 40 feet, the practical sustained yield of the well would be about 1,100 gpm (28 gpm/ft $\times$ 40 feet). Taking into consideration interference from the nearby production well, which will be pumped at the same rate as the production well in question, and the limited areal extent of the aquifer, it is assumed as a first approximation that the practical sustained yield of a single production well under existing hydrogeologic and

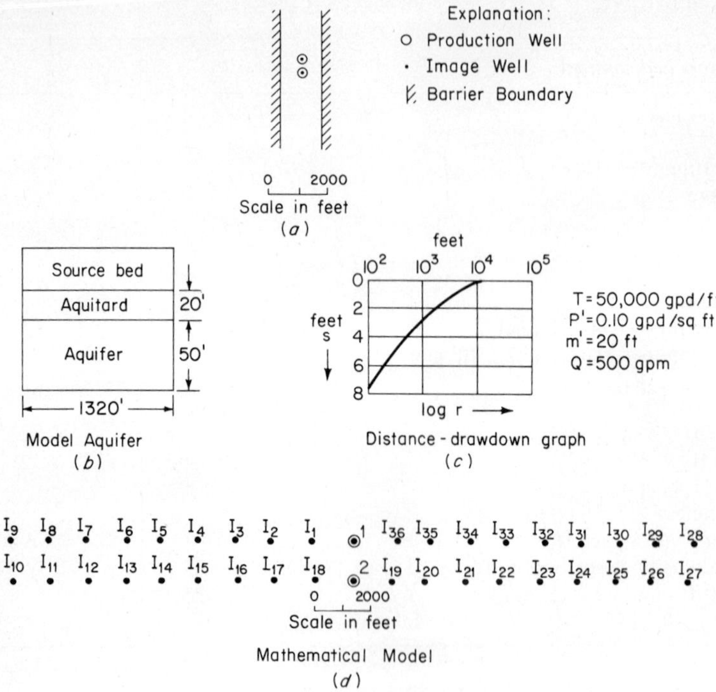

Fig. 8.1 Model aquifer (*a* and *b*) and mathematical model (*c* and *d*) for hypothetical groundwater development problem. (*Drawn by W. C. Walton.*)

development conditions would be about half (500 gpm) of the yield of a single production well in an infinite aquifer.

A distance-drawdown graph (Fig. 8.1*c*) is constructed using leaky artesian equations, assuming $T = 50{,}000$ gpd/ft, $P' = 0.10$ gpd/sq ft, $m' = 20$ feet, and $Q = 500$ gpm. It is noted that drawdown is negligible beyond a distance of about 13,000 feet. The mathematical model (Fig. 8.1*d*), consisting of the two production wells and the image wells associated with the two barrier boundaries within a distance of 13,000 feet from the production wells, is constructed based on the multiple-boundary image-well theory. Since both production wells are identical, the individual yields of the wells will be the same.

Considering production well 1, the distances from production well 1 to production well 2 and the 36 image wells are determined and tabulated as shown in Table 8.2. Components of drawdown due to pumping production well 1 itself at a rate of 500 gpm, interference from production well 2 pumping at a rate of 500 gpm, and the effects of the barrier boundaries simulated by the 36 image wells each pumping at a rate of 500 gpm are estimated with Fig. 8.1*c*. The sum of the components of drawdown, 39.95 feet, is approximately equal to the available drawdown, 40 feet. Thus, 1,000 gpm, 500 gpm from each production well, can be continuously withdrawn from

TABLE 8.2 DATA FOR HYPOTHETICAL GROUNDWATER DEVELOPMENT PROBLEM

Well no.	*Distance, ft*	*Drawdown, ft*
P 1	1	18.00
2	500	4.40
I 1	1,200	2.50
2	2,700	1.00
3	4,000	0.50
4	5,500	0.25
5	6,700	0.15
6	8,200	0.10
7	9,500	0.08
8	10,800	0.04
9	12,200	0.02
10	12,300	0.02
11	10,800	0.04
12	9,500	0.08
13	8,200	0.10
14	6,700	0.15
15	5,500	0.25
16	4,000	0.50
17	2,700	1.00
18	1,300	2.30
19	1,500	2.10
20	2,700	1.00
21	4,200	0.50
22	5,500	0.25
23	7,000	0.14
24	8,300	0.10
25	9,700	0.08
26	11,000	0.04
27	12,300	0.02
28	12,300	0.02
29	11,000	0.04
30	9,600	0.08
31	8,300	0.10
32	6,900	0.15
33	5,500	0.25
34	4,100	0.50
35	2,700	1.00
36	1,500	2.10
	Total	39.95

the two production wells. If the total drawdown associated with the 500-gpm pumping rate was not approximately equal to the available drawdown, other pumping rates would be selected and the process would be repeated until the computed total drawdown was equal to the available drawdown.

It should be pointed out that most hydrogeologic systems analysis problems are much more complex than the simple problem described. Many problems involve components of drawdown or buildup due to well loss, effects of partial penetration of production wells, effects of dewatering a portion of the aquifer, effects of complex recharge conditions, and effects of variations in hydrogeologic properties.

8.2 Case Histories of Hydrogeologic Systems Analysis with Analytical Methods

As will be demonstrated, it is often possible to evaluate the response of wells and aquifers to heavy development with analytical methods by devising approximate methods of analysis based on idealized models of aquifer situations. Case histories of groundwater development indicate that aquifer behavior actually does coincide rather closely with what may be predicted theoretically with model aquifers and mathematical models. It is recognized that methods of analysis described provide only approximate answers. To quantitatively describe in detail hydrogeologic systems having highly complex geometry and great variations in hydraulic properties, the hydrologist may have to turn to computer methods which are more versatile in simulating aquifer conditions. Quantitative answers depend primarily upon the accurate description of hydrogeologic conditions. In the future, as the techniques of groundwater resource evaluation are refined, a need for more precise, quantitative data concerning requisite hydrogeologic information will develop. In evaluating aquifers the hydrologist is forced to take a good hard look at the forest as well as the trees and in doing so his understanding of the major factors governing the yields of wells and aquifers is greatly enriched. Too often in the past hydrologists have overemphasized the complexity of conditions and have spent most of their time on details, at the expense of understanding the overall problem. Careful study often greatly reduces the importance of complexities and indicates that stress should be placed on more dominant factors.

For demonstrations of the applicability of model aquifers and mathematical models under more complex conditions, several case histories of the use of analytical methods in solving groundwater development problems will now be presented.

Arcola, Illinois

The village of Arcola is located in the southern part of Douglas County, Illinois, 23 miles south of the city of Champaign and 45 miles east-southeast of the city of Decatur. The municipal water supply in 1961 was obtained from production wells in the unconsolidated deposits within and near the city corporate limits (Walton and

Walker, 1961). The municipal water supply had been obtained from three pumping centers shown in Fig. 8.3*a*. Average daily groundwater withdrawal increased from 20,000 gallons in 1891 to 146,000 gallons in 1959.

Based on geologic studies made by the Illinois State Geological Survey, the unconsolidated glacial deposits in the Arcola area are mainly Wisconsinan and Illinoian in age and range in thickness from 80 to 125 feet. As shown in Fig. 8.2, these deposits consist primarily of ice-laid till with some permeable water-laid silt, sand, and gravel outwash. The thicker sections of glacial material are contained in a narrow bedrock valley cut in Pennsylvanian bedrock, which consists mainly of shale. The thicker and more permeable outwash materials, hereafter referred to as the aquifer, are generally found in the lower parts of the drift and are Illinoian in age. The thick upper unit of the Wisconsinan glacial till, which occurs from the surface to an average depth of 60 feet, contains a high percentage of silt and clay. The lower Wisconsinan unit and the Illinoian deposits immediately overlying the aquifer, hereafter referred to as the aquitard, contain sand and lenses within sandy till. The aquifer contains a large amount of fine sand and silt, and its permeability is not great.

The geologic cross section and aquifer-thickness map shown in Fig. 8.2 were drawn from the few available drillers' logs of wells and test holes. As is often the case, data are not sufficient for a rigorous description of the areal extent of the aquifer. Analysis of existing geologic information suggests that the aquifer occurs as a thin and narrow strip of permeable sand and gravel exceeding 20 feet in thickness in many places and trending from northeast to southwest through Arcola. The more permeable part of the aquifer suitable for development ranges in width from about 800 to less than 200 feet. Water occurs under leaky artesian conditions, and recharge is derived from the vertical leakage of water through the aquitard into the aquifer.

During the period 1940 to 1955, five aquifer tests were made at Arcola to determine the hydrogeologic properties of the aquifer. Based on aquifer test data the coefficient of permeability of the aquifer ranges from 280 to 660 gpd/sq ft and the coefficient of transmissibility ranges from 2,200 to 18,000 gpd/ft. The smaller values of T and P reflect thinner and less permeable deposits near the edge of the aquifer; the large values reflect much thicker and more permeable deposits near the center of the aquifer. The average coefficient of storage of the aquifer is 0.001.

Past records of water-level decline and pumpage and analytical methods were used to determine the practical sustained yield of the aquifer. An idealized model aquifer that duplicates hydrogeologic conditions in the Arcola area was created. Major factors considered in the model aquifer were (1) the external barrier boundaries (bedrock walls) are tapered and irregular in shape; (2) the bedrock walls are not entirely impervious and some subsurface flow will occur from the bedrock into the aquifer, the effective boundaries do not coincide exactly with physical boundaries; (3) the hydrogeologic properties of the aquifer vary from place to place and are highly variable in the proximity of boundaries but are fairly uniform on a gross basis; and (4) the permeability and thickness of the aquitard vary from place to place but are fairly uniform on a gross basis. With these factors taken into account, the results of

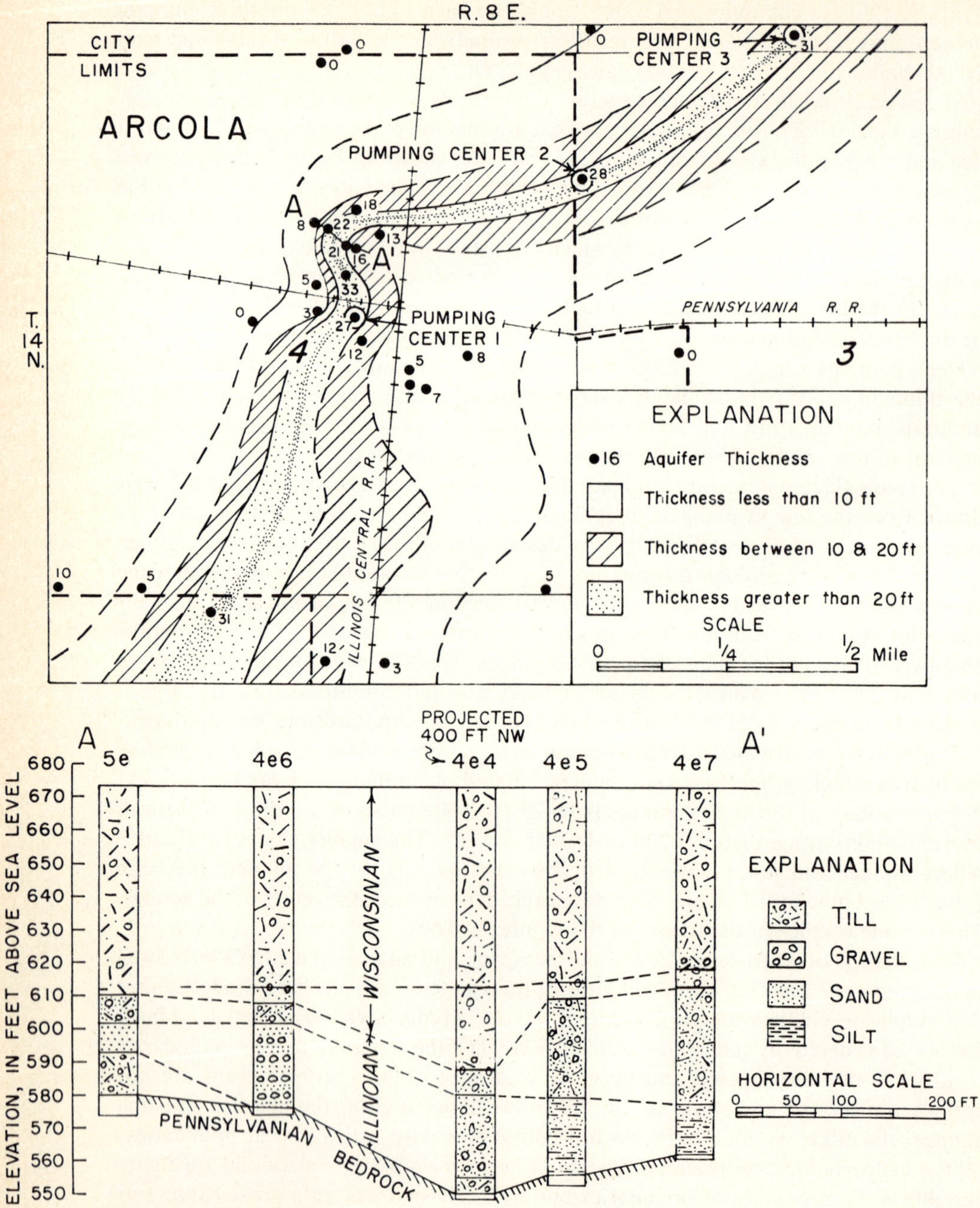

Fig. 8.2 Map and geologic cross section showing thickness and areal extent of aquifer at Arcola, Illinois. (*From Walton and Walker, 1961.*)

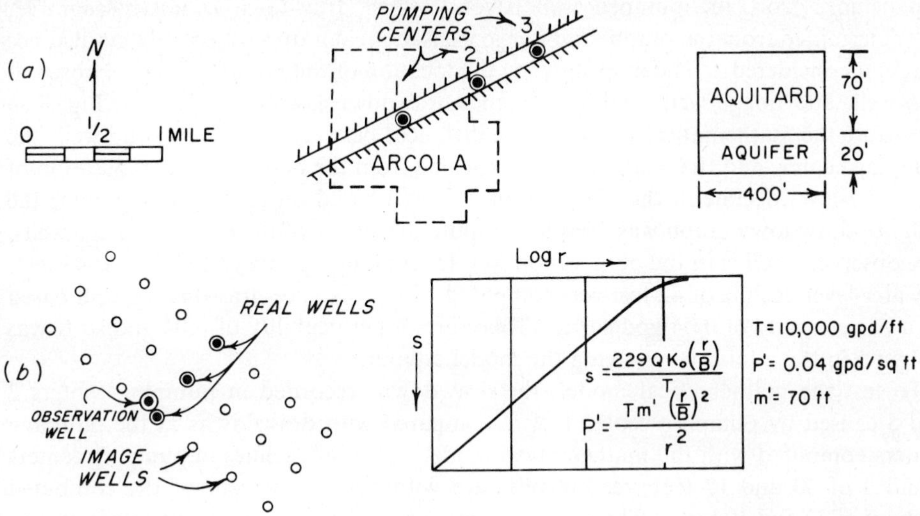

Fig. 8.3 Model aquifer (*a*) and mathematical model (*b*) for Arcola area, Illinois. (*Walton and Walker, 1961.*)

hydrogeologic studies indicate that it is possible to simulate complex aquifer conditions with an infinite strip of sand and gravel which is 400 feet wide, 20 feet thick, bounded on the sides and bottom by impermeable material, and overlain by an aquitard 70 feet thick. The model aquifer and its orientation with respect to Arcola are shown in Fig. 8.3*a*. The average coefficients of transmissibility and storage of the model aquifer are 10,000 gpd/ft and 0.001, respectively.

Most drawdown data collected during aquifer tests are affected by barrier boundaries, and it is impossible to isolate the effects of the leakage through the aquitard. Although the permeability of the aquitard cannot be determined from aquifer test data, it can be estimated with the model aquifer. The water-level decline in an observation well near pumping center 1 was computed using a mathematical model based on the model aquifer, calculated hydrogeologic properties of the aquifer, the image-well theory, the steady-state leaky artesian equation, estimated pumpage data, and several assumed values of the vertical permeability of the aquitard. The computed decline was then compared with the actual decline and that permeability which gave a computed decline equal to actual observed decline was assigned to the aquitard.

The pumping center, the observation well, and the image wells associated with the boundaries of the model aquifer were drawn to scale on a map. The boundaries are parallel; therefore an image-well system extending to infinity is required. However, in practice it is only necessary to add pairs of image wells until the effect of the next pair has no measurable influence. A semilog distance-drawdown graph based on assumed hydrogeologic properties of the model aquifer and its aquitard and past ground-water withdrawals was constructed. The virtual radius of the cone of depression, the distance from the pumped well beyond which drawdown is not measurable, was determined from the graph, and image wells at greater distances than that distance were not considered. The map showing the location of the pumping center, observation well, and image wells and the distance-drawdown graph as shown in Fig. 8.3*b* constitute the mathematical model. The distances between the observation well, the pumping center, and the image wells were scaled from the map. The computation of the water-level decline in the observation well was based on past pumping data; the distance-drawdown graph was used to compute the effects of the real and image wells. The observed decline in the observation well for a pumping rate of 115 gpm is 42 feet. A water-level decline of 42 feet was computed with a distance-drawdown graph based on a permeability of 0.04 gpd/sq ft. Therefore, a permeability of 0.04 gpd/sq ft was assigned to the aquitard overlying the model aquifer.

To test the mathematical model, the drawdowns recorded in pumping centers 2 and 3 caused by pumping center 1 were compared with drawdowns at the pumping centers computed with the mathematical model. Actual declines in pumping centers 2 and 3 of 30 and 17 feet, respectively, are within a few percent of the computed declines of 32 and 19 feet. The close agreement between computed and actual decline indicates that the model aquifer and mathematical model closely describe the hydrogeologic conditions at Arcola. It is reasonable to assume that the model aquifer and the mathematical model may be used to predict with reasonable accuracy the effects of future groundwater development and the practical sustained yield of the aquifer. The mathematical model is based on a particular combination of aquifer boundaries and properties. There are probably other mathematical models involving several slightly different combinations of parameters which would also duplicate aquifer conditions.

In 1959, the multiple-well system at Arcola consisted of three production wells ranging in depth from 106 to 122 feet and spaced 2,500 feet apart in the more permeable parts of the aquifer as shown in Fig. 8.3*a*. Available drawdowns in the production wells, assuming pumping levels above the screens, range from 66 to 75 feet. Computations made with the mathematical model indicate that the practical sustained yield of the existing three-well system is about 137 gpm or 200,000 gpd. The practical sustained yield can be developed by pumping production wells 1 and 3 at 57 gpm and production well 2 at 25 gpm. Variations in the hydrogeologic properties of the aquifer were considered in computations of drawdowns due to pumping the production wells themselves.

The rate of groundwater withdrawal increased from 115,000 gpd in 1957 to 146,000 gpd in 1959. If pumpage continues to increase at this rate in the future, it is estimated that the practical sustained yield of the existing three-well system will be exceeded within a three-year period.

Rio Grande de San Miguel Basin, El Salvador

An intensive hydrogeologic study was made to determine the physical feasibility of developing the groundwater resources of the lower basin of the Rio Grande de San

Miguel, El Salvador, for irrigation purposes (see FAO, 1964). The potential area for irrigation is approximately 10,000 hectares in the valley of the Rio San Miguel just before the river empties into the Pacific Ocean. The valley is bounded on the north by the Usulatan–San Miguel chain of volcanoes, on the south by the Jucuaran Range, on the east by the El Delirio Hills, and on the west by the Rio Grande Hills. Average annual rainfall is 2,040 millimeters; 90 percent of the rains fall during the period between May and November. The average air temperature is 26°C.

Slopes of the volcanoes are underlain by pyroclastic materials and talus debris, grading into reworked pyroclastics and lake and river sediments toward the lower valley floor. The center of the valley is underlain by lacustrine tuffaceous silts with thin interbeds of coarser pyroclastics. The unconsolidated pyroclastics are underlain by impermeable basement rocks including cemented lahars, intrusive rocks, old extrusive lava flows, and consolidated pyroclastics. The logs of test holes indicate that the permeable and saturated pyroclastic, lake, and river sediments constituting the aquifer in the valley exceed 300 feet thick and are highly stratified and heterogeneous.

In uplands, the water ranges from 50 to 105 meters below land surface; in the valley the water table ranges from 3 to 30 meters below land surface. The average annual fluctuation of the water table is about 0.6 meter. Based on aquifer and well-production test data, the average coefficients of transmissibility and storage of the aquifer are 250,000 gpd/ft and 0.15, respectively. The average permeability of the pyroclastic materials falls into the class commonly assigned to sand and gravel deposits having fine to medium grain sizes.

The results of hydrogeologic studies suggest that it is possible to simulate the complex aquifer conditions with a finite strip of pyroclastic, lake, and river materials which exceeds 300 feet in thickness and 25 kilometers wide in the proposed area for groundwater development, bounded on the bottom by impermeable material, and bounded partially on the south, east, and west by impermeable materials (barrier boundaries) as shown in Fig. 8.4. With regard to the possible cones of depression associated with the proposed area for groundwater development, the model aquifer extends to infinity in northeastern, northern, and northwestern directions. The following factors were taken into account in creating the model aquifer: (1) the boundaries are not clear-cut straight-line features but are irregular in shape and extent; (2) during the dry season and under heavy pumping conditions some water in the San Miguel River will be induced to flow from the stream into the aquifer; (3) the coefficient of transmissibility of the aquifer decreases rapidly in a southernly direction; and (4) the existence of generally impermeable tuffaceous lahars, massive lava flows, crystallized tuffs, domes of intrusive basalts, and other basement rocks serves to limit the continuity of the aquifer at places within the proposed area for groundwater development and south, east, and west of the area.

The following propsoed scheme of groundwater development was tested, using the model aquifer and a mathematical model (Walton, 1965b). During a five-month period, November through April, 2.1×10^{10} gallons of groundwater will be withdrawn

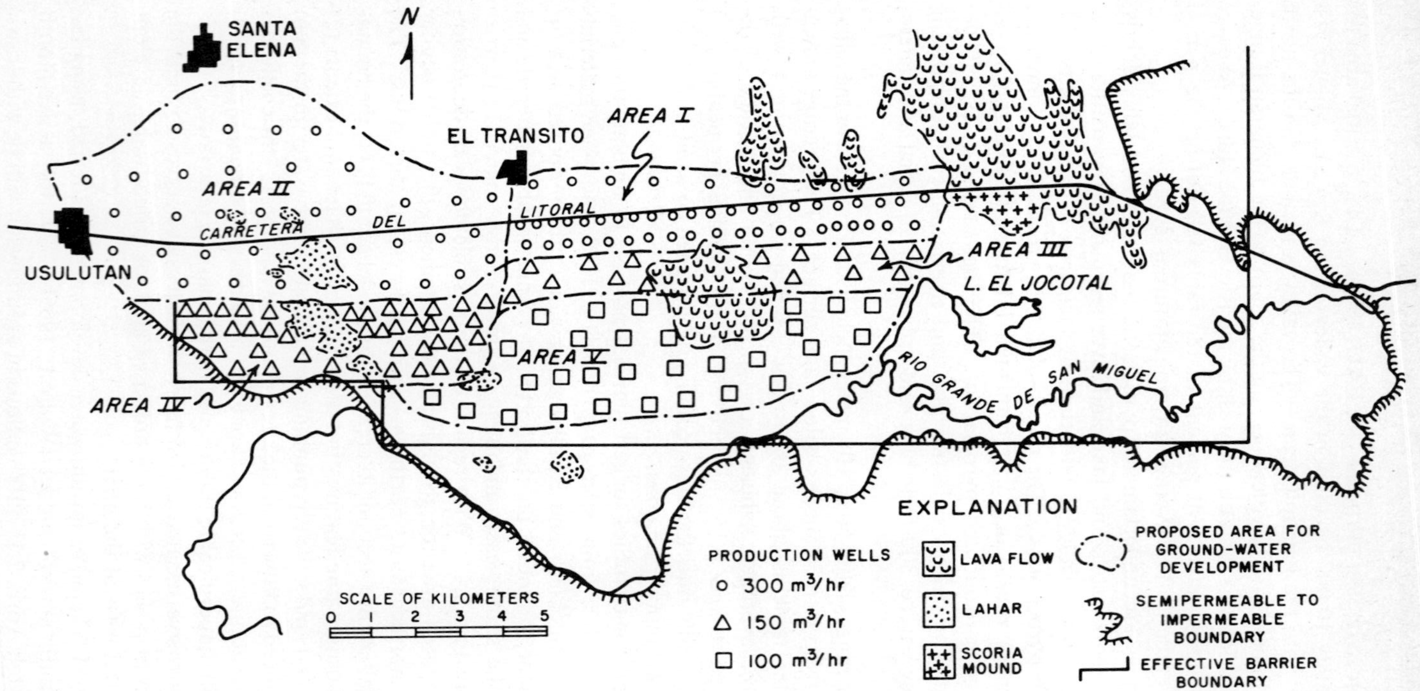

Fig. 8.4 Model aquifer and selected groundwater development plan for lower basin of Rio Grande de San Miguel, El Salvador. (*From Walton, 1965b.*)

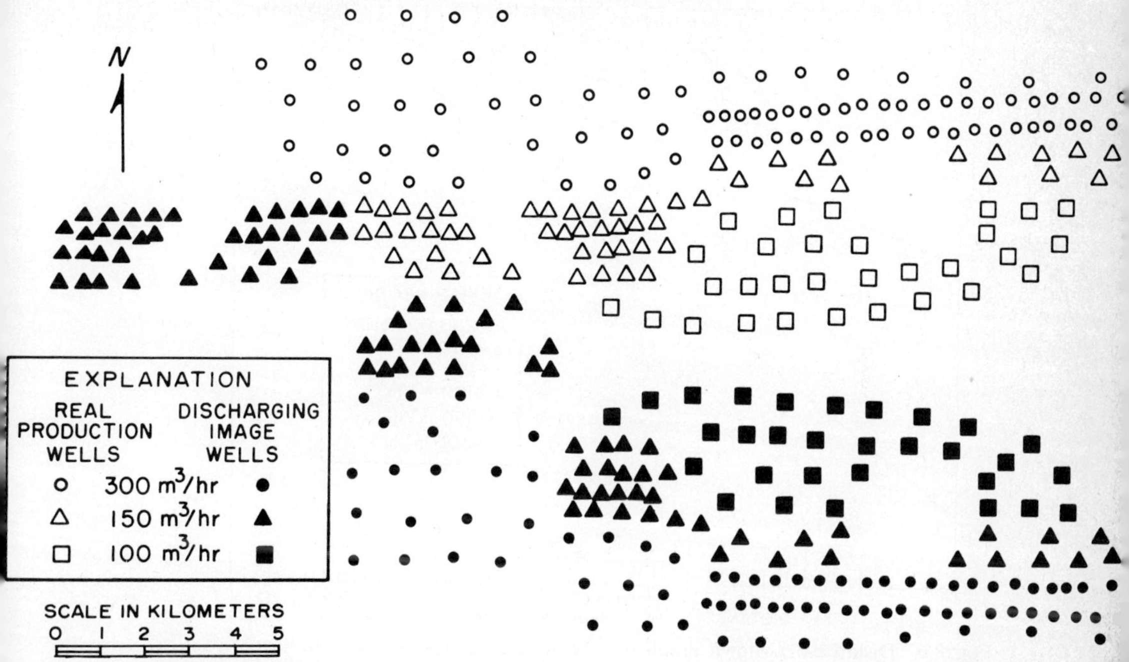

Fig. 8.5 Mathematical model for lower basin of Rio Grande de San Miguel, El Salvador. (*From Walton, 1965b.*)

from 170 production wells which will be pumped 14 hours per day at rates varying from 1,320 to 440 gpm for a total daily discharge of 140 mgd. Each production well will be drilled to a depth not to exceed 300 feet. Casing size will vary from 12 to 14 inches. The casing will be perforated throughout the lower 80 to 90 percent of the saturated zone. Production wells will be spaced at least 1,000 feet apart.

The mathematical model in Fig. 8.5 was constructed based on the model aquifer and the proposed scheme of development. The distance-drawdown graph in Fig. 8.6 was prepared and the distances between each production well and other real and image wells were scaled from the mathematical model. Components of drawdown in each production well due to pumping each production well itself, the effects of dewatering a portion of the aquifer, and the effects of well losses were estimated with Fig. 8.6 and appropriate groundwater equations. Values of C of 2 sec^2/ft^5 were assigned to production wells with pumping rates of 660 and 440 gpm. Components of drawdown were summed and added to existing nonpumping water levels to obtain pumping levels in the production wells. The results of the hydrogeologic systems analysis are summarized in Table 8.3.

It is concluded that it is physically feasible to develop the groundwater supplies mentioned earlier without creating critical water levels in production wells. The mathematical model assumes that recharge from precipitation will balance discharge. According to the results of groundwater recharge and hydrologic budget studies made

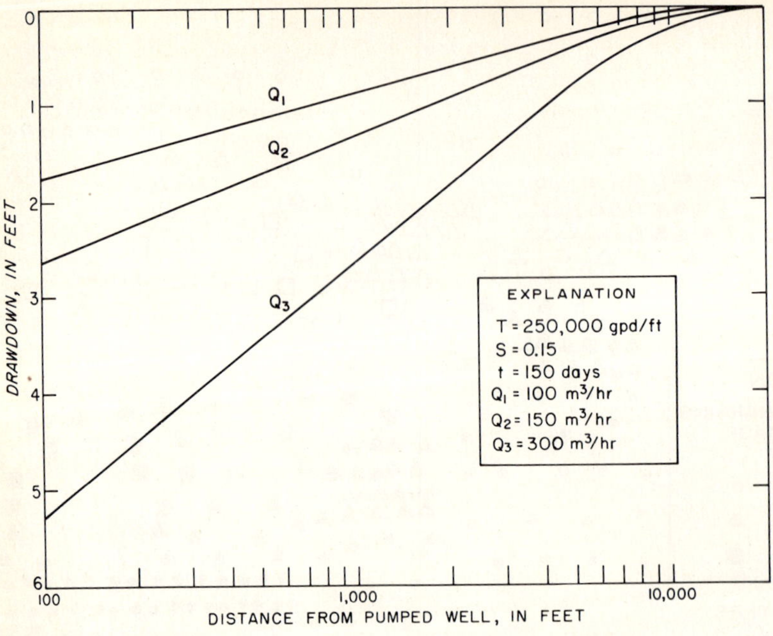

Fig. 8.6 Distance-drawdown graph for lower basin of Rio Grande de San Miguel, El Salvador. (*From Walton, 1965b.*)

TABLE 8.3 RESULTS OF HYDROGEOLOGIC SYSTEMS ANALYSIS FOR LOWER BASIN OF RIO GRANDE DE SAN MIGUEL, EL SALVADOR. (*From Walton, 1965b.*)

No. of wells	*Capacity, gpm*	*Average depth to static water level, ft below land surface*	*Drawdown, ft*	*Average pumping level, ft below land surface*
Area I				
8	1,320	132	58	190
22	1,320	98	73	171
20	1,320	83	65	148
Area II				
10	1,320	132	51	183
25	1,320	66	58	124
Area III				
7	660	66	41	107
8	660	33	47	80
Area IV				
30	660	33	44	77
10	660	17	43	60
Area V				
10	440	50	62	112
10	440	33	56	89
10	440	17	52	69

in the area, recharge from precipitation will balance total withdrawals. The average rate of recharge from precipitation in the area is 1.65×10^6 gpd/sq mi, whereas a recharge rate of about 5.0×10^5 gpd/sq mi is required to balance proposed withdrawals assuming a reasonable area of influence (110 square miles) of the production wells.

Assumption, Illinois

A well field was developed 2.5 miles southeast of the city of Assumption, Illinois, to supplement the municipal water supply (Walton and Walker, 1961). Geologic studies suggest that the aquifer underlying the well field is mainly poorly sorted sand, ranging in thickness from 5 to 13 feet. The aquifer is of limited size; rectangular dimensions are 600 feet wide by 2,200 feet long. Clayey materials (aquitard) with an average saturated thickness of 8 feet overlie the aquifer. Water occurs under leaky artesian conditions, and recharge is received chiefly from precipitation by the vertical leakage of water through the aquitard. Based on the results of two aquifer tests, the coefficients of transmissibility and permeability of the aquifer are 4,900 gpd/ft and 410 gpd/sq ft, respectively. The permeability of the aquitard is 0.10 gpd/sq ft.

The idealized model aquifer for the Assumption area, shown in Fig. 8.7*a*, is a box of sand 600 feet wide, 2,200 feet long, and 10 feet thick. The model aquifer is bounded on the sides and bottom by impermeable material and is overlain by an aquitard with an average saturated thickness of 8 feet. The mathematical model for the model aquifer is shown in Fig. 8.7*b*.

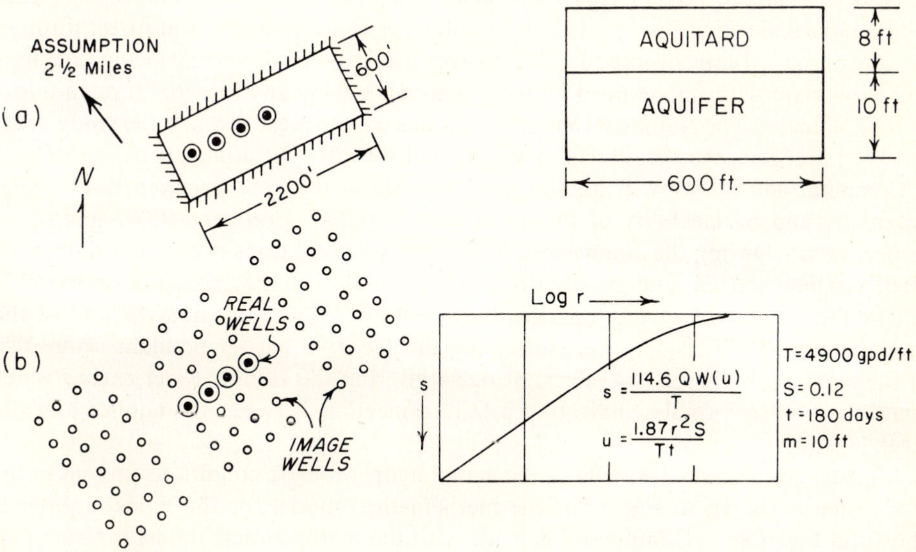

Fig. 8.7 Model aquifer (*a*) and mathematical model (*b*) for Assumption, Illinois, area. (*From Walton and Walker, 1961.*)

The response of the aquifer to long-term pumping was studied by means of the mathematical model. Computations based on an average available drawdown of 12 feet indicate that the practical sustained yield of a four-well system, consisting of production wells 6 inches in diameter, 24 feet deep, with 5 feet of screen, and spaced 300 feet apart, is 38 gpm or 55,000 gpd. Gravity drainage of the aquitard and part of the aquifer during six-month dry periods was taken into account in estimating the yield of the aquifer.

Pekin, Illinois

An intensive hydrogeologic study was made to determine the feasibility of developing for industrial use a large groundwater supply from unconsolidated deposits in an area along the Illinois River, about 3 miles southwest of the city of Pekin, Illinois (Walton and Walker, 1961). The study included a test-hole drilling program and a controlled aquifer test. The data thus obtained together with information from other sources led to an evaluation of the practical sustained yield of the aquifer in the study area and to the design of a multiple-well system capable of meeting the demand of the industry.

The unconsolidated deposits in the study area consist of recent silty alluvial materials and glacial outwash of Wisconsinan age. These deposits are contained in a buried valley cut into the underlying bedrock of Pennsylvanian age. The permeable outwash forming the aquifer ranges in thickness from less than 5 feet to more than 33 feet and consists of stratified beds of gravel and sand. Logs of wells and test holes show that the boundary marking the limits of the aquifer trends northeast to southwest through the study area. In the proposed well field area the aquifer averages 20 feet in thickness and is overlain with fine-grained alluvial materials having an average saturated thickness of 50 feet. The Illinois River, which trends east to west through the study area, has been dredged into the alluvial materials but not into the aquifer.

Computations made with aquifer test data show that the coefficients of transmissibility and permeability of the aquifer are 180,000 gpd/ft and 9,000 gpd/sq ft, respectively. During the aquifer test, leaky artesian conditions occurred a short time after pumping started, and gravity drainage of the alluvial materials was appreciable during the latter part of the test period, especially in the immediate vicinity of the production well. Test data were adversely affected by a barrier boundary (the edge of the aquifer); however, the effects of recharge from the Illinois River caused water levels to stabilize rapidly, indicating a fair connection between the aquifer and the river.

A model aquifer which simulates the actual hydrogeologic conditions present in the study area is shown in Fig. 8.8*a*; the mathematical model for the model aquifer is shown in Fig. 8.8*b*. Computations made with the mathematical model indicate that 4,000 gpm or 5.76 mgd can be obtained with maximum drawdown above the top of the aquifer from five production wells spaced about 325 feet apart.

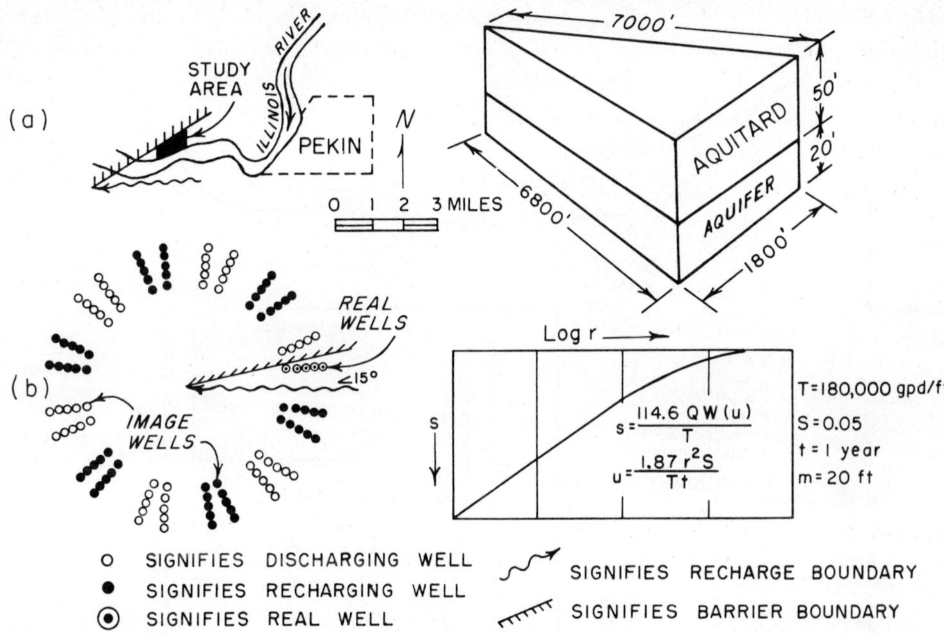

Fig. 8.8 Model aquifer (*a*) and mathematical model (*b*) for Pekin, Illinois, area. (*From Walton and Walker, 1961.*)

Bracebridge, Ontario, Canada

The bedrock in the Bracebridge, Ontario, Canada, area is composed of Precambrian crystalline rocks of the Canadian Shield. The bedrock surface is scored by numerous deep and narrow valleys and the bedrock-surface topography is rugged with more relief than the land surface. Bedrock crops out at the land surface in many places within the area. The bedrock is compact and for practical purposes impermeable.

Where bedrock is not exposed, the Bracebridge area is covered mostly with fine-grained unconsolidated deposits which at places exceed 300 feet thick. The most prominent constituents of the unconsolidated deposits are clay and silt. Lacustrine deposits of clay and silt are largely associated with extensive postglacial lakes formed on the retreat of an ice front. Beach deposits of fine sand and gravel often occur. The unconsolidated deposits in general have a low permeability. The fine-grained character of existing sand and gravel deposits makes development of high-capacity production wells most difficult. Hydrogeologic studies were made to determine the practical sustained yields of well fields in the Peake Springs area 1 mile northeast of Bracebridge and the Monck Township area 0.5 mile southeast of Bracebridge (see Walton, 1964b).

Logs of test holes indicate that the unconsolidated deposits in the Peake Springs area range in thickness from 130 feet in the center of the bedrock valley to 40 feet on the sides of the bedrock valley as shown in Fig. 8.9. The bedrock-valley walls are

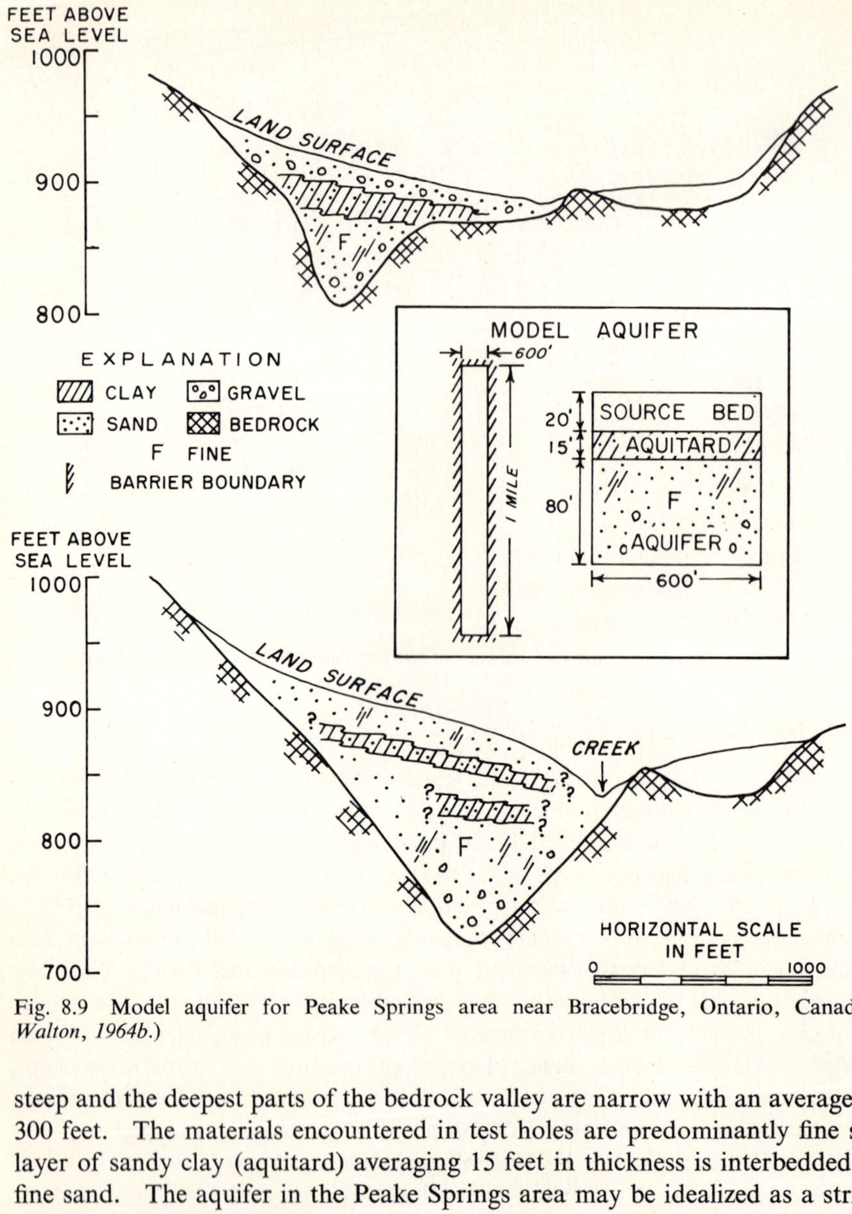

Fig. 8.9 Model aquifer for Peake Springs area near Bracebridge, Ontario, Canada. (*From Walton, 1964b.*)

steep and the deepest parts of the bedrock valley are narrow with an average width of 300 feet. The materials encountered in test holes are predominantly fine sand. A layer of sandy clay (aquitard) averaging 15 feet in thickness is interbedded with the fine sand. The aquifer in the Peake Springs area may be idealized as a strip of fine sand and some gravel, 600 feet wide, 1 mile long, and 80 feet thick. The model aquifer shown in Fig. 8.9 is overlain by an aquitard (sand clay) averaging 15 feet thick which is in turn overlain by a layer of sand (source bed) averaging 20 feet thick. The model aquifer is underlain by impervious bedrock.

Based on aquifer test results the coefficient of transmissibility of the model aquifer is

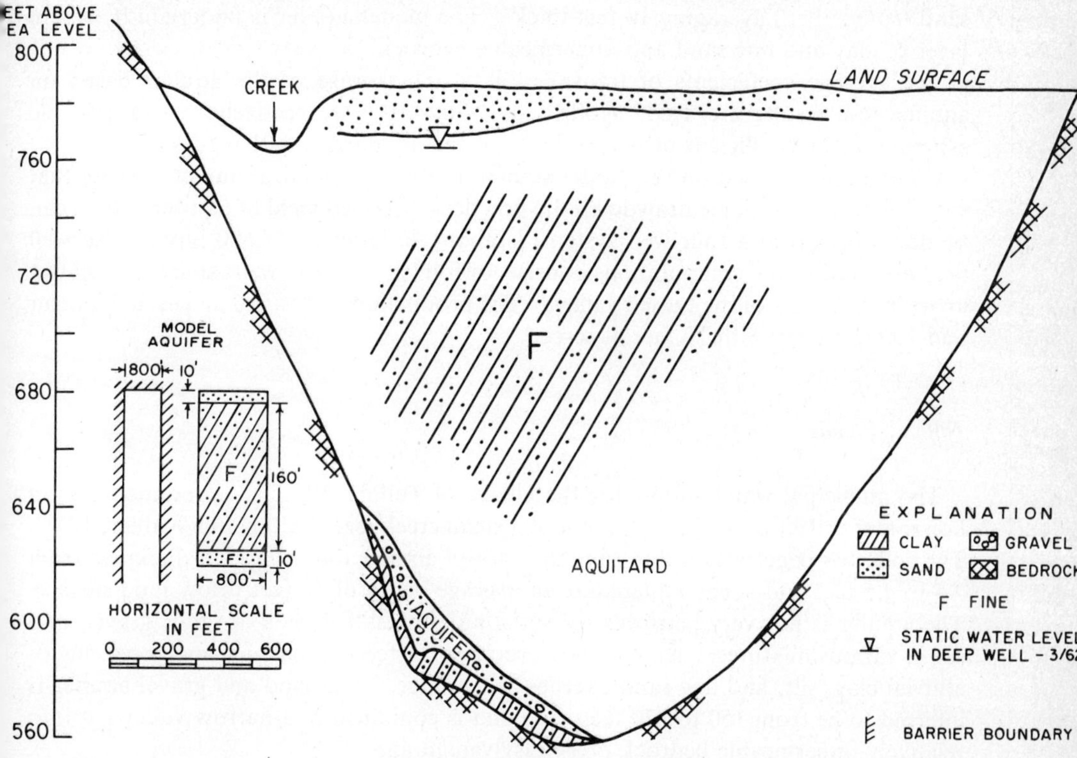

Fig. 8.10 Model aquifer for Monck Township area near Bracebridge, Ontario, Canada. (*From Walton, 1964b.*)

34,400 igpd/ft. The long-term coefficient of storage is 0.15. The coefficient of permeability of the aquitard is 0.43 igpd/sq ft. Computations made with the model aquifer and a mathematical model, taking into consideration gravity drainage of the aquitard and part of the aquifer during extended dry periods, indicate that 38,000 igpd could be obtained from a production well without creating critical water levels or exceeding recharge from precipitation. The production well is 6 inches in diameter and 160 feet deep with 10 feet of screen and it is located near the center of the buried valley. Recharge from precipitation (39.37 inches annually) will balance the discharge.

The materials encountered in the Monck township area are predominantly clay and fine sand. The thickness and areal extent of the unconsolidated deposits are shown in Fig. 8.10. The deposits are about 230 feet thick in the center of the bedrock valley. The deepest parts of the bedrock valley have an average width of 800 feet. The aquifer in the Monck Township area may be idealized as a semi-infinite strip of cemented sand and gravel 800 feet wide and 10 feet thick. The model aquifer is overlain by an aquitard (clay and fine sand) averaging 160 feet thick which in turn is overlain by a layer of fine sand (source bed) averaging 10 feet thick. The model aquifer is underlain by a thin layer of clay and fine sand and impermeable bedrock.

The average coefficients of transmissibility and storage of the aquifer based on aquifer test results are 1,650 igpd/ft and 4.6×10^{-5}, respectively. A reasonable estimate of the coefficient of permeability of the aquitard is 0.01 igpd/sq ft.

Computations based on the model aquifer and a mathematical model indicate that with 160 feet of available drawdown the practical sustained yield of aquifer, which can be developed from a four-well system, is about 40 igpm or 57,600 igpd. The well field assumed in the computations consists of four production wells spaced 1,500 feet apart in the center of the buried valley. The production wells are 6 inches in diameter and 200 feet deep with 10 feet of screen.

Tallula, Illinois

The municipal water supply for the village of Tallula, Illinois, is obtained from a horizontal collector on the floodplain of a small creek (see Walton and Walker, 1961). The collector penetrates a thin sand and gravel aquifer that ranges in thickness from 2.5 to 4.5 feet and is encountered at an average depth of 16 feet below land surface. The aquifer is not very permeable, consisting of stratified beds of sand, gravel, and silt in various mixtures. An aquitard averaging 10 feet in thickness and consisting of alluvial clay, silt, and fine sand overlies the aquifer. The sand and gravel aquifer is inferred to be from 150 to 370 feet wide and is contained in a narrow valley cut into relatively impermeable bedrock of Pennsylvanian age.

Recharge is derived chiefly by the vertical leakage of water through the aquitard. Because of the small area and silted condition of the bed of the creek, low streamflow, and the presence of silty materials beneath the streambed, very little recharge occurs by the induced infiltration of surface water, especially during summer, fall, and winter months. Large amounts of water enter the aquifer through a recharge well connected to a lagoon and located 60 feet from the end of one of the laterals in the collector.

The coefficients of transmissibility and permeability of the aquifer determined from the results of two aquifer tests are 2,750 gpd/ft and 790 gpd/sq ft, respectively. Under natural conditions, leaky artesian conditions exist; however, under heavy pumping conditions and during prolonged dry periods the aquitard is partially drained.

Pumpage from the horizontal collector increased from 9,000 gpd in 1955 to 29,000 gpd in 1959. The horizontal collector consists of a 6-foot-diameter concrete caisson 26 feet deep from which two horizontal 8-inch-diameter vitrified perforated clay pipe laterals are projected into the aquifer. One horizontal lateral (upper lateral) projects from the caisson at a depth of 18 feet and is 478 feet long. The other horizontal lateral (lower lateral) projects from the caisson at a depth of 21 feet and is 310 feet long. As the result of pumping at a rate of 37,000 gpd during the summer months of 1959, the water levels declined below the top of the upper lateral in the caisson.

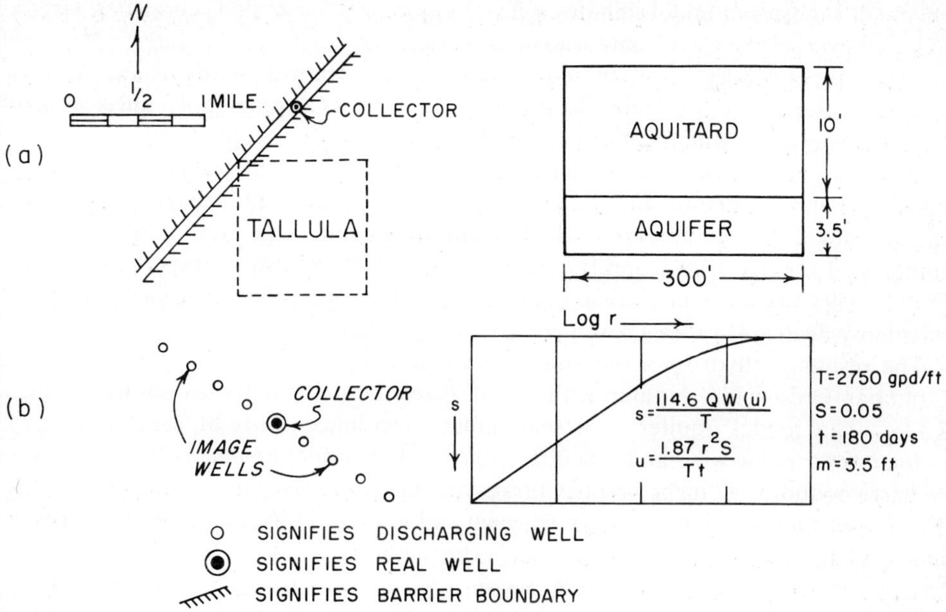

Fig. 8.11 Model aquifer (*a*) and mathematical model (*b*) for Tallula, Illinois, area. (*From Walton and Walker, 1961.*)

The idealized model aquifer for the Tallula area as shown in Fig. 8.11*a* is a semi-infinite rectilinear strip of sand and gravel 300 feet wide and 3.5 feet thick. It is bounded on the sides and bottom by impermeable material and is overlain by an aquitard 10 feet thick. The mathematical model for the model aquifer is shown in Fig. 8.11*b*.

Aquifer test data and the mathematical model were used to compute the radius of a vertical well that would have the same specific capacity as the horizontal collector. Computations made simulating the collector with a vertical well having a radius of 66 feet and using the mathematical model indicate that the practical sustained yield of the collector is 11 gpm or 16,000 gpd during extended dry periods and 20 gpm or 25,000 gpd during years of normal precipitation.

Northeastern Illinois

The Cambrian-Ordovician Aquifer (Suter et al., 1959) is the most highly developed aquifer for large groundwater supplies in the Chicago, Illinois, region. The Cambrian Ordovician Aquifer is encountered at an average depth of about 500 feet below the land surface at Chicago; it has an average thickness of 1,000 feet and is composed chiefly of sandstones and dolomites. The Maquoketa Formation, consisting largely of shale, overlies the Cambrian-Ordovician Aquifer and confines the water in the deep aquifer under leaky artesian conditions. The Cambrian-Ordovician Aquifer receives water from overlying glacial deposits in areas averaging 47 miles west of Chicago where the Maquoketa Formation is absent.

Based on the results of 63 aquifer tests and other studies, the coefficients of transmissibility and storage of the Cambrian-Ordovician Aquifer and the permeability of the Maquoketa Formation are fairly uniform throughout large areas in northeastern Illinois and average 17,000 gpd/ft, 0.00035, and 0.00005 gpd/sq ft, respectively (Suter et al., 1959; Walton and Csallany, 1962; Walton, 1960). The coefficient of transmissibility decreases rapidly south and east of Chicago.

The results of hydrogeologic studies indicate that it is possible to simulate the Cambrian-Ordovician Aquifer with an idealized model aquifer as shown in Fig. 8.12*a*. The model aquifer is a semi-infinite rectilinear strip of sandstones and dolomites 84 miles wide and 1,000 feet thick. The model aquifer is bounded by a recharge boundary 47 miles west of Chicago and by two intersecting barrier boundaries 37 miles east and 60 miles south of Chicago and is overlain by an aquitard consisting mostly of shale averaging 200 feet thick. The mathematical model for the Cambrian-Ordovician Aquifer is shown in Fig. 8.12*b*.

Pumpage of groundwater from the Cambrian-Ordovician Aquifer increased gradually from 200,000 gpd in 1864 to 50 mgd in 1959. Pumpage is concentrated in six

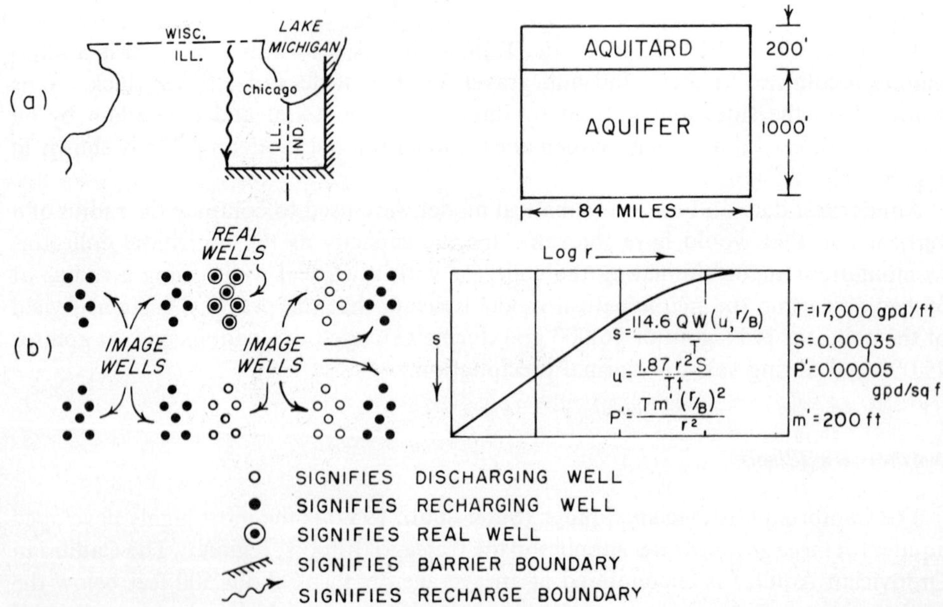

Fig. 8.12 Model aquifer (*a*) and mathematical model (*b*) for Chicago region, Illinois. (*From Walton and Walker, 1961.*)

centers as shown in Fig. 8.12*b*: the Chicago, Joliet, Elmhurst, Des Plaines, Aurora, and Elgin areas. As a result of heavy pumping, artesian pressure in deep wells declined more than 600 feet at Chicago between 1864 and 1959.

Studies made with the mathematical model show that the practical sustained yield of the Cambrian-Ordovician Aquifer is about 46 mgd and is largely limited by the rate at which water can move eastward through the aquifer from recharge areas. The practical sustained yield of the aquifer is here defined as the maximum amount of water that can be continuously withdrawn with the present distribution of pumping centers without eventually dewatering the most productive and basal water-yielding formation of the Cambrian-Ordovician Aquifer. The practical sustained yield of the aquifer was exceeded in 1959.

Declines in nonpumping water levels that may be expected between 1958 and 1980 at pumping centers were computed by using the mathematical model and assuming that the distribution of pumping remains the same as it was in 1958. Computed declines ranged from 300 feet in the Chicago area to 190 feet at Elgin and averaged 250 feet.

Camas Prairie, Idaho

Camas Prairie is an eastward trending intermontane basin along the north flank of the Snake River plain in southern Idaho (Walton, 1962). The basin is about 40 miles long and averages about 8 miles wide. It was formed as a structural depression in which a considerable thickness of alluvial and lake deposits have accumulated behind basalt flows which at times blocked the outlet to the east. Intrusive and extrusive rocks of Cretaceous to Quaternary age enclose the basin on the north, west, and east. The enclosing rocks yield small amounts of water to springs and wells from the weathered mantle and fractures.

The principal aquifers in Camas Prairie are the sand and gravel deposits (valley fill) and the basalts at the eastern outlet of the prairie. The valley-fill deposits are about 550 feet thick; 188 feet of basalt was penetrated above clay in an irrigation well near the east margin of the prairie. An extensive clay bed averaging 90 feet in thickness, penetrated between the average depths of 120 and 210 feet below the land surface, separates the shallow and deep valley-fill deposits. Permeable sand and gravel deposits are found in two zones below the clay bed; above the clay are alternating beds of sand, sandy silt, silt, and clay which are not very permeable. Fine- to medium-grained sand and some gravel lie immediately below the clay bed. The average thickness of this zone (upper artesian aquifer) is about 50 feet. Permeable sand and gravel deposits (lower artesian aquifer) averaging 85 feet in thickness occur at the base of the valley fill. The upper and lower aquifers are separated by beds of sandy and clayey materials which are relatively impermeable. The bulk of the sediments of the two aquifers are rather fine grained and their permeability is not great.

Confined water is encountered by wells extending to depths greater than 40 feet in the valley fill and by wells in the basalt aquifer. The confined water is under sufficient pressure to cause it to rise above the land surface over much of the prairie, and flowing artesian wells are common. The head above land surface in the flowing wells seldom exceeds 10 feet. Water occurs under water-table conditions in the shallow deposits above the clay bed. The water table lies at shallow depths, less than 10 feet in most places.

An unbroken unit of basalt is relatively impermeable, but porous permeable zones along contacts between separate flows, joints, and other crevices yield large amounts of groundwater to wells. The Snake River basalt yields large amounts of water near the east margin of the prairie.

The transmissibility of a 126-foot interval in the lower alluvial aquifer was determined from the results of an aquifer test to be about 30,000 gpd/ft. The coefficient of transmissibility of the upper and lower alluvial aquifers together was estimated to be about 70,000 gpd/ft. The coefficient of storage for purposes of computation was assumed to be 0.001. Specific-capacity data suggest that the coefficient of transmissibility of the upper 200 feet of the basalt aquifer is about 200,000 gpd/ft, or roughly three times that of the upper and lower alluvial aquifers together. The coefficient of permeability of aquitards averages about 0.2 gpd/sq ft.

The results of hydrogeologic studies made in Camas Prairie indicate that the effects of barrier and recharge boundaries and other hydrogeologic features on the response of the valley-fill deposits to development by wells can be determined roughly by mathematical analysis of a rectangular model aquifer. The model aquifer is 7 miles wide and 30 miles long and is bounded by two parallel barrier boundaries (on the north and south) intersected at right angles by a third barrier boundary on the west and a recharge boundary on the east.

A study was made of the theoretical drawdowns associated with the development of 9,000 acre feet per year (acre ft/yr) from production wells in the artesian aquifers and 3,000 acre ft/yr from production wells in the basalt aquifer west of Camas Creek. Computations, using the estimated and assumed hydrogeologic properties and boundaries of the aquifer in Camas Prairie, suggest that water levels in production wells in the artesian aquifers throughout the prairie would decline to a position a few feet above the top of the upper artesian aquifer and 50 feet below the top of the basalt aquifer at the end of a 60-day pumping season as the result of such development. Additional recharge that might result from lowering water levels in the shallow water-table aquifer and the decrease in groundwater in storage in the alluvial deposits around the margin of the area where water-table conditions occur were not considered in this study.

Assuming an average duty of water of 1.75 acre ft/acre, withdrawals of 12,000 acre ft/yr by the development of the aquifers is sufficient to irrigate about 7,000 acres. It is emphasized that this discussion contemplates withdrawing the entire 12,000 acre feet during a 60-day pumping season. Hydrologic budget studies suggest that recharge from precipitation and streamflow will balance the discharge. The average annual precipitation on the area under study is estimated to be about 17 inches.

The postulated large withdrawals from production wells on the prairie would be supplied in part by a reduction in underflow from the prairie and in part by a decrease in or cessation of leakage from the upper and lower artesian aquifers. Large-scale development would reduce the artesian pressure considerably, and production wells on the prairie would cease to flow. Under present conditions, the leakage from the deeply buried deposits upward into the shallow valley-fill deposits partially supports the shallow unconfined water body. The postulated large drawdown of water levels accompanying heavy development would reduce or stop leakage to the water table and cause it to decline.

Snake River plain, Idaho

The Snake River plain extends in a broad arc, concave to the north, from eastern Idaho near Yellowstone Park into Oregon. The plain is from 30 to 75 miles wide. It is bordered on both its north and south flanks by high plateaus and mountain ranges. The Snake River flows variously in shallow valleys and deep canyons along the southern (convex) margin of the plain. About 15 major valleys are tributary to the Snake River plain. The Snake River serves as the trunk drainageway for all water moving out of the basin and receives all surface and groundwater discharge from the Snake River plain and tributary valleys. Average annual precipitation throughout much of the Snake River plain and in many of the intermontane valleys is less than 10 inches; in the mountains and high plateaus, annual precipitation at places exceeds 50 inches.

The rocks forming the mountains are largely consolidated rocks of low permeability. The plain and plateau areas are underlain chiefly by rocks of high permeability and porosity. These rocks, mostly basaltic lava flows (Snake River basalt) and alluvial materials, are the principal aquifer in the area. In the mountainous areas, aquifers mostly are recharged directly by precipitation. In the plains, lowlands, and valleys, which contain the principal aquifers, far more water is evaporated than falls on the area, and direct precipitation is a minor part of the recharge. Most recharge to these aquifers is by percolation from streams, canals, and irrigated lands.

Several large developments of groundwater for irrigation are in progress on the Snake River plain, and other large developments are expected. These groundwater withdrawals will ultimately affect water levels in the National Reactor Testing Station (NRTS) area in the north-central part of the plain because the irrigation wells and wells at the NRTS tap the same regional aquifer—the Snake River basalt. The development of groundwater at the NRTS will also affect water levels in irrigation wells. The regional effects from competitive pumping at NRTS and at other places on the Snake River plain were appraised with a model aquifer and mathematical model (Nace et al., 1959).

The hydrogeologic boundaries of the Snake River basalt are not clear-cut straight-line features but are irregular in shape and extent. The hydrogeologic properties of the aquifer are not uniform and vary from place to place on the plain. The regional effects of pumping, however, are not affected to any great degree by the irregularities of hydrogeologic boundaries or the variability of hydrogeologic properties, but are controlled for the most part by the average hydrogeologic properties of the Snake River basalt within the large areas of influence of pumping, and by the general shape and extent of boundaries. The regional effects of pumping can be determined by treating hydrogeologic boundaries as straight-line demarcations. The results of hydrogeologic studies made in the Snake River plain indicate that the effects of hydrogeologic boundaries on the response of the Snake River basalt to development by production wells can be simulated by mathematical analysis of a model aquifer consisting of a rectangular aquifer 190 and 65 miles in length and width, respectively, enclosed by two intersecting barrier and two intersecting recharging boundaries. The barrier boundaries are the borders of the Snake River basalt about 10 and 80 miles northwest and northeast, respectively, from the NRTS. These boundaries coincide approximately with the boundaries of mountainous or hilly country that limit the Snake River plain on the north and on the east (see Fig. 8.13).

The recharging boundaries are the Snake River. The Snake River has influent sections in alternation with effluent sections. Springs and seeps discharge groundwater to the river in some areas, and water percolates from the river into the aquifer in areas where the water table is below stream level. Lowering of water levels along reaches of the river where influent or effluent conditions exist and where the water table is above the bed of the river will result in decreased discharge of groundwater to the river or increased percolation of water from the river into the aquifer. Thus, the reaches of the river where the water table is above the bed of the river are recharge boundaries which limit the spread of the cone of depression. In certain reaches of the river the water table lies considerably below the bed of the river and further lowering of water levels would not increase river losses. These stretches of the river are not recharge boundaries. Thus, the Snake River is not an infinite boundary but is a finite boundary. This condition and the fact that the river is not a straight-line demarcation were taken into consideration in the model aquifer by increasing the distances from heavily pumped areas to the line of recharge and by assuming a rectangular aquifer. The influence of the hydrogeologic boundaries on the regional effects of pumping was determined with a mathematical model involving image wells.

The results of aquifer tests and specific-capacity data indicate that average coefficients of transmissibility and storage of the Snake River basalt are 4×10^6 gpd/ft and 0.025, respectively. These values were computed from wells which only partially penetrate the aquifer and from data collected during relatively short periods of pumping. A study of the gradient of the water table and underflow beneath the Snake River plain indicates that the average coefficient of transmissibility of the entire saturated thickness of the Snake River basalt is at least twice the value of T computed from aquifer tests and specific-capacity data. Further, because some basal drains very slowly, the coefficient of storage computed from test data is probably much less than the long-term S of the aquifer. As time passes water readily derived from storage by the gravity drainage of relatively large voids is augmented by the slow drainage

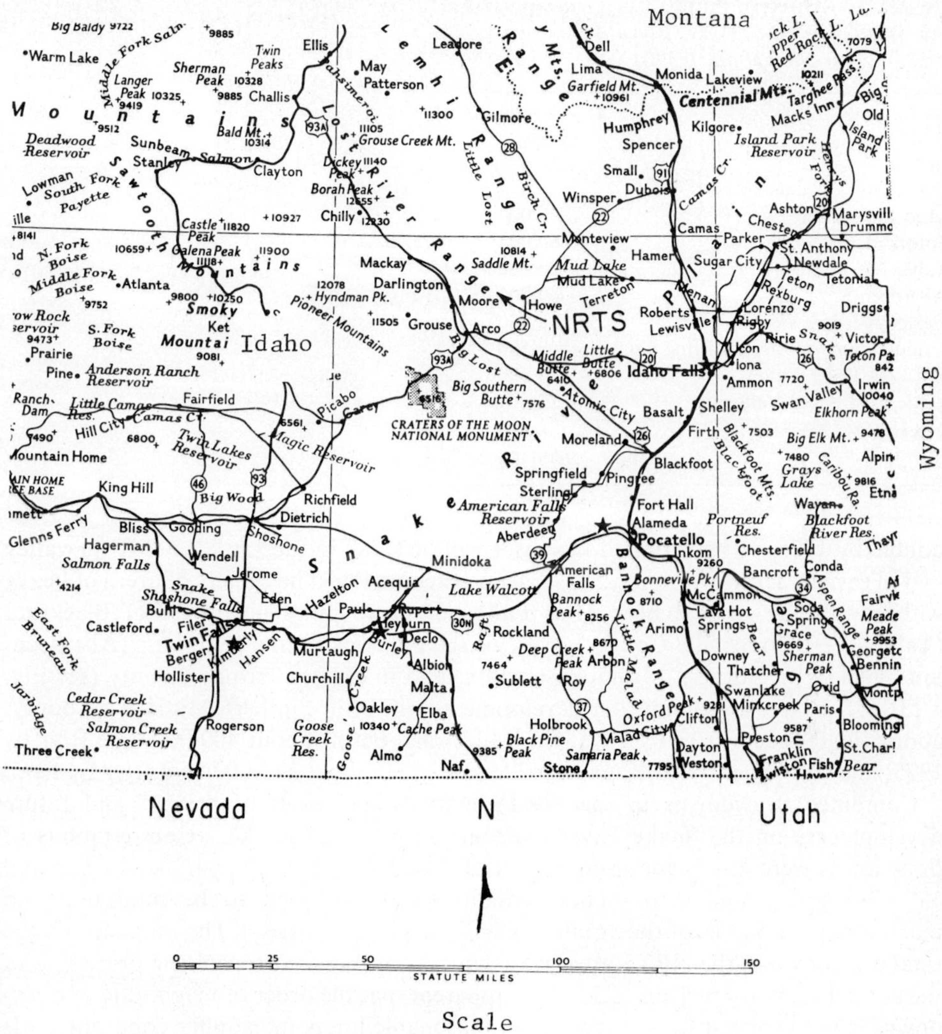

Fig. 8.13 Location of Snake River plain, Idaho. (*Drawn by W. C. Walton.*)

of small pores (vesicles, etc.). It is believed that the long-term average coefficient of storage of the Snake River basalt is of the magnitude of 5 percent (0.05). Values of T and S of 8×10^6 gpd/ft and 0.05, respectively, were substituted in the nonleaky artesian equation to determine the effects of real and image wells associated with the mathematical model. A time interval of 100 years was used in computations because it takes that long for the full effects of pumping to be felt.

Estimated withdrawals of groundwater for irrigation on the Snake River plain are given Table 8.4. Pumpage was about 635,000 acre ft/yr in 1955 and is expected to

TABLE 8.4 ESTIMATED PUMPAGE OF GROUNDWATER FOR IRRIGATION ON SNAKE RIVER PLAIN, IDAHO, 1955. (*After Nace et al., 1959.*)

Area	*Water pumped in 1955, acre ft/yr*
Mud Lake basin	270,000
Southern Minidoka Co.	160,000
Taber–W. Bonneville Co.	80,000
Aberdeen-Springfield	60,000
Eastern Jerome Co.	28,000
Gooding and Western Jerome Co.	9,500
Big Lost River Valley	14,000
Little Lost River Valley	11,000
NRTS	2,900
Total	635,400

double by 1975. The expected increases will be small in the Little Lost River valley and in eastern Jerome County, but large in other areas. The centers of areas of heavy withdrawals are located near Mud Lake, about 10 miles northwest of Blackfoot (Taber–West Bonneville County), west of the American Falls Reservoir (Aberdeen-Springfield), in southern Minidoka County, and in eastern Jerome County (see Fig. 8.13). The largest groundwater developments will be in southern Minidoka County, about 500,000 acre ft/yr by 1975, the Mud Lake basin, about 400,000 acre ft/yr by 1975, and the Taber–West Bonneville County area, about 240,000 acre ft/yr by 1975.

Computed drawdowns in the NRTS area as the result of present and future developments on the Snake River plain are given in Table 8.5. Computations of drawdowns were also made using values of T and S of 4×10^6 gpd/ft and 0.025 and 1.2×10^7 gpd/ft and 0.10. These conditions are believed to be minimum and maximum possibilities of the aquifer hydrogeologic properties. The influence on the total drawdown in the NRTS area of varying the average hydrogeologic properties of the aquifer is shown in Table 8.5. It is apparent that the order of magnitude of drawdowns is not changed by varying within reasonable limits the aquifer conditions. It is believed that the computed values of drawdown based on a T and S of 8×10^6 gpd/ft and 0.05 are realistic.

Groundwater pumpage from production wells on the NRTS was about 4 cfs or 2,900 acre ft/yr in 1955. Much of the pumped water was returned to ground storage through disposal wells and infiltration pits; the net depletion of groundwater was low, probably ranging from 15 to 30 percent of the amount pumped at different facilities. The estimated consumptive use of water on the NRTS in 1955 was 1.3 cfs. The withdrawal of such a small amount of water will theoretically effect the water levels in large irrigation well fields on the Snake River plain only a few hundredths of a foot or less.

TABLE 8.5 RESULTS OF HYDROGEOLOGIC SYSTEMS ANALYSIS FOR SNAKE RIVER PLAIN, IDAHO. (*After Nace et al., 1959.*)

	Computed effects on water levels in NRTS, drawdown in ft	
Assumed values of T and S	*Based on 1955 consumptive use*	*Based on 1975 consumptive use*
$T = 4 \times 10^6$ gpd/ft, $S = 0.025$	7	16
$T = 8 \times 10^6$ gpd/ft, $S = 0.05$	5	11
$T = 1.2 \times 10^7$ gpd/ft, $S = 0.10$	2	6

Point of observation	*Effects on water levels due to consumptive use of 13.3 cfs (10,000 acre ft/yr) at NRTS, drawdown in ft*
Mud Lake	0.5
Midway between Taber and Blackfoot	0.2
South Minidoka County	0.1

Suppose that the consumptive use of groundwater on the NRTS increases tenfold by 1975 to 13.3 cfs (10,000 acre/ft yr). Computations similar to those previously described were made to determine the effects of withdrawing 13.3 cfs on water levels in irrigated areas on the Snake River plain. The results of the computations given in Table 8.5 indicate that the withdrawals would not noticeably lower levels in large irrigation well fields in the Mud Lake, Taber, and Minidoka areas.

8.3 Mathematical-model Computation by Electronic Digital Computer

Frequently, mathematical-model computations are too laborious and costly to be practical. The time-consuming computations associated with mathematical models can in most instances be reduced to simple standardized procedures which can be inexpensively handled by means of digital computing machines (Walton and Neil, 1960). A description of machine programming is presented to show how a digital computer can expedite the solution of a mathematical model using the nonleaky artesian equation.

A complete set of instructions for the University of Illinois electronic digital computer (Illiac) was devised for the solution of the nonleaky artesian equation. Data enter the Illiac in the form of a pattern of holes punched in paper tape. Arrangement of data on the input paper tape is important to repeated automatic solution of the nonleaky artesian equation. Many of the same data items may be used repeatedly in various combinations; consequently, the data input format and the instructions to

Number of values in T-table

$T_1, T_2, T_3, \ldots\ldots T_N$ $\quad 1 \leq N \leq 24$

$-T$ Last entry in table is a negative number

S= Coefficient of storage

Number of values in r-table

$r_1, r_2, r_3, \ldots\ldots r_n$ $\quad 1 \leq n \leq 325$

$(Q,t)_{11}, (Q,t)_{12}, \ldots\ldots (Q,t)_{1m_1}$ $\quad 1 \leq m_1 \leq 46$

$-Q$ This entry in table must be a negative number

$ID_{111}, ID_{112}, \ldots\ldots ID_{11k}$ $\quad k = m_1$

$-ID$ This entry in ID set must be a negative number

$+ID$ When this entry is positive, a new ID-table follows

$ID_{211}, ID_{212}, \ldots\ldots ID_{21k}$

$-ID$

$-ID$ When this entry is negative, new Q,t and new ID follow

$(Q,t)_{21}, (Q,t)_{22}, \ldots\ldots (Q,t)_{2m_2}$ $\quad 1 \leq m_2 \leq 46$

$-Q$

$ID_{321}, ID_{322}, \ldots\ldots ID_{32k}$ $\quad k = m_2$

$-ID$

$+ID$ When this entry is positive, a new ID-table follows

$ID_{421}, ID_{422}, \ldots\ldots ID_{42k}$

$ID_{ij1} \ldots\ldots ID_{ijk}$ $\quad i = n$; j = no. of Q,t-tables

$-ID$

$-ID$

$\pm ID$ When this entry is +, the format can be repeated in a similar manner beginning with the r-table. Computer stops when this entry is negative.

Fig. 8.14 Illustrative data input format for digital computer analysis of a mathematical model involving nonleaky artesian equation. (*From Walton and Neill, 1960.*)

the Illiac were designed to minimize repetitious labor in data tape preparation. An illustrative order of items on the data tape is shown in Fig. 8.14. The *T* and *S* portion is located at the leading end of the data tape since only one set of these values is required for a single mathematical model. Either all or portions of the remainder of the format may be repeated many times.

An outline or a flow diagram, Fig. 8.15, and the data input format, Fig. 8.14, will be used as a basis for a brief discussion of steps in the solution of a mathematical model. Each block in Fig. 8.15 may be represented either an individual instruction to the computer or many individual instructions. The first step after starting the Illiac, block 1 of Fig. 8.15, involves arrangement of the complete program of instructions in designated locations in the memory. Block 2 inserts values of T and S as input data and blocks 3 and 4 insert values or r and tables of Q and t, respectively. Instruction blocks 3 and 4 are used several times during analysis, with block 4 functioning more frequently than block 3.

Increments of drawdown and buildup for given values of Q and t and particular real and image wells are given identification numbers (ID numbers). Instructions given in blocks 5, 9, 17, and 18 ensure proper identification of the many drawdowns and buildups. Block 5 functions more times than the other input blocks in providing sets of identification numbers for use during analysis. The printing of T, block 7, also provides identification and segregates groups of values in the output format.

The first five blocks of Fig. 8.15 provide the Illiac with a complete program of instructions and one set of data with identification numbers, which are required before the machine can proceed to obey arithmetic-type instructions. During the computational phase, the programmer must arrange for arithmetic instructions to function

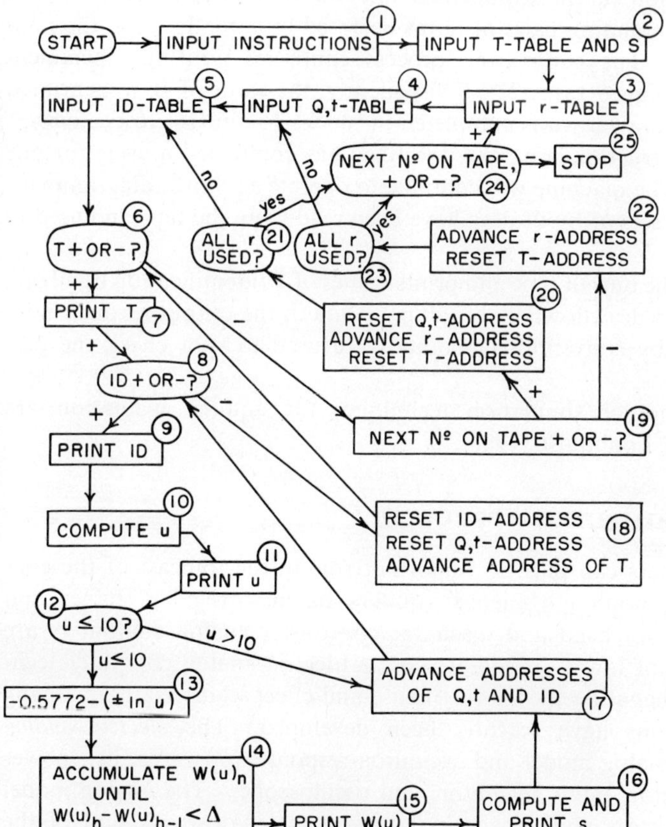

Fig. 8.15 Abbreviated flow diagram for digital computer analysis of a mathematical model involving nonleaky artesian equation. (*From Walton and Neill, 1960.*)

in harmony with instructions involving movement of data in the computer, subsequent insertion of other sets of data from the input tape, and the output of results. The algebraic sign of a number is frequently used as a basis for synchronizing the operation of the different types of instructions.. For example, quantities conveniently referred to as $-T$, $-Q$, and ID are punched at the end of their respective tables (see Fig. 8.14) to enable the Illiac to make pertinent decisions in flow diagram blocks 6, 8, 19, and 24. An alternate cycle from block 12, when u is greater than 10, was also arranged because drawdowns associated with these values of u are too small to be of practical significance. The alternative saves computer time and omits several insignificant entries in the output record.

A table of values for $W(u)$ and u has been published and could be stored in the Illiac. However, a table would use valuable memory space and it is quicker and easier for the Illiac to compute the required values of $W(u)$ when needed than to locate them in a table stored in the computer memory. Instructions presented by block 14 are concerned with the solution of the exponential integral $W(u)$ that is handled as a convergent series by accumulating pairs of terms enclosed in parentheses in the nonleaky artesian equation. The summation process continues until the increment becomes less than a preassigned value Δ. Subsequently, the value of $W(u)$ is printed and the computation for drawdown is completed in blocks 15 and 16, respectively.

At this point in the instruction program, the Illiac has completed a cycle for one drawdown or buildup. The machine will continue to operate as shown diagrammatically in Fig. 8.15 until all sets of input data have been read from the tape and used in computations.

A teletypewriter reads the output tape and prints values of T, identification numbers, values of u and $W(u)$, and drawdown or buildup. Data in the output format and a few computations made by conventional methods are used to spot check the performance of the Illiac.

Advanced digital computer simulation techniques for aquifer evaluation are described by Prickett and Lonnquist (1968).

ELECTRIC ANALOG COMPUTER APPROACH

Electric analog computers can play an important role in the forecast of the consequences of developing nonhomogeneous aquifers having irregular shapes and boundaries and a variety of head and discharge controls. Analog computers are versatile and simple and of low to moderate cost. Electric analog computer techniques (Skibitzke, 1961) that allow the study of cause-and-effect relationships involving complex aquifer conditions have recently been developed. The *electric analog computer* consists of an analog model and excitation-response apparatus, i.e., power supply, waveform generator, pulse generator, and oscilloscope. The analog model is a regular array of resistors and capacitors and is a scaled-down version of the aquifer. Resistors are approximately inversely proportional to the hydraulic conductivity of the aquifer, and capacitors store electrostatic energy in a manner analogous to the storage of water in an aquifer. The behavior of the electrical network is described by an equation that has the same form as the finite-difference equation for non-steady-state, two- or three-dimensional flow of groundwater. Electrical units (voltage, coulombs, amperes, and seconds) and corresponding hydraulic units (feet, gallons, gallons per day, and days) are connected by four scale factors.

Excitation-response equipment forces electrical energy in the proper time phase into the analog model and measures energy levels within the energy-dissipative resistor-capacitor network. Ocilloscope traces, i.e., time-voltage graphs, are analogous to time-drawdown graphs that would result after a step function type change in discharge or head. A catalog of time-voltage graphs provides data for construction of a series of water-level-change maps.

The analog computer can deal with problems of much greater complexity than is practicable with analytical methods. The accuracy, flexibility, and speed of the analog computer give the hydrologist the ability to analyze rapidly almost any conceivable hydrogeologic system with a degree of accuracy compatible with the quality of existing data. Transient and complex conditions, forbiddingly complicated for analytical study, are easily handled by the analog computer. The analog model is reusable, easy to assemble, and bears a close correspondence to the hydrogeologic system, thus giving rise to comparatively simple problem formulation and computer setup for relatively complicated situations. Unlike the digital computer, an analog computer does not require instruction in the proper mathematical method of solving a problem.

Methodology available for analyzing groundwater problems has been founded almost entirely on analytical solutions and in the past has been inadequate in coping with complex aquifer conditions. With the advent of analog computers, quantitative studies of groundwater resources are no longer hampered by inadequate methods of analysis, but are limited by the magnitude and cost of the field effort in the collection of sufficient hydrogeologic data to properly describe aquifers. Analog computers have not caused analytical methods to become obsolete, but they have, when used in conjunction with these other tools available to the hydrologist, greatly improved the analysis of groundwater problems.

8.4 Analog Model

Analog models for simulating aquifers consist of regular arrays of resistors and capacitors. In the resistor-capacitor network, resistors are approximately inversely proportional to the hydraulic conductivity of the aquifer, and the capacitors store electrostatic energy in a manner analogous to the storage of water within an aquifer. The electrical network bears a close resemblance in form to the aquifer and is a scaled-down version of the aquifer.

A resistor-capacitor network is characterized by "junctions" and discrete branches. A masonite pegboard, $\frac{1}{8}$-inch thick and perforated with holes on a 1-inch square pattern, is used in the construction of networks. The holes function as junctions or nodes; brass shoe eyelets are inserted in the holes to provide terminals. For two-dimensional flow problems and in internal parts of the analog model, four resistors and a capacitor are connected to each terminal. The capacitor also is secured to the ground connection of the electrical system. The boundary of the network is adjusted in a step fashion to approximate the actual boundary of the aquifer. To do this, boundary terminals are connected to two or three resistors and a capacitor, depending on the geometry of the boundary Exact duplication of the actual boundary of the aquifer, when the boundary does not coincide with the lines of the network grid, is possible through the use of the "vector volume" technique (Karplus, 1958). In this method the magnitudes of resistors and capacitors in the proximity of the boundary are modified to take into consideration an irregularly shaped boundary, and resistors and capacitors are selected so that there is correspondence between network parameters and aquifer parameters.

Since aquifers are continuous phenomena, while a resistor-capacitor network consists of many discrete branches, the network is only an approximation of a true analog However, it can be demonstrated mathematically that if the mesh size of the network is small in comparison to the size of the aquifer, the behavior of the network closely describes the response of an aquifer to pumping. To obtain solutions of uniform accuracy, it is often necessary to vary the net spacing from fine to coarse in areas where aquifer characteristics undergo sharp variations. Methods for effecting a variation in net spacing are available (Karplus, 1958).

Homogeneous and isotropic aquifers

The partial differential equation governing the non-steady-state, two-dimensional flow of groundwater in a homogeneous and isotropic aquifer is (Jacob, 1950)

$$\frac{\partial^2 h}{\partial x^2} + \frac{\partial^2 h}{\partial y^2} = \frac{S}{T}\frac{\partial h}{\partial t} \tag{8.1}$$

Equations of the same form occur in the theories of non-steady-state flow of heat and electricity. Consider a finite-difference approximation of Eq. (8.1). The aquifer is subdivided into squares of equal area, $\partial x\, \partial y$. The sides of the squares, ∂x and ∂y, are equal and are of finite length, Δx and Δy, respectively. The square grid is shown in Fig. 8.16*a*. The intersections of grid lines are called nodes. The infinitesmal $\Delta x\, \Delta y$ is approximated by $a_g{}^2$, in which a_g is the width of the grid interval. The area, $a_g{}^2$, is small compared with the area of the aquifer.

Examination of node 1 and Eq. (8.1) shows that the second differentials of head can be approximated by (Stallman, 1956)

$$\frac{\partial^2 h}{\partial x^2} = \frac{h_2 + h_4 - 2h_1}{a_g{}^2} \tag{8.2}$$

$$\frac{\partial^2 h}{\partial y^2} = \frac{h_3 + h_5 - 2h_1}{a_g{}^2} \tag{8.3}$$

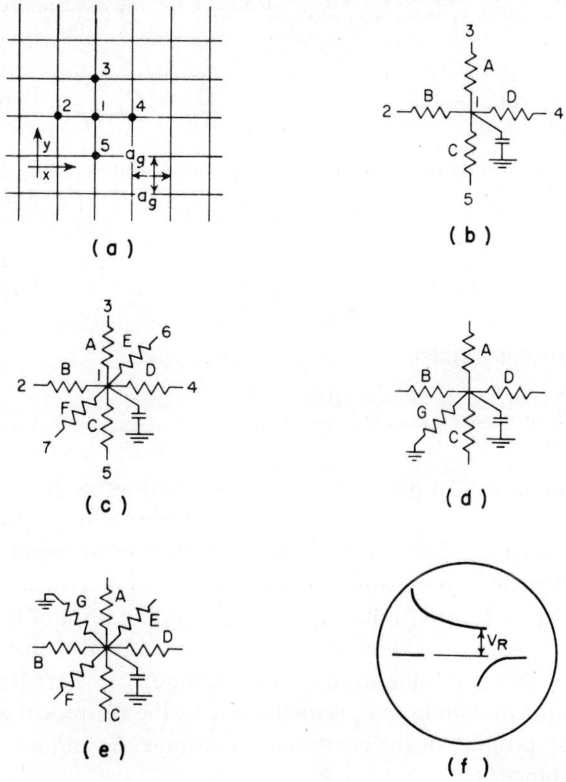

Fig. 8.16 Finite-difference grid (*a*), resistor-capacitor nets (*b*–*e*), and pumping-rate oscilloscope trace (*f*) associated with electric analog computer. (*From Walton and Prickett, 1963.*)

Substitution of Eqs. (8.2) and (8.3) in Eq. (8.1) results in

$$\frac{h_2 + h_3 + h_4 + h_5 - 4h_1}{a_g^2} = \frac{S}{T}\frac{\partial h}{\partial t} \tag{8.4}$$

which is the finite-difference form of the partial differential equation governing the non-steady-state, two-dimensional flow of groundwater in an infinite aquifer. Equation (8.4) may be rewritten in abbreviated form as

$$T\left(\sum_2^5 h_i - 4h_1\right) = a_g^2 S \frac{\partial h}{\partial t} \tag{8.5}$$

where h_1 is the head at node 1; h_i ($i = 2, 3, 4,$ and 5) represents heads at nodes 2–5; a_g denotes the width of grid interval; T stands for the coefficient of transmissibility; and S equals the coefficient of storage.

A resistor-capacitor network with a square pattern as shown in Fig. 8.16*a* and network junctions at nodes as defined in Fig. 8.16 is set up. The junctions consist of four resistors and one capacitor connected to a common terminal; the capacitor is also connected to ground. The relation of electrical potentials in the vicinity of the junction, according to Kirchhoff's current law, can be expressed as (Millman and Seely, 1941)

$$(V_3 - V_1)\frac{1}{R_A} + (V_2 - V_1)\frac{1}{R_B} + (V_5 - V_1)\frac{1}{R_C} + (V_4 - V_1)\frac{1}{R_D} = C_p\frac{\partial V}{\partial t} \qquad (8.6)$$

where V_{1-5} is the electrical potential at ends of resistors; R_{A-D} denotes the resistance; and C_p represents the capacitance. If the values of the resistors are equal, Eq. (8.6) may be written as

$$(V_2 + V_3 + V_4 + V_5 - 4V_1)\frac{1}{R} = C_p\frac{\partial V}{\partial t} \qquad (8.7)$$

Equation (8.7) may be rewritten in abbreviated form as

$$\frac{1}{R}\left(\sum_2^5 V_i - 4V_1\right) = C_p\frac{\partial V}{\partial t} \qquad (8.8)$$

where V_i ($i = 2, 3, 4$, and 5) is the electrical potential at ends of resistors A–D.

Comparison of Eqs. (8.5) and (8.8) shows that the finite-difference equation governing the non-steady-state, two-dimensional flow of groundwater in an aquifer is of the same form as the equation governing the flow of electrical current in a resistor-capacitor network. For every term in Eq. (8.5) there is a corresponding term of the same order of differentiation in Eq. (8.8). The analogy between electrical systems and aquifer systems is apparent. The hydraulic heads h are analogous to electrical potentials V. The coefficient of transmissibility, T, is analogous to the reciprocal of the electrical resistance, $1/R$. The product of the coefficient of storage, S, and $a_g{}^2$ is analogous to the electrical capacitance, C_p.

Continuing the comparison, water moves in an aquifer just as charges move in an electrical circuit. The quantity of water is calculated in gallons while the charge is calculated in coulombs. The rate of flow of water through a given cross section of the aquifer is expressed in gallons per day, while the flow of electricity is expressed in coulombs per second, or amperes. The hydraulic head losses between two points in an aquifer are expressed in feet, while the potential drop across a part of the electrical circuit is in volts.

Thus, there are four units that are analogous. There is necessarily a scale factor connecting each unit in one system to the analogous unit in the other system. Knowing the four scale factors, the hydrologist is able to relate electrical units associated with the analog model to hydraulic units associated with the aquifer. The four scale factors, K_1, K_2, K_3, and K_4, were defined as follows (Bermes, 1960)

$$q = K_1 Q \qquad (8.9)$$

$$h = K_2 V_0 \qquad (8.10)$$

$$Q = K_3 A \qquad (8.11)$$

$$t_d = K_4 t_f \qquad (8.12)$$

where q is gallons; $\mathcal{Q}$ equals coulombs; Q denotes gpd; A represents amperes; h indicates feet; V_0 corresponds to volts; t_d is equivalent to days; t_f describes seconds; K_1 refers to gallons per coulomb (gal/coulomb); K_2 stands for feet per volt (ft/volt); K_3 is gallons per day per ampere (gal/day amp); and K_4 equals days per second (days/sec). The relation between scale factors K_1, K_3, and K_4 is expressed by

$$\frac{K_3K_4}{K_1} = 1 \tag{8.13}$$

The values of resistors and capacitors required to model a given aquifer may be determined from Eqs. (8.9) to (8.13). To minimize computer costs, scale factors are chosen that allow the use of standard tolerance, low wattage, fixed carbon resistors with values from 10^2 to 10^7 ohms, and capacitors of low voltage rating with values from 10^1 to 10^5 $\mu\mu$f.

Substitution of Ohm's law and Darcy's law in Eq. (8.11) yields

$$R_R = \frac{K_3}{K_2T} \tag{8.14}$$

where R_R is the resistance, in ohms; and T refers to the coefficient of transmissibility, in gpd/ft. Equation (8.14) may be used to determine the value of the resistors of the interior parts of the analog model (see Walton and Prickett, 1963).

Equation (8.15), which may be used to determine the value of the capacitors of the interior parts of the analog model, is derived by taking into consideration the definitions of the storage and capacitance coefficients and the analogy between $(a_g{}^2S)$ and C_p.

$$C_p = 7.48a_g{}^2S\frac{K_2}{K_1} \tag{8.15}$$

where C_p is the capacitance, in farads; a_g denotes the network spacing, in feet; and S represents the coefficient of storage, fraction.

Network spacings are selected to minimize the errors due to finite-difference approximation. For regional studies on a gross basis the network spacing can be 10,000 feet; for local studies involving details, a network spacing of 100 feet is adequate.

Thus far, only two-dimensional flow through homogeneous and isotropic aquifers has been considered. The hydrogeologic properties of the aquifer were assumed to be constant everywhere in space. In many problems, however, the variations of hydrogeologic properties with space are large and flow is three-dimensional. These variations and vertical components of flow can also be taken into account. A generalization of Eqs. (8.5) and (8.8) permits accounting for differences in space of the coefficients of transmissibility and storage by varying resistors and capacitors (Skibitzke, 1961).

Resistance elements of an analog model of two-dimensional groundwater flow through nonhomogeneous aquifers may be assigned values inversely proportional to the aquifer transmissibility if variations of transmissibility in space can be represented accurately by linear or quadratic equations. According to Stallman (1963) the solution error is independent of the analogic distance between resistor units, except for errors due to truncation of the finite-difference approximations of head, in models of aquifers with variations of transmissibility of first or second degree. Where transmissibility changes in space are of third or higher degree, model size for a given solution accuracy can be made small by selecting resistance values inversely proportional to an integral form of transmissibility. For some types of aquifers, the latter design procedure may be preferred for solutions of adequate accuracy from a comparatively small number of components. A procedure for computing local error at each point in the analog solution has been proposed by Stallman (1963), using finite-difference techniques. An approximation of the total error in the analog solution can be obtained by injecting currents proportional to the calculated local error at each point on the model.

The finite-difference equation governing the three-dimensional flow of groundwater in an anisotropic and homogeneous aquifer is

$$\frac{h_2 + h_4 - 2h_1}{a_{g_1}{}^2/P_1} + \frac{h_3 + h_5 - 2h_1}{a_{g_2}{}^2/P_2} + \frac{h_6 + h_7 - 2h_1}{a_{g_3}{}^2/P_3} = S_0 \frac{\partial h}{\partial t} \tag{8.16}$$

in which a_{g_1}, a_{g_2}, and a_{g_3} are the sides of a rectangular network of junctions corresponding to the three principal directions; P_1, P_2, and P_3 are the three principal coefficients of permeability; $h_1 - h_7$ are heads at nodes; and S_0 is the specific storativity which is defined as the volume of water released from (or added to) a unit volume of aquifer when the head is lowered (or raised) by a unit head.

A resistor-capacitor network with a square pattern and network junctions at nodes as defined in Fig. 8.16*c* is constructed. The junctions consist of six resistors and one capacitor connected to a common terminal, with the capacitor connected to ground. If R_1, R_2, and R_3 are the corresponding resistors in the 1, 2, and 3 principal directions, then from Eqs. (8.6) and (8.16) the analogy can be derived between S_0 and C_p, $a_{gi}{}^2/P_i$ and R_i $(i = 1, 2, 3)$. If $P_1 = P_2 \neq P_3$ and $a_g = a_{g1} = a_{g2} \neq a_{g3}$ is chosen, then the analogous terms are $a_g{}^2S_0$ and C_p, $1/P_1 = 1/P_1 = 1/P_2$ and $R_1 = R_2$, $a_{g3}{}^2/a_{g2}{}^2P$ and R_3.

Leaky artesian conditions exist where aquifers are overlain by aquitards through which leakage occurs. Leakage through the aquitard into the aquifer is vertical and proportional to drawdown, and the hydraulic head in the deposits supplying leakage remains more or less uniform. Under these conditions leakage can be simulated in an analog model by the addition of resistors connected to ground and to each node of the network. Network junctions for two-dimensional and three-dimensional flow in a leaky aquifer are shown in Fig. 8.16*d* and *e*, respectively.

Substitution of Ohm's and Darcy's laws in Eq. (8.11) results in

$$R_G = \frac{K_3}{K_2(P'/m')a_g{}^2} \tag{8.17}$$

where P'/m' is the coefficient of leakage, in gpd/cu ft; P' is the coefficient of permeability of aquitard, in gpd/sq ft; and m' denotes the saturated thickness of aquitard in feet. Equation (8.17) may be used to determine the value of the resistor G of the interior parts of the analog model. Variations in the coefficient of vertical permeability of the aquitard may be simulated by varying the value of resistors connected to ground.

Barrier boundaries (lines across which there is no flow) can be simulated by an open circuit. Resistors connected to the nodes along the edge of the analog model and to the ground simulate horizontal leakage through boundaries of the aquifer.

Leakage of water through a streambed can be simulated with a resistor connected to appropriate nodes and to ground. The magnitude of resistors (R_S) simulating the streambed are inversely proportional to the areas of the portions of the streambed they represent. The following equation, derived by substituting Ohm's and Darcy's laws into Eq. (8.11), may be used to compute the values of resistors simulating the streambed (Walton and Ackroyd, 1966)

$$R_S = \frac{K_3}{K_2(I_h)A_s} \tag{8.18}$$

where A_s is the area of the streambed represented by the resistor, in square feet, and I_h is the infiltration rate of the streambed, in gpd/sq ft/ft.

The values of resistors and capacitors adjacent to boundaries can be computed based on the "vector volume" technique described by Karplus (1958). Variation in net spacing also can be accomplished with techniques described by Karplus (1958). The following equations (see Karplus, 1958) may be used to compute values of resistors adjacent to boundaries:

$$R_x = R_b \frac{\Delta x}{\Delta y} \tag{8.19}$$

$$R_y = R_b \frac{\Delta y}{\Delta x} \tag{8.20}$$

where R_x is the value of the x-direction resistor adjacent to a boundary, in ohms; R_y is the value of the y-direction resistor adjacent to a boundary, in ohms; R_b is the value of resistors in interior portions of the model, in ohms; x is the x dimension of the portion of the aquifer represented by the resistor, in feet; and y is the y dimension of the portion of the aquifer represented by the resistor, in feet.

Suppose that a barrier or recharge boundary does not coincide with the lines of part of the grid as shown in Fig. 8.17*a*. The vector volumes corresponding to R_5 and R_6 have the same width Δy (see Fig. 8.17*b*) as those of interior resistors R_b and a length of 0.5 Δx (see Fig. 8.17*b*), so that from Eq. (8.19), $R_5 = R_6 = 0.5R_b$. The vector volumes corresponding to R_7 and R_8 have the same width Δx (see Fig. 8.17*c*) as those of the interior resistors R_b, but a length of 0.5 Δy (see Fig. 8.17*c*), so that from Eq. (8.20), $R_7 = R_8 = 0.5R_{14}$. The vector volume corresponding to R_4 has an average

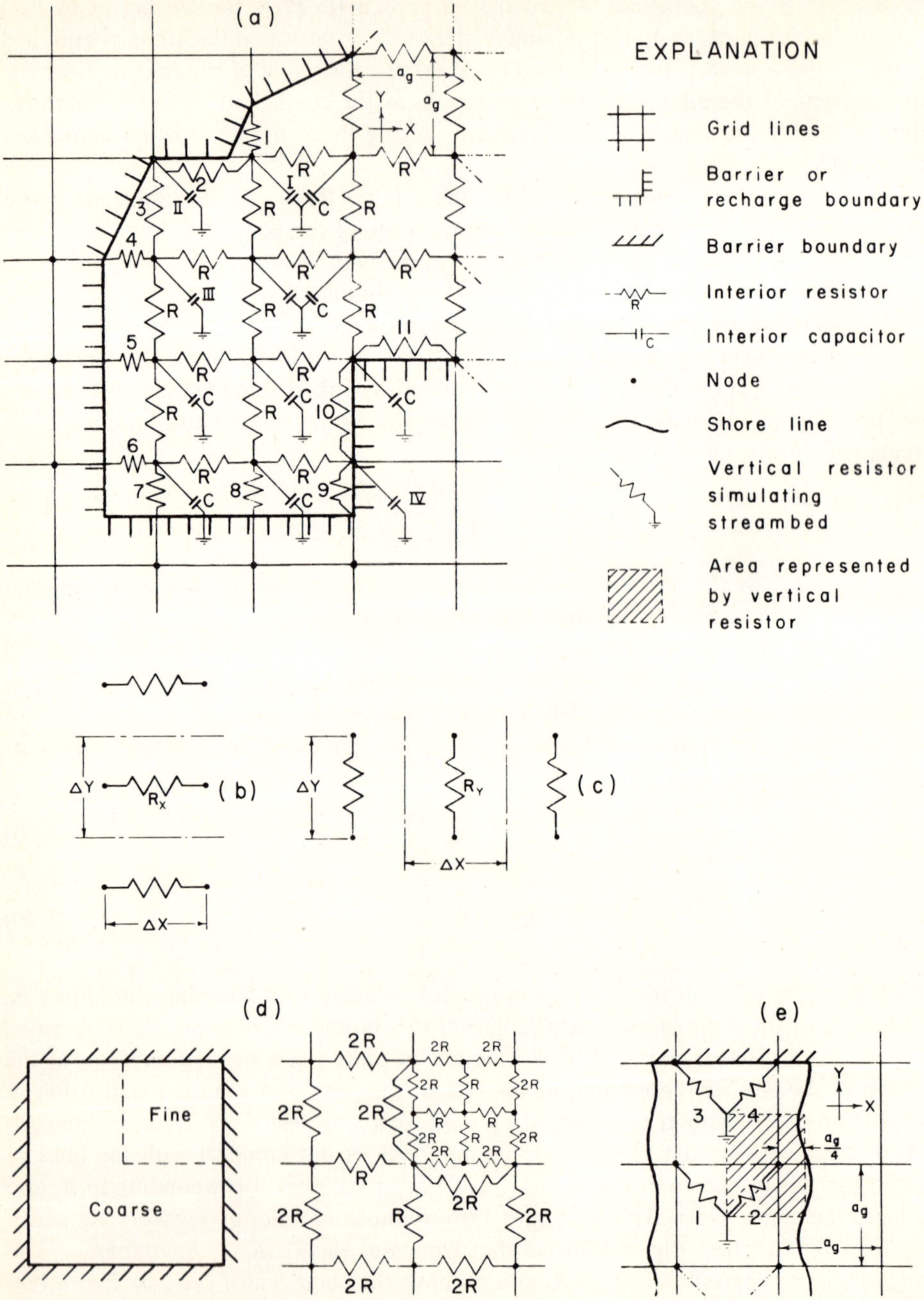

Fig. 8.17 Vector volume techniques for computing values of resistors and capacitors adjacent to boundaries of analog model. (*From Walton and Ackroyd, 1966.*)

width (Δy) of 0.9 Δy compared to an interior resistor R_b and an average length (Δx) of 0.45 Δx compared to an interior resistor R_b, so that from Eq. (8.19), $R_4 = 0.5R_b$.

Suppose that a barrier or recharge boundary coincides with the lines of part of the grid as shown in Fig. 8.17*a*. The vector volume corresponding to R_{11} has the same length Δx (see Fig. 8.17*b*) as an interior resistor R_b, but a width of 0.5 Δy (see Fig. 8.17*b*), so that from Eq. (8.19), $R_{11} = 2R$. The vector volume corresponding to R_{10} has the same length Δy (see Fig. 8.17*c*) as an interior resistor R, but a width of 0.5 Δx (see Fig. 8.17*c*), so that from Eq. (9.20), $R_{10} = 2R_b$. The vector volume corresponding to R_9 has a length of 0.5 Δy (see Fig. 8.17*c*), compared to an interior resistor R_b, and a width of 0.5 Δx (see Fig. 8.17*c*), so that from Eq. (8.20), $R_{10} = R_b$. Values of resistors in Fig. 8.17*a* are $R_1 = 1.1R_5$, $R_2 = 1.9R_b$, $R_3 = 1.5R_b$, $R_4 = 0.5R_b$, $R_5 = 0.5R_b$, $R_6 = 0.5R_b$, $R_7 = 0.5R_b$, $R_8 = 0.5R_b$, $R_9 = R_b$, $R_{10} = 2R_b$, and $R_{11} = 2R_b$.

The vector volumes (A_s) adjacent to boundaries corresponding to R_{s1} and R_{s2} in Fig. 8.17*e* are $\frac{3}{4}a_g^2$ ($\frac{3}{4}a_g \times a_g$); thus, $R_{s1} = R_{s2} = (K_3/K_2)(I_h)(\frac{3}{4}a_g^2)$. The vector areas ($A_s$) corresponding to R_{s3} and R_{s4} in Fig. 8.17*e* are $\frac{3}{8}a_g^2$ ($\frac{3}{4}a_g \times \frac{1}{2}a_g$); thus, $R_{s3} = R_{s4} = (K_3/K_2)(I_h)(\frac{3}{8}a_g^2)$

The magnitudes of capacitors are directly proportional to the areas of the portions of the aquifer they represent. The values of the capacitors (C_b) adjacent to boundaries may be computed with the following equation

$$C_b = 7.48A_vS\frac{K_2}{K_1} \tag{8.21}$$

where A_v is the area of the aquifer represented by the capacitor, in square feet.

The vector area corresponding to C_{III} (see Fig. 8.17*a*) is $0.94b^2$, so that $C_{\mathrm{III}} = 0.94C_G$. Values of capacitors in Fig. 8.17*a* are $C_{\mathrm{I}} = 0.80C_p$, $C_{\mathrm{II}} = 0.38C_p$, $C_{\mathrm{III}} = 0.94C_p$, and $C_{\mathrm{IV}} = 0.25C_p$.

In order to obtain satisfactory solutions to groundwater flow problems with a minimum number of electrical components, it is often advantageous to employ coarse grids in some sections of the model and fine grids in other sections. A simple method for effecting a variation in net spacing is to regard the "coarse" and "fine" sections as separate portions of the model and to treat the interface between the two sections as a barrier boundary (Karplus, 1958). The two grids are then jointed at appropriate nodes. Figure 8.17*d* illustrates how a variation in the ratio of 1:2 may be accomplished.

If a production well is simulated by a junction of an analog model, its radius will be as indicated in the following equation (Prickett, 1967)

$$r_w = \frac{a_g}{4.81} \tag{8.22}$$

where r_w is the radius of the production well, in feet, and a_g is the grid spacing, in feet.

A fully penetrating production well with any given radius can be simulated by addition of a single resistor to the junction. The value of the resistor, R_w, may be computed from the following equation (Prickett, 1967)

$$R_w = \frac{K_3}{K_2 T} 0.3665 \log \left(\frac{a_g}{4.810 r_w} \right) \tag{8.23}$$

The effects of partial penetration of the production well, well losses, and gravel packs can be simulated with methods developed by Prickett (1967).

The cost of analog models largely depends on the price of resistors and capacitors embodied in the model and on the labor charge for the design and construction of the analog model. Fixed carbon resistors with tolerances of ± 10 percent can be purchased (1968) in bulk at a cost of approximately \$0.04 each; ceramic capacitors with tolerances of ± 10 percent are available in bulk for approximately \$0.05 each. Analog models of average size (4 by 8 feet or smaller) and complexity can usually be constructed in less than 80 man-hours. Computations involved in determining the values of resistors and capacitors of complex analog models usually can be performed in less than 40 man-hours.

8.5 Excitation-response Apparatus

Excitation-response apparatus (Korn and Korn, 1952; Wass, 1955) forces electrical energy in the proper time phase into the analog model and measures energy levels

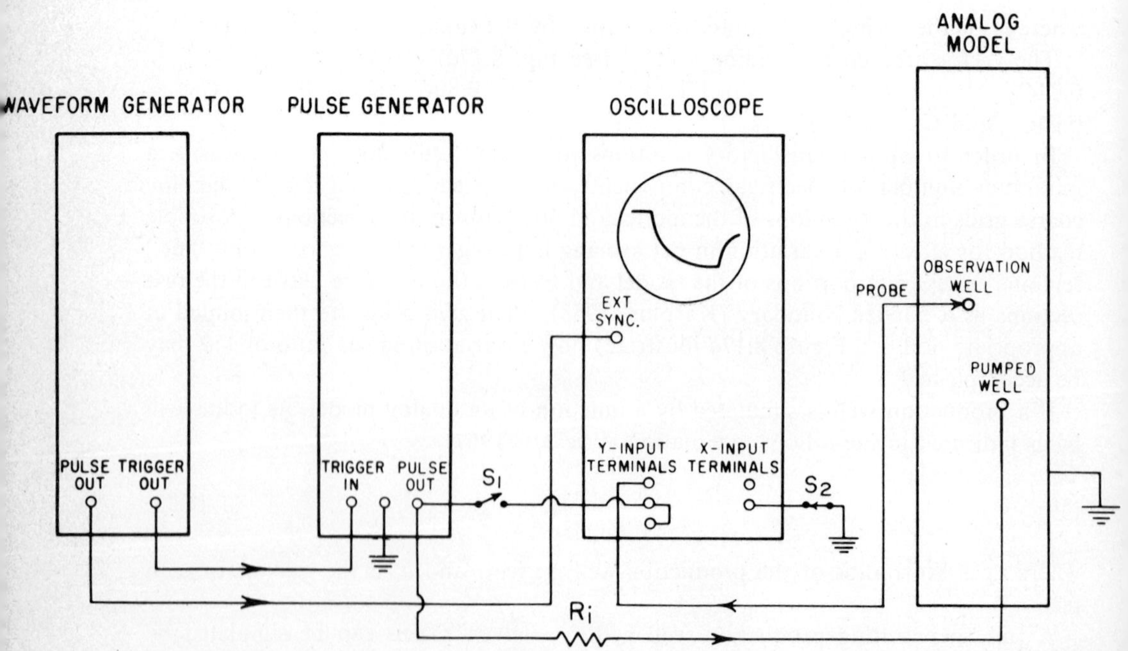

Fig. 8.18 Block diagram describing analog computer excitation-response apparatus. (*From Walton and Prickett, 1963.*)

within the energy-dissipative, resistor-capacitor network. The excitation-response apparatus is comprised of four major parts, as shown in Fig. 8.18: a power supply, a waveform generator, a pulse generator, and an oscilloscope. The waveform generator, which produces sawtooth pulses, is connected to the trigger circuits of the pulse generator and oscilloscope, thereby controlling the repetition rate of computation and synchronizing the oscilloscope's horizontal sweep and the output of the pulse generator. The pulse generator, which produces rectangular-shaped pulses (Gottlieb, 1958) of various duration and amplitude on command from the waveform generator, is coupled to that junction in the analog model representing the production well. The oscilloscope is connected to junctions of the analog model in which it is desired to determine the response of the analog model to excitation. An electron beam is swept across the cathode-ray tube of the oscilloscope, providing a time-voltage graph that is analogous to the time-drawdown graph for an observation well.

Figure 8.19 shows the simultaneous sequences of events that take place in the excitation-response apparatus during two computational cycles. The waveform generator sends a positive pulse to the oscilloscope (Fig. 8.19*a*) to start a horizontal sweep; at the same time, it sends a negative sawtooth waveform to the pulse generator (Fig. 8.19*b*). At a point along the sawtooth waveform (arbitrarily chosen as 25 percent of the waveform duration in Fig. 8.19*a*) the pulse generator is triggered to produce a negative, rectangular-shaped pulse as shown in Fig. 8.19*c*. The duration of this pulse is analogous to the pumping rate Q. This pulse is sensed by the oscilloscope as a function of the analog-model components, boundary conditions, and node position of the junction connected to the oscilloscope. Thus, the oscilloscope trace (Fig. 8.19*d* and *e*) is analogous to the water-level fluctuation that would result after a step-function type pumpage change of known duration and amplitude. To provide data independent of the pulse repetition rate, the time interval between pulses is set at a multiple of the longest time constant in the analog model. The time constant is the product of the capacitance at a point and the resistance in its discharge path.

A method of computing the pumping rate is incorporated in the circuit between the pulse generator and the analog model by the small resistor, R_i, in series, shown in Fig. 8.18. Substitution of Ohm's law in Eq. (8.11) results in

$$Q = \frac{V_R}{1.44 \times 10^3 R_i} K_3 \tag{8.24}$$

where Q denotes the pumping rate, in gpm; V_R represents the voltage drop across the resistor R_i, in volts; and R_i describes the calibrated resistance, in ohms. Equation (8.24) may be used to compute the pumping rate.

The voltage drop across the calibrated resistor is measured with the oscilloscope. Switches S_1 and S_2 in Fig. 8.18 are closed and opened, respectively, and the oscilloscope is connected to the pumped well junction. The waveform in Fig. 8.16*f* will

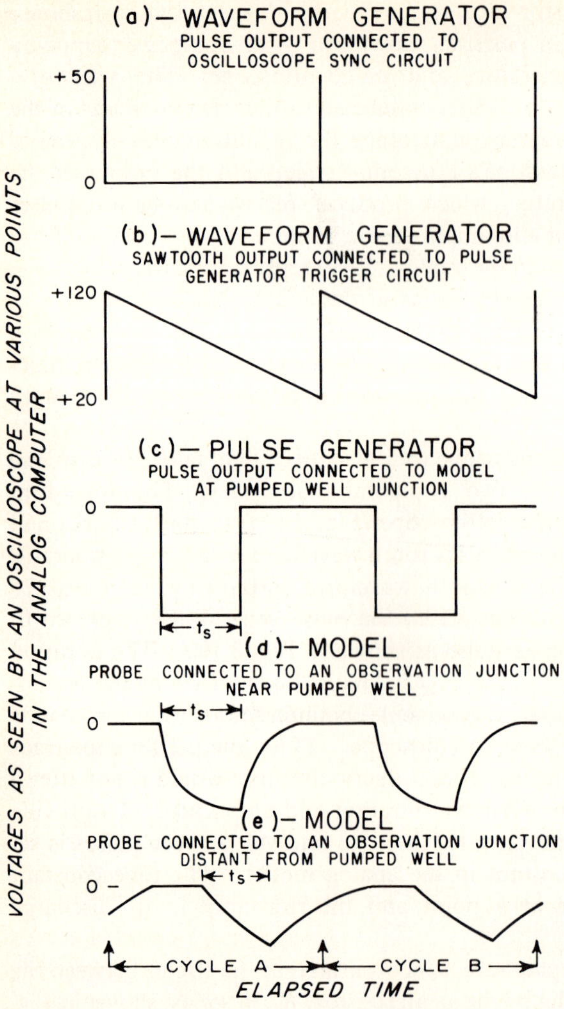

Fig. 8.19 Diagram showing sequence of events in two computational cycles of analog computer. (*From Walton and Prickett, 1963.*)

appear on the cathode-ray tube; the vertical distance as shown is the desired voltage drop V_R.

The switches S_1 and S_2 are returned to their original positions. The oscilloscope is then connected to all junctions of the analog model representing observation wells. Time-voltage curves for junctions can be recorded by photographing oscilloscope traces. The screen of the oscilloscope is accurately calibrated so that voltage and time may be read on the vertical and horizontal axes, respectively. The time is in seconds; the value of each horizontal division on the screen is determined by noting the duration of the rectangular-shaped pulse and the number of divisions covered by the time-voltage trace for a junction adjacent to the pumped well. The time-voltage graphs obtained from the oscilloscope can be converted into time-drawdown graphs by means of Eqs. (8.10) and (8.12), relating electrical units to hydraulic units. A catalog of time-drawdown graphs provides data for the construction of a series of water-level-change contour maps. Thus, water-level changes are described everywhere in the aquifer for any desired pumping period. The pulse generator can be connected to many junctions, and a variety of pumping conditions can be studied.

The pulse generator can be coupled to junctions in the analog model representing a stream recharge boundary, and water-level fluctuations due to a step-function type change in stream stage can be determined with the oscilloscope. Factors such as well discharge and recharge, stream-stage changes, recharge from precipitation, and tidal fluctuations may be imposed on the analog model.

The effects of complex pumpage changes on water levels may be determined by approximating the pumpage graph by a group of step functions and analyzing the effect of each step function separately. The total water-level change, based on the superposition theorem, is obtained by summation of individual step-function water-level changes. Costly variable-function generators are operational conveniences for imposing complex pumpage or boundary fluctuations on the analog model.

The pulse generator commonly used has a maximum output of 50 volts and 20 ma; the pulse generator and oscilloscope have rise times of less than 1 μsec and waveform durations ranging from less than 10 μsec to 100 msec. The performance specifications of the waveform generator, pulse generator, and oscilloscope are compatible with the following criteria desired for analog computers: low power requirements, repetitive calculation at variable rates, and fast computing speeds. Excitation-response equipment can be purchased (1968) for less than $1,500, depending chiefly on the quality of oscilloscope desired. Special apparatus also may be coupled to the excitation-response apparatus to reject some of the unwanted noise pick-up, thus permitting the measurement of smaller voltages.

The accuracy and reliability of the electric analog computer were assessed by a study of selected idealized aquifer situations that could be modeled and accurately analyzed both analytically and electrically. Analytical solutions involving mathematical models (Walton and Prickett, 1963) were compared with results obtained with the analog computer. In general, the accuracy of the analog solutions is a matter of the quality of resistors and capacitors, the effects of finite-difference approximation, and the signal-noise ratio. Inaccuracies due to the quality of resistors and capacitors are small, owing to the error-equalizing property of the resistor-capacitor network. Errors due to finite-difference approximation are minimum when the network spacing is small in comparison to the size of the aquifer; errors depend in part on the nature of the potential field being analyzed. Examples of analysis indicate that if the excitation-response apparatus in Fig. 8.18, inexpensive resistors and capacitors, and a reasonable network spacing are used, differences between analog and analytical solutions are not excessive. The overall accuracy of the analog solutions can be

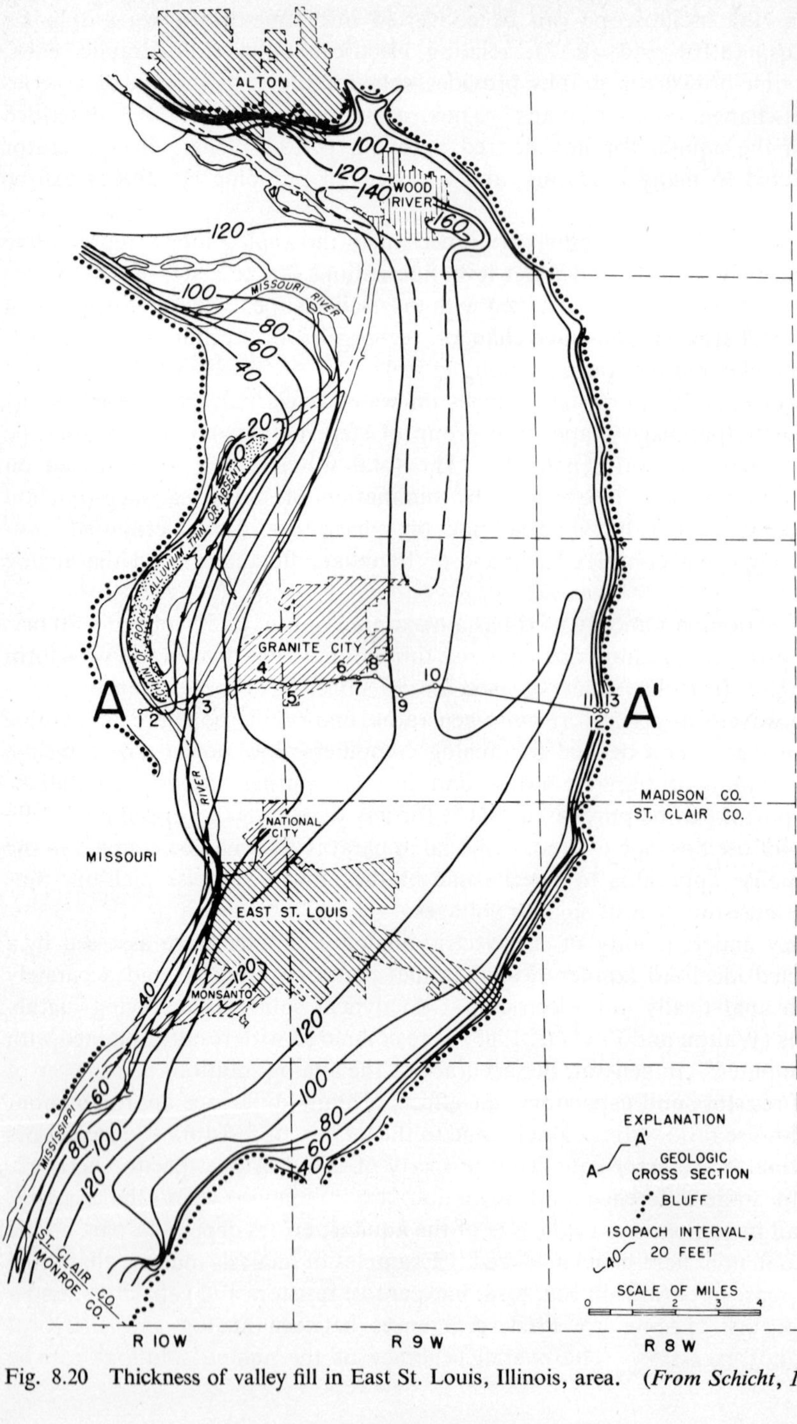

Fig. 8.20 Thickness of valley fill in East St. Louis, Illinois, area. (*From Schicht, 1965.*)

increased to 1 or 2 percent by using precision resistors and capacitors and refined excitation-response apparatus; however, the benefit-cost ratio of the increased accuracy is low, so that in most cases the precision is not warranted.

8.7 Case Histories of Hydrogeologic Systems with Electric Analog Computers

East St. Louis, Illinois

An electric analog computer (Schicht, 1965) for the East St. Louis, Illinois, area was constructed in order that the consequences of further development of a sand and gravel aquifer in the area could be forecast, the practical sustained yield of existing pumping centers could be evaluated, and the potential yield of the aquifer with a selected scheme of development could be appraised. The East St. Louis area is in southwestern Illinois and extends along the valley lowlands of the Mississippi River from the city of Alton south to the village of Dupo. The area covers about 175 square miles and is approximately 30 miles long and 11 miles wide at the widest point. It is one of the most heavily populated and industrialized areas in Illinois.

The groundwater resources of a sand and gravel aquifer underlying the East St. Louis area have been extensively developed. It is estimated that 105 mgd was withdrawn during 1962 chiefly from industrial and municipal wells. Pumpage from wells increased from 2.1 mgd in 1900 to 111.0 mgd in 1956 and declined sharply from 111.0 gmd in 1956 to 92.0 mgd in 1958 (Schicht and Jones, 1962). Pumpage is largely concentrated in five major pumping centers: Alton, Wood River, Granite City, National City, and Monsanto areas. During 1962 total pumpage from the Alton, Wood River, and Monsanto areas was 74.8 mgd, of which 31.2 mgd was withdrawn from wells near the Mississippi River.

As the result of heavy pumping, water levels declined about 50 feet in the Monsanto area, 40 feet in the Wood River area, 20 feet in the Alton area, 15 feet in the National City area, and 10 feet in the Granite City area from 1900 to 1962. In general, water levels within pumpage centers fluctuate in response to precipitation and changes in river stage and pumpage. Taking these factors into consideration, water-level declines are directly proportional to pumping rates. Recharge within the area is from precipitation, induced infiltration of surface water from the Mississippi River and small streams traversing the area, and subsurface flow from the bluffs bordering the area. Average annual precipitation is 38 inches.

According to Bergstrom and Walker (1956), unconsolidated valley fill in the area is composed of recent alluvium and glacial valley-train material and is underlain by Mississippian and Pennsylvanian rocks consisting of limestone and dolomite with subordinate amounts of sandstone and shale. Because of the low permeability of the bedrock formations and poor water quality with depth, the rocks do not constitute an important aquifer in the area. The valley fill has an average thickness of 120 feet and ranges in thickness from a feather's edge, near the bluff boundaries of the area and along the reach of the Mississippi River known as the "Chain of Rocks," to more than 170 feet near the city of Wood River (Fig. 8.20). The thickness of the valley fill is

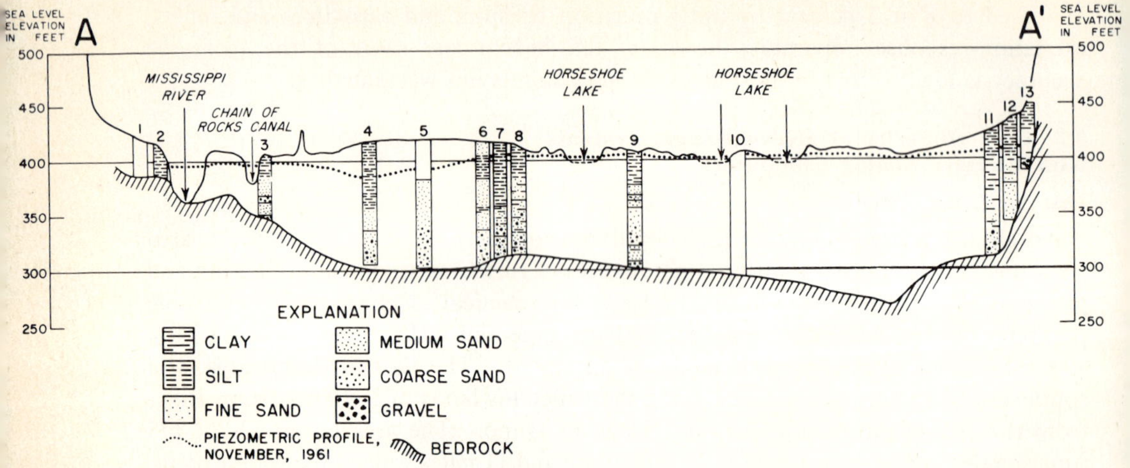

Fig. 8.21 Geologic cross section and piezometric profile of valley fill in East St. Louis, Illinois, area. (*After Bergstrom and Walker, 1956.*)

generally greatest and exceeds 120 feet in places near the center of a buried-bedrock valley that bisects the area.

Recent alluvium comprises the major portion of the valley fill in most of the area (Bergstrom and Walker, 1956). The alluvium is composed of fine-grained materials with a low permeability; the grain size increases from the surface down. Recent alluvium rests on older deposits including in many places valley-train materials. The valley-train materials are predominantly medium-to-coarse sand and gravel which increase in grain size with depth. The coarsest deposits most favorable for development are commonly encountered near bedrock and often average 30 to 40 feet thick. Logs of wells in cross section *A–A′* in Fig. 8.21 show that the valley fill grades with depth from clay to silt to sand and gravel interbedded with layers of silt and clay.

Groundwater in the valley fill occurs under leaky artesian and water-table conditions. Leaky artesian conditions exist at places where fine-grained alluvium overlies valley-train deposits, and water in the valley-train deposits is under artesian pressure. Water-table conditions prevail at many places where alluvium is missing and the upper surface of the zone of saturation is in valley-train deposits, and at places within deep cones of depression, where water is unconfined.

A coefficient-of-transmissibility map was constructed (Fig. 8.22) based on aquifer test and well-production data and information on the saturated thickness of the aquifer. The coefficient of transmissibility ranges from less than 50,000 gpd/ft near the bluffs and the Chain of Rocks canal to greater than 300,000 gpd/ft near Monsanto. The long-term coefficient of storage of the aquifer is estimated from aquifer test data to be 0.15.

The analog model for simulating the aquifer in the East St. Louis area consists of a regular array of 2,800 resistors and 1,350 capacitors and is shown in Fig. 8.23. The

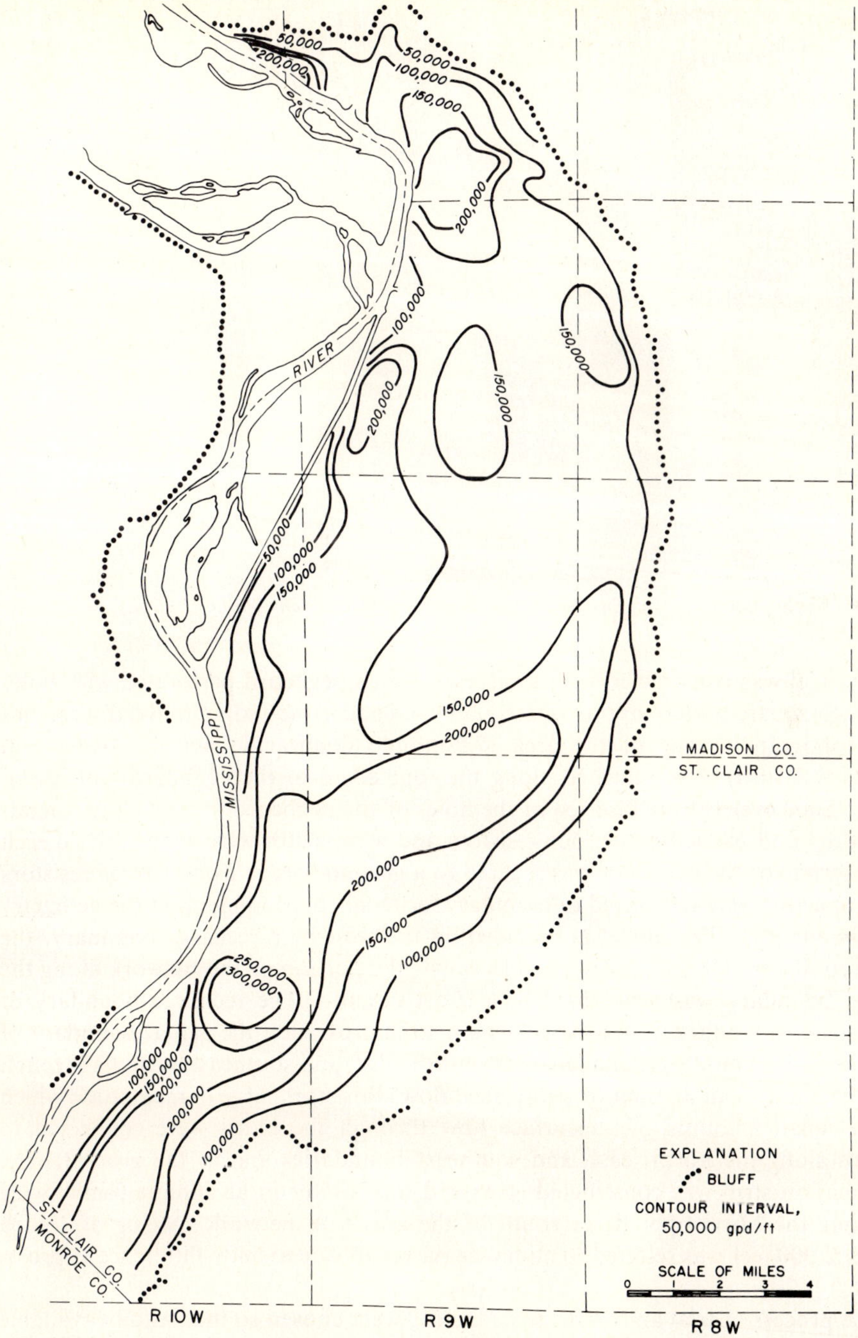

Fig. 8.22 Coefficient of transmissibility of aquifer in East St. Louis, Illinois, area. (*From Schicht, 1965.*)

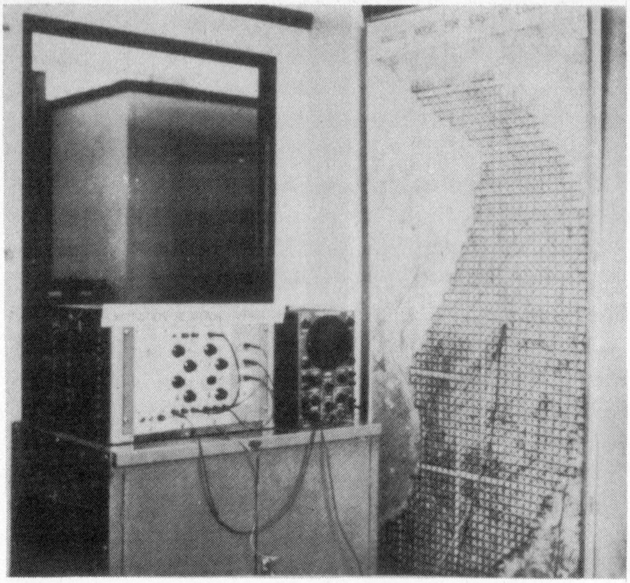

Fig. 8.23 Analog computer for East St. Louis, Illinois, area. (*From Walton, 1964d.*)

analog model was constructed with a piece of $\frac{1}{8}$-inch pegboard perforated with holes on a 1-inch-square pattern approximately 3 by 6 feet corresponding to the dimensions of the topographic map of the area (7.5-minute quadrangle maps). Aluminum angles (1×1 inch) were attached along the four edges of the pegboard with metal screws. Shoe eyelets were inserted in the holes of the pegboard to provide terminals for resistors and capacitors. Four resistors and a capacitor were connected to each interior terminal; the capacitor was secured to a ground wire. Two or three resistors and a capacitor were connected to boundary terminals depending upon the geometry of the boundary. The model is bounded on the west by a recharge boundary, the Mississippi River and the Chain of Rocks canal; the portion of the network along the recharge boundary was terminated in a short circuit. The recharge boundary of the network was adjusted in a step fashion to approximate the actual boundary of the aquifer. The model is bounded on the north, east, and southeast by bluffs through which there is a small amount of subsurface flow. Resistors large in magnitude which simulate small amounts of subsurface flow through the bluffs were connected to terminals along the north, east, and southeast boundaries and to the ground wire. A termination strip was constructed to extend the aquifer in an infinite 3-mile-wide strip along the Mississippi River south of the area. A network spacing of 1 inch equaling 2,000 feet was selected to minimize the errors due to finite-difference approximation.

By the process of trial and error, scale factors were chosen so that readily available and inexpensive resistors and capacitors and existing excitation-response apparatus could be used. Selected scale factors are given below:

$$K_1 = 1.826 \times 10^{15} \text{ gal/coulomb}$$

$$K_2 = 1 \text{ ft/volt}$$

$$K_3 = 1 \times 10^{10} \text{ gal/day amp}$$

$$K_4 = 1.826 \times 10^{5} \text{ days/sec}$$

A maximum pumping period t_a of 5 years was chosen which is a sufficient period for water levels to stabilize under the influence of recharge from the Mississippi River. Fixed carbon resistors with tolerances of ± 10 percent and disk ceramic capacitors with tolerances of ± 10 percent were used in constructing the analog model.

Values of resistors were computed with data on the coefficient of transmissibility given in Fig. 8.22. Values of resistors in the internal parts of the model range in magnitude from 470,000 ohms near the bluff to 33,000 ohms near Monsanto. Resistors are greatest in magnitude, 2,200,000 ohms, along the bluff where the coefficient of transmissibility is about 5,000 gpd/ft. Values of the capacitors of the interior portions of the model are 2,500 micromicrofarads ($\mu\mu$f).

The accuracy and reliability of the electric analog computer were assessed by a study of records of past pumpage and water levels (Schicht, 1965). Water-level declines and a piezometric-surface map obtained with the electric analog computer were compared with actual water-level declines and the piezometric-surface map for December 1956 (see Fig. 8.24).

Hydrographs of observation wells indicate that stabilization of the piezometric surface during 1956 was mostly due to the effects of the Mississippi River. During 1956 there were long periods when little water was in the small streams and lakes in the interior portion of the East St. Louis area and these hydrologic features had for practical purposes negligible influence on water levels.

Production wells were grouped into centers of pumping and the average discharges during the period 1952 to 1956 (Fig. 8.25) for each pumping center were determined. The analog model was coupled to the excitation-response apparatus and the pulse generator was connected to junctions at locations of pumping centers. The output of the pulse generator was adjusted in accordance with discharge data. The oscilloscope was connected to terminals representing observation wells, and water-level declines were computed. Thus, water-level declines everywhere in the aquifer were described. The total water-level decline, based on the superposition theorem, at each terminal was obtained by summation of individual effects of each pumping center. The water-level declines were superposed on the piezometric map for 1900 (see Fig. 8.26) and a piezometric-surface map for 1956 was constructed. This map takes into account only the effects of pumping centers and assumes that the average stage of the Mississippi River was the same in 1956 as it was in 1900. However, records show that the average stage of the Mississippi River was about 11 feet lower in 1956 than in 1900. The effect of the change in the average stage of the river on

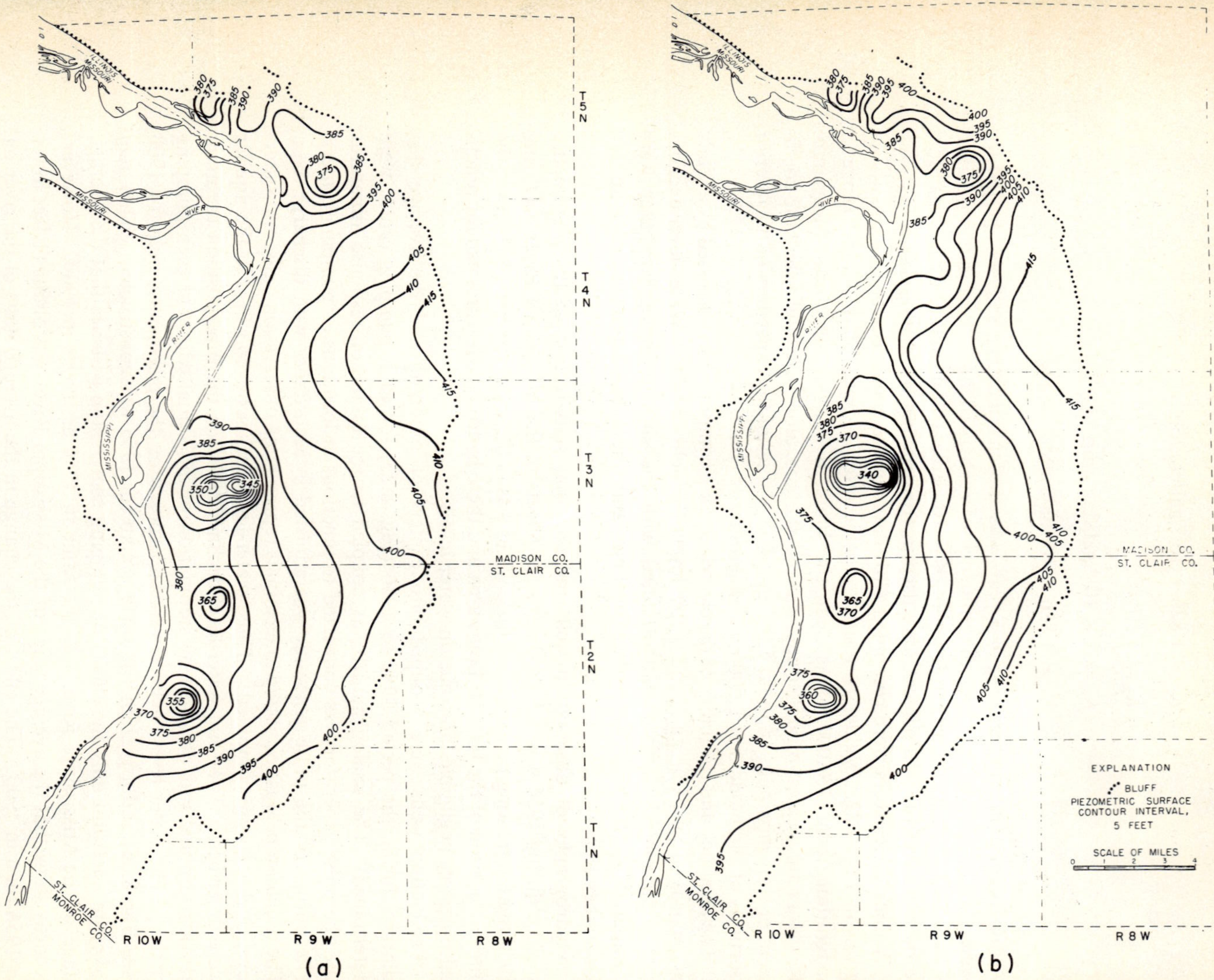

Fig. 8.24 Elevation of piezometric surface, December 1959, actual (*a*) and based on analog computer results (*b*) in East St. Louis, Illinois,

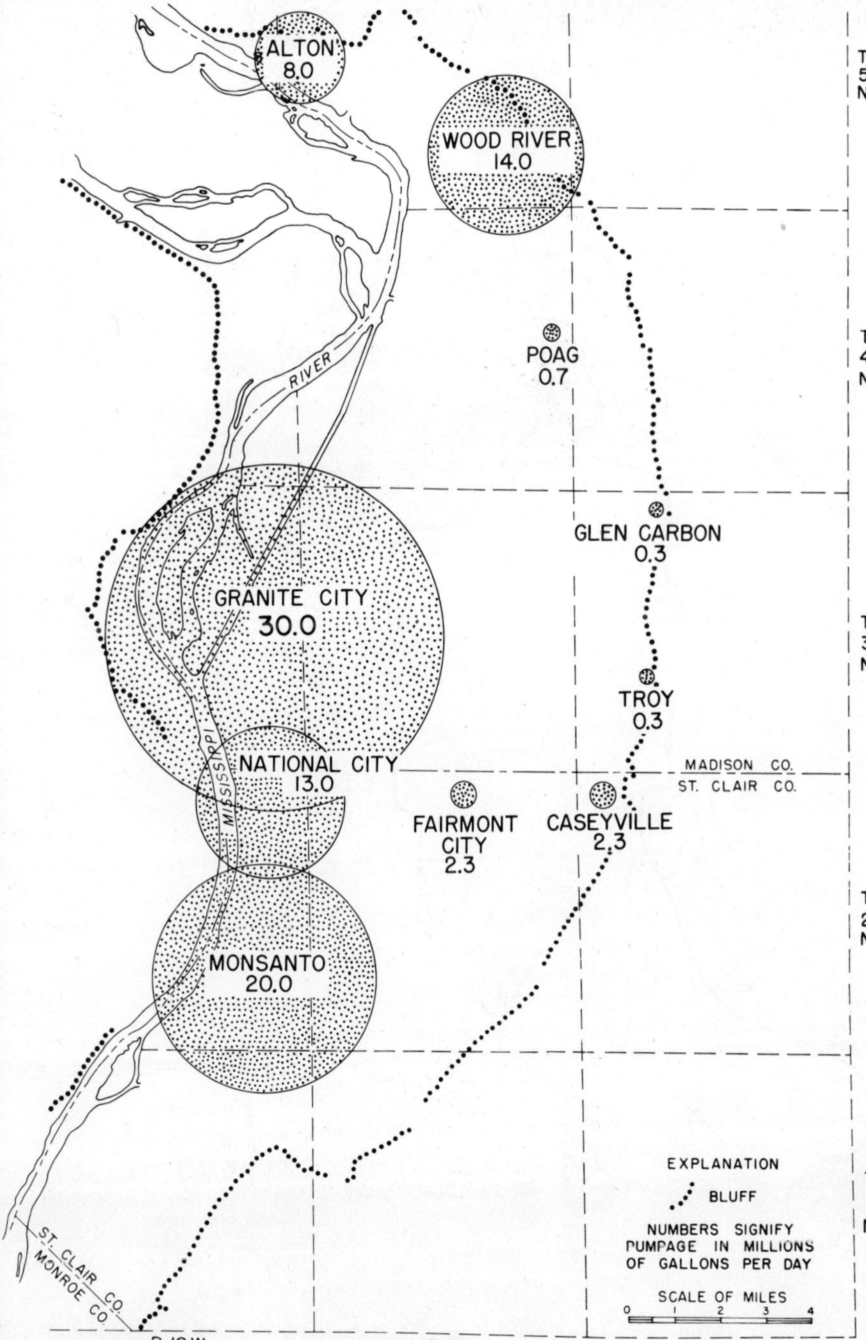

Fig. 8.25 Distribution of average pumpage 1952 to 1956 in East St. Louis, Illinois, area. (*After Schicht, 1965.*)

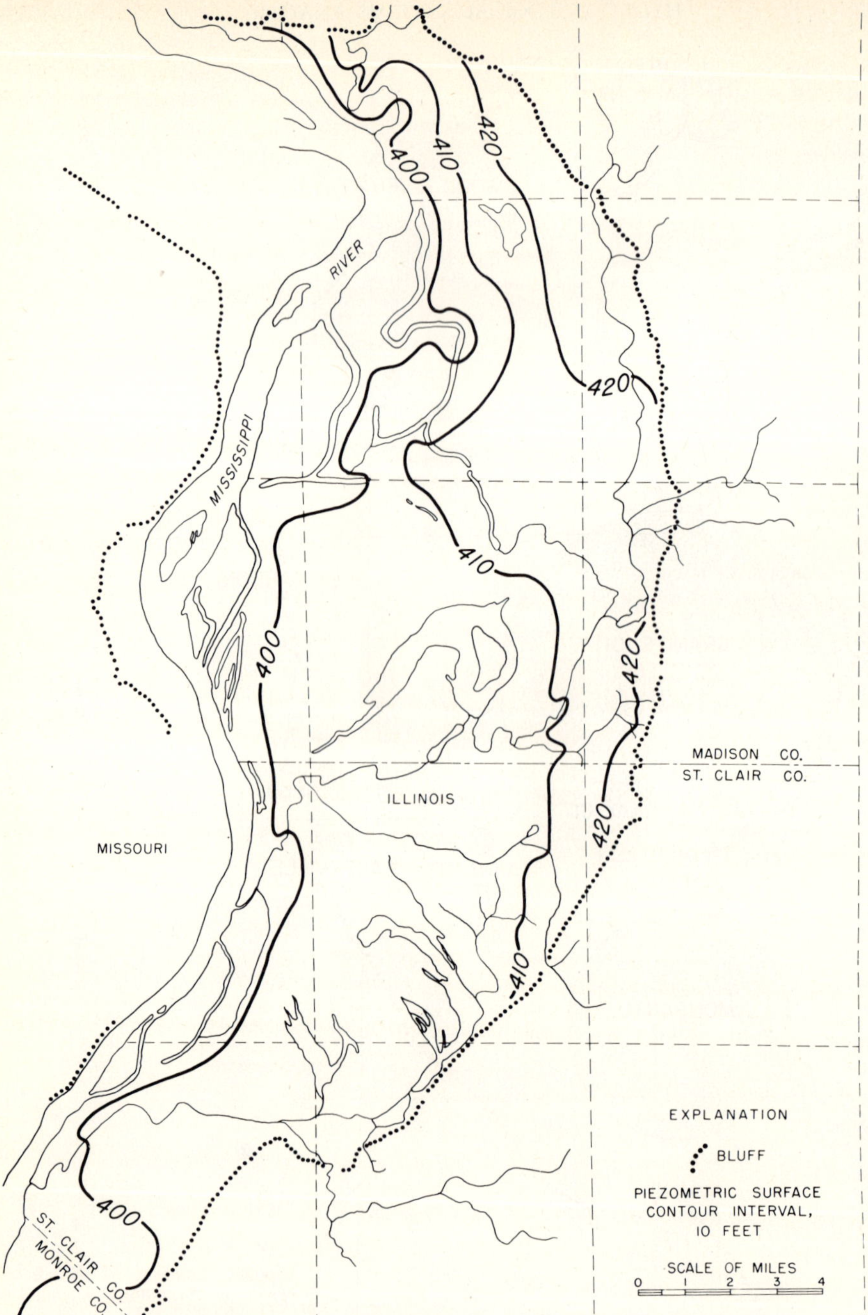

Fig. 8.26 Drainage system and estimated elevation of piezometric surface about 1900 in East St. Louis, Illinois, area. (*From Schicht and Jones, 1962.*)

water levels was estimated by coupling the pulse generator to the ground wire connecting junctions in the analog model along the river and measuring water-level changes due to the given change of the stage of the river with the oscilloscope connected to junctions in the interior portions of the analog model. Water-level declines due to dewatering portions of the aquifer were computed with analytical methods. The above water-level declines due to the decline in river stage and dewatering were superimposed upon water-level changes due to pumpage, and a total water-level-change map covering the period 1900 to December 1956 was prepared. A piezometric-surface map (Fig. 8.24*b*) was constructed by superimposing the total water-level-change map on the piezometric-surface map for 1900.

Features of the piezometric-surface map prepared with data from the analog computer and the piezometric-surface map prepared from actual water-level data are generally the same as shown in Fig. 8.24. Differences in analog computer and actual piezometric-surface maps are not significant when considered in relation to the accuracy and adequacy of hydrogeologic data. The close agreement between analog computer and actual piezometric maps indicates that the analog computer may be used to predict with reasonable accuracy the effects of future groundwater development and the practical sustained yield of existing pumping centers.

In 1962 water levels were not at critical stages in any pumping center and there were areas of the aquifer unaffected by pumping. Thus, the practical sustained yield of existing pumping centers exceeds total withdrawals in 1962 (Schicht, 1965). The practical sustained yield is here defined as the rate at which groundwater can be continuously withdrawn from wells in existing pumping centers without lowering water levels to critical stages or exceeding recharge. Critical water levels occur when pumping water levels are below tops of screens and/or more than one-half of the aquifer is dewatered. The electric analog computer with a pumping period of 5 years was used to determine pumping center discharge rates which would cause water levels in all major pumping centers to decline to the critical stages. The practical sustained yields of the existing major pumping centers are given in Table 8.6.

Pumpage in the Monsanto area during 1962 (22.6 mgd) was near the estimated practical sustained yield of 23 mgd. Extrapolation of pumpage graphs suggests that

TABLE 8.6 PRACTICAL SUSTAINED YIELDS OF EXISTING MAJOR PUMPING CENTERS IN EAST ST. LOUIS, ILLINOIS, AREA. (*From Schicht, 1965.*)

Pumping center	*1962 pumping rate, mgd*	*Additional possible withdrawal, mgd*	*Practical sustained yield, mgd*	*Year after which practical sustained yield may be exceeded*
Alton area	6.3	9.7	16	2000
Wood River area	14.1	5.9	20	1990
Granite City area	9.5	5.5	15	1980
National City area	11.6	6.4	18	2000
Monsanto area	22.6	0.4	23	1965
Total	64.1	27.9	92	

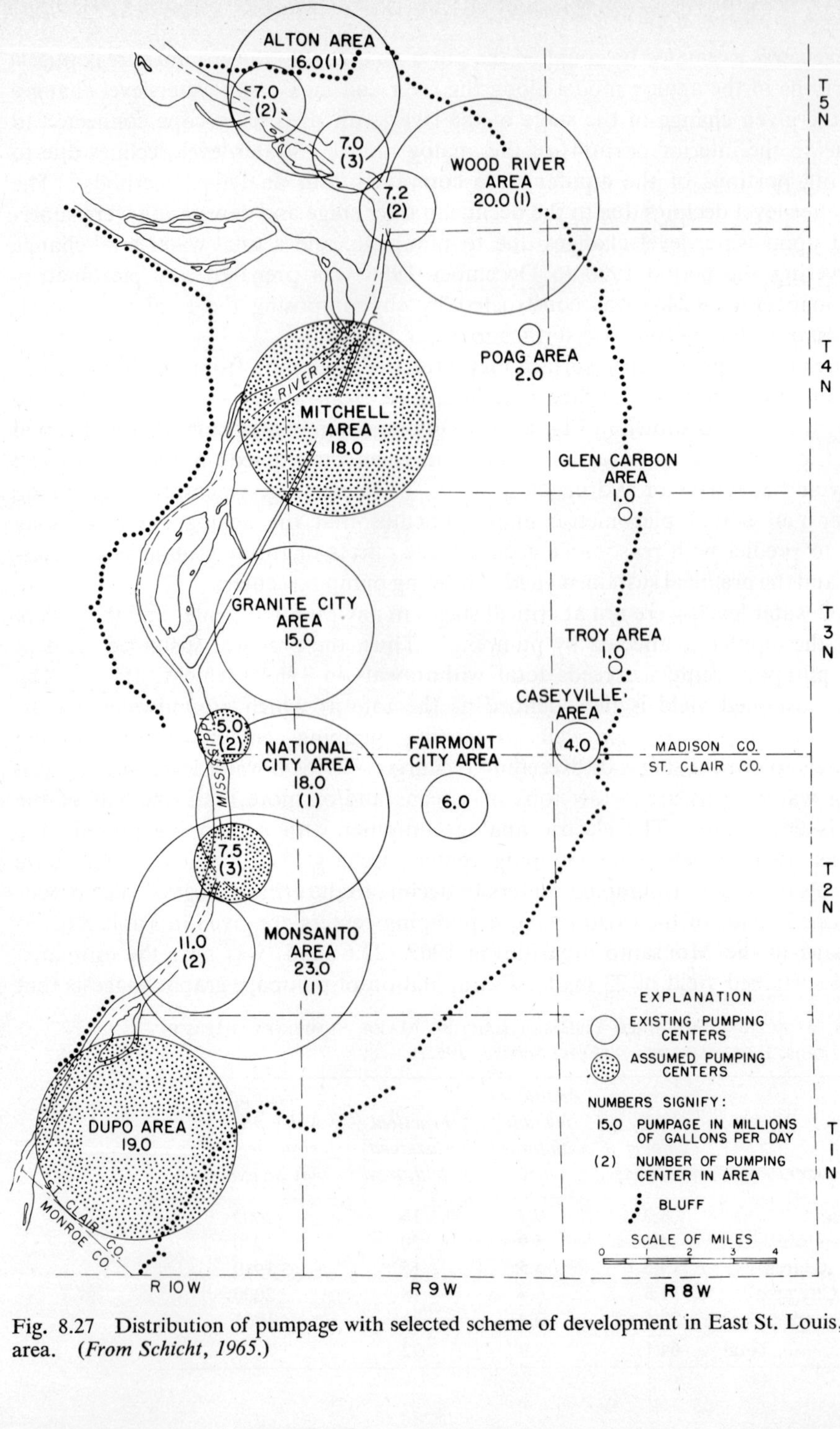

Fig. 8.27 Distribution of pumpage with selected scheme of development in East St. Louis, Illinois, area. (*From Schicht, 1965.*)

total groundwater withdrawals in other existing major pumping centers will exceed the practical sustained yields starting in 1980.

The electric analog computer and analytical methods were used to describe the effects of a selected scheme of development and to determine the potential yield of the aquifer under assumed pumping conditions (Schicht, 1965). The potential yield of the aquifer is here defined as the maximum amount of water that can be continuously withdrawn from a selected system of well fields without creating critical water levels or exceeding recharge.

The distribution of pumpage with the selected scheme of development is shown in Fig. 8.27. The electric analog computer was used to determine pumping center discharge rates which would cause water levels to decline in all major pumping centers to critical stages. Anticipated discharge rates for minor pumping centers in the year 2015 were assumed and water-level declines throughout the East St. Louis area were determined. Model aquifers and mathematical models based on available geohydrologic data and information on induced infiltration rates of the Mississippi River were used to determine the local effects of withdrawals in pumping centers near the river. The effects of dewatering portions of the aquifer were determined with analytical methods. The approximate elevation of the piezometric surface with the selected scheme of development is shown in Fig. 8.28.

Assuming that pumpage will continue to grow in the future as it has in the past, total pumpage in the East St. Louis area will exceed the potential yield with the selected scheme of development (188 mgd) after about 45 years or by 2015. There are sites near the river where additional pumping centers could be developed. Thus, the potential yield of the aquifer with other possible schemes of development exceeds 188 mgd.

Estevan, Saskatchewan, Canada

An investigation was made to determine the nature and extent of sand and gravel aquifers in the buried Missouri River valley system in the vicinity of Estevan, Saskatchewan, Canada, and to estimate the potential yield of the aquifers with selected schemes of development (see Walton, 1965a). The reach of the buried Missouri River valley system west and north of Estevan was considered. The investigation included collection and analysis of water-level data to define the piezometric surfaces of the aquifers and associated deposits and data on the chemical character of the groundwaters, an extensive test-hole drilling program, controlled aquifer and well-production tests, and an electric analog computer study.

Hydrogeologic conditions in the Estevan area are very complex and available data are not sufficient to permit a precise quantitative description of the aquifers and associated deposits (see case histories of groundwater exploration in Chap. 2). Aquifer conditions had to be simplified to some extent because of the incompleteness of data. For practical purposes, the stratified interbedded sand and gravel and silt

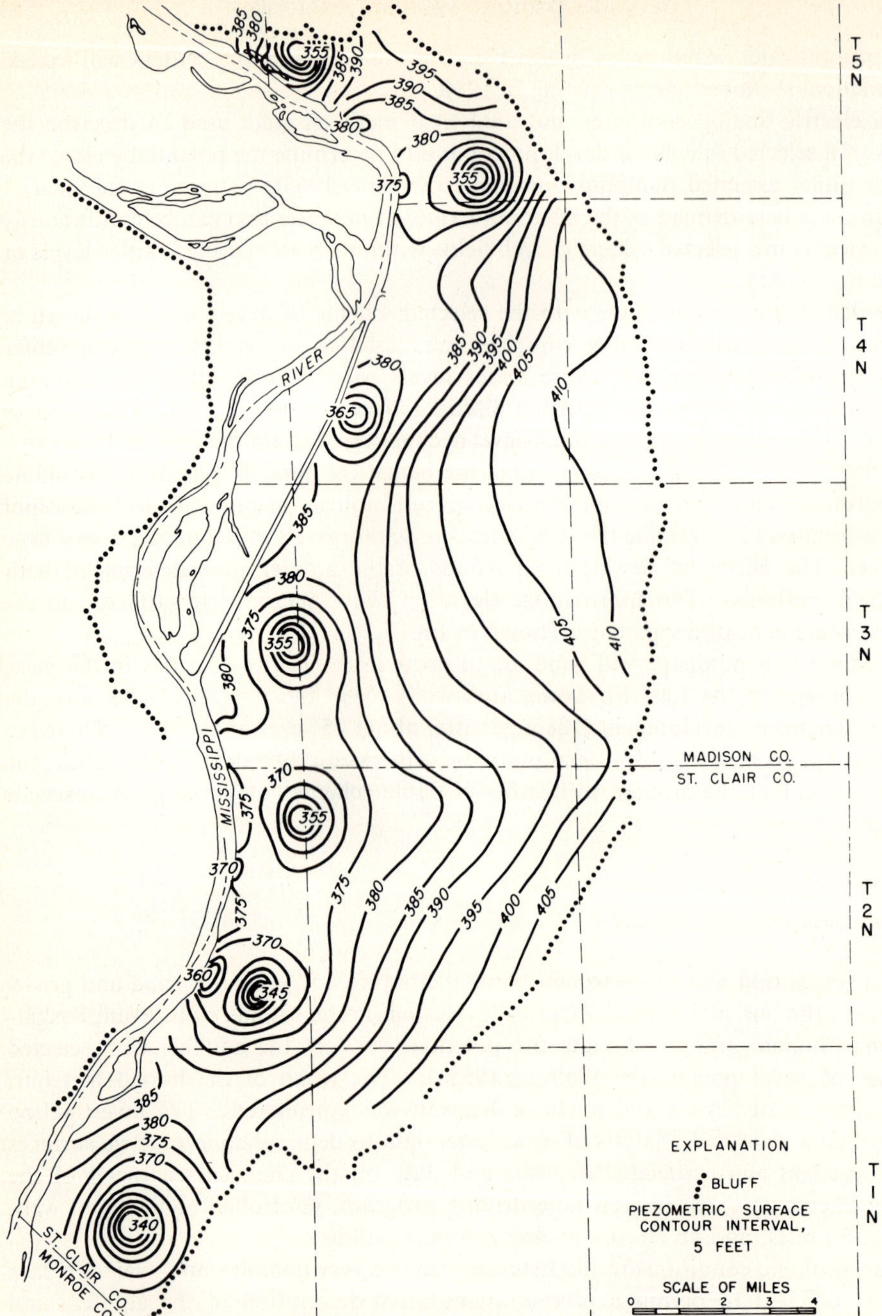

Fig. 8.28 Approximate elevation of piezometric surface with a selected scheme of development in East St. Louis, Illinois, area. (*From Schicht, 1965.*)

and clay glacial deposits below an average depth of about 300 feet were grouped as a single aquifer with an average width of 8,000 feet. The aquifer occurs in a valley cut into the Ravenscrag Formation (poorly consolidated thin sandstone, shale, and coal beds) which has some permeability. The Ravenscrag Formation was grouped with the stratified glacial sediments as a lower aquifer. The Ravenscrag Formation rests on shales which have negligible permeability and were considered as an aquiclude. Most of the till deposits overlying the lower aquifer were considered as a single aquitard. The stratified sediments interbedded in the till near the land surface and penetrated by shallow water-table wells were considered as an upper aquifer. Thus, hydrogeologic conditions in the Estevan area were idealized as a three-layer situation (upper aquifer–aquitard–lower aquifer).

All available hydrogeologic data for the Estevan area were studied and hydrogeologic maps were prepared which describe the following factors: (1) hydrogeologic properties, location and extent of the upper aquifer, (2) leakage coefficient, location and extent of aquitard hydrogeologic properties, location and extent of the lower aquifer. These maps encompass all nonhomogeneous and irregular hydrogeologic conditions including the boundaries of aquifers and the aquitard. The average coefficient of transmissibility of the upper aquifer was assumed to be 100 igpd/ft based on the general performance characteristics of existing shallow farm wells. The average coefficient of storage of the upper aquifer was assumed to be 0.05 based on data on the specific yields of sediments given by Todd (1959).

The saturated thickness of the aquitard (see Fig. 8.29) was estimated by comparing the bedrock-surface map of the Estevan area and the water-table map and by taking into consideration the saturated thickness of till encountered in test holes drilled in the buried Missouri River valley. The average coefficient of permeability of the aquitard was assumed to be 0.01 igpd/sq ft based on data given by Walton (1960b) and Prickett et al. (1964), and the map in Fig. 8.30 was constructed.

The saturated thicknesses of the sand and gravel aquifers and the Ravenscrag Formation were estimated by studying the logs of test holes and by comparing the bedrock-surface map, the water-table map, and the map showing contours on the base of the Ravenscrag Formation (Fig. 8.31). The coefficients of permeability of the individual zones of sand and gravel were estimated based on the results of the aquifer test, well logs, and the summary of mechanical analyses of test-hole drilling cutting samples. The coefficients of transmissibility of individual zones of sand and gravel were computed and added together to determine the total coefficient of transmissibility of the sand and gravel aquifer. The average coefficient of permeability of the Ravenscrag Formation was assumed to be 2 igpd/sq ft based largely on field-permeability tests conducted in connection with an investigation of foundation conditions at a nearby Boundary Dam site. The permeability is mostly concentrated in coal and sandstone beds. The coefficient of transmissibility of the Ravenscrag Formation was estimated by multiplying the average coefficient of permeability by the saturated thickness of the formation (Fig. 8.32). The coefficient of transmissibility of the sand and gravel aquifer was added to the coefficient of transmissibility of the

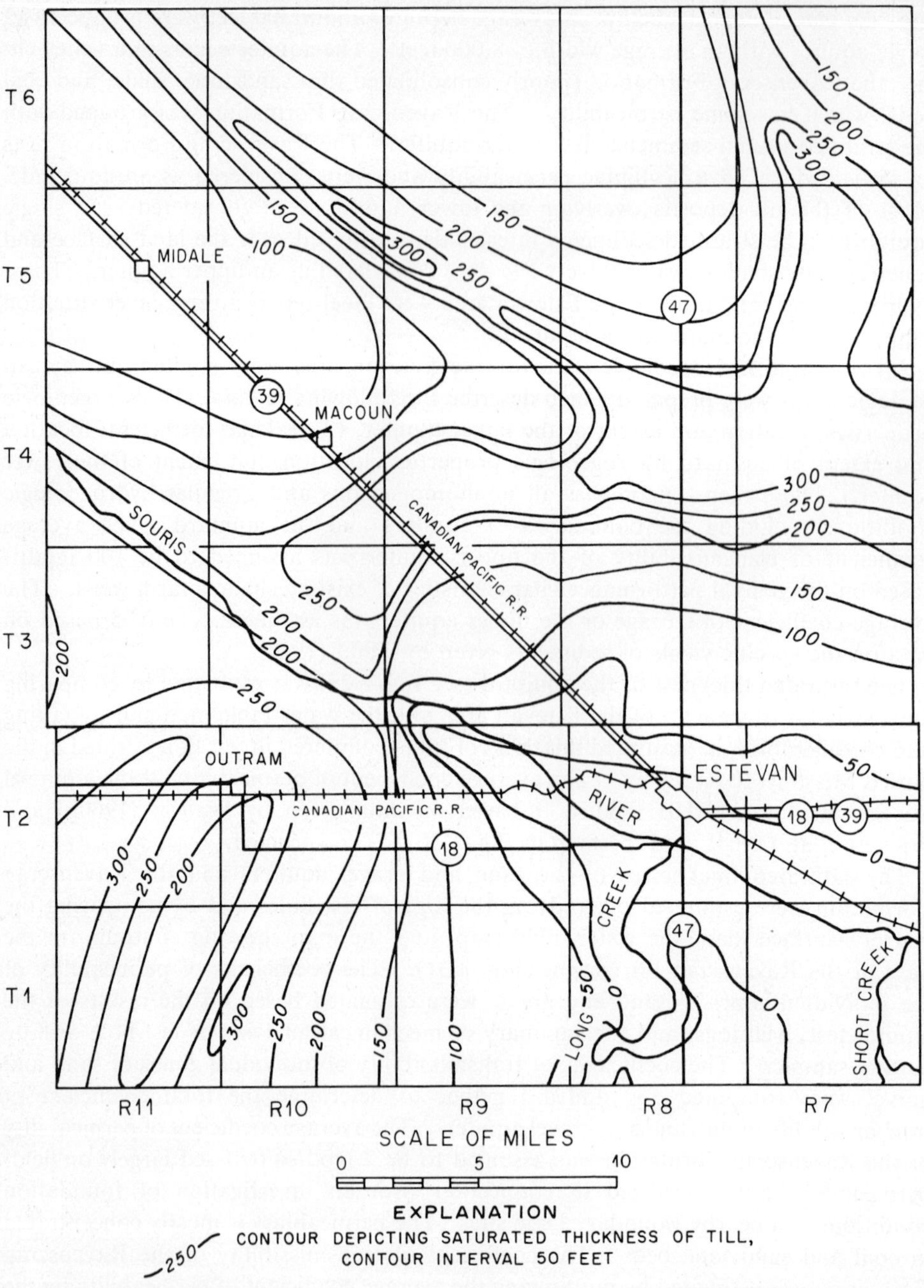

Fig. 8.29 Saturated thickness of till in Estevan, Saskatchewan, Canada, area. (*From Walton, 1965.*)

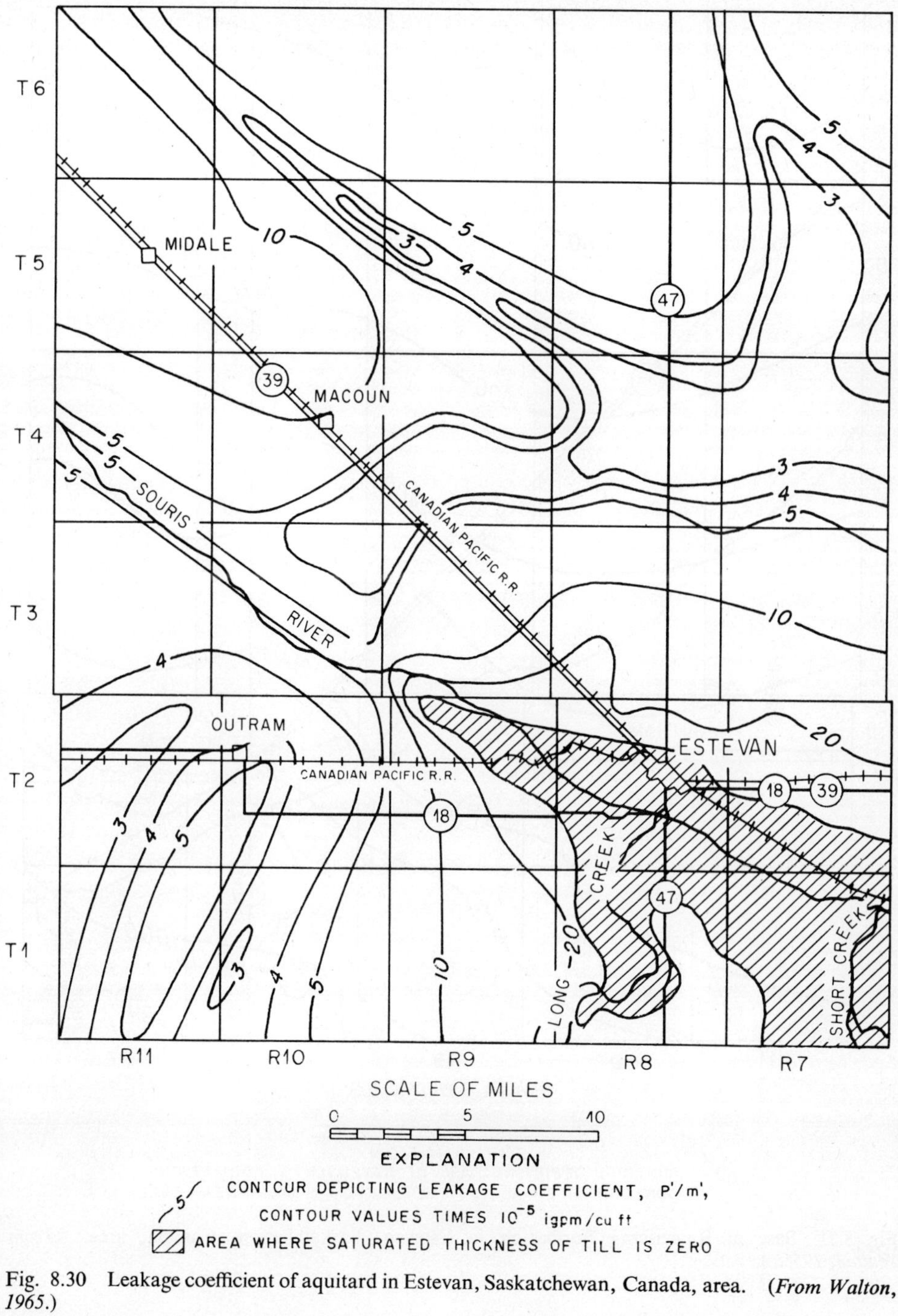

Fig. 8.30 Leakage coefficient of aquitard in Estevan, Saskatchewan, Canada, area. (*From Walton, 1965.*)

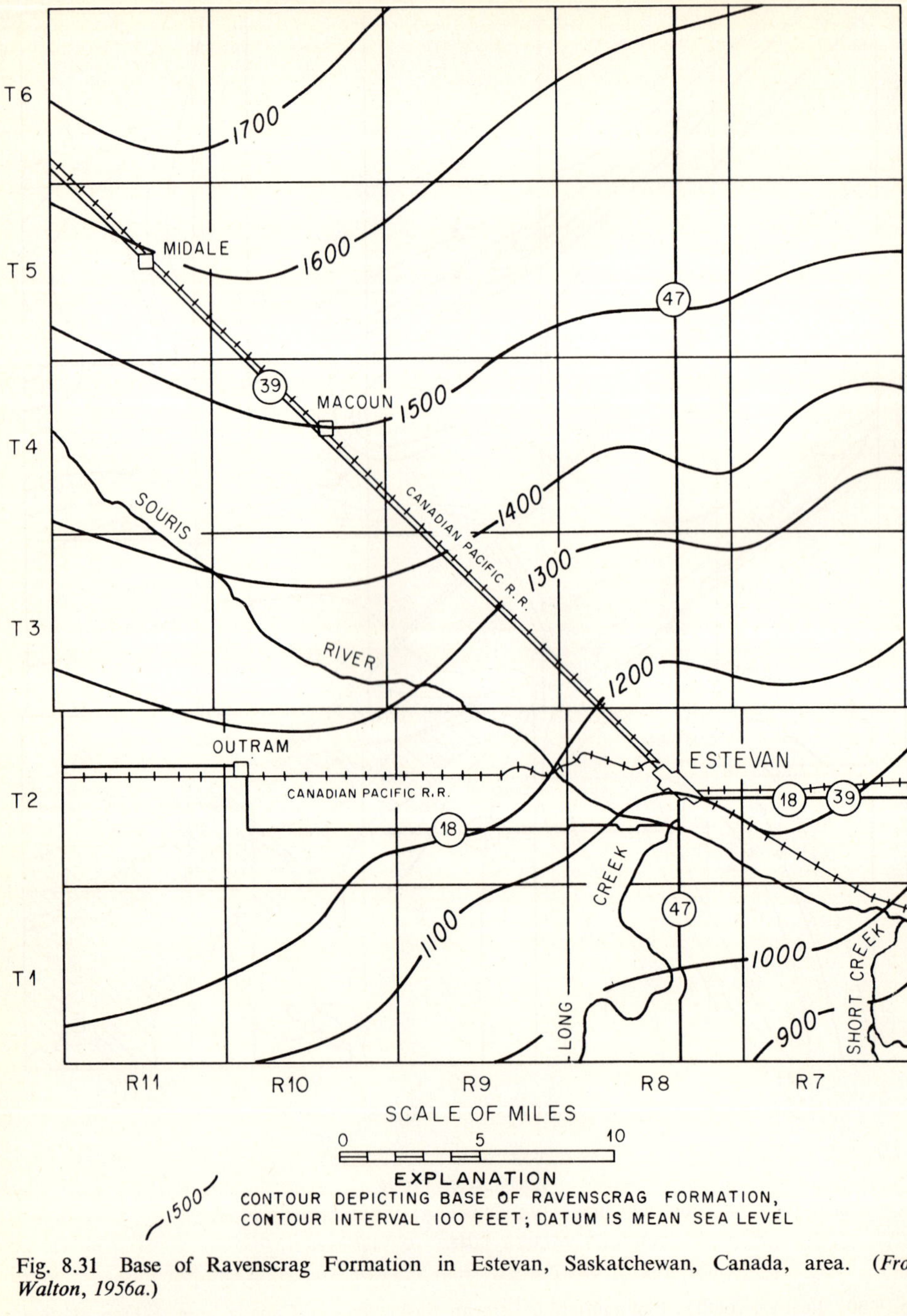

Fig. 8.31 Base of Ravenscrag Formation in Estevan, Saskatchewan, Canada, area. (*From Walton, 1956a.*)

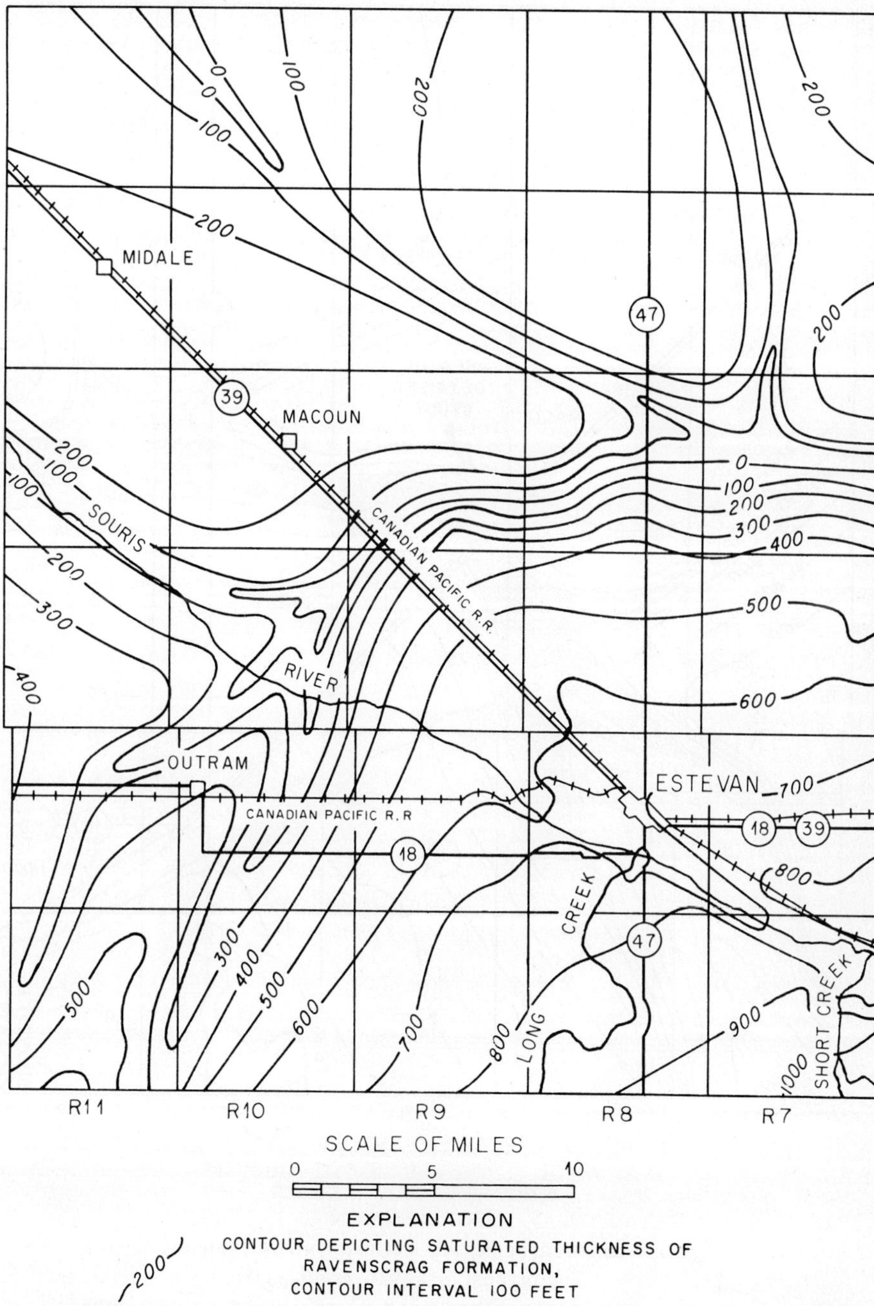

Fig. 8.32 Saturated thickness of Ravenscrag Formation in Estevan, Saskatchewan, Canada, area. (*From Walton, 1965a.*)

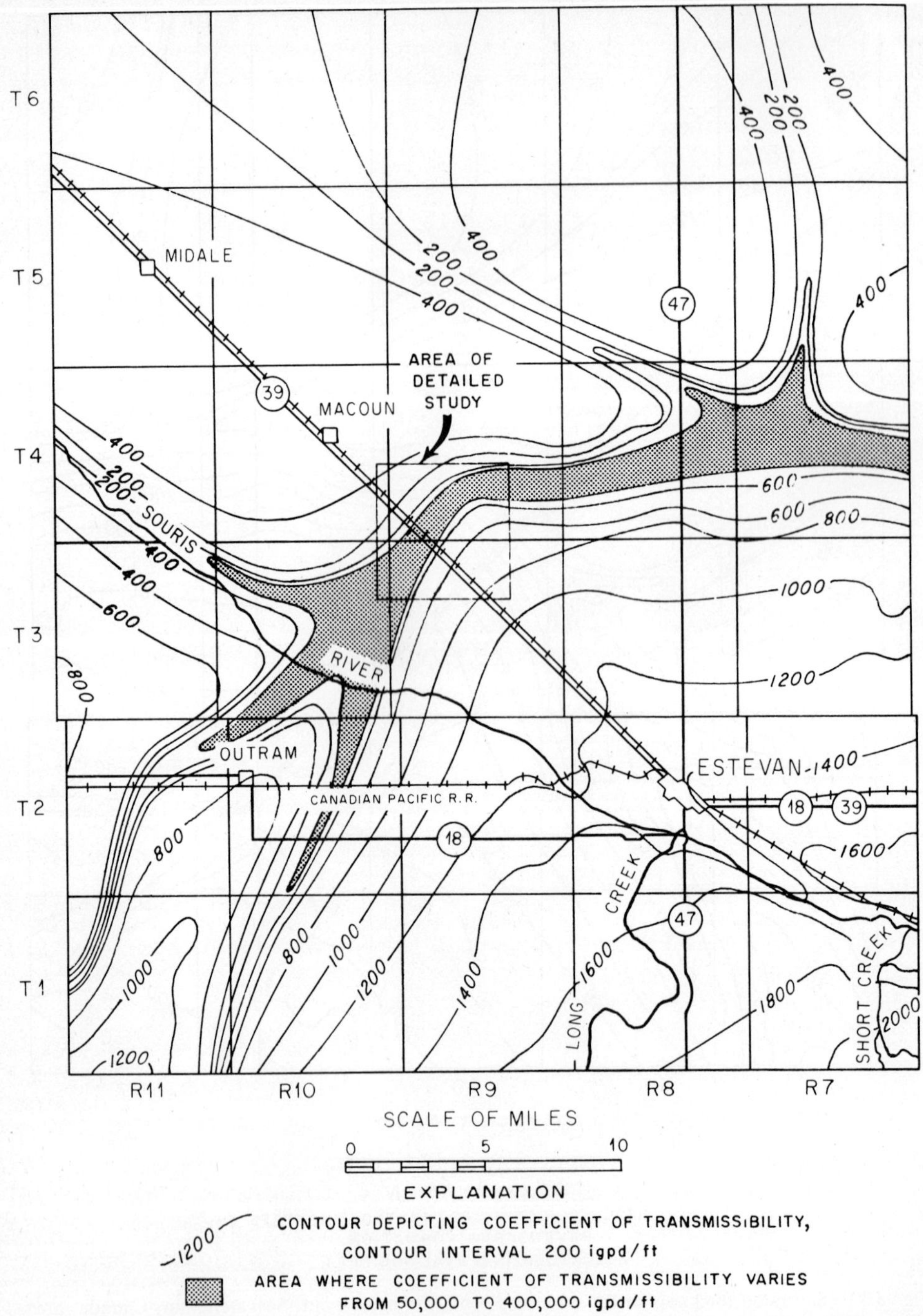

Fig. 8.33 Coefficient of transmissibility of lower aquifer in Estevan, Saskatchewan, Canada, area. (*From Walton, 1965a.*)

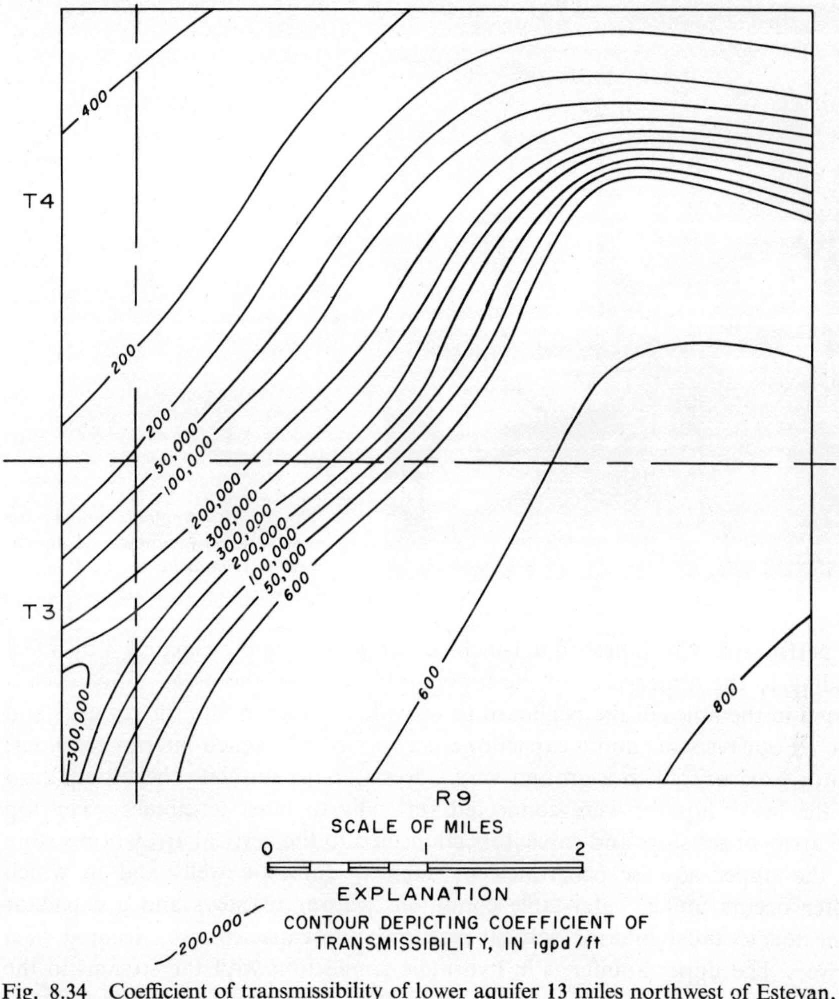

Fig. 8.34 Coefficient of transmissibility of lower aquifer 13 miles northwest of Estevan, Saskatchewan, Canada. (*From Walton, 1965a.*)

Ravenscrag Formation to obtain the total coefficient of transmissibility of the lower aquifer (see Figs. 8.33 and 8.34).

The analog model (Fig. 8.35) for simulating hydrogeologic conditions in the Estevan area consists of two horizontal regular arrays of resistors and capacitors interconnected by a vertical regular array of resistors. The model contains approximately 17,000 resistors and capacitors. The bottom horizontal array of resistors and capacitors simulates the lower aquifer in the deeper parts of the buried Missouri River valley and the adjacent Ravenscrag Formation. It was constructed on a piece of $\frac{1}{8}$-inch

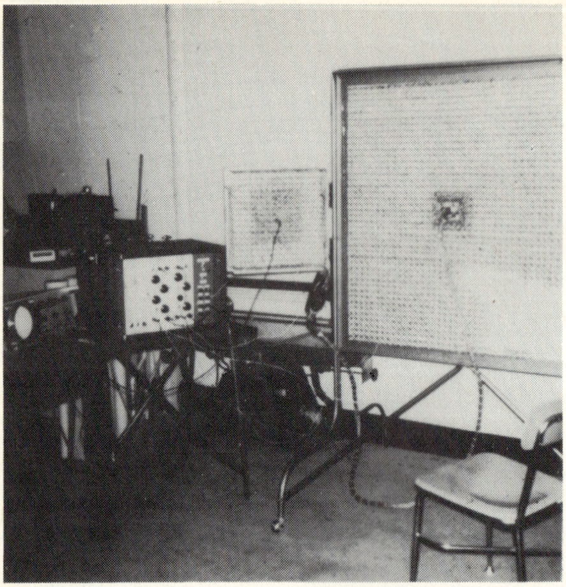

Fig. 8.35 Analog model for Estevan, Saskatchewan, Canada, area. *(Photo by W. C. Walton.)*

pegboard perforated with holes on a 1-inch-square pattern approximately 4 by 4 feet corresponding to the dimensions of the topographic map of the area. Shoe eyelets were inserted in the holes of the pegboard to provide terminal nodes for resistors and capacitors. Four resistors and a capacitor were connected to each interior terminal; the capacitor was secured to a ground wire. Resistors to simulate the till aquitard overlying the lower aquifer were connected vertically to most terminals. The top horizontal array of resistors and capacitors connected to the vertical array of resistors simulates the upper aquifer penetrated by shallow domestic wells and in which groundwater occurs under water-table conditions. Four resistors and a capacitor were connected to most interior vertical resistors; the capacitor was secured to a ground wire. The upper aquifer is in hydraulic connection with the streams in the study area (Souris River, Long Creek, Short Creek). Nodes of the upper array of resistors and capacitors along these recharge boundaries were connected to ground to simulate the recharge boundaries which distort cones of depression.

A network spacing of 1 inch equals 4,166 feet was selected for the major portion of the analog model to minimize the errors due to finite-difference approximation. A network spacing of 1 inch equals 1,041 feet was used in a small area centered about 13 miles northwest of Estevan where heavy groundwater development is likely to occur so that the cones of depression in the immediate vicinity of production wells could be mapped in detail. Equations given by Karplus (1958) suggest that the selected network spacings are adequate.

By the process of trial and error, scale factors were chosen to utilize inexpensive resistors and capacitors and existing excitation-response apparatus. Selected scale factors are as follows:

$$K_1 = 1.61 \times 10^{15} \text{ igal/coulomb}$$
$$K_2 = 2 \text{ ft/volt}$$
$$K_3 = 4.4 \times 10^9 \text{ igpd/amp}$$
$$K_4 = 3.65 \times 10^5 \text{ days/sec}$$

A maximum pumping period of 100 years was chosen. The pulse generator (see Fig. 8.35) has a maximum pulse duration of 10^{-1} second. A scale factor K_2 of 2 ft/volt was selected for ease in reading the oscilloscope graph.

Values of resistors of the lower aquifer were computed using data on the coefficient of transmissibility given in Figs. 8.33 and 8.34. Values of resistors range in magnitude from 2.2×10^7 ohms in areas where the Ravenscrag Formation is thin and T is about 100 igpd/ft to 4.8×10^3 ohms in the center of the buried Missouri River valley where T is about 400,000 igpd/ft. Values of capacitors of the network having a spacing of 1 inch equals 4,166 feet were computed to be 20 $\mu\mu$f; values of capacitors of the network having a spacing of 1 inch equals 1,041 feet were computed to be 3 $\mu\mu$f. The average coefficient of storage of the lower aquifer was assumed to be 0.00015.

Values of resistors of the aquitard were computed using data on the leakage coefficient given in Fig. 8.30. Values of resistors range from 3.9×10^5 ohms where the aquitard is thin and P'/m' is 4×10^{-4} igpd/cu ft to 5.6×10^6 ohms in areas where the aquitard is thick and P'/m' is 2.9×10^{-5} igpd/cu ft.

Values of resistors of the upper aquifer were computed to be 2.2×10^7 ohms based on an average T of 100 igpd/ft. Values of capacitors of the network having a spacing of 1 inch equals 4,166 feet were computed to be 6.8×10^3 $\mu\mu$f; values of capacitors of the network having a spacing of 1 inch equals 1,041 feet were computed to be 1.0×10^2 $\mu\mu$f. The long-term coefficient of storage of the upper aquifer was assumed to be 0.05.

Techniques described by Karplus (1958) were used to determine the values of resistors and capacitors connecting the two networks having different spacings. A termination strip was constructed to extend the analog model a distance of 60 miles east of the map shown in illustrations.

The electric analog computer was used to determine the effects of two selected schemes of development, i.e., a 5-imgd supply from a well field consisting of 5 production wells and located in an area about 13 miles northwest of Estevan and a 10-imgd supply from a well field consisting of 12 production wells and located in an area about 13 miles northwest of Estevan.

The analog model was coupled to the excitation-response apparatus and the pulse generator was connected to junctions at locations of production wells. The output of the pulse generator was adjusted in accordance with discharge data and a maximum time period of 50 years. The oscilloscope was connected to terminals representing observation wells in both the upper and lower aquifers and voltage readings were converted into water-level declines. Thus, water-level declines everywhere in the upper and lower aquifers were described for the two schemes of development and a

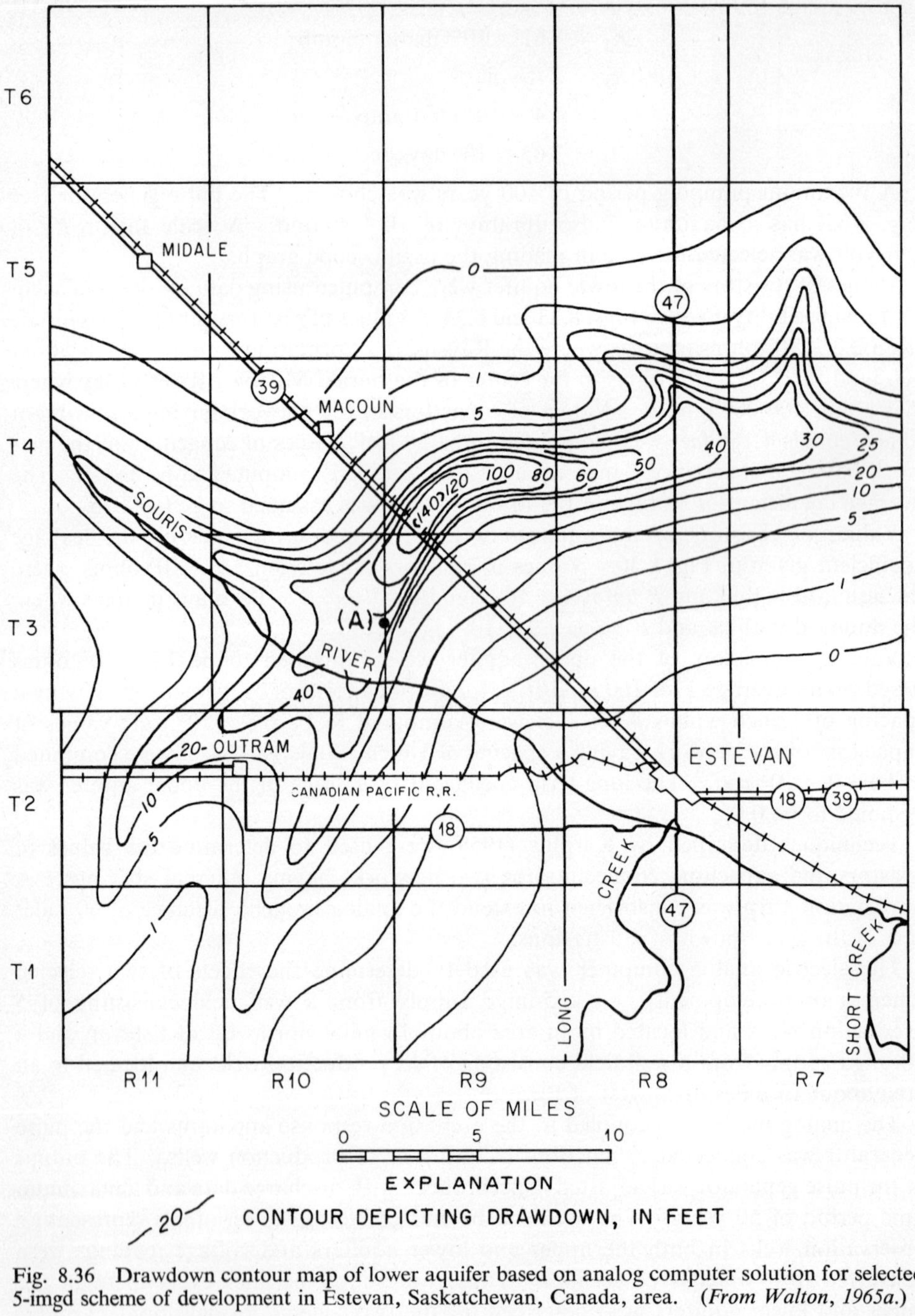

Fig. 8.36 Drawdown contour map of lower aquifer based on analog computer solution for selected 5-imgd scheme of development in Estevan, Saskatchewan, Canada, area. (*From Walton, 1965a.*)

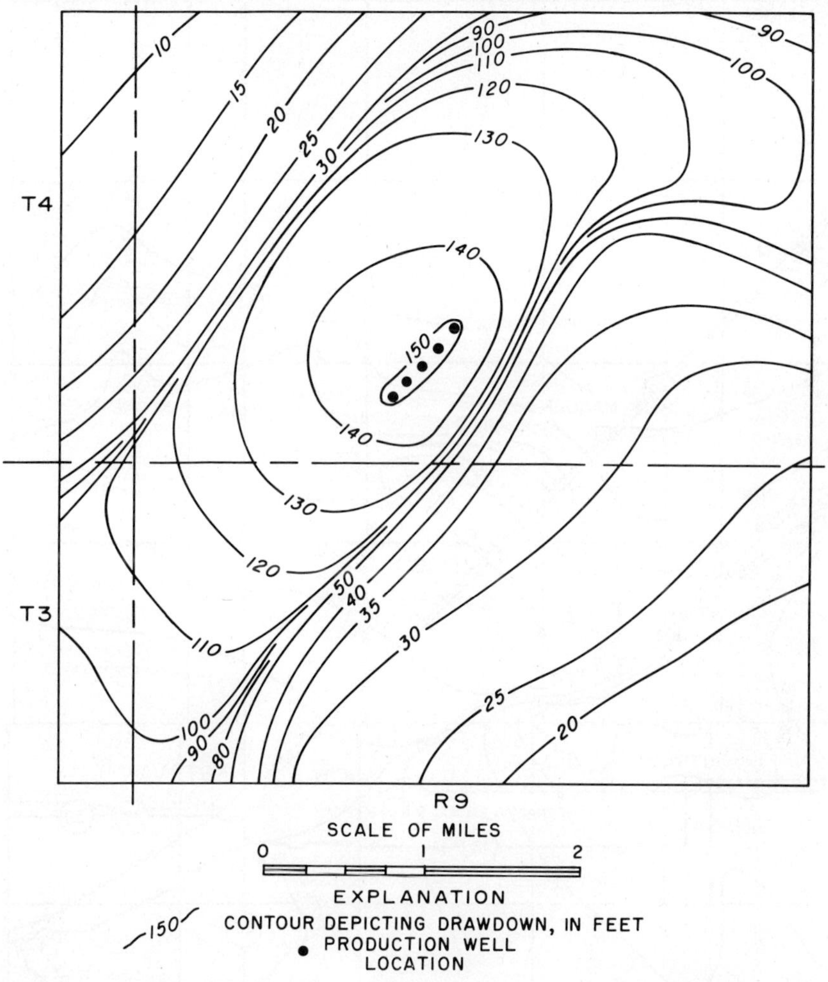

Fig. 8.37 Drawdown contour map of lower aquifer based on analog computer solution for selected 5-imgd scheme of development in area 13 miles northwest of Estevan, Saskatchewan, Canada. (*From Walton, 1965a.*)

pumping period of 50 years. Water-level-decline contour maps are shown in Figs. 8.36 to 8.43. The total water-level decline for each scheme of development, based on the superposition theorem, at each terminal was obtained by summation of individual effects of each production well. The water-level-decline maps were prepared taking into account contour deflection theory at contrasting T boundaries.

Natural-pack production wells with 12-inch casings and 50 feet of 12-inch screen were assumed in computations of drawdowns in the well field. The production wells were designed to have screens opposite highly permeable zones in the lower part of the

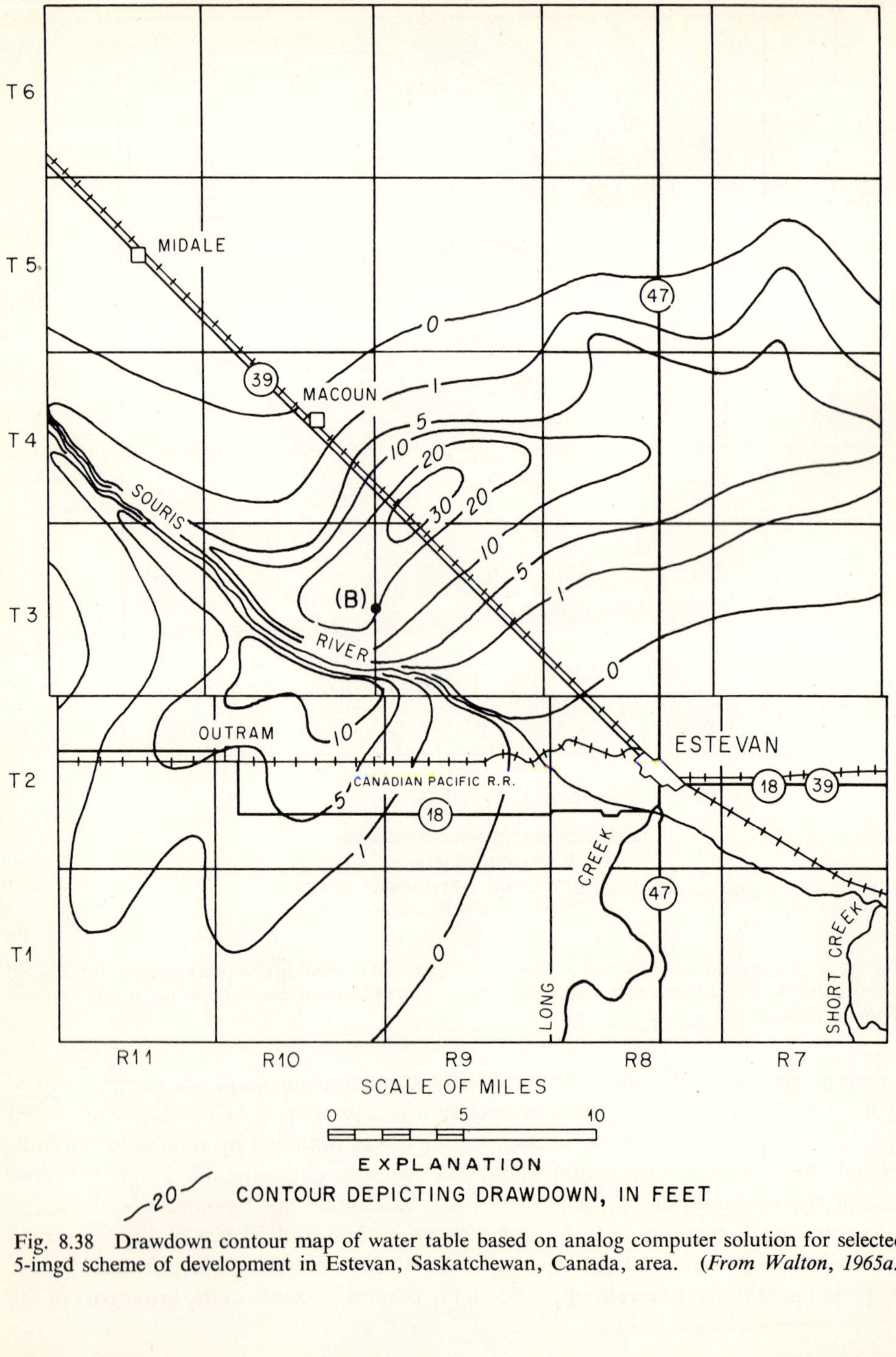

Fig. 8.38 Drawdown contour map of water table based on analog computer solution for selected 5-imgd scheme of development in Estevan, Saskatchewan, Canada, area. (*From Walton, 1965a.*)

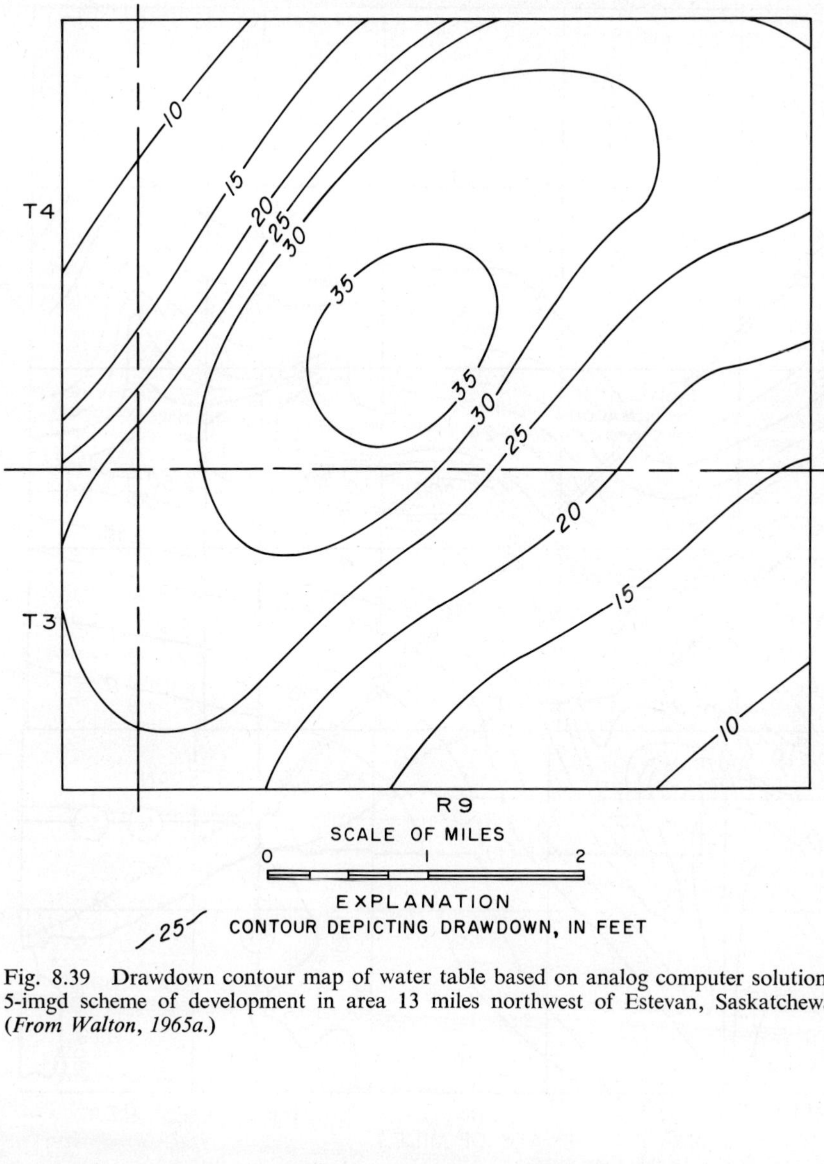

Fig. 8.39 Drawdown contour map of water table based on analog computer solution for selected 5-imgd scheme of development in area 13 miles northwest of Estevan, Saskatchewan, Canada. (*From Walton, 1965a.*)

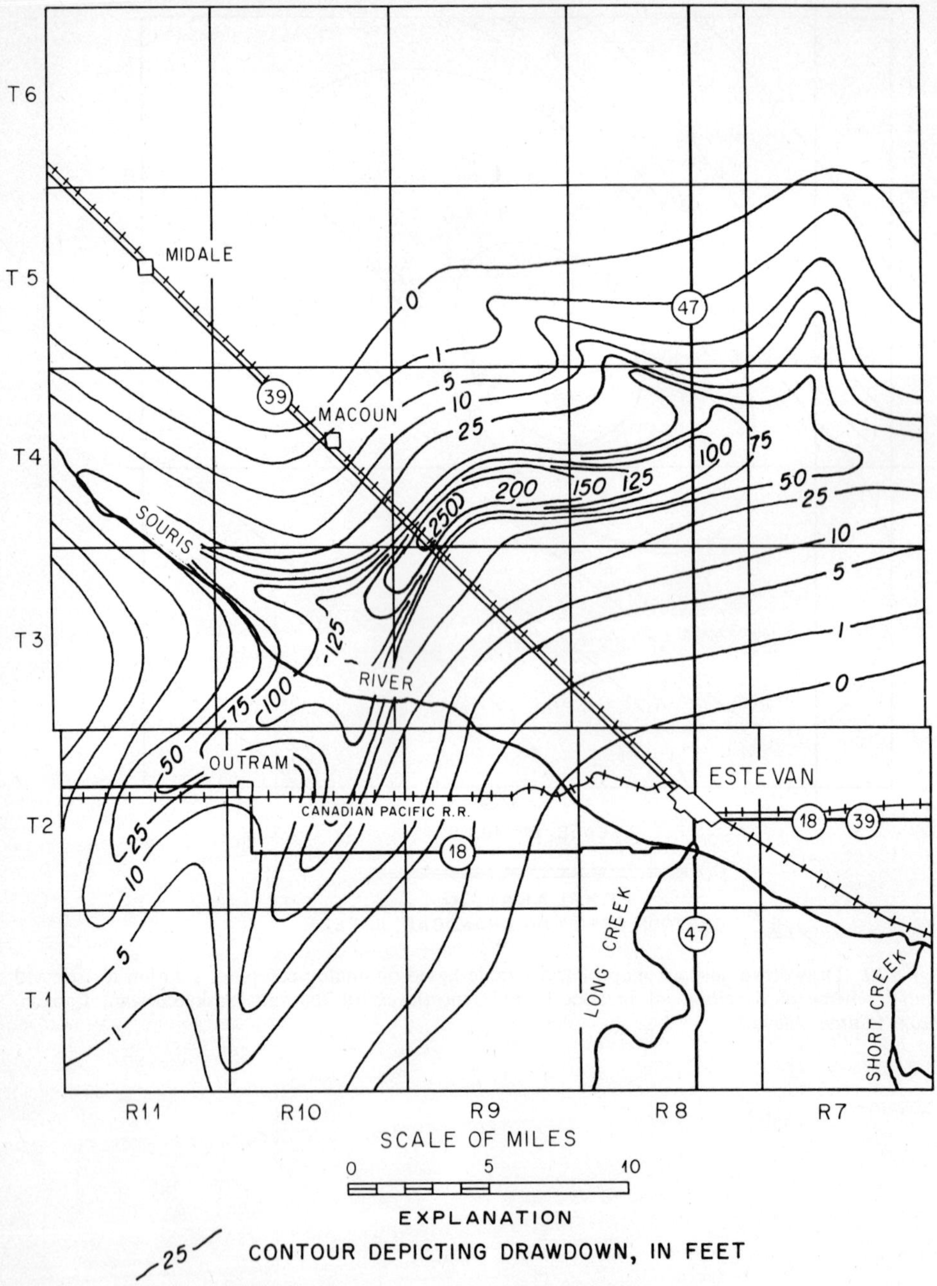

Fig. 8.40 Drawdown contour map of lower aquifer based on analog computer solution for selected 10-imgd scheme of development in Estevan, Saskatchewan, Canada, area. (*From Walton, 1965a.*)

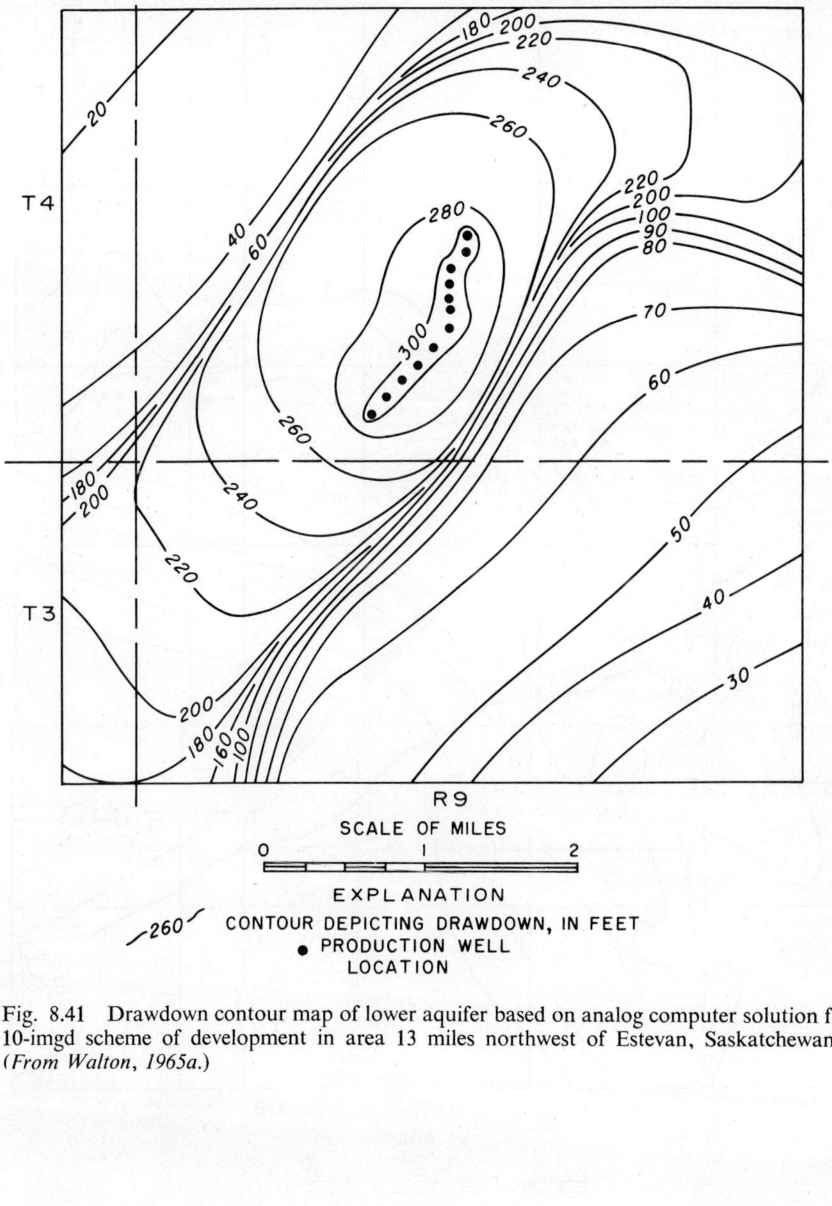

Fig. 8.41 Drawdown contour map of lower aquifer based on analog computer solution for selected 10-imgd scheme of development in area 13 miles northwest of Estevan, Saskatchewan, Canada. *(From Walton, 1965a.)*

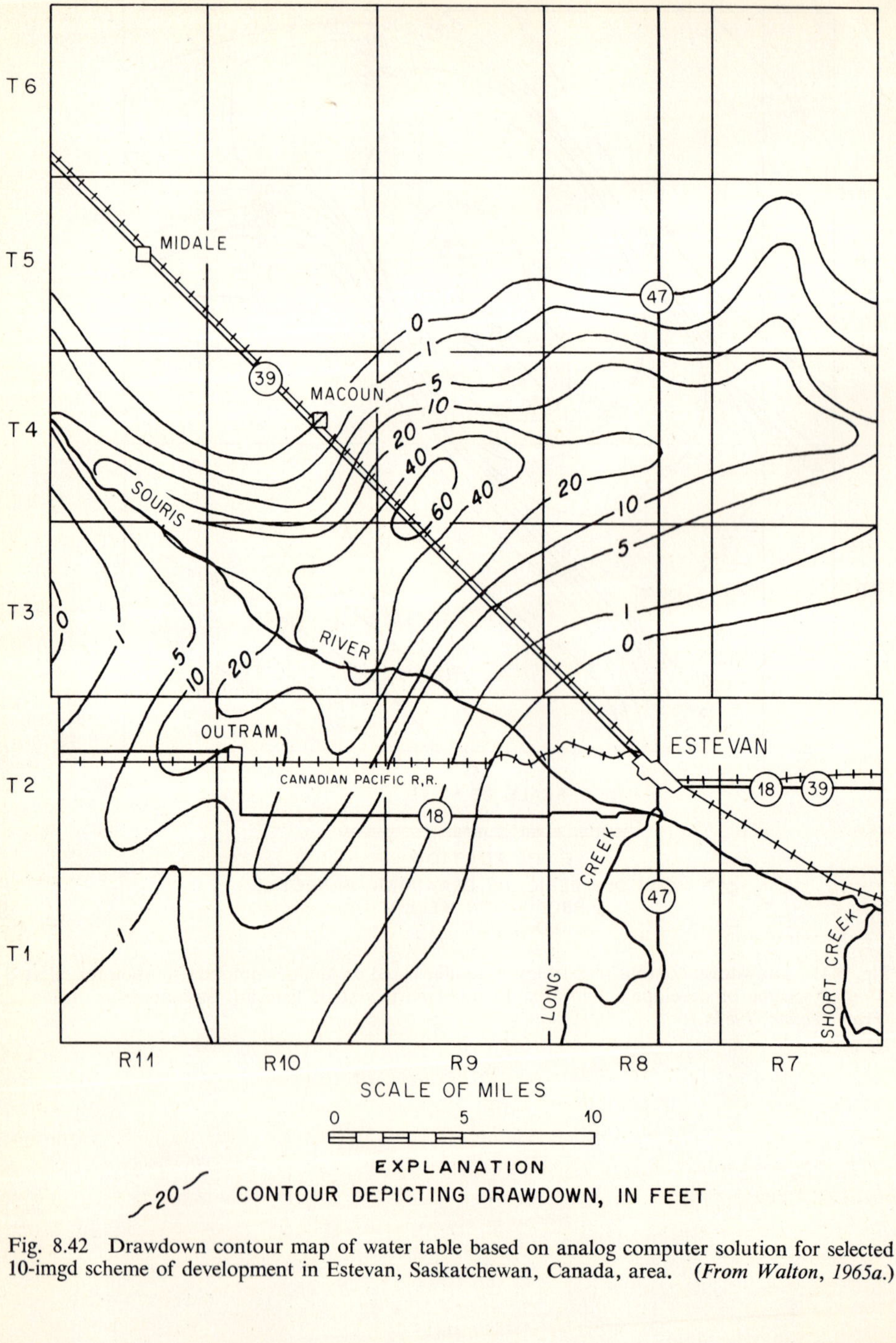

Fig. 8.42 Drawdown contour map of water table based on analog computer solution for selected 10-imgd scheme of development in Estevan, Saskatchewan, Canada, area. (*From Walton, 1965a.*)

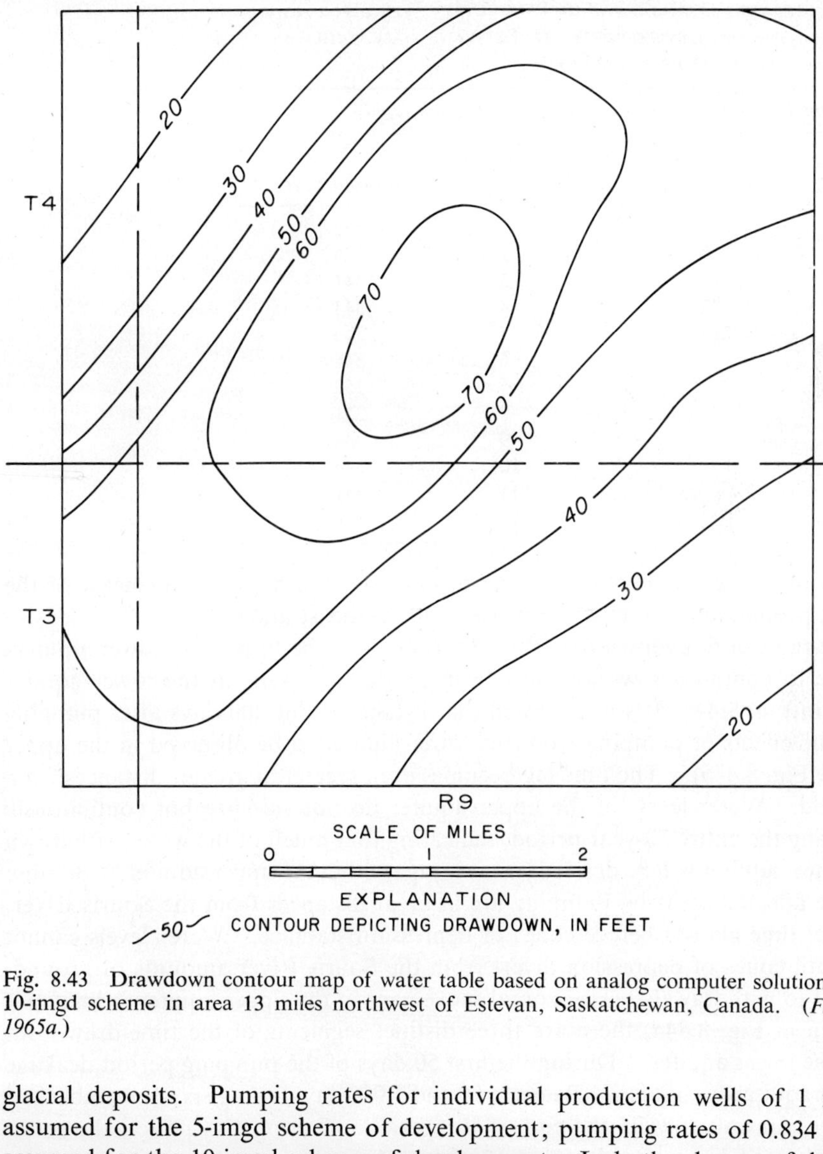

Fig. 8.43 Drawdown contour map of water table based on analog computer solution for selected 10-imgd scheme in area 13 miles northwest of Estevan, Saskatchewan, Canada. (*From Walton, 1965a.*)

glacial deposits. Pumping rates for individual production wells of 1 imgd were assumed for the 5-imgd scheme of development; pumping rates of 0.834 imgd were assumed for the 10-imgd scheme of development. In both schemes of development the spacing of production wells was assumed to be 500 feet with the first production well located about 500 feet northeast of the production well used during the aquifer test. In general, the well fields trend southwest to northeast. A well-loss constant of 2.0 sec^2/ft^5 and a 0.6 fraction penetration were assumed in computing well losses and drawdowns due to partial penetration in production wells. Estimated drawdowns in production wells for the two schemes of development are given in Table 8.7. The

TABLE 8.7 ESTIMATED DRAWDOWNS IN PRODUCTION WELLS FOR TWO SELECTED SCHEMES OF DEVELOPMENT IN ESTEVAN, SASKATCHEWAN, CANADA, AREA. (*From Walton, 1965a.*)

5-imgd supply		*10-imgd supply*	
Production well no.	*Drawdown, ft*	*Production well no.*	*Drawdown, ft*
1	216	1	332
2	217	2	336
3	219	3	341
4	222	4	343
5	219	5	346
		6	348
		7	352
		8	351
		9	348
		10	344
		11	341
		12	334

production wells are numbered from the production well 500 feet northeast of the aquifer test production well in order proceeding northeast and north.

The time-rates of drawdown for selected terminals in the upper and lower aquifers in response to continuous withdrawals from production wells in the lower aquifer are shown in Fig. 8.44. It will be noted that it takes about 200 days after pumping starts for the effects of pumping from the lower aquifer to be observed in the upper aquifer (see Fig. 8.44*b*). The time lag becomes even greater at greater distances from the well field. Water levels in the upper aquifer do not stabilize but continuously decline during the entire 50-year period, indicating that much of the water withdrawn from the lower aquifer will be derived from storage within the upper aquifer. Because most of the affected areas by pumping are at great distances from the Souris River, considerable time elapses before cones of depression stabilize. Water levels cannot stabilize until cones of depression divert from the Souris River amounts of groundwater equal to withdrawals. Horizontal movement in the upper aquifer is limited.

As shown in Fig. 8.44*a*, there are three distinct segments of the time-drawdown curve for the lower aquifer. During the first 50 days of the pumping period, leakage through the aquitard is limited. Between 50 and 3,000 days leakage is appreciable and balances withdrawals toward the end of the period. During this period the water table has declined but a few feet. After 3,000 days the drawdown curve deflects downward in response to significant declines in the water table and changing vertical hydraulic gradients.

As shown in Figs. 8.36 and 8.37, water-level declines in the lower aquifer of 150 feet or more are restricted to the immediate vicinity of the well field in the case of the 5-imgd scheme of development. Water-level declines of 50 feet or more occur in a belt averaging 2 miles wide and extending about 7 miles northeast and southwest

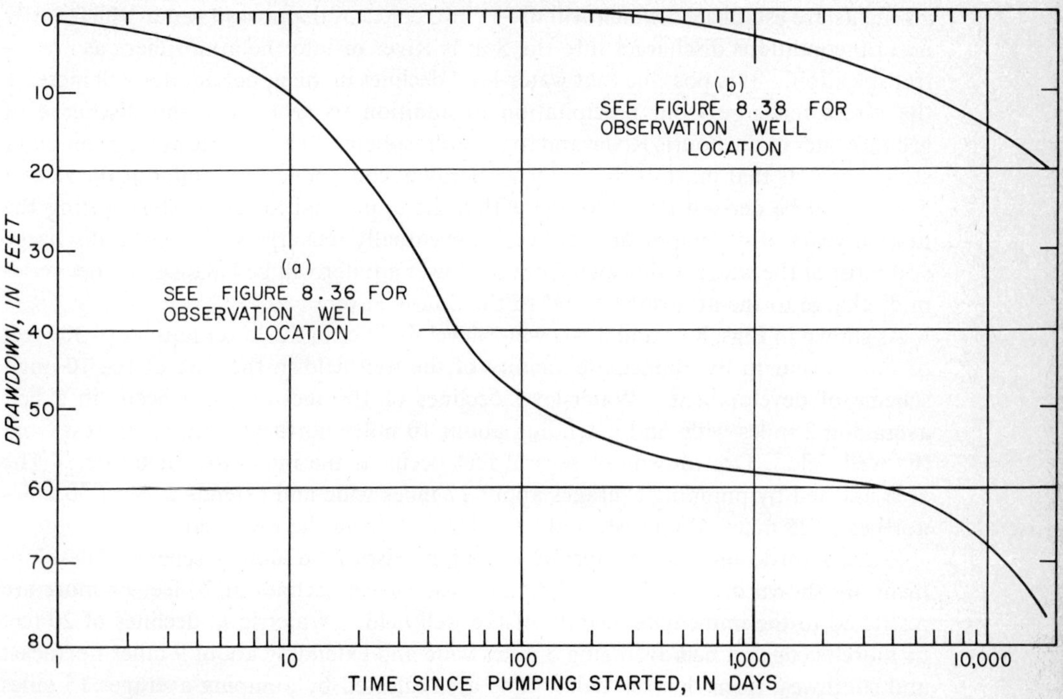

Fig. 8.44 Time-drawdown graph for observation wells located in lower aquifer (*a*) and in water-table aquifer (*b*) in Estevan, Saskatchewan, Canada, area. (*From Walton, 1965a.*)

from the well field. The area affected by pumping averages about 15 miles wide and extends at least 80 miles northeast, 25 miles southwest, and 28 miles west from the well field. Drawdowns of a few feet occur at the international border, indicating that very small amounts of groundwater in the lower aquifer may be diverted across the border.

Water-level declines in the upper aquifer in the case of the 5-imgd scheme of development are shown in Figs. 8.38 and 8.39. Water-level declines of 35 feet or more are restricted to the immediate vicinity of the well field. Water-level declines of 10 feet or more occur in a belt averaging 4 miles wide and extending about 7 miles northeast and southwest from the well field. The area affected by pumping averages about 14 miles wide and extends at least 25 miles northeast, southwest, and west from the well field. Drawdowns of a fraction of a foot occur at the international border, indicating that very small amounts of groundwater in the upper aquifer may be diverted across the border. Drawdown along the Souris River is negligible.

In the case of the 5-imgd scheme of development, the upper aquifer area of influence of pumping is over 1,000 square miles. Taking into account a direct recharge from precipitation of 0.4 inch per year (see Farvolden et al., 1963), recharge to the upper aquifer under natural conditions in the area of influence exceeds 5 imgd. However, large quantities of water must be mined from the upper aquifer before hydraulic gradients are established which will divert into cones of depression water which under natural conditions discharges into the Souris River or into the atmosphere as evapotranspiration. It is possible that water-level declines in the upper aquifer will increase the direct recharge from precipitation in addition to decreasing the discharge of groundwater to the Souris River and to the atmosphere. The electric analog computer study suggests that most of the water withdrawn from the lower aquifer at the rate of 5 imgd will be derived from storage within the upper and lower aquifers during the first 50 years of development. However, eventually recharge will balance discharge and most of the water withdrawn from the lower aquifer will be balanced by decreases in discharge to the atmosphere and to the Souris River.

As shown in Figs. 8.40 and 8.41, water-level declines in the lower aquifer of 300 feet or more occur in the immediate vicinity of the well field in the case of the 10-imgd scheme of development. Water-level declines of 100 feet or more occur in a belt averaging 2 miles wide and extending about 10 miles northeast and southwest from the well field. Drawdowns of several feet occur at the international border. The area affected by pumping averages about 17 miles wide and extends at least 80 miles northeast, 25 miles southwest, and 28 miles west from the well field.

Water-level declines in the upper aquifer in the case of the 10-imgd scheme of development are shown in Figs. 8.42 and 8.43. Water-level declines of 70 feet or more are restricted to the immediate vicinity of the well field. Water-level declines of 20 feet or more occur in a belt averaging 5 miles wide and extending about 9 miles northeast and southwest from the well field. The area affected by pumping averages 15 miles wide and extends at least 25 miles northeast, southwest, and west from the well field. Drawdowns of 1 foot or more occur at the international border. There is substantial drawdown along the Souris River.

In the case of the 10-imgd scheme of development, the upper aquifer area of influence of pumping exceeds 1,200 square miles. Taking into account a direct recharge from precipitation of 0.4 inch per year, recharge to the upper aquifer under natural conditions in the area of influence exceeds 10 imgd. During the first 50 years of development most of the water withdrawn will be derived from storage within the upper and lower aquifers. However, eventually recharge will equal discharge and water withdrawn will be balanced by decreases in discharge to the atmosphere and to the Souris River.

Pumping levels in production wells in the case of the 10-imgd scheme of development are at critical stages near the tops of screens in the wells. Thus, further development is not possible and the potential yield of the lower aquifer with development concentrated in the area 13 miles northwest of Estevan and a pumping period of 50 years is about 10 imgd.

Lake Calumet area, Illinois

A hydrogeologic systems analysis of the Chicago region was made in connection with a preliminary evaluation of the feasibility of a tunnel drainage project (see

Walton, 1966). The project consists of a system of tunnels excavated beneath the Lake Calumet area of Chicago into which runoff from the most severe storms could be quickly drawn from the surface and stored temporarily underground. Large pump-turbine installations, normally used for hydroelectric generation, would discharge the stored floodwaters at a gradual rate into the Illinois River system. Storm waters and sewage effluent could be carried through the tunnel system without contaminating the surface waterways within the area. The tunnel drainage project could give full drainage protection to the Lake Calumet area at a cost comparable to presently contemplated systems, which can only protect against much smaller rainstorms and which cannot prevent waterway pollution (*Engineering News Record*, July 22, 1965).

The project would utilize the existing network of lateral sewers and main sewers. To this system would be added new main sewers having sufficient grade to induce higher velocities. Both the existing and new main sewers would be intercepted at appropriate intervals by vertical shafts extending to a depth of 600 feet or more below the ground surface. The bottom of these shafts would be connected to large galleries excavated in rock at locations shown in Fig. 8.45. The galleries in turn would drain into a tunnel leading away from the area to a point near the Sanitary and Ship Canal. There a reversible pump-turbine plant would draw water from the underground system and discharge into the canal.

The tunnel drainage project could overcome the present limitations of a shallow drainage system by the provision of adequate storage space for floodwaters in underground galleries beneath the urban areas. Sewer velocities could be substantially increased due to the greater slope, and storm runoff would be drawn rapidly into the underground storage space before water could accumulate on the surface. Water with an intensity occurring once every 100 years could be accommodated. After passage of the rainstorm, the waters stored in the galleries could be pumped at controlled rates into the Sanitary and Ship Canal, downstream of the urban areas.

Total underground storage in the Lake Calumet project would be about 2,500 acre feet, sufficient for a storm of almost 6 inches in 24 hours which might be expected to occur not more than once in 100 years. A conveyance tunnel, about 20 miles long, would connect the galleries to a pump-turbine plant in the Palos Hills area to the southwest. This plant could evacuate the 100-year flood in 72 hours during the time of surplus power availability. Maximum discharge into the Sanitary and Ship Canal would be about 500 cfs, or only one-tenth of the discharge from a shallow drainage system.

The pump-turbine plant would add 200,000 kw to the capacity of the electric utility system. An artificial reservoir covering about 375 acres in Palos Hills would supply water for daily use of the generating plant. While generating, the plant would fill about 40 percent of the underground storage. During the nongenerating hours, this water would be pumped back into the surface reservoir.

Studies were made to determine the rate of seepage into the drainage tunnels, the extent of effects on the aquifer system, and the measures necessary to assure protection

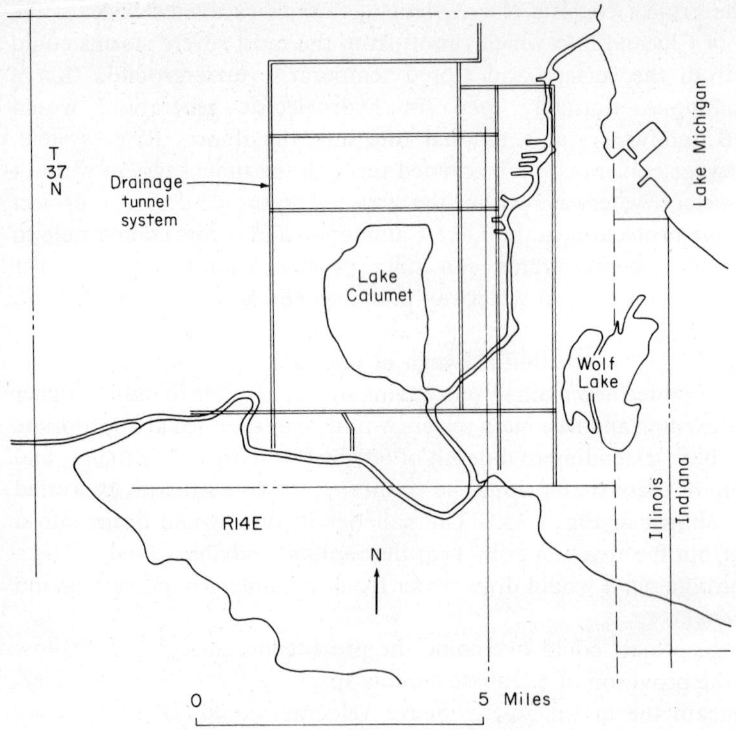

Fig. 8.45 Location of proposed subsurface drainage tunnels in Lake Calumet area, Chicago. (*From Walton, 1966.*)

of the groundwater resource from any ill effects resulting from the project. It was proposed that the drainage tunnel be located in the Galena-Platteville Formation, which constitutes a part of the principal regional aquifer, the Cambrian-Ordovician Aquifer. The extensive development of this aquifer requires that it be protected from any possible pollution as a result of the project.

An electric analog computer was constructed to simulate the Cambrian-Ordovician Aquifer in northeastern Illinois. The computer was used to determine: (1) the probable magnitude of seepage into the tunnel drainage system during periods between storms; (2) the amount of artificial recharge which will be required to eliminate the pollution hazard associated with the subsurface temporary storage of storm water; and (3) the effects of the tunnel drainage and a selected artificial recharge scheme on water levels in the Cambrian-Ordovician Aquifer.

The analog model (Fig. 8.46) for simulating the Cambrian-Ordovician Aquifer in northeastern Illinois consists of three horizontal regular arrays of resistors and capacitors interconnected by two vertical arrays of resistors. The model contains approximately 9,500 resistors and 4,000 capacitors. The bottom horizontal array of resistors

Fig. 8.46 Analog computer for Lake Calumet area, Chicago, tunnel drainage project. (*Photo by W. C. Walton.*)

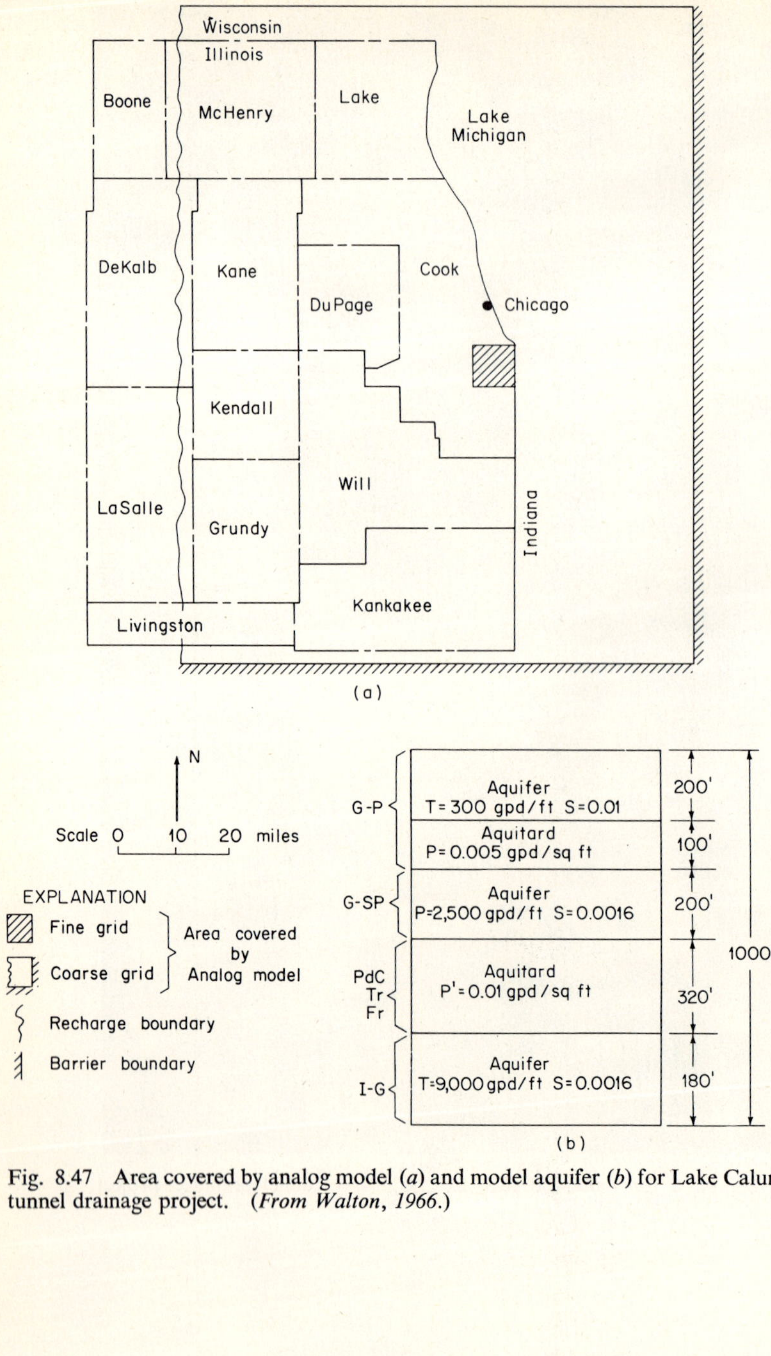

Fig. 8.47 Area covered by analog model (*a*) and model aquifer (*b*) for Lake Calumet area, Chicago, tunnel drainage project. (*From Walton, 1966.*)

and capacitors simulates the aquifer unit consisting of the Ironton-Galesville Sandstone (see Suter et al., 1959). Resistors to simulate the aquitard units consisting of the Prairie-du-Chien Series, Trempealeau Dolomite, and Franconia Formation were connected vertically to terminals of the bottom horizontal array. The second horizontal array of resistors and capacitors simulates the aquifer unit consisting of the Glenwood–St. Peter Sandstone. Resistors to simulate the semipermeable aquitard portions of the Galena-Platteville Dolomite were connected vertically to terminals of the second horizontal array. The third horizontal array of resistors and capacitors simulates the permeable aquifer portions of the Galena-Platteville Dolomite. Appropriate nodes of the third horizontal array were connected to a ground wire to simulate the tunnels in the Lake Calumet area. The Maquoketa Formation, Silurian rocks, and glacial deposits overlying the Cambrian-Ordovician Aquifer were not included in the model because water levels by 1970 will be below these rocks in most areas of northeastern Illinois (Walton, 1964c) and the tunnel drainage and recharge scheme will not materially affect movement of water in these rocks.

The model was constructed with terminals on a 1-inch-square pattern and is approximately 46 inches high and 48 inches wide. The area covered by the model and the locations of hydrogeologic boundaries (see Walton, 1964c) are shown in Fig. 8.47. The hydrogeologic properties and dimensions of aquifer and aquitard units of the model were estimated from data given by Suter et al. (1959) and Walton and Csallany (1962). Hydrogeologic conditions had to be simplified and modified to some extent because of model-component limitations. Four resistors and a capacitor were connected to each interior terminal of the horizontal arrays; the capacitors were secured to a ground wire. Barrier boundaries were simulated with open circuits; the recharge boundary was simulated by connecting appropriate nodes of the Galena-Platteville Dolomite to ground wires.

By the process of trial and error, scale factors were chosen to utilize inexpensive resistors and capacitors and existing excitation-response apparatus. Selected scale factors are:

$$K_1 = 1 \times 10^{16} \text{ gal/coulomb}$$
$$K_2 = 1 \text{ ft/volt}$$
$$K_3 = 1 \times 10^{9} \text{ gal/day/amp}$$
$$K_4 = 1 \times 10^{7} \text{ days/sec}$$

A maximum pumping period of 100 years was chosen.

Network spacings of 1 inch equals 20,000 feet and 1 inch equals 40,000 feet were selected for the major portion of the analog model to minimize the errors due to finite difference approximation and the cost of the model. A network spacing of 1 inch equals 2,000 feet was used in the 3.75-square-mile area centered over Lake Calumet and the tunnel drainage system so that cones of depression and recharge mounds in the immediate vicinity of the tunnels could be mapped in detail.

Fixed carbon resistors and disk ceramic capacitors were used in constructing the model. Values of resistors range from 180 to 6.8×10^6 ohms; values of capacitors range from 5 $\mu\mu$f to 3×10^{-3} μf. The analog model was constructed with $\frac{1}{8}$-inch pegboard perforated with holes on a 1-inch-square pattern. Aluminum angles (1×1 inch) were attached along the edges of the model with metal screws. Shoe eyelets were inserted in the holes of the pegboard to provide terminals for resistors and capacitors.

The excitation-response apparatus used is shown in Fig. 8.46. The analog model was coupled to the excitation-response apparatus and the pulse generator was connected to the wires simulating the tunnels. The output of the pulse generator was adjusted in accordance with water-level data (see Walton, 1964c) shown in Fig. 8.48. Seepage into the tunnels with the varying artesian pressure conditions in the Cambrian-Ordovician Aquifer was estimated. It was assumed that the sloping tunnels would be at an average depth of 150 feet below the top of the Galena-Platteville Dolomite.

As shown in Fig. 8.48, the artesian pressure is expected to decline from a position of about 200 feet above the top of the Galena-Platteville Dolomite in 1966 to the top of the Galena-Platteville Dolomite in 1980 as the result of estimated regional water-level declines and withdrawals from the tunnels. Artesian-pressure declines in the vicinity of the tunnels resulting from withdrawals from the tunnels were estimated by connecting the oscilloscope to terminals representing observation points.

The average seepage into the tunnels will probably decline from about 4.5 mgd in 1966 to 2.0 mgd in 1980 as shown in Fig. 8.49. The withdrawals from the tunnel necessary to keep the tunnels dewatered will decline as the artesian pressure recedes. Estimated withdrawals assume that the permeability of the Galena-Platteville Dolomite is uniform and that seepage into the tunnels is uniform throughout the tunnel systems. Lower portions of the tunnels may experience average seepage rates twice those shown

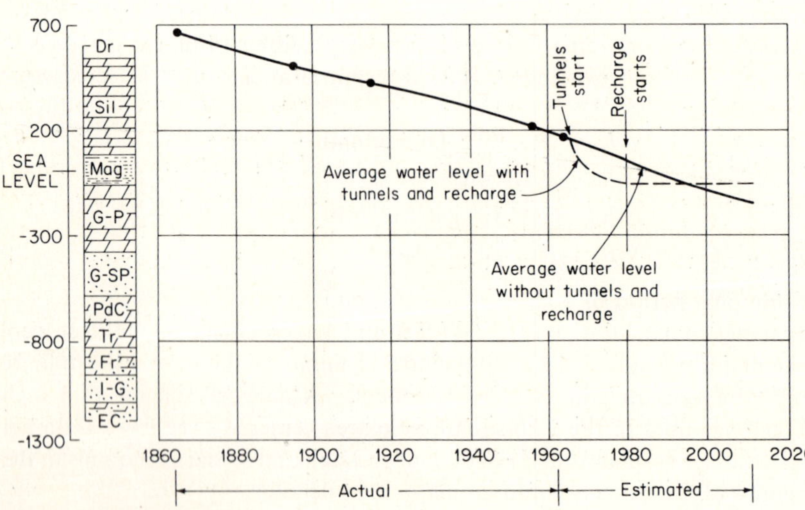

Fig. 8.48 Water levels in Cambrian-Ordovician Aquifer in Lake Calumet area, Chicago, 1864 to 2010. (*From Walton, 1966.*)

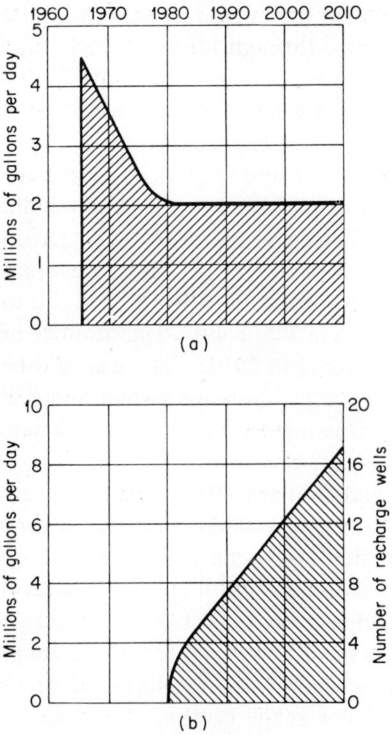

Fig. 8.49 Estimated pumpage from tunnels, 1966 to 2010 (*a*), and recharge required to stabilize water levels at top of Galena-Platteville 1980 to 2010 (*b*), Lake Calumet area, Chicago, tunnel drainage project. (*From Walton, 1966.*)

in Fig. 8.49; however, upper portions may experience average seepage rates one-half those shown. Because of the irregular occurrence and interconnection of openings in the Galena-Platteville Dolomite, it is probable that some portions of the tunnels will intercept large openings, which will result in local seepage rates several times the average seepage rate, and some portions of the tunnels will intercept small or practically no openings, which will result in local seepage rates much less than the average seepage rate. The larger openings should be grouted. No great accuracy for the estimated average seepage rates should be inferred, because of complex hydrogeologic conditions and data on the yield of the Galena-Platteville Dolomite in the project area and in northeastern Illinois are meager, and there are no case histories of seepage rates associated with such an extensive tunnel system as is contemplated. Seepage into portions of the tunnels locally could exceed five times the average estimated seepage rate based on well-production data given by Csallany and Walton (1963).

If withdrawals from the tunnels were continued at an average rate of 2 mgd beyond 1980 without artificial recharge, water levels in the Galena-Platteville Dolomite would recede below upper parts of the tunnel and storm water diverted into the tunnel drainage system could migrate into the Cambrian-Ordovician Aquifer and pollute existing water supplies. To prevent pollution of the Cambrian-Ordovician Aquifer, water levels in the Galena-Platteville Dolomite must be kept at the top of the Dolomite above the tunnels. This will require artificial recharge through injection wells near the tunnels starting in 1980.

The analog computer was used to determine the most efficient scheme of artificial recharge which would maintain water levels at the top of the Galena-Platteville Dolomite despite regional water-level declines shown in Fig. 8.48 and withdrawals from the tunnels. By the process of trial and error, it was determined that water levels could be maintained above tunnels most efficiently by recharging 17 injection wells terminating in the Glenwood–St. Peter Sandstone. Analog computer studies suggest that the required amount of recharge water will increase from less than 1 mgd in 1980 to about 8.5 mgd in 2010 as water levels recede as shown in Fig. 8.49. The number of recharge wells will probably increase from 1 in 1980 to 17 in 2010. There would be four injection wells at each of the four corners of the tunnel drainage system and one injection well near the center of the tunnel system. The injection wells at the corners could be spaced about 2,000 feet apart.

Based on well-production data given by Walton and Csallany (1962), the long-term specific capacity of injection wells terminating in the Glenwood–St. Peter Sandstone will average about 0.5 gpm/ft. As indicated in Fig. 8.48, the available buildup (difference between the top of the Galena-Platteville Dolomite and the ground surface) is about 650 feet. Thus, it will not be possible to recharge each injection well at a rate much in excess of $\frac{1}{2}$ mgd, because the recharge rate is equal to the product of the long-term specific capacity and the available drawdown. The injection wells could be 12 inches in diameter at the surface, 10 inches in diameter at the bottom, and open to

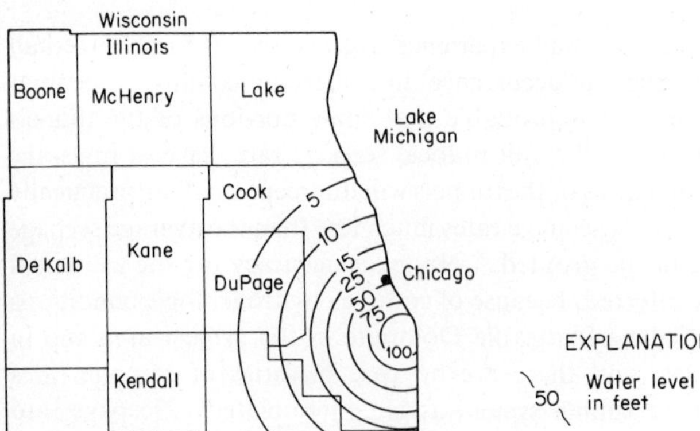

Fig. 8.50 Theoretical water-level rises in Cambrian-Ordovician Aquifer as the result of artificial recharge and tunnel drainage in Lake Calumet area, Chicago. (*From Walton, 1966.*)

noncaving portions of the Galena-Platteville Dolomite and the Glenwood–St Peter Sandstone.

Some of the recharge water will not move vertically upward into the Galena-Platteville Dolomite but will move vertically downward to the Ironton-Galesville Sandstone and be lost to the cause of maintaining water levels in the Galena-Platteville Dolomite (see Fig. 8.50). However, if injection wells were completed in the Ironton-Galesville Sandstone, water levels in the Galena-Platteville would be even less affected by artificial recharge. It is impossible to terminate the injection wells in the Galena-Platteville Dolomite because the formation has a very low average permeability and the interconnection of openings is poor.

Based on analog computer studies, it is concluded that physically the pollution hazard can be eliminated by artificially recharging the Cambrian-Ordovician Aquifer in the vicinity of the tunnel drainage system. Recharge rates are not physically excessive. The average seepage rate into the tunnel system also may not be excessive.

Springfield, Ohio, and Monsanto, Illinois

The accuracy and reliability of induced infiltration rates and the effects of induced infiltration on the water levels in wells estimated by use of the image-well theory were appraised with electric analog computers (Walton and Ackroyd, 1966). In the electric analog computers, the streambed is simulated as an area of recharge in accordance with natural conditions instead of as a recharging image well. Induced infiltration rates and water-level declines for two aquifer-stream situations determined with electric analog computers and by analysis of aquifer test data with the image-well theory were compared. In addition, the effects of a stream-stage change on water levels in a selected aquifer were determined with electric analog computers.

The results of two aquifer tests were studied to determine the relation between the effects of induced infiltration on water levels estimated with the image-well theory and electric analog computers. The selected aquifer tests involve the two extreme aquifer-stream situations, i.e., the case where the cone of depression spreads beneath and beyond the entire streambed and the case where the cone of depression spreads only part way beneath the streambed.

During 1955 and 1956 an intensive hydrogeologic study was made to determine the feasibility of developing a large water supply for the city of Springfield, Ohio, from a sand and gravel aquifer in an area along the Mad River, about 4 miles northwest of the city (see Walton and Schaefer, 1956a). The study included a test-hole drilling program and a controlled aquifer test (aquifer test 1). The unconsolidated deposits in the study area consist of recent silty alluvial materials and glacial outwash sand and gravel. These deposits are contained in a buried valley cut into the underlying bedrock of Ordovician and Silurian ages. The permeable outwash forming the aquifer averages about 100 feet thick and consists of permeable sand and gravel and occasional thin lenses of clayey materials. Logs of wells and test holes show that the

TABLE 8.8 SUMMARY OF INDUCED STREAMBED INFILTRATION AQUIFER TEST RESULTS. (*From Walton and Ackroyd, 1966.*)

Owner of wells	*Location of test site*	*Date of test*	*Pumping period, days*	*Pumping rate, gpm*	*Hydrogeologic properties* T, *gpd/ft*	P, *gpd/sq ft*	S	I_a, *gpd/acre*	h_r, *ft*	I_h, *gpd/acre/ft*	*Temperature of surface water, °F*
City of Springfield, Ohio	Along Mad River, about 4 miles northwest of Springfield, Ohio	2/20 to 23/56	3	1,034	547,000	5,470	0.01	92,000	0.09	1,000,000	39
Monsanto Chemical Corp.	Along Mississippi River, near East St. Louis, Illinois	8/4 to 8/52	4	1,100	290,000	3,900	0.08	15,500	0.17	91,000	83

aquifer is about 4,000 feet wide and trends southwest to northeast through the study area. The Mad River has been dredged into the aquifer.

Aquifer test 2 was made during the period August 4 to 8, 1952, by Ranney Method Water Supplies, Inc., for the Monsanto Chemical Corp. (see Schicht, 1965). The test site is located east of Monsanto, Illinois, along the Mississippi River in section 27, T2N, R10W. The permeable aquifer in the area consists of alluvial materials and glacial outwash sand and gravel. Impermeable bedrock is encountered at a depth of 120 feet. The aquifer is hydraulically connected to the Mississippi River.

Aquifer test data were analyzed with analytical methods. A summary of the results of the aquifer tests is given in Table 8.8.

Electric analog computers for the two aquifer test sites were constructed using the computed values of the coefficients of transmissibility and storage of the aquifers and infiltration rates of the streambeds. The analog model for aquifer test site 1 consists of regular arrays of 6,100 resistors and 3,100 capacitors and is shown in Fig. 8.51. The values of resistors range from 1.2×10^4 to 5.6×10^6 ohms and the values of capacitors range from 5 $\mu\mu$f to 4×10^{-3} μf. The fine grid of the model is 5 by 4 feet and has a scale of 1 inch = 40 feet; the coarse grid of the model is 1 by 2 feet and has a scale of 1 inch = 800 feet. Scale factors are: $K_1 = 1.19 \times 10^{13}$ gal/coulomb, $K_2 =$ 1 ft/volt, $K_3 = 5.47 \times 10^{10}$ gal/day amp, and $K_4 = 2.18 \times 10^2$ days/sec. The analog model for aquifer test site 2 consists of regular arrays of 2,150 resistors and 1,060 capacitors and is shown in Fig. 8.52. The values of resistors range from 4.7×10^4 to 1.0×10^6 ohms and the values of capacitors range from 5.0×10^2 $\mu\mu$f to 1×10^{-2} μf. The model is 3 by 4 feet. The fine grid of the model has a scale of 1 inch = 150 feet; the coarse grid of the model has a scale of 1 inch = 450 feet. Scale factors are: $K_1 = 1.38 \times 10^{13}$ gal/coulomb, $K_2 = 1$ ft/volt, $K_3 = 2.1 \times 10^{10}$ gal/day amp, and $K_4 = 6.58 \times 10^2$ days/sec.

The analog models were constructed with $\frac{1}{8}$-inch pegboard perforated with holes on a 1-inch-square pattern. Aluminum angles (1 $\times$ 1 inch) were attached along the edges of the models with metal screws. Shoe eyelets were inserted in the holes in the pegboard to provide terminals for resistors and capacitors. Horizontal arrays of resistors and capacitors simulate aquifers and vertical arrays of resistors simulate streambeds.

The excitation-response apparatus used in this investigation is shown in Fig. 8.53. The analog models were separately coupled to the excitation-response apparatus and the pulse generator was connected to junctions at the locations of production wells used during aquifer tests 1 and 2. The output of the pulse generator was adjusted in accordance with the aquifer test pumping rates and periods. The oscilloscope was connected to terminals representing observation wells, and water-level declines were computed from time-voltage graphs. Thus, water-level declines everywhere in the aquifers were described.

The drawdown contours for aquifer test 1 (pumping period 4,310 minutes and pumping rate 1,034 gpm) based on the analog computer solution are shown in Fig. 8.54. The drawdown contours for the same pumping period and rate computed with

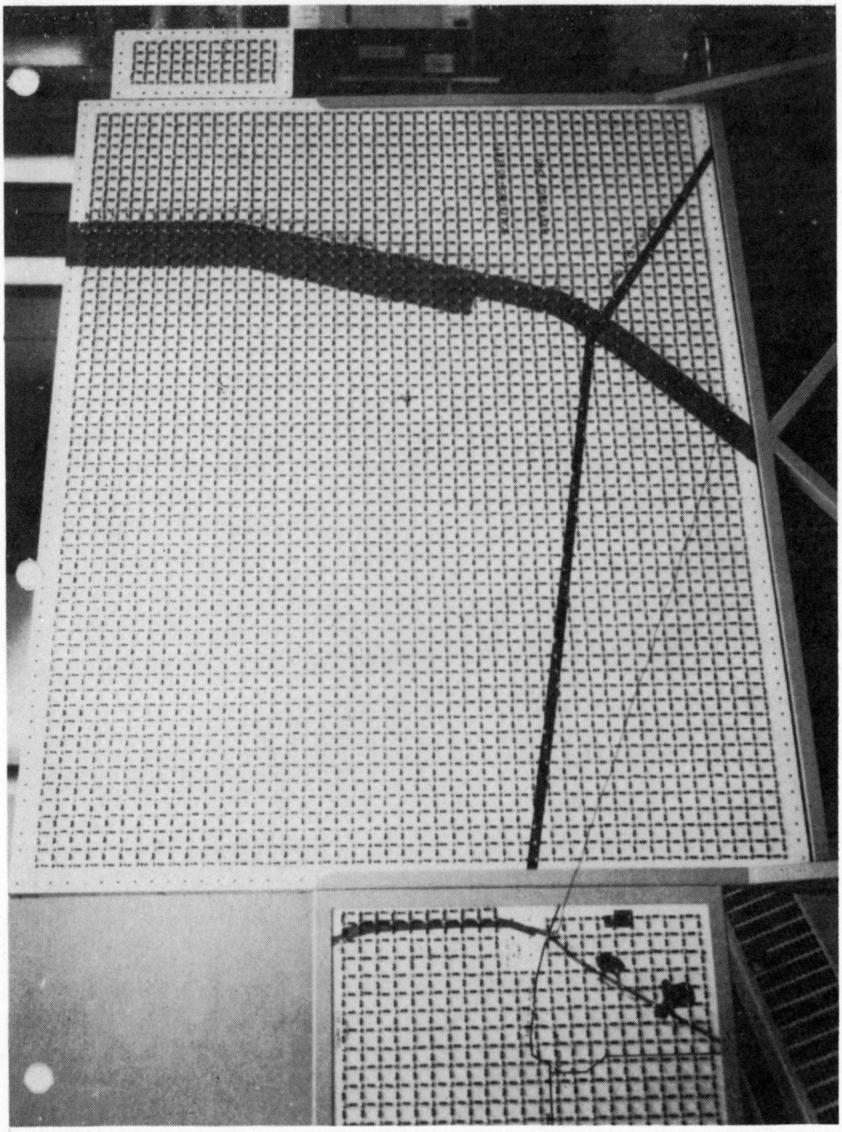

Fig. 8.51 Analog model for aquifer test site 1 near Springfield, Ohio. (*From Walton and Ackroyd, 1966.*)

Fig. 8.52 Analog model for aquifer test site 2 near Monsanto, Illinois. (*From Walton and Ackroyd, 1966.*)

the results of aquifer test 1 and based on an analytical solution involving the image-well theory are shown in Fig. 8.55.

The drawdown contours for aquifer test 2 (pumping period four days and pumping rate 1,100 gpm) based on the analog computer solution are shown in Fig. 8.56. The drawdown contours for the same pumping period and rate computed with the results of aquifer test 1 and based on an analytical solution involving the image-well theory are given in Fig. 8.57.

In the case of aquifer test 1, close agreement between analog and analytical drawdown contours on the land side of the river is noted if the irregular course of the river is taken into consideration. Differences in individual drawdowns at observation points west of the river based on analog computer and analytical solutions are small.

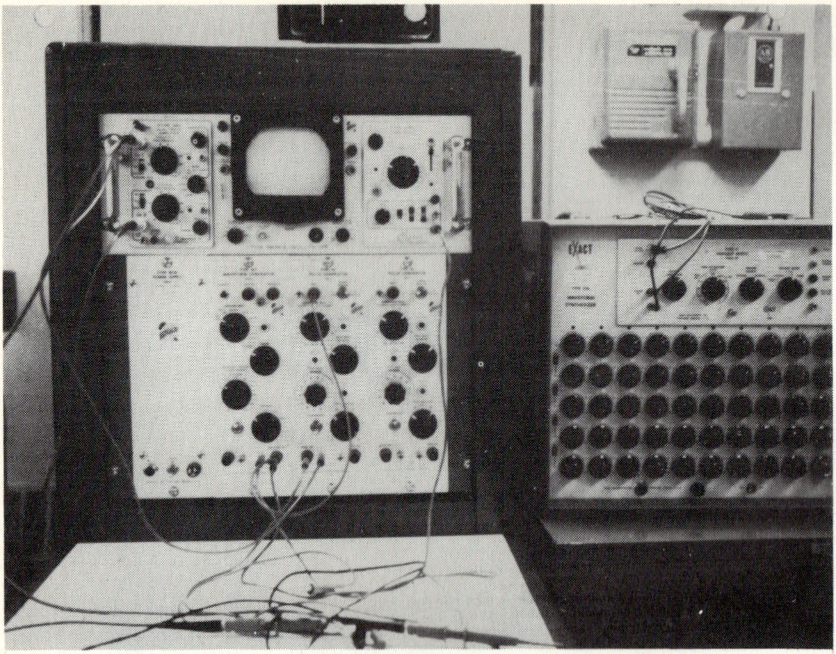

Fig. 8.53 Excitation-response apparatus used in induced streambed infiltration studies. *(From Walton and Ackroyd, 1966.)*

Analog and analytical drawdown contour solutions beneath the river and east of the river differ appreciably. According to the analog computer solution, the cone of depression spreads a much greater distance beyond the river than suggested by the analytical solution. Also, as indicated in Table 8.9, the I_a and h_r values based on analog computer and analytical solutions do not agree. The analog computer solution indicates that the streambed area of infiltration is greater but the average head loss is less than suggested by the analytical solution. In the case of aquifer test 2, close agreement between analog and analytical drawdown contours on the land side of the river is noted. Differences in individual drawdowns at observation points east of the river based on analog computer and analytical solutions are small. Analog and analytical drawdown contours beneath the river differ appreciably. According to the analog computer solution, the cone of depression spread a much greater distance beyond the shoreline than suggested by the analytical solution and the effective line of recharge is curved. Also, as indicated in Table 8.9, the I_a and h_r values based on analog computer and analytical solutions do not agree. The analog computer solution indicates that the streambed area of infiltration is greater but the average head loss is less than suggested by the analytical solution.

The results of the studies indicate that during induced infiltration aquifer tests, the image-well theory closely describes drawdowns on the land side of the streams with a

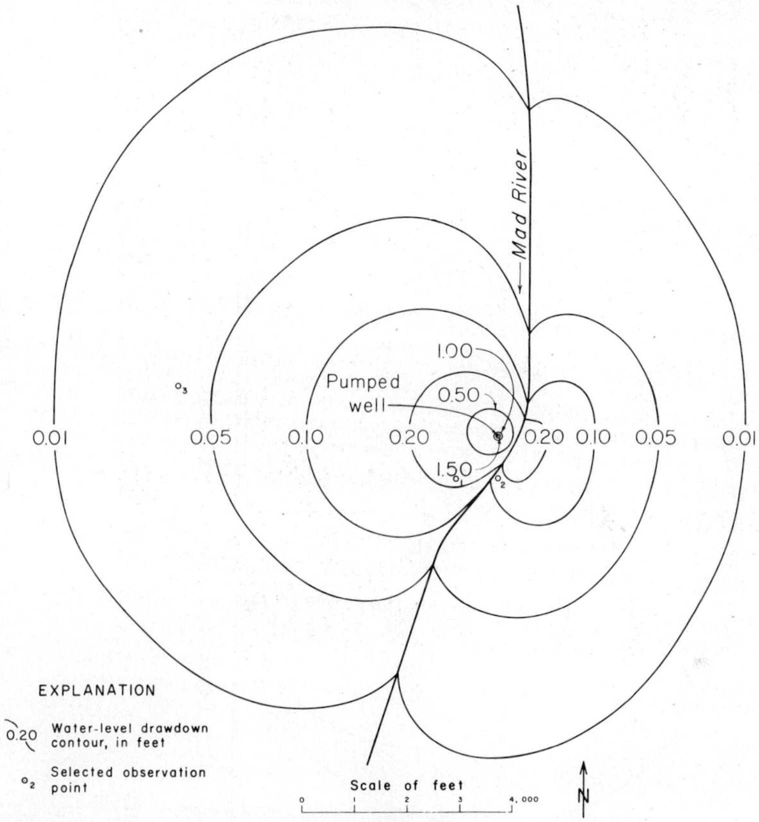

Fig. 8.54 Drawdown contours for aquifer test 1 near Springfield, Ohio, based on analog computer solution. *(From Walton and Ackroyd, 1966.)*

high degree of accuracy, whether the cone of depression spreads beneath and beyond or only part way beneath the streambeds. Drawdowns beneath or beyond the streambeds and the streambed areas of infiltration based on the image-well theory are not those which would be observed in the field (Walton and Ackroyd, 1966). Analog computer studies indicate that the streambed areas of infiltration are greater but the average head losses beneath the streambeds are less than suggested by analytical solutions. Thus, it would appear that analytical methods involving the image-well theory can be used to estimate streambed infiltration rate per unit area per foot of head loss. The streambeds may be narrow or wide and have high or low permeabilities. The image-well theory may also be used to estimate drawdowns in wells on the land side of the river. However, analytical methods cannot be used to estimate streambed areas of infiltration and drawdowns beneath and beyond streambeds which would occur as the result of pumping a well adjacent to a stream.

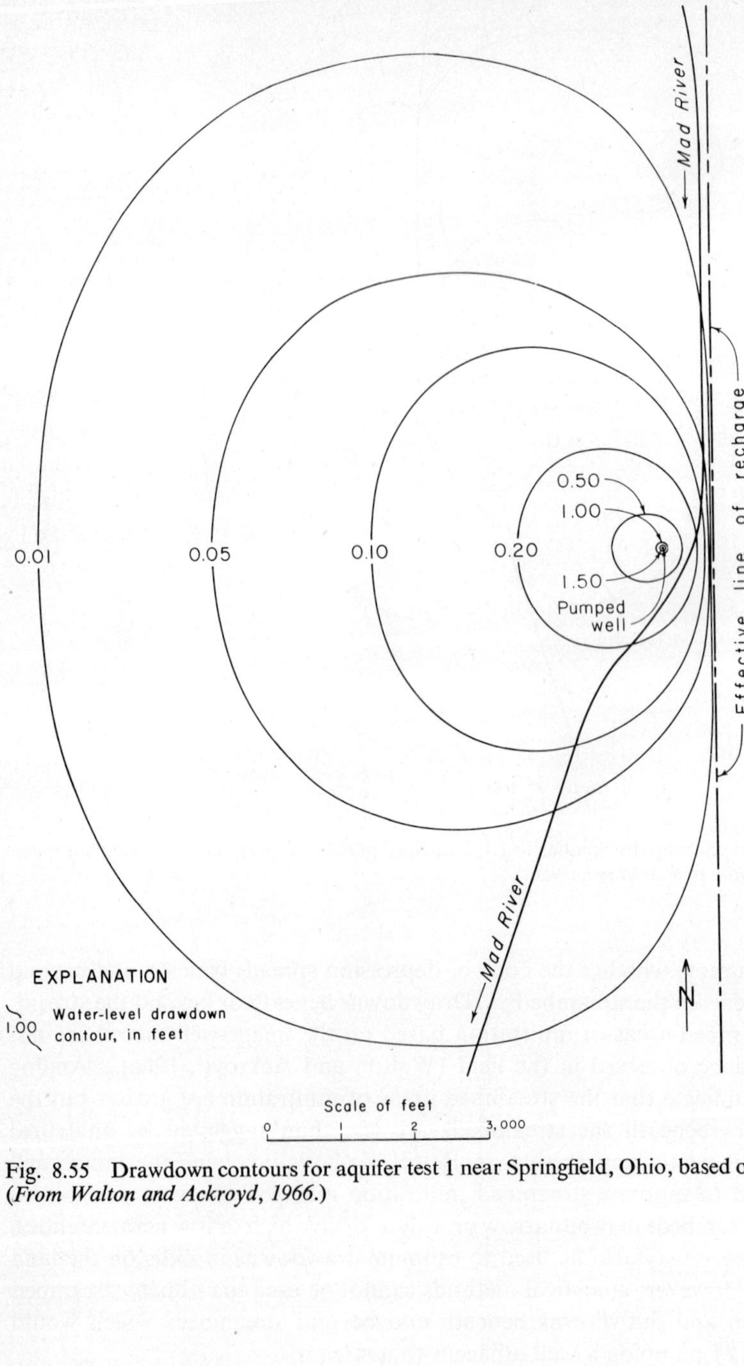

Fig. 8.55 Drawdown contours for aquifer test 1 near Springfield, Ohio, based on analytical solution. (*From Walton and Ackroyd, 1966.*)

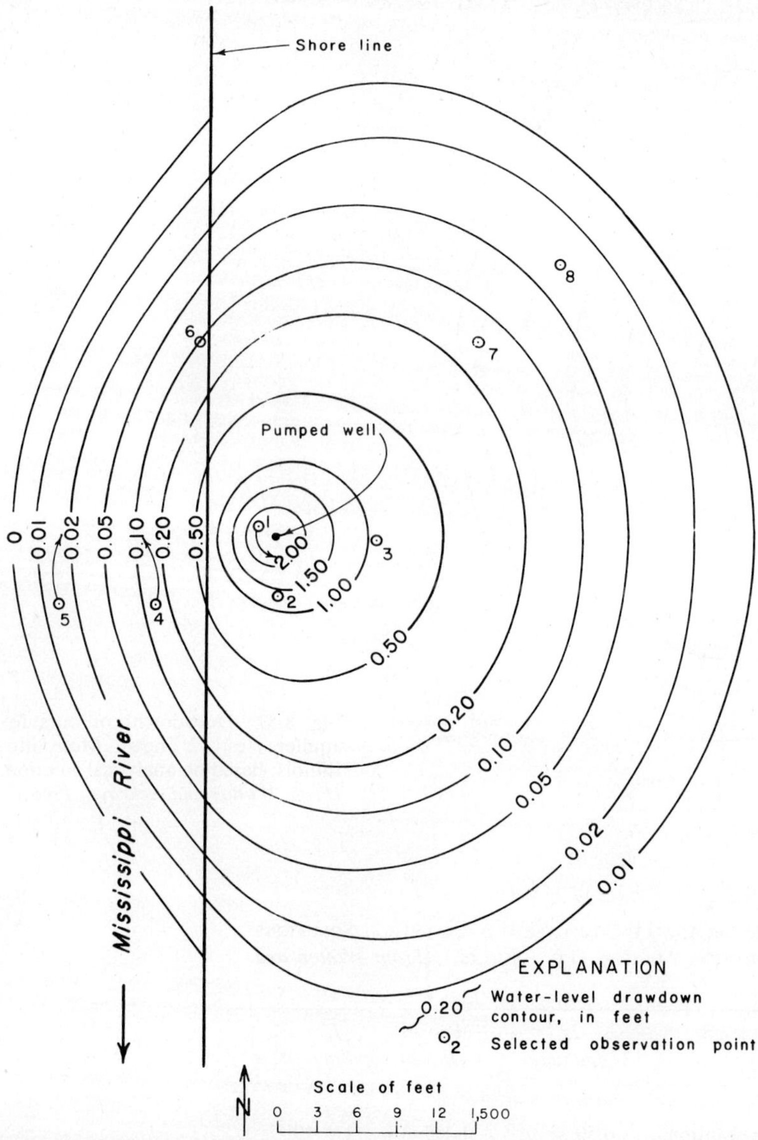

Fig. 8.56 Drawdown contours for aquifer test 2 near Monsanto, Illinois, based on analog computer solution. (*From Walton and Ackroyd, 1966.*)

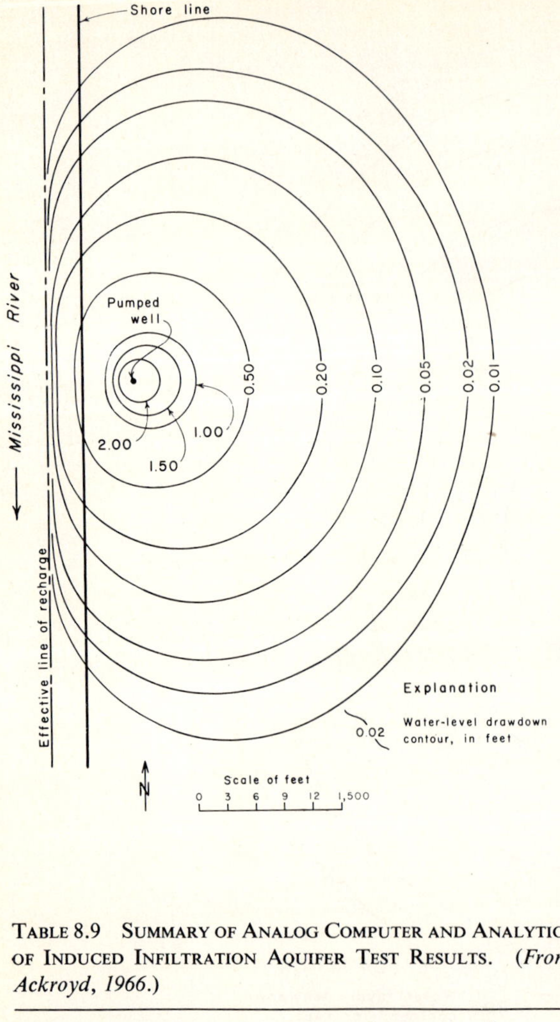

Fig. 8.57 Drawdown contours for aquifer test 2 near Monsanto, Illinois, based on analytical solution. (*From Walton and Ackroyd, 1966.*)

TABLE 8.9 SUMMARY OF ANALOG COMPUTER AND ANALYTICAL SOLUTIONS OF INDUCED INFILTRATION AQUIFER TEST RESULTS. (*From Walton and Ackroyd, 1966.*)

	I_a, *gpd/acre*	h_r, *ft*	I_h, *gpd/acre/ft*
Aquifer test 1			
Analog computer solution	6.50×10^4	0.065	1×10^6
Analytical solution	9.20×10^4	0.090	1×10^6
Aquifer test 2			
Analog computer solution	6.5×10^3	0.072	9.1×10^4
Analytical solution	1.55×10^4	0.170	9.1×10^4

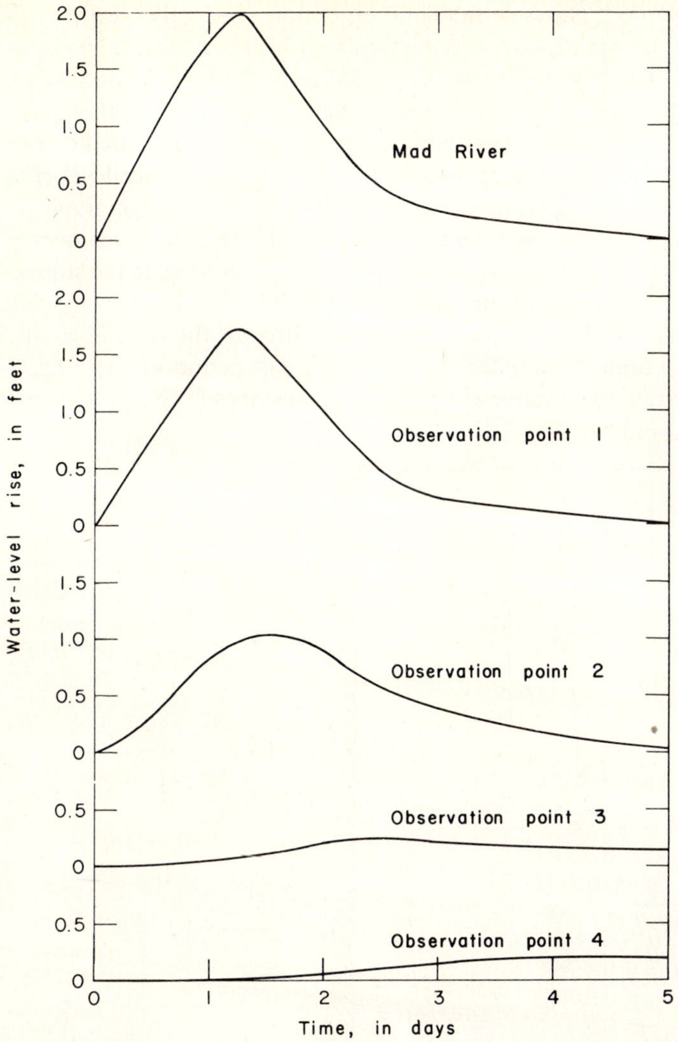

Fig. 8.58 Water-level rises at observation points due to a selected stream-stage rise in aquifer test site 1 near Springfield, Ohio. Location of observable points shown in Fig. 8.59. (*From Walton and Ackroyd, 1966.*)

In the past, little if any attention has been given water-level declines beneath and beyond streambeds. During future tests, observation wells should be installed within the streambeds at depths just below the streambed so that the average head losses due to the vertical percolation of water through the streambeds and the streambed areas of infiltration can be estimated.

The analog model for aquifer test site 1 was coupled to a waveform synthesizer with a variable-width plug-in unit (variable-function generator) and other excitation-response apparatus. The output of the variable-function generator was adjusted in accordance with a selected small stream rise (Fig. 8.58) recorded for the Mad River. The variable-function generator was connected to the vertical resistors simulating the streambed; the vertical resistors were disconnected from ground. The oscilloscope was connected to terminals representing observation points, and water-level rises caused by the complex stream-stage rise were computed from time-voltage graphs.

The water-level-rise contours representing the maximum effects of the stream-stage rise are shown in Fig. 8.59. Water-level rises varied from 1.75 feet beneath the streambed to 0.20 foot about $1\frac{1}{2}$ miles west of the river due to the 2.0-foot stream-stage rise. Approximately 15.5 million gallons of water percolated through the streambed and into temporary storage within the aquifer during the $1\frac{1}{4}$-day period of the stream-stage rise. The average rate of recharge (bank storage) was about 3.7 mgd per mile

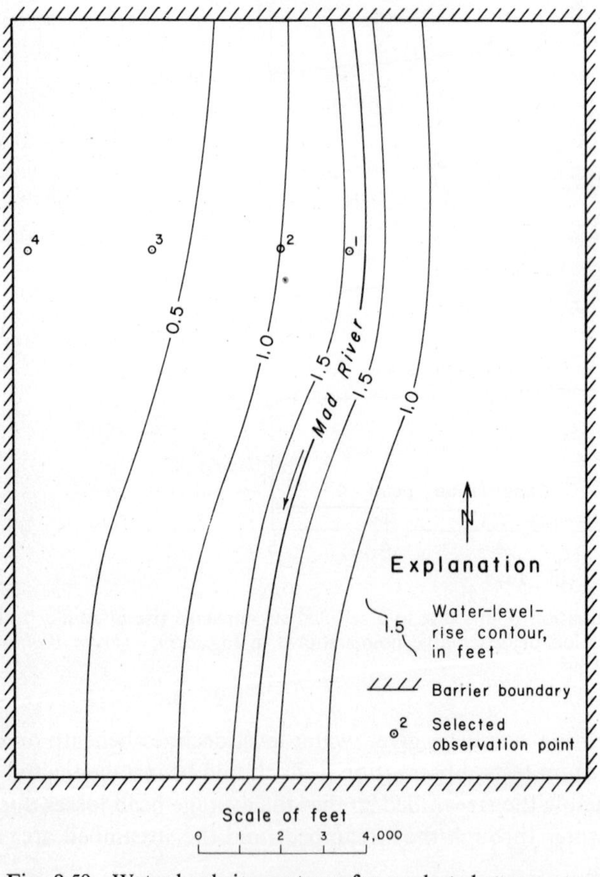

Fig. 8.59 Water-level-rise contours for a selected stream-stage rise at aquifer test site 1 near Springfield, Ohio. (*From Walton and Ackroyd, 1966.*)

length of the streambed. Approximately 2.6 million gallons or 17 percent of the recharge remained in storage within the aquifer $3\frac{3}{4}$ days after the stream-stage peak even though the stream stage had returned to its original stage. Thus, about 12.9 million gallons or 83 percent of recharge returned to the stream during the $3\frac{3}{4}$-day period following the peak.

Venice, Ohio

The aquifer-stream system described in Chap. 4 in the Venice area, Ohio, for which hydrogeologic data are available, was studied to gain insight into the magnitude of recharge by induced streambed infiltration under complex stream-stage and ground-water level conditions (Walton et al., 1967). Electric analog computers based on the aquifer-stream system in the Venice, Ohio, area were constructed. The analog computers consist of analog models and excitation-response apparatus.

Analog model 1 is a regular array of resistors and capacitors. In the resistor-capacitor network, resistors are inversely proportional to the coefficients of transmissibility of the outwash deposits and the infiltration rate of the Miami River streambed; capacitors store electrostatic energy in a manner analogous to the storage of water within the outwash deposits. The electrical network is a scaled-down version of the Venice, Ohio, aquifer-stream system assuming that leakage through the Miami River streambed is directly proportional to declines in water levels beneath the streambed regardless of the stage of the water table.

Analog model 1 consists of a regular array of 1,985 resistors and 1,073 capacitors. The values of resistors range from 15,000 to 10,000 ohms and the values of capacitors range from 100 to 4,700 $\mu\mu$f. The fine grid of the model is 20 by 20 inches and has a scale of 1 inch = 200 feet; the coarse grids of the model have scales of 1 inch = 400 feet, 1 inch = 800 feet, and 1 inch = 8,000 feet. Scale factors are: $K_1 = 9.95 \times 10^{13}$ gal/coulomb, $K_2 = 1$ ft/volt, $K_3 = 2.77 \times 10^7$ gal/day amp and $K_4 = 7.30 \times 10^6$ days/sec. The analog model was constructed with $\frac{1}{8}$-inch pegboard perforated with holes in a 1-inch-square pattern. Aluminum angles (1×1 inch) were attached along the edges of the models with metal screws. Shoe eyelets were inserted in the holes of the pegboard to provide terminals for resistors and capacitors. A horizontal array of resistors and capacitors simulates the outwash deposits and a vertical array of resistors simulates the Miami River streambed.

The analog model was designed to have a low impedance of about 50 ohms. This low impedance made it possible to quickly discharge the model to equilibrium. For analog model 1, equilibrium is reached in less than 0.5 μsec (a microsecond is 10^{-6} second) and therefore the time-voltage curve appears as a stable waveform on the oscilloscope. The time-voltage waveform with high-speed excitation-response apparatus repeats itself every 100 μsec. The low-impedance analog model utilizes to better advantage available excitation-response apparatus than do the higher-impedance (50,000 ohms or more) models commonly designed. With a low-impedance model, currents through components of the model are large enough to be measured with a current probe capable of measuring from 10 ma to 5 amp.

Analog model 1 was designed in part to permit the later use of transistors in streambed elements of analog model 2. Transistors become useful in terms of their operating characteristics at impedances of typically 500 ohms. To be economical, the transistors in the streambed element should have the following current, response time, and impedance capabilities:

Current	5 to 150 ma, typical
Response time	10^{8} sec [10 nonoseconds (nsec)] or less
Forward impedance	20 ohms at 50 ma
Reverse impedance	20,000 ohms at 0.1 ma

The upper limit of the value of the aquifer element resistor of analog model 1 must be at least 10 times smaller than the value of the transistor reverse impedance (about 20,000 ohms). To avoid exceeding transistor current ratings, the value of the aquifer elements resistor must exceed 50 ohms.

The transistor response time must be 10 times faster than the analog model 1 aquifer element circuit response time. Since the transistor responds relatively fast, small values of capacitance [less than 10,000 picofarads (pf)] should be used in aquifer element construction. To assist in capacitor computation, 1 picofarad $= 10^{-12}$ farad. The capacitance must exceed 100 pf per aquifer node to avoid unwanted secondary capacitive effects from excess wire lengths, resistor leads, and transistors.

Analog model 1 was designed taking into consideration all the above-mentioned requirements associated with the use of transistors in streambed elements. In addition, to eliminate high-current problems associated with high collector discharge rates (0.24 amp for $Q = 6.65$ mgd), a fine grid scale was selected so that collector nodes would be three or more nodes away from streambed element nodes and power transistors were used in simulating collectors.

Excitation-response apparatus forces electrical energy in the proper time phase into analog model 1 and measures energy levels within the energy-dissipative resistor-capacitor network. The excitation-response apparatus consists of two power supplies, a waveform generator, a pulse generator, two power amplifiers, and an oscilloscope as shown in Fig. 8.60.

The repetition rate of the pulse generator was adjusted to a value approximately 10 times the value of the pulse width taking into consideration scale factors and a desired maximum pumping period of 1 year. Under these conditions, a low duty cycle (10 percent) allows the use of only 0.5-watt resistors in analog model 1 aquifer elements. If the duty cycle were 100 percent, each of the four resistors of the collector node would have to dissipate 5 watts of heat.

Two transistorized power amplifiers were designed to make the commercial pulse

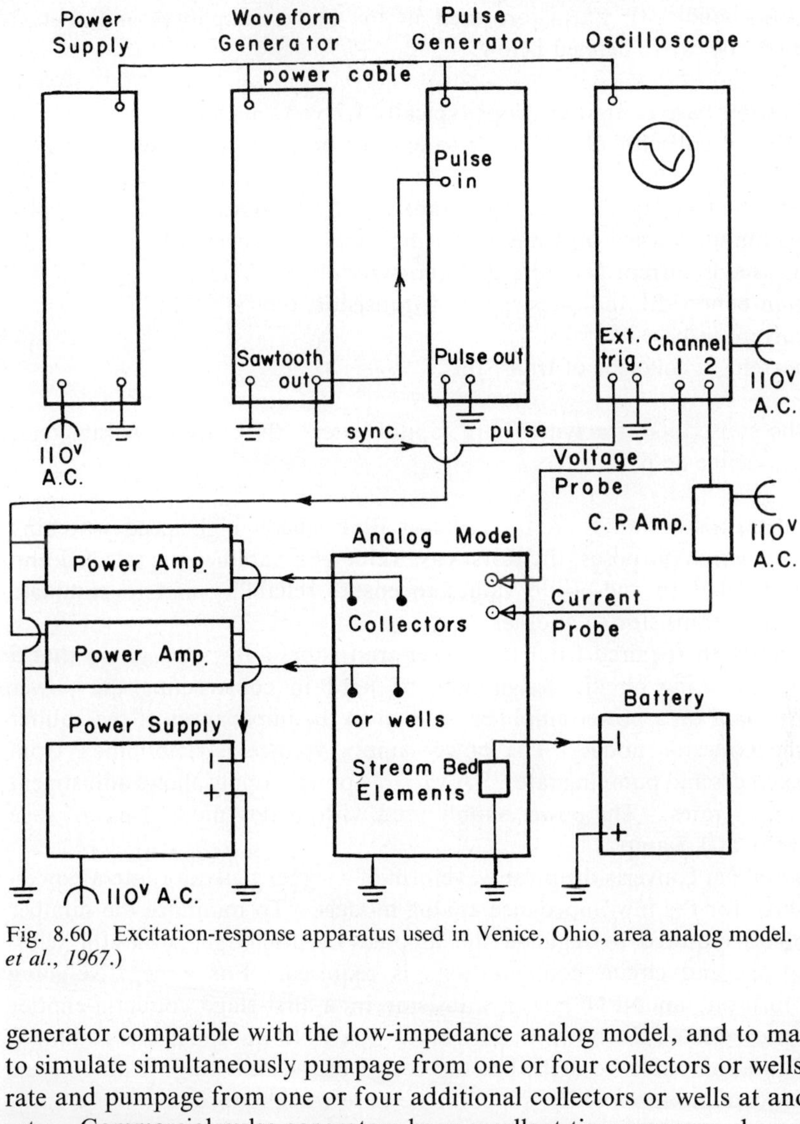

Fig. 8.60 Excitation-response apparatus used in Venice, Ohio, area analog model. (*From Walton et al., 1967.*)

generator compatible with the low-impedance analog model, and to make it possible to simulate simultaneously pumpage from one or four collectors or wells at a selected rate and pumpage from one or four additional collectors or wells at another selected rate. Commercial pulse generators have excellent time-response characteristics for a wide range of pulse amplitudes, width, and repetition rates, but do not meet the power output requirement of analog model 1. The pulse width with the power amplifiers does not exceed 50 μsec and is compatible with a current probe's response characteristics. The speed of the pulse rise time limits how soon time-voltage data can be accurately interpreted. The power amplifiers have rise times less than 1.0 μsec and therefore only the first 2 percent of the total pumping period (pulse width) is obscured by the use of the power amplifiers.

The terms associated with transistors used in the power amplifiers and typical transistor characteristics are defined below.

V_{BE} = voltage, base to emitter drop; typically 1.2 volts maximum
V_{CE} = voltage, collector of transistor to emitter drop; normally limited to 30 volts maximum
V_{CES} = saturated voltage, collector of transistor to emitter, typically 1.3 volts maximum (turned ON transistor using excess base current)
B or h_{FE} = transistor current gain at a dc level, typically 30 minutes
f_t = gain bandwidth indicates speed of transistor, typically 300 MHz
I_B = current in base of transistor
I_c = current in collector of transistor

Note: When the subscripts of a symbol are capital letters, these are dc parameters; lowercase letters define ac parameters.

These typical transistor values are for NPN or PNP silicon high-speed switching transistors. For design purposes, the worst-case value of a parameter is selected and then normally derated to end-of-life values to ensure reliability and to eliminate variations from one transistor to another.

The gain bandwidth required from the power transistor is low enough so that a saturated-mode switching circuit design may be used in constructing the power amplifiers. The load on a power amplifier is equal to the impedances of the aquifer elements at the collector nodes. The power-supply voltage is determined from estimated impedances and pumping rates. A variable power supply allows adjustment to desired pumping rates. The power supply used with analog model 1 has a range from 1 to 20 volts at 1.5 amp.

The power amplifier converts the pulse waveform of a signal generator into a powerful driving source for the low-impedance analog models. To minimize the number of power supplies required to regulate the amount of discharge, a careful selection of transistors and circuit configurations is required. For a negative-going pulse during turn-on, an NPN power transistor in a first-stage common-emitter configuration was selected as shown in Fig. 8.61*a*. A PNP transistor in a common-collector configuration could be used, but the design of the next stage is an emitter-follower circuit configuration which supplies the base current to Q_1. Figure 8.61*a* shows the first-stage power transistor Q_1 directly coupled to the emitter-follower Q_2. When Q_2 is turned on, resistor R_1 supplies sufficient base current to Q_1 to keep both transistors in saturation. To ensure turn-off of Q_1, resistor R_2 depletes the base current of Q_1 when Q_2 is turned off. A positive-going signal referenced to the -12 volts variable power supply controls the switching of Q_2. Notice that emitter followers do not invert the signal; they have current gain but not voltage gain. Another stage of inversion conveniently references a negative-going signal to ground for connection to the commercial pulse generator. Figure 8.61*b* shows the complete

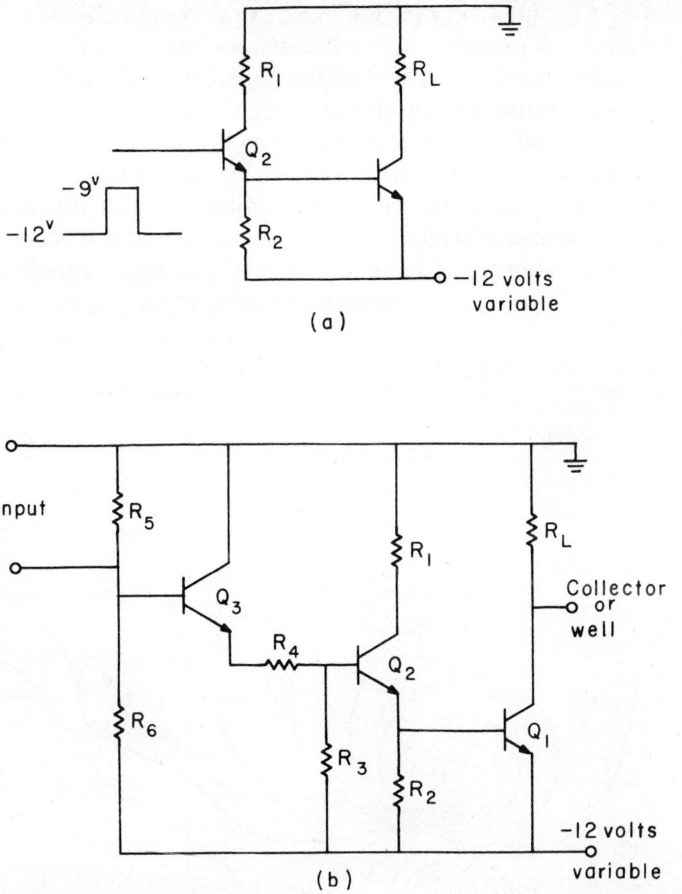

Fig. 8.61 Wiring diagrams for first two stages of power amplifier (*a*) and complete power amplifier (*b*) used in Venice, Ohio, area analog model. (*From Walton et al., 1967.*)

power amplifier using a PNP transistor for the third stage. The voltage divider network R_3 and R_4 controls the voltage level required to keep Q_2 on. The input of the X_3 resistor network biases the base of Q_3. R_L is the estimated impedance seen by the power amplifiers.

Analog model 1 was coupled to the excitation-response apparatus and the power amplifiers were connected to junctions at the locations of existing and hypothetical collectors or production wells. The output of the pulse generator was adjusted in accordance with selected pumping rates and periods. The oscilloscope was connected to terminals representing observation wells, and water-level declines were computed from time-voltage graphs.

Drawdown contours for a two-collector system are shown in Fig. 8.62. Pumping rates of collectors 1 and 2 are 6.2 and 8.5 mgd, respectively. The drawdown contours are for a pumping period of 6 months. Thus, Fig. 8.62 shows the distribution of water-level decline as the result of pumping two collectors at constant rates for a period of 6 months with no recharge directly from precipitation but with recharge from induced streambed infiltration. Although not shown in Fig. 8.62, drawdown was measurable to distances of 32,000 and 33,000 feet up and down the buried valley from collector 2. Water-level declines beneath the streambed are less than 4 feet.

The minimum average depth of water in the Miami River during August 1956 was about 4 feet. If it is assumed that stream stages remained at levels recorded in August 1956 during the selected 6-month pumping period and the water table before pumping started was level with the surface of the stream, then the water table was not lowered

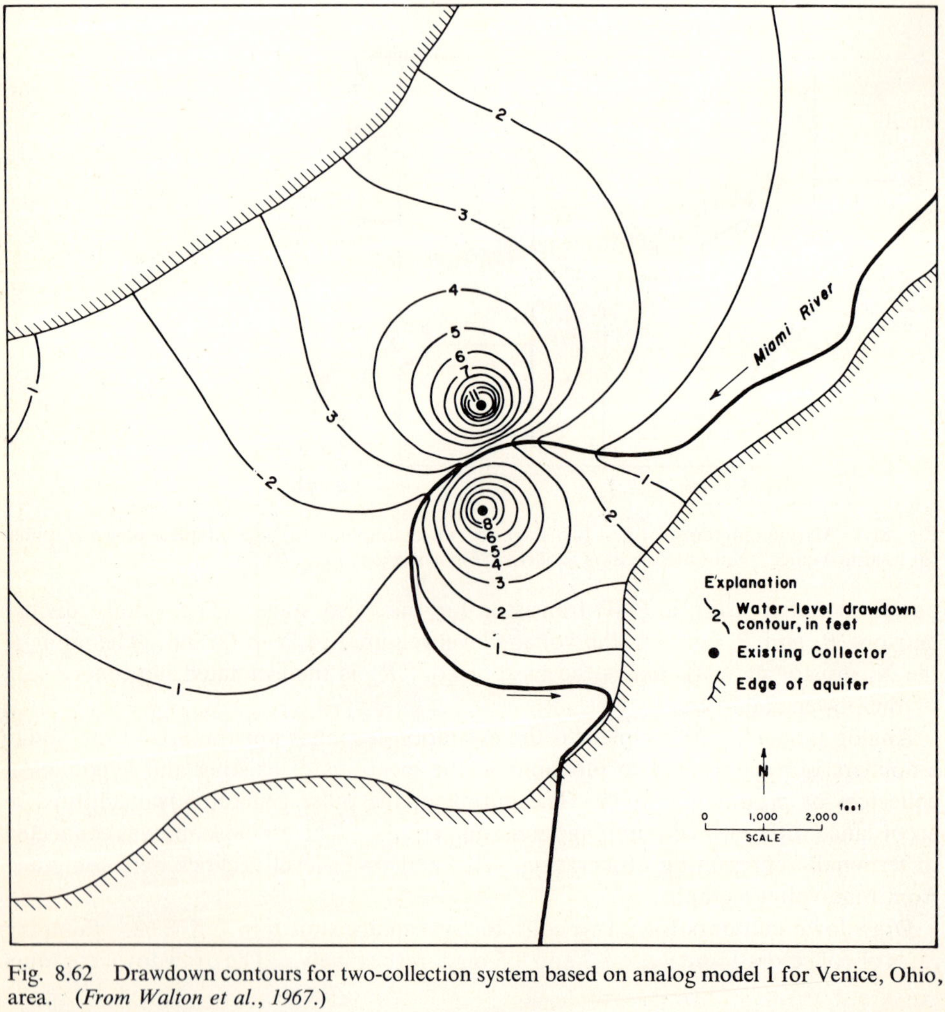

Fig. 8.62 Drawdown contours for two-collection system based on analog model 1 for Venice, Ohio, area. (*From Walton et al., 1967.*)

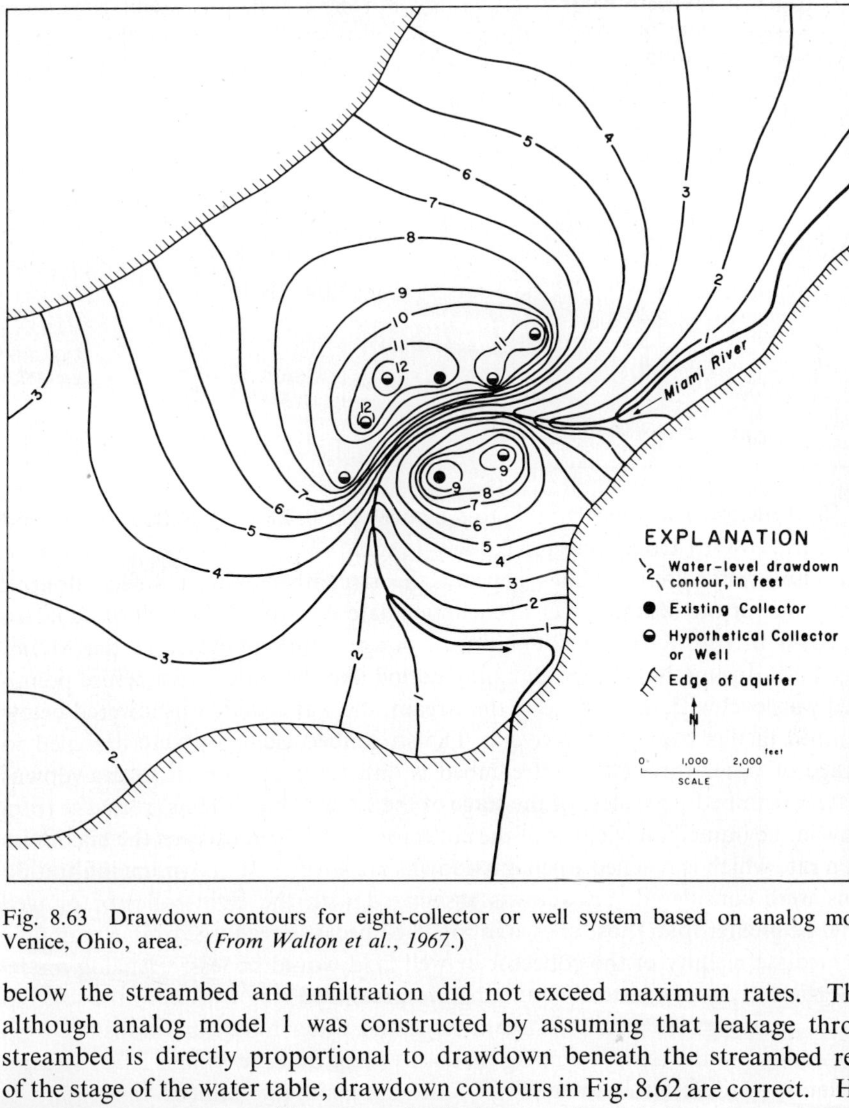

Fig. 8.63 Drawdown contours for eight-collector or well system based on analog model 1 for Venice, Ohio, area. (*From Walton et al., 1967.*)

below the streambed and infiltration did not exceed maximum rates. Therefore, although analog model 1 was constructed by assuming that leakage through the streambed is directly proportional to drawdown beneath the streambed regardless of the stage of the water table, drawdown contours in Fig. 8.62 are correct. However, drawdown contours would not be valid if the water-level declines beneath the streambed exceeded 4 feet.

Water-level declines in Fig. 8.62 are approximately those that were observed in the Venice, Ohio, area during the summer and late fall months of 1956. Except in short reaches of the Miami River where the stream was shallow and narrow, the water table in August 1956 was everywhere above the streambed.

Drawdown contours for a hypothetical eight-collector or well system are shown in Fig. 8.63. Assumed pumping rates for collectors or wells are given in Table 8.10. The drawdown contours are for a pumping period of 6 months. Although not shown

TABLE 8.10 ASSUMED PUMPING RATES FOR AN EIGHT-COLLECTOR OR WELL SYSTEM IN VENICE AREA, OHIO. (*From Walton et al., 1967.*)

Collector or well no.		*Pumping rate, mgd*
1		4.9
2		3.2
3		4.7
4		4.8
5		4.2
6		4.8
7		2.9
8		3.5
	Total	33.0

in Fig. 8.63, drawdown was measurable to distances of 50,000 and 44,000 feet up and down the buried valley from collector 2.

In this case, water-level declines beneath the streambed exceed 4 feet along a 3,200-foot reach of the Miami River in the immediate vicinity of the well or collector field as shown in Fig. 8.63. Assuming the average depth of water in the Miami River was 4 feet during the selected pumping period and the water table before pumping started was level with the surface of the stream, the water table was lowered below the streambed in this reach of the river. The streambed elements were designed so that leakage of water through the streambed is directly proportional to drawdowns beneath the streambed regardless of the stage of the water table. Thus, recharge from streamflow in the immediate vicinity of the collector or well field exceeds the maximum infiltration rate which is reached when drawdowns are 4 feet. If maximum infiltration conditions were considered, drawdowns associated with the eight-collector or well field would be greater than those shown in Fig. 8.63 because recharge from streamflow in the immediate vicinity of the collector or well field would be less. Analog model 1 with a vertical array of resistors simulating the streambed is not valid for the selected high rate of pumping and assumed stream-stage and water-table conditions.

Analog model 2 is the same as analog model 1 except that the Miami River streambed simulation is different. In analog model 2 maximum induced infiltration is taken into consideration by simulating the streambed with vertical arrays of resistors, transistors, and diodes instead of simple vertical arrays of resistors as in analog model 1. Analog model 2 and excitation-response apparatus are shown in Fig. 8.64.

The streambed elements were designed so that leakage through the streambed reaches a maximum rate when the water-level decline is 4 feet and thereafter remains constant. Therefore, analog model 2 is valid for any pumping rate whereas analog model 1 is valid only for pumping rates low enough so that the water table is not lowered below the streambed.

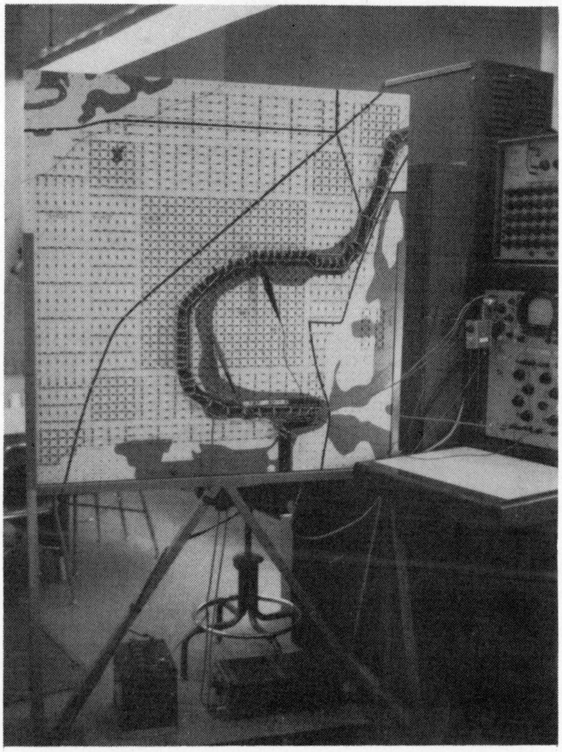

Fig. 8.64 Analog model 2 and excitation-response apparatus for Venice, Ohio, area. (*From Walton et al., 1967.*)

The elements simulating the streambed in analog model 2 were designed with the following conditions in mind: (1) the current through a streambed element must increase directly proportional to the associated aquifer element voltage drop until the voltage is 4.0 volts (drawdown beneath the streambed is 4.0 feet); (2) thereafter the current through a streambed element and the voltage drop across the element must remain constant; and (3) the aquifer elements must at all times be affected by current moving through the streambed elements, but the voltage of the aquifer elements must continuously drop after the current through the streambed element becomes constant.

Figure 8.65*a* shows aquifer elements at a streambed node connected to a vertical streambed resistor (R_G) which is in turn connected to ground and a diode connected to a negative 4.0 volts. As the voltage in the aquifer elements decreases to −4.0 volts, the current through the streambed resistor increases in direct proportion to the voltage drop and the leakage of water through the streambed is correctly simulated. When the voltage in the aquifer elements decreases below −4.0 volts, the diode becomes forward-biased (turned on). The low forward impedance of the diode prevents the voltage in the aquifer elements from going further negative. One node away from the streambed, the voltage of the aquifer elements is still decreasing. Therefore, the current from the streambed node to the adjacent aquifer node must be

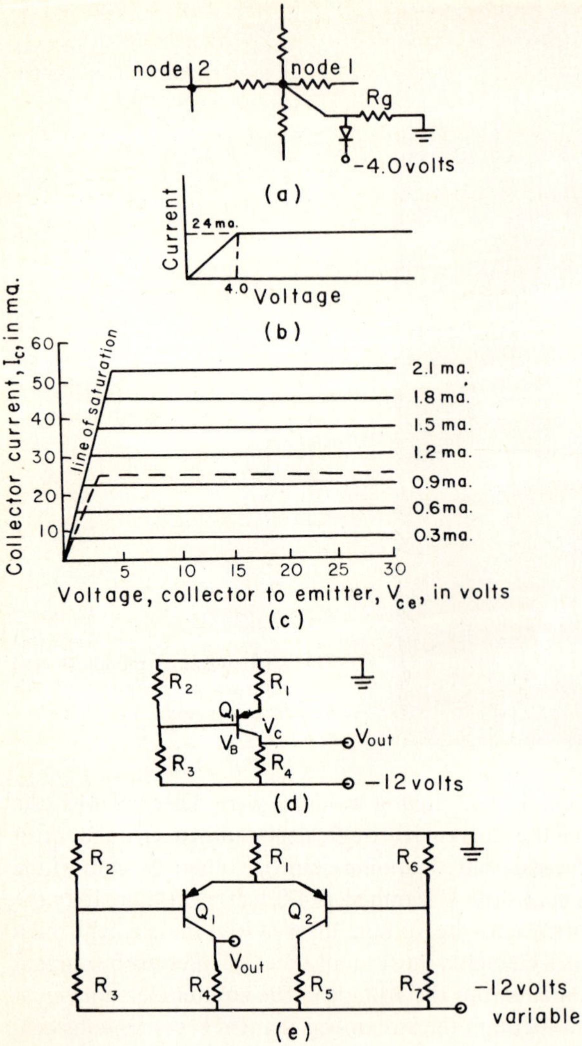

Fig. 8.65 Resistor-diode streambed element wiring diagram (*a*), current-voltage diagram for streambed element (*b*), graph showing transistor characteristics (*c*), streambed elements basic current sources wiring diagram (*d*), and complete streambed element wiring diagram (*e*) used in Venice, Ohio, area analog model. (*From Walton et al., 1967.*)

increasing. Thus, the current through the streambed element increases after the diode is turned on and the voltage at the streambed node remains constant. For this reason, the streambed element consisting of a resistor and a diode does not correctly simulate the leakage of water through the streambed. The streambed element must be a current source independent of the voltage of the aquifer elements after the voltage of the aquifer elements decreases below −0.4 volts.

Figure 8.65*b* shows the required electrical conditions at the streambed node with respect to the current through the streambed element resistor. Since the current is nonlinear, an active device (transistors have nonlinear characteristics) must be used.

With two transistors the streambed leakage can be simulated as a piecewise linear current in accordance with Fig. 8.65*b*. The slope of the current-vs.-voltage graph is controlled by the streambed element resistor; after the voltage decreases to -4.0 volts the slope of the current-vs.-voltage graph is ideally zero. The streambed element acts as either a simple resistive element or a constant current element, depending upon the state of the applied voltage.

The streambed element is a current diverter network changing with voltage. A constant-current source must be independent of the applied voltage. A transistor inherently has these characteristics by putting a constant-current source on the emitter (holding the base current constant) and allowing the transistor collector voltage to vary. Figure 8.65*c* is a typical transistor curve characteristic. The voltage V_{CE} is plotted as a function of transistor collector current I_c for different values of base current I_B. For practical purposes, the transistor collector current equals the emitter current. If the base current is held constant (I_b in Fig. 8.65*c*) and the transistor collector current I_c (which is the emitter current I_E) is also held constant at 25 ma, the base current would be about 1.1 ma. However, the values I_c and I_b would be independent of V_{CE} over the range indicated by the dashed horizontal line shown in Fig. 8.65*b*, thus verifying that a constant-current source can be connected to the emitter while the transistor collector voltage is allowed to vary.

The basic current source for the streambed element is shown in Fig. 8.65*d* with R_1 controlling the magnitude of the current available at V_{out}. By adjusting the ratio of R_2 to R_3, V_C (or V_{out}) can be made 4.0 volts negative with Q_1 in saturation. V_{out} is the output of the streambed element and, therefore, is a function of the analog model. Any resistors in the streambed element except the current source must be 10 times larger than the analog model impedances. For V_{out} less than 4.0 volts, Q_1 is in saturation (symbolized V_{CES}) which results in a low impedance (about 10 ohms) between the transistor collector and emitter (see Fig. 8.65*c*, line labeled "line of saturation"). The streambed element then acts as a simple resistor, its value being R_1. For V_{out} greater than 4.0 volts, the transistor comes out of saturation and the emitter resistor R_1 becomes a constant-current source.

The basic current source of Fig. 8.65*d* must be modified to allow for the analog model impedance as seen from the streambed aquifer node. Since the coefficient of transmissibility is known at this point, the impedance is also known. Figure 8.65 shows the modified current source which is compatible with the analog model impedance. The additional transistor Q_2 acts as a current diverter when V_{out} is less than 4.0 volts. The diverter maintains the current through R_1 until Q_2 reaches its threshold for turn-on at 4.0 volts. To maintain a balanced current diversion, R_5 must be the analog model impedance at the streambed node. For design purposes, R_5 must be estimated and later verified when it is connected to the analog model. The dummy load R_5 resistor value is critical because it determines how sensitive the streambed element is to the threshold voltage, 4.0 volts. For a perfect representation of the piecewise linear simulation in Fig. 8.65*b* the breakpoint (threshold) must be discontinuous.

Streambed elements were assembled and attached to the streambed-aquifer nodes of the analog model. The excitation-response apparatus was coupled to the analog model and the power amplifiers were connected to junctions at the locations of collectors or wells. The excitation-response apparatus was turned on and the performance of streambed elements was checked. It was found that due to the low impedances between streambed elements, an offset voltage with equilibrium conditions was developed along the streambed. A diode was added to each streambed element so that the transistor bias current would not leak into the stream. Each streambed element by itself is not capable of affecting the voltage distribution within the analog model, but 50 streambed elements caused appreciable voltage drops with equilibrium conditions. After the diodes were added to the streambed elements, the performance (voltage versus current at connection to analog model) of each streambed element was measured and compared with the idealized piecewise linear streambed element described earlier. The streambed elements functioned properly and correctly simulated leakage through the streambed.

In the initial design of the streambed elements, it was necessary to estimate the impedance at streambed element nodes so that the value of the load-balancing resistor (R_5) could be determined. The impedance at streambed nodes of completed analog model was measured and the value of the associated load-balancing resistor (R_5) in the streambed element was defined. The impedance was calculated by plotting voltage (relative to ground) versus current into the streambed element node from the streambed element. The slope of the graph is the impedance. The value of R_5 based on completed analog measurements was within 30 percent of the original estimated value of R_5 (50 ohms). The value of R_5 was increased to 250 ohms to reduce heat dissipation in the transistor. Degradation in system performance from this resistor change was found to be negligible.

Because the analog model used pulses at high frequencies, the ground connection between the excitation-response apparatus and the model is very important. Time-voltage curves are degraded if voltage filtering and grounding are improper. By properly grounding the analog computer, the effects of electrical noise radiating from normal laboratory environments (fluorescent lights, electric motors, etc.) were practically eliminated.

Analog model 2 was coupled to the excitation-response apparatus and the power amplifiers were connected to junctions at the locations of the eight hypothetical collectors or wells shown in Fig. 8.63. The output of the pulse generator was adjusted in accordance with the assumed pumping rates in Table 8.10 and a pumping period of 6 months.

Drawdown contours for the eight-collector or well system are shown in Fig. 8.66. Although not shown in Fig. 8.66, drawdown was measurable to distances of 50,000 and 44,000 feet up and down the buried valley from collector 2. A comparison of Figs. 8.63 and 8.66 shows that much greater drawdowns occur when maximum infiltration conditions are taken into account than when maximum infiltration conditions are ignored. With the streambed correctly simulated, water-level declines

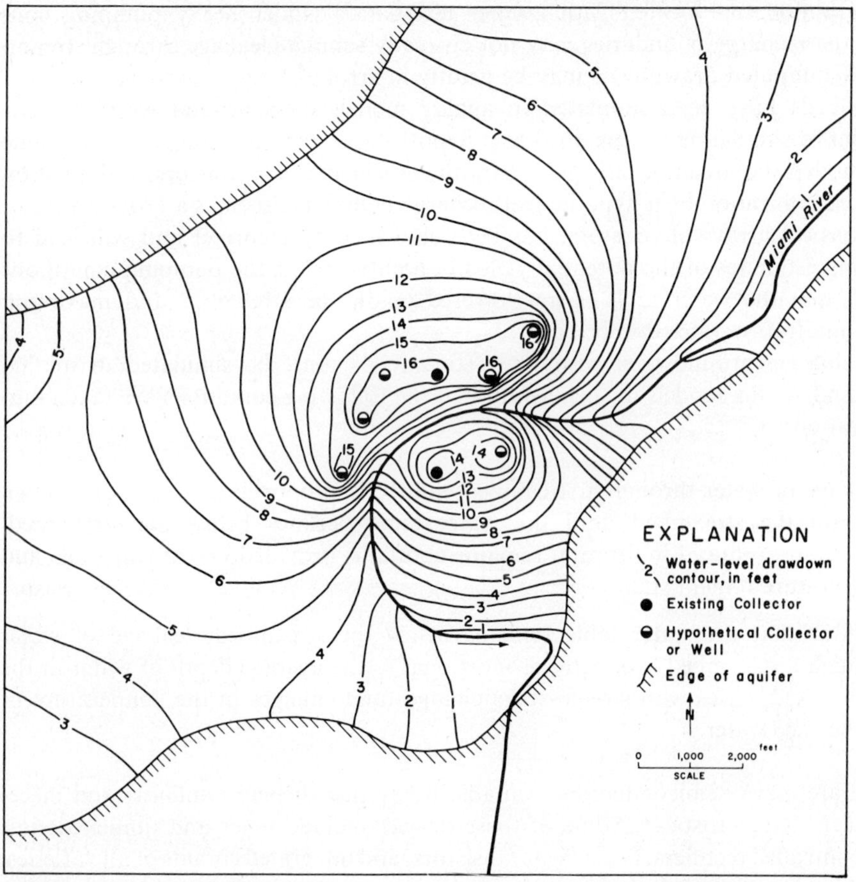

Fig. 8.66 Drawdown contours for eight-collector or well system based on analog model 2 for Venice, Ohio, area. (*From Walton et al., 1967.*)

beneath the streambed exceed 4 feet along a 5,500-foot reach of the Miami River as shown in Fig. 8.66; whereas, with the streambed incorrectly simulated, water-level declines beneath the streambed exceed 4 feet along only a 3,200-foot reach of the Miami River (see Fig. 8.63). With the streambed correctly simulated, water-level declines near the collectors or wells north and south of the river average about 16 and 14 feet, respectively, as shown in Fig. 8.66; whereas, with the streambed incorrectly simulated, water-level declines near the collectors or wells north and south of the river average about 11 and 9 feet, respectively (see Fig. 8.66). The importance of correctly simulating the streambed is apparent.

In the past, little if any attention has been given to the complex nature of recharge from induced streambed infiltration under varying groundwater level and stream-stage conditions. Streambeds are often simulated in mathematical models by hypothetical recharge boundaries located with aquifer test data. Under heavy pumping conditions, the recharge boundaries may not correctly simulate leakage through streambeds and computed drawdowns may be greatly in error.

Streambeds have been simulated in analog models by a ground wire which is equivalent to simulating a streambed as a hypothetical recharge boundary. In some analog models the streambed is simulated with vertical arrays of resistors. Simulation of induced infiltration by a hypothetical recharge boundary, such as a ground wire or simple vertical arrays of resistors, has been shown to be incorrect and will lead to misleading estimates of the potential yields of aquifer unless the pumping conditions are such that the water table is not lowered below the streambed and maximum induced infiltration rates do not occur.

In evaluating groundwater resources, streambeds must be simulated in mathematical and analog models in such a way that the following conditions are taken into consideration:

1. Leakage of water through a streambed is directly proportional to the drawdown beneath the streambed until the water table declines below the streambed. Thereafter, induced infiltration remains constant, provided the stream stage and temperature remain stationary.

2. Provided that the water table remains below the streambed, leakage of water through a streambed is directly proportional to the average depth of water in the stream and varies with stream-stage changes and changes in the temperature of the surface water.

There are many semiconductors available today that display nonlinear and piecewise linear characteristics. Some of these devices include zener and tunnel diodes, silicon controlled rectifiers, field effect transistors, and integrated circuits of all varieties. The application of these devices to hydrogeologic systems analysis using electric analog computers has unlimited research possibilities. In addition, excitation-response apparatus could be improved with use of recent equipment technological advances.

Problems

8.1 An aquifer has the following average hydrogeologic properties: $T = 86{,}000$ gpd/ft and $S = 0.20$. The aquifer is an infinite strip of sand and gravel bounded on one side by a recharge boundary and on the other side by a parallel barrier boundary. Groundwater is withdrawn from the four-well system located as shown in diagram. The fully penetrating production wells are 18 inches in diameter. Available drawdown is 19 feet. The time required for water levels

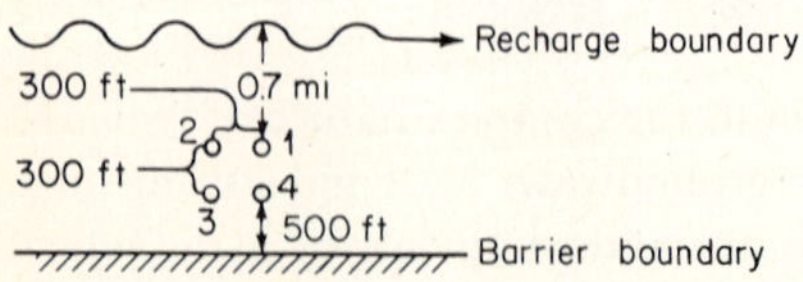

Fig. P8.1

to stabilize is 3 years. The original saturated thickness of the aquifer is 34 feet. Well loss is negligible. Estimate the practical sustained yield of the four-well system.
Answer: 800 gpm.

8.2 A wedge-shaped aquifer (see following diagram) has the following average hydrogeologic properties: $T = 547{,}000$ gpd/ft and $S = 0.20$. Each of five fully penetrating production wells

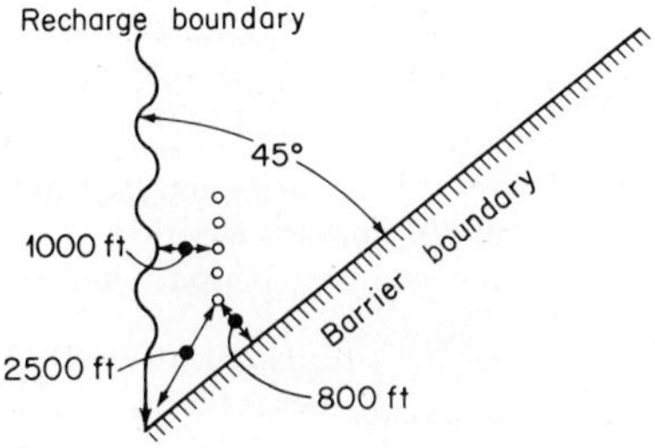

Fig. P8.2

30 inches in diameter and spaced 600 feet apart as shown in the diagram above are pumped continuously at a rate of 4 mgd or 2,780 gpm (a 20-mgd supply). The time required for water levels to stabilize is 365 days. The effects of well loss and dewatering are negligible. Compute the drawdown in the center well of the five-well system.
Answer: 15 ft.

8.3 Consider the following model aquifer.

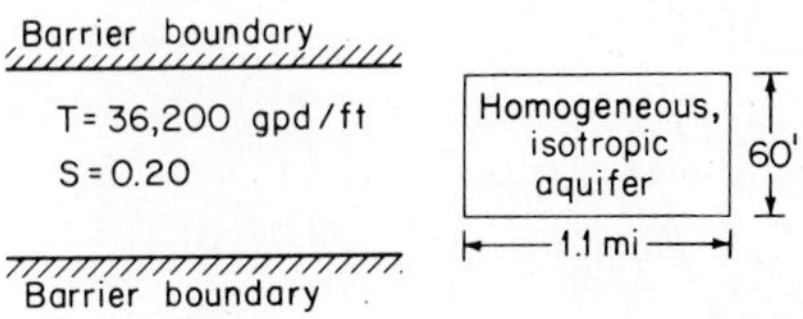

Fig. 8.3(A)

A production well located in the center of the aquifer is pumped continuously at a rate of 772 gpm for a period of 180 days. Determine the drawdown distribution at the end of the pumping period. Do not consider the effects of dewatering.
Answer:

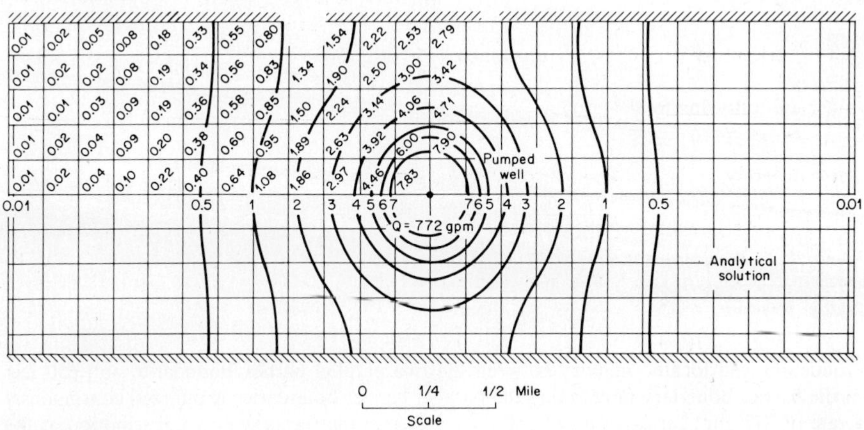

Fig. P8.3(B)

8.4 Consider the following model aquifer.

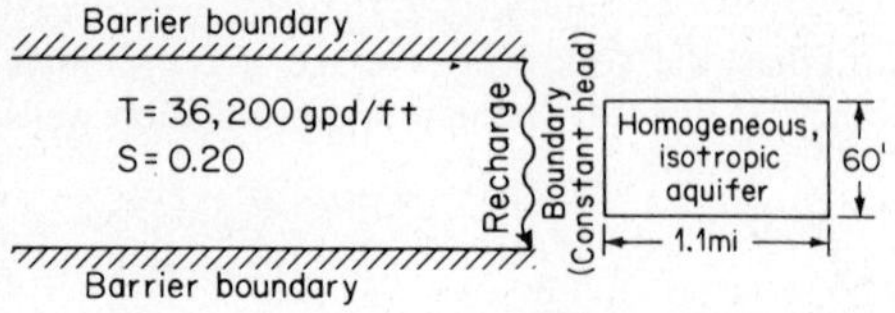

Fig. P8.4(A)

A production well located midway between the two parallel barrier boundaries and 0.25 mile from the recharge boundary is pumped continuously at a rate of 772 gpm for a period of 180 days. Determine the drawdown distribution at the end of the pumping period. Do not consider the effects of dewatering.

Answer:

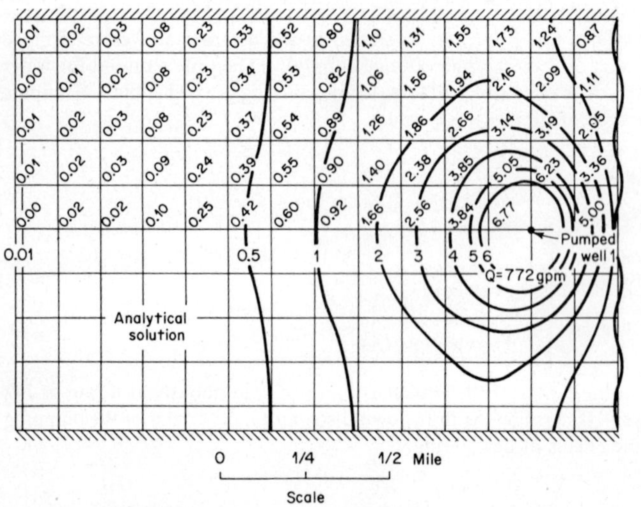

Fig. P8.4(B)

8.5 Consider the following model aquifer.

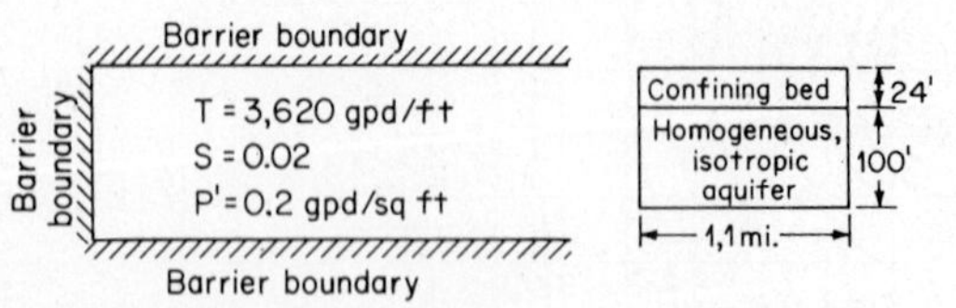

Fig. 8.5(A)

A production well located midway between the two parallel barrier boundaries and 650 feet from the barrier boundary intersecting the parallel barrier boundaries is pumped continuously at a rate of 772 gpm for a period of 180 days. Determine the drawdown distribution at the end of the pumping period. Do not consider the effects of dewatering.

Answer:

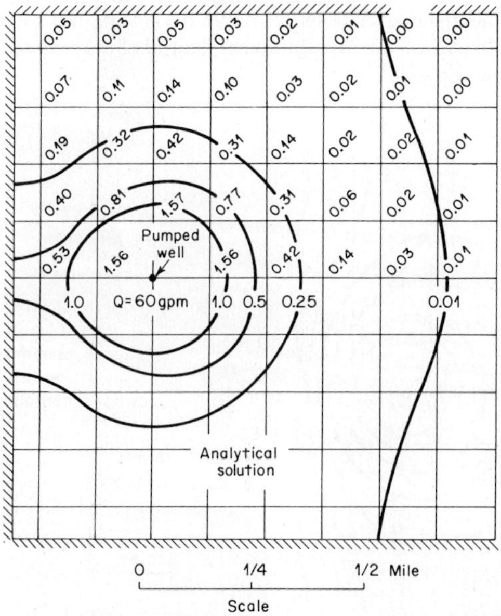

Fig. P8.5(B)

8.6 Consider the model aquifer and pumping conditions in Problem 8.3. Let $a_g = 660$ feet, $K_1 = 6.52 \times 10^{13}$ gal/coulomb, $K_2 = 1$ ft/volt, $K_3 = 3.62 \times 10^{11}$ gal/day amp, and $K_4 = 1.8 \times 10^3$ days/sec. Construct an analog model for the given conditions. Couple the analog model to excitation-response apparatus. Determine the drawdown distribution at the end of the pumping period. Do not consider the effects of dewatering.
Answer:

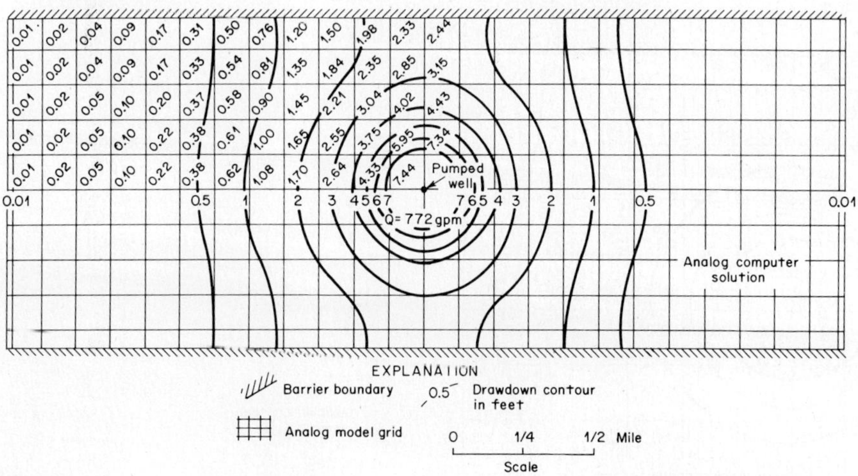

Fig. P8.6

8.7 Consider the model aquifer and pumping conditions in Problem 8.4. Let a_g = 660 feet, K_1 = 6.52 × 10^{13} gal/coulomb, K_2 = 1 ft/volt, K_3 = 3.62 × 10^{10} gal/day amp, and K_4 = 1.8 × 10^3 days/sec. Construct an analog model for the given conditions. Couple the analog model to excitation-response apparatus. Determine the drawdown distribution at the end of the pumping period. Do not consider the effects of dewatering.
Answer:

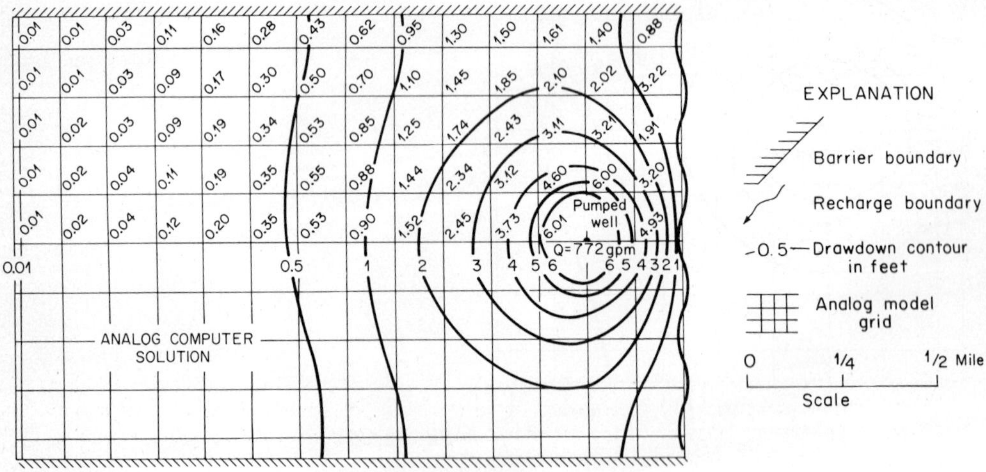

Fig. P8.7

8.8 Consider the model aquifer and pumping conditions in Problem 8.5. Let a_g = 660 feet, K_1 = 6.52 × 10^{13} gal/coulomb, K_2 = 1 ft/volt, K_3 = 3.62 × 10^{10} gal/day amp, and K_4 = 1.8 × 10^3 days/sec. Construct an analog model for the given conditions. Couple the analog model to excitation-response apparatus. Determine the drawdown distribution at the end of the pumping period. Do not consider the effects of dewatering.
Answer:

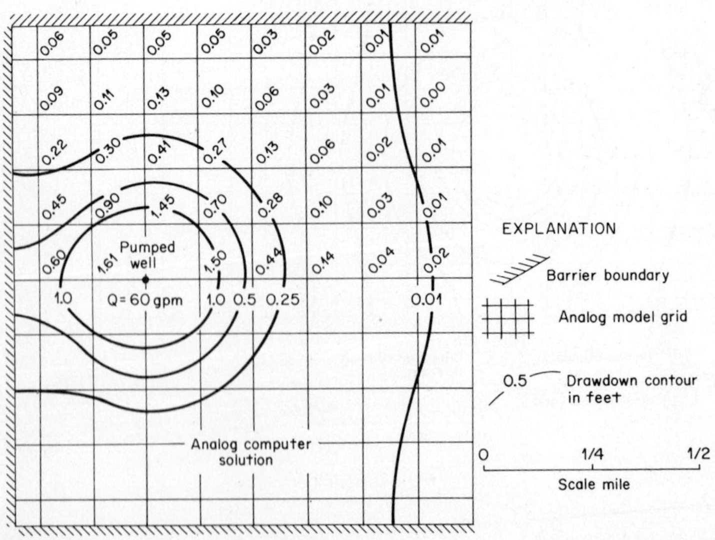

Fig. P8.8

8.9 Compare the results of Problems 8.3 to 8.8 to determine the accuracy of the analog computers.

8.10 Consider the following model aquifer.

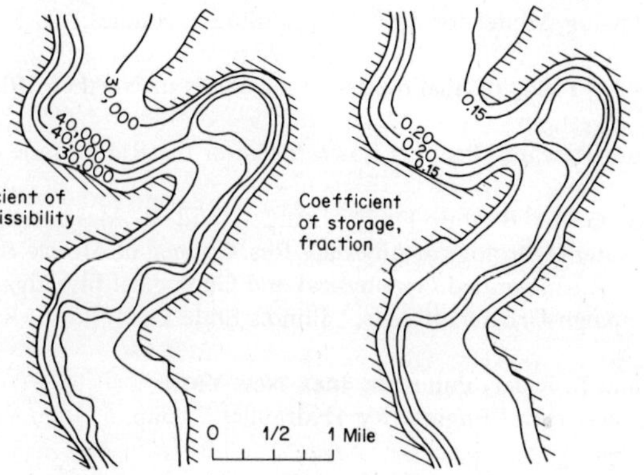

Fig. P8.10(A)

The production well is pumped continuously at a rate of 772 gpm for a period of 180 days. Construct an analog model for the given conditions. Couple the analog model to excitation-response apparatus. Determine the drawdown distribution at the end of the pumping period. Do not consider the effects of dewatering.

Answer:

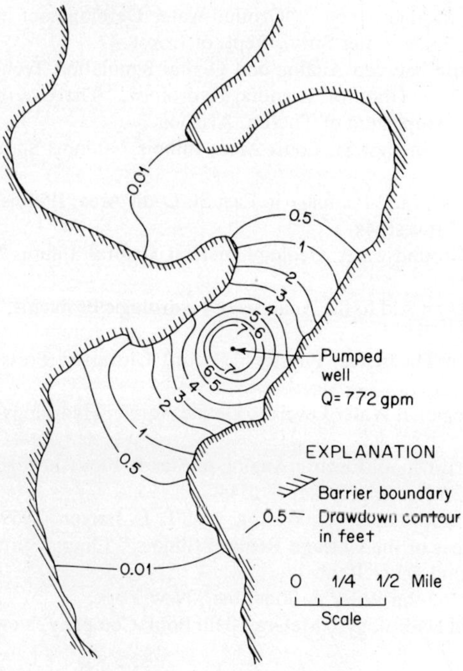

Fig. P8.10(B)

References

Bergstrom, R. E., and T. R. Walker, 1956. "Ground-water Geology of the East St. Louis Area, Illinois," Illinois State Geol. Surv., Rept. of Invest. 191.

Bermes, B. J., 1960. "An Electric Analog Model for Use in Quantitative Studies," U.S. Geol. Surv., mimeographed report.

Csallany, S., and W. C. Walton, 1963. "Yields of Shallow Dolomite Wells in Northern Illinois," Illinois State Water Surv., Rept. of Invest. 46.

FAO/SF: 5/ELS, 1964. "Ground-water Research Project: Lower Basin of the Rio Grande de San Miguel, El Salvador," vols. 1–5.

Farvolden, R. N., W. A. Meneley, E. G. Le Breton, D. D. Lennox, and P. Meyboom, 1963. "Early Contributions to the Groundwater Hydrology of Alberta," Res. Council of Alberta Bull. 12.

Foster, J. W., and M. B. Buhle, 1951. An Integrated Geophysical and Geological Investigation of Aquifers in Glacial Drift near Champaign-Urbana, Illinois," Illinois State Geol. Surv., Rept. of Invest. 155.

Gottlieb, I., 1958. "Basic Pulses," John F. Rider, Publisher, Inc., New York.

Jacob, C. E., 1950. Flow of Ground Water, in "Engineering Hydraulics," chap. 5, John Wiley & Sons, Inc., New York.

Karplus, W. J., 1958. "Analog Simulation," McGraw-Hill Book Company, New York.

Korn, G. A., and T. M. Korn, 1952. "Electronic Analog Computers," McGraw-Hill Book Company, New York.

Millman, J., and S. Seely, 1941. "Electronics," McGraw-Hill Book Company, New York.

Nace, R. H., et al., 1959. "Geography, Geology and Water Resources of the National Reactor Testing Station, Idaho," pt. 3, Hydrology and Water Resources, U.S. Geol. Surv. Admin. Rept.

Prickett, T. A., 1967. Designing Pumped Well Characteristics into Electric Analog Models, *Ground Water*, vol. 5, no. 4.

———, L. R. Hoover, W. H. Baker, and R. T. Sasman, 1964. "Ground-water Development in Several Areas of Northeastern Illinois," Illinois State Water Surv., Rept. of Invest. 47.

———, and C. G. Lonnquist, 1968. "Comparison between Analog and Digital Simulation Techniques for Aquifer Evaluation," International Association of Scientific Hydrology, "The Use of Analog and Digital Computers in Hydrology," Symposium of Tucson, Arizona.

Schicht, R. J., 1965. "Ground-water Development in East St. Louis Area, Illinois," Illinois State Water Surv., Rept. of Invest. 51.

——— and E. G. Jones, 1962. "Ground-water Levels and Pumpage in East St. Louis Area, Illinois, 1890–1961," Illinois State Water Surv., Rept. of Invest. 44.

Selkregg, L. F., and J. P. Kempton, 1958. "Ground-water Geology in East Central Illinois," Illinois State Geol. Surv. Circ. 248.

Skibitzke, H. E., 1961. "Electronic Computers As an Aid to the Analysis of Hydrologic Problems," Intern. Assoc. Sci. Hydrology Publ. 52.

Southwell, R. V., 1948. "Relaxation Methods in Theoretical Physics," Oxford University Press, London.

Stallman, R. W., 1956. Numerical Analysis of Regional Water Levels to Define Aquifer Hydrology, *Trans. Am. Geophys. Union*, vol. 37, no. 4.

———, 1963. "Calculation of Resistance and Error in an Electric Analog of Steady Flow through Nonhomogeneous Aquifers," U.S. Geol. Surv. Water Supply Paper 1544G.

Suter, M., R. E. Bergstrom, H. F. Smith, G. H. Emrich, W. C. Walton, and T. E. Larson, 1959. "Preliminary Report on Ground-water Resources of the Chicago Region, Illinois," Illinois State Water Surv. and Geol. Surv. Cooperative Ground-water Rept. 1.

Todd, D. K., 1959. "Ground Water Hydrology," John Wiley & Sons, Inc., New York.

Ven Te Chow (ed.), 1964. "Handbook of Applied Hydrology," McGraw-Hill Book Company, New York.

Walker, W. H., and W. C. Walton, 1961. "Ground-water Development in Three Areas of Central Illinois," Illinois State Water Surv., Rept. of Invest. 41.

Walton, W. C., 1960. "Leaky Artesian Aquifer Conditions in Illinois," Illinois State Water Surv., Rept. of Invest. 39.

———, 1962. "Ground-water Resources of Camas Prairie, Camas and Elmore Counties, Idaho," U.S. Geol. Surv. Water Supply Paper 1609.

———, 1964a. Electric Analog Computers and Hydrogeologic System Analysis in Illinois, *Ground Water*, vol. 2, no. 4.

———, 1964b. "Ground-water Conditions at Bracebridge, Ontario," typewritten report prepared for James F. MacLaren Limited, Toronto, Ontario, Canada.

———, 1964c. Future Water-level Declines in Deep Sandstone Wells in Chicago Region, *Ground Water*, vol. 2, no. 1.

———, 1964d. Electric Analog Computers and Hydrogeologic System Analysis in Illinois, *Ground Water*, vol. 2, no. 4.

———, 1965a. "Potential Yield of a Sand and Gravel Aquifer in a Buried Valley near Estevan, Saskatchewan—Results of Aquifer and Well-production Tests and Evaluation of Ground-water Potential," typewritten report prepared for Saskatchewan Res. Council.

———, 1965b. "Feasibility of Scheme of Ground-water Development in Lower Basin of Rio Grande de San Miguel, El Salvador," typewritten report prepared for Harza Engineering Co. International, Chicago.

———, 1966. "Pre-feasibility Report on Chicago Tunnel Drainage Project," typewritten report prepared for Harza Engineering Co. International, Chicago.

——— and E. A. Ackroyd, 1966. "Effects of Induced Streambed Infiltration on Water Levels in Wells during Aquifer Tests," Minn. Water Resources Res. Center Bull. 2.

——— and S. Csallany, 1962. "Yields of Deep Sandstone Wells in Northern Illinois," Illinois State Water Surv., Rept. of Invest. 43.

——— and J. C. Neill, 1960. "Analyzing Groundwater Problems with Mathematical Models and a Digital Computer," Intern. Assoc. Sci. Hydrology Publ. 52.

——— and T. A. Prickett, 1963. Hydrogeologic Electric Analog Computers, *J. Hydraulics Div., Am. Soc. Civil Engrs.*, vol. 89, no. HY6.

——— and E. J. Schaefer, 1956a. "Report on New Ground Water Supply in the Eagle City Area for the City of Springfield, Ohio," typewritten report prepared for Black and Veatch, Consulting Engineers, Kansas City, Missouri.

——— and ———, 1956b. "Report to the Southwestern Ohio Water Co. on Evaluation of the Potential Yield of the Aquifer in the Big Bend Area, Ohio," mimeographed rept.

——— and W. H. Walker, 1961. Evaluation Wells and Aquifers by Analytical Methods, *J. Geophys. Res.*, vol. 66, no. 10.

———, D. L. Hills, and G. M. Grundeen, 1967. "Recharge from Induced Streambed Infiltration under Varying Ground-water Level and Stream-stage Conditions," Minn. Water Resources Res. Center Bull. 6.

Wass, C. A. A., 1955. "Introduction to Electronic Analog Computers," McGraw-Hill Book Company, New York.

CHAPTER 9

DEVELOPMENT AND MANAGEMENT OF AQUIFERS

Scientific analysis of available information for many aquifers in the conterminous United States suggests that on a gross basis considerably more development can be made of the groundwater supplies without exhausting the supplies or ruining them for others. Only a small proportion of the nation's important groundwater reservoirs can be considered "overdeveloped," and the rest of the country still offers vast opportunities for development. However, problems associated with the erratic distribution of groundwater resources and development in space and time do exist in most parts of the nation. Past development of some aquifers has been in ignorance of their potentialities and especially their limitations. Realization of the limits of surface supplies came early, particularly in the arid states, where deficient precipitation sets a low ceiling on all water supplies. The limitations in groundwater supplies in those regions became apparent much later and have just began to be apparent in humid regions.

Groundwater is a critical resource in much of the semiarid or more arid parts of the country. It may set the limit on the use that man can make of the other resources, and thus ultimately on the density of population and the standards of living that can be sustained in those regions. In the regions of abundant rainfall, however, water is not generally the first of the natural resources to falter. Nature generally provides a surplus of water in those regions where water flows to the sea, and this surplus can be put to many uses. Even here local shortages of water may develop, because of the failure to pattern the development and management in accordance with the availability of groundwater.

The arid regions have groundwater reservoirs in almost every stage of development, from those practically unused and capable of considerable development to those capable of very little sustained perennial use and seriously overdeveloped. Many reservoirs are in areas where the land and other resources are eminently suitable for the enterprises of civilization, and those enterprises are limited only by the amount of water that can be made available perennially for their use. Eventually, every groundwater reservoir in the arid regions will doubtless be scrutinized to determine its capabilities in this respect, although attention to date has been concentrated on those areas where needs have been most urgent.

The groundwater reservoirs in humid regions having adequate replenishment are by no means devoid of groundwater problems. As the number of producing wells has increased, there has inevitably been a lowering of the water table or artesian pressure in many areas, with resulting decline or cessation of yield from production wells, and increased cost of obtaining water by pumping. Climatic cycles have caused concern in several areas because the subnormal replenishment in a series of dry years has caused declining water tables and artesian pressures and suggested the possibility of overdevelopment. Pumping from closely spaced production wells has caused significant declines of water levels in parts of nearly every state, chiefly in municipal or industrial areas that use large quantities of groundwater. Water levels have reached approximate equilibrium in some of those areas, indicating that the pumped water is now being replaced by water transmitted through aquifers. In other areas the water levels are still declining each year.

Most of the analysis of groundwater problems to date has been restricted to the groundwater phase of the hydrologic cycle, and recommendations for optimum utilization have been based largely on the rate of natural replenishment. Few groundwater reservoirs are yet utilized to store flood streamflows for later use, and they are generally not even considered in river-basin flood-control and storage programs. Yet many have capacity far greater than the largest artificial surface reservoirs. They can provide holdover storage with minimum loss of water by evaporation and with minimum loss of productive land. The utility of the existing groundwater storage capability may be enhanced, and the water supply required for the support of a rapidly growing economy may be provided, by recognition of the possibilities inherent in the coordinate, or conjunctive, use of both surface and groundwater storage capacity for the maximum conservation and regulation of the available water supply. In order to obtain increased supplies without excessive pumping lifts, the natural recharge to aquifers has been increased in several places by the use of spreading areas, ponds, and recharge wells.

A major factor in the successful production of groundwater for use is the "know-how" of the technicians who are primarily responsible for the development and maintenance of production. In the past two decades there have been great advances in the technical knowledge, procedures, and equipment used to get water out of the ground.

Optimum development and management are achieved in most cases when the average replenishment is balanced by the average discharge and when the discharge for nonbeneficial purposes has been reduced to minimum. In most individual years, however, such a balance will not be achieved. When recharge is large, the precipitation and streamflow are likely to be above normal, so that the need for groundwater may be less; and the need for groundwater may be greatest in dry years, as a supplement to low precipitation and streamflow. As with surface reservoirs, full use of groundwater reservoirs cannot be achieved by attempting to hold the storage at a constant level. Maximum use is achieved if some of the inflow of wet years is held in storage until dry years when there is need for it.

A significant management effort lies in providing for better natural and artificial recharge of important groundwater reservoirs. This can only be accomplished by a comprehensive understanding of the hydrogeology. Another management effort lies in importing or supplementing by other means water for recharging groundwater reservoirs.

Many activities unrelated to pumping of groundwater have modified the storage of water below the land surface. Drainage projects and irrigation projects have proved that it is possible to manipulate the storage in groundwater reservoirs. Unfortunately, groundwater storage has been increased by irrigation in some places until good agricultural lands have been waterlogged and abandoned; it has been decreased in other localities by drainage to the detriment of agricultural use of the land or municipal use of the water. Man has changed the quantity of water stored in aquifers by his structures for storage of surface water or for protection against floods, and by improving channels for navigation. He has damaged some water supplies by discharging contaminated water into the ground or into streams from which it enters groundwater reservoirs, and also by puncturing protective layers, thus permitting entry of sea water or other mineralized water into aquifers. Urban areas cover a relatively small part of the total area of the country, but they include many of the areas where man has changed the land surface most profoundly. The topography has been modified by extensive cutting and filling, and natural drainage patterns have been revised materially. So far as the hydrologic cycle is concerned, some of the most significant changes have resulted from construction of buildings, pavements, and other impermeable surfaces upon the land. These have prevented infiltration of precipitation into the ground in extensive areas.

To maintain groundwater resources indefinitely, recharge must balance discharge. Economic, legal, and quality aspects must be considered. The cost of pumping groundwater may become excessive. Withdrawals may produce a groundwater of inferior quality. Legal considerations may limit safe yield if there is interference with prior water rights.

The practical sustained yield is the amount of water which can be withdrawn annually without producing undesirable effects. The practical sustained yield may be limited to an amount less than recharge but cannot exceed the long-term mean annual recharge. The practical sustained yield can be limited either by the physical size of the groundwater reservoir or by the rate at which water moves through the reservoir from the recharge area to the withdrawal area. Withdrawals exceeding this amount must come from storage within the aquifer. Storage depletion is often referred to as mining of groundwater. In most groundwater reservoirs, the quantity of storage is many times the annual recharge or withdrawal; therefore in any one year the withdrawals can exceed recharge without causing permanent depletion. But on a long-term basis, when series of wet and dry years tend to average out, the withdrawals become on overdraft if the mean recharge is exceeded.

It is seldom that any single value of practical sustained yield can be correct for an extended time, in part due to changing economic conditions. Any determination of practical sustained yield is based upon specified conditions, either existing or assumed, and any changes in these conditions will change the yield. This fact applies to the degree and pattern of groundwater development. Lowering groundwater levels may increase water supplies by increasing recharge and decreasing natural discharge. Rearranging pumping patterns and changes in vegetation and crops may affect recharge. Urbanization of an area may reduce recharge. Changes in the use of groundwater, such as from irrigation to municipal or industrial use, can change the practical sustained yield.

There is a real problem in development and management in arid regions: Should the water be extracted for maximum benefit of the present generation as minerals and other nonrenewable resources are mined or should the pumping be limited to the negligible quantity that can be supplied perennially? Evaluation of mining yield is primarily a matter of determining the amount of extractable, but nonrenewable, water in a groundwater reservoir. The maximum mining yield is the total quantity of nonreplenishable water stored in the groundwater reservoir. The permissive mining yield is that part of the maximum mining yield that is of economic quality, within economic pumping lifts, obtainable through economic pumping patterns, and legally available.

9.1 Groundwater Development Problems

Groundwater is generally a renewable resource. Precipitation periodically replenishes aquifers which supply wells and springs. However, the natural supply of groundwater is usually limited as to time and place. In recent years our use of this limited supply has grown enormously. Because of this tremendous growth in the use of groundwater, many startling news reports have appeared in the press and on the radio about the groundwater situation in certain cities and other areas. In some parts of the nation the water-level trend in most wells has been downward for several years. These areas are sufficiently numerous and widely scattered over the nation that they have given rise to statements that our groundwater resources are being depleted quite generally and may be exhausted within the lifetime of the present generation. However, in many localities groundwater supplies are as abundant as in the past. The popular notion is that there has been a general and progressive decline of water tables. There are enough records to show also that this condition is not universal; water tables in many areas are higher today than in past decades and in some instances higher than at any time in history (Thomas, 1951).

It is true that in many areas water is pumped from production wells faster than it is replaced by nature. Such overdevelopment affects the considerable part of our population dependent upon groundwater for agricultural, industrial, and municipal uses, but the aggregate of these areas is only a few percent of the nation. In a much larger proportion of the nation the water levels are not declining progressively, and in some areas groundwater storage has increased appreciably in past decades. Throughout many parts of the country there has been little development of groundwater. There are extensive areas where adequate supplies of good-quality groundwater cannot be obtained from wells and large areas where very little is known about the groundwater possibilities. These are areas where large, undeveloped supplies of groundwater are known to be available.

Historical development of many groundwater reservoirs has resulted in conflict between those pumping from groundwater and those diverting from surface-water supplies. Wherever groundwater discharges into or is recharged from a stream, the utilization of groundwater from the basin is likely to modify the natural regimen of streamflow. Utilization of groundwater storage capacity has been confined principally to haphazard local development and individual or community use of the water thus made available.

The difficulties created by pumping from production wells are of several types. Some problems pertain to entire groundwater reservoirs where the rate of replenishment is inadequate to meet the continuing demand. Other problems pertain to the inability of groundwater to move rapidly enough through aquifers to supply the demand. The most serious problem of groundwater shortage occurs in areas where water is pumped out faster than the entire groundwater reservoir is replenished. Under these conditions the reservoir is being emptied of water that may have taken decades or centuries to accumulate and there is no possibility of a continuous supply unless present conditions are changed. Many groundwater reservoirs in arid regions are not replenished in dry years. In the driest basins, the groundwater reservoir may receive no recharge in a year of normal precipitation because all the moisture is dissipated from the soil zone into the atmosphere. Recharge to these reservoirs occurs only during years of exceptionally abundant rainfall, which may mean only 1 in 5 or 10 years or longer. Nearly all these excessively pumped reservoirs are in arid regions. Generally the groundwater users are aware that they are using more than the perennial supply and that the supply will be exhausted unless corrective action is taken. Corrective measures already applied in certain areas include: prevention of waste, pro-rata reduction of pumping from production wells, prohibition of further development, reclaiming of used water, artificial groundwater replenishment by surplus stream water, and importation of water from other areas.

For many groundwater reservoirs, it has been possible to reduce the disparity between draft and natural replenishment by increasing the replenishment artificially, which is a more desirable solution to the problem than the alternative of restricting the withdrawal of groundwater. Several of California's coastal valleys are traversed by streams responsible for much of the groundwater recharge but which also discharge considerable quantities of water into the ocean.

Since 1934, the Santa Clara Valley Conservation District has constructed storage dams and percolating works to divert surplus stream water into the groundwater reservoir. Similarly, spreading basins along the Santa Clara River in southern California have put nearly 10,000 acre-feet of water in aquifers beneath the Oxnard Plain, thus reducing the overdraft. The Santa Cruz Valley in Arizona, Antelope Valley in California, and Mimbres Valley in New Mexico are typical of areas with appreciable groundwater replenishment but where current pumping is so great that emptying of the reservoir is proceeding apace. In many groundwater reservoirs of the arid regions the average annual recharge is adequate for present uses. Indeed the number of such reservoirs is far greater and their total areal extent is larger than the reservoirs in which current draft from production wells now exceeds the estimated average rate of replenishment.

Several groundwater reservoirs contain appreciable amounts of water in storage but are prevented from receiving more than negligible recharge by enclosing or overlying materials. It is inevitable that the reservoirs will be depleted when production wells pump from them. Such conditions, though rare, occur in humid as well as arid regions.

There are problems associated with the slow movement of water within an aquifer, rather than of replenishment of the aquifer. All developed aquifers serve not only as reservoirs to hold water but also as pipelines to carry water to the production wells from the areas where it enters the aquifer. The perennial yield of a production well or group of wells is determined by the quantity of water that can move through the aquifer from the recharge areas. If the transmissivity of the aquifer is inadequate, the water levels in production wells will decline whether the aquifer as a whole is adequately recharged or not.

These problems most commonly afflict cities and industrial areas that use large quantities of groundwater, whether in humid or arid climate. Problems of inadequate transmissivity of aquifers have developed in every part of the country. The greatest recorded declines of water levels have been in areas of concentrated draft from artesian aquifers. One of the outstanding areas of concentrated draft from artesian aquifers is that around Houston and Galveston, Texas, where total pumpage from an artesian groundwater reservoir exceeds 300 mgd. Corrective measures already taken include a wider spacing of pumping centers, moving production wells closer to recharge areas, artificial recharge practices, and reduction of withdrawals.

There are very few aquifers in which fresh water is known to have traveled as much as 100 miles from the recharge area, and most commonly the travel in artesian aquifers is 50 miles or less. Some areas of concentrated development are at no great distance from the recharge area but the aquifers have been cut by barriers that prevent the free movement of water into the area of development. A general lowering of water levels over a broad area has been recorded in some aquifers where production wells have been widely dispersed, resulting from a coalescing of the cones of depression formed by the individual pumping centers remote from recharge areas.

Pumping itself increases the rate of movement from the recharge areas to production wells, by increasing the hydraulic gradient. In several areas the water levels in production wells have reached approximate equilibrium after years of pumping, indicating that the pumped water is now replaced by inflow of water through the aquifer.

Declining water levels are inevitable if water is taken from a production well, and if the draft is increased the water levels will decline further, whether the increase is obtained by drawing more water from one well or by drawing from more wells. If the creation of a cone of depression results in diversion of water from seeps or other points of natural discharge, the draft may be sustained perennially. If water is unable to move toward a production well rapidly enough to replace the water pumped out, the pump inevitably takes water from progressively greater depths in the immediate vicinity of the production well, the pumping lift and cost of pumping increase, and the production well eventually may be pumped dry. This indicates a shortage of water at the production well but not necessarily an insufficiency of water in the groundwater reservoir. Indeed, water levels have declined markedly in small portions of some reservoirs whose recharge areas are annually filled to overflowing but the recharging water moves too slowly to the areas where production wells are pumped. In groundwater reservoirs being emptied because the total replenishment of the aquifer is less than the draft, pumping from closely spaced production wells accelerates the declining trend of water levels in some areas.

In several California valleys open to the ocean, pumping from groundwater storage has progressed to the point where the water table or piezometric surface is now below sea level at the coast. Unless there is an impermeable barrier to prevent it, the ocean water can move into the groundwater reservoirs of those coastal valleys. Intrusion of saline water into reservoirs once occupied by fresh water has been reported along the shores of San Francisco Bay and in the lower part of the adjacent Santa Clara Valley; in the lower end of Salinas Valley; along the Los Angeles Coastal Plain from Venice to Newport Beach; and near Ventura, San Juan Capistrano, Oceanside, and San Diego. Saline water has been drawn from adjacent brine-bearing formations into the overdeveloped portions of some groundwater reservoirs. In the Lake Charles industrial area and adjacent rice-growing areas of southwest Louisiana, salinity is greatest in the areas where water levels have been lowered most by pumping, and it appears that the pumping is drawing up water from underlying beds that carry saline waters.

In some cases, the contaminated water has come in through production wells perforated in several water-bearing formations. At the other places there is as yet no evidence that saline waters are entering the fresh-water reservoir. However, wherever the cone of depression drops below sea level, there is some cause for fear that the reversed gradient may eventually induce inflow of saline water from the ocean. Maintenance of groundwater quality at acceptable levels is one of the major requirements for successful utilization of groundwater reservoirs.

Problems result from pumping along streams, where the groundwater is so closely related to water in the stream that pumping from production wells depletes the streamflow. Problems arise because the pumped water is not readily replaced from the streams or it is readily replaced but the stream water is either needed downstream or is unsuitable for use. Some production wells close to streams produce water free from harmful bacteria even though the stream is highly polluted. The dumping of brines from oil fields into streams at places has rendered the streams temporarily unfit for use, but appears to have brought more permanent damage on the groundwater reservoirs of the watercourse.

The valley groundwater reservoirs of the west are typically bordered on two sides and sometimes completely enclosed by mountain ranges, from which both the water and the water-yielding material are generally derived. In many instances one range is far higher than the other and furnishes the great bulk of the water to the valley. As a result, the water from one side may saturate the valley materials clear to the base of the unproductive range on the other side, as is the case in Utah Valley in Utah. Commonly, however, the sediments in the central part of the valley are too fine or other conditions are unfavorable for movement of water across the valley, and the "dry" side thus receives negligible recharge. This is the situation in San Joaquin Valley, California, where the pumping for irrigation of the west slope is almost entirely from storage, and water levels in wells have declined markedly as a result.

The reduction of artesian pressure in thick valley deposits has induced compaction of aquifers and adjacent silts and clays. Many areas of subsidence due to groundwater pumping are known in the United States as well as in Japan, Mexico, and England. The most spectacular subsidence areas, those of the southern Great Valley of California and the Mexico City area, are in large valleys of tectonic origin. In both areas the maximum subsidence has probably exceeded 15 feet and most certainly has exceeded 10 feet over relatively large areas. Extensive problems can be caused by subsidence. Groundwater storage capacity is permanently destroyed. Elevations determined by expensive surveys must be reestablished periodically; topographic maps must be redrawn. Water flow in canals and rivers may become sluggish because of the reduction of the already low gradients by subsidence. If production wells extend below major zones of subsidence, the bottoms of casings will remain stationary while the overlying material settles downward and exerts a drag along the casing. The resulting stress transmitted to the casing will commonly collapse the casing. Piles for buildings and other structures may also extend into or below zones of subsidence. The structures then tend to remain stationary as the ground surface subsides. Any connections with the ground such as ramps and stairs will fail if they are not flexible.

The problems resulting from groundwater development, by means of production wells, are alike in one respect: there is not enough usable water where it is needed. In most river basins the groundwater data are so meager that proper evaluation cannot be made of the potentialities of groundwater reservoirs, and those reservoirs are therefore often ignored in development plans. The lack of adequate hydrologic data is recognized by technicians in all fields related to development and utilization of water resources. Many have stressed the need for far more basic data to ensure that these developments will be sound and economical.

From previous discussions, some guides may be offered for the effective development of aquifers for perennial use. The practical sustained yield of a groundwater reservoir cannot be greater than the average recharge, whether that recharge occurs by natural or artificial means; and generally the quantity recovered from the entire reservoir for use will be appreciably less than that inflow, because of the practical limitations of the recovery techniques. In any part of the reservoir, even at the particular production well, a further limitation is imposed by the rate at which water can be transmitted through the aquifer; this limitation applies least to production wells within the recharge area, where they can reap the benefit of any inflow in short order. These limitations are an oversimplification of a complex problem, and are only the hydrologic limitations. In many areas economic, social, or legal factors limit the yield of groundwater reservoirs to a level far below their physical capabilities.

There is increasing awareness that wherever the nation hopes to make full use of its resources to meet its progressively increasing requirements for water, it must develop and maintain a continuing inventory that will show the quantities of water passing through each phase of the hydrologic cycle (precipitation, infiltration, runoff, evapotranspiration, and groundwater), and permit a periodic accounting of the water storage in the soil, in surface reservoirs, and in groundwater reservoirs. With this quantitative information, and the application of sound hydrologic principles, full development of the water resources can be achieved.

For effective groundwater management, it is essential that the unit controlled be a complete hydrologic unit, including the places of natural and artificial recharge, the places of natural and artificial discharge, and all the intervening rocks through which water may move. Such a prerequisite for effective control is readily apparent in the analogous operation of a surface reservoir, which could not be effective unless there were some inflow that could be counted on and unless all diversions from the reservoir were subject to the control of the operator.

The distribution of the facilities determining the amount of groundwater that can be brought under management for its controlled uses is as variable as the geological conditions that have provided mineral deposits that are feasible of development. While it is possible with groundwater facilities, as with coal or iron-ore deposits, to mark out the general regions of their occurrence and after preliminary exploration to visualize something of their volume and qualities, it is difficult in either case to determine all the necessary factors for the wise use without thorough mapping and adequate exploration, and in many cases careful checking and observation after the facility is in operation. Manageable water supplies can be enlarged to serve high-value controlled uses through reduction of wasteful evapotranspiration, suppression of low-valued uses, and increase of the efficiency of facilities of control.

9.2 Groundwater Use

Groundwater is an important source of water supply. The use of groundwater in irrigation, industries, municipalities, and rural homes continues to increase. Estimates of groundwater use in the United States in 1960 were made by MacKichan and Kammerer (1961). Of the total water withdrawn in 1960 (267,000 mgd), 46,539 mgd came from groundwater; in 1955 groundwater use was 46,300 mgd. In 1960, irrigation used the largest part of the groundwater withdrawn (30,162 mgd); industry used the next greatest amount (7,312 mgd). Groundwater withdrawals for public supplies amounted to 6,304 mgd; rural, domestic, and stock use was the smallest (2,761 mgd). The withdrawal of groundwater was the greatest in western areas where irrigation is practiced extensively and in the eastern industrial areas. Withdrawals in the midcontinent region were the smallest. Data concerning groundwater withdrawals in 1960 are given in Table 9.1.

Uses of groundwater may be classified in several ways including withdrawal, consumptive, and nonconsumptive uses. *Withdrawal uses* require that the water be removed from the ground; public, industrial, irrigation, domestic, and stock uses are of this type. The quantity of water withdrawn at a place is often called pumpage. *Consumptive use* is the quantity of water discharged to the atmosphere or incorporated in the products of the process in connection with vegetative growth, food processing, or incidental to the industrial process.

Water used for public supplies includes all water pumped into the system. This water may be used for fire protection, street flushing, irrigation of lawns and gardens, and by industry and commerce, as well as for domestic supply. Because the water is measured at the source, leakage is also included.

Water used for rural supplies includes water for livestock and for domestic use. Rural use is often estimated by multiplying per capita uses by human and livestock population. The following human per capita rates were given by MacKichan and Kammerer (1961):

	With running water	*Without running water*
Arizona	100 gpd	10 gpd
California	80	20
Florida	60	5
Hawaii	160	70
Idaho	100	20
Illinois	60	10
Kentucky	35	5
Minnesota	50	5
New Jersey	60	10
New York	75	10
Puerto Rico	25	10
All other states	50	10

The quantity of water used by livestock ranges widely, depending on the kind and age of the animal and the temperature of the air. The following livestock rates per head were given by MacKichan and Kammerer (1961):

Milk cows	20 gpd
Horses and mules	10
Beef cattle	10
Hogs	3
Sheep	2
Goats	2
Chickens	0.04
Turkeys	0.06

TABLE 9.1 GROUNDWATER WITHDRAWALS IN 1960. (*After MacKichan and Kammerer, 1961.*)

State	*Public supplies, mgd*	*Domestic and livestock, mgd*	*Irrigation, mgd*	*Self-supplied industries, mgd*	*Total, mgd*
Alabama	84	50	2	80	216
Alaska	8	5	0.05	12	25
Arizona	100	36	3,000	80	3,216
Arkansas	48	33	770	147	998
California	1,200	260	8,500	740	10,700
Colorado	41	29	1,800	47	1,917
Connecticut	14	24	0.05	40	78
Delaware	11	8	1	29	49
Florida	500	100	270	688	1,558
Georgia	110	68	16	233	427
Hawaii	74	7	380	145	606
Idaho	91	29	2,300	91	2,511
Illinois	200	120	1	328	649
Indiana	130	110	3	161	404
Iowa	91	120	38	74	323
Kansas	120	66	900	144	1,230
Kentucky	18	21	0.5	69	109
Louisiana	91	49	340	377	857
Maine	11	8	0	12	31
Maryland	21	39	1	49	110
Massachusetts	120	6	0.19	66	192
Michigan	170	130	8	113	421
Minnesota	100	110	4	120	334
Mississippi	87	50	320	166	623
Missouri	49	69	20	59	197
Montana	24	15	34	36	109
Nebraska	120	86	1,300	12	1,518
Nevada	43	6	270	37	356
New Hampshire	18	5	0.16	8	31
New Jersey	230	85	26	198	539
New Mexico	84	22	810	23	1,039
New York	330	140	10	166	646
North Carolina	34	100	7	33	174
North Dakota	14	17	0.13	11	42
Ohio	230	110	3	288	631
Oklahoma	56	20	190	75	341
Oregon	170	24	240	140	574
Pennsylvania	110	110	0.25	300	520
Rhode Island	10	1	0.04	15	26
South Carolina	35	33	19	54	141
South Dakota	46	33	34	11	124
Tennessee	150	37	2	420	609
Texas	530	110	7,700	811	9,151
Utah	100	11	240	61	412
Vermont	8	10	0.03	9	27
Virginia	43	74	12	51	180
Washington	200	7	420	170	797
West Virginia	37	20	0.06	73	130
Wisconsin	160	130	12	230	532
Wyoming	33	8	57	9	107
District of Columbia	0	0	0	1	1
U.S. excluding Alaska, Hawaii, Puerto Rico, and Virgin Islands	6,240	840	30,000	7,163	44,243
Total for United States	6,304	2,761	30,162	7,312	46,539

Major exceptions to the above rates per head given by MacKichan and Kammerer (1961) are as follows:

Arizona	Milk cows	30 gpd
Arkansas	Milk cows	30
	Beef cattle	15
	Horses and mules	15
California	Milk cows	25
	Beef cattle	15
	Horses and mules	15
Colorado	Milk cows	15
Kentucky	Beef cattle	8
Maryland	Milk cows	35
Missouri	Milk cows	30
Nevada	Beef cattle	6
Virginia	Milk cows	25

Irrigation water is applied during only part of each year and at variable rates; therefore, the actual rate of application is much greater than the average daily rate. Irrigation varies greatly throughout the nation; the quantity of water used for irrigation in the western states is much greater than in the remaining states. The water used for irrigation in the East is usually conveyed to the fields in pipes, so that the conveyance losses are very small compared to those losses in the western states where the water is usually transported to the fields in ditches. About 60 percent of the irrigation water applied in 1960 was consumed by evaporation and transpiration. This percentage is more than 25 times the percent consumed by self-industry, and about 4 times the percent consumed by public supplies.

Water has many industrial uses, including processing, cooling, washing, conveyance of material, boiler feed, and sanitation. About 94 percent of the self-supplied industrial water is used for cooling. The amount of water used by public utilities for fuel-electric power was more than twice the amount of self-supplied industrial water used by other industries.

GROUNDWATER RIGHTS

Development and management of groundwater resources involve the legal control of groundwater (see Piper and Thomas, 1958). The laws of groundwater rights in particular and water rights in general must be considered in addition to hydrogeologic environments. A *groundwater right* is a right, granted by law, to take possession of groundwater and to put it to beneficial use.

In the conterminous United States, groundwater rights are derived from two contrasting doctrines of law: the common-law or English doctrine of riparian rights, under the basic premise that the water (unless in "underground streams") is the absolute property of the owner of the overlying land, in perpetuity; and the doctrine of prior appropriation, under the basic premise that the groundwater is the property of the state or of the "public"—the individual appropriates a right to use a specified quantity of groundwater provided the use is "beneficial"—also in perpetuity. Under the common-law doctrine, the groundwater is appurtenant to the land, is privately owned and available for use by the landowner, and the right is not lost by nonuse. Under the prior appropriation doctrine the right is not dependent upon land ownership and it is forfeited after a specified period of nonuse. The common-law doctrine with variations dominates in the humid or subhumid East (most states east of the 97th meridian); western semiarid or arid states adhere to the prior-appropriation doctrine. The 31 eastern states constitute a humid region where annual precipitation exceeds 20 inches and is greater than evapotranspiration and there is a relative abundance of water. The 17 western states include areas where the average precipitation is less than potential evapotranspiration and there is a perennial water deficiency. In some states both doctrines are followed. For example, in California most rights to groundwater are under a variant of the common-law doctrine, although appropriative rights can be acquired to water that is surplus to the needs of the owners of overlying land.

9.3 Common-law Doctrine

The specific basis for the common-law doctrine of groundwater rights in the United States usually is cited as the case of *Acton* v. *Blundell* in England in 1843. Its rigid concepts, expressed forcefully in a Connecticut decision of 1850, an Ohio decision of 1861, and an Illinois decision of 1899, follow:

The laws of its (percolating groundwater) existence and progress cannot be known or regulated. It rises to great heights, and moves collaterally, by influences beyond our apprehension. These influences are so secret, changeable and uncontrollable, we cannot subject them to the regulations of law, nor build them a system of rules, as has been done with streams upon the surface.

The reasoning is briefly this: In the absence of express contract, and of positive authorized legislation, as between proprietors of adjoining land, the law recognizes no correlative rights in respect to underground waters percolating, oozing or filtrating through the earth; and this mainly from considerations of public policy. 1. Because the existence, origin, movement and course of such waters, and the causes which govern and direct their movements, are so secret, occult, and concealed that an attempt to administer any set of legal rules in respect to them would be, therefore, practically impossible. 2. Because any such recognition of correlative rights would interfere to the material detriment of the common wealth, with drainage and agriculture, mining, the construction of highways and railroads, with sanitary regulations, building and the general progress of improvement in works of embellishment and utility.

Water which is the result of natural and ordinary percolation through the soil is part of the land itself, and belongs absolutely to the owner of the land, and, in the absence of any grant, he may intercept or impede such underground percolations, though the result be to interfere with the source of supply of springs or wells on adjoining premises.

The street common-law doctrine applies only to "percolating" groundwater, not to water in a "defined underground stream." (The latter is governed by the companion riparian doctrine of surface watercourses.) In general, the common-law doctrine tends to foster, rather than restrain, excessive competition for a common groundwater source (Piper, 1960).

An early variant of the common law of groundwater softens the basic doctrine; this is the so-called American doctrine or principle of reasonable use. The concept is expressed in an early decision by the Supreme Court of New Hampshire to the effect that, as between two neighbors, the right to withdraw "percolating groundwater" restricts each to a reasonable exercise of his own right, a reasonable use of his own property, in view of the similar rights of others. The same principle has been expressed on other decisions in terms such as "reasonable beneficial use," "reasonable economic use," or "best reasonable use."

In the end, like the strict common-law doctrine, the principle of reasonable use may accelerate rather than prevent competitive withdrawal under which the total groundwater supply of a given area might in effect be confiscated by a few users (Piper, 1960).

Another variant of the common-law doctrine evolved in California and Utah. This variant, the principle of correlative rights, holds "that the rights of all landowners over a common basin, saturated strata, or underground reservoir, are coequal or correlative, and that one landowner cannot extract more than his share even for use on his own lands where the rights of others are injured thereby; nor can he claim more than his share on the ground of peculiar benefit to him from its use" (Piper, 1960). Under this principle, in a time of water shortage, all the landowners would share the shortage proportionately. Apportionment of a total supply is reasonably straightforward if all the landowners use water for the same purpose, as for irrigation. Complexities arise if unlike uses of the water are involved.

To cope with excessive competition for limited supplies of groundwater or to resolve other troubles, several states adhering to the common-law doctrine have resorted to regulation of groundwater use. This is done under statutes that invoke the police power to assure the "public welfare, safety, and health." New Jersey probably has gone farthest in an effort to limit groundwater withdrawals to perennial yield through regulation. The scope of its statute is indicated by the following quotation:

The Division of Water Policy and Supply of the State Department of Conservation shall delineate from time to time such areas of the State where diversion of subsurface and percolating waters exceeds, or threatens to exceed, or otherwise threatens or impairs, the natural replenishment of such waters. In areas delineated by the Division of Water Policy and Supply no person, corporation or agency of the public shall hereafter divert or obtain water from subsurface or percolating sources in excess of one hundred thousand (100,000) gallons per day for any purpose unless such person, corporation or agency of the public shall first obtain a permit for such withdrawal from the Division of Water Policy and Supply. Such permit may be refused, or if granted, may include such stipulations as may be necessary to conserve the subsurface and percolating waters of the State and prevent their exhaustion.

Under this statute, two "protected areas" have been designated as of 1950; one comprised parts of Middlesex and Monmouth Counties, the other, parts of Burlington, Camden, Gloucester, and Salem Counties. In such areas, control is exercised through a system of term licenses that are issued to users and that have many of the effects of the prior-appropriation doctrine. Licenses are renewable if the Division of Water Policy and Supply determines that the total water supply remains adequate. However, because the licenses do not run indefinitely, at least one manufacturer declined to locate a new plant in the state.

9.4 Prior-appropriation Doctrine

All groundwater statutes under the prior-appropriation doctrine recognize and validate "prior vested rights" that may have accrued to landowners under the common law. Some, but not all, specifically limit such prior rights to the quantity of water used beneficially. Nearly all the statutes exempt, from the transitional application-and-permit procedure, uses of water by individuals for domestic purposes or for watering livestock and small uses for any purpose (see Bliss, 1951). Statutes invoking the doctrine of prior-appropriation have been applied to groundwaters in Colorado, Idaho, Kansas, Nevada, New Mexico, Oklahoma, Oregon, South Dakota, Utah, Washington, and Wyoming.

The essential common point is that the waters and the lands are declared to be separate entities, with no inherent attachment of one to the other. At least in the public-land states, the legal basis for such legislative declarations usually has been cited as the Desert Land Act of March 1877, which reserved the waters of the concerned lands to be disposed of by the several states under their respective laws and court procedures. It is held widely that no reasonable basis exists for a legislative declaration separating land and groundwater, under the principles that the water is absolute property of the landowner and that the owner cannot be deprived of that property, even by legislation, without just compensation. These principles are invoked commonly, even though no use is made of the property, that is, of the water.

All groundwater statutes that invoke the prior-appropriation doctrine embody phraseology equivalent to the principle that "beneficial use shall be the basis, the measure and the limit of the right to the use of water." Essentially all the statutes provide for the right to be forfeited to the extent that it may not be exercised for some specified number of years, commonly from three to five years.

The New Mexico appropriative groundwater statute was the first to be put into wide administrative operation (Hutchins, 1955), and it has established a general pattern for much of the subsequent legislation of that sort. That statute was enacted first in 1927. Basic provisions of the current statute follow:

75-11-1. The water of underground streams, channels, artesian basins, reservoirs, or lakes, having reasonably ascertainable boundaries, are hereby declared to be public waters and to belong to the public and to be subject to appropriation for beneficial use [*modified by 75-11-9 and 75-11-21*].

75-11-2. Beneficial use is the basis, the measure, and the limit of the right to the use of the [*ground*] *waters.*

75-11-9. All underground waters of the State of New Mexico are hereby declared to be public waters and to belong to the public of the State of New Mexico and to be subject to appropriation for beneficial use within the State of New Mexico. All existing rights to the beneficial use of such waters are hereby recognized.

75-11-20. No person shall withdraw water from any underground source in the State of New Mexico for use in any other State by drilling a well in New Mexico and transporting the water outside the State or by drilling a well outside the boundaries of the State and pumping water from under lands lying within the territorial boundaries of the State of New Mexico.

75-11-21. No permit and license to appropriate underground waters shall be required except in basins declared by the State Engineer to have reasonably ascertainable boundaries.

75-11-22. The State Engineer and the Attorney General or the various District Attorneys are authorized and directed to use any and all legal means neccessary to enforce the provisions of [*the act*].

An intending appropriator within a designated basin applies to the State Engineer for a permit. If no protests are filed, and if the State Engineer finds that unappropriated waters exist in the groundwater reservoir designated in the application, or that the proposed appropriation would not impair prior rights to that source, the State Engineer issues a permit to appropriate all or part of the waters applied for, subject to the rights of prior appropriators. If protests are filed, the State Engineer holds a hearing before granting or denying the permit.

A variety of regulations have been established to facilitate groundwater development and management. In many states, water well contractors must be licensed; installation, deepening, or reperforation of a well must be reported on prescribed forms. Some states require well-drilling permits and drillers must submit geologic logs for all new and deepened wells. Flowing artesian wells must be capped or regulated by suitable devices to prevent waste. Abandoned wells must be sealed to prevent waste, contamination, and accidents. In many states, water pumped for cooling or air conditioning must be returned to the ground through injection wells. The disposal of any pollutants to the subsurface is often restricted (Todd, 1959).

The inadequacy of present doctrines to cope with problems arising from the increasing demand for groundwater has been noted by several study groups. Although the common-law doctrine appears to grant perpetual security in water rights to each landowner, whether he uses the water or not, there is insecurity in any investment for developing, storing, or using groundwater, because there is no assurance that other landowners will not some day undertake to develop and use the same water (Thomas, 1958).

When it applies exclusively, the prior-appropriation doctrine affords a practical mechanism for curtailing the use of groundwater. However, the mechanism is far from effective for reasons including the following:

1. Extensive records and investigations are prerequisite to monitoring groundwater reservoirs. Few areas have the required records available. Overdraft may therefore escape recognition until it is too late because the effects of development commonly are obscure and develop slowly.
2. The causes and effects of groundwater overdraft are not reversible immediately or fully. Consequently, cutting off withdrawals in the reverse order of their priorities of appropriation does not assure that the overdraft trend will be reversed. Under these circumstances, here stated all too briefly, responsible officials very commonly are reluctant to invoke statutory provisions for reducing use of groundwater. They may have either of two reasonable doubts: that the available facts would suffice to sustain them against any appeal from an order for reduction, or that the statutory procedure would in fact recapture the status of the earlier appropriators.
3. Even in a groundwater basin in which perennial yield and appropriations for use are about equal in total, some further development may be feasible. Such is a common feature of irrigated areas, in which development tends to concentrate around the most productive wells and the most fertile lands, whereas the less fertile lands or the areas of small yield from wells remain virtually untouched.

Strict application of the prior-appropriation doctrine would preclude the potential further development. Declaration of "critical" or "restricted" subareas does not offer a satisfactory solution because the boundaries of such areas would be arbitrary, at least in part, and almost certainly would be indefensible in detail.

Effectiveness of the prior-appropriation doctrine is further and substantially limited when, as is generally the case, the groundwater code is administered under a register of priorities that is separate and distinct from the register of surface-water priorities. Combined registers would not be universally appropriate.

Strict adherence to current water-right doctrines may create many obstacles to adequate future management of groundwater resources. Some resources may go unused or may be used at less than capacity, even while others are under needless and excessive competition. Substantial changes in water-right doctrine and law are inevitable. The laws of water rights in general and of groundwater rights in particular should: afford flexibility whereunder the nation as a whole may effectively manage its water destiny, under reasonable but adequate checks and balances; in defining rights as between individuals, prescribe only essential and minimum restrictions; and afford to groups of water users all reasonable freedom to act concertedly, mutually conceding some part of their individual "rights" to their common advantage (Thomas, 1958).

LAND SUBSIDENCE DUE TO GROUNDWATER WITHDRAWALS

In areas where groundwater reservoirs are composed of unconsolidated and semi-consolidated alluvial sediments confined or partially confined by thick fine-grained beds, subsidence of the land surface is likely to result as artesian pressure is reduced by groundwater pumpage. A reduction of artesian pressure results in an increase of grain-to-grain load on the aquifer skeleton system which may cause compaction of deposits. The subsidence at any location depends upon the subsurface lithology and upon the magnitude and duration of artesian pressure decline, and it is mostly inelastic and permanent. Subsidence can be detected by careful releveling of surface bench marks. Compaction of deposits may extend to depths exceeding 1,000 feet and is directly related to declines in artesian pressure.

A definite period of time is required for fine-grained sediments to compact and adjust completely to artesian pressure declines. The time required for complete compaction is proportional to the square of the thickness of the sediments (Terzaghi, 1943). In beds of silty clay and clay, the rate of compaction depends upon the rate at which water can be squeezed from the pore spaces. Thick clay beds may require hundreds of years to adjust, but thinner beds may need only a few tens of years or less (Green, 1964). Most compaction in sands and silts occurs within a few years after the reduction of artesian pressure.

Land subsidence of several feet has been recorded in the San Joaquin Valley, California (Poland, 1961); the upper Gulf coastal region, Texas (Winslow and Wood, 1959); the Savannah area, Georgia (Davis et al., 1963); Las Vegas Valley, Nevada (Domenico and Clark, 1964); Mexico City and London (Wilson and Grace, 1942). Land subsidence affects an area of more than 2,500 square miles in San Joaquin Valley, California. In areas of maximum subsidence, groundwater levels show a downward trend, and subsidence rates range from 0.4 to 1.5 feet per year. In these areas, 1 foot of subsidence has been observed for each 10 to 25 feet of artesian pressure decline.

A special type of recorder (Lofgren, 1961) has been developed to measure the rate and magnitude of compaction occurring at depths in areas of land subsidence. The assembly (Fig. 9.1) consists of a heavy weight placed in the formation below the bottom of a well casing with an attached cable stretched upward in the casing and counter-weighted at the land surface to maintain constant tension. An automatic water-stage recorder mounted over the open casing is used to measure directly the amount of cable that appears above the casing as subsidence occurs. At the land surface it appears as if the bottom-hole weight is rising; actually, the land surface is settling with respect to the bottom-hole weight. A specially manufactured $\frac{1}{8}$-inch, stainless steel, 7 by 7 stranded, plastic-coated cable is used. Ball-bearing sheaves are used to reduce the frictional drag of the system. The bottom weight is placed in an open

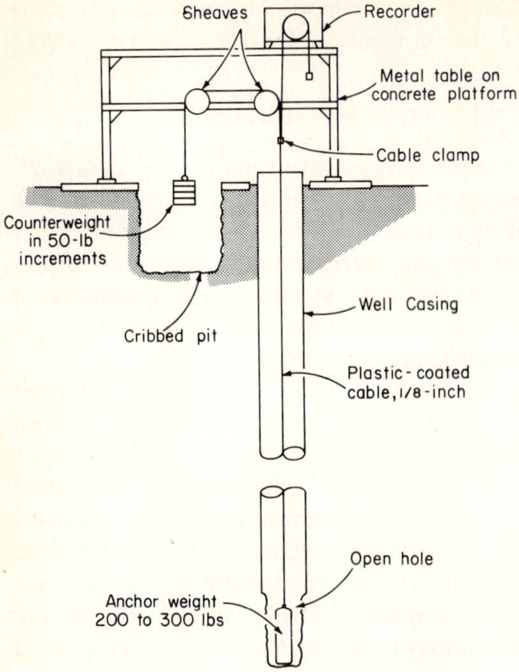

Fig. 9.1 Diagram of compaction-recorder installation for measuring land subsidence due to groundwater withdrawals. (*From Lofgren, 1961.*)

hole 15 to 25 feet below the bottom of the well casing so that measurements of vertical shortening are independent of the casing.

A method for estimating inelastic compaction in artesian aquifers and therefore land subsidence using the results of one-dimensional consolidation tests made upon core samples of the aquifers and data on the decline in artesian head was described by Miller (1961). An extension of this method can be used to predict future subsidence. Aquifers and overburdens are divided into sufficient segments so that each segment can be represented by a single consolidation test typical for that segment. Declines in artesian pressure are estimated from data on nonpumping water levels in aquifers and converted into pounds per square inch. The overburden loads on aquifers in pounds per square inch are computed from the wet unit weights of core samples. The effective loads on the tops of aquifers are computed by subtracting the aquifer pressures from the bulk weights of the aquifer overburdens.

One-dimensional consolidation tests (see Terzaghi, 1943) are made on the core samples of the aquifers for the maximum load ranges that could occur in the aquifers. An increase in loading due to a decline in artesian pressure results in a decrease in the void ratio of the sample tested. Clays tend to consolidate more under loads than do sands, but not so rapidly. The compaction due to the load change on each aquifer is computed from the void-ratio change determined graphically from the extension of the straight-line part of the one-dimensional consolidation curves. Compaction is computed by the equation (Miller, 1961)

$$\Delta h_L = \frac{l_1 - l_2}{l + l_1} m \tag{9.1}$$

where Δh_L = inelastic compaction or land subsidence, in feet
l_1 = initial void ratio
l_2 = void ratio after loading
m = thickness of aquifer, in feet

In segments of aquifers that have very low permeability, years or even decades may be required before enough groundwater is displaced so that all of the computed compaction can occur. The time required for the computed compaction to be completed can be estimated from the coefficient of consolidation which is determined as part of the one-dimensional consolidation test. The equation given by Terzaghi and Peck (1948) for computing the compaction time is

$$t_L = \frac{T_L m^2}{C_L} \tag{9.2}$$

where t_L = compaction time, in years
T_L = time factor (1.0 for 93 percent compaction, 0.2 for 50 percent compaction, and 0.0076 for 10 percent compaction)
m = thickness of aquifer, in feet
C_L = coefficient of consolidation, in sq ft/yr

If drainage can take place from both top and bottom of the aquifer, $(m/2)^2$ is used in place of m^2. Equation (9.2) is valid only when artesian pressures are lowered abruptly and then maintained at a constant level.

Unconsolidated granular deposits reduced in volume by compaction due to artesian pressure decline are subjected to rearrangement and deformation of grains and a reduction in porosity. If the deposits experience more than a single stage of artesian pressure decline through the same range, the storage coefficient during the first decline greatly exceeds that during subsequent declines (Green, 1964).

The approximate porosity decrease of any zone is computed from the following equation (Green, 1964)

$$\Delta\theta = \frac{l_1}{l + l_1} - \frac{l_2}{l + l_2} \tag{9.3}$$

where $\Delta\theta$ = porosity change
l_1 = void ratio prior to load change
l_2 = void ratio after load change

The average porosity decrease of the entire aquifer system is the weighted average of the porosity decreases for all compacting zones.

The relation of void ratio to load, on a semilogarithmic plot, is approximately linear after the preconsolidation load is exceeded. As loading progresses, more load is required to produce an equal decrease in void ratio and in porosity. Deeply buried sediment, under greater natural overburden load than that at the top, requires a greater change in effective load to produce uniform porosity decrease through the aquifer system.

The elastic subsidence of the land surface can be computed from the following equation (see Lohman, 1961)

$$\Delta m = \Delta p \left(\frac{S}{\gamma} - \theta m \beta \right) \tag{9.4}$$

where Δm = subsidence, in feet
Δp = reduction in artesian pressure, in psf
S = coefficient of storage, fraction
γ = specific weight of water (0.434 lb/in.²/ft)
θ = porosity of formation, fraction
m = thickness of formation, in feet
β = reciprocal of bulk modulus of elasticity of water (3.3×10^{-6} in.²/lb)

The specific storage S_s' is defined as the volume of water that a unit volume of aquitard releases from storage, owing to its compression when the average excess pressure within the unit volume undergoes a unit decline, and can be expressed (Domenico and Mifflin, 1965)

$$S_s' = \alpha\gamma = \frac{a_v \gamma}{1 + e} = \frac{\gamma}{E_s} \tag{9.5}$$

where α = vertical compressibility of deposits
γ = unit weight of water
a_v = slope of line obtained by plotting void ratio versus pressure for test specimens
e = void ratio
E_s = bulk modulus of compression of deposits

To reduce the number of variables, the coefficient of consolidation has been introduced to include combined effects of compressibility and permeability, and its value is determined in the laboratory. The coefficient of consolidation C_L is given as $P'(1 + e)/a_v\gamma$. The specific storage can be stated as $S_s' = P'/C_L$, where P' = coefficient of permeability. Variations in the bulk modulus of compressibility and specific storage based on compressibility alone are shown in Table 9.2.

Decreases in porosity unquestionably affect permeability, but not to the extent permitted in the consolidation apparatus. Therefore, the values of specific storage calculated from the conventional consolidation test data are too small.

The coefficient of storage of a compressible layer, defined as the volume of water removed from a prism with a 1-square-foot base per unit lowering of excess pore-water pressure throughout the prism, is expressed as the product of specific storage and thickness and depends upon E_s. The pressure change within each element is not constant but is variable depending mainly on the depth of the element. When dealing with clay beds subjected to recurrent loading the bulk modulus of compression is not

TABLE 9.2 VARIATIONS IN THE BULK MODULUS OF COMPRESSION AND SPECIFIC STORAGE. (*Modified after Jumikis, 1962.*)

Material	E_s, *lb/sq ft*	γ/E_c, *lb/ft*
Plastic clay	1×10^4 to 8×10^4	6.2×10^{-3} to 7.8×10^{-4}
Stiff clay	8×10^4 to 1.6×10^5	7.8×10^{-4} to 3.9×10^{-4}
Medium-hard clay	1.6×10^5 to 3×10^5	3.9×10^{-4} to 2.8×10^{-4}
Loose sand	2×10^5 to 4×10^5	3.1×10^{-4} to 1.5×10^{-4}
Dense sand	1×10^6 to 1.6×10^6	6.2×10^{-5} to 3.9×10^{-5}
Dense sandy gravel	2×10^6 to 4×10^6	3.1×10^{-5} to 1.5×10^{-5}
Rock, fissured, jointed	3×10^6 to 6.25×10^7	2.1×10^{-5} to 1×10^{-6}
Rock, sound	Greater than 6.25×10^7	Less than 1×10^{-6}

the same for each cycle of loading (Domenico and Mifflin, 1965). The change in height per unit original height decreases with recurrent cycles of equal load, resulting in an apparent increase in the bulk modulus and decrease in S.

In areas where land subsidence occurs, the total coefficient of storage under pumping conditions is equal to the coefficient of storage due to the expansion of the groundwater, plus the coefficient of storage due to the elastic compaction of deposits, plus the coefficient of storage due to the inelastic compaction of deposits. Electric analogs for solving problems associated with time-settlement of sediments were described by Domenico and Clark (1964).

The amount of compaction represents an equivalent reduction in porosity and in stored water. At some localities the component of storage derived from the inelastic compaction of the artesian aquifer is equal to the amount of land subsidence divided by the reduction of artesian head in feet (see Poland, 1961). Because compaction is almost wholly inelastic and rebound is negligible, the coefficient of storage for the first drawdown phase greatly exceeds that of the second. At places in the Santa Clara Valley, California, the coefficient of storage during the first drawdown phase was 0.04 and during the second drawdown phase was 0.001.

The volume of water removed from compressible aquitards equals the product of their specific storage and the effective-pressure area developed across them. Often, pressure levels in groundwater reservoirs undergoing subsidence remain in a state of progressive decline for years or even decades. Hence, the effective-pressure area is continually expanding. For aquitards of equal specific storage and thickness, equal changes in seepage pressures will produce at equilibrium the same amount of vertical shortening and recovery of stored water.

CASE HISTORIES OF GROUNDWATER DEVELOPMENT

9.5 Northeastern Illinois

Groundwater resources in the Chicago region, northeastern Illinois, are developed from four water-yielding systems: (1) sand and gravel aquifers of the glacial drift;

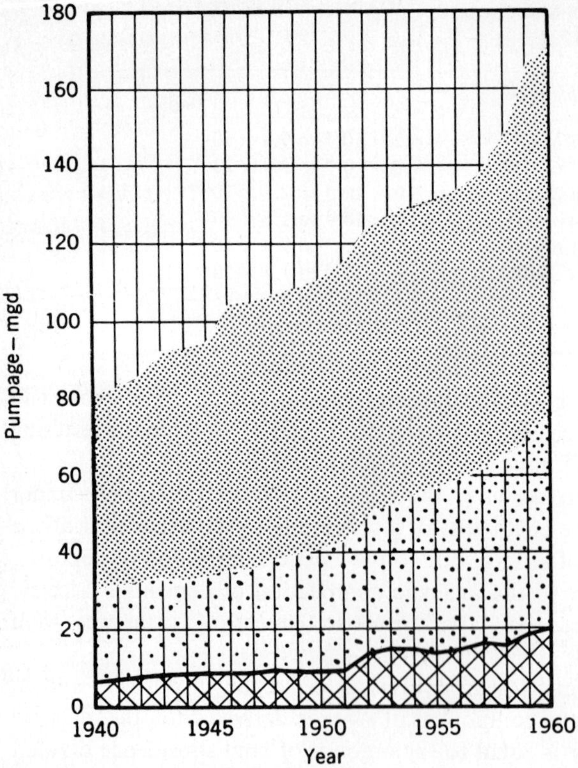

Fig. 9.2 Estimated rate and apparent source of groundwater pumpage in northeastern Illinois, 1940 to 1960. The topmost curve is for the total groundwater pumpage; the solid area is for the Cambrian-Ordovician and Mt. Simon Aquifers; the stippled area is for the shallow dolomite aquifers; and the crosshatched area is for the glacial-drift aquifers. (*From Walton, 1963b.*)

(2) shallow dolomite aquifers of Silurian and Ordovician age; (3) the Cambrian-Ordovician Aquifer, of which the Ironton-Galesville and Glenwood–St. Peter Sandstones are the most productive formations; and (4) the Mt. Simon Aquifer, consisting of sandstone beds of the Mt. Simon and Eau Claire Formations of Cambrian age (see Suter et al., 1959). Groundwater pumpage in northeastern Illinois increased from 80 mgd in 1940 to 171 mgd in 1960, as shown in Fig. 9.2. In Fig. 9.2, the term "apparent source" is used because the pumpage from a well is assigned only to the aquifer in which the production well is finished, although many production wells are open to more than one aquifer. It is estimated that in 1960 about 88 mgd or 51 percent of the total pumpage was from shallow dolomite aquifers; 63 mgd or 37 percent of the total was from the Cambrian-Ordovician Aquifer and Mt. Simon Aquifer; and 20 mgd or 12 percent of the total was from sand and gravel aquifers.

The Cambrian-Ordovician Aquifer is encountered at an average depth of 500 feet below the land surface at Chicago. It has an average thickness of 1,000 feet, is composed chiefly of sandstones and dolomites, and is considered the best bedrock aquifer in Illinois because of its consistently high yield. The sand and gravel and the shallow dolomite aquifers are hydraulically interconnected and are separated from the

Cambrian-Ordovician Aquifer in most of northeastern Illinois by shale beds of the Maquoketa Formation. Shale beds of the Eau Claire Formation separate the Cambrian-Ordovician Aquifer from the Mt. Simon Aquifer.

The first deep sandstone well completed in the Cambrian-Ordovician Aquifer at Chicago was drilled at the corner of Chicago and Western Avenues in 1864, and had an artesian flow estimated at about 150 gpm or about 200,000 gpd. The estimated pumpage from deep sandstone wells in northeastern Illinois increased gradually from 200,000 gpd in 1864 to more than 76 mgd in 1958. Figure 9.3 shows estimated withdrawal rates for the period 1864 through 1958 in the six major pumping centers in northeastern Illinois. Figure 9.4 and Table 9.3 show the distribution of estimated pumpage from deep sandstone wells in 1958. The greatest quantities of water were withdrawn from deep sandstone wells in the Chicago, Joliet, and Aurora areas (Suter et al., 1959).

Many deep sandstone wells in northeastern Illinois are uncased in the Mt. Simon Aquifer as well as in the Cambrian-Ordovician Aquifer and the Silurian-age dolomite. Thus, a large portion of the water pumped from deep sandstone wells is obtained from the aquifers above and below the Cambrian-Ordovician Aquifer. Figure 9.4 shows the distribution of estimated pumpage from the Silurian-age dolomite and Mt. Simon and Cambrian-Ordovician Aquifers in 1958. In 1958 about 33.3 mgd or 43 percent of the total amount of 76 mgd pumped from deep sandstone wells was derived from the Silurian-age dolomite and the Mt. Simon Aquifer. Based on an inventory of production-well construction and quality of water studies, it is estimated that in 1958 about 20.5 mgd were derived from the Silurian-age dolomite and about 12.8 mgd were obtained from the Mt. Simon Aquifer. A large part of the water derived from the Mt. Simon Aquifer was pumped from deep sandstone wells in the Fox Valley (Suter et al., 1959). Figure 9.3 gives approximate amounts of water derived from the Silurian-age dolomite and the Mt. Simon and Cambrian-Ordovician Aquifers in the pumping centers prior to 1958.

In 1908, there was very little pumpage outside Chicago except at Aurora and Elgin (Table 9.3). Since that time the net increase in pumpage at Chicago has been only about 10 percent although in the 1920s and 1940s much higher pumping rates were obtained. Pumpage in areas near Chicago (Elmhurst and Des Plaines) has increased to 15 times that recorded in 1908. Pumpage at Joliet increased from 1.8 to 14.0 mgd and pumpage along the Fox Valley in the Aurora and Elgin areas has more than tripled. Prior to about 1940, pumpage in the Elmhurst and Des Plaines areas increased fairly uniformly at an average rate of about 50,000 gpd per year. The total pumpage in 1940 was 2.3 mgd in the Elmhurst area and 1.7 mgd in the Des Plaines area. During and after World War II, pumpage in these two areas increased very rapidly at an average rate of 400,000 gpd per year, or at a rate 8 times that recorded for the period prior to 1940. The rate of increase in withdrawal in the Aurora area has been very uniform since 1890 and averaged about 110,000 gpd per year. Pumpage in the Elgin area increased gradually during the period 1904 through 1928 after which it increased rapidly until about 1952. From 1952 to 1958 pumpage at this center

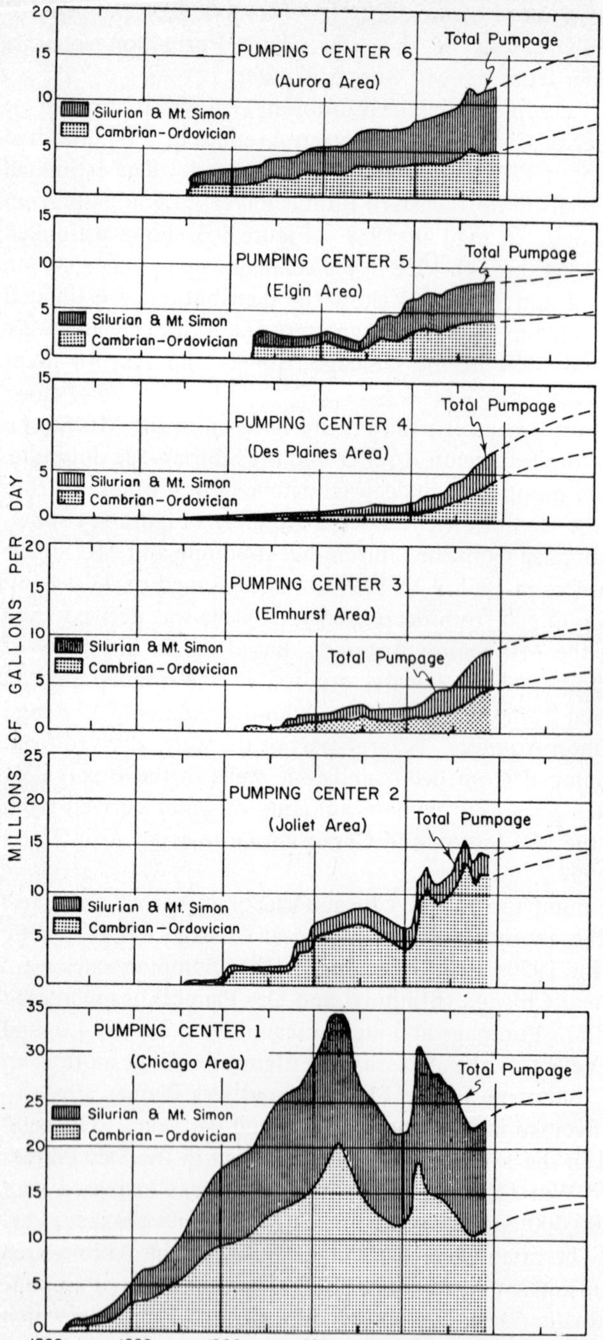

Fig. 9.3 Pumpage from deep sandstone wells in northeastern Illinois, 1864 through 1958. (*From Suter et al., 1959.*)

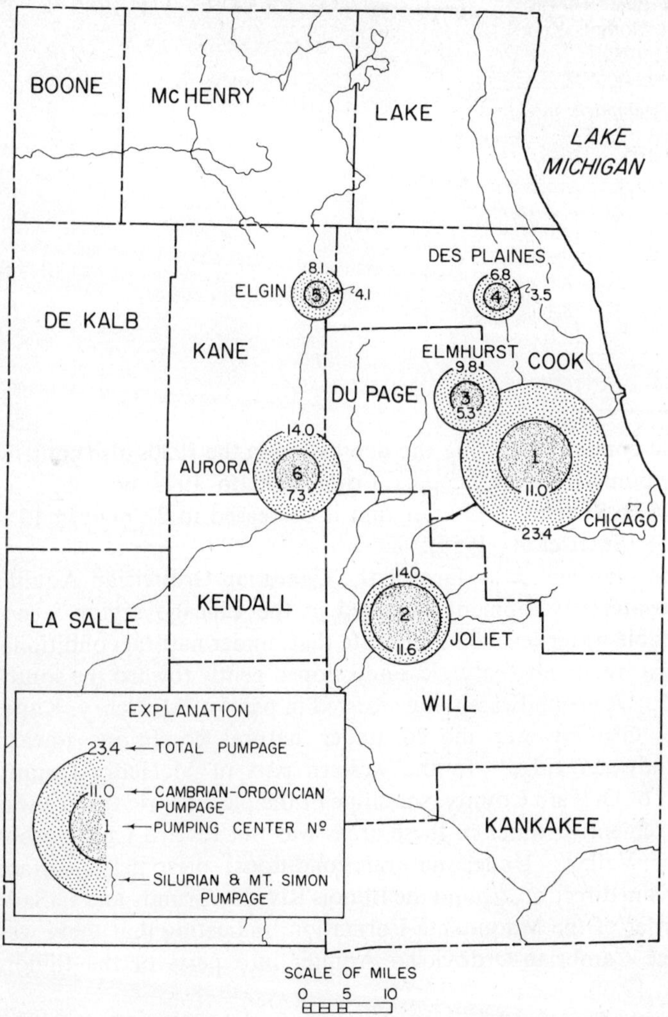

Fig. 9.4 Distribution of estimated pumpage from deep sandstone wells in northeastern Illinois in 1958. (*From Suter et al., 1959.*)

remained fairly constant at about 8 mgd. Pumpage in the Joliet area greatly increased during World Wars I and II and the Korean War in response to water demands of manufacturing industries. Pumpage at Joliet also fluctuated with economic conditions as indicated by the decrease in pumpage during the 1930s. Pumpage in the Chicago area increased at a rapid rate from about 200,000 gpd in 1864 to a maximum of 34.4 mgd in 1924. During the period 1925 through 1937, pumpage decreased rapidly as many industries, including the stockyards, abandoned deep sandstone wells owing to the great decline in artesian pressure and began using the city of Chicago

TABLE 9.3 DISTRIBUTION OF PUMPAGE FROM DEEP SANDSTONE WELLS IN NORTHEASTERN ILLINOIS. (*From Suter et al., 1959.*)

	Total pumpage, mgd	
Pumping center	*1908*	*1958*
Chicago area	21.3	23.4
Joliet area	1.8	14.0
Elmhurst area	0.4	9.8
Des Plaines area	0.4	6.8
Elgin area	2.6	8.1
Aurora area	3.9	14.0
Total	30.4	76.1

water supply. Economic conditions during the depression in the 1930s also contributed to the decrease in pumpage at Chicago. From 1940 to 1942 pumpage increased rapidly from 21.6 to 31 mgd, and after that it decreased to 22 mgd in 1954 and was 23.4 mgd in 1958 (Suter et al., 1959).

The exact shape of the piezometric surface of the Cambrian-Ordovician Aquifer before extensive groundwater development occurred in the Chicago region is not known. However, available water-level data indicate that, under natural conditions, the piezometric surface was relatively featureless and sloped gently toward the southeast, as shown in Fig. 9.5. A groundwater ridge existed in parts of McHenry, Kane, and DeKalb Counties. Groundwater moved under natural conditions toward Chicago from the groundwater ridge. In the western part of McHenry County and in the northwest part of DeKalb County, the slope of the piezometric surface was to the west, and the movement of water in these areas was not toward Chicago but toward the Rock River Valley. Under natural conditions, piezometric-surface contours bent in an upstream direction around the Illinois River in Grundy and LaSalle Counties west of the border of the Maquoketa Formation, indicating that there was upward leakage from the Cambrian-Ordovician Aquifer into parts of the Illinois River valley (Suter et al., 1959).

In 1864 the piezometric surface at Chicago had an average elevation of about 710 feet. The elevation of the piezometric surface was between 750 and 850 feet in McHenry, Kane, Boone, and DeKalb Counties, between 40 and 140 feet higher than at Chicago. In 1864, the 650-foot piezometric-surface contour bent southward in Will County and some water moved toward Grundy, Livingston, and Kankakee Counties south of Chicago. Hundreds of deep sandstone wells were drilled in Chicago and Joliet after 1864, and the pumping of large quantities of groundwater changed considerably the direction and rate of groundwater movement. The artesian pressure was lowered and cones of depression were produced. As early as 1895 the piezometric surface was greatly modified owing to heavy withdrawals of groundwater. The hydraulic gradients from the west were steepened and the slopes of the piezometric

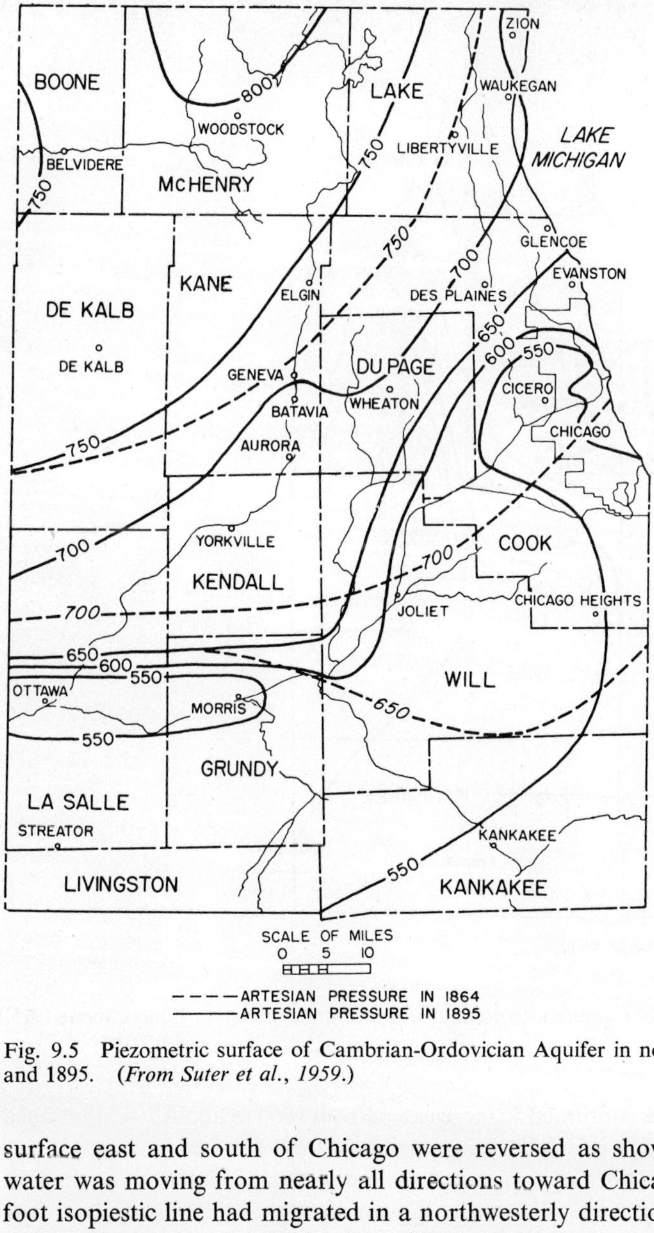

Fig. 9.5 Piezometric surface of Cambrian-Ordovician Aquifer in northeastern Illinois about 1864 and 1895. (*From Suter et al., 1959.*)

surface east and south of Chicago were reversed as shown in Fig. 9.5. By 1895 water was moving from nearly all directions toward Chicago and Joliet. The 700-foot isopiestic line had migrated in a northwesterly direction about 22 miles from its estimated position in 1864. The 550-foot line had moved northwestward from northern Indiana and south of Kankakee, Illinois, to a position west of the centers of pumpage in the vicinity of Chicago, as large quantities of water were taken from storage within the aquifer (Suter et al., 1959).

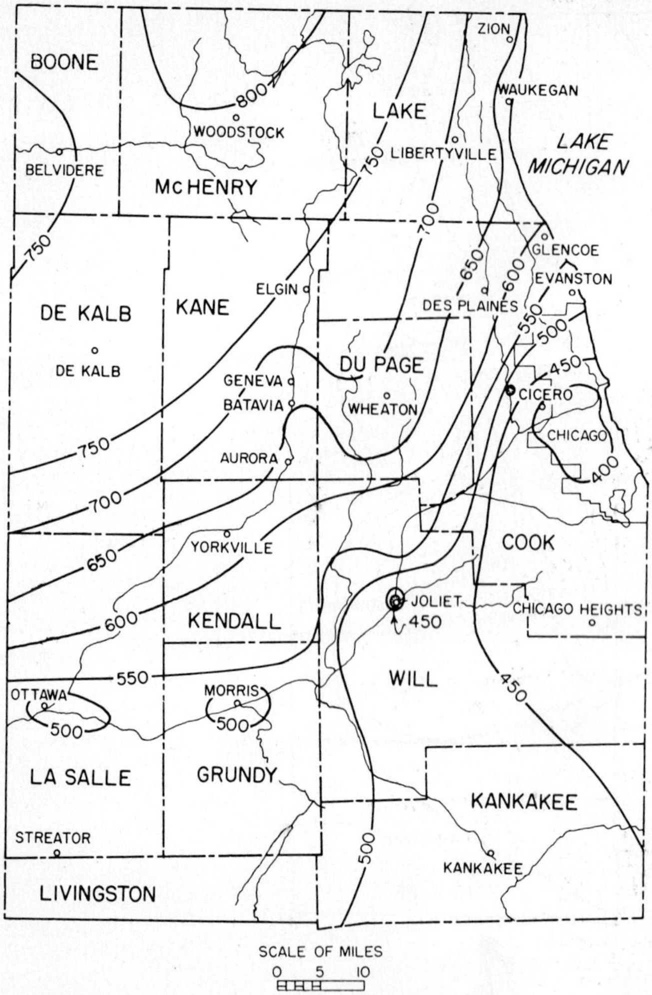

Fig. 9.6 Piezometric surface of Cambrian-Ordovician Aquifer in northeastern Illinois about 1915. (*From Suter et al., 1959.*)

Pumpage of groundwater continued to increase. From 1864 to about 1915 the cones of depression around Chicago and Joliet had deepened about 300 feet to elevations of about 400 feet and isopiestic lines were distorted around pumpage centers along the Fox Valley in Kane County (Fig. 9.6). Discharge of groundwater to the Illinois River was decreased as the cone of depression moved into Kendall County. The 700-foot isopiestic line had migrated 6 miles from its estimated position in 1895 (Suter et al., 1959).

Figure 9.7 shows the piezometric surface of the Cambrian-Ordovician Aquifer in

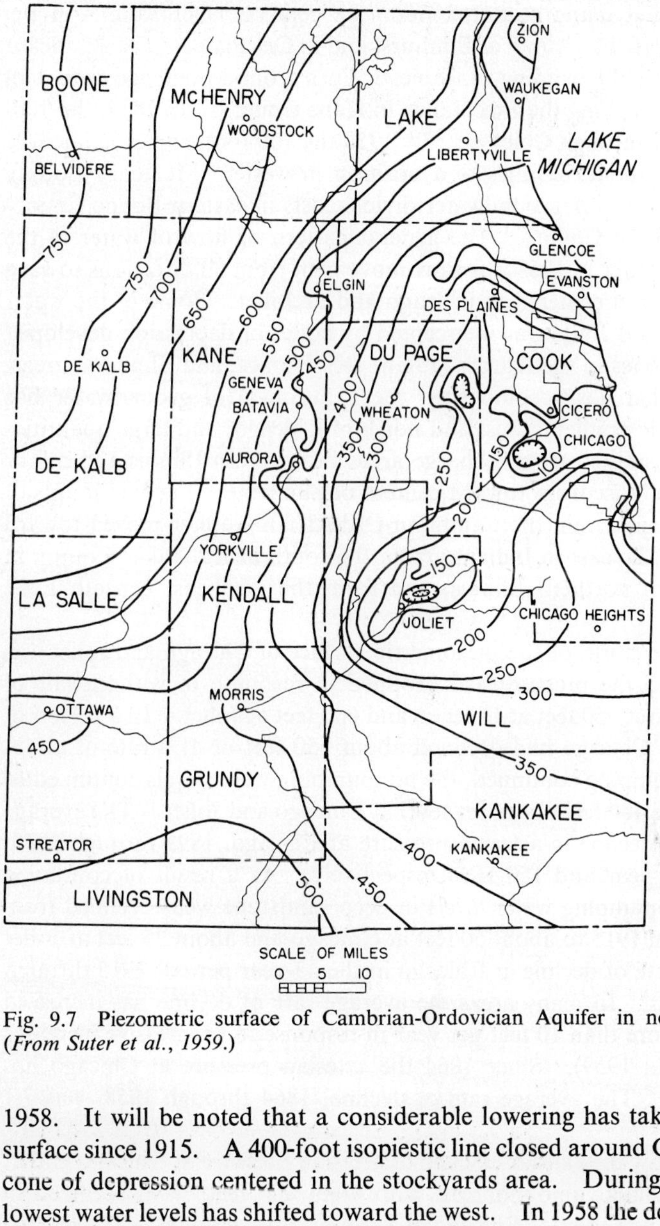

Fig. 9.7 Piezometric surface of Cambrian-Ordovician Aquifer in northeastern Illinois in 1958. (*From Suter et al., 1959.*)

1958. It will be noted that a considerable lowering has taken place in the pressure surface since 1915. A 400-foot isopiestic line closed around Chicago in 1915, with the cone of depression centered in the stockyards area. During recent years the area of lowest water levels has shifted toward the west. In 1958 the deepest cone of depression in Cook County (50 feet above sea level) was in the vicinity of Summit, which is southwest of Chicago. Another pronounced cone is centered at Joliet where the artesian pressure was at an elevation of about 25 feet above sea level in 1958. Several small cones of depression have developed and distort the isopiestic lines, causing them to bend irregularly around many scattered pumpage centers. Depressions in the piezometric surface occur in the vicinity of Elmhurst and at Des Plaines. A significant feature shown in Fig. 9.7 is the bending of isopiestic lines around pumpage centers in the Elgin and Aurora areas along the Fox Valley in Kane County. In 1864, the 700-foot isopiestic line passed through Chicago. By 1958 the 700-foot isopiestic line had migrated northwestward about 52 miles to a position in western McHenry County and eastern DeKalb County. A groundwater divide exists in eastern Boone County and in northeastern DeKalb County. The general pattern of flow of water in the Cambrian-Ordovician Aquifer in 1958 was slow movement from all directions toward the deep cones of depression centered at Chicago and at Joliet. Some of the water flowing toward Chicago and Joliet was intercepted by cones of depression developed locally within the large cones in the Aurora, Elgin, Des Plaines, and Elmhurst areas. The lowering of head that has accomplished the withdrawal of groundwater has established steep hydraulic gradients west and north of Chicago, and large quantities of water were being transmitted from recharge areas in northern Illinois and minor quantities from southern Wisconsin toward centers of pumpage. Large amounts of water derived from storage within the Cambrian-Ordovician Aquifer moved toward Chicago and Joliet from the east in Indiana, from the south in Kankakee County in Illinois, from the west in northern Illinois, and from the northeast beneath Lake Michigan.

In 1864 the average elevation of the piezometric surface at Chicago and Joliet was about 700 feet. By 1895 the pressure had dropped in response to withdrawals of water to elevations of about 550 feet at Chicago and 600 feet at Joliet. In a period of 31 years, water levels at Chicago had declined about 150 feet or at a rate of about five feet per year. As pumping continued, the nonpumping water levels continued to decline, and by 1915 were 400 feet above sea level at Chicago and Joliet. The average rate of decline and total decline in artesian pressure at Chicago, 1895 through 1915, were about 7.5 feet per year and 150 feet, respectively. As a result of continued heavy pumping, the nonpumping water levels in deep sandstone wells declined from an elevation of 400 feet in 1915 to about 50 feet at Chicago and about 25 feet at Joliet in 1958. The average rate of decline at Chicago in the 43-year period, 1915 through 1958, was 8 feet per year. In many areas the average rate of decline has increased during recent years to more than 10 feet per year in response to progressive increases in pumping (Suter et al., 1959). Since 1864 the artesian pressure at Chicago has declined about 660 feet. The average rate of decline, 1864 through 1958, was 7.1 feet per year.

Figures 9.8 and 9.9 show the rates of decline of artesian pressure at Chicago and at Joliet. The greatest declines, amounting to more than 600 feet, have occurred in areas of heavy pumpage just west of Chicago at Summit, and at Joliet. The decline was 10 feet or less west of the margin of the Maquoketa Formation in recharge areas in DeKalb, Boone, and McHenry Counties. There has been fairly widespread decline of more than 100 feet west of the margin of the Maquoketa Formation in southwestern Kendall County, northwestern Grundy County, and LaSalle County.

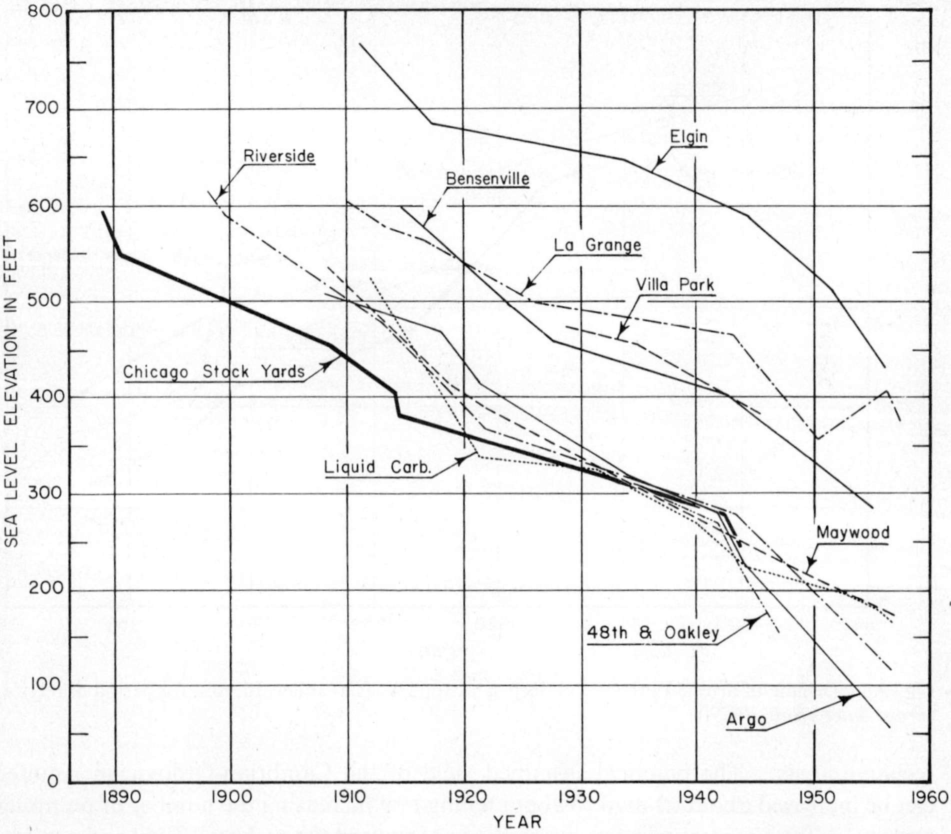

Fig. 9.8 Decline of artesian pressure in deep sandstone wells in Chicago region, Illinois, for period of record. (*From Suter et al., 1959.*)

The hydraulic gradients in these areas have been readjusted to divert large quantities of water from the recharge areas into the Joliet and Chicago cones of depression.

Pumping of groundwater from deep sandstone wells in the Chicago region was 96.5 mgd in 1961. It is estimated that of the 96.5 mgd pumped from deep wells in 1961, 55.0 mgd came from the Cambrian-Ordovician Aquifer, and 41.5 mgd came from the Silurian-age dolomite and Mt. Simon Aquifer.

The practical sustained yield of the Cambrian-Ordovician Aquifer has been estimated (Walton, 1964a) to be about 46 mgd. The practical sustained yield is here defined as the maximum amount of water that can be continuously withdrawn from existing pumping centers without eventually dewatering the most productive water-yielding formation, the Ironton-Galesville Sandstone, or exceeding recharge. The practical sustained yield of the Cambrian-Ordovician Aquifer, with the present distribution of pumping centers, is limited not by the rate of replenishment in recharge areas but by the rate at which water can move eastward through the aquifer from

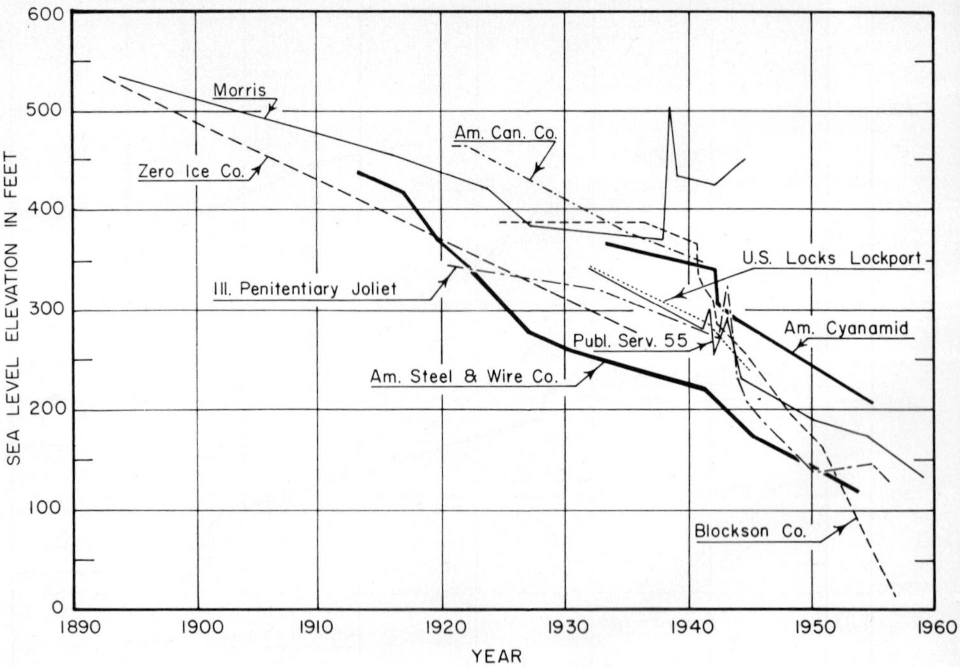

Fig. 9.9 Decline of artesian pressure in deep sandstone wells at Joliet, Illinois, for period of record. (*From Suter et al., 1959.*)

recharge areas. The practical sustained yield of the Cambrian-Ordovician Aquifer can be increased about 19 mgd to about 65 mgd by increasing the number of pumping centers, shifting centers of pumping to the west toward the recharge area, and spacing wells at greater distances (Walton, 1964a).

Pumpage from 1961 to 2010 was extrapolated as shown in Fig. 9.10 taking into consideration past rates of growth of pumpage. Assuming that the total pumpage from deep sandstone wells will increase at a uniform rate from about 96.5 mgd in 1961 to about 243 mgd in 2010, the daily withdrawal from the Cambrian-Ordovician Aquifer in 2010 will be about 190 mgd. The distribution of pumpage in 2010, subdivided by pumping center, will be as shown in Table 9.4.

Declines in water levels from 1961 to 2010 which may reasonably be expected were calculated (Walton, 1964a) for these pumping centers. The following factors were considered in computations: (1) in the future it can be expected that the rate of water-level decline will decrease even though pumping rates increase as upper units of the Cambrian-Ordovician Aquifer are partially dewatered and the coefficient of storage increases; (2) after about 1970 gravity drainage of upper units will occur but water in the lower units will still occur under artesian conditions; and (3) leakage through the Maquoketa Formation was almost at a maximum in 1958.

Hydrographs showing estimated future water-level declines in pumping centers are shown in Figs. 9.11 through 9.13. It should be emphasized that the water levels

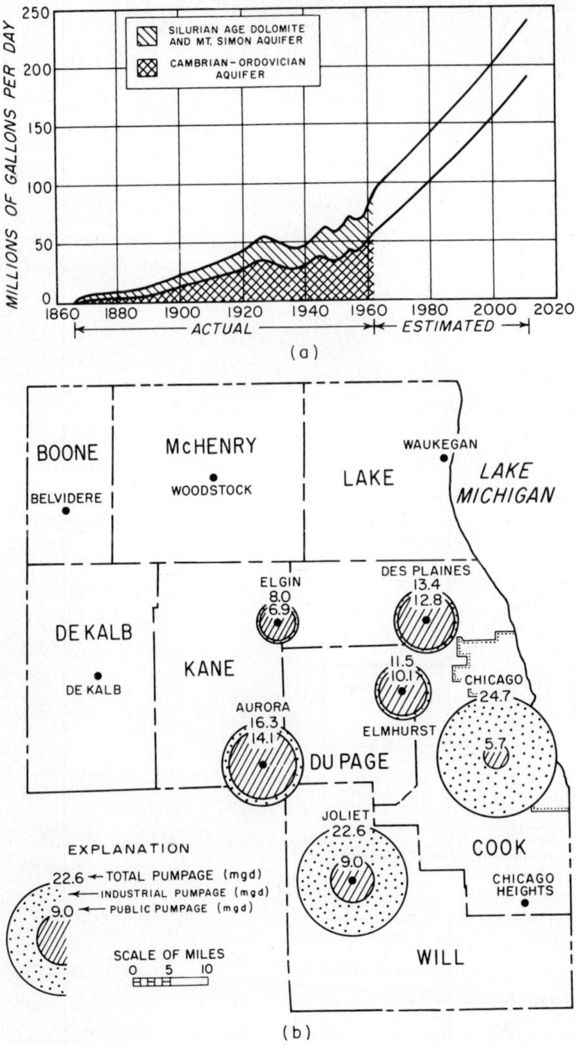

Fig. 9.10 Pumpage, 1864 to 1961 (*a*), and distribution of estimated pumpage from deep sandstone wells in 1961 (*b*) in Chicago region. (*From Walton, 1964a.*)

cited are nonpumping water levels. Pumping levels in production wells in most of the pumping centers, if the present rates of pumping from individual wells are maintained, will at critical stages within a few feet of the Ironton-Galesville Sandstone in 2010. Pumping levels exceeding 1,000 feet below the surface will be common by 1980.

It has been estimated (Walton, 1964a) that about 1.6×10^{13} gallons of water were in storage within the Cambrian-Ordovician Aquifer in 1961 and about 1.5×10^{13} gallons of water will be in storage in 2010. It would take on the order of magnitude of 4,500 additional production wells in upper units of the Cambrian-Ordovician Aquifer to mine water in storage during a 50-year period.

Table 9.4 Estimated Withdrawals from Cambrian-Ordovician Aquifer in Chicago Region, 1961 and 2010. (*From Walton, 1964a.*)

Pumping center	*Total pumpage, mgd* 1961	*Total pumpage, mgd* 2010
Chicago area	24.7	39
Joliet area	22.6	52
Elmhurst area	11.5	37
Des Plaines area	13.4	56
Elgin area	8.0	17
Aurora area	16.3	42
Total	96.5	243

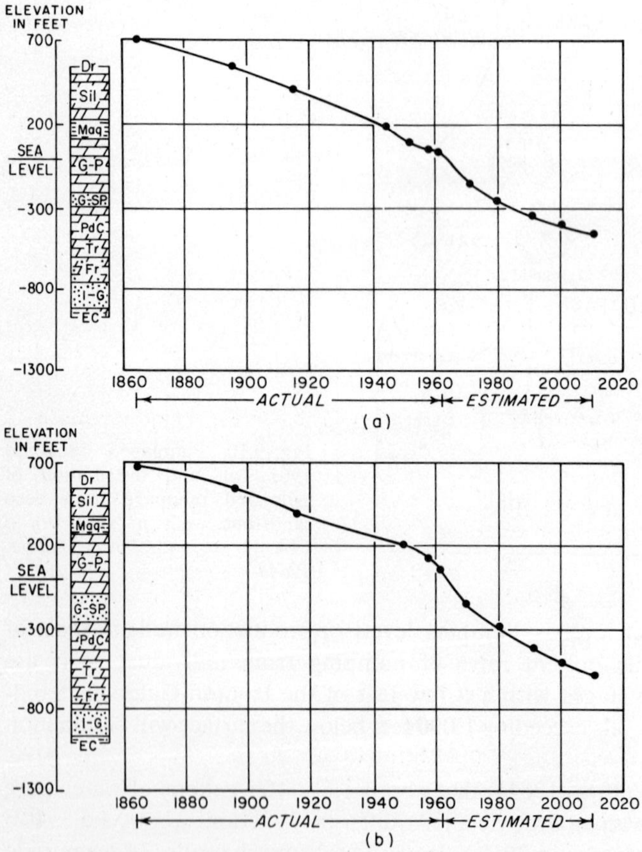

Fig. 9.11 Water levels in deep sandstone wells in Chicago (*a*) and Joliet (*b*) pumping centers in northeastern Illinois. (*From Walton, 1964a.*)

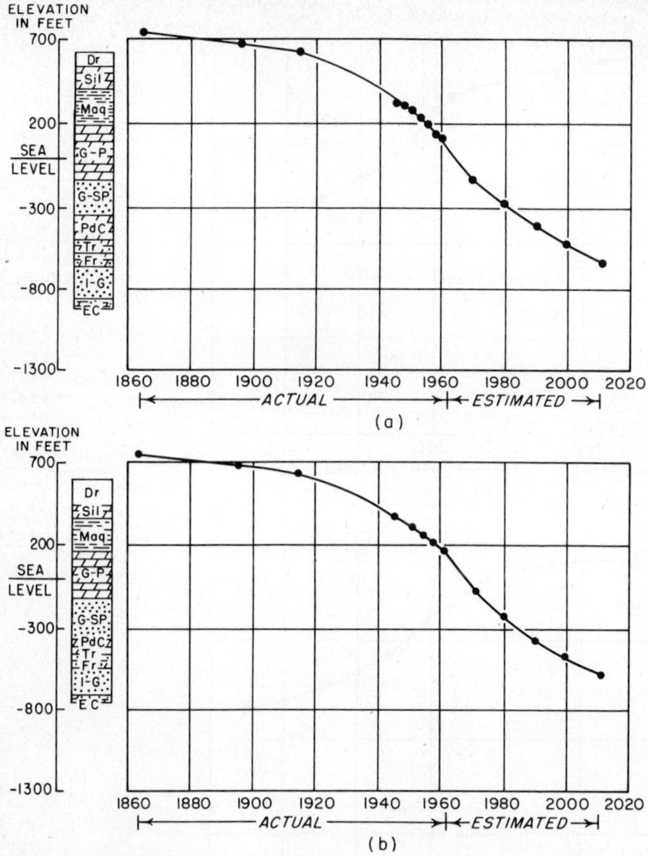

Fig. 9.12 Water levels in deep sandstone wells in Elmhurst (*a*) and Des Plaines (*b*) pumping centers in northeastern Illinois. (*From Walton, 1964a.*)

Withdrawals since 1959 exceed the practical sustained yield of the Cambrian-Ordovician Aquifer, but critical water levels will not exist until 2010 if the rate of pumpage increases in the future as it has in the past. Thus, groundwater users in northeastern Illinois have perhaps 50 years at some places in which to reduce pumpage from the Cambrian-Ordovician Aquifer and take full advantage of the groundwater resources of shallow sand and gravel and dolomite aquifers.

The Mt. Simon Sandstone and lower sandstones of the Eau Claire Formation are hydrologically interconnected and collectively are called the Mt. Simon Aquifer. These sandstones yield moderate to large quantities of water to production wells especially along the Fox River valley in Kane County. Water encountered in the Chicago region in the Mt. Simon Aquifer at less than 1,300 feet below sea level is commonly too salty for municipal uses. Present withdrawals probably do not exceed, but are about equal to, the practical sustained yield of the Mt. Simon Aquifer.

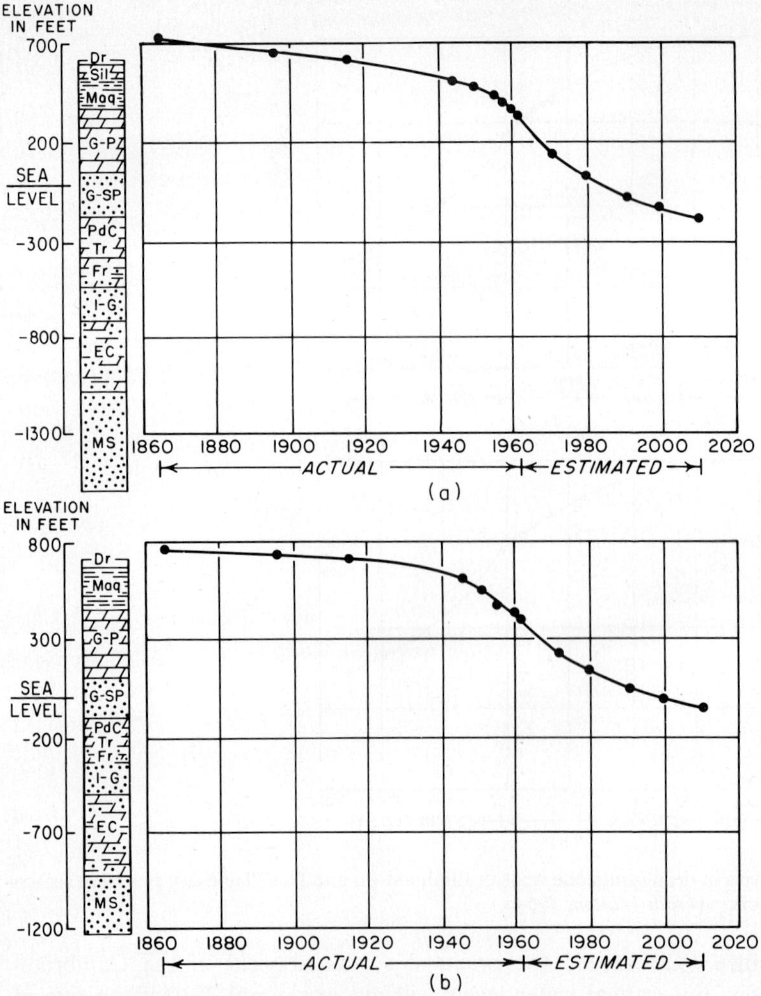

Fig. 9.13 Water levels in deep sandstone wells in Aurora (*a*) and Elgin (*b*) pumping centers in northeastern Illinois. (*From Walton, 1964a.*)

Sand and gravel aquifers are known to exist in scattered areas throughout much of northeastern Illinois. According to Bergstrom et al. (1955) and Suter et al. (1959), groundwater development possibilities in deeply buried sand and gravel beds are best in central and western Lake County; McHenry County; northwestern, south-central, and southern Cook County; northwestern DuPage County; central and western Kane County; along the Illinois River in Grundy County; northwestern and northeastern Kendall County; and central Will County. Sand and gravel beds are more widespread in Lake, McHenry, and DuPage Counties than they are in other counties.

The poorest possibilities for water-yielding sand and gravel beds are along reaches of the Fox, Des Plaines, Illinois, and Kankakee Rivers, where bedrock is encountered at shallow depths; in southern Kendall County, northern Grundy County, southern Will County, and western and southwestern Cook County, where the glacial drift is thin; and in eastern Cook County, where sands are scarce in the lake beds of the Chicago plain. Fairly extensive surficial sand and gravel deposits occur along reaches of the Fox, Kishwaukee, Des Plaines, and DuPage Rivers in Lake, McHenry, Kane, Cook, DuPage, Kendall, and Will Counties.

Sand and gravel aquifers in large areas in northeastern Illinois are in hydraulic connection with shallow dolomite aquifers. The shallow dolomite aquifers receive water from the sand and gravel aquifers. Thus, in some areas, pumping from sand and gravel aquifers overlying dolomite aquifers reduces recharge to, and, therefore, the potential yield of, the shallow dolomite aquifers. The potential yields of sand and gravel and overlying shallow dolomite aquifers are interdependent and cannot be considered separately. In some areas, shaly dolomite interbedded with shallow dolomite aquifers retards the vertical movement of water from sand and gravel aquifers into permeable zones within the shallow dolomite aquifers. More water can be developed by pumping both sand and gravel and dolomite wells than by pumping from only shallow dolomite aquifers.

In many areas in northeastern Illinois, the shallow dolomite aquifers have much greater areal extent than sand and gravel aquifers and about the same water-yielding potentials. Dolomite aquifers, being more deeply buried than sand and gravel aquifers, often yield more water than do sand and gravel aquifers, because available drawdown and, therefore, yields of production wells depend in part upon the depth of tops of aquifers.

Because of the larger areal extent and greater available drawdowns, it is wise to develop the shallow dolomite aquifers fully, where conditions permit, and to use existing sand and gravel aquifers to augment the shallow dolomite aquifers in areas where shaly dolomite beds retard recharge, or the shallow dolomite aquifers yield very little water to production wells. In many places, sand and gravel aquifers are much more permeable and more readily recharged than shallow dolomite aquifers. In these places, it is advantageous to develop the sand and gravel aquifers fully, instead of the shallow dolomite aquifers.

It is estimated (Walton, 1964b) that, under heavy pumping conditions, recharge rates for surficial sand and gravel aquifers in northeastern Illinois will range from 200,000 to 300,000 gpd/sq mi. Recharge rates for several sand and gravel aquifers deeply buried beneath clayey materials in Illinois have been estimated to range from 50,000 to 150,000 gpd/sq mi under heavy pumping conditions.

Specific capacities of existing sand and gravel wells in northeastern Illinois range from 1 to 300 gpm/ft of drawdown and average about 25 gpm/ft. In contrast, the specific capacities of production wells open to the Cambrian-Ordovician Aquifer range on the average from 4 to 7.5 gpm/ft. Thus, in many parts of northeastern Illinois, sand and gravel and dolomite production wells have greater specific capacities

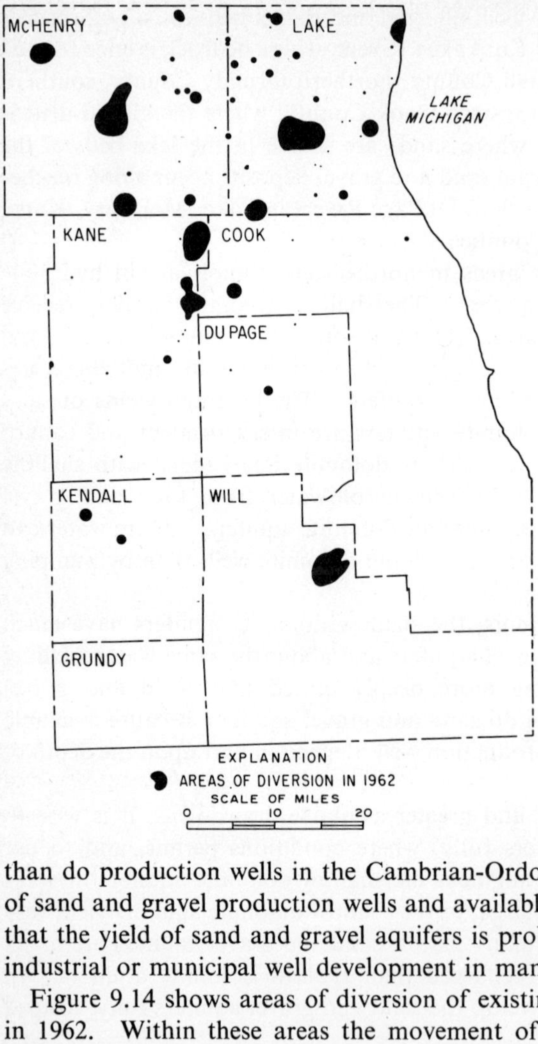

Fig. 9.14 Areas of diversion of sand and gravel production wells in northeastern Illinois in 1962. The solid areas represent the areas of diversion in 1962. (*From Walton, 1964b.*)

than do production wells in the Cambrian-Ordovician Aquifer. Data on the yields of sand and gravel production wells and available hydrogeologic information suggest that the yield of sand and gravel aquifers is probably high enough to support heavy industrial or municipal well development in many areas.

Figure 9.14 shows areas of diversion of existing sand and gravel production wells in 1962. Within these areas the movement of groundwater is toward production wells in pumping centers. There are large areas not influenced by present pumpage where heavy production-well development is possible. This suggests that the potential yield of the sand and gravel aquifers is much greater than present withdrawals.

That large quantities of water can be successfully obtained from sand and gravel aquifers is further indicated by water-level and pumpage data for existing pumping centers. Pumping center specific capacities range from 900 to 380,000 gpd/ft and average 45,000 gpd/ft. Pumping center specific capacity, as herein defined, is the total pumpage from production wells within a pumping center per foot of average drawdown within the pumping center. In contrast, pumping center specific capacities for shallow dolomite aquifers range from 6,000 to 120,000 gpd/ft and average 35,000 gpd/ft. On the basis of aquifer distribution, the potential yield of sand and gravel aquifers is greatest in McHenry, Lake, Kane, Kendall, and DuPage Counties and least in Grundy, Cook, and Will Counties.

It is estimated (Walton, 1964b) that under heavy pumping conditions recharge rates for the shallow dolomite aquifers will average 140,000 gpd/sq mi (about 100 gpm/sq mi) in much of the eastern two-thirds of northeastern Illinois and 60,000 gpd/sq mi (40 gpm/sq mi) in large areas of the western third.

The productivity of the shallow dolomite aquifer is highest in the eastern part of Will County and least in western Will County. Specific capacities of production wells in all areas except eastern Will County are less than the specific capacities of production wells in DuPage County. With the exception of McHenry County, wells in Silurian rocks in all counties have average specific capacities that exceed 10 gpm/ft and are greater than the average specific capacity of deep sandstone wells penetrating the Cambrian-Ordovician Aquifer in northeastern Illinois. It is possible to drill what is essentially a dry hole at any location; on the basis of data for 50 percent of existing production wells, the chances of obtaining a dolomite production well with a yield of 250 gpm or more are good in most areas where Silurian rocks and the Galena-Platteville Dolomite immediately underlie the glacial drift. These areas cover most of the eastern two-thirds of northeastern Illinois and western Kendall County. Thus, the yield of the shallow dolomite aquifers is probably large enough to support heavy industrial or municipal well development in wide areas. The chances of obtaining a production well with a yield of 250 gpm or more are very poor in areas in southwestern McHenry, western Kane, eastern Kendall, and western Grundy Counties where the Maquoketa Formation immediately underlies the glacial drift.

The Cambrian-Ordovician Aquifer commonly yields 1 mgd or more to deep sandstone wells, mostly because of the large available drawdowns. As a result, water users in northeastern Illinois have come to frown upon production wells with lesser yields. In the future, as greater development of shallow aquifers becomes necessary, groundwater users will have to readjust their thinking and accept more freely production wells with yields of 250 gpm. In large part because of available drawdown the number of production wells in well fields will increase as shallow aquifer groundwater resources are more fully developed.

Because the shallow dolomite aquifers are thick, fairly deeply buried, and, on a regional basis, have a high-to-moderate permeability and great areal extent, the areas of diversion of production wells extend for considerable distances, and available groundwater resources can be developed with a reasonably small number of production wells and well fields.

Figure 9.15 shows the areas of diversion of pumping centers in 1962. There are large areas outside areas of diversion where heavy production-well development is possible. This suggests that the potential yield of the shallow dolomite aquifers is much greater than present withdrawals. Detailed quantitative studies suggest that the potential yield of shallow dolomite aquifers within areas of diversion is also greater than present withdrawals.

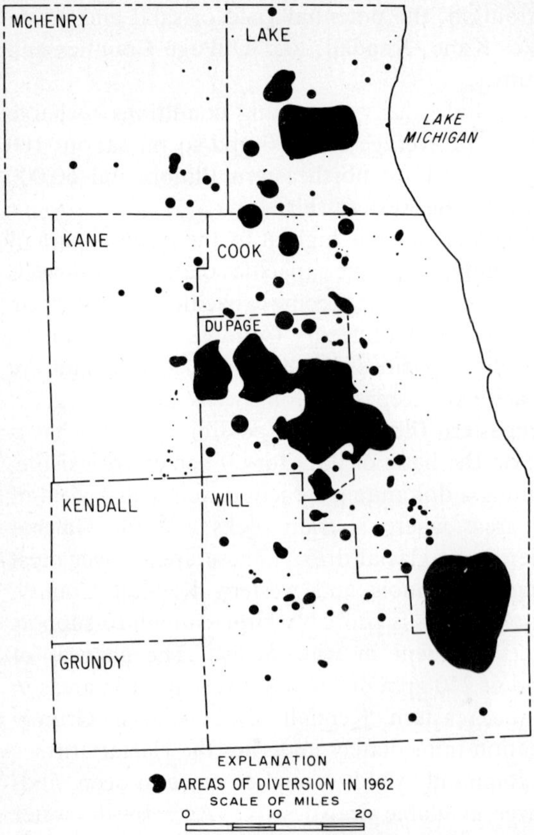

Fig. 9.15 Areas of diversion of dolomite production wells in northeastern Illinois in 1962. The solid areas represent the areas of diversion in 1962. (*From Walton, 1964b.*)

Areas of diversion include sites where the shallow dolomite aquifers yield very little water to individual production wells. In addition, water-level maps for the aquifers are regular in appearance. These facts indicate that, except in areas where the Maquoketa Formation immediately underlies the glacial drift, the inconsistency in production of wells in the shallow dolomite aquifers has little effect on the regional response of the aquifers to pumping and should not seriously deter full development of available groundwater resources. Areas of diversion and cones of depression of individual pumping centers overlap one another and do not respect political boundaries. This suggests that individual water users cannot develop groundwater resources independently of one another.

That large quantities of water can be successfully obtained from the shallow dolomite aquifers is further indicated by pumping center specfiic capacities. The shallow dolomite aquifers commonly yield more than 35,000 gpd/ft in heavily pumped areas; pumping center specific capacities exceed 10,000 gpd/ft in most places where data are available.

A careful study of the water-yielding possibilities of the shallow dolomite and sand

and gravel aquifers indicates that, on a gross basis, these aquifers are capable of yielding far more water to production wells with proper development of both aquifers than will be recharged under heavy pumping conditions. Thus, the potential yield of the shallow aquifers depends upon recharge rates rather than on the water-yielding possibilities. It should be kept in mind that recharge rates for sand and gravel aquifers range between 40 and 250 gpm/sq mi.

The potential yields of shallow dolomite and sand and gravel aquifers are grouped together because of the hydraulic interconnection of these aquifers. The potential yield of shallow aquifers in areas outside the Chicago Sanitary District in northeastern Illinois, on the basis of estimated recharge rates and water-yielding possibilities of aquifers, is of the order of magnitude of 440 mgd (Walton, 1964b).

A study was made of available data for high-capacity shallow aquifer production wells in service in 1962 to determine the probable number of production wells needed for full development of the potential yield. Including necessary standby production wells, in 1962, an average of four wells were in service for each 1 mgd withdrawn. This would mean that less than 2,000 production wells would be required to develop the potential yield of shallow aquifers fully, not an unreasonably large number of production wells. It is true that a few pipelines up to 10 miles long might be required.

The total potential and practical sustained yields of shallow and deep aquifers in northeastern Illinois is of the order of magnitude of 500 mgd, and the total groundwater pumpage in 1960 was 171 mgd. The potential yield of shallow aquifers (dolomite and sand and gravel) is 440 mgd, or about 88 percent of the total. In the future, emphasis must be placed on the development of the groundwater resources of the shallow aquifers.

The potential yield of the shallow aquifers, 440 mgd, amounts to 2.5 inches of precipitation over northeastern Illinois and is less than 10 percent of the mean annual precipitation. Average annual groundwater runoff to streams is estimated at 4.5 inches; minimum average groundwater runoff during extended dry periods, at 2 inches. Streamflow will probably not be diminished to any great extent by full development of shallow aquifers, because most water pumped from production wells and diverted from streamflow will be returned to the streams after use. Artificial recharge by induced infiltration of surface water or through wells, pits, and so forth is not included in computations of potential yield.

9.6 Taylorville, Illinois

Water for municipal and industrial use at Taylorville in central Illinois in 1960 was obtained from production wells in a sand and gravel aquifer approximately $\frac{3}{4}$ mile wide and 50 feet thick that receives recharge directly from precipitation (Walker and Walton, 1961). Since 1888, when the aquifer was first tapped for the municipal water supply, the daily withdrawal increased from about 27,000 gallons to over 3 million gallons in 1953. Exploitation of groundwater resources caused a general

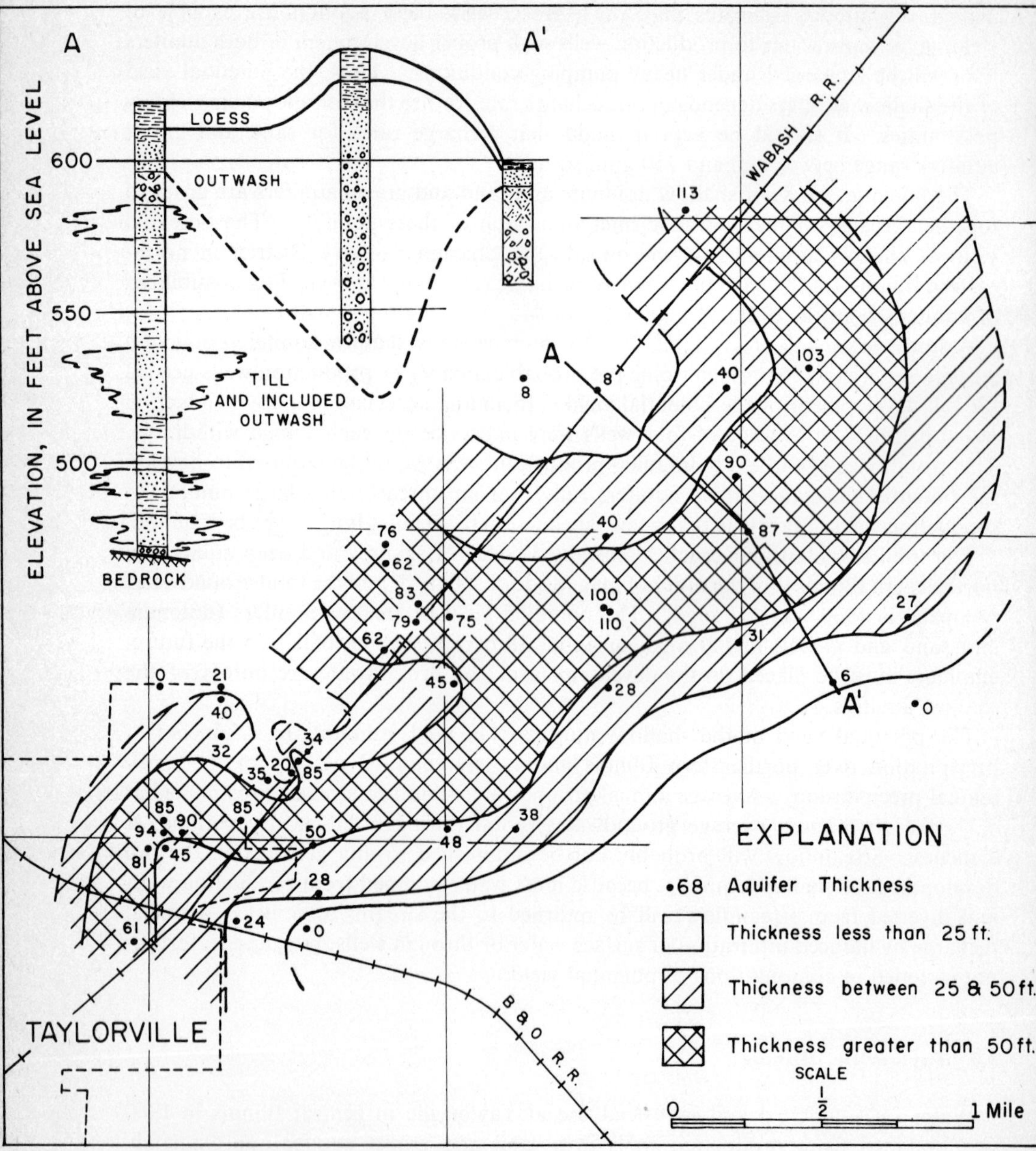

Fig. 9.16 Map and geologic section showing thickness and areal extent of aquifer at Taylorville, Illinois. (*From Walker and Walton, 1961.*)

water-level decline in the Taylorville area and the water table receded to near critical levels in some of the major pumping centers.

Except for a low narrow ridge extending southwest to northeast through the center of the Taylorville area, the land surface is a level to gently undulating plain. Drainage is largely southeastward to Flat Branch and to South Fork Sangamon River, which flows in courses about $\frac{1}{2}$ mile southeast of the area.

The mean annual precipitation is 36.18 inches. The months of greatest precipitation are April, May, June, and August, each having more than 3.5 inches; February is the month of least precipitation, having less than 2 inches. A large part of central and southern Illinois including the Taylorville area experienced a severe drought beginning in the latter half of 1952. For the period 1952 through 1956, cumulative deficiency of precipitation at Taylorville was about 23 inches. Recharge from precipitation was much below normal during the dry years and large quantities of water were taken from storage within the aquifer to balance pumpage. The mean annual temperature is 53.9°F. June, July, and August are the hottest months with mean temperatures of 73.2, 77.2, and 75.2°F, respectively; January is the coldest month with a mean temperature of 30.7°F. The mean length of the growing season is about 185 days.

The unconsolidated glacial deposits in the Taylorville area are chiefly Illinoian in age and range in thickness from about 50 to 180 feet. These deposits vary in thickness in part because of a buried valley cut into the underlying bedrock of Pennsylvanian age, and in part because of the uneven topography of the land surface.

The glacial deposits are a complex of ice-laid till, water-laid silt, sand and gravel outwash, and wind-deposited silt and fine sand. The sand and gravel outwash is permeable and yields water in large quantities to production wells. Both the glacial till, which contains considerable silt and clay, and the underlying shale bedrock of Pennsylvanian age are relatively impermeable. The loess is of Wisconsinan age and has been deposited as a blanket, 5 to 10 feet thick, on the older tills and outwash deposits.

Logs of wells, test holes, and coal test borings indicate that the bedrock valley in the Taylorville area was at one time in the past largely filled with glacial till containing only a few thin, probably discontinuous beds or lenses of glacial outwash. As shown by cross section A–A′ in Fig. 9.16, the aquifer occurs as a fill of outwash sand and gravel in a relatively narrow valley cut into the upper part of the till deposits. The map in Fig. 9.16 shows that the outwash deposits range in width from about $\frac{1}{2}$ to 1 mile and trend northeast to southwest. The permeable outwash forming the aquifer ranges in thickness from 0 to 113 feet and consists of stratified beds of sand and gravel with silt in various mixtures. Generally, the upper part of the aquifer contains finer-grained material than the lower part.

The first municipal production wells at Taylorville were constructed in 1888 at the city water plant near Cherokee and Vine Streets. In 1900 about 28,000 gpd were pumped from eight tubular production wells to satisfy the commercial and domestic water needs of the city. As shown in Fig. 9.17 water use has steadily increased since 1888 as the city has grown in population from 4,248 to 9,750, and in 1958

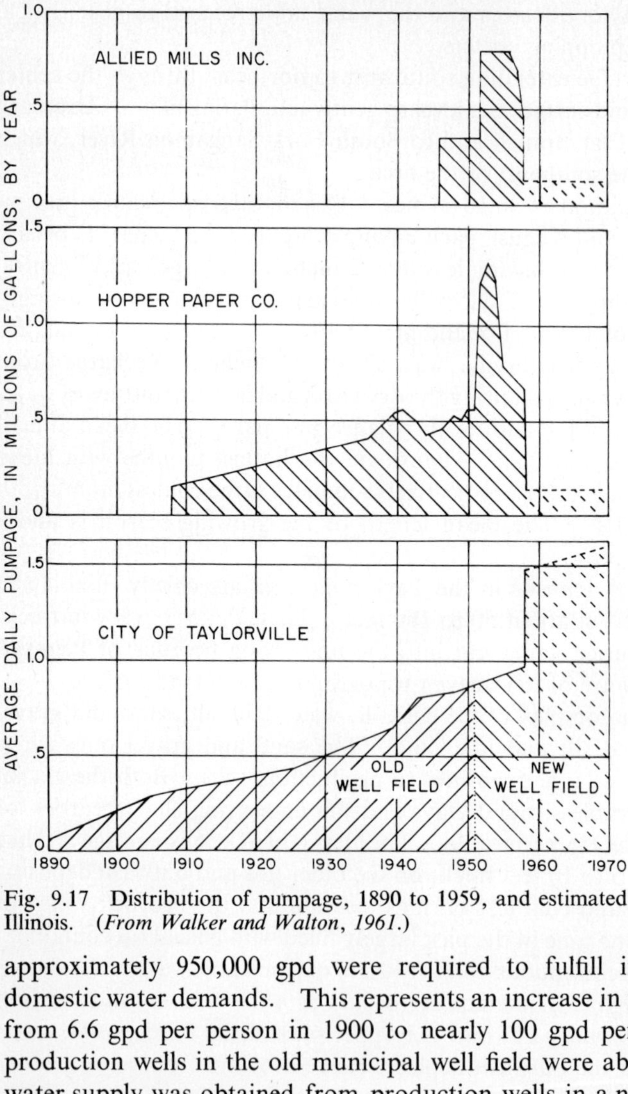

Fig. 9.17 Distribution of pumpage, 1890 to 1959, and estimated pumpage to 1970 at Taylorville, Illinois. (*From Walker and Walton, 1961.*)

approximately 950,000 gpd were required to fulfill industrial, commercial, and domestic water demands. This represents an increase in total per capita consumption from 6.6 gpd per person in 1900 to nearly 100 gpd per person in 1958. In 1951, production wells in the old municipal well field were abandoned and the municipal water supply was obtained from production wells in a new field northeast of the city limits. Quality of water and production-well-deterioration problems were responsible for the change in pumping centers.

Two industries, Hopper Paper Company and Allied Mills, Inc., also pump large quantities of groundwater from the aquifer at Taylorville. The water demand at Hopper Paper Company increased as shown in Fig. 9.17 at a relatively uniform rate from 150,000 gpd in 1908 to 550,000 gpd in 1952. In 1953, the rate of groundwater withdrawal was increased to 1,320,000 gpd. This increase in pumpage caused such a rapid rate of water-level decline that the pumpage had to be reduced to 770,000 gpd in 1955, and to 650,000 gpd in 1958. A further reduction to 144,000 gpd was made in April 1958, and since that time Hopper Paper Company has purchased an average of 632,000 gpd from the city to supplement the amount pumped from their production well field.

The soybean processing plant owned by Allied Mills, Inc., pumped about 350,000 gpd from 1946 until 1952. Plant expansion resulted in an increase of the water demand to 800,000 gpd in 1952. Groundwater levels declined rapidly to critical stages and in 1957 cooling towers were installed to conserve water. Conservation practices resulted in a reduction in groundwater withdrawal at the plant of about 675,000 gpd. Since 1958 about 125,000 gpd have been pumped from production wells at Allied Mills, Inc.

The distribution of groundwater pumpage in May 1959 at Taylorville was 1,500,000 gpd from municipal production wells; 144,000 gpd from the Hopper Paper Company production well field; and 125,000 gpd from production wells at Allied Mills, Inc. (see Fig. 9.18). Total groundwater withdrawals increased progressively from about 28,000 gpd in 1890 to a maximum of about 3,000,000 gpd in 1953 as shown in Fig. 9.19. Pumpage decreased rapidly from 3,000,000 gpd in 1953 to about 1,750,000 gpd in 1957 owing to the great decline in the water table. In May 1959, total groundwater withdrawal was about 1,770,000 gpd.

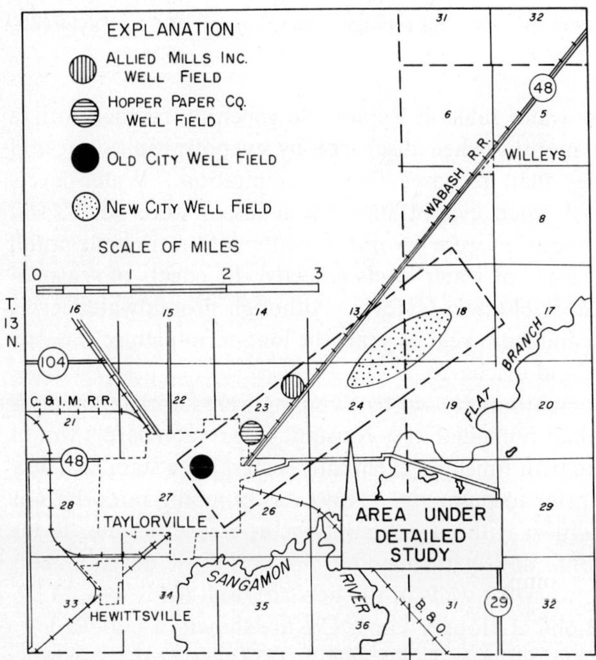

Fig. 9.18 Map showing location of study area and pumpage centers at Taylorville, Illinois. (*From Walker and Walton, 1961.*)

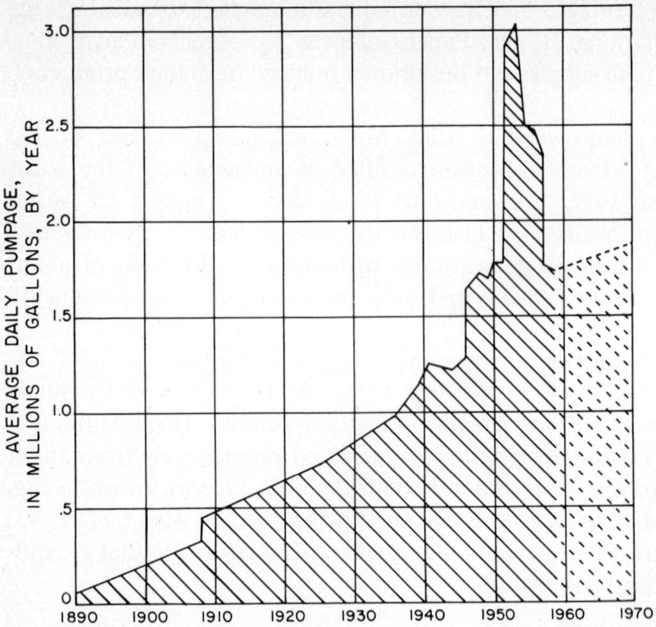

Fig. 9.19 Groundwater pumpage, 1890 to 1959, and estimated pumpage to 1970 at Taylorville, Illinois. (*From Walker and Walton, 1961.*)

Under natural conditions, the water table at Taylorville generally receded in late spring, summer, and early fall months when discharge by evapotranspiration and groundwater runoff were greater than recharge from precipitation. Water levels began to recover late in the fall when evapotranspiration losses were small, soil moisture was replenished, and conditions were favorable for the infiltration of rainfall to the water table. The annual cycle of water levels reflected the effects of seasonal variation in precipitation and other climatic factors. Although groundwater levels fluctuated from season to season and from year to year, the long-term picture was one of equilibrium between recharge and discharge.

Starting in 1888 water-level fluctuations caused by pumping were superimposed on the natural cycle. If pumpage had remained at a constant rate, water levels would have declined at a decreasing rate with time and eventually would have stabilized at a stage lower than that measured prior to pumping. However, pumping rates did not remain constant but increased almost without interruption; as a result, water levels never stabilized but declined continuously throughout the period of development. Nonpumping water-level hydrographs for wells in the new city well field, well 23.2f6 at Allied Mills, Inc., and well 23.6b6 at Hopper Paper Co. are shown in Fig. 9.20.

The average elevation of the water table at Taylorville in 1888 was probably about 615 feet. By 1951 the water table had declined in response to continual withdrawals of water to an average elevation of 595 feet. Thus, in a period of 63 years, water levels declined 20 feet or at a rate of about 0.3 ft/yr. As the result of progressive increases in pumpage and drought conditions, water levels declined from an average elevation of 595 feet in 1951 to an average elevation of 575 feet in 1956. The average rate and total decline of water levels, 1951 through 1956, were about 3.3 ft/yr and 20 feet, respectively. Between 1956 and 1958 water levels recovered slightly due to reduced rates of pumpage and above normal precipitation. Recovery terminated in production wells in the new city well field early in 1958, and since that time water levels in those wells have declined continuously.

A combination of several effects is reflected in the hydrographs presented in Fig. 9.20. The graphs of water-level fluctuations in wells 24.4g, 18.8a, and 18.6d in the new city well field reflect alternate pumping from each of the four municipal wells until about October 1952, when pumping was shifted to wells 24.4g and 18.8a. Alternate pumping of these two wells continued until October 1953, when well 18.8a was placed on a standby basis and most of the municipal water supply was pumped from well 24.4g. From about September 1 through October 15, 1956, well 18.8a was used exclusively and a distinct recovery of water levels in wells 24.4g and 23.2f6 occurred while the water levels in wells 18.8a and 18.8c declined as the water table adjusted to the shift in pumping centers. The recovery of water levels in all wells from May through December 1957 was the result of above-average rainfall during this period and a reduction of pumpage at the Allied Mills, Inc., plant.

The hydrographs for well 23.2f6, located between production wells at the Allied Mills, Inc., plant, and for well 23.6b6 at the Hopper Paper Company plant, correlate in general with the hydrographs of wells in the new city well field given in Fig. 9.20. The hydrographs for wells 23.2f6 and 23.6b6 are greatly affected by shifts in pumping

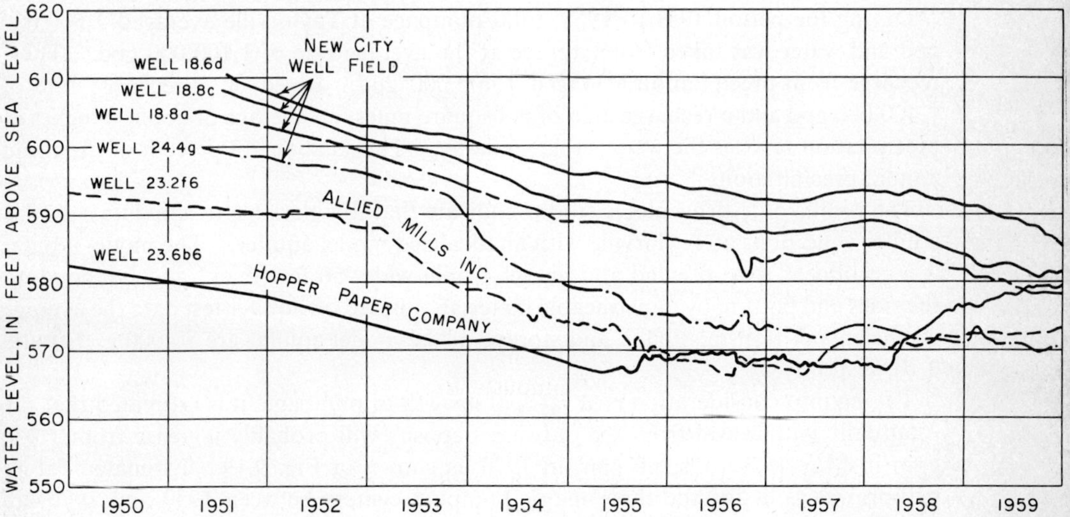

Fig. 9.20 Water levels in wells at Taylorville, Illinois, 1950 to 1959. (*From Walker and Walton, 1961.*)

centers and changes in pumping rates within the Allied Mills, Inc., and Hopper Paper Company well fields. From January 1952 to September 1956, for example, the withdrawal rate at Allied Mills, Inc., was increased to 800,000 gpd, twice the original pumping rate. Water levels in the area declined 24 feet as a result of this increase in pumpage. Pumpage was reduced in 1956 and as a result the water table recovered 7 feet from 1956 to 1959. The hydrograph of well 23.6b6 at Hopper Paper Company indicates that the 632,000-gpd reduction in pumpage from the Hopper well field on April 1, 1958 caused the water table to rise 8 feet in less than a year.

It is estimated that 7.7 billion gallons of water were taken from storage within the aquifer during the period 1888 to 1958. The average volumetric rate of decrease of groundwater storage during that period was about 300,000 gpd. Computations show that about 2.7 billion gallons of water were taken from storage at an average rate of 1,000,000 gpd during the period 1951 to 1958.

The groundwater in storage in the aquifer permits pumping at rates greater than recharge for limited periods. Eventually, however, discharge from wells must be balanced by recharge if pumping is to be continued without eventually dewatering the aquifer below tops of screens in production wells.

The source of recharge to the aquifer is precipitation. Groundwater recharge is greatest in spring months of heavy rainfall and least in the summer, fall, and winter months. Most recharge occurs during spring months when evapotranspiration is small and soil moisture is maintained at or above field capacity by frequent rains. During summer and fall months evapotranspiration and soil moisture requirements have first priority on precipitation and are so great that little precipitation percolates to the water table except during periods of excessive rainfall. Recharge during the winter months when the ground is frozen is negligible.

During the period 1951 to 1958, total pumpage at Taylorville averaged 2,500,000 gpd and water was taken from storage at the average rate of 1,100,000 gpd. Thus, recharge from precipitation averaged 1,400,000 gpd. Based on a recharge rate of 1,400,000 gpd and a recharge area of 6.3 square miles, an average of about 6 inches of precipitation reaches the water table annually. This is about 17 percent of the mean annual precipitation.

The results of hydrogeologic studies indicate that it is possible to simulate complex aquifer conditions at Taylorville with an idealized model aquifer. The model aquifer is a rectilinear strip of sand and gravel $\frac{3}{4}$ mile wide, 50 feet thick, and bounded on the sides and bottom by impermeable material. Based on aquifer test data the average coefficients of transmissibility and storage of the model aquifer are 70,000 gpd/ft and 0.20, respectively.

Taking into consideration past rates of growth of pumpage, it is estimated that the total daily withdrawal from the outwash deposits will probably increase from 1,230 gpm in May 1959 to 1,300 gpm in 1970 as shown in Fig. 9.19. Estimated future withdrawal rates for the three major pumping centers between 1959 and 1970 are shown in Fig. 9.21.

Water-level declines in wells 24.4g, 18.8c, 23.6b9, and 23.2f4 were calculated using

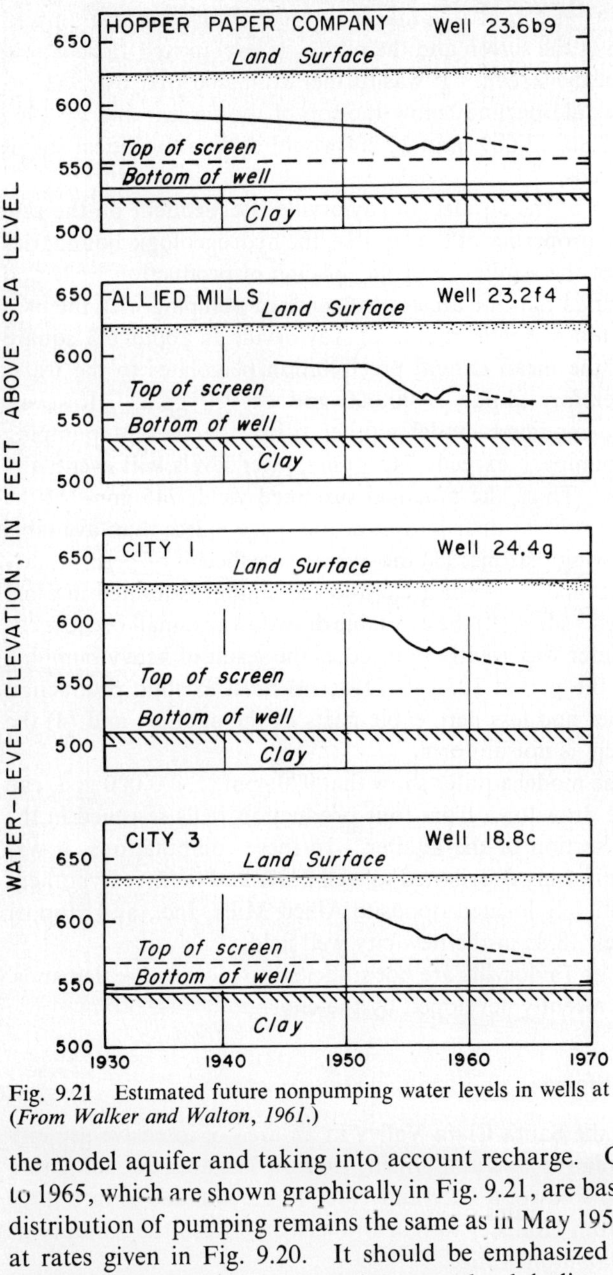

Fig. 9.21 Estimated future nonpumping water levels in wells at Taylorville, Illinois, 1959 to 1965. (*From Walker and Walton, 1961.*)

the model aquifer and taking into account recharge. Computed declines from 1959 to 1965, which are shown graphically in Fig. 9.21, are based on the assumption that the distribution of pumping remains the same as in May 1959 and that pumpage increases at rates given in Fig. 9.20. It should be emphasized that the water-level declines shown in Fig. 9.21 are nonpumping water-level declines. Pumping levels will decline at about the same rate as the nonpumping water levels if the May 1959 rates of pumping from individual wells are maintained.

Based on computations, it seemed probable that in 1961 the pumping level in well 23.2f4 would be below the top of the screen and the pumping level in well 18.8c would be only 2 feet above the top of the screen. It was further estimated that by 1963 the pumping level in well 18.8c would decline below the top of the screen, and by 1965 the pumping levels in wells 18.8c, 23.6b9, and 23.2f4 would decline to critical stages several feet below tops of screens.

The practical sustained yield of the aquifer at Taylorville is dependent on the rate of recharge, the hydrogeologic properties of the aquifer, the hydrogeologic boundaries of the aquifer, the thickness of the aquifer, and the spacing of production wells and well fields. It has been estimated that the area of influence of pumping with the production wells distributed within a 3-mile radius of Taylorville is about 6.3 square miles and that 17 percent of the mean annual precipitation percolates to the water table. Total recharge is therefore about 1,400,000 gpd or 970 gpm. However, computations made with the modified model aquifer using the present pumping distribution indicate that if pumpage exceeds 745 gpm, water levels will eventually decline below tops of screens. Thus, the practical sustained yield, 745 gpm, of the aquifer at Taylorville with the present distribution of pumpage is less than available recharge, largely for the following reasons: (1) the average coefficient of permeability of the outwash deposits, 1,400 gpd/ft, is not great and the aquifer is limited in areal extent by the walls of the buried valley; (2) the available drawdown is small because the saturated thickness of the aquifer was greatly reduced as the result of heavy pumping in excess of recharge during the period 1951 to 1958; (3) most existing production wells are located in the thinner and less permeable parts of the aquifer; and (4) the distribution of pumping centers is not uniform.

Computations made with the model aquifer show that 970 gpm (1,400,000 gpd) can be obtained without excessive drawdown from four production wells screened in the thicker and more permeable section of the aquifer. In these computations, it was assumed that pumpage was uniformly distributed to well 24.4g and to three hypothetical production wells approximately located opposite Allied Mills, Inc., and Hopper Paper Company, and near well 18.8c in the new city well field.

The groundwater resources at Taylorville are not sufficient to meet future demands. Surface-water resources were recently developed by the city.

9.7 Santa Clara Valley, California

Land subsidence occurs in the Santa Clara Valley in an area of intensive groundwater development in California (Poland and Green, 1962). The area extends southeastward about 30 miles from Redwood City and Niles to Coyote and is in a large structural trough extending from Hollister to San Francisco between the Santa Cruz Mountains and the Mount Hamilton Range.

The bedrock underlying the area ranges in age from Jurassic to Pliocene and consists mainly of consolidated sedimentary rocks with minor areas of metamorphic and igneous rocks. The Santa Clara Formation of Pliocene and Pleistocene age overlies the bedrock. It consists of semiconsolidated conglomerate, sandstone, siltstone, and claystone with low permeabilities. Unconsolidated alluvial and bay deposits of clay, sand, and gravel (valley fill) reaching a thickness of 1,000 feet or more rest on the Santa Clara Formation.

Fine-grained materials such as clay, silt, and sandy clay constitute the bulk of the valley fill; sand and gravel occur in lesser amounts. None of these deposits can be traced laterally for more than a short distance. The valley fill yields moderate to large quantities of water to production wells especially for irrigation purposes.

A 1932 releveling of a 1912 first-order level line indicated that several feet of settlement had occurred in the San Jose area. To determine the extent and magnitude of subsidence, the U.S. Coast and Geodetic Survey established an extensive level network in 1933. The network was leveled in the spring of 1934 and releveled seven times from November 1934 to April 1940, in 1948, and in 1954. The elevations of certain bench marks at the valley margins were found to have ramained stable; these bench marks were assumed to be fixed, and elevations of the valley-floor bench marks were adjusted accordingly by the least-square method.

Subsidence from 1934 to 1954 has occurred in an area exceeding 230 square miles. Control is not available to define the line of zero subsidence. Maximum subsidence was 4.06 feet; subsidence exceeded 3 feet in many places. The area of subsidence lies wholly within the area of the valley fill where groundwater occurs under leaky artesian conditions and large quantities of water are withdrawn from production wells.

During the period from the spring of 1948 to the summer of 1954 subsidence in or near San Jose exceeded 2 feet at an average rate of 0.4 foot per year. Subsidence at San Jose from 1912 to 1959 was 9.04 feet. No appreciable subsidence has occurred in areas where groundwater is unconfined and coarse-grained materials predominate.

Figure 9.22 shows the change in altitude at a bench mark in San Jose and the change in artesian pressure at a nearby well, 1912 to 1959. Groundwater levels, which were above the land surface in 1915, continuously declined 100 feet or more from 1916 to 1934, rose 74 feet to a peak in 1943, and then declined 95 feet by 1951. Subsidence was 0.39 foot from 1912 to 1919 and 4.56 feet from 1919 to 1934. From

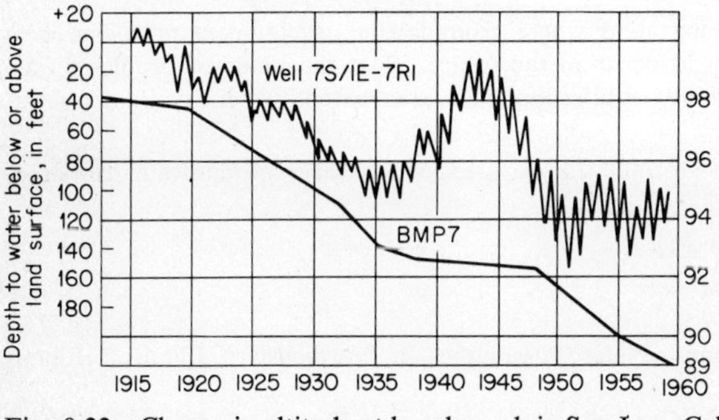

Fig. 9.22 Change in altitude at bench mark in San Jose, California, and change in artesian pressure at nearby well. (*After Poland and Green, 1962.*)

November 1934 to November 1937 the rate of subsidence decreased; from 1937 to 1940 subsidence was only 0.03 foot. From 1938 to 1940 recovery, if any, was small. Subsidence of 2.72 feet occurred from 1948 to 1958. From April 1932 to November 1934, subsidence averaged 0.5 foot per year; from 1948 to 1954 the average rate was 0.32 foot per year. The ratio of subsidence to water-level decline from 1920 to 1935 was about 1 to 13.

Subsidence caused by the compaction of the valley fill due to artesian-pressure decline occurred rapidly in the coarse-grained sediments and more slowly in the fine-grained beds. The cessation of subsidence in 1938 suggests that all compaction attributable to the net artesian-pressure decline from 1916 to 1938 had taken place. Data indicate that a substantial decline of the water table may cause only a small amount of subsidence.

The decline of artesian pressure is the principal cause of land subsidence in the Santa Clara Valley. The area of intensive subsidences is limited to the area where leaky artesian conditions exist; subsidence is not appreciable in areas of unconfined water. The rate of subsidence generally corresponded to the rate of artesian pressure.

Problems

9.1 Visit the offices of state water agencies and the U.S. Geological Survey Water Resources Division in your state. Obtain available information concerning development and management of the major aquifers in your state. Prepare a report describing the problems encountered in developing and managing groundwater resources.

9.2 Obtain available information on pumpage and associated water-level declines in the major pumping centers in your state. Prepare a report describing the relations between pumping center specific capacities and hydrogeologic conditions and well-field spacings. Using available pumping center specific-capacity data and information concerning hydrogeologic conditions, estimate the practical sustained yields of existing pumping centers.

9.3 Assemble all laws related to groundwater development and management in your state. Prepare a report summarizing the effects groundwater laws have had on the development and management of groundwater resources.

9.4 Prepare a report describing case histories of the conjunctive management of groundwater and surface water resources in your state.

9.5 Prepare a map of your state showing areas where groundwater development problems have occurred in the past and are likely to occur in the future. List the kinds of problems (i.e., insufficient recharge, low transmissibility, aquifer limited in areal extent, water-quality deterioration, etc.).

9.6 Prepare a map of your state showing undeveloped areas where large groundwater resources are available.

References

Bergstrom, R. E. et al., 1955. "Ground Water Possibilities in Northeastern Illinois," Illinois State Geol. Surv. Circ. 198.

Bliss, J. H., 1951. Administration of the Ground-water Law of New Mexico, *J. Am. Water Works Assoc.*, vol. 43.

Davis, G. H., J. B. Small, and H. B. Counts, 1963. Land Subsidence Related to Decline of Artesian Pressure in the Ocala Limestone at Savannah, Georgia, Eng. Geol. Case Histories, *Geol. Soc. Am.*, vol. 4.

Domenico, P. A., and G. Clark, 1964. Electric Analogs in Time-settlement Problems, *Proc. Am. Soc. Civil Engrs.* 90 (SM3).

——— and M. D. Mifflin, 1965. "Water from Low-permeability Sediments and Land Subsidence," Res. Associates, Desert Res. Inst., Univ. Nevada Publications, vol. 1, no. 4.

Gibbs, H. J., 1960. A Laboratory Testing Study of Land Subsidence, *Proc. Pan Am. Conf. Soil Mech. Foundation Eng.*, 1st, Mexico City, 1959, vol. 1.

Green, J. H., 1964. "The Effect of Artesian-pressure Decline on Confined Aquifer Systems and Its Relation to Land Subsidence," U.S. Geol. Surv. Water Supply Paper 1779-T.

Hutchins, W. A., 1955. "The New Mexico Law of Water Rights," State Engineer of New Mexico (in cooperation with U.S. Dept. of Agr.).

Jumikis, A. R., 1962. "Soil Mechanics," D. Van Nostrand Company, Inc., Princeton, N.J.

Lofgren, B. E., 1961. "Measurement of Compaction of Aquifer Systems in Areas of Land Subsidence," U.S. Geol. Surv. Profess. Paper 424-B, art. 24.

Lohman, S. W., 1961. "Compression of an Elastic Artesian Aquifer," U.S. Geol. Surv. Profess. Paper 424-B.

MacKichan, K. A., and L. C. Kammerer, 1961. "Estimated Use of Water in the United States, 1960," U.S. Geol. Surv. Circ. 456.

Miller, R. E., 1961. "Compaction of an Aquifer System Computed from Consolidated Tests and Decline in Artesian Head," U.S. Geol. Surv. Profess. Paper 424-B, B54-B58.

Piper, A. M., 1959. Requirements of a Model Water Law, *J. Am. Water Works Assoc.*, vol. 51.

———, 1960. "Interpretation and Current Status of Ground-water Rights," U.S. Geol. Surv. Circ. 432.

——— and H. E. Thomas, 1958. Hydrology and Water Law—What Is Their Future Common Ground? in "Water Resources and the Law," Univ. Mich. Law School.

Poland, J. F., 1961. "The Coefficient of Storage in a Region of Major Subsidence Caused by Compaction of an Aquifer System," U.S. Geol. Surv. Profess. Paper 424-B.

——— and J. H. Green, 1962. "Subsidence in the Santa Clara Valley, California, a Progress Report," U.S. Geol Surv. Water Supply Paper 1619-C.

Suter, M., et al., 1959. "Preliminary Report on Ground Water Resources of the Chicago Region," Illinois State Water Surv. and Geol. Surv. Cooperative Ground-water Rept. 1.

Terzaghi, K., 1943. "Theoretical Soil Mechanics," John Wiley & Sons, Inc., New York.

——— and R. B. Peck, 1948. "Soil Mechanics in Engineering Practice," John Wiley & Sons, Inc., New York.

Thomas, H. E., 1951. "The Conservation of Ground Water," McGraw-Hill Book Company, New York.

———, 1958. Hydrology vs. Water Allocation in the Eastern United States, in "The Law of Water Allocation in the Eastern United States," The Ronald Press Company, New York.

Todd, D. K., 1959. "Ground Water Hydrology," John Wiley & Sons, Inc., New York.

Walker, W. H., and W. C. Walton, 1961. "Ground-water Development in Three Areas of Central Illinois," Illinois State Water Surv., Rept. of Invest. 41.

Walton, W. C., 1964a. Future Water-level Declines in Deep Sandstone Wells in Chicago Region, *Ground Water*, vol. 2, no. 1.

———, 1964b. Potential Yield of Aquifers and Ground Water Pumpage in Northeastern Illinois, *J. Am. Water Works Assoc.*, February.

Wilson, G., and H. Grace, 1942. The Settlement of London due to Under-drainage of the London Clay, *J. Inst. Civil Engrs.*, vol. 19, no. 2.

Winslow, A. G., and L. A. Wood, 1959. Relation of Land Subsidence to Groundwater Withdrawals in the Upper Gulf Coastal Region of Texas, AIME Mining Sec., *Mining Engr.*

INDEX